D0919471

Guide to Identification of Marine and Estuarine Invertebrates

CAPE HATTERAS TO THE BAY OF FUNDY

KENNETH L. GOSNER

Curator of Zoology
The Newark Museum
Newark, New Jersey

Wiley-Interscience, a Division of John Wiley & Sons, Inc.

New York • London • Sydney • Toronto

Library of Congress Catalog Card Number: 70–149771

ISBN 0 471 31897 3

Printed in the United States of America

10 9 8 7 6 5 4 3 2 1

To LDG and PWG

Preface

The sole aim of this book is to acquaint the reader with the scope of the invertebrate fauna of the shore and shallow sea bordering the northeastern United States and to provide a primary tool for the identification of its species.

There is, at present, no single comprehensive guide to the identification of invertebrate animals from this coast. Nor is there even a full check list of the species that are likely to be encountered here. An enormous and widely scattered literature exists that deals with these animals, but no full regional guide to this literature. At the same time there is growing interest in the sea and an expanding base for the study of it. A relatively small part of this new work is concerned with the systematics of invertebrates, and progress in revising and monographing individual groups proceeds at a slow pace. The mechanics for applying names to these animals lies mainly in the knowledge of experts whose time for routine identifications and assistance is necessarily limited. This book is offered in the hope that it may prove useful to nonexperts.

The composition of a text of this sort is primarily a job of selection and compromise. At the start there is the choice between a single author and a committee of experts. The lone worker of necessity becomes a jack-of-all-phyla with considerably less immunity to error than the specialist. On the positive side this less specialized experience is advantageous in favoring an approach that assumes no special knowledge on the part of the reader. There is substantial advantage, too, in having to reach a working accord with fewer points of view in matters of content and style; and there is, indeed, a range of opinion about what a manual of this sort should do and how it should go about doing it.

Quite early it was decided to sacrifice descriptive detail to attain greater scope in species coverage. To restrict coverage to supposedly common species with no indication of the total fauna seemed quite unsatisfactory, particularly where the boundary between "common" and "uncommon" is so uncertain. Compromise in this area includes the elimination of protists and endoparasites, and only restricted coverage is given microscopic forms. I have attempted to give some indication of the occurrence of freshwater species in estuaries, and the existence of distinctive larval stages is acknowledged.

The choice of geographic limits is reasonably natural; to extend coverage south of Hatteras even so far as Beaufort, North Carolina, would increase beyond reason the number of species to be dealt with. The choice of a bathymetric limit is more difficult and is especially aggravated by imperfections in current knowledge. Pelagic, oceanic forms present additional problems in selection. Some deep water or otherwise rarely encountered species are listed but not keyed.

The decision to attempt to include virtually all species recorded from the inner and middle parts of the Continental Shelf proved to be the main determinant of style and format, chiefly in requiring a shorthand treatment of distribution and taxonomic criteria. Consequently the text does not claim to do more than *assist* in making preliminary identifications and to provide a guide through the maze of technical literature. The conscientious student will corroborate or refute his initial findings through reference to these sources.

The first two chapters provide background on the physical and biotic peculiarities of this coast. Chapter 3 is principally a guide to the use of the systematic text and includes the means for the preliminary identification of material to phylum. The chapter also lays the "ground rules" according to which the guide was written, and the student, however well grounded he may be, is urged to read it lest by not doing so he misinform himself of the text's and his own responsibilities in pursuing the goal of making species determinations. The succeeding chapters follow the sequence of formal systematics and in general adhere to a standard format.

The reader should realize that evaluations in systematics are frequently, indeed usually, a matter of personal judgment rather than of hard, provable fact. No attempt can be made here to penetrate the esoteric realm of phylogenetic discourse, and I have not felt obliged to involve the reader in taxonomic controversies. Where a diversity of published opinion exists, I have felt free to accept one or another of divergent views, without comment, and without, I hope, allowing the personal inclinations of a taxonomic "lumper" to affect unduly the selection.

It would be impossible to exaggerate the extent of my indebtedness to specialists in the subjects covered. Every bit of the original manuscript was submitted to one or more specialists for criticism, and I have been greatly aided by their responses, both in the deletion of errors and in the addition of information and opinion. This has been particularly helpful in bringing systematic names and classifications up to date, in refining the check lists, and in making me aware of current literature. The graciousness, patience, and generosity of these readers is gratefully and most humbly acknowledged. I must quickly add that none of the persons whose names are listed in the Acknowledgments saw final script. While they can justly claim credit for substantially improving the book, they are in no way responsible for any errors or inadequacies that remain. The publisher and I welcome, and in fact solicit, contributions aimed at removing such shortcomings as inevitably will be found.

KENNETH L. GOSNER

March, 1971
Newark, New Jersey

Acknowledgments

It is a pleasure to thank the following specialists who have read and commented on the sections indicated.

Chapters 1 and 2: Alfred C. Redfield, Woods Hole, Mass.

Chapter 3: Norman Meinkoth, Swarthmore College; James D. Anderson, Rutgers University.

Chapter 4, Porifera: Willard D. Hartman, Yale University.

Chapter, Cnidaria: Sears Crowell, Indiana University (polyps); Kay W. Petersen, Zoologiske Museum, Copenhagen (polyps and both hydrozoan and scyphozoan medusae); G. F. Gwilliam, Reed College (scyphozoan and hydrozoan medusae); Mary Sears, Woods Hole, Mass. (siphonophores and chondrophores); Cadet Hand, University of California, Berkeley (anemones); Frederick M. Bayer, University of Miami, Institute of Marine Science (octocorals); John W. Wells, Cornell University (corals).

Chapter 6, Ctenophora: G. F. Gwilliam, Reed College.

Chapter 7, Platyhelminthes: Louise Bush, Drew University.

Chapter 8, Rhynchocoela: William E. McCaul, Harrisburg College.

Chapter 9, Aschelminthes: W. D. Hummon, Ohio University (Gastrotricha); W. D. Hope, United States National Museum (Nematoda); Robert Higgins, Smithsonian Institution (Kinorhyncha, Priapula); W. T. Edmondson, University of Washington (Rotifera).

Chapter 10, Entoprocta: J. D. Soule and Dorothy Soule, Allan Hancock Foundation.

Chapter 11, Tardigrada: Robert Higgins, Smithsonian Institution.

Chapter 12, Chaetognatha: George C. Grant, Virginia Institute of Marine Science.

Chapter 13, Bryozoa: Roger Cuffey, Pennsylvania State University; Neil A. Powell, National Museum of Natural Sciences, Ottawa; Frank Maturo, University of Florida; Thomas Schopf, Leigh University.

Chapter 14, Phoronida: Joan Rattenbury Marsden, McGill University, Montreal.

Chapter 15, Brachiopoda: Alwyn Williams, The Queen's University, Belfast.

Chapter 16, Mollusca: William K. Emerson, American Museum of Natural History (Introduction and Scaphopoda); William Old, American Museum of Natural History (Aplacophora and Polyplacophora); Kenneth J. Boss, Museum of Comparative Zoology (shell-less gastropods and pelecypods); Ruth Turner, Museum of Comparative Zoology (gastropods).

Chapter 17, Annelida: Ernst Kirsteuer, American Museum of Natural History (archiannelids); Marian H. Pettibone, United States National Museum (polychaetes); David G. Cook, National Museum of Natural Sciences, Ottawa (oligochaetes); Kenneth H. Mann, Bedford Institute, Dartmouth, Nova Scotia (Hirudinea).

Chapter 18, Sipunculida: Mary Rice, United States National Museum.

Chapter 19, Echiurida: Mary Rice, United States National Museum.

Chapter 20, Arthropoda: Joel W. Hedgpeth, Oregon State University, Marine Science Laboratory (pycnogonids and introduction); Lorus J. Milne and Margery Milne, University of New Hampshire (Merostomata); William B. Muchmore, University of Rochester (pseudoscorpions); Clarence C. Hoff, University of New Mexico (pseudoscorpions); Irwin M. Newell, University of California, Riverside (mites;) Vincent Roth, American Museum of Natural History, Southwestern Research Station (spiders); Hugo Jamnback, New York State Museum (insects). For information regarding estuarine occurrences of arthropods: Ralph E. Crabill, Jr., United States National Museum (Chilopoda, Diplopoda); George F. Edmunds, Jr., University of Utah (Ephemeroptera); Olive S. Flint, United States Department of Agriculture, Systematic Entomology Laboratory (Neuroptera, Megaloptera, Trichoptera); Lenora K. Gloyd, Illinois State Natural History Survey (Odonata); R.L. Hoffman, Radford College (millipeds); W.E. Ricker, Fisheries Research Board of Canada (Plecoptera); Herbert H. Ross, Illinois State Natural History Survey (Trichoptera); J.L. Tuxen, Zoologisk Museum, Copenhagen (Pauroptera, Symphyla).

Chapter 21, Crustacea: Howard L. Sanders, Woods Hole Oceanographic Institution (Mystacocarida and Cephalocarida); Georgiana B. Deevey (introduction, copepods, Branchiura); Edward Ferguson, Jr., Lincoln University of Missouri (ostracoda); E. L. Bousfield, National Museum of Canada, Ottawa (crustacean checklists, Amphipoda); Roland L. Wigley,

United States Bureau of Commercial Fisheries (Mysidacea); Thomas E. Bowman, United States National Museum (Isopoda, Tanaidacea); William D. Burbanck, Emory University (Cumacea); John C. McCain, Oregon State University, Marine Science Laboratory (caprellid amphipods); Eric L. Mills, Dalhousie University, Nova Scotia (Amphipoda); N. S. Jones, Marine Biological Station, Isle of Man (Cumacea); Arnold Ross, San Diego Natural History Society (introduction, Cirripedia); Hilary B. Moore, University of Miami (Euphausiacea); Raymond B. Manning, United States National Museum (Stomatopoda); Janet Haig, Allan Hancock Foundation (Anomura); Marvin L. Wass, Virginia Institute of Marine Science (Anomura); Anthony J. Provenzano, University of Miami, Institute of Marine Science (Anomura); Fenner Chace, Jr., Smithsonian Institution (introduction, Thalassinidea, Caridea); Austin B. Williams, University of North Carolina (Penaeidea, Caridea); Henry Roberts, United States National Museum(Brachyura).

Chapter 22, Echinodermata: H. Barraclough Fell, Museum of Comparative Zoology (all echinoderms); David L. Pawson, United States National Museum (holothurians); Porter M. Kier, United States National Museum (Echinoidea); Lowell Thomas, University of Miami, Institute of Marine Science (Ophiuroidea); Maureen Downey, United States National Museum (Asteroidea); E. H. Grainger, Fisheries Research Board of Canada (Asteroidea).

Chapter 23, Hemichordata: Dixie Lee Ray, University of Washington.

Chapter 24, Chordata: N. J. Berrill, Swarthmore, Pa. (tunicates); H. H. Plough, Amherst College (whole chapter).

Several additional sources of help should be acknowledged. Pursuit of the literature would have been impossible without the kind assistance of the library staff of the American Museum of Natural History and particularly of Miss Mary Wissler. Special thanks are due Dawn Hassinger who most valiantly dealt with my tortured revisions in typing the final script and also made substantial editorial contributions.

And especially I thank my wife, Pamela, who sustained and participated in the somewhat exotic, nocturnal schedule required to produce this manual. In addition she made the supreme sacrifice of reading the entire script with a librarian's keen eye and shared the arduous chores of proofreading and indexing. Finally, my thanks to the staff of John Wiley and their printers for their efforts in making this a book.

K.L.G.

Contents

GUIDE TO IDENTIFICATION OF MARINE AND ESTUARINE INVERTEBRATES

ONE
The Physical Environment

The straight-line distance from Cape Hatteras to the Bay of Fundy is somewhat more than 800 miles—a span of less than 9° of latitude. Biologically this coast is touched by both the polar and the tropical Atlantic, exhibiting a diversity of life that occupies a range of forty geographical degrees on the opposite shore of the same ocean.

This diversity reflects the climatic extremes of the adjacent continent whose harsh temperature contrasts are transmitted to the coastal waters, producing the greatest range in temperature in the world's oceans. In addition, the places where animals live—the habitats and the substrata on which they are developed—are unusually varied. Within the scope of these primary influences, of which temperature is the single most important one, local distributions are chiefly molded by changes in salinity, by varying exposures to light and to the rise and fall of the tide, and by the movement of water in currents and waves.

The most important result of the interaction of these factors is the division of the coast into two distinct regions. The dividing line is at Cape Cod. The regions north and south of this cape differ climatically, physiographically, and hydrographically with consequent effects on the distribution of life in coastal waters. Differences between the regions north and south of Cape Hatteras are chiefly climatic, but zoogeographically this cape is also an important change point. No such climatic boundary exists on the coast of Nova Scotia; the limit of our discussion is drawn at the Bay of Fundy as a matter of convenience.

Physiography

The seaward edge of the continent as it appears on hydrographic charts is bounded by a broad ledge or Continental Shelf that dips gently seaward until at about 70 to 100 fathoms it plunges into the abyss of the deep ocean. At Cape Hatteras the edge of the Continental Shelf is only 20 miles offshore, but northward the submerged shelf widens gradually to about 70 miles off New Jersey and 100 miles off Cape Cod.

South of Cape Cod the inner part of the shelf rises above sea level to form a low Coastal Plain. Here the coastline is formed along most of its length by a band of barrier beaches breached now and again by inlets that admit the sea to shallow, marsh-bordered sounds and lagoons. Between Cape Henry and Cape Cod the plain is dissected by the deep estuaries of the Chesapeake and Delaware bays, New York harbor, and Long Island Sound, which are the drowned lower valleys of more ancient drainage systems. East of the Hudson River, only fragments of a largely submerged continental border are exposed; the farthest exposures are in coastal New Hampshire, according to Johnson (1925).

In terms of biological habitats the outer shore south of Cape Cod consists almost entirely of unconsolidated material—sand beaches on the ocean front and, sublittorally, more sand with varying mixtures of shell, gravel, and mud. The bays and estuaries allow a greater accumulation of silt, and the textural quality of the bay shore littoral ranges from somewhat muddier sand beaches to mud flats and eroding banks of silty peat. Reefs of shell, both dead and in living beds of oysters and mussels, provide an anchorage for algae and sessile invertebrates, and such organisms find additional footholds on submerged structures of rock, wood, and concrete, and on the abundance of lost and discarded junk that litters the shallow sea bottom.

In contrast to the soft landforms of the shore south of Cape Cod, the coastline to the north is famed for its ruggedness; its features are those of a drowned upland of glacier-scoured, rocky hills. The ancient rocks—granites, schists, and gneiss—which at Hatteras lie buried beneath 10,000 feet of unconsolidated sediment here stand against the sea, and offshore thrust up in submerged ledges and pinnacles through a bottom laden with glacial debris. Competing stresses have warped and folded the Paleozoic and older strata, and the present shore owes its character to varying structural alignments and rock textures. From Casco Bay east the main trend of the coast is "across the grain"; this is a shoreline strewn with islands and with valleys deeply penetrated by the sea. Westward the coast runs more nearly parallel to the structure and is by contrast straight and shelterless. The largest embayments, Boston and Narragansett bays, intrude on lowlands of soft, easily weathered Carboniferous rock.

The predominantly rocky quality of the New England coast extends along the north shore of Long Island Sound to the Hudson River, but the outer coast, including Long Island, the peninsula of Cape Cod, and the smaller islands between, mark the terminus of at least two glacial advances. Eastward from Nantucket the shoals and fishing banks follow the path of the glacial moraine beneath the sea. Elsewhere east of the Hudson, heaps and pockets of glacial debris provide the raw material for bay head beaches, thinly bedded river marshes, and for a variety of spits, tombolos, and other classic constructional features. The morainal materials, easily cut by the sea, are deeply eroded in steep, falling cliffs and bluffs, exposing layers of sand, gravel, and unsorted mixtures with cobbles and glacial boulders.

As a consequence of the recurring cycles of glaciation the level of the sea has fluctuated, falling during periods when water was withdrawn and frozen in the glaciers and rising again with their decline. During periods of glacial maxima the sea receded, exposing bottom down to 100 meters (m) below the current level. As recently as 11,000 years ago peat was deposited on Georges Bank at what is now 59 m below sea level. The dating of this deposit is in agreement with other records placing the retreat of the last glacier from southern New England at about 13,000 years before the present.

In New England, interpretations of sea level fluctuation are complicated by changes in the level of the land itself. During the time of glacial maxima the weight of ice depressed the earth's crust hundreds of feet. With the removal of this load the crust recovered, rebounded higher than before, then sank again. At Boston the rebound is calculated at about 290 feet during the period from 14,000 to 6000 years before the present, followed in the next 3000 years by subsidence to nearly the present level. Crustal movements beyond the glacier's edge were minimal but, where the burden of ice was greater, as on the coast of Maine, the fluctuations were presumably even greater than at Boston.

Despite the seemingly inevitable recession of the coast before a rising sea level, constructive effects on unconsolidated coasts may locally override destructive ones; in southern New Jersey, for example, the shore in most places receded from 200 to 1200 ft in an 80-year period beginning in 1839, but locally the beaches built out by similar amounts. Such coastlines are subject to continual change in detail. Inlets and other features shift with the prevailing coastal currents as, for example, East Rockaway Inlet, which migrated 2¾ miles in 35 years. Inlets are conspicuously impermanent if left to their own devices. Of a score of breaches in the Carolina bar with sufficient longevity to have acquired names, only Ocracoke Inlet predates the historic period. Such changes profoundly affect biotic conditions in the coastal lagoons to which these inlets give access.

The physiographic contrast between the tough-grained, rocky, northern part of this coast and the labile, loose-sedimented southern shore is only one element of contrast that divides the region into two dissimilar segments. Another element is climate.

Hydrographic Climate

The coast between Cape Cod and Cape Hatteras is, in effect, sandwiched between two regions of relative thermal stability, one cold, the other warm. This stability is indicated by an annual range of extreme water temperatures on the Maine and Carolina coasts amounting to about 20°F compared to almost 48° at New York.

The Gulf of Maine is a cold water bight, a "backwater of the Atlantic" with a deep central basin nearly sealed off by the shallow fishing banks across its mouth. Its waters are fed by a chill current that drifts down past Nova Scotia and enters the gulf by way of the Eastern Channel and Browns Bank. Circulation within the gulf is counterclockwise, and water exits by way of the Great South Channel and Nantucket Shoals. Its refrigerated condition is sustained by additional hydrographic influences, including the upwelling of cold bottom water inshore, and by climatic effects due to the coast's leeward position in the circulation pattern of continental air masses and their attendant cyclonic storms.

The coast south of Cape Cod is influenced by such storms in winter, and the Gulf Stream is also famed as a "weather breeder" in this season. In the warmer season, cyclogenesis offshore is minimal and the paths of cyclonic storms shift northward out of the area. The diminished winds reduce vertical mixing and encourage thermal stratification. Surface warming trends are intensified by summer weather patterns dominated by the Bermuda High with a sluggish circulation of warm, moist air from the south. Under these conditions, inshore coastal waters between Cape Hatteras and Cape Cod become subtropical in temperature.

In Figure 1.1 the annual thermal regime for surface waters is shown graphically. The seasonal bunching of isotherms at Cape Cod and Cape Hatteras indicates the alternating development of thermal barriers at each cape (see Parr, 1933). Thus in winter the east coast as far south as Hatteras is "open" to the southward migration of cold water species. Such migrations do occur, in species such as the cod, for example, which is fished commercially at least as far south as Cape May. Northern plankters occur on the middle coast as winter species, carried as passive migrants by the southward flow of coastal waters inshore. The pattern of this countercurrent of the Gulf Stream has been characterized as "chiefly a series of great eddies" with a

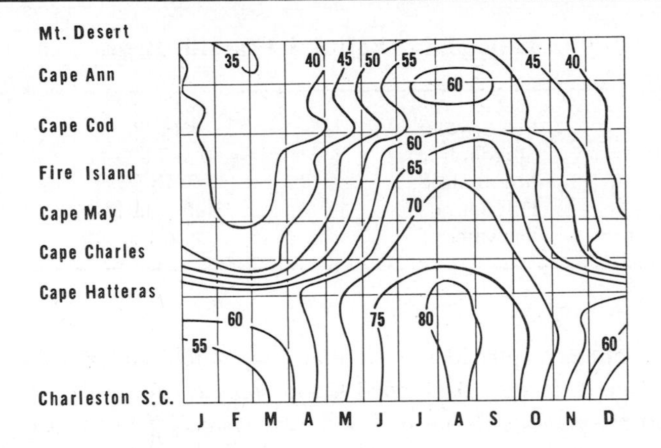

Fig. 1.1. Annual thermal regime for surface waters, redrawn from Parr (1933). Temperature in degrees Fahrenheit.

net drift to the southwest of about 0.5 knot or less. With the coming of warm weather, boreal fish move back into colder water. The plankters presumably perish. Northern, benthic species that might tend to expand their ranges southward in winter are weeded out when the water warms up.

The reverse situation is found in summer. Then warm water species are free to migrate north of Cape Hatteras. Prevailing southerly breezes in summer favor shoreward movements in certain areas—at Cape Hatteras where the Gulf Stream is close inshore and at Cape Cod where the stream swings eastward. The "cold wall" off Cape Cod usually stops such migrations, but under suitable conditions Gulf Stream forms are found in the Gulf of Maine.

Owing mainly to the peculiarities of their thermal regimes, Cape Cod and Cape Hatteras are biogeographical change points setting the broad pattern of distributions on this coast. As such, they operate as more or less selective filters rather than as absolute barriers. For the determination of more local aspects of distribution we must look to additional influences of which salinity and substratum are most important in coastal waters.

Salinity

The substances that make seawater salty are found in nearly constant proportions in water from widely separated parts of the world's oceans, but salinity (the total amount of these salts measured in parts per thousand and symbolized by ‰) is an important variable in the coastal environment. The classification in Table 1.1, based on Hedgpeth's adaptation of earlier

Table 1.1. Salinity Classification

Fresh water	<0.5‰
Oligohaline	0.5–3.0‰
Mesohaline	3.0–16.5‰
Polyhaline	16.5–30.0‰
Seawater	>30.0‰

European schemes, provides a standard to which biological distributions may be referred. Hedgpeth divided the mesohaline into *meiomesohaline*, 3.0 to 8.0‰, and *pleiomesohaline*, 8.0 to 16.5‰; *metahaline* refers to situations where excessive evaporation produces saltier conditions than those usually found in ocean waters, i.e., in excess of about 37 to 40‰. High salinities are associated with salt pans and lagoons on hot, dry coasts, but locally in salt marshes the soil water may reach concentrations at least as high as 42 to 46‰. Extensive salinity variations in our coastal waters generally result from the relatively heavy rainfall on the area east of the Appalachian Mountains and the confinement of diluted seawater in lagoons, estuaries, and inshore waters.

A few plants and animals, termed *euryhaline*, can tolerate a wide salinity range; such adaptability is almost a prerequisite for life in estuaries. Species with limited tolerance are *stenohaline;* strictly oceanic forms that cannot stand even the reduced and varying salt concentrations of neritic waters are an example. The more adaptable freshwater plants and animals range through the oligohaline, and the more tolerant marine species invade polyhaline waters. Endemic "brackish water" types are characteristic of the mesohaline, particularly in concentrations between 7 and 10‰. The total biota is theoretically at a minimum, in terms of numbers of species, at about 7‰ and increases progressively in more and less salty waters.

The average salinity of seawater is usually given as 35‰. Along the main flow lines of the Gulf Stream, salt concentrations usually exceed 36.5‰, but salinities as high as this are exceptional inshore. From the Gulf of Maine to Hatteras the usual values are near 32‰ close to shore, increasing to 34 to 35‰ near the edge of the shelf. Lightship records indicate even lower salinities immediately adjacent to the coast and especially at the mouth of large rivers. The lowest monthly means at the Chesapeake and Ambrose lightship stations were near 27‰ with daily lows of 19 to 22‰.

Estuarine situations bridge the full salinity range from fresh water to the somewhat dilute seawaters just indicated. The influx of fresh waters from river flow and runoff from the adjacent upland are the chief factors promoting dilution. In the north, minimum salinities are usually recorded in

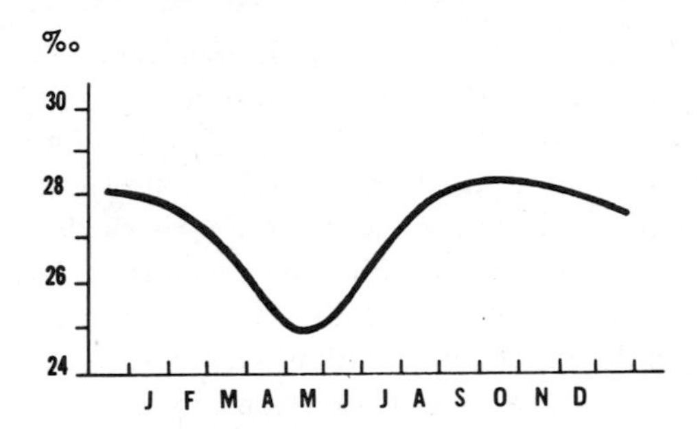

Fig. 1.2. Annual cycle of surface water salinity in Long Island Sound.

April and May when melting snow in the distant uplands fills the rivers, Figure 1.2. Rainfall is fairly evenly distributed throughout the year along much of this coast, but hurricanes and heavier than usual cyclonic storms may bring rapid and wide departures from the normal curve.

Some of the factors favoring a rise in salinity are less obvious. During the winter, ice formation may lead to a rise in salinity, partly by reducing the flow of river waters. In summer, on the other hand, high salinity values often coincide with an excess of evaporation. In southeastern Maryland, for example, summer evaporation is about five times the winter rate. Mixing in the water column by winter storms tends to raise the surface salinity by adding deeper, saltier waters to the more dilute surface layer.

Tides and Currents

Tides are the prime movers in determining the characteristics of the estuarine environment, transporting sediment, mixing fresh and salt waters, renewing nutrients, removing wastes, etc. They also create the intertidal zone with its characteristic biota, and their range is important in fixing the extent of this zone.

Tides on the Atlantic coast are *semidiurnal;* that is, there are two high and two low tides each day. The moon is the chief controlling influence, and since the lunar day is longer than the solar day the tides change accordingly. In practice this means that today's tides will peak or dip to their lowest an hour later than yesterday's. Tide tables, published annually by the Coast and Geodetic Survey, give daily predictions of high and low tides at a series of key stations along the coast with corrections for numerous secondary localities.

Quite commonly one of the daily high tides is somewhat higher than the other, and the low tides are also unequal. During the lunar cycle the range is at a maximum at about the time of the full moon and again at new moon; these are the *spring* tides, and they are both higher and lower than the average. The tidal range is smallest during the intervening *neap* tide periods. These pulsations are superimposed on a broader pattern of seasonal change.

Stations separated by very few miles may differ markedly in both time and range of tides. Coastal tides are oscillations engendered by the tides of the deep ocean beyond the Continental Shelf; their range at Bermuda, for example, is 2.6 feet. When these oceanic tidal waves reach the open coast, the range is increased as a result of "reflection" by the coastal barrier. (The term tidal wave here should not be confused with the so-called tidal waves or *tsunamis* that are caused by submarine earthquakes or volcanic eruptions.) The tidal range increases with the width of the Continental Shelf: it is 3.6 feet at Cape Hatteras, 4.1 feet at Atlantic City, and 9.0 feet on the coast of Maine. In large, open embayments the range may increase still more toward the head owing to reflection and resonance. The maximum increase is about fourfold in large embayments where a quarter of the tidal period (about 3 hours) is required for the tidal wave to travel the length of the basin. Long Island Sound fulfills this requirement, and the tidal range increases from 2.0 feet at Montauk Point to 7.3 feet at Glen Cove near the western end. In the Bay of Fundy the range increases from about 9 feet in the Gulf of Maine to 36 feet or more in the upper bay. When large sounds are entered at either end by tidal waves of different range and timing, interference may reduce the range within the sound. This happens in Vineyard and Nantucket sounds, where the range is about 3.0 feet or more at Monomoy Point or Cuttyhunk Island, and drops to about 1.8 feet or less at Woods Hole and Vineyard Haven. In lagoons the range also decreases, the change being relative to the size of the inlet and the area of the lagoon. The greatest reduction is found in large lagoons with restricted inlets. Examples are Great South Bay on Long Island, Barnegat Bay, New Jersey, and Pamlico Sound, North Carolina, where the range diminishes from 3 or 4 feet on the open coast to a foot or less in the lagoon.

Tidal range is also affected by the weather. Offshore winds generally produce tides lower than predicted, and onshore winds produce higher tides. In the shallow, almost tideless lagoons of the middle and southern coast persistent winds may pile up water on the lee shore and cause flooding in areas normally above the tide. Coastal flooding reaches an extreme in the storm surges accompanying hurricanes and powerful extratropical cyclones.

Tides are vertical movements; transverse movements accompanying them are *tidal currents*. Tidal currents, like the tides themselves, vary in intensity and are at their strongest during new and full moon periods. They are also influenced by the weather. Offshore tidal currents are usually rotary; that is, the direction of the current flow shifts around the compass during the tidal cycle, though the force of the current may vary on different headings. Inshore, where the movement of water is directed by channels in bays and sounds, the current reverses, flowing first one way and then the other.

The timing relationship of tides and tidal currents is often complicated. In open embayments where the tide is a standing wave, slack water occurs at high and low water, and maximum currents flow at about half tide. In rivers the maximum current occurs at about high and low water, and slack water at half tide. Annual tables predicting tidal currents are published by the Coast and Geodetic Survey as companion volumes to the tide tables. Maximum currents occur at the entrance to embayments and decrease upstream; for example, in Long Island Sound, currents are generally under 1 knot, but at the eastern end the current through the Race reaches more than 5 knots. Very strong currents also occur in narrow passages separating waters in which the range and time of high water differ substantially. Hydraulic currents of this kind account for the famous currents in Hell Gate between Long Island Sound and the East River in New York, and in the Cape Cod Canal and Woods Hole Passage.

Another aspect of the circulation in estuaries is of special importance ecologically. The principles of this circulation were worked out in detail in the Chesapeake but appear to be of general occurrence on this coast, Figure 1.3. Current measurements show that, although the surface flow in such systems is essentially downstream, the net movement along the bottom is upstream. The bottom current is small but sufficient, for example, to carry plankters 100 miles and more up the Chesapeake in less than 3 weeks. This bottom current also maintains the salinity gradient in such estuaries, balancing the downstream flow of fresh water; thus there is usually an increase of salinity with depth. This circulation is obviously of vital importance in maintaining populations of sessile or sedentary benthic animals, most of which have planktonic dispersive stages.

Having considered very briefly certain aspects of the physical environment important to life in the shallow waters of this coast, we may pass on to an

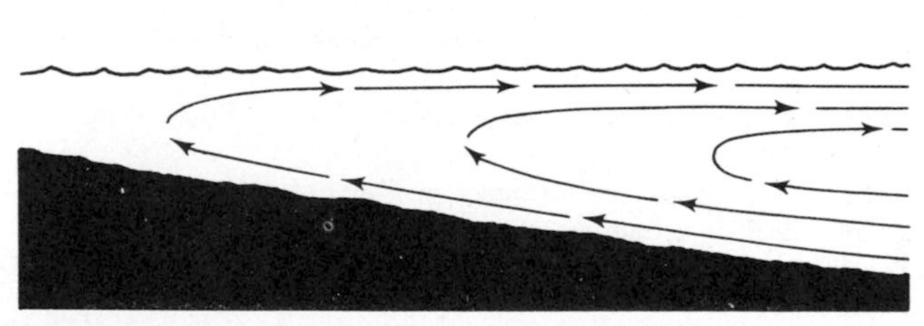

Fig. 1.3. General scheme of net movement of waters in estuaries in the northeastern United States, redrawn from Pritchard (1951). Heavy barred arrow points downstream.

equally general consideration of the effects of these factors on local biotic distributions.

References

Bigelow, H. P. 1928. Physical oceanography of the Gulf of Maine. Bull. Bur. Fish. Wash. Vol. 40.

———, and Mary Sears. 1935. Studies of the waters of the Continental Shelf, Cape Cod to Chesapeake Bay. II. Salinity. Pap. Phys. Oceanogr. Met.

Haight, J. J. 1942. Coastal currents along the Atlantic Coast of the United States. Spec. Pub. 230, Coast and Geodetic Surv. U.S. Dept. Comm. iv + 73 pp.

Haupt, L. M. 1905. Changes along the New Jersey coast. Ann. Rept. State Geol. (N.J.).

Hicks, S. D., and W. Shofnos. 1965. Yearly sea level variations for the United States. Am. Soc. civ. Engrs, J. Hydraul. Div. 91(4468):23–32.

Hedgpeth, J. 1951. The classification of estuarine and brackish waters and the hydrographic climate. Rept. Comm. Treatise mar. Ecol. and Paleoecol. Geol. Soc. Am.

Johnson, D. 1925. The New England Acadian Shoreline. John Wiley and Sons, New York.

Ketchum, B. H., A. C. Redfield, and J. C. Ayers. 1951. The oceanography of the New York Bight. Pap. Phys. Oceanog. Met. 13(1): 2–46.

Lauff, G. H., ed. 1967. Estuaries. Am. Ass. Advancement Sci. Publ. 83. See especially articles by various authors on Basic Considerations, Physical Factors, Geomorphology, and Sediments and Sedimentation.

Parr, A. E. 1933. A geographic-ecological analysis of the seasonal changes in temperature conditions in shallow waters along the Atlantic coast of the United States. Bull. Bingham oceanogr. Coll. 4(3):1–90.

Pritchard, D. W. 1951. The physical hydrography of estuaries and some applications to biological problems. Trans. Sixteenth N. Am. Wild. Conf. 368–376.

———. 1952. The physical structure, circulation and mixing in a coastal plain estuary. Chesapeake Bay Inst. tech. rept. 3:1–56.

———. 1959. Computation of the longitudinal salinity distribution in the Delaware estuary for various degrees of river inflow regulation. Chesapeake Bay Inst. tech. rept. 18:1–72.

———. 1960. Salt balance and exchange rate for Chincoteague Bay. Chesapeake Sci. 1(1): 48–57.

Redfield, A. C. 1953. Interference phenomena in the tides of the Woods Hole Region. Sears Fnd.: J. mar. Res. 12(1):121–140.

———. 1958. The influence of the Continental Shelf on tides of the Atlantic coast of the United States. Sears Fnd. J. mar. Res. 17:432–448.

———, and A. R. Miller. 1957. Water levels accompanying Atlantic coast hurricanes. Met. Monogr. 2(10):1–23.

Riley, G. A. 1956. Oceanography of Long Island Sound II. Physical oceanography Bull. Bingham. oceanogr. Coll. 15:15–46.

———, and S. A. M. Conover. Oceanography of Long Island Sound III. Chemical oceanography Bull. Bingham oceanogr. Coll. 15:47–61.

Wright, H. E., and D. G. Frey, eds. 1965.The Quarternary of the United States. Princeton Univ. Press.

U.S. Naval Oceanogr. Office. 1965. Pub. 700, Oceanographic atlas of the North Atlantic Ocean. Sect. 1. Tides and currents.

TWO
The Biotic Environment

Biotic distributions can be categorized in several ways that stress different aspects of cause and relationship. Thus, in a broad geographical view, we may delineate *regions* such as Arctic, Boreal, and Temperate. This scheme stresses temperature as the controlling factor, and the regions are defined in terms of characteristic assemblages of species. The *biome* concept is more emphatically ecological in stressing biotic communities and habitat. Although particular plants and animals may be used to identify a biome, it is the type of animal or plant and its role in the community that is most important.

Biogeographical Regions

As indicated in Chapter 1, the coastal waters north and south of Cape Cod have markedly different thermal regimes. There are also substantial differences in biotic distribution, sustaining the recognition of two distinct biogeographical regions.

The *American Atlantic Boreal Region* extends from Cape Cod north to the coast of Labrador. It is populated by many species that are typical of the European Boreal Region, which is a biogeographical unit extending from the English Channel to the North Cape of Norway. Sixty to eighty percent of the forms in such unrelated taxa as laminarian seaweeds, mollusks, various crustacean orders, and teleost fish are common to both the eastern and western Atlantic. As might be expected, many boreal species range north into the Arctic, but the fauna of the polar region is sufficiently unique to merit recognition as a separate entity. According to Ekman (1953), nearly half of the genera of arctic fish are endemic; the others are common to the arctic and boreal regions. A relatively few invertebrates with a center of abundance in the high arctic range as far south as the boreal coastal waters covered by this text.

South of Cape Cod the proportion of species shared with northern Europe immediately drops to about 7 or 8%. Such characteristic Boreal species as do range south of the Cape tend to be local or scarce or they are confined to deeper waters. Boreal plankters ranging south of Cape Cod usually appear only in the cooler seasons.

The fauna of the coast from Cape Cod to southern Florida, which constitutes the *American Atlantic Temperate Region*, is essentially transitional to the warmer waters of subtropical and tropical America. It has few endemic species, less than 2% of the decapods for example, and as a zoogeographic region it is perhaps most clearly delimited by the species that fail to enter it. There are, for example, 385 species of crabs in the Atlantic American Warm Water Region according to Ekman (1953); the recent and complete summary of Williams (1965) shows 120 of these species in the southern part of the temperate region. In the northern part of the region about 20% of decapods is derived from the boreal region. The tropical Atlantic is faunistically the most important contributor to the temperate region, but the fauna is considerably impoverished.

The temperate region can be divided into two subregions or provinces of which only the northern or *Virginian Province* concerns us in detail. The dividing point is Cape Hatteras. Using decapods as an example, less than 30% of the fauna of the Carolinian Province ranges northward into the Virginian Province. If the coverage of this manual were extended even as far as Beaufort, North Carolina, the additional species to be considered would be enormously greater.

The categorization of animal distributions as Boreal or Temperate is useful in the present text chiefly as a shorthand notation of geographical range. The ecological distribution of animals in terms of biomes reflects the species response to the environment in a more intimate sense.

Marine Biomes

Major biotic assemblages, such as those forming the tropical rain forest, are called *biomes.* (They may also be called *biome types*, in which case "biome" becomes a subheading with a more specific connotation—such as the African rain forest biome.) Three marine biomes are represented on this coast: Pelagic, Balanoid-Thallophyte, and Pelecypod-Annelid.

Pelagic plants and animals live within the water column; those that are dependent on the Continental Shelf for the completion of some aspect of their life history are termed *neritic*. Such species are tolerant of the environmental vagaries of coastal waters. *Oceanic* organisms, on the other hand, belong to the high seas beyond the shelf. Pelagic animals are either planktonic or nektonic, depending on whether they float freely and more or less passively in the *plankton* or propel themselves actively as members of the *nekton*. The principal distinction is the plankter's inability to alter its distribution in any wide geographic sense; it goes where it is carried. Few invertebrates except squids qualify as nekton. Some, such as prawns and shrimp, nereid worms, portunid crabs, jellyfish, and others swim for short distances but have a limited capacity for migration compared to fishes or marine mammals.

Microscopic plants in the *phytoplankton*, diatoms and flagellates mostly, are at the base of marine food pyramids. The contribution of macroscopic algae and marine seed plants to primary production in the ocean is insignificant by comparison. Herbivores in the *zooplankton*, notably calanoid copepods, graze on these pelagic blooms and are eaten in turn by carnivorous plankters, including comb-jellies, arrow worms, and jellyfish, and by menhaden, mackerel, and other filter-feeding fish. Both plant and animal plankters feed a host of benthic animals including bivalved mollusks, barnacles, bryozoans, tubicolous worms and amphipods, hydroids, sponges, and ascidians.

Nomenclatural distinctions categorize plankters according to spatial distribution and life history. Plankters that live near the bottom belong to the *hypoplankton;* surface forms are *epiplankton*. Some species are planktonic throughout their lives; these typically oceanic forms belong to the *holoplankton*. Members of the *meroplankton* have planktonic and benthic phases in their life history and most have neritic distributions; hydromedusans with alternately free medusoid and sessile polypoid stages and the host of benthic animals with planktonic larvae belong in this category. Finally, the plankton of coastal waters usually contains plants and animals that are not properly planktonic at all but get swept up by turbulence; typical examples of this *tychoplankton* are benthic foraminiferans, harpacticoid copepods, and fragments of delicate filamentous algae and chains of sessile diatoms. Classic subdivisions of the pelagic realm have equivalents in the plankton. The *epiplankton* lives in the euphotic zone, which is usually stated to extent about 80 m below the surface; the *mesoplankton* occupies the disphotic zone, whose base is at about the outer rim of the Continental Shelf; the *bathyplankton* belongs to the aphotic zone of the Continental Slope and Abyss.

Though many plankters are widely distributed or even cosmopolitan, others have restricted tolerances and are particularly useful as hydrographic indicators; for example, neritic or oceanic plankters frequently accompany the intrusion of seawater into coastal sounds and estuaries. Periodic invasions of Gulf Stream water in the Gulf of Maine have been traced by the appearance of normally extralimital plankters.

More predictable changes in plankton composition accompany the cycle of the seasons. Primary production reaches its lowest level in winter when light and temperature values are down; during this period, nutrients, chiefly nitrates and phosphates, accumulate and reach annual maxima. In the absence of phytoplankton the dependent zooplankters are also scarce. A vernal "flowering" period dominated by diatoms usually begins in shallow waters on the southern part of the coast as early as January. Everywhere the initial bloom is commonly dominated by the minute diatom *Skeletonema costatum* with smaller numbers of a few additional genera. The net plankton at this time may reach forty million cells per cubic meter (in Long Island Sound); this is an eightyfold increase over the winter low. Mass estimates for nannoplankton blooms suggest that these ultra small phytoplankters may exceed microplankters 100 to 1000 times.

As the season advances, dominance in the phytoplankton frequently shifts from diatoms to flagellates. While flagellate blooms are usually quantitatively inferior to those of diatoms, the dinoflagellates assume a major ecological role in producing "red tides" that are destructive to fish and other organisms. Occurrences of this sort have been observed on this coast usually in the shallow waters of confined estuaries and usually in warm weather.

Although individual phytoplankters may still outnumber zooplankters two to one, the plankton of late spring and summer is usually more conspicuously zoological. Species composition changes seasonally, but at all seasons copepods are usually the most important zooplankters. These, in turn, sustain predator populations including ctenophores, arrow worms, crustacean larvae, hydromedusae, and larger jellyfish such as *Aurelia*. During the summer the larvae of benthic organisms make a significant contribution to the total zooplankton, attaining values of the order of 50,000 to 100,000 individuals per cubic meter.

The cyclic pattern so briefly outlined above is one of the chief traits of the shallow water plankton on this coast. Frequently the broad pattern is bimodal with renewed peaks of both phyto- and zooplankton in the fall of the year. Virtually all animal phyla are represented in the plankton, and members of Ctenophora and Chaetognatha are exclusively planktonic. The pelagic larval forms of many benthic marine animals are quite different in appearance from their parents and present special problems in identification.

Free-living animals that are not pelagic are *benthic*. They creep, crawl, burrow, or attach themselves to the "bottom," be it sand or rock, wharf piling, or floating buoy. The *Balanoid-Thallophyte Biome* is typical of the rocky coast benthos, while the *Pelecypod-Annelid Biome* occupies unconsolidated sediments. Tidal marshes present some difficulty in categorization, and the term *ecotone* can be applied to them in the sense that they are marginal areas between communities of different biomes, in this case between marine and terrestrial communities and between salt and fresh waters. They are zones of transition and are highly zonated in the same sense as rocky shores. Shallow water benthic situations can also be described as either *littoral* or *sublittoral*. In sheltered places the littoral is essentially equivalent to the intertidal zone. On more exposed coasts, wave action and spray make any precise correlation with tide levels impractical, and the zones are best delineated biologically in terms of typical communities occupying different levels. This definition conforms to the view of Stephenson and Stephenson (1954) and Lewis (1964). Unfortunately there are differences in the use of the word "littoral." An alternative opinion to the one espoused here uses littoral for the whole Continental Shelf or some ill-defined inner part of it.

Subzones in the shelf benthos depend on differences in substratum. Water depth in this comparatively shallow section influences biotic distribution chiefly through its effects on light and temperature; water pressure is probably a relatively minor factor in comparison. The bathymetric range of individual species changes with latitude, and both northern and southern species tend to seek deeper, more stable levels when they intrude on the middle coast.

A conspicuous pattern of littoral zonation is one of the foremost characteristics of the *Balanoid-Thallophyte Biome*. Usually one or a few plants or animals dominate each community and give it a characteristic appearance. The broad scope of environmental variables produces a considerable diversity on different shores.

Typically the sequence of biotic zones on a rocky coast occurs as a series of horizontal color bands. Lichens dapple the bare rock of the supralittoral shore with vivid splashes of orange and yellow. Land animals live here—a few insects, land snails, terrestrial mites, isopods, and spiders. Below, in the *littoral fringe*, the rocks are greenish black with a smooth, paper-thin mat of microscopic blue-green algae. The rocks are extraordinarily slippery when wet. Semiaquatic lichens of the genus *Verrucaria* are often present, and equally characteristic of this zone are the periwinkles, *Littorina littorea* and *L. saxatilis*, that abound in shaded places and in humid crevices. In protected coves this "black zone" is only wetted by high spring tides. On the open coast the entire zone may lie above the reach of all but extreme storm tides. In such places, spray provides the small amount of moisture necessary for life.

The vertical limit of barnacles defines the top of the *midlittoral* zone. Typically, *Balanus balanoides* forms a conspicuous white band; locally, a smaller, more delicate, grayish barnacle, *Chthamalus fragilis*, is found with *balanoides* or is segregated at a higher level. On the open coast especially, barnacles may dominate nearly the whole midlittoral, but frequently they share the substratum with the blue mussel, *Mytilus edulis*. In such cases the two may have a patchy or mosaic distribution at the same level or they may be segregated in upper and lower bands.

In many localities the midlittoral is dominated in its lower levels by seaweeds. These are distributed in a characteristic sequence—with numerous local variations. The topmost level is often a fucoid subzone belonging to the rockweeds. At a distance it is golden brown in color. *Fucus vesiculosus* is the most widespread species, but other members of the same genus are common. *F. spiralis* sometimes dominates the highest fucoid level, and another exposure-tolerant species, *F. edentatus*, is either found with *vesiculosus* or segregated at a lower level. Rockweeds range through the littoral and even into the sublittoral, but other seaweeds commonly take over dominance at lower midlittoral levels. The olive-green knotted wrack, *Ascophyllum nodosus*, frequently forms almost pure "stands" just below the rockweeds. Both *Ascophyllum* and *F. veisiculosus* tolerate brackish waters, and both assume a wide range of ecological variant forms.

The lowest level of the midlittoral is covered in many localities by species of red algae, actually they are frequently brownish or purple rather than red. *Chondrus* predominates in some places; elsewhere *Gigartina* forms a tough spongy mat with a dense stubble of coralline algae filling every available interstice. *Rhodymenia* and *Halosaccion* are additional red genera important in this subzone along with the brown algae *Chordaria* and *Leathesia*. These associations are subject to considerable local variation.

The upper limit of the *sublittoral* is exposed only by very low tides. Here the rubbery fronds of laminarian kelp cling tenaciously to rocks, horse mussels, or any other promise of firm anchorage against the pull of the surf. The common species, in some places at least, are *Alaria esculentia* and *Laminaria digitata*. The undergrowth at this level consists primarily of the weeds mentioned in the preceding paragraph.

The seaweeds of the rocky littoral shelter a rich invertebrate fauna that increases in abundance, both of individuals and of variety at lower levels. The only conspicuous animals of the high barnacle-mussel level, aside from the dominants themselves, are dog whelks, *Nucella lapilla*, and the limpet, *Acmaea*. The fucoids and other weeds, which provide both a hiding place and protection from desiccation, harbor numerous small invertebrates. Colman (1940) found a dozen species of copepods with *F. vesiculosus* and thirty with *Ascophyllum*, but mites are the most abundant animals in some of these associations. Amphipods and worms also occur in great profusion. Among the latter are turbellarian flatworms, nemertines, the white, coiled tubes of *Spirorbis*, oligochaetes, and other annelids, plus the ubiquitous nematodes. With these are encrusting bryozoans, hydroids, and small anemones.

An even richer fauna is found at lower levels wherever there is shelter in rubble-strewn gullies and in dark, eroded grottoes that do not quite dry out between tides. Larger animals are found here—the crabs *Cancer* and *Carcinus*, brittle stars, starfish, sea urchins and tunicates, luxurious patches of the green sponge *Halichondria*, and vivid, enamel-like washes of encrusting red algae. The small animals increase, too. In the holdfasts of the kelp, Colman (1940) found 51 kinds of polychaete worms, 12 or more genera of encrusting bryozoans, an equal assortment of amphipods, 24 species of copepods, and a great assortment of small sponges, hydroids, and other forms.

Tide pools present a special situation in that they shelter plants and animals at higher levels than those at which they would otherwise occur. Many of the animals found in tide pools in the lower littoral do not occur elsewhere in the intertidal zone. The highest pools, on the other hand, are often stagnant and relatively poor in plant and animal life; conditions here are too severe—too likely to become too hot or too cold, too fresh with rainwater, or too salty from evaporation.

South of Cape Cod the littoral changes as rock gives way to sand, and higher temperatures inhibit cold water species, driving them into deeper water, if they do not disappear entirely. Nevertheless an impoverished, ever-diminishing boreal fauna is found wherever there are hard substrata along the south shore of Cape Cod and into Long Island Sound. South of New York the fauna of the rock jetties and bulkheads has a more southern flavor. The common barnacle of the northern littoral, *B. balanoides*, reaches its southern limit in the region of Delaware Bay at about the latitude where oysters become important littoral animals.

The general disappearance of northern algae is the most conspicuous change in balanoid-thallophyte assemblages south of Cape Cod; *Fucus vesiculosus* and, to a lesser extent, *Ascophyllum* alone remain of the northern dominants. The common, red algal genera of the lower littoral from New Jersey south are *Gracilaria*, *Agardhiella*, *Dasya*, and *Polysiphona*. On sea beach jetties the lower littoral is usually sterile because the constant agitation of the sand alternately buries the lower parts of the rocks or scours them clean.

The unconsolidated substrata of the region south of Cape Cod support a subdivision of the *Pelecypod-Annelid Biome* whose membership is predominantly southern in affinity; however, this fauna also forms the core of soft-ground communities north of Cape Cod.

Biotic distribution in the pelecypod-annelid biome is determined to a large extent by the textural quality of the substratum, and the application of quantitative techniques of soil analysis to describe bottom samples gives precision to evaluations that previously were largely subjective. This procedure describes sediments in terms of particle size, and the proportion of different fractions in a sample can be related to a standard nomenclature, Table 2.1.

Table 2.1. Nomenclature; Particle Sizes

	Designation	Size, mm	Sieve Opening Passes
Gravel	Boulders	>500	—
	Cobbles	25–500	—
	Pebbles	10–25	—
	Fine gravel	2–10	—
Sand	Very coarse sand	1–2	1.19
	Coarse sand	0.5–1.0	—
	Medium sand	0.250–0.500	0.50
	Fine sand	0.100–0.250	0.25
	Very fine sand	0.050–0.100	0.105
Silt	Coarse silt	0.020–0.050	0.046
	Medium silt	0.005–0.020	—
	Fine silt	0.002–0.005	—
Clay	Clay	0.002	—

Most substrata contain a mixture of particle sizes, and the system of nomenclature diagrammed in Fig. 2.1 is based on the three "dimensions" of silt, clay, and sand; a pyramidal scheme can be devised to include gravel.

Differences in bottom texture lead to a segregation of animal feeding types. Coarse sediments, composed mainly of sand or sand with gravel or shell fragments, shelter filter feeders that sort planktonic and other organic materials from the water. Bottoms of this sort, charted as "sandy" or "hard," are common on the open coast and in estuaries where tidal currents are sufficiently strong to carry away or inhibit the deposition of finer particles.

Silt and clay particles settle out in quiet waters, producing substrata that are described on charts as "mud" or "soft." The attachment of organic debris to fine particles is a major factor in the distribution of the infauna of soft bottoms, since this organic matter, held mainly by the clay component, forms the principal food of deposit feeders. Excess of organic matter, however, may lead to a deficiency of oxygen and a reduction in the total fauna.

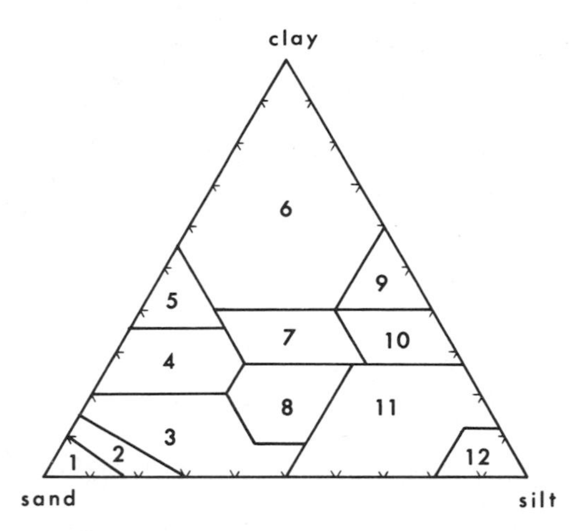

Fig. 2.1. A textural classification of sediments according to percentage composition of silt, clay, and sand. Apices of the triangle are 100% for the components indicated. The classes are: 1, sand; 2, loamy sand; 3, sandy loam; 4, sandy clay loam; 5, sandy clay; 6, clay; 7, clay loam; 8, loam; 9, silty clay; 10, silty clay loam; 11, silt loam; 12, silt.

Such conditions are often indicated by the blackening of subsurface layers, which is taken to indicate the presence of anaerobic bacteria. In Long Island Sound, Sanders (1956) found that bottoms with a silt-clay fraction in excess of 50% had fewer individuals and species than coarser sediments. Substrata with between 25 and 50% silt-clay are likely to have an abundance of deposit feeders. The most productive grounds are those with coarser sediments—less than 25% silt-clay—and a high concentration of suspension feeders.

The definition of discrete communities in the Pelecypod-Annelid biome is complicated by continuous variation in textural combinations and by the different requirements of individual kinds of organisms. More precise studies of the textural qualities of different substrata coupled with inventories of their fauna are needed.

Littoral members of the pelecypod-annelid biome are, for the most part, burrowers. The perpetual instability of surf-beaten sand beaches imposes a Spartan regime, and the resident macrofauna is extremely limited. Most of the animals migrate with the tides. From the seaward side the hippid *Emerita*, and in the south, wedge clams, *Donax*, move up and down the beaches in the swash zone of breaking waves, alternately burrowing and being disinterred. Amphipods of the family Haustoridae are especially characteristic of beach sand habitats. The microfauna of the interstitial waters of beach sand (psammo-littoral) has attracted increasing attention in recent years because of the abundance there of animals previously thought rare; they include various amphipods, gastrotrichs, tardigrades, and platyhelminthes; see Swedmark (1964) and references for the groups mentioned. From the land side, ghost crabs, *Ocypode*, and beach fleas, Talitridae, range down to the water's edge when foraging but have their burrows in the supralittoral. Almost sterile conditions may prevail on shingle beaches—shores covered with large pebbles and cobbles. These fragments are too coarse to permit burrowing, and the tumbling of such stuff by the surf constitutes a mill capable of grinding up the thickest armor. On beaches with mixed sediments a belt of shingle sometimes accumulates at the upper level reached only by storm waves. Beaches entirely of shingle occur in New England, where subtidal deposits of glacial debris provide the raw material.

The Tidal Marsh Ecotone and Estuarine Transition

Estuarine marshes are found along the whole coast. Most New England marshes and those of the middle coast that are developed on rivers and near inlets experience tidal changes in water level. The extensive wetlands that fringe the almost tideless coastal lagoons have a horizontal circulation of water caused by tidal currents, winds, and freshwater runoff, but changes in water level occur primarily in connection with storms or the prolonged press of winds from any one direction.

Marshes are, above all, transitional situations. Their initial formation requires a substratum bare for about half the tidal period; water calm enough to prevent the uprooting of the plants, and a sufficient supply of sediment to enable the upward growth of the marsh to keep pace with or exceed the rise in sea level due to glacier melt. The ecotonal qualities of tidal marshes are indicated by the dynamic sequence of communities that develops in the transition from wetland to dry land and from salt to fresh water. Variation in substratum, drainage, aeration, and tidal cycling in addition to salinity produce a wealth of community subdivisions. The dominant species are usually botanical. The sequence is not obligatory, and the zonation depends on elevation.

In quiet, submerged shallows the eel grass, *Zostera*, and another seed plant, *Ruppia*, take root, grow, and gather round them an ever-increasing accumulation of sediment. When the bottom has been raised to a sufficient elevation, whether through the effects of subtidal plants or otherwise, it becomes suitable for colonization by littoral marsh pioneers. The most common of these is *Spartina alterniflora*, though glasswort, *Salicornia*, or other submersion-tolerant plants may be the first to take hold. Between the submergent eel grass and the first marsh community there is often an expanse of bare sandy or muddy ground. In a continuing sequence of marsh development the first plants maintain the process of sediment gathering and raise the level to an elevation at which other plant associations may succeed them. *S. alterniflora* grows in the upper two-thirds of the littoral, but is supplanted at elevations equal to or exceeding mean high water by high marsh communities. Plants growing at this level tolerate short-term daily submergence but require only a few wettings per month. This higher zone is dominated by short cord grass, *Spartina patens*, or by species of *Distichlis* or *Salicornia* in relatively pure stands or in combinations one with another or with *Sueda* or *Limonium*.

Most of the plants of the littoral are distributed along the whole coast and have circumboreal or, in some cases, almost cosmopolitan distributions. The vegetation above the *patens* level is more varied, and geographical differences in community composition are more prominent. Here the flora is more conspicuously American and southern.

The salt to fresh water transition characteristically passes from salt marsh to either reed swamp, with almost pure stands of the tall grass *Phragmites*, or to cattail swamps with a somewhat more diverse flora, or to associations dominated by rushes (*Juncus*) or sedges (*Scirpus* and others). The marshes of the upper Bay of Fundy more often merge with freshwater bogs.

Tidal marshes contain a diverse fauna. The eelgrass community has attracted special attention because it is an extraordinarily rich assemblage and also because of the periodic "epidemics" that attack *Zostera* and decimate its associates. Allee (1923b) listed 140 animal species in eleven phyla as members of this community at Woods Hole; of these species 55 were considered "characteristic"; and a third of the fauna disappeared after the epidemic that began in 1931. Animals are associated with *Zostera* in much the same way that they occur with algal communities on rocky coasts: they attach themselves to the plant, hide among its stems and foliage, or dwell in the holdfasts, roots, or substratum below. The attached forms in Allee's list include 2 sponges, 14 hydroids, and 8 bryozoans. Among the swimmers and crawlers are several amphipods, the ubiquitous shore shrimp *Palaemonetes* and *Crangon* plus an eel grass "endemic," *Hippolyte zostericola*, the snails *Bittium alternatum*, *Littorina littorea*, and *Lacuna vincta*. In the long list of forms inhabiting the substratum the six most abundant were annelid worms, plus the bivalve *Tellina agilis*.

Most of the invertebrates of the *Zostera* community actually are common on soft-ground habitats generally, but the population density is probably greater in the presence of eel grass. *Ruppia*, which is also a spermatophyte, provides a cover somewhat equivalent to that of *Zostera* and also harbors a fauna rich in species and individuals. As of 1954 the recovery of eelgrass in different sections varied from "good" to "poor"; in 1964 its recovery was still incomplete.

A typical assemblage at the lower *alterniflora* level illustrates the ecotonal qualities of the marsh as a transition between hard- and soft-ground biomes. The ribbed mussel *Modiolus* is a dominant form here providing a base of attachment for estuarine barnacles, bryozoans, anemones, and the fucoids *Ascophyllum* and *Fucus vesiculosus*. From Delaware Bay south the oyster becomes an important member of this community. Throughout its range, which extends north through New England, the *Modiolus* assemblage includes a variety of small invertebrates.

Higher marsh associations harbor periwinkles and a characteristic snail, *Melampus bidentatus*, plus an assortment of salt marsh beetles, dipterans, and other essentially terrestrial arthropods. Fiddler crabs form extensive colonies at the same level on bare flats or in eroding banks adjacent to marsh areas. A terrestrial fauna with crickets, earwigs, oniscoid isopods, termites, and other small arthropods shelters under debris at the edge of the high marsh where there are patches of goldenrod and such xerophytic plants as *Cakile*, *Salsola*, and *Atriplex*. The "marine" contingent is represented here by salt marsh beach fleas, which are intolerant of prolonged or frequent submersion in water.

The transition to fresh water has been studied in connection with the distribution of the intertidal, estuarine isopod *Cyathura polita*, which is most common where the *alterniflora* and cattail marshes meet. The diminution of animals in lower mesohaline situations is demonstrated by the absence of *any* macroscopic animals except *Cyathura* in some localities.

References

Allee, W. C. 1923a. Some physical factors related to the distribution of littoral invertebrates. Biol. Bull. mar. Biol. Lab. Woods Hole 44:205–253.

———. 1923b. The distribution of common littoral invertebrates of the Woods Hole region. Biol. Bull. mar. Biol. Lab. Woods Hole 44:167–191.

Bigelow, H. B. 1928. Plankton of the offshore waters of the Gulf of Maine. Bull. U.S. Bur. Fish. 40:1–509. Doc. 968.

Blum, J. L. 1968. Salt marsh Spartinas and associated algae. Ecol. Monogr. 38:199–221.

Burbanck, W. D., M. E. Pierce, and G. C. Whiteley, Jr. 1956. A Study of the bottom fauna of Rand's Harbor, Massachusetts: an application of the ecotone concept. Ecol. Monogr. 26(3):213–243.

Colman, J. 1940. On the fauna inhabiting intertidal seaweeds. J. mar. biol. Ass. U.K. 24(1):129–183.

Conover, S. A. M. 1956. Oceanography of long Island Sound 1952–1954. IV. Phytoplankton. Bull. Bingham oceanogr. Coll. 15:62–112.

Cronin, L. E., J. C. Daiber, and E. M. Mulbert. 1962. Quantitative seasonal aspects of zooplankton in the Delaware River estuary. Chesapeake Sci. (2):63–93.

Deevey, G. B. 1956. Oceanography of Long Island Sound 1952–1954. V. Zooplankton. Bull. Bingham oceanogr. Coll. 15:113–155.
———. 1960. The zooplankton of the surface waters of the Delaware Bay region. Bull. Bingham oceanogr. Coll. 17:5–52.
Dexter, R. 1944. The bottom community of Ipswich Bay, Massachusetts. Ecol. 25(3):352–359.
———. 1947. The marine communities of a tidal inlet at Cape Ann, Massachusetts, a study in bio-ecology. Ecol. Monogr. 17(3):262–294.
Ekman, S. 1953. Zoogeography of the Sea. Sidgwick and Jackson Ltd. London.
Fish, C. J. 1925. Seasonal distribution of the plankton of the Woods Hole region. Bull. U.S. Bur. Fish. 41:91–179.
Johnson, D. S., and A. F. Skutch. 1928. Littoral vegetation of a headland of Mt. Desert Island, Maine, Ecology 9(3): 188–215, 307–338, 429–448.
Lee, R. E. 1944. A quantitative survey of the invertebrate bottom fauna of Menemsha Bight. Biol. Bull. mar. biol. Lab Woods Hole 86(2):83–97.
Lewis, J. R. 1964. The ecology of rocky shores. English Universities Press Ltd., London.
Miner, R. W. 1950. Field Book of Seashore Life. G. P. Putnam's Sons, New York.
Patten, B. C., R. A. Mulford, and J. E. Warinner. 1963. An annual phytoplankton cycle in the lower Chesapeake Bay. Chesapeake Sci. 4(1):1–20.
Pearse, A. S., H. J. Humm, and G. W. Wharton. 1942. Ecology of sand beaches at Beaufort, N.C. Ecol. Monogr. 12(12):137–190.
Pfitzenmeyer, H. T. 1961. Benthic shoal water invertebrates from tidewaters of Somerset County, Maryland. Chesapeake Sci. 2(1–2):89–94.
Proctor, W. 1933. Biological survey of the Mount Desert region. Part V. Marine fauna.
Redfield, A. C. 1965. Ontogeny of a salt marsh estuary. Science 147(3653):50–55.
Sanders, H. L. 1956. Oceanography of Long Island Sound 1952–1954. X. The biology of marine bottom communities. Bull. Bingham oceanogr. Coll. 15:345–414.
———, D. C. Mangelsdorf, Jr., and G. R. Hampson. 1965. Salinity and faunal distribution in the Pocasset River, Massachusetts. Limn. and Oceanogr. 10 (supplement):219–229.
Southward, A. J. 1967. Life on the Seashore. Harvard Univ. Press.
Stauffer, R. C. 1937. Changes in the invertebrate community of a lagoon after disappearance of the eel grass Ecol. 18(3):427–431.
Stephenson, T. A., and Anne Stephenson. 1954. Life between tide-marks in North America. IIIA. Nova Scotia and Prince Edward Island: description of the region. IIIB. The geographical features of the region. J. Ecol. 42:14–45,46–70.
Stickman, A. P., and L. D. Stringer. 1957. A study of the bottom fauna of Greenwich Bay, Rhode Island, Ecol. 38(1):111–122.
Sumner, F. B., R. C. Osburn, and L. J. Cole. 1911 (1913). A biological survey of the waters of Woods Hole and vicinity. Part 1, Sect. 1. Physical and zoological. Bull. U.S. Bur. Fish. 31:1–442.
Swedmark, B. 1964. The interstitial fauna of marine sand. Biol. Rev. 39:1–42.
Taylor, W. R. 1957. Marine algae of the northeastern coast of North America. Univ. Michigan Press.
Wass, M. L. 1963. Check list of the marine invertebrates of Virginia. Va. Inst. mar. Sci., Spec. sci. rept. 24:1–56.
Williams, A. B. 1965. Marine decapods of the Carolinas. Fishery Bull. 65(1):1–298.

THREE
Systematic Introduction

Method and Organization

The primary goal of this work is to *assist* in making identifications. This section, in effect, establishes "ground rules" for the pursuit of this goal. Only experience with the book can tell where it is to be accepted as providing a final identification and where it only leads in this direction. This experience involves the verification of identifications by comparison with published material or with accurately identified museum specimens.

Of the following chapters, one is devoted to each phylum regardless of size. The only exception is the very large phylum, Arthropoda, with one chapter for Crustacea and another for all other arthropod classes. The basic classification used is that of Rothschild (1965), but I have departed from this source to accept more recent classifications of specialists in individual taxa. In general, I have preferred to follow authors of the *Treatise on Invertebrate Paleontology*, which includes both recent and extinct animals. No attempt is made to involve the reader in theoretical controversy, but he should be aware that differences of opinion regarding nomenclature may be encountered at any level of classification.

A standard format is generally adhered to in each chapter, and each element of this format will be briefly examined. Most accounts begin with a *diagnosis* which is a brief statement, telegraphic in style, of traits that distinguish the taxon from similar or related ones. For the most part the traits are ones that are readily observed; they may not be the most important ones theoretically. The diagnosis and all generalizations in this text apply to the local fauna and may not cover extralimital forms.

The *general descriptive section* that follows amplifies and qualifies the diagnosis. It is a broad statement of the characteristics of the taxon, but always with the problem of species identification in mind. Morphological description is largely confined to that necessary to work the keys and is not intended to fill the needs of a course in invertebrate anatomy. Plain language is used as much as possible, but special terms cannot be escaped entirely. Generally applicable terms are defined in the descriptive section; terms of limited use are defined as needed in the keys. A brief account is given of reproductive habits and life history, mainly as an indication of the possible presence of polymorphism due to sexual dimorphism and ontogenetic change, including metamorphosis through distinctive larval stages. An even briefer statement is given of the scope of ecological adaptation, including mention of the existence of planktonic and parasitic forms that are especially likely to depart from the norm.

A *systematic list and distribution* section follows the general descriptive text. The list is also a phyletic classification, limited where possible to the primary Linnaean categories. The original hope of including *all* forms reported from the Continental Shelf proved somewhat impractical, particularly in requiring the extension of the length and complexity of the keys to include poorly known, deep water forms the general worker has little chance of encountering. Nevertheless, omissions are chiefly from the list of oceanic plankters and benthic species that have been reported only from the outer limits of the Shelf. The extent of these omissions is indicated in general terms at appropriate points in the text. Some of the rarer species are listed but not keyed.

Distribution is given in a conventionalized form. Arctic species, usually occurring as scarce or erratically distributed forms within the range of this text, are indicated as A+. Boreal species generally distributed between Cape Cod and the Bay of Fundy are designated B. When a Boreal species ranges south of Cape Cod, the range is given in one of two ways. B+ indicates a slight extension, for example, into Vineyard, Block Island, or Long Island sound; when the species has a more substantial southward range, the southern limit is given in degrees North Latitude, as B,39°. Virginian species generally distributed between Cape Cod and Cape Hatteras are designated V. The distribution of such species north of Cape Cod is also given in one of two ways. When the range extends into Cape Cod Bay, the distribution is given as V+; for greater extensions the northern limit is given in degrees North Latitude. Southern species that range slightly north of Cape Hatteras are indicated by a C+. Extralimital distributions for local species generally are not given. (See Chapters 1 and 2 for a discussion of biotic distributions.)

Bathymetric range is stated in meters, and literature records in fathoms have been converted directly. In cases where the original data were rounded off or generalized, this practice may lead to a fallacious impression of precision; for example, a literature statement of 10 to 50 fathoms converts to 18 to 91 m. The reader should also realize that published distributional data, however expressed, often tell more about the distribution of collectors than about the animals described. Bathymetric ranges for pelagic species in particular should be viewed with caution, because they are often based on vertical tows with open nets; the species may actually occupy only a limited part of the total water column sampled. "Lit." indicates that the species is littoral or intertidal in the sense of Stephenson or Lewis, as discussed in Chapter 2. A dash indicates that no bathymetric datum was found.

Brackish water occurrences are indicated by appending "eury." to the bathymetric range, sometimes with an indication of the minimum salinity in which the species has been found, given in parts per thousand, ‰. This information is not readily available for all estuarine species.

Sources of check lists and distributional data include the references cited in the systematic sections and chapters. The general faunal reports and lists cited in Chapter 2 were also consulted, but these sources are generally not repeated in the systematic accounts.

When a species has been tentatively identified, the student should look up its distribution in the check list. If the recorded distribution is wildly inconsistent with the history of the specimen being identified, it may be worth running the specimen through the keys again. Such a lack of agreement constitutes grounds for suspicion but is not justification for rejecting a carefully made identification.

Identification is accomplished mainly with keys which have been constructed with this as their only goal. The reader should know that some systematists argue against a purely utilitarian approach in favor of rigorously phylogenetic keys that may serve as a means of taxonomic analysis. This alternative has been rejected, taking heed of the statement by Mayr (1969) that "he who tries to include classification and identification in a single operation is bound to become confused and thwarted in both endeavors." This should not be taken as a guarantee that the keys offered will entirely spare the student exasperation in dealing with the problem of species identification.

The familiar bracket key is generally used, but I have departed at times from a strictly dichotomous arrangement, and the indented form is also used in combination with the bracket format. See Mayr (1969) for a discussion of types of taxonomic keys. Elsewhere, diagnostic traits are presented in tabular form rather than keys. The choices have necessarily been somewhat idiosyncratic. The student inevitably will find himself on the wrong track at times, and backtracking in long keys can be especially frustrating and confusing. The solution offered is to divide what might be handled in one long key into subkeys. This in effect is what happens in a long key, and the points of departure are simply made more obvious by separating the sections entirely. To facilitate scanning, few keys in this volume have more than 12 couplets. Each couplet presents characters in diminishing order of importance, and clues are often included that are not strictly diagnostic. Thus size is seldom a strong diagnostic trait, but it is useful to have some idea of scale.

The illustrations are usually diagrammatic and are intended to clarify and supplement points made in the keys and tables. Several reviewers have indicated a preference for fully rendered representations. Aside from practical considerations making this impossible, there is a valid objection to presenting detailed drawings without an equally detailed descriptive text to aid the reader in evaluation. It is also desirable that the reader approach the problem of identification systematically and not by haphazard browsing.

References complete the systematic accounts. Abbreviations are those used in Butterworth's *World List of Scientific Periodicals* (Fourth Edition, 1963). The references do not constitute an exhaustive bibliography but are a source of additional descriptive detail, distributional data, and bibliographic citations. In some cases, references are given for individual species in the keys, indicated by an author and date set in parentheses and preceding species names in the couplets. This convention is followed especially in large groups with a complex literature. In small groups or where a few publications cover the group, this aid is omitted. It is impractical in any event to give detailed literature references for each species. The author and date of publication of original species descriptions are given in the systematic index, which also serves as a cross reference to the more common synonyms for generic and species names.

General manuals for this coast include those by Miner (1950), Smith *et al.* (1964), Richards (1938), and Arnold (1901). One-volume textbooks such as those of Barnes (1963), Kerkut (1959), Storer and Usinger (1957), Brown *et al.* (1950), and Bullough (1960) are useful introductory manuals but are necessarily more limited than such multivolumed works as Hyman's *The Invertebrates*, the *Traité de Zoologie* edited by Grasse, or the *Treatise on Invertebrate Paleontology* edited by R. C. Moore. (Specific volumes are cited in the systematic chapters.) Jargon is a special problem for the general worker, since familiarity with a term as used in one animal group is no guarantee that it will make any sense when encountered elsewhere. The *Treatise* is especially useful in this connection because special terms are presented in glossary form. Of the several biological dictionaries available, those compiled by Kenneth (1963) and Jaeger (1955) have proved most useful.

Taxonomic catalogs and revisions are points of departure in tracing references and, for later works than such compilations, the most valuable reference aid is the *Zoological Record*. Published annually since 1864, this extremely useful work lists virtually all publications in systematic zoology and is exhaustively cross-indexed to cover aspects other than systematics. The *Zoological Record* is quite complete, but it is usually several years behind. *Biological Abstracts* is more current but is less complete for taxonomic papers.

The systematic literature is scattered through an extraordinary range of journals. Libraries holding individual periodicals can be located through the *Union List of Serials*, and books may be located in the *National Union Catalog*. Most good libraries have these references and can arrange for interlibrary loans or photocopies of literature needed.

For the locations of the taxonomists themselves, see Blackwelder (1961), *American Men of Science*, and similar directories for foreign scientists. It is difficult to keep these publications up to date, however, and a better indication of specialists active in individual groups may be had by examining the current literature.

Preparation of Material for Study

Most of the systematic literature is based, necessarily, on the study of preserved specimens, and methods of preparing such material affect the ease of subsequent analysis. (Specialists, and particularly those working with delicate forms, stress the value of examining live material.)

Fixing and preservation are consecutive steps in the preparation of material for study. In addition, narcotization prior to fixation is required for some animals. The ideal fixing agent should preserve the morphological structure of the animal in as nearly natural a posture as possible. The agent may also prepare the material for staining, sectioning, or other subsequent treatment. A good preservative should maintain fixed material in this condition indefinitely. None of the available fixatives or preservatives fully qualifies as an ideal agent.

The most commonly used general fixative for marine invertebrates is a formalin solution made by diluting one part of commercial formalin (which is itself a 40% solution of formaldehyde) in 8 to 10 parts of water, preferably seawater. Three characteristics of formalin limit its usefulness. It is obnoxious to work with because of the vapor and of irritating effects on the worker's skin. Secondly, formalin is usually acid and should not be used without modification as a fixative or permanent preservative for specimens containing structural elements of calcium carbonate. Formalin can be neutralized temporarily by adding borax or "Boraxo" in the proportion of 1 tablespoon per quart. For long-term use, a buffered, neutral solution can be made by adding about 1 pound of hexamine to 1 gallon of solution. Formalin also has a corrosive effect on iron and should be kept in glass or plastic containers with similar covers. If formalin must be kept in steel drums, Rochelle salt may be added in the proportion of 40 grams (about two heaping tablespoons) to the gallon to prevent dissolved iron from precipitating out of solution.

Formalin, buffered or not, should never be used, even as a fixing agent, for sponges or most ctenophores, since rapid maceration or total destruction of the animal may occur. Formalin is generally a better fixative than alcohol in microscopy, but other reagents such as FAA or Bouin's are preferable. Their formulae are:

Bouin's

75 parts picric acid (saturated aqueous solution made by adding 1 gram of crystals to about 75 cc of water)
25 parts formalin
5 parts glacial acetic acid

FAA

10 parts formalin
50 parts 95% alcohol
2 parts glacial acetic acid
40 parts water

Workers in different groups frequently develop special formulae or techniques, and such preferences are indicated in the "identification" sections of the systematic chapters. General collectors may be obliged to use only one or two reagents in the field, and Table 3.1 may serve as a guide for this purpose.

For permanent preservation the concentration of formalin can usually be reduced to about one-half to two-thirds that used for fixing. In a few cases, for medusae, ctenophores, and some microscopic forms, formalin is actually recommended for permanent preservation; 70% isopropyl or ethyl alcohol are more common general preservatives. Common modifications include the addition of formalin or glycerin in the proportion of 1 part per 19 parts of 70% alcohol. Glycerin is especially recommended because it prevents total dessication if the alcohol evaporates. Glycerin also serves as a clearing agent; it renders specimens more transparent. (The term "clearing" is also used in microscopy as a synonym for dealcoholization.)

Narcotization is used to avoid the extreme distortion, including *autotomy* or self-fragmentation, that fixing agents induce in many animals. Animals that are contracted or have contracted parts must be in an extended condition before narcotization is begun. This is frequently a tedious business, particularly when there is no guarantee of success with all material. The variety

Table 3.1.

	Taxon	Fixative	Preservative
	Porifera	Alcohol	Alcohol
N(1)	Cnidaria, except	Formalin	Formalin or alcohol
	Octocorallia	Alcohol	Alcohol or dry
	Scleractinia	Alcohol	Alcohol or dry
	Ctenophora	Special	Formalin
N	Platyhelminthes	Special	Formalin
N	Rhynchocoela	Special	Alcohol
	Aschelminthes		
N	Rotifera	Formalin, special	Formalin
N	Gastrotricha	Formalin, special	Formalin
	Kinorhyncha	Formalin, special	Alcohol
	Priapulida	Alcohol, special	Alcohol
	Nematoda	Formalin	Formalin
N(2)	Entoprocta	Formalin	Alcohol or dry
	Tardigrada	Formalin	Formalin or alcohol
	Chaetognatha	Formalin	Formalin or alcohol
N(2)	Bryozoa	Formalin	Alcohol or dry
	Phoronida	Formalin	Formalin or alcohol
	Brachiopoda	Formalin	Alcohol
N(3)	Mollusca	Formalin	Alcohol
N	Annelida	Formalin	Formalin or alcohol
N	Sipunculida	Formalin	Formalin or alcohol
N	Echiurida	Formalin	Formalin or alcohol
	Arthropoda	Formalin	Alcohol
N(4)	Echinodermata	Formalin	Alcohol or dry
	Hemichordata	Formalin	Formalin or alcohol
N(5)	Chordata	Formalin	Formalin or alcohol

N = narcotization recommended, (1) for polypoid forms, (2) for extended zooids, (3) for naked gastropods, (4) for holothurians and brittle stars. Formalin is diluted in the proportion 1:8–10 in water for fixing and about 1:19 for permanent preservation. Alcohol for fixing is 90–95%, and for preservation is 70–75%. See text.

of opinion and individual preferences for relaxants and procedures gives some indication of the difficulties involved. Patience is one of the chief ingredients in most formulae. Chloretone, menthol crystals, clove oil, MS-222 (Sandor Pharmaceuticals, Hanover, N.J.), or an 8% stock solution of magnesium chloride in tap water works well with a wide variety of animals; the usual technique is to add the relaxant to the surface of water containing the animals in the proportion of a few crystals or drops per large finger bowl. Magnesium chloride is especially recommended for Cnidaria, Polychaeta, Echinodermata, and shell-less mollusks. Propylene phenoxytol is used for shelled mollusks and can be obtained from Goldschmidt Chemical Corp., New York, N.Y.; the stock solution is 10% in tap water. For hemichordates, annelids, and their relatives, alcohol is added, drop by drop, until a 10% solution is obtained. Otherwise reagents are usually added at hourly or half hourly intervals over a period of 3 or 4 hours, the container being closed during the intervals.

Narcotization technique must avoid unduly disturbing the subject, but the animal must be gently prodded now and then to determine the effect, if any, of the procedure. When the specimen has become insensitive, it is usually left for an additional period that may equal the time required to induce insensitivity; tunicates, anemones, and annelids are left for periods of 8 to 12 hours. The specimen may then be removed to a fixing agent, or the water siphoned off and a fixative added. Suitably relaxed material can often be obtained by allowing specimens to expire in foul or deoxygenated water or by placing them in a freezer. One of the problems in all these procedures is that of choosing the right moment to add the fixing agent. Too soon, and the whole process is defeated by the animal's response; too late, and decay may have already started. An alternative technique to narcotization that works for some material is to flood the live animal with hot (50 to 60°C) FAA, Bouin's, or corrosive sublimate.

Knudsen (1966) has reviewed the literature on these procedures and lists numerous references for methodology; see also Edmondson (1959) and Pennak (1953). Emerson and Ross (1965) have reviewed procedures for curating invertebrate collections.

The Practice of Systematics

The systematist's primary chore is to assemble specimen materials from which he can name and describe taxa. In the process he must determine criteria that define and differentiate related or similar forms, and he must then arrange these forms in a sound classification. The basic Linnaean hierarchy of categories or taxa is:

Kingdom
Phylum
Class
Order
Family
Genus
Species

Additional categories, such as suborder, tribe, cohort, division, appear in the classification of the larger phyla. An attempt has been made to standardize these secondary categories (see Mayr, 1969), but their usage varies, especially in the older literature. In this text the higher categories of classification chiefly serve a utilitarian function in facilitating the presentation of material, and categorization is generally limited to the basic Linnaean series.

For our purposes the species is the most important category in the Linnaean hierarchy. By definition, biological species are "groups of interbreeding natural populations that are reproductively isolated from other such groups" (Mayr, 1969). Taxonomists have traditionally relied almost exclusively on morphological criteria in defining individual species, but the method of the "new systematics" attempts to integrate all aspects of the organism's biology in making this evaluation. The modern approach is much concerned with causal interpretations. With sufficient data the limits of most species can be determined with some assurance. Unfortunately, but through practical necessity, much taxonomic work is still based on the assumption of reproductive isolation as evidenced by morphological discontinuity usually without supporting evidence from other aspects of biology. As a result, taxonomic decisions, even at the species level, may be quite arbitrary, and their validity depends on the judgment of individual taxonomists. (It is considerably more difficult to obtain a consensus on the definition of categories above the species level.)

Taxonomic judgment is influenced by the philosophical approach of the individual student. Thus the taxonomist may be a "splitter" who recognizes minor shades of difference or relationship by formally naming them. The "lumper" is inclined to regard such procedures as essentially typological and scientifically medieval. Practically, the complexity of the splitter's taxonomy tends to obscure relationship rather than clarify it, and the multiplicity of names adds to the befuddlement of the nontaxonomist. Mayr (1969) notes that the classification of individual groups tends to follow a cycle in which taxa are split ever more precisely until the taxonomy becomes overburdened with names; then a lumping reaction is initiated. Invertebrate taxonomy appears to be in an actively splitting phase, as evidenced by the tendency to elevate taxa to higher levels. These activities result in name changes, all of which become a permanent burden on the systematic literature.

One of the main sources of contention among taxonomists and a major practical concern for users of the nomenclature is the chronic instability of species names. Quite legitimate name changes result from revisions in the conception of individual species, such as when named species prove to be mere variants in a natural interbreeding population. Conversely, what was once assumed to be a single species may on close study turn out to include morphologically similar but reproductively isolated populations that qualify as separate species.

Different names for the same taxon are called synonyms. Revisionary texts commonly include synonymies, and such lists are invaluable for anyone attempting to use the literature of species with a complex history. The present manual cannot possibly include complete synonymies, but to ignore the problem completely is also undesirable. Therefore common synonyms for names in current use are cross-indexed in the systematic index.

A further bit of information that is given only in the systematic index is the author and date of publication of current species names. In systematic works this information is usually appended to the species name and is considered by some systematists to be an integral part of the name. By convention, if the generic name has been ammended, the author's name is put in parentheses. Thus the name *Leiochone dispar* (Verrill) 1873 indicates that the species *dispar* was described by Verrill in 1873. The original description placed *dispar* in the genus *Nichomache*, but someone (it could have been Verrill himself) later moved *dispar* to *Leiochone*. This convention does not tell us when the revision was made, and it says nothing about the validity of *Nichomache* as a generic name for other species. The convention above is followed in most (but not all) modern texts, but cannot be relied on in 19th century literature.

The naming of species is controlled by a complex set of rules fully expounded in the *International Code of Zoological Nomenclature*. The code governs almost every conceivable aspect of the subject, but as in most legalistic productions there is ample scope for contention. The International Commission on Zoological Nomenclature is chiefly concerned with refereeing disputes in the interpretation of the code. Somewhat aside from this function is the following recommendation which gives an amusing insight into the activities of systematists (as quoted by Smith *et al.*, 1964):

OPINIONS AND DECLARATIONS RENDERED
BY THE INTERNATIONAL COMMISSION
ON ZOOLOGICAL NOMENCLATURE

Volume I. Part 4. Pp. 23–30.
1943

DECLARATION 4

On the need for avoiding intemperate language
in discussions on zoological nomenclature

DECLARATION.—In the opinion of the Commission the tendency to enter into public polemics over matters which educated and refined professional gentlemen might so easily settle in refined and diplomatic correspondence is distinctly unfavorable to a settlement of the nomenclatorial cases for which a solution is sought. It may be assumed that the vast majority of zoologists agree with the Commission in desiring results rather than polemics, and the Commission ventures to suggest that results may be obtained more easily by the utmost consideration for the usual rules of courtesy when discussing the views of others.

The dangers attending the use of sarcasm and intemperate language in discussions on zoological nomenclature were specially considered by the International Commission on Zoological Nomenclature at their Session held at Monaco in March 1913 during the Ninth International Congress of Zoology. The Commission considered that this question was sufficiently pressing to require special treatment in their report to the Congress. In framing that report the Commission accordingly devoted paragraphs (68) and (69) to this subject.

2. Paragraph (68) of that report reads as follows:

(68) Intemperate Language.—Whether or not it be an actual fact, appearances to that effect exist that if one author changes or corrects the names used by another writer, the latter seems inclined to take the change as a personal offense. The explanation of this fact (or appearance, as the case may be) is not entirely clear. If one person corrects the grammar of another, this action seems to be interpreted as a criticism upon the good breeding or education of the latter person. Nomenclature has been called "the grammar of science" and possibly there is some inborn feeling that changes in nomenclature involve a reflection upon one's education, culture and breeding. Too frequently there follows a discussion in which one or the other author so far departs from the paths of diplomatic discussion, that he seems to give more or less foundation to the view that there is something in his culture subject to criticism. It is with distinct regret that the Commission notices the tendency to sarcasm and intemperate language so noticeable in discussions which should be not only of the most friendly nature, especially since a thorough mutual understanding is so valuable to an agreement, but which are complicated and rendered more difficult of results by every little departure from those methods adopted by professional gentlemen.

As a final comment on the practice of systematics, we may note the tendency of some nontaxonomists to regard the need for taxonomic nomenclature as obsolete—an extraordinary notion actually espoused by some biologists. For many types of investigation, particularly in ecological and behavioral fields, accurate species identification is fundamental. There are numerous examples in the experimental literature where inattention to this truth has rendered otherwise sound work virtually worthless. The taxonomy of most marine invertebrate groups is in a relatively primitive state compared with the refined studies that have been carried out by students of terrestrial vertebrates and of a few groups of terrestrial invertebrates. Much work, indeed, remains to be done in this field.

Introduction to Invertebrate Phyla

Faced with the "mixed bag" of a general collecting expedition, the student can easily find himself in possession of representatives of a dozen or more phyla. The first step in attempting to identify individual species is to sort the material to higher taxa. The essence of species identification is the tracking of a specimen down through the hierarchy of Linnaean categories.

The phyletic position of such familiar forms as starfish, annelid worms, bivalve mollusks, and crabs may be immediately obvious, and the student may proceed via the index to the appropriate parts of the text without bothering with preliminary keys. There is some danger in this technique, however. In each large taxon there are species or groups of species that do not conform to "type"; this often means that they simulate, superficially at least, the "type" of some other taxon. For this reason many of the keys in this book are "artificial," and they group species that are superficially similar but not closely related phylogenetically.

Many of the theoretically important criteria that distinguish higher taxa are unfortunately not very useful for evaluating specimens in the hand. A useful synopsis of the formal criteria separating the phyla is given in Hyman (1940, pp. 32–34). The following artificial keys are intended to lead the reader directly to sections of the text in which species identifications can be made. The introductory keys are developed in some detail in an attempt to cover atypical as well as typical forms. Animals are grouped in three categories; larval forms are considered in an addendum.

A. *Sessile forms* in which the animal is so tenacious of its hold on the substratum that it usually can be removed only with difficulty and frequently not without damage to the specimen. The category includes related immotile species that are anchored in or simply lie loose on a soft substratum. (See Table 3.2.)

B. *Wormlike forms* including not only elongate, cylindrical, thread- or ribbonlike animals but also shorter, stouter, cucumberlike or peanut-shaped forms, soft discoid or filmlike types, and others that most students are accustomed to accept as "wormlike."

C. *All others* including shelled forms and animals with a more or less rigid exoskeleton, as well as benthic and pelagic animals that are not wormlike or sessile as broadly defined above.

Sessile Forms: Invertebrates in several phyla are "plantlike" to the extent of being firmly fixed to a substratum and in developing branching and sometimes distinctly bushy growth forms. Others included here are closely encrusting animals that develop as films, networks, or wandering creeperlike *stolons*. And still others grow in more massive encrustations that are thicker and may have erect lobes or fanlike projections. The majority of these animals are colonial.

Table 3.2. Habits of Marine Invertebrates

Taxon	Number of Species in Local Fauna	Sessile	Pelagic	Benthic (Free-Moving)
Porifera	46	+	—	—
Cnidaria				
Hydrozoa	250	+	+	—
Scyphozoa	19	+	+	—
Anthozoa	56	+	Rarely	+ w
Ctenophora	7	—	+	—
Platyhelminthes	ca.30 (Turbellaria only)	—	+	+ w
Rhynchocoela	47	—	+	+ w
Aschelminthes				
Rotifera	Numerous	+ m	+ m	+ m
Gastrotricha	30	—	—	+ m
Kinorhyncha	6	—	—	+ m
Priapulida	1	—	—	+ w
Nematoda	Numerous	—	—	+ w, usually m
Nematomorpha	1	—	—	+ w
Entoprocta	7	+	—	—
Tardigrada	7	—	—	+ m
Chaetognatha	16	—	+ w	—
Bryozoa	127	+	—	—
Phoronida	1	—	—	+ w
Brachiopoda	4	+	—	—
Mollusca				
Polyplacophora	10	—	—	+ w
Aplachophora	2	—	—	+ w
Gastropoda	ca. 270	+ w	+	+ some w
Scaphopoda	14	—	—	+
Bivalvia	196	+	—	+
Cephalopoda	18	—	+	+
Annelida				
Polychaeta	327	+	+ w	+ w
Myzostomaria	2	—	—	Parasitic
Oligochaeta	ca.30	—	—	+ w
Hirudinea	14 genera	—	+ w	+ w
Sipunculida	7	—	—	+ w
Echiurida	4	—	—	+ w
Arthropoda				
Merostomata	1	—	+	+
Arachnida	Numerous	—	+	+
Pantopoda	16	—	—	+
Crustacea				
Cephalocarida	1	—	—	+
Branchiopoda	ca.12	—	+ m	—

(Continued)

Table 3.2. (*Continued*)

Taxon	Number of Species in Local Fauna	Sessile	Pelagic	Benthic (Free-Moving)
Ostracoda	Numerous	—	+ m	+ m
Copepoda	ca.400	—	+ m	+ m
Mystacocarida	1	—	—	+ m, w
Branchiura	6	—	+	Parasitic
Cirripedia	ca.20	+	—	—
Malacostraca	372	—	+	+
Insecta	Numerous	—	+	+ some w
Echinodermata				
Crinoidea	3	+	+	+
Holothuroidea	21	—	—	+ w
Echinoidea	9	—	—	+
Asteroidea	42	—	—	+
Ophiuroidea	21	—	—	+
Hemichordata	3	—	—	+ w
Chordata				
Ascidiacea	48	+	—	+ w
Thaliacea	ca.15	—	+	—
Larvacea	ca.7	—	+	—

w = wormlike, m = exclusively microscopic, i.e., smaller than about 2 mm in greatest dimension. Terms are defined in introductory statements to keys. Parasitic forms occur in taxa other than those indicated, but those above are exclusively parasitic.

1. Growing chiefly on snail shells occupied by hermit crabs:

- Minute (about 5 mm tall or less), slender, stalked, contractile, polyplike animals in colonies emerging from a chitinous crust; Fig. 5.3. Cnidaria: Hydrozoa: Hydractiniidae.
- Larger (about 10 mm or more tall), thick, anemonelike animals, solitary or in aggregations of no more than about 18 individuals. Cnidaria: Anthozoa: Zoanthidea or Actiniaria.

1. Not "endemic" on shells occupied by hermit crabs—2.

2. Hard forms armored at least in part with calcareous shells, plates, corallites, or boxlike encasements (calcareous skeletal elements "fizz" when treated with dilute hydrochloric acid)—7.

2. Soft or flexible forms without a calcareous skeleton except sometimes as embedded spicules or deposits—3.

3. Body of individual animal or matrix of colony with embedded spicules or supported by massed calcareous deposits; to determine, remove a block of material a few millimeters square to a microscope slide, add a drop or two of Clorox; minute deposits 0.01–3.0 mm remain after soft parts have dissolved away—4.

3. Without embedded spicules but sometimes coated externally with foreign debris—5.

4. Spongy in structure with internal cavities and fibrous network of spicules; growth form extremely diverse (see Fig. 4.1), spicules siliceous as in Fig. 4.2 or Fig. 4.3 or, if calcareous, then tri- or quadriradiate as in Fig. 4.4. Porifera.

4. Colonial forms consisting of minute zooids embedded in a common matrix; deposits calcareous as in Fig. 5.29 or Fig. 24.3:
- Animals polyplike with a single gut opening at the center of a whorl of 8 tentacles; zooids may be completely retractile in matrix; internal structure of matrix frequently with concentric layers, with a hollow, solid, or chambered core; colony form diverse; Fig. 5.29. Cnidaria: Octocorallia.
- Zooids saclike, without tentacles, frequently in oval, circular, or starlike "systems" with minute incurrent and excurrent openings; internal structure not as above; colony form diverse; Fig. 24.3. Chordata: Ascidiacea.

5. Solitary animals, sometimes aggregated but without structural connections; form polyplike, pedunculate, or saclike:
- Pedunculate forms with peduncle expanded to an urnlike, 8-lobed calyx; Fig. 5.27. Cnidaria: Scyphozoa: Stauromedusae.
- Polyplike forms consisting of a pedal disc, column (a few to many millimeters thick), and oral disc with a simple or complex whorl of tentacles; partitioned internally with longitudinal septa; more or less contractile to soft, hemispherical or amorphous lumps; Fig. 5.30. Cnidaria: Anthozoa: Actiniaria.
- Saclike animals, frequently encrusted with foreign debris; with incurrent and excurrent, more or less contractile siphons and complex internal structure; Figs. 24.4–24.7. Chordata: Ascidiacea.

Note that quite a few hydroids, entoprocts, and naked bryozoans have minute colonies in which single or sparsely branching zooids arise from a stolon or stoloniferous network or holdfast; the zooids are usually minute (<5 mm). These are not interpreted as "solitary animals"; see next part of couplet.

5. Colonial forms or, less often, solitary with individual zooids usually <5 mm:
- Zooids embedded in a gelatinous or rubbery matrix; encrusting or partly erect, more or less amorphous, stoloniferous, pedunculate, lobular, or hemispherical. Bryozoa (Group A), Fig. 13.1, or Chordata: Ascidiacea, Figs. 24.2, 24.3.
- Zooids not embedded in gelatinous or rubbery matrix, colonies more delicate, sometimes solitary, or with individuals or sparsely branching stalks 10 mm or less tall, or more profusely branching in bushy or mossy colonies up to several hundred millimeters in extent, rising from a stolen or from a holdfast—6.

6. Zooids with a single whorl of tentacles; gut with two openings:
- Both gut openings within circlet of tentacles; minute forms, 2–10 mm tall; Fig. 10.1. Entoprocta.
- With mouth inside circlet of tentacles, anus outside; some colonies reach several hundred millimeters in extent; Figs. 13.1, 13.2. Bryozoa (Groups A, B).

6. Zooids with a single or multiple whorls of tentacles or tentacles scattered on hydranth; gut simple with one opening at center of whorl of tentacles; Figs. 5.3–5.16. Cnidaria: Hydrozoa.

7. Colonial forms, the individual animals 5 or 6 mm or less in height or diameter—9.

7. Solitary forms of larger size—8.

8. Stalked forms:

- Stalk scaly or rubbery; body compressed, armored with calcareous plates; Fig. 21.13. Crustacea: Cirripedia.
- Stalk segmented, calcareous; crown with 4–10 arms; oral surface with 5 branching grooves; Fig. 22.2. Echinodermata: Crinoidea.

8. Conical or bivalved forms:

- Shell consisting of closely interlocking plates; opening sealed by two bipartite, plated valves; Fig. 21.14. Crustacea: Cirripedia.
- Inverted cup- or caplike shell closely adherent to substratum but not cemented; Fig. 16.3. Mollusca: Gastropoda (Group 1).
- With two valves, lower valve cemented to substratum or attached by a fleshy peduncle, or by byssus threads. Brachiopoda and Bivalvia. See second couplet of "other forms" key.

9. Colonies consisting of up to about 30 cuplike or short cylindrical corallia 5 or 6 mm in diameter with calcareous septa; united by a thin crust or in a more or less branching form; the live animal anemonelike; Fig. 5.1A. Cnidaria: Anthozoa: Scleractinia.

9. Colonies consisting of recumbent, stoloniferous, lacy, sheetlike, or more massive aggregations of minute zooids, (<0.5 mm) more or less encased in individual calcareous shells that are frequently boxlike, coffin-shaped, or tubular; or colonies of tubular or boxlike individuals in branching or fanlike growths; Figs. 13.3–13.8. Bryozoa (Groups C and D).

Wormlike Forms: As an evolutionary adaptation, *the worm* has obviously been extremely successful. Most of the invertebrate phyla have evolved forms that can be broadly described as wormlike, and they are a likely source of confusion for the novice. Most wormlike animals should be narcotized. They frequently are highly contractile or have diagnostic structures that are retractile. Also, some fragment easily of their own accord or through careless handling.

Omitted from the key are chitons, Fig. 16.1. Mollusca: Polyplacophora.

---

1. Body distinctly segmented, or annulated, or regularly ringed with papillae or spines, or faintly annulated but with a sucker at each end; without tube feet—2. Annulations may be superficial or they may reflect a serial repetition of internal structures. Faint annulations or folds may result from contractions in preservative causing some difficulty of interpretation.

1. Body not segmented, or if faintly annulated then without a sucker at each end, or with tube feet; smooth or with papillae, bristles, or other ornamentation—4.

2. With a sucker at each end, anterior one frequently less strongly developed; Fig. 17.25. Annelida: Hirudinea. Note also that some free-living platyhelminths have adhesive organs, but not as above.

2. Not so—3.

3. With 20–40 rings of papillae or spines:

- With 30–40 rings of papillae; rear end with terminal clusters of contractile appendages, which in life may be greatly elongated but are usually short and thick in preserved material; front end with completely retractile, bulbous, spiny proboscis and short collar; Fig. 9.5. Aschelminthes: Priapulida.
- With about 22 spinous rings; front end with nonretractile proboscis with a ventral gutter; Fig. 19.1. Echiurida.

Note that some nematodes have rings of papillae or spines; these are mostly microforms and have an anterior, terminal mouth surrounded by several rings of short papillae, Fig. 9.7. Nematoda. Also, annelids in the family Sphaerodoridae have rings of capsules and are faintly annulated, Fig. 17.9. These worms do not have a proboscis.

3. Segmentation usually distinct, or not as above; some with dorsal scales or "furry" covering:

- Lateral body appendages (parapodia) unsegmented, simple, or lobed, with chitinous setae; head usually with tentacles, cirri, or other appendages; dorsum sometimes with scales or furry; diverse group with many species; Figs. 17.2–17.23. Annelida: Polychaeta.
- Without parapodia; most segments with setae in four bundles; head end without appendages; Fig. 17.24. Annelida: Oligochaeta.
- Distinctly annulated, or soft and maggotlike; usually with soft prolegs or segmented legs; some with filamentous caudal appendages or a breathing tube; head distinctly developed or not, or retractile; Fig. 20.6. Arthropda: Insecta.

Note that some nematodes, mostly micro-forms, are distinctly annulated; see note in first half of couplet.

4. Body usually more or less sluglike with a solelike creeping foot; head usually with one or more pairs of tentacular appendages; back frequently with a lateral fold (mantle), with soft, clublike dorsal appendages in rows or scattered, or with a cluster of retractile gills posteriorly; with or without an internal shell; Figs. 16.12–16.14. Mollusca: Gastropoda: naked gastropods.

4. Not so—5.

5. Anterior end with a circlet of retractile tentacles or lobes and with body regionated or not, or with a nonretractile proboscis or single long tentacle and with body regionated—6.

5. Anterior end without a proboscis or circlet of tentacles or, if with a retractile proboscis, then with body regions ill-defined or with a short caudal cirrus—7.

6. With a U-shaped *lophophore* with a single row of tentacles along both edges; Fig. 14.1. Phoronida.

6. With a nonretractile proboscis or single tentacle:

- With a stalked proboscis, short collar, and trunk, the trunk sometimes regionated; Fig. 23.1. Hemichordata.
- With a proboscis and trunk, the proboscis with a ventral gutter; Fig. 19.1. Echiurida.
- With a long, coiled tentacle with lateral pinnules; body with middle section consisting of a cephalic lobe and a longer *protomesosome* and long trunk; Fig. 3.1A. Pogonophora.

Members of this phylum are mainly confined to abyssal or Continental Slope depths. Wigley (1963) reported *Siboglinum* in 366 and 567 m off New England.

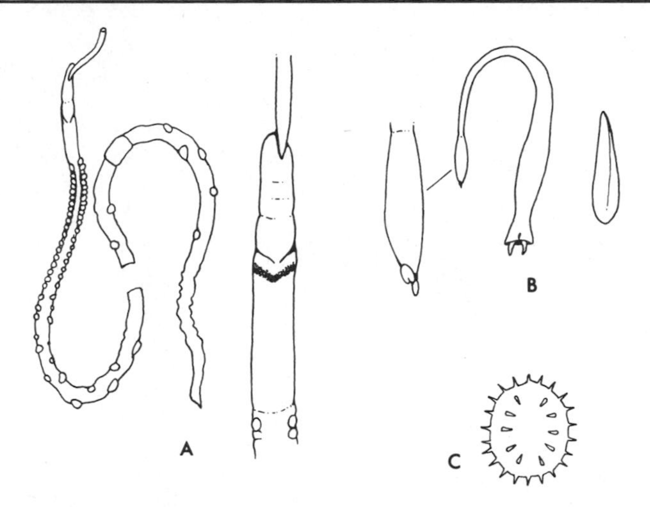

Fig. 3.1. Miscellaneous worm-like forms. A, typical pogonophore, whole animal and detail of anterior part of *Siboglinum longicollum;* B, *Crystallophrisson nitidulum*, an aplacophoran mollusk, whole animal and spicule of *C. vadorum;* C, myzostomarian annelid.

Southward and Brattegard (1968) reviewed western North Atlantic distributions for this group, reporting tubes of *Siboglinum longicollum* and *S. holmei* north of Hatteras in as little as 43 m; tubes of *S. holmei* were also reported in the Hatteras area along with the actual animal for *S. longicollum* and *S. lepida*, all in 70 m. See also Southward and Southward (1967) and Hyman (1959). The phylum is not treated elsewhere in this manual.

6. With a retractile proboscis with terminal mouth more or less surrounded by tentacles or lobes; anus opens dorsally and anteriorly on trunk; Fig. 18.1. Sipunculida.

6. With a retractile circlet of tentacles:

- Skin smooth, pseudo-annulate, or wrinkled, usually with minute, embedded calcareous deposits; with or without tube feet; with anterior mouth and posterior anus; Figs. 22.3–22.4. Echinodermata: Holothuroidea.
- Skin smooth, pseudo-annulate, without embedded calcareous bodies but sometimes with foreign particles adhering; without tube feet but sometimes with adhesive papillae; gut with only one opening, the anterior mouth; Figs. 5.1, 5.31. Cnidaria: Anthozoa: Actiniaria and Ceriantharia.

7. With one or another of the following combinations of characters:

- Somewhat regionated, head spindle-shaped and set off by a constriction, body expanding posteriorly; with a covering of minute spicules increasing in size posteriorly; a pair of short pinnate gills posteriorly; Fig. 3.1B. Mollusca: Aplacophora.
- Disclike with 5 pairs of parapodia with chitinous setae; parasitic on crinoids and basket stars; Fig. 3.1C. Annelida: Myzostomaria.
- With a U-shaped gut with 2 tubular siphons opening near each other. Chordata: Ascidiacea: *Pelonaia*, simple ascidian.

7. Not so—8

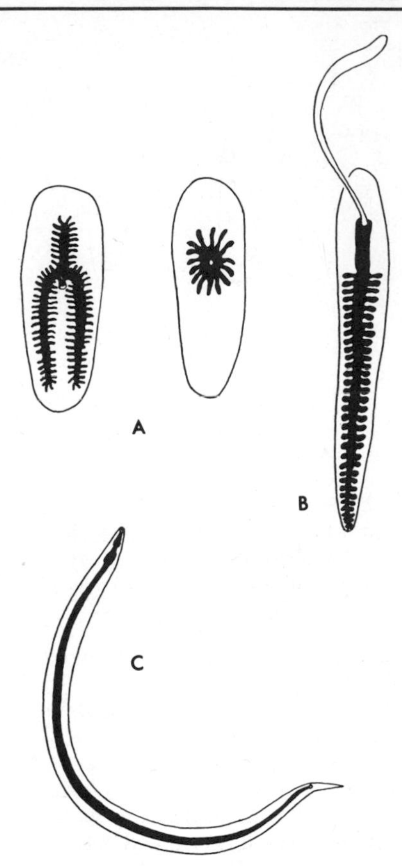

Fig. 3.2. Schematic representation of acoelomate and pseudocoelomate worms with digestive system in solid black: A, triclad (left) and polyclad (right) platyhelminths; B, rhynchocoel; C, nematode.

8. Body usually flat or oval in cross-section; without a coelom:
- With a retractile proboscis opening through mouth or in front of mouth; anus terminal but obscure; Figs. 8.1, 8.2, Fig. 3.2B. Rhynchocoela.
- Without a proboscis except in microforms; gut saclike, or with radiating diverticula, or absent; mouth usually ventral, no anus; Fig. 7.1, Fig. 3.2A. Platyhelminthes.

8. Body usually round in cross-section; pseudocoelomate:
- Mouth and anus usually terminal; mouth surrounded by circlets of papillae; mostly very small forms; Fig. 9.7, Fig. 3.2C. Aschelminthes: Nematoda.
- Mouth parts degenerate in adults; with a double row of bristles laterally; young parasitic in brachyuran or hermit crabs; anus terminal; Fig. 9.6. Aschelminthes: Nematomorphia.

Other Forms.

1. With 1 or 2 calcareous shells, the shell thin and transparent in many pelagic forms—2.
1. With or without a calcareous or chitinous exoskeleton but not as above—3.

2. With two shells:
- Fixed to substratum by a fleshy pedicel that passes through a hole or slot in one valve; with a lophophore fringed with tentacles; shells dorsal and ventral; Fig. 15.1. Brachiopoda. Note that *Crania anomala* is a brachiopod without a pedicel and with one valve cemented to the substratum, while the bivalue *Anomia* has one valve perforated for the passage of a bundle of byssal threads.
- Without pedicel; burrowing forms, or anchored by byssus threads, or shell cemented to substratum; shells lateral; Figs. 16.16–16.26. Mollusca: Bivalvia.

2. With one shell:
- Shell elongate, tusklike, not strongly coiled; Fig. 16.15. Mollusca: Scaphopoda, tusk shells. Note also the shells of some pelagic gastropods, Fig. 16.5.
- Shell distinctly coiled, broadly conical, or cuplike. Mollusca: Gastropoda. Note also the flattened coiled tubes of serpulid worms, Figs. 17.6, 17.19, and the chambered shells or paper-thin "egg shells" of some cephalopod mollusks, Fig. 16.28.

3. Forms with radial symmetry—4.
3. Forms with bilateral symmetry—5.
3. Pelagic colonies consisting of a gelatinous, cartilaginous, or leathery, thick-walled tube, closed at one end with small (ca. 7 mm) zooids embedded in tube wall; Fig. 24.8. Chordata: Thaliacea: Pyrosomida.

4. Pelagic forms; soft and gelatinous:
- Umbrella- or bell-like, usually with marginal or submarginal tentacles; with simple radial symmetry; Figs. 5.2, 5.17–5.24, 5.28. Cnidaria: medusoid forms.
- Spheroidal, saclike, or lobed with 8 rows of comblike ciliary plates; with or without two retractile tentacles; with biradial symmetry; Fig. 6.2. Ctenophora. Note also the atypical, ribbonlike *Cestum*, Fig. 6.1.
- Disclike with a horny, chambered float and suspended, more or less polypoid bodies below; Fig. 5.25. Cnidaria: Hydrozoa: Chondrophora.

Note that fragmented parts of siphonophore colonies may also key out here; Figs. 5.25, 5.26.

4. Benthic forms; with an exoskeleton of calcareous ossicles embedded in skin:
- Oral surface leathery with 5 branching, radial grooves; with 10 slender, pinnate arms; Fig. 22.2. Echinodermata: Crinoidea.
- With a central disc and 5 sharply differentiated arms; tube feet reduced, papillalike, in pores on ventral surface of arms; Figs. 22.10, 22.11. Echinodermata: Ophiuroidea.
- With 5 or more arms merging with central disc; tube feet in 2 or 4 rows in grooves on ventral surface of arms; Figs. 22.6–22.8. Echinodermata: Asteroidea.
- With a globular or disclike test of close-fitting plates, with movable spines; Fig. 22.5. Echinodermata: Echinoidea.

5. With a segmented exoskeleton and jointed legs; body usually regionated; a large and extremely diverse group. Arthropoda.
5. Torpedo- or saclike animals; with 2 well-developed, complex eyes; with 8 or 10 arms with rows of suckers; Fig. 16.27. Mollusca: Cephalopoda.
5. Body spindle- or sluglike with winglike parapodia; pelagic; Fig. 16.12. Mollusca: Gastropoda: Gymnosomata.
5. Barrel-shaped, prismatic, or spindlelike, transparent tests open at both ends; with complete or incomplete hooplike muscle rings; sometimes joined in chains; Figs. 24.9, 24.10. Chordata: Thaliacea: Doliolida and Salpida.
5. With an elongate, finned, chordate tail attached at right angles to a trunk about ¼ to ⅓ as long; whole animal sometimes enclosed in transparent, more or less ovoid "house"; Figs. 24.1, 24.11. Chordata: Larvacea: appendicularians.

Small Plankters and Larval Forms. The most abundant small animals in the plankton, except for short-term "swarms" of special forms, are species of Copepoda. Adult members of other crustacean classes, such as cladocerans and euphausiids, are also planktonic, and larval crustaceans in even greater variety may be encountered. The young of other invertebrate phyla add to the diversity of small zooplankters.

Table 3.3 summarizes developmental data for local marine invertebrate phyla. *Direct development* implies that the hatchling or earliest free-living young is essentially like the adult in form, at least to the extent that its phyletic affinities can be determined with the general keys. *Metamorphic* species pass through one or more developmental stages that bear little if any resemblance to the adult. Most of these free planktonic forms are minute (<0.5 mm to several millimeters) and transparent; their primary function is to bring about the dispersal of species that for the most part are sessile or sedentary when adult. Pelagic larvae may be considered under four main headings: planulae, trochophores, dipleurulae, and crustacean larvae. The list is not exhaustive, and in some groups exceptional larvae are found that do not conform strictly to one or other of these types. The larvae of invertebrate chordates fall outside this list and may be described as superficially tadpole-like (Fig. 24.1A). In most of the larger groups some species have abbreviated or direct development or are otherwise atypical.

The *planula* is the basic larval type of the phylum Cnidaria. Somewhat planulalike larvae have also been reported in sponges, Fig. 3.3B,C. Planulae are more or less cylindrical, multicellular, mouthless, ciliated, and motile, Fig. 3.3A. In some cnidarians, tentacled larvae are produced either as the initial phase or as a transitional stage in metamorphosis to the adult. The most common of these tentacled forms is the *actinula*, Fig. 3.3D, but other types of larvae have been reported in Ceriantharia and Zoantharia.

Table 3.3. Developmental Patterns

Taxon	Development	Principal Larva Types
Porifera	Metamorphic; see text, Chap. 4	Amphiblastula, stereogastrula
Cnidaria	Metamorphic	Planula, actinula
Ctenophora	Direct	—
Platyhelminthes	Direct or metamorphic	Muller's larva and others
Rhynchocoela	Direct or metamorphic	Pilidium and others
Aschelminthes		
Rotifera	Direct or metamorphic	See text, Chap. 9
Gastrotricha	Direct (little known)	—
Kinorhyncha	Direct or metamorphic	See text, Chap. 9
Priapulida	Metamorphic	Rotiferlike
Nematoda	Direct, but see Chap. 9	—
Nematomorpha	Direct	—
Entoprocta	Metamorphic	Trochophore- or rotiferlike
Tardigrada	Direct	—
Chaetognatha	Direct	—
Bryozoa	Metamorphic	Cyphonautes or rotiferlike
Phoronida	Metamorphic	Actinotroch
Branchiopoda	Metamorphic	Trochophorelike
Mollusca	Direct or metamorphic	Trochophore, veliger
Annelida		
Polychaeta	Direct or metamorphic	Trochophore, polytrochula, nectochaeta, mitraria, and others
Myzostomaria	Metamorphic	Trochophore
Oligochaeta	Direct	—
Hirudinea	Direct	—
Sipunculida	Direct or metamorphic	Trochophore
Echiurida	Metamorphic	Trochophore
Arthropoda		
Merostomata	Slightly metamorphic	See Chap. 20
Pantopoda	Slightly metamorphic	See Chap. 20
Arachnida	Slightly metamorphic	See Chap. 20
Insecta	Direct or metamorphic	See Chap. 20
Crustacea	Direct or metamorphic	See Chap. 21
Echinodermata		
Crinoidea	Metamorphic	Auricularia, doliolaria
Holothuroidea	Metamorphic	Auricularia, doliolaria
Echinoidea	Metamorphic	Echinopluteus
Ophiuroidea	Metamorphic	Ophiopluteus
Asteroidea	Metamorphic	Bipinnaria, brachiolaria
Hemichordata	Direct or metamorphic	Tornaria
Chordata		
Ascidiacea	Direct or metamorphic	Larvacealike
Thaliacea	Direct or metamorphic	Larvacea-tadpolelike
Larvacea	Direct	—

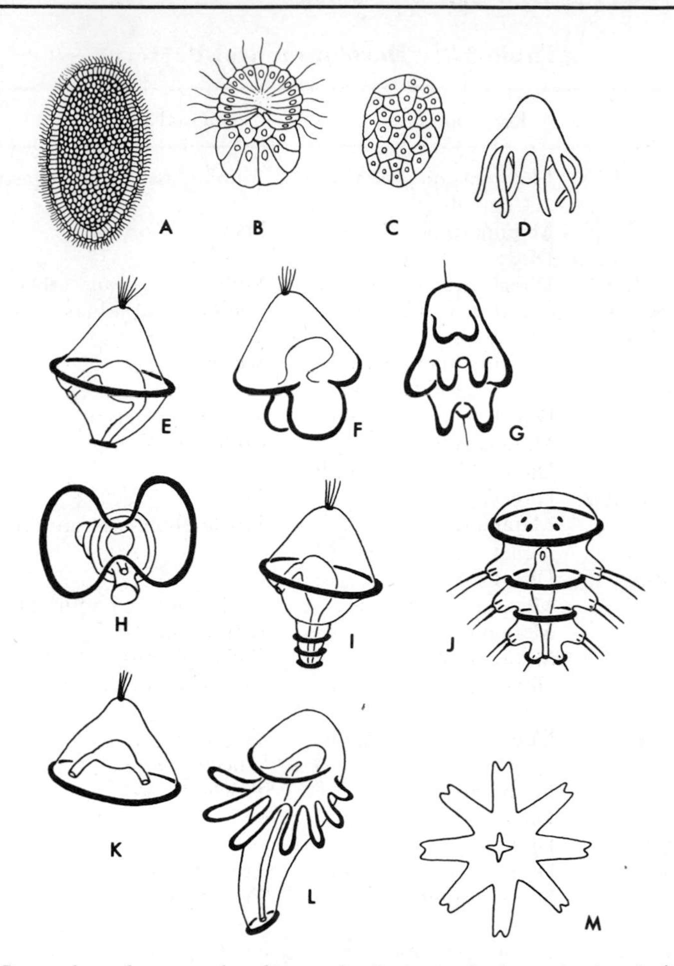

Fig. 3.3. Invertebrate larvae—planulae, trochophores, and ephyra. A, planula (Cnidaria); B. amphiblastula (Porifera); C, stereoblastula (Porifera); D, actinula (Cnidaria); in E–L, ciliary bands shown as a heavy line; E, basic trochophore (Annelida, Mollusca); F, pilidium (Rhynchocoela); G, Muller's larva (Platyhelminthes); H, velliger (Mollusca); I, polytroch (Annelida); J. nectochaeta (Annelida); K, cyphonautes (Bryozoa); L, actinotroch (Phoronida); M, ephyra (Cnidaria).

Many invertebrate phyla have *trochophore* larvae or larvae that can be derived from the trochophore type. Hardy (1956) presents a very lucid resume of these relationships; Fig. 3.3E–L is a diagrammatic representation of trochophore types based on Hardy's sketches and on figures in several volumes of Hyman's *The Invertebrates*. The basic trochophore is a top-shaped or spherical, microscopic larva with a band of cilia around the middle (called the *prototroch*), a smaller circlet of cilia about the anus (*telotroch*), and a tuft of cilia at the opposite pole. The basic form is often modified by flattening or is otherwise distorted, and the prototroch is frequently extended in lobes or armlike projections. The occurrence of trochophore types may be summarized briefly for the principal phyla:

- Rhynchocoela and Platyhelminthes. Trochophorelike larvae in these phyla have a single opening to the digestive track. The basic form may be highly modified. One rhynchocoelan larval type, the *pilidium*, is helmet-shaped with the prototroch drawn down in two "ear flaps" Fig. 3.3F. In metamorphosis the bulk of the pilidium acts as a carrier; the larva becomes encapsuled or separated within the helmet and finally drops out to continue its metamorphosis as a benthic form. A type of platyhelminthine trochophore, called a *Muller's larva*, has the prototroch modified into 8 lobes, and this larva changes gradually into the definitive flat worm Fig. 3.3G.
- A few primitive annelids and mollusks have a simple trochophore larva, but the larva shortly becomes substantially modified. In some annelids the trochophore is a prehatching phase. Molluscan larvae quickly develop a shell or shells, as do the cyphonautes larvae of Bryozoa. Most gastropods are univalved in larval as well as adult stages, but some have a larval shell for flotation in addition to the shell that will be carried through metamorphosis to the adult snaillike form. The molluscan prototroch is expanded as a supporting *velum*, and in gastropods this is usually drawn out in two or more lobes Fig. 3.3H. From this structure the molluscan trochophore-derived larva is called a *veliger*. The anelid trochophore becomes a *polytroch* by the addition of segments at the anal pole, and with transformation to the *nectochaeta* becomes more obviously wormlike; setae are developed early in these larvae Fig. 3.3I,J. A more elaborate trochophorelike modification is found in members of the annelid family Oweniidae; these *mitraria* larvae (not illustrated) also have the prototroch drawn out in lobes, and in metamorphosis the young worm separates from the larval carrier.
- The trochophorelike, bivalved larva of bryozoans is called a *cyphonautes* Fig. 3.3K. This type is often common in coastal plankton. The bryozoan trochophore begins life as a helmet-shaped form, becoming flattened and flanked by two triangular shells.
- As Table 3.3 indicates, several other phyla with few or little-known species on this coast have trochophore larvae. Among sipunculids the larva of *Golfingia* is recognizably trochophorelike. Sipunculid larvae, so far as known, metamorphose without any indication of segmentation; see Hyman (1959). The larvae of *Phoronis* as ultimately elaborated are very distinctive and are called *actinotrochs* Fig. 3.3L.

Hemichordates link the phyla Echinodermata and Chordata in a complex network of evidence. The affinities of Hemichordata and Chordata are discussed by Hyman (1959). Hemichordata and Echinodermata are allied in having distinctive larvae that can be derived from a somewhat hypo-

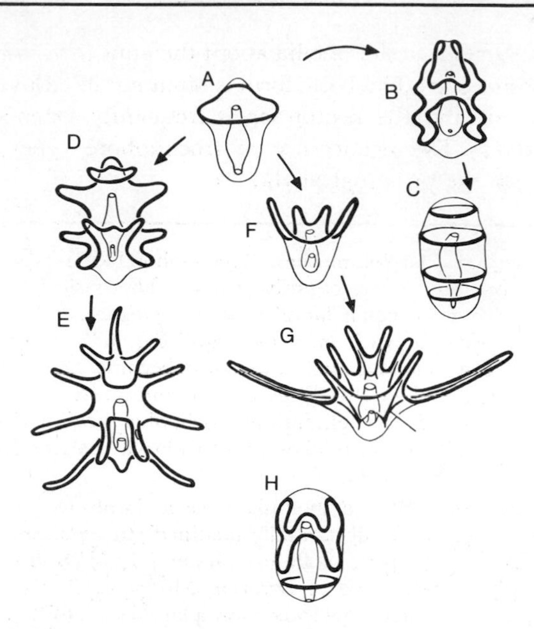

Fig. 3.4. Invertebrate larvae—echinoderm and hemichordate types. Ciliary bands are shown as a heavy line; A, dipleurula; B, auricularia; C, doliolaria; D, bipinnaria; E, brachiolaria; F, early ophiopluteus; G, later ophiopluteus; H, tornaria larva of hemichordate.

thetical *dipleurula* plan (Fig. 3.4). The derived larvae are called *tornaria* and *bipinnaria*, respectively.

The basic dipleurula is a short, sausage- or bean-shaped form with a U-shaped gut and a sinuous band of cilia that passes in front of both the mouth and the anus, i.e., it passes *between* the mouth and anus. (Note that the prototroch of trochophorelike larvae lies on one side of both anus and mouth.) There is also a postoral band of cilia in the dipleurula. From this initial form various larvae, including the bipinnaria of starfish and the tornaria of hemichordates, are developed. In tornaria there is a third band of cilia, the *telotroch*, which is missing in bipinnaria.

In the subsequent development of the tornaria the pattern of ciliary bands regresses; the larva develops a constriction that will become the boundary between the collar and trunk, the body elongates, and the tornaria is gradually transformed into the wormlike hemichordate. The starfish bipinnaria becomes a *brachiolaria* as the ciliary bands develop lobes or arms and a median sucker. The brachiolaria, in turn, assumes a benthic existence and metamorphoses into the definitive starfish. In other echinoderms the dipleurula plan is developed along one of two lines. One of them leads to larval forms called *auricularia* and *doliolaria* found in crinoids and holothurians. The other line leads to several types of pluteus larvae in Ophiuroids and echinoids. See Chapter 22 for a more extended account of echinoderm larvae.

Adult and larval crustaceans are keyed in the introductory part of Chapter 21. In microcrustaceans the segmentation of thorax and abdomen may be suppressed, but the locomotor appendages are segmented in contrast to the ciliated lobes or arms that support and propel other invertebrate larval types.

References

Arnold, Augusta F. 1901 (1968 reprint). The sea-beach at ebb-tide. Dover, New York.
Barnes, H. 1959. Apparatus and methods of Oceanography. Allen and Unwin.
Barnes, R. D. 1963. Invertebrate zoology. W. B. Saunders, Philadelphia.
Blackwelder, R. 1961. Directory of zoological taxonomists of the world.
Brown, F. A., ed. 1950. Selected invertebrate types. John Wiley and Sons, New York.
Bullough, W. S. 1960. Practical invertebrate anatomy. Macmillan, New York.
Dawydoff, C. 1928. Traité d'embryologie comparée des invertébrés. Masson et Cie, Paris.
Edmondson, W. T., ed. 1959. Fresh-water biology. Second ed. John Wiley and Sons.
Emerson, W. K., and A. Ross. 1965. Invertebrate collections: trash or treasure. Curator 8(4): 333–346.
Hardy, A. C. 1956. The open sea, its natural history: the world of plankton. Houghton Mifflin Co.
Hyman, Libbie H. 1940. Invertebrates, Protozoa through Ctenophora. McGraw-Hill Book Company, New York.
———. 1959. The Invertebrates: small coelomate groups. McGraw-Hill.
Jaeger, E. C. 1955. A source-book of biological names and terms. Charles C. Thomas.
Kenneth, J. H. 1963. A dictionary of biological terms. D. Van Nostrand Co., Princeton, N.J.
Kerkut, G. A. 1959. The invertebrata: a manual for the use of students. Cambridge Univ. Press.
Knudsen, J. W. 1966. Biological Techniques: Collecting, Preserving, and Illustrating plants and animals. Harper and Row, New York.
Mayr, E. 1969. Principles of Systematic Zoology. McGraw-Hill.
———. E. G. Linsley, and R. L. Usinger. 1953. Methods and principles of systematic zoology. McGraw-Hill.
Miner, R. W. 1950. Field Book of seashore life. G. P. Putnam's Sons, New York.
Moore, R. C., ed. Treatise on Invertebrate Paleontology. Geol. Soc. Am. and Univ. Kansas. A continuing series of volumes, cited in subsequent sections by the authors of individual parts.
Needham, J. G., P. Galtsoff, F. Lutz, and P. Welch. Culture methods for invertebrate animals. Dover.
Pennak, R. W. 1953. Fresh-water invertebrates of the United States. Ronald Press Co. New York.
Pratt, H. S. 1935. A manual of the common invertebrate animals. McGraw-Hill Book Co.

Richards, H. G. 1938. Animals of the seashore. Bruce Humphries, Inc., Boston.
Rothschild, Lord. 1965. A classification of living animals. Second ed. Longmans.
Smith, R. I., ed. 1964. Keys to marine invertebrates of Woods Hole region. Marine Biological Laboratory, Woods Hole, Mass.
Southward, Eve C., and A. J. Southward. 1967. The distribution of Pogonophora in the Atlantic Ocean. *In* Aspects of Marine Zoology, N. B. Marshall, ed. Symposia of Zool. Soc. Lond. 19.
———. and T. Brattegard. 1968. Pogonophora of the Northwest Atlantic: North Carolina Region. Bull. mar. Sci. U. Miami 18(4):836–875.
Storer, T. I., and R. L. Usinger. 1957. General Zoology. McGraw-Hill Book Co.
Wigley, R. L. 1963. Occurrence of Pogonophora on the New England Continental Slope. Sci. 141:358–359.

FOUR
Phylum Porifera

SPONGES

Diagnosis. *Porous, multicellular animals with no indication of formed organs or organ systems; invariably sessile and ranging from simple encrustations to amorphous masses or more definite forms, e.g., fanlike, cuplike, or branching; a few species tubular with simple radial symmetry. Sponges range in size from a few to many hundred millimeters.*

Members of the same species tend to be similar in form and color, but environmental influences may alter normal developmental inclinations. Figure 4.1 shows the principal growth forms to which descriptions are referred. The presence of zoochlorellae imparts a greenish tinge to the normal body color of many species.

Surface openings are of two types and are under muscular control; in some species the openings are highly contractile, in others, less so. The incurrent pores or *ostia* are small, even microscopic, while the excurrent *oscula* vary in diameter in different species from 1 mm or less to 8 or 9 mm. The oscula may be few or numerous; they may be scattered or regularly placed. Frequently they are raised, the elevations varying from low, conelike mounds or papillae to extended tubes or *fistulae*. The surface of the colony may be slimy; it may be smooth and leathery, rough, shaggy (*hispid*), or raised in low warts or tentlike projections (*conulose*) without conspicuous pore openings. Most sponges have a "skin" or *ectosome*. This may be a relatively thick *cortex*, perhaps a millimeter in thickness, or a microscopically thin *dermis* about $^{1}/_{50}$ mm thick.

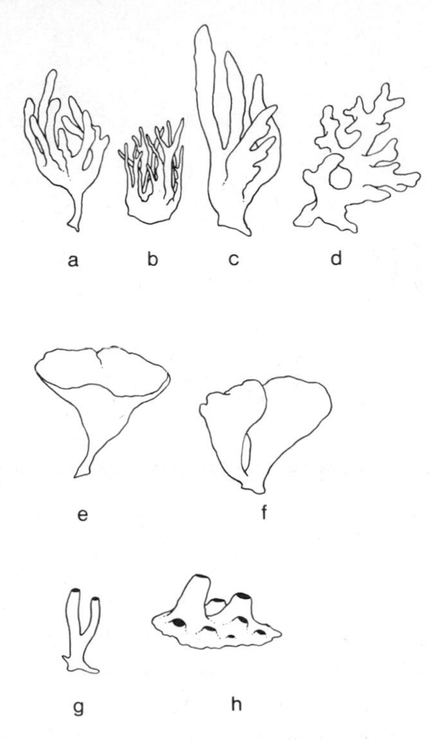

Fig. 4.1. Sponge growth forms. a–d, ramose types; e, vasiform; f, flabellate; g, tubular; h, with tubular oscula.

Most sponges have a "skeleton" of calcareous *spicules*, or of siliceous spicules and/or a protein material, *spongin*. The spicules are variously aligned or fused as support structures; in section, sponges frequently show a more or less open network of hard elements with or without spongin. Structural variations are of taxonomic importance, as are the complements of different spicule types. The latter are classified initially as *megascleres* and *microscleres*. The implied size difference is not absolute.

Microscleres are distributed more or less haphazardly in the gelatinous matrix of the sponge, and there is an evolutionary tendency toward the loss or reduction of spicules, particularly microscleres. In some species these are variably scarce or absent, and the failure to find particular elements in individual sponges has led to much taxonomic confusion. The spicule types of interest to the present account are illustrated in Figures 4.2–4.4; extralimital spicule types including those found in the deep water hexactinellid or glass sponges are omitted. Particular types of spicules are sometimes segregated in endo- or ectosome. Further comments on spicules are reserved for the section on identification.

Sponges reproduce both sexually and asexually. The regenerative powers of poriferans are well known, but growth from accidently displaced fragments is evidently slow; budding and deliberate fragmentation are known, in Calcarea for example. It is a somewhat moot point whether most Desmospongiae are single animals or colonies developed by vegetative growth. Adverse environmental conditions induce the formation of *reduction bodies* in some marine sponges. These may be roughly described as minimal packets of representative cells from which the beginning of a new colony may be reconstituted when favorable conditions return. *Gemmules* provide another reproductive device sometimes equated with periods of stress but evidently occurring as a reproductive process during nonstress periods as well. In some marine forms, gemmules are special cell aggregations that are released as motile, so-called larval forms that subsequently settle and metamorphose into new colonies. Gemmules are especially characteristic of freshwater sponges—Spongillidae—in which case development is direct, without a motile stage. Such nonmotile gemmules are also found among some marine sponges.

The details of sexual reproduction in Porifera are only vaguely understood. Most species are presumed to be hermaphroditic but may not produce male and female gametes simultaneously. In the few forms in which sexual reproduction has been studied, fertilization is internal, the offspring developing into blastuloid, flagellated larvae that are released as dispersive agents. These larvae approximate the *planulae* of cnidarians and also resemble the vegetatively produced motile type of gemmule.

Sponges are filter feeders; they produce through currents by means of the flagellated *choanocytes* or collar cells that line internal cavities. The food presumably consists of minute plankters, detritus, and possibly even nutrients in solution. Zoochlorellae are common in species inhabiting lighted situations. While such algae are not considered vital to the survival of their hosts, it appears that sponges may derive commensal benefits from them equivalent to those conveyed to corals by the presence of zooxanthellae.

Sponges, particularly the larger cavernous types, frequently harbor a variety of invertebrate commensals. Poriferans, in turn, may be found growing on other living animals. Several species, e.g., *Suberites ficus*, *Myxilla incrustans*, *Iophon nigricans*, and *Pellina sitiens*, are particularly associated with bivalve mollusks, barnacles, or, in the case of *Iophon*, the brachiopod *Terebratulina septentrionalis*. *Mycalecarmia ovulum* is found chiefly on hydroids, bushy bryozoans, and algae of similar form.

The best-known association of sponges and living hosts involves the boring sponge *Cliona* and the Virginia oyster *Crassostrea*, although *Cliona* may also utilize other calcareous substrata. Severe infestations have definitely deleterious effects on oysters, in which case the relationship can be described as parasitic rather than commensal. The association of sponges with crabs and other invertebrates may benefit the latter to the extent that most predators find sponges distasteful, either because of a presumably offensive odor or because of their spicules. Among relatively few known predators on sponges are littorinid snails and nudibranchs.

Most sponges require a firm substratum, but they are otherwise widely distributed at all depths. They do not withstand drying and, except for occurrences in rock pools, are poorly represented in the littoral. A single family, Spongillidae, has invaded fresh waters, where its species occur in a wide range of ecological situations; the extent of penetration of freshwater species in estuarine situations on this coast is not well known, but some, e.g., *Spongilla lacustris*, are euryhaline and presumably can be expected. Spongillids are reported to be intolerant of polluted waters. Among marine forms the common red-beard sponge, *Microciona*, survives in polluted estuarine situations, e.g., in the Hudson River estuary in salinities near 15‰. *Lissodendoryx carolinensis* and species of *Halichondria*, *Haliclona*, and *Cliona* penetrate waters in the polyhaline or mesohaline range.

Systematic List and Distribution

Sponges have relatively few structural features amenable to precise description or measurement and, as a result, they present especially difficult problems in taxonomy below the class level. One of the three recognized poriferan classes, the Hexactinella or glass sponges, is excluded from consideration here as being bathymetrically extralimital. The class Calcarea has only a few species that occur within the range of this text; the most recent revisor of the group (Burton 1963), whose treatment is followed here, has lumped most of the names that will be encountered in the local literature under a very few species; this extreme view has met with opposition from other sponge systematists (see Hartman, 1964).

The classification of Desmospongiae follows the very thorough generic survey of deLaubenfels (1936) with modifications from more recent literature; deLaubenfels subsequently revised this classification somewhat (1955), but the later treatment is too incomplete in placing recent genera to be suitable for the present account. The basic distribution list was compiled from deLaubenfels's paper (1949), which includes his detailed review of the literature of this coast; additions to this list were made after reference to Lambe (1896), Proctor (1933), and Burton (1930) for boreal forms and to Wells *et al.* (1960) for Carolinian species. For freshwater sponges see Pennak

(1953) and Jewell (1959). Problems of synonymy are particularly difficult in this phylum. No thorough study of the sponges of the Continental Shelf north of Cape Hatteras has yet been made. Consequently, dredging on the shelf may be expected to turn up additional species.

Class Calcarea (for an alternative arrangement of this class see Hartman, 1958b)		
Family Homocoelidea		
Clathrina coriacea	BV	Lit. to 33 m
Leucosolenia botryoides	B	18 to 117 m
Family Heterocoelidae		
Scypha ciliata	BV	Lit. to 102 m
Class Desmospongiae		
Order Keratosa		
Family Aplysillidae		
Pleraplysilla minchini	C	Lit. to 55 m
Family Halisarcidae		
Halisarca sp.	B	Shallow water
Order Haplosclerida		
Family Spongillidae		
Spongilla lacustris		Cosmopolitan, fresh-brackish water
Meyenia fluviatilis		Cosmopolitan, fresh-brackish water
Family Desmacidonidae		
Isodictya palmata	B	7 to 14 m
Isodictya deichmannae	B	20 to 40 m
Family Haliclonidae		
Haliclona urceola	AB	17 to 20 m
Haliclona palmata	B	Lit. to ?
Haliclona oculata	B	Lit. to 150 m
Haliclona loosanoffi	V	Lit. to ?
Haliclona canaliculata	V	Lit. to?
Callyspongia vaginalis	C+	—
Family Adociidae		
Sigmadocia flagellifer	B	69 to 146 m
Order Poecilosclerida		
Family Tedaniidae		
Tedania suctoria	AB	9 to 1462 m
Lissodendoryx carolinensis	V	Shallow water, eury
Family Microcionidae		
Microciona prolifera	BV	Lit. to 26 m, eury.
Dictyociona adioristica	C+	Shallow water
Family Myxillidae		
Myxilla incrustans	AB+	Lit. to 366 m
Myxilla fimbriata	AB	55 to 146 m
Iophon nigricans	B	29 to 740 m
Family Mycalidae		
Mycale lingua	A+	18 to 160 m
Mycale fibrexilis	V	Shallow water
Mycalecarmia ovulum	A+	6 to 200 m

Order Halichondrida		
Family Halichondridae		
Halichondria panicea	AB	Lit. to 88 m
Halichondria bowerbanki	V	Lit. to 18 m
Pellina sitiens	AB	15 to 183 m
Order Axinellida		
Family Axinellidae		
Axinella sp.	C+	—
Cladocroce ventilabrum	B	64 to 1894 m
Higginsia strigilata	C+	3 to 50 m
Hymeniacidon heliophila	BV	5 to 50 m
Order Hadromerida		
Family Suberitidae		
Suberites ficus	BV	18 to 55 m, on bivalves
Subertechinus hispidus	B	15 to 388 m
Prosuberites sp.	BV	Shallow water
Tentorium semisuberites	AB	37 to 176 m
Polymastia robusta	BV	15 to 55 m
Family Clionidae		
Cliona vastifica	BV	Lit. to 128 m
Cliona truitti	V	Estuarine
Cliona lobata	V	Lit. to 20 m
Cliona celata	BV	Lit. to 40 m
Order Epipolasida		
Family Jaspidae		
Topsentia genitrix	AB	27 to 201 m
Order Choristida		
Family Craniellidae		
Craniella gravida	CV	27 to 1126 m
Family Geodiidae		
Geodia gibberosa	C+	Lit. to 100+ m (North Carolina records are from deep water)

Identification

Barrett and Yonge (1958), in their pocket guide to the British seashore, succinctly evaluate the prospects for field identification of sponges with the statement, "the last shadow of doubt seldom disappears." As previously indicated, most features of external morphology are subject to modification by environmental effects and to changes during growth. Color notes from life are useful, however, and the spicule complements are reasonably constant in size and type, but it must be borne in mind that some spicule types are scarce and that mature spicules are accompanied by developmental forms.

Formalin, even when buffered, is not a satisfactory preservative; 95% alcohol is recommended for fixing with a change to about 75% after a few days.

In the following account the primary emphasis is on spicule complements. This requires use of the higher powers of a compound microscope; microscleres are frequently between 10 and 20 μ in length. (With a combination of 10× eyepiece and 43× objective a spicule 12 μ long extends only $\frac{1}{30}$ of the field diameter.) The preparation of simple spicule examinations is not difficult. Remove a block of sponge several millimeters square and include both endosome and ectosome, or preferably take separate samples of each because the spicule complements may differ. Place the sample on a microscope slide with a drop or so of Clorox. Solution of the soft parts may be accelerated by heating. For a temporary mount dissolve any residue with a drop or two of water, add a cover slip, and examine microscopically. For more elaborate preparations, including the use of stains, see deLaubenfels (1953), Pennak (1953), or Hartman (1964). More detailed study makes use of thin sections.

Spicules are present in all local sponges except two species of doubtful occurrence which are listed but not keyed. They are *Halisarca* sp. and *Pleraplysilla* sp. Another southern species, listed but not keyed, *Geodia gibberosa*, is the only local form in which asters are the characteristic spicule type. Tri- or quadriradiate spicules immediately identify a calcareous sponge in the local fauna, although this type (in siliceous form) is also characteristic of the deep water hexactinellids; among local forms, the desmosponge *Craniella gravida* also contains a type of tetraxonid spicule (triaenes, 100 to 2000μ), which in this case is siliceous rather than calcareous (Fig. 4.4D). *Craniella* also has spiny, caterpillarlike microscleres called spirasters, and oxeas equal in length to the triaenes; the oxeas are the predominant megasclere type in this species.

To identify local species, begin with the following preliminary key. Note that some taxa are keyed in more than one subgroup. Note also the tabular index to color (Table 4.1). Dimensions, other than those for spicules, are approximate maximum colony sizes. A logical procedure is to prepare a spicule sample; make a list of spicule types and their sizes; then proceed with the Keys.

A. Solitary or branching assemblages of sponges of simple tubular form with terminal oscula; spicules calcareous and include tri- or quadriradiates and oxeas; Fig. 4.4A,B. Calcarea.

B. Boring forms; often yellow; spicules siliceous; megascleres are monactinals only (tylostyles 139–440 μ); microscleres are diactinals (microoxeas and spirasters or spirasters only, 16–64 μ); Fig. 4.5. Clionidae.

C. Not as above: sponges of various forms, habits, colors, and spicule complements; spicules siliceous.

- Group One. Megascleres exclusively monactinal, consisting of styles, tylostyles, and acanthostyles, the largest usually >200 μ (range of mature megascleres 60–2600 μ); microscleres absent or including palmate

Table 4.1. Color Index to Common Sponges.

Brown or black sponges		Orange sponges	
Isodictya	One, 3	*Hymeniacidon*	One, 2
Mycale	One, 5	*Pellina*	Three, 2
Yellowish sponges		*Haliclona*	Two, 7
Isodictya	One, 3	Red sponges	
Mycale	One, 5	*Microciona*	One, 5
Cliona	B, One, 10	*Polymastia*	One, 8
Subertechinus	One, 9	*Isodictya*	Two, 9
Suberites	One, 10	Lavender sponges	
Tedania	Two, 2	*Haliclona*	Two, 7
Myxilla	Two, 3		
Cladocroce	Two, 6		
Haliclona	Two, 7		
Lissodendoryx	Two, 9		
Halichondria	Three, 5		
Greenish or olive sponges			
Hymeniacidon	One, 2		
Mycale	One, 5		
Cladocroce	Two, 6		
Lissodendoryx	Two, 9		
Halichondria	Three, 5		

Numbers indicate group key and couplet in which the genera occur.

anisochelae, toxas, sigmas, or microstrongyles, -styles, -oxeas, or -centrostrongyles <60 μ.

- Group Two. Megascleres include both monactinal and diactinal types (range of mature megascleres 140–1500 μ); microscleres absent or as in Group One.
- Group Three. Megascleres exclusively diactinal, consisting of oxeas plus, in a few species, strongyles and centrotylote oxeas (range of mature megascleres 55–900 μ); microscleres absent or with palmate isocheles, or in mesohaline species, birotulates, 18–80 μ. (Note comments on the spicules of *Craniella* on page 57.)

A. Calcarea

1. Colony a network of anastomosing tubules, the colony seldom more than 100 mm in diameter; whitish or pale yellowish; Fig. 4.4D. *Clathrina coriacea.*

1. Tubular or vaselike, solitary or in branching colonies but not forming a network except basally—2.

2. Tubular colonies, sometimes branching, rising from a stoloniferous base which may form a network; individual tubes >1.5 mm in diameter, 10–20 mm tall; Fig. 4.4C. *Leucosolenia botryoides.*

2. Solitary or clustered, unbranched vase or urnlike individuals with a fringe of spicules 1–2 mm long surrounding the osculum; individual sponges reach a diameter of 3 mm and a height of 12 mm; Fig. 4.4E. *Scypha ciliata.* The

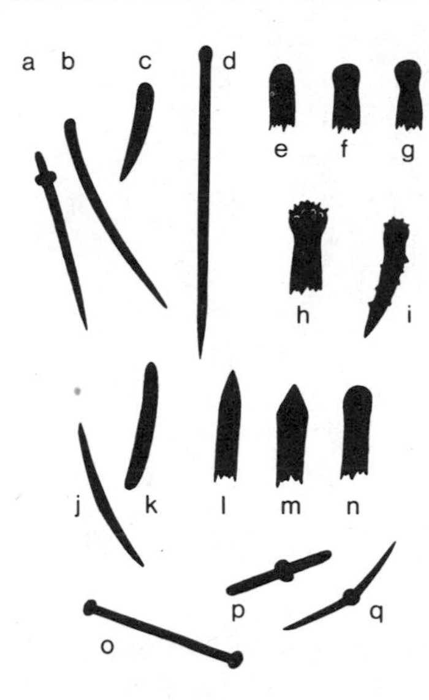

Fig. 4.2. Megasclere spicules. When these types also occur as microscleres the prefix micro is added. a–i, *monactinal types*, with two ends dissimilar; a, b, c, *styles*, with one end rounded the other pointed; a, *tylostyle with subterminal head;* the position of the swelling along the shaft varies, and the addition of the prefix "centro" to a more specific designation indicates a spicule with a nonterminal head; d–g, variations with rounded ends; e, simple *style;* d, g, *tylostyles* with terminal head; f, *subtylostyle* with less pronounced head; h, i, *acanthostyle* and detail of head; when any part of a spicule is spiny, the prefix "acantho" is added. j–q, *diactinal types*, with two ends similar (diactines or rhabds); j, *oxea*, with both ends pointed; k, *strongyle*, with both ends rounded; l–o, variants: l, simple oxea; m, *tornote*, with lance-headed ending; n, *tylote*, with knobbed ending; p, q, with central swellings.

latter species was mistaken for *Grantia* sp. and has been represented as such by dealers in biological specimens. However, the only authentic American record for this genus, according to deLaubenfels (1949), is *Grantia canadensis* from the St. Lawrence estuary. *Grantia* has a well-developed cortex which is absent in *Scypha*.

B. Clionidae. Boring sponges riddle oyster and other shells, the excavations consisting of shallow, branching tunnels that open to the surface periodically to expose the incurrent and excurrent pores. These openings in the surface of the shell range in size from 0.8 to 4.5 mm in different species. The substance of the living sponge, usually yellow in the species listed, fills these cavities. This condition represents the *alpha* stage of development, Figure 4.5. In the *beta* stage the sponge grows over the host shell as a surface encrustation in addition to its boring activities. In the *gamma* form the sponge is a massive growth and is "free-living." The beta and gamma forms are

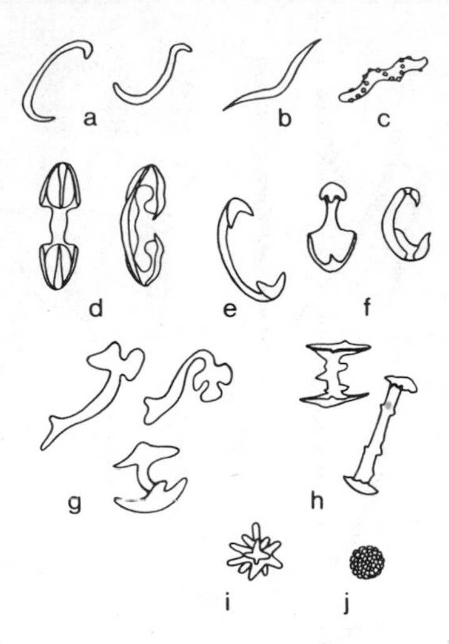

Fig. 4.3. Microsclere spicules. *Sigmas*, a, are variously C– or S-shaped and *toxas*, b, are bow-shaped. *Spirasters*, c, are a type of twisted, spiny microsclere; *streptaster* is a generic term including special spirasters and other forms not found in the local fauna. *Chelas*, may have the two ends similar, in which case they are *isochelas*, d, e, or, when dissimilar, *anisochelas*, f; palmate chelas are the common form in the local fauna, d. *Bipocilli*, g, are regarded as highly modified chelas and are only found locally in *Iophon*. *Birotulates*, h, are found in certain genera of fresh or brackish water sponges (Spongillidae). *Asters*, i, j, are polyaxial microscleres with radial projections in three dimensions; they are common in only a few local forms, chiefly as short-rayed *sterrasters*, j.

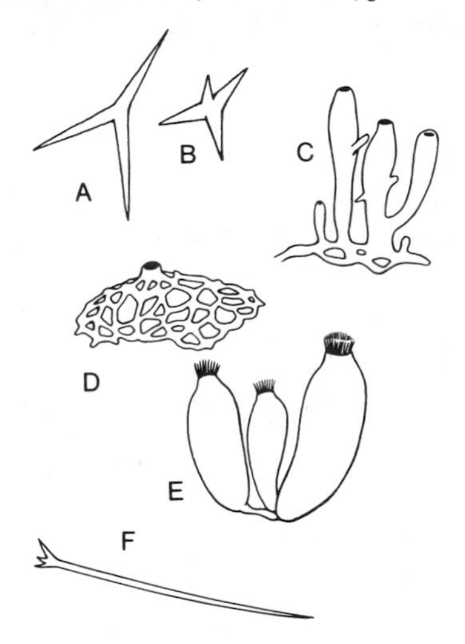

Fig. 4.4. Calcareous sponges. A, triradiate spicule; B, quadriradiate spicule; C, *Leucosolenia botryoides*, whole colony; D, *Clathrina coriacea*, whole colony; E, *Scypha ciliata*, whole colony; F, triaene spicule (Desmospongiae).

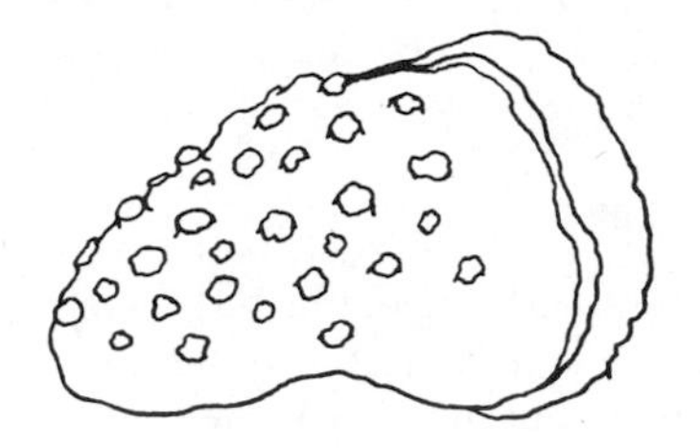

Fig. 4.5. Calcareous sponge, *Cliona*.

found in *Cliona celata*. The gamma form results from the complete dissolution of the host.

Local species may be differentiated according to the data in Table 4.2; *truitti* may be no more than an ecophenotype of *vastifica* (see Hartman, 1958a).

Table 4.2.

Cliona	Largest Perforations in Substratum	Tylostyles, length	Oxeas, length	Spirasters, length
celata	>2.0 mm	200–400 μ	None	None
lobata	<1.6 mm	170–210 μ	None	8–25, 25–50 μ
truitti	<1.6 mm	170–230 μ	75–130 μ*	7–14 μ
vastifica	<1.6 mm	260–330 μ	50–130 μ*	6–15 μ

* Some oxeas, at least, are spiny (acanthoxeas).

C. Group One. Megascleres exclusively monactinal, the largest usually >200 μ.

---

1. Mature megascleres, 98–490 μ, predominantly or exclusively styles—2.
1. Mature megascleres, up to 2600 μ, predominantly or exclusively tylostyles or subtylostyles—4.

2. Isochelae or anisochelae present—3.
2. Chelae absent; megascleres straight, 212–490 μ; form encrusting, lobular, ramose; orange, olive-tinted; surface conulose; 100 mm. *Hymeniacidon heliophila.*

3. Microscleres include palmate isochelae, ca. 25 μ); megascleres straight or curved, 98–316 μ; ramose; yellow-brown; surface minutely lumpy; 80 by 40 mm. *Isodictya deichmannae.*

3. Microscleres include palmate *an*isochelae; megascleres curved, 166–310 μ; egg-shaped or elongate globular, growing on hydroids, bryozoans, or algae; light to dark yellowish; surface finely hispid; 15 mm. *Mycalecarmia ovulum.*

4. Microscleres include palmate chelae—5.

4. Microscleres scarce or absent, no chelae. Suberitidae—7.

5. At least some megascleres with spiny heads; microscleres include toxas 12–58 μ; usually bright red; encrusting, lobular, ramose; to 150 mm at least. *Microciona prolifera.*

5. Without spiny or spiny-headed megascleres; microscleres include sigmas; usually dull grayish tinted with either yellow, brown, or olive; encrusting, lobular, globular; 55 mm. *Mycale* spp.—6.

6. Megascleres straight, 530–1150 μ. *Mycale lingua.*

6. Megascleres curved. *Mycale fibrexilis.*

7. Largest megascleres >700 μ—8.

7. Largest megascleres <600 μ—10.

8. Megascleres reach 1360 μ; encrusting with raised fingerlike processes; yellow-orange or salmon with light yellow fistules; 52 mm. *Polymastia robusta.*

8. Megascleres reach 2000 μ or more—9.

9. Subhemispherical with central, depressed osculum; surface hispid, with large cortical tylostyli up to 2600 μ; yellowish; 37 mm. *Subertechinus hispidus.*

9. Tall, mushroom-shaped with 1–6 tubular oscula on top; larger tylostyli to 2400 μ; 28 mm tall. *Tentorium semisuberites.*

10. Microscleres present, 14–56 μ, especially near the surface but easily overlooked; encrusting to globular, figlike; yellowish; 36 mm. *Suberites ficus.*

10. Microscleres absent; emergent from riddled mollusk shells, forming massive crusts with scattered, wartlike papillae ca. 4 mm diameter, or massive, pumpkin-like with no trace of host shell; bright sulfur yellow; 200 mm. *Cliona celata.* See Table 4.2 and B. Clionidae section above.

C. Group Two. Megascleres include both monactinal and diactinal types.

---

1. With spiny, monactinal megascleres (acanthostyli), 190–430 μ, or spiny diactinal microscleres (rhaphides) equally long—2.

1. Spicules in this size range not spiny—4.

2. With acanthostyli; dermis with diactinals including tylotes, tornotes, strongyles 17–260 μ—3.

2. With spiny rhaphides ca. 500 μ; dermis with tylotes 250–500 μ; endosome with smooth styles ca. 400 μ; encrusting to massive, ramose; yellowish, surface raised in warts with pores or oscula to 1.5 mm; 40 mm. *Tedania suctoria.*

3. Microscleres include highly modified chelae (bipocilli), 13–18 μ; encrusting to lobate; brownish, turning black in alcohol; 80 mm. *Iophon nigricans*

3. Microscleres include anchorate isochelae and sigmas 60–90 μ. *Myxilla: M. incrustans* is cushion-shaped, lumpy, or lobular; surface smooth; light yellowish in alcohol; 115 mm. *M. fimbriata* reaches 80 mm and is dark brown to black in alcohol.

4. Microscleres scarce or absent, or consisting of chelae—5.
4. Microscleres easily overlooked but present especially near surface, consisting of centrotylote microoxeas, strongyles, and styles, 14–56μ; megascleres predominantly tylostyles but sometimes 10–20% oxeas 144–426 μ; encrusting to massive, figlike; yellowish; 36 mm. *Suberites ficus.*

5. Microscleres absent—6.
5. Microscleres present ca. 20–25 μ—8.

6. Megascleres are curved or twisted strongyles ca. 1500 μ; styles ca. 440 μ; fan- to vase-shaped; pale yellowish, greenish tinted; 150 mm. *Cladocroce ventilabra.*
6. Megascleres are oxeas—7.

7. Megascleres ca. 85–156 μ, plus a few somewhat smaller styles; thick crusts with raised tubules; tan, gold, yellowish, or lavender-tinted; 150 mm wide by 15 mm thick. *Haliclona loosanoffi.* Note also *Cladocroce ventilabra*, sometimes with oxeas 145–186 μ.
7. Megascleres ca. 102–125 μ, plus a few somewhat smaller styles; thin crusts without raised tubules; color similar to above; 100 mm wide by 8 mm thick. *Haliclona canaliculata.*

8. Microscleres ca. 25 μ, palmate isochelae; megascleres <250 μ; flabellate to ramose; yellowish to dark red—9.
8. Microscleres ca. 20 μ, isochelae and modified sigmas; dermal tylotes ca. 175 μ, ectosomal styles ca. 145 μ; encrusting to massive with small oscula raised on short tubules; yellow or greenish; 110 mm. *Lissodendoryx carolinensis.*

9. Megascleres mostly diactinals (oxeas ca. 210 μ) plus 5% or less styles; 600 mm. *Isodictya palmata.*
9. Megascleres mostly monactinals (styles ca. 177 μ) plus a few oxeas; 80 mm. *Isodictya deichmannae.*

C. Group Three. Megascleres exclusively diactinal.

---

1. With palmate isochelae ca. 27 μ; oxeas ca. 210 μ; ramose, 600 mm (see Group Two, couplet 8). *Isodictya palmata.*
1. Without palmate isochelae—2.

2. Largest oxeas >600 μ (698–876 μ); no microscleres; tubular, ramose to globular; yellow-tan, orange-tinted; surface smooth; 150 mm. *Pellina sitiens.*
2. Largest oxeas <400 μ—3.

3. Largest oxeas <200 μ; without dermal skeleton. *Haliclona* spp.—6.
3. Largest oxeas >200 μ; with or without dermal skeleton—4.

4. Freshwater species, possibly entering low salinity, estuarine situations; without dermal skeleton. Family Spongillidae. Subfamily Spongillinae with smooth oxeas only; subfamily Meyeninae with some spiny oxeas and with birotulate microscleres 18–70 μ in the gemmules only.
4. Marine species some of which enter polyhaline waters—5.

5. Without dermal skeleton; microscleres present, consisting of deformed sigmas 26–65 μ; globular, yellowish; surface hispid; 55 mm. *Sigmadocia flagellifer.*

5. With dermal skeleton; microscleres absent; encrusting to globular or ramose; buff or cinnamon, frequently green-tinted; 300 mm. *Halichondria* spp. The two local species differ chiefly in the form of the dermal spiculation, for which see Hartman (1958a).

6. Largest oxeas <100 μ (ca. 55–60 μ); encrusting to ramose; dull yellow-orange; 80 mm. *Haliclona palmata.*

6. Largest oxeas >100 μ. Identification of individual species of *Haliclona* is greatly complicated by overlap in criteria and uncertainty about the identity of species whose types have been lost and which were inadequately described originally. In addition to the species in the check list there are others of uncertain validity. If a dermal skeleton is present, some of these forms may belong to *Adocia* (see also Group Two, couplet 6.).

References

M. W. deLaubenfels (1949) thoroughly reviewed the literature on this coast. Of the 60 titles cited, 47 predate 1900 and 7 predate 1800.

Barrett, J. H., and C. M. Yonge. 1958. Collins Pocket Guide to the Sea Shore. Collins, London.

Burton, M. 1930. Norwegian sponges from the Norman Collection. Proc. zool. Soc. Lond. 487–546.

———. 1963. A revision of the classification of the calcareous sponges. British Museum of Natural History.

deLaubenfels, M. W. 1936. A discussion of the sponge fauna of the Dry Tortugas in particular and the West Indies in general, with material for a revision of the families and orders of the Porifera. Pap. Tortugas Lab. 30:1–225.

———. 1947. Ecology of the sponges of a brackish water environment at Beaufort, N.C. Ecol. Monogr. 17(1): 31–46.

———. 1949. The sponges of Woods Hole and adjacent waters. Bull. Mus. comp. Zool. Harv. 103(1):3–55.

———. 1953. A guide to the sponges of eastern North America. Spec. Publ. mar. Lab. Univ. Miami.

———. 1955. Porifera. *In* Treatise on Invertebrate Paleontology. R. C. Moore. ed. Part E:21–122., Geol. Soc. Amer. and Univ. of Kansas.

Dendy, A., and R. W. H. Row. 1913. The classification and phylogeny of the calcareous sponges; with a reference list of all the described species, systematically arranged. Proc. zool. Soc., Lond. 704–813.

George, W. C., and H. V. Wilson. 1919. Sponges of Beaufort (N.C.) Harbor and Vicinity. Bull. Bur. Fish., Wash. 36:130–179.

Hartman, W. D. 1958a. Natural history of the marine sponges of southern New England. Bull. Peabody Mus. nat. Hist., 12:x + 155 pp.

———. 1958b. A re-examination of Bidder's classification of the Calcarea. Syst. Zool. 7:97–110.

———. 1964. Taxonomy of calcareous sponges. Science 144:711–712.

Hechtel, G. J. 1965. A systematic study of the Desmospongiae of Port Royal, Jamaica. Bull. Peabody Mus. nat. Hist., 20:vi + 103 pp.

Hopkins, S. H. 1962. Distribution of species of *Cliona* (boring sponges) on the eastern shore of Virginia in relation to salinity. Chesapeake Sci. 3(2):121–124.

Jewell, Mina. 1959. Porifera. *In* Fresh-Water Biology, W. T. Edmondson, ed. John Wiley and Sons.

Lambe, L. M. 1896. Sponges from the Atlantic Coast of Canada. Trans. R. Soc. Can. (2):ii + 181–211.

Lundbeck, W. 1902–1910. Porifera, Parts I–III. Dan. Ingolf-Exped. 6:1–108, 1–219, 1–124.

Old, M. C. 1941. The taxonomy and distribution of the boring sponges (Clionidae) along the Atlantic Coast of North America. St. Md. Dep. Res. Educ. 44:3–16.

Pearse, A. S., and L. G. Williams. 1951. The biota of the reefs off the Carolinas. J. Elisha Mitchell scient. Soc. 67:133–161.

Pennak, R. W. 1953. Fresh-water invertebrates of the United States. Ronald Press, New York.

Proctor, W. 1933. Biological Survey of Mt. Desert Region. Part V. Marine Fauna. Wistar Inst. Press.

Wells, H. W., M. J. Wells, and I. E. Gray. 1960. Marine sponges of North Carolina. J. Elisha Mitchell scient. Soc. 76:200–245.

FIVE

Phylum Cnidaria—Coelenterata

HYDROIDS, ANEMONES, MEDUSAE

Introduction

The larger phyla do not lend themselves to definitions that are at once brief and sufficiently descriptive to encompass all morphological possibilities. A most succinct definition, borrowed with modifications from Hyman (1940), describes the members of this phylum as tentacle-bearing animals with at least superficial radial symmetry and with intrinsic nematocysts. The body plan is relatively simple, lacks complex organs, and is chiefly characterized by a single *gastrovascular cavity* with but one opening, the mouth. The essential construction is based on two tissue layers separated by a *mesoglea* which may exhibit some cellular organization or may function chiefly as a gelatinous or fibrous "connective tissue."

Nematocysts, literally "thread bladders," are microscopic structures that function defensively to repel enemies and offensively to poison and ensnare prey. They are uniquely a development of this phylum, but some flatworms and nudibranchs are able to ingest cnidarians without discharging their nematocysts, later incorporating the nematocysts for their own use.

Halstead (1965) has reviewed the literature on medical aspects of cnidarian poisoning and presents a useful bibliography. The virulence of stings by the more notorious species has frequently been exaggerated, but sound evaluations are difficult to make because of suspect information as to species identity, conditions under which the "attack" occurred, symptomatic responses, etc. Probably the Portuguese man-of-war, *Physalia*, is the most powerfully armed species to be encountered locally, but Totton and Mackie (1960) were unable to verify episodes of fatal encounters with this species in the Atlantic. Pacific fatalities attributed to *Physalia* may have been caused by the cubomedusan *Chironex fleckeri*, which is known to be capable of lethal stings (see Barnes, 1966; Cleland and Southcott, 1965). The larger jellyfish, cubomedusae, and siphonophores should be treated with care, however, since severe stings are, at the very least, discomforting.

Many members of the phylum Cnidaria are polymorphic with two major morphological types, *polyp* and *medusa*. Anemones typify the solitary polyp which is essentially cylindrical, basally attached or adherent, with a central mouth and peripheral whorl of tentacles. The medusoid form, typified by scyphozoan jellyfish, is frequently represented as a bell or umbrella with, again, a central mouth and peripheral tentacles. While the polypoid or medusoid form of most postlarval cnidarians is fairly evident, the orders Siphonophora and Chondrophora have developed complex and atypical forms at variance with the usual pattern (Figs. 5.25, 5.26). Additional exceptions occur in the order Stauromedusae, class Scyphozoa, and rarely in Hydrozoa where a few atypical polypoids are pelagic. In Stauromedusae the medusae are sessile, adhering by an aboral peduncle, the umbrella being calyxlike with eight lobes bearing short knobbed tentacles (Fig. 5.1G).

Polymorphism within successive stages of the same ontogeny greatly complicates the problem of identification. Cnidarian life histories are inordinately varied, and it is desirable to have a general view of details relevant to problems of identification.

Major Patterns of Cnidarian Life History

It is not pertinent to this account to enter into a discussion of the complex and controversial philosophical implications of this subject, for which see Rees (1966), Hyman (1940), and Dougherty (1963). The abbreviated classification in Table 5.1. provides a frame of reference for the discussion that follows. Extralimital groups are omitted. The classification is mainly that of Rothschild (1965).

A common feature of cnidarian life histories is the alternation of polypoid and medusoid phases seen in Table 5.2. (*Metagenesis* is a term used by some students in preference to "alternation of generations" to avoid confusion with the alternation of haploid and diploid generations in botanical and

Table 5:1 Classification of Cnidaria

Class Hydrozoa	Hydroid polyps and medusae, etc.
Order Athecata	
Order Thecata	
Order Limnomedusae	
Order Narcomedusae	
Order Trachymedusae	
Order Siphonophora	
Order Chondrophora	
Class Scyphozoa	Scyphozoan jellyfish
Order Stauromedusae	
Order Cubomedusae	
Order Coronatae	
Order Semaeostomeae	
Order Rhizostomeae	
Class Anthozoa	
Subclass Octocorallia	Sea whips, gorgonians, etc.
Order Gorgonacea	
Order Pennatulacea	
Order Telestacea	
Order Alcyonaceae	
Subclass Zoantharia	Anemones, corals, etc.
Order Zoanthidea	
Order Actiniaria	
Order Scleractinia (Madreporaria)	
Subclass Ceriantipatharia	Ceriantharian anemones
Order Ceriantharia	

protist life histories.) Sexual and asexual reproduction both occur in all major groups. The sexes are frequently separate in both solitary and colonial species. Hermaphroditism also occurs in all classes. Asexual reproduction takes place by budding and fragmentation. The principal patterns of cnidarian ontogeny may be summarized in tabular form. Table 5.2 is subject to philosophical dispute at several points; its main purpose is to relate sequential stages in the various major taxa in a logical order.

It will be seen that in nearly all groups, except freshwater hydroids and the brackish water *Protohydra*, the zygote develops into a nearly microscopic, free-living larval form called a *planula*. Typically the planula (length ca. 0.5 mm) is somewhat elongated and cylindrical, multicellular, mouthless, ciliated, and motile, Figure 3.3A. Planulae may develop from eggs that have been freely shed in water, but in nearly all major groups "marsupialism" occurs in which development to the planula or a later stage takes place within the parental body or in some structure developed for the purpose.

Table 5.2. Summary of Cnidarian Life Histories

	Systematic Group	Egg	Larvae	Polypoids	Medusoids
1.	*Protohydra*	Egg	None	Polyp	None
2.	Thecata, Athecata	Egg	Planula-actinula	Polyp	Sessile gonophore
3.	Thecata, Athecata, Limnomedusae	Egg	Planula	Polyp	Free medusa
4.	Trachymedusae, Narcomedusae	Egg	Planula	None*	Free medusa
5.	Siphonophora, Chondrophora	Egg	Planula	Polypoid-medusoid colonies†	
6.	Semaeostomeae (*Pelagia*)	Egg	Planula	Polypoid larva	Free medusa
7.	Stauromedusae	Egg	Planula	Polypoid larva	Sessile medusa
8.	Most Scyphozoa except 6 and 7 above	Egg	Planula	Scyphistoma	Free medusa
9.	Ceriantharia, Zoanthidea	Egg	Special larva	Polyp	None
10.	Most Anthozoa except 9 above	Egg	Planula	Polyp	None

* A parasitic polypoid stage may occur in Narcomedusae.

† The "colonial" status of chondrophorans is disputed, as is the homology of individual elements to polyps or medusae in Siphonophora.

In some forms the initial free phase is a more advanced, tentacle-bearing larva called an *actinula*, Figure 3.3D. Free planulae, in metamorphosis, may or may not pass through a similar phase. Ceriantharians and some zoanthid anemones (Anthozoa) produce distinctive larval forms; most zoanthids have a simple planula.

Metamorphosis in those groups with a sessile polypoid phase is accomplished with the abandonment of a pelagic existence; the larva selects a substratum, settles or attaches itself, and changes into or buds a polypoid form. A period of asexual reproduction commonly ensues. The initial polyp may bud additional *zooids* which remain attached to form a colony, as occurs in hydroids, corals, and octocorallians. When asexual reproduction occurs in anemones—it does not in most species—the newly budded individual separates from the "parental" anemone. In Scyphozoa the polypoid stage or *scyphistoma* is primarily a device for budding off juvenile jellyfish. In exclusively medusoid taxa, i.e., Trachymedusae, Narcomedusae, and some Scyphozoa, the planulae metamorphose directly into medusae or, in siphonophores, into a postlarval stage from which both medusoid and polypoid "persons" will be budded to form a pelagic, polymorphic colony, the Portuguese man-of-war, *Physalia*, being an example.

In summary, we are concerned only with polypoid forms in Anthozoa, and with medusoid forms almost exclusively in Scyphozoa. In Hydrozoa the orders Trachymedusae and Narcomedusae are exclusively medusoid or very nearly so. The Siphonophora and Chondrophora are highly aberrant, colonial types less obviously medusoid or polypoid; whether these are indeed colonies or individual animals, as well as the question of homology between various structural elements or whole individuals are subjects for academic dispute—happily beyond the concern of this text. Nonfeeding medusoid, sexual forms are released by the chondrophores *Velella* and *Porpita*. In the other hydrozoan orders both medusae and polyps must be considered.

The matching of the polyps and medusae of individual species of Thecata and Athecata poses some difficulty. Each body type in a metagenetic species is subject to its own selective pressures and rate of evolution. Species that are quite divergent and morphologically distinct in one stage may be indistinguishable in the other. As a result of these ontogenetic idiosyncrasies a dichotomy of names and family level phylogenies exists, and it is more practical at the present, primitive stage of understanding to examine the polyps and medusae separately.

Preliminary Identification

As a first step toward species identification it is necessary to sort cnidarian types to order. The orders will then be dealt with in the sequence of Table 5.1. Polypoid forms are considered first. (Siphonophores and chondrophores are not considered in this summary; see Figures 5.25, 5.26, and associated text.)

Polypoids. The animals are usually attached to or adherent on a substratum. Many of the species are more or less plantlike and as such are perhaps likely to be confused with filamentous algae or, as is more likely, with bryozoans or ramose sponges. (Differences between these "zoophytes" are discussed in Chapter 3.)

The fundamental differences separating species of colonial polypoids involve internal structure, life history, and other criteria that may not be immediately evident to the student. The following informal summary should suffice to separate the main groups.

Colonial Polypoids. The taxa to be considered are the orders Thecata and Athecata in Hydrozoa with some 200 species plus the anthozoan orders Zoanthidea and Scleractinia with but a single species each (locally), and some half-dozen species of octocorallians. The hydrozoan order Limnomedusae, whose members are chiefly known as medusae, may be ignored for all practical purposes. The groups separate as follows.

- Colonies of typical corals with an external, calcareous skeleton; Fig. 5.1A. Scleractinia.
- Colonies of widely varying appearance, e.g., whiplike, arborescent, plume- or lily padlike, or sparsely branching from a stolon, the individual polyps minute; with an internal skeleton of embedded spicules; Fig. 5.29. Octocorallia.
- Colonies without a skeletal support but frequently encrusted with sand, the polyps short, thick, macroscopic; local species usually encrust shells occupied by hermit crabs in deeper waters; Fig. 5.1B. Zoanthidea.
- Colonies without a skeletal support, usually not encrusted with debris, the colonies diverse in form and ranging from solitary or minute stolonate forms to extensive bushy growths, the usually minute polyps either naked or armored

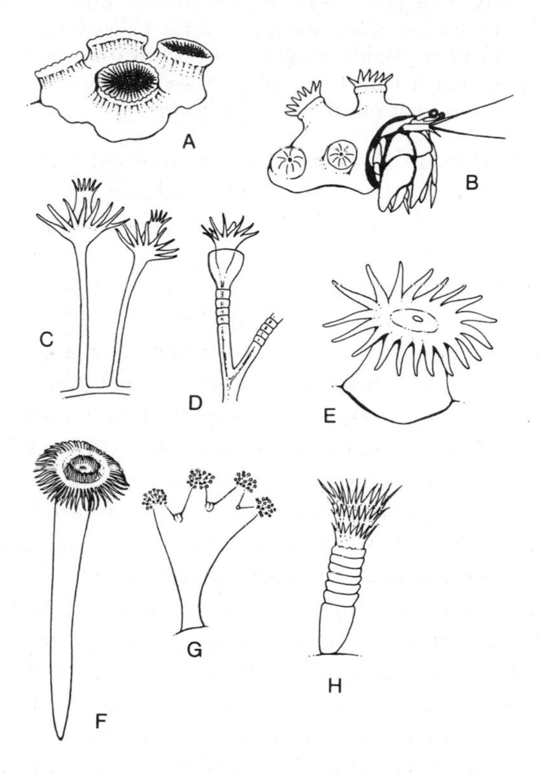

Fig. 5.1. Polypoid types. A, *Astrangia danae*, a stony coral, order Scleractinia; B, *Epizoanthus incrustatus*, a zoanthid anemone, order Zoanthidea; C, athecate hydroid polyps, order Athecata; D, thecate hydroid polyp, order Thecata; E, sea anemone, order Actiniaria; F, *Cerianthus borealis*, a ceriantharian anemone, order Ceriantharia; G, sessile jellyfish, order Stauromedusae; H, scyphistoma or polypoid stage of jellyfish, order Semaeostomeae.

wholly or in part with a relatively thin chitinous sheath; Fig. 5.1C,D. Thecata and Athecata.

Solitary Polypoids. Actinaria (anemones), the anemonelike cerianthrians, and the relatively few solitary hydroids are to be considered here as well as the superficially polypoid Stauromedusae. A typical scyphozoan scyphistoma is figured, but no attempt will be made to differentiate species (Fig. 5.1H).

- Polyps with a differentiated *base* (a holdfast or stolon usually), *stem*, and flowerlike *hydranth*, the solitary forms frequently minute or, in *Tubularia*, with elongate, slender stems. Thecata and Athecata.
- Polyps usually columnar with an expanded, aboral *pedal disc* and tentacular *oral disc*; burrowing forms have the aboral end tapered, rounded, and the oral disc with only one whorl of tentacles; Fig. 5.1E. *Actiniaria.*
- Burrowing polyps; elongate, columnar, aboral end tapered, rounded, and the oral disc with *two* whorls of tentacles; Fig. 5.1F. *Ceriantharia.*
- Superficially polypoid, highly modified sessile medusoids, with an adherent, aboral peduncle and a calyxlike umbrella bearing eight lobes with knobbed tentacles; Fig. 5.1G. Stauromedusae (Scyphozoa).

Medusae. A substantial range of morphological variation is found in the two classes and seven orders in which medusae occur. The differential traits overlap and are otherwise liable to uncertainty of interpretation.

The typical medusa consists of a swimming bell or *umbrella* with the mouth at the center of the concave, subumbrellar surface; marginal tentacles and other appended parts are usually present (Fig. 5.2). The umbrella may be shallow or deep, ranging in shape from subspherical to an almost flat saucer. In both Scyphozoa and Hydrozoa the umbrella exhibits tetraradiate symmetry, and a body plan of four identical quadrants is usually apparent. In hydrozoans this symmetry is diagrammatically expressed by a system of *radial canals* extending from the central stomach to the marginal *ring canal*, Figure 5.2A. In local anthomedusae (Athecata) there are four simple radial canals in all genera except *Niobia*, where two of the canals are bifurcated, and in *Calycopsis*, where additional *centripetal canals* extend from the ring canal toward and sometimes reaching the stomach. In the remaining hydrozoan orders some genera follow the simple anthomedusan pattern, but frequently there are 6, 8, or as many as about 100 radial canals with or without additional centripetal canals. In Narcomedusae, radial canals are absent.

The subumbrellar cavity in hydromedusae is partly sealed by a marginal, veillike partition called the *velum*. This structure is characteristic of hydromedusae, generally, though varying in width in different genera; in *Obelia* it is rudimentary. The velum is absent in scyphozoan medusae, but in this class in the order Cubomedusae and in *Aurelia* an analogous structure, the *velarium*, is present.

The umbrella in profile usually presents a smoothly contoured outline, but in the scyphozoan order Coronatae the umbrella is divided horizontally by a deep *coronal groove*, Figure 5.2H. In several genera of anthomedusae the apex of the umbrella bears a cone- or domelike *apical process*. In hydromedusae the umbrella has a continuous margin except in Narcomedusae, where it is lobed. The lobed condition is typical of scyphomedusae except Cubomedusae; in Scyphozoa the individual lobes are called *lappets*.

In local hydromedusae marginal tentacles are only absent in *Pennaria* and *Stylactis*, where they are reduced to bumps; in Trachymedusae, however, the tentacles are liable to break off basally when the animals are captured, making it difficult to determine tentacle complements. Narcomedusae are also fragile. The tentacles are well armed with nematocysts, which may be evenly distributed or concentrated in clumps, giving the tentacle a nodular appearance. Additional structures are found along the margin of the hydromedusan umbrella including *cirri* and various sense organs. Cirri resemble tentacles but are decidedly smaller; they are "lateral" when closely associated with tentacular bulbs, or "marginal" when they emerge from the interspaces between tentacles. The chief sense organs are *ocelli* and *statocysts* which, like cirri, are variably present or absent in different genera. The ocelli are usually pigmented but may fade in preservative. In some genera they are conspicuous, in some, obscure. They are usually located on the tentacular bulbs. Statocysts, also known as marginal vesicles, lithocysts, etc., are minute structures of several types, presumably concerned with equilibrium. They are located along the rim of the bell. The distinction between "open," pitlike statocysts and "closed," vesicular statocysts is of importance to formal systematics. *Sensory clubs* and *cordyli* are marginal, clublike sense organs that differ histologically and are also taxonomically important.

Tentacles are absent in the scyphomedusan order Rhizostomeae. They are essentially marginal in the other orders, but actually emerge either from the lappets, or from just below the coronal groove in Coronatae, or in the niches between lappets, or on the subumbrella. In Cubomedusae the tentacles have two distinct segments, a basal, flattened bladelike *pedalium*, and a flexible distal portion, Figure 5.2C. The basal expansions from which tentacles arise in Coronatae are also called pedalia. The marginal sense organs of Scyphozoa are called *rhopalia* and are found either on the side of the umbrella in Cubomedusae, on pedalia in most Coronatae, or in niches between lappets in most other species. The rhopalia contain statocysts and, sometimes, ocelli. The ocelli of Cubomedusae are especially well developed and can be described as eyelike, an appropriate development in these active jellyfish.

Variations in mouth structure also provide important identification criteria. In Rhizostomeae the mouth is a single opening initially, but the original mouth closes, and multiple mouth openings develop on the *mouth-arms* and their appurtenances, Figure 5.2F. The mouth may have a simple tubular opening, e.g., in some Anthomedusae, but more commonly four or more lobes or lips are developed, and they may be modified by special features including nodules or nematocysts, frills or crennelations, and tentacular extensions, Figure 5.17. In a few genera, *oral tentacles* emerge above the mouth. The mouth opens into a simple saclike stomach in hydromedusae; in some cases the stomach is attached to the subumbrellar surface by a *peduncle* which may be short and inconspicuous (or absent), or it may be elongate. The pendulous, clapperlike portion of the stomach (or stomach plus peduncle) which is suspended in the subumbrellar cavity is called the *manubrium*. In Scyphozoa, except Rhizostomae, the mouth is central, single, and usually four-lobed. The scyphozoan stomach is more complex than that of Hydromedusae and includes a central stomach and four lateral gastric pouches. In a very loose sense these are comparable to the radial canals of Hydrozoa, radial canals being otherwise absent in Scyphozoa.

Among the simpler anthomedusae there is relatively little ontogenetic change except in size and maturation of the gonads. In more complex genera and in leptomedusae and scyphozoan medusae the changes may be substantial, involving especially the form of the umbrella and the complement of tentacles. Frequently there is a rapid increase in the number of tentacles during the early growth stages. The initial medusoid stage of scyphozoans is called an *ephyra;* ranging in size from one to a few millimeters, it is eight-armed, the arms or lobes being bifurcated terminally (Fig. 3.3M). The metamorphosis of the ephyra involves the filling in of the bell, addition of tentacles, and elaboration of other details of the adult form.

The generative organs are conspicuous in mature medusae and are of major importance in taxonomy. The position of the gonads on either the stomach or radial canals is usually consistent within orders except in Limnomedusae. Further details of form and location are also significant at lower systematic levels. Medusae of both Hydrozoa and Scyphozoa are usually dioecious, but hermaphroditism occurs among local forms in the scyphozoan *Chrysaora*. "Marsupialism" occurs in genera of both classes. Asexual reproduction in Scyphozoa is confined to larval stages, but in Hydrozoa, and chiefly in Athecata, medusae are budded from medusae, e.g., in *Rathkea*, *Lizzia*, *Sarsia*, and *Bougainvillia* on the stomach wall, and on the tentacular bulbs in *Hybocodon* and *Niobia*.

Anthomedusae are generally small, ranging from 1 or 2 mm in *Lizzia*, *Pennaria*, and *Margelopsis* to about 40 mm in *Calycopsis;* the latter is unusual, however, and only a few species exceed 15 to 20 mm in bell diameter or height. Local Narcomedusae are equally small, but Trachymedusae and Leptomedusae average substantially larger, and in several genera, e.g.,

Staurophora and *Rhacostoma*, the bell diameter exceeds 200 mm. The larger and generally more conspicuous jellyfish belong in Scyphozoa. In higher latitudes, *Cyanea* is reported to reach a bell diameter of 2.3 m, but the New England form seldom exceeds 750 mm; south of Cape Cod it is even smaller.

The preceding sketch may be summarized in key form.

---

1. Subumbrellar cavity partly sealed by a velum or velarium, Fig. 5.2A; velum rudimentary in *Obelia*, Fig. 5.23A; the scyphozoan medusa *Aurelia* has a pseudovelum, Fig. 5.28A—2.

1. Velum or velarium lacking—4.

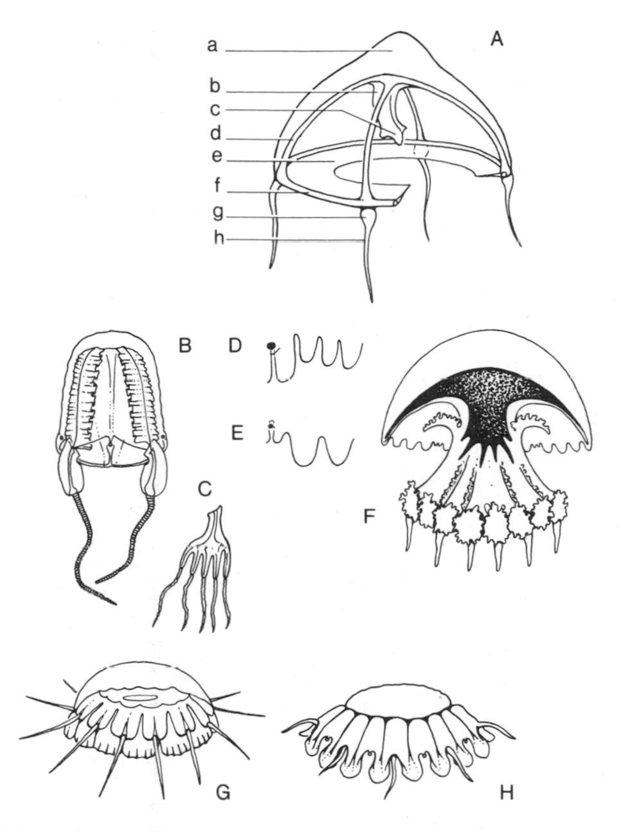

Fig. 5.2. Medusoid types. A, hydromedusa with the following parts named: a, apical process; b, peduncle; c, manubrium; d, radial canal; e, velum; f, ring canal; g, tentacular bulb; h, tentacle; B, *Tamoya haplonema*, a cubomedusan; C, *Chiropsalmus* sp., single tentacle; D, *Stomolophus* sp., detail of bell margin; E, *Rhopilema* sp., a rhizostomeaen scyphozoan, detail of bell margin; F, same, with bell shown diagrammatically in cross-section to indicate multiple mouth openings; G, *Cunina* sp., a narcomedusan; H, *Nausithoe* sp., a coronataean scyphozoan.

2. Tentacles 4 in number, with a basal pedalium and a distal, more typically tentacular part; Fig. 5.2B,C. Scyphozoa, Cubomedusae.
2. Tentacles 1 to >4000, not separated into two dissimilar parts (tentacles rudimentary in *Pennaria* and *Stylactis*, Fig. 5.18J)—3.

3. Umbrellar margin lobed; radial canals lacking; tentacles emerge above the margin; gonads on stomach; Fig. 5.2G. Hydrozoa, Narcomedusae.
3. Not so; for the remaining hydrozoan orders see Table 5.11 p. 108.

4. Tentacles absent; mouth-arms well developed, with multiple mouth openings; Fig. 5.2F. Scyphozoa, Rhizostomeae.
4. Tentacles present; with a single, central mouth—5.

5. Umbrella divided horizontally by a coronal groove; Fig. 5.2H. Scyphozoa, Coronatae.
5. Coronal groove lacking, Fig. 5.2E. Scyphozoa, Semaeostomeae.

References

See also references following individual taxa.

Barnes, J. H. 1966. Studies on three venemous cubomedusae. *In* The Cnidaria and their Evolution. W. J. Rees, ed. Symposia of the Zoological Society of London, No. 16. Academic Press, New York.

Bayer, F. M., *et al.* 1956. Coelenterata. Part F. Treatise on Invertebrate Paleontology. R. C. Moore, ed. Geol. Soc. Amer. and Univ. Kansas.

Cleland, J. B., and R. V. Southcott. 1965. Injuries to man from marine invertebrates in the Australian Region. Nat. Health and Med. Council, Spec. Rept. Ser. 12, Canberra.

Dougherty, E. C., ed. 1963. The Lower Metazoa; Comparative Biology and Phylogeny. University of California Press. (See especially articles by Hand, Uchida, and Mackie.)

Halstead, B. W. 1965. Poisonous and Venomous Marine Animals of the World. Vol. I, Invertebrates. U.S. Government Printing Office.

Hand, C. 1958. On the origin and phylogeny of the coelenterates. Syst. Zool. 8(4):191–202.

Hyman, Libbie H. 1940. The Invertebrates: Protozoa through Ctenophora. McGraw-Hill Book Co., New York.

Rees, W. 1957. Evolutionary trends in the classification of capitate hydroids and medusae. Bull. Br. Mus. nat. Hist. Zool. 4:1–60.

———, ed. 1966. The Cnidaria and their Evolution. Symposia Zool. Soc. London.

Rothschild, Lord. 1965. A classification of Living Animals. Longmans.

Totton, A. 1954. Siphonophora of the Indian Ocean. "Discovery" Rept. 27:1–162.

———, and H. E. Bargmann. 1965. A synopsis of the Siphonophora. British Museum.

———, and G. O. Mackie. 1960. Studies on *Physalia physalis* (I): Natural History and Morphology. Part 2: Behavior and Histology. "Discovery" Rept. 30:301–408.

Class Hydrozoa, Part One
SESSILE POLYPOIDS

While pelagic polypoids occur in *Margelopsis* and (questionably) in *Hypolytus*, the polypoid phase in other genera of the orders Athecata, Limnomedusa, and Thecata is sessile, and most species are colonial. Detached hydranths of a few species may survive long enough in the plankton to bud new members and assume somewhat aberrant growth forms.

Variations in polyp structure and growth habit necessitate a somewhat extended "diagnosis."

Each zooid usually has two structural elements, a stem or *pedicel* and a *hydranth* typically with tentacles and a central mouth. Zooids may arise from a rootlike holdfast, a *stolon* or network of stolons, or a laminated crust; such colonies are termed rhizocaulous. The colonies of some species are branched (hydrocaulous) with individual zooids developing from a common stem or stem branches; profusely branching colonies may be bushy or plantlike. The stems or pedicels may be "naked" or armored with a chitinous *periderm;* they may be annulated, and two or more stems may be bound together in a *fascicled* condition. The hydranth may be naked, or it may be protected by a cuplike *hydrotheca* which is a continuation of the periderm of the pedicel. In most such forms the hydranth can withdraw into the hydrotheca, and hydrotheca and pedicel are clearly differentiated. Colonies may be polymorphic in which case the basic member is a feeding polyp or *gastrozooid;* zooids bearing sexual buds may be differentiated as *gonozooids* and additional modified zooids may be present.

The production of sexual buds or *gonophores* is seasonal in most species. Sexual arrangements in Hydrozoa are extremely diverse, and the variations have greatly complicated the systematics of the group. The gonophores vary in structure and in their location in the colony. In about a third of the hydroid genera gonophores produce free medusae. Medusae are degenerate in some species, however, and lack functional mouth parts, tentacles, or other features of the typical motile form. In some genera, more or less medusoid buds are produced, but they do not become free. Medusae may mature, produce gametes, and even have their ova fertilized while still attached to the colony. In *Stylactis* the medusae are sometimes released, sometimes not.

In the forty-odd genera listed by Fraser (1944) that do not produce free medusae, the degree to which medusoid features have been obliterated varies. In all Sertularidae and Plumularidae as well as in numerous other families, the gonophores are reduced to a *sporosac* that produces eggs or sperm. The sporosac has, at best, histological traces of a medusoid origin. In *Dicoryne*, however, the sporosacs are described as "free-swimming." In a few athecate genera and in most thecate ones the gonophores are budded laterally along a modified hydranth called a *blastostyle*. In Thecata the gonophores are encased by a perisarc, the receptacle being called a *gonotheca*. Colonies are dioecious and may exhibit sexual dimorphism.

In forms producing free medusae the eggs may be dispersed in the water, or they may be brooded; special structures are developed in some genera for this purpose. Forms without free medusae retain the eggs, and the larvae are released as planulae or in a few genera, including *Tubularia* and *Myriothela*, as actinulae. In some genera a brood chamber called an *acrocyst* extends from the gonophore; *Calycella* and *Opercularella* have this habit, and in *Gonothyraea* the acrocyst is described as medusiform. The larva, after a motile period, settles down and either metamorphoses into the initial polyp of a new colony or produces a growing stem from which polyps emerge by budding. Where pertinent to problems of identification, further details of reproductive structure or habits will be indicated.

Hydroid colonies range in size from minute, rhizocaulous forms with polyps <5.0 mm tall to bushy colonies between 500 and 1000 mm in extent. They are found on a wide variety of substrata, animate as well as inanimate. More than two score species have been reported on *Sargassum*, and hydroids are common on other algae, eelgrass, and on a variety of invertebrates, including other hydroids. Some of these preferences are fairly specific; e.g., *Hydractinia* and *Podocoryne* are found on gastropod shells occupied by hermit crabs, and *Clava* is usually found on *Fucus* and other thallophytic algae. Most species are fixed to relatively firm substrata, but a few genera, e.g., *Hypolytus*, *Dahlgrenella*, *Corymorpha*, and, sometimes, species of *Tubularia* live on unconsolidated material.

Plankters and minute benthic forms are the principal food items. *Corymorpha* has been reported to feed on detritus, and a single local species, *Hydrichthys mirus*, is parasitic on fish. Its only reported host is the rudderfish *Seriola*, a subtropical species that strays northward to Nova Scotia. The hydroid is reported to send out rootlike processes that penetrate to the muscle layer.

Hydroids are intolerant of drying, and they range from the lower littoral to abyssal depths. They are often important fouling organisms on buoys and fixed structures, in estuaries especially. The different species occur in a wide range of salinities, and *Cordylophora* and *Protohydra* may be found in waters that are essentially fresh.

Systematic List and Distribution

The scheme of classification used is based on that of Rothschild (1965) with modifications from other sources, particularly Russell (1953), Kramp (1959, 1961), and Rees (1957). Although progress is being made toward a unified classification for polyps and medusae, this happy goal is yet to be achieved. The provisional status of the taxonomy is particularly apparent in Thecata. Fraser (1944) and Kramp (1959, 1961) provide the basic lists of polyps and medusae, respectively.

In the lists below, the annotations are: (mp) medusoid and polypoid phases known under the same name; (m) presumably with a free medusoid phase but medusa unknown; (p) polypoid phase unknown or questionably identified; (s) no free medusa; numbers in parentheses refer to taxa in which medusa and polyp are questionably correlated, the notes indicating medusoid name: (1) equals *Hybocodon pendulus*. (2) equals *Bougainvillia ramosa?* (3) equals *Leuckartiara octona; jonesi* and *repens* conspecific? Bathymetric data are omitted where the polyp is unknown, and such information is also missing for several other forms.

Order Athecata (Anthomedusae, Gymnoblastea)			
Family Hydridae			
Protohydra leuckarti	(s)	BV	Fresh-brackish water.
Family Corymorphiidae			
Corymorpha pendula	(1)	B+	Lit. to 274 m
Heteractis aurata	(mp)	B	73 to 110 m
Hypolytus peregrinus	(s)	V	Shallow, eury.
Euphysora gracilis	(p)	BV	—
Euphysa flammea	(p)	A+	—
Family Tubulariidae			
Tubularia couthouyi	(s)	B+	Lit. to 132 m
Tubularia crassa	(s)	V	55 m
Tubularia crocea	(s)	BV	Lit. to 60 m
Tubularia indivisa	(s)	B+	7 to 787 m
Tubularia larynx	(s)	B+	13 to 46 m
Tubularia spectabilis	(s)	B+	Lit. to 335 m
Tubularia tennella	(s)	B+	Lit. to 73 m
Hybocodon prolifer	(mp)	B+	Lit., tide pools
Ectopleura dumortieri	(mp)	V	18 m
Family Margelopsidae			
Margelopsis gibbesi	(mp)	C+	Both phases pelagic
Family Acaulidae			
Acaulis primarius	(s)	B	9 to 91 m
Blastothela rosea	(s)	B	13 m
Family Myriothelidae			
Myriothela phrygia	(s)	A+	—
Family Cladocorynidae			
Cladocoryne pelagica	(m)	C+	Usually on *Sargassum*

Family Corynidae			
Dipurena strangulata	(p)	V	—
Sarsia tubulosa	(mp)	BV	0 to 25 m
Sarsia princeps	(mp)	A+	Hydroid unknown in W. Atlantic
Sarsia hargitti	(p)	V	—
Linvillea agassizi	(p)	V	May equal *Corynitis agassizii* or *Zanclea costata*
Family Zancleidae			
Zanclea costata	(mp)	V	—
Family Pennariidae			
Pennaria tiarella	(mp)	V+	Lit. to 29 m
Family Hydrichthyidae			
Hydrichthys mirus	(m)	C+	Parasitic on fish
Family Clavidae			
Turritopsis nutricola	(mp)	V	Lit. to 9 m
Clava leptostyla	(s)	B+	Lit. to 37 m
Rhizogeton fusiformis	(s)	BV	Lit.
Tubiclava cornucopiae	(s)	BV	Lit.
Cordylophora lacustris	(s)	BV	Fresh-brackish water
Family Hydractiniidae			
Podocoryne carnea	(mp)	B+	Lit. to 51 m
Podocoryne borealis	(mp)	B+	Polyp unknown in W. Atlantic
Hydractinia echinata	(s)	BV	Lit. to 662 m
Hydractinia valens	(s)	B	57 m
Stylactis hooperi	(mp)	V	Shallow
Stylactis arge	(m)	V	Shallow
Family Bougainvilliidae			
Bougainvillia carolinensis	(mp)	BV	Shallow, eury
Bougainvilla superciliaris	(mp)	B+	—
Bougainvillia rugosa	(mp)	V	Lit. to 4 m
Bougainvillia britannica	(2)	B+	—
Dicoryne conferta	(s)	B	15 to 368 m
Dicoryne flexuosa	(s)	B	91 to 229 m
Bimeria brevis	(s)	B	0 to 24 m
Bimeria tunicata	(s)	V	Shallow
Calyptospadix cerulea	(s)	B, 37°	Lit. to 4 m
Garveia groenlandica	(s)	B+	0 to 110 m
Nemopsis bachei	(p)	V	—
Lizzia fulgurans	(p)	V	Genus frequently confounded with *Podocoryne*
Lizzia blondina	(p)	V	—
Family Rathkeidae			
Rathkea octopuncata	(p)	B+	—
Family Calycopsidae			
Calycopsis typa	(p)	V	—
Family Pandeidae			
Catablema vesicarium	(p)	B+	—

Amphinema rugosum	(mp)	V	Polyp unknown in W. Atlantic
Amphinema dinema	(mp)	V	Polyp indistinguishable from preceding
Perigonimus jonesi	(3)	V	Shallow
Perigonimus repens	(3)	B+	26 to 84 m
Stomotoca pterophylla	(p)	BV	—
Leuckartiara nobilis	(p)	B	—
Niobia dendrotentaculata	(p)	V	—
Halitholus cirratus	(p)	A+	—
Family Eudendriidae			
Eudendrium album	(s)	BV	2 to 132 m
Eudendrium capillare	(s)	B+	0 to 82 m
Eudendrium carneum	(s)	V	0 to 137 m
Eudendrium dispar	(s)	B+	0 to 82 m
Eudendrium insigne	(s)	B	0 to 18 m
Eudendrium rameum	(s)	B	9 to 183 m
Eudendrium ramosum	(2)	BV	3 to 2296 m
Eudendrium tenellum	(s)	BV	0 to 479 m
Eudendrium tenue	(s)	BV	Lit. to 137 m
Eudendrium vaginatum	(s)	B	0 to 37 m

Order Limnomedusae

Family Proboscidactylidae			
Proboscidacytla ornata	(p)	V	Polyp may equal the European *Lar.* sp.
Family Olindiidae			
Gonionemus vertens	(p)	BV	—

Order Thecata (Leptomedusae, Calyptoblastea)

The taxonomy of the following species is particularly troubled; the allocation of polypoid genera to family follows Fraser (1944), and medusae names have simply been integrated without attempting to reconcile systematic contradictions.

Family Campanularidae			
Campanularia amphora	(s)	B, 40.5°	Shallow water
Campanularia angulata	(s)	B, 37°	0 to 327 m
Campanularia calceolifera	(s)	BV	Shallow, eury.
Campanularia fasciculata	(s)	B	60 m
Campanularia flexuosa	(s)	B+	Lit. to 55 m
Campanularia fragilis	(s)	B	15 to 55 m
Campanularia gelatinosa	(s)	BV	Lit. to 640 m
Campanularia gigantea	(s)	B+	9 to 132 m
Campanularia groenlandica	(s)	B+	46 to 91 m
Campanularia hincksii	(s)	BV	0 to 110 m
Campanularia integra	(s)	B+	0 to 132 m
Campnaularia neglecta	(s)	B+	Lit. to 132 m
Campanularia pygmaea	(s)	B	—
Campanularia verticillata	(s)	B+	13 to 957 m

Campanularia volubilis	(s)	B+	Lit. to 109 m, on *Sargassum*
Gonothyraea gracilis	(s)	BV	Lit. to 137 m, on *Sargassum*
Gonothyraea integra	(s)	V	24 m
Gonothyraea loveni	(s)	B+	Lit. to 957 m
Eucopella caliculata	(m)	B+	Lit. to 73 m
Eucopella sargassicola	(m)	C+	on *Sargassum*

Obelia has free medusae but the medusae are not distinguishable to species.

Obelia articulata	—	B+	Lit. to 82 m
Obelia bicuspidata	—	BV	Lit. to 35 m, on *Sargassum*
Obelia commissuralis	—	BV	Lit. to 46 m
Obelia dichotoma	—	BV	0 to 73 m, on *Sargassum*
Obelia flabellata	—	B+	Lit. to 82 m
Obelia geniculata	—	BV	Lit. to 82 m, on *Sargassum*
Obelia hyalina	—	C+	Gulf Stream (on *Sargassum?*)
Obelia longissima	—	*B*+	Lit. to 69 m, eury.

Clytia has free medusae which are generally known under the name *Phialidium;* no attempt is made here to correlate medusae with polyps.

Clytia coronata	(m)	V	0 to 46 m
Clytia cylindrica	(m)	BV	0 to 119 m, on *Sargassum*
Clytia edwardsi	(m)	B+	Lit. to 112 m
Clytia fragilis	(m)	C+	13 to 26 m, on *Sargassum*
Clytia johnstoni	(m)	BV	Lit. to 439 m
Clytia longicyatha	(m)	V	9 to 165 m
Clytia minuta	(m)	B+	Shallow
Clytia noliformis	(m)	C, 38°	Pelagic, on *Sargassum*
Clytia raridentata	(m)	V	9 to 82 m
Phialidium bicophorum	(p)	BV	—
Phialidium singularis	(p)	V	—
Phialidium languidum	(p)	BV	—
Phialidium folleatum	(p)	V	—
Family Campanulinidae			
Calycella syringa	(s)	B+	Lit. to 274 m
Opercularella lacerata	(s)	B+	Lit. to 71 m
Opercularella pumila	(s)	B+	9 to 27 m
Stegopoma plicatile	(s)	BV	17 to 1941 m
Cuspidella costata	(m)	V	16 to 20 m
Cuspidella humilis	(m)	V	0 to 110 m
Lovenella gracilis	(m)	V	3 to 29 m
Lovenella grandis	(m)	V	0 to 22 m
Lovenella producta	(m)	B	18 to 787 m
Family Mitrocomidae			
Tiaropsis multicirrata	(p)	B+	Polyp equals *Campanulina* sp.

Halopsis ocellata	(p)	B+	Polyp presumably "*Cuspidella-like*"
Mitrocomella polydiademata	(p)	B+	Polyp presumably "*Cuspidella-like*"
Family Laodiceidae			
Laodicea undulata	(p)	BV	Polyp presumably "*Cuspidella-like*
Staurophora mertensi	(p)	B+	Polyp presumably "*Cuspidella-like*"
Toxorchis kellneri	(p)	BV	—
Ptychogena lactea	(p)	B+	—
Family Melicertidae			
Melicertum octocostatum	(p)	B+	—
Orchistomella tentaculata	(p)	V	—
Family Lovenellidae			
Blackfordia virginica	(p)	V	—
Blackfordia manhattensis	(p)	V	—
Eucheilota ventricularis	(p)	BV	—
Eucheilota duodecimalis	(p)	V	Polyp presumably *Lovenella*-like
Family Dipleurosomidae			
Dipleurosoma typicum	(p)	B	Polyp presumably *Cuspidella*-like.
Family Eutimidae			
Tima formosa	(p)	BV	Polyp presumably *Campanopsis*-like.
Eutima mira	(p)	V	Polyp presumably *Campanopsis*-like.
Family Aequoreidae			
Rhacostoma atlanticum	(p)	BV	—
Aequorea albida	(p)	B+	—
Aequorea aequorea	(p)	V	—
Aequorea tenuis	(p)	V	—
Family Lafoeidae			
Hebella calcarata	(m)	V	Lit. to 18 m, on *Sargassum*
Hebella pocillum	(m)	B	0 to 37 m
Lafoea dumosa	(s)	BV	7 to 823 m
Lafoea fruticosa	(s)	B+	5 to 957 m
Lafoea gracillima	(s)	B+	5 to 957 m
Lafoea pygmaea	(s)	B+	26 to 46 m
Lafoea symmetrica	(s)	B+	13 to 37 m
Grammaria abietina	(s)	B+	Lit. to 165 m
Grammaria gracilis	(s)	B	Lit. to 73 m
Lictorella crassitheca	(s)	B	31 to 327 m
Filellum serpens	(s)	BV	9 to 476 m
Family Halecidae			
Halecium articulosum	(s)	B+	14 to 245 m
Halecium beani	(s)	B, 39°	0 to 1135 m
Halecium curvicaule	(s)	B	46 to 71 m

Halecium diminutivum	(s)	B	13 m
Halecium flexile	(s)	V	9 to 201 m
Halecium gracile	(s)	BV	Lit. to 285 m, eury.
Halecium halecium	(s)	B+	13 to 370 m
Halecium labrosum	(s)	B+	27 to 73 m
Halecium minutum	(s)	B+	27 to 132 m
Halecium muricatum	(s)	B+	15 to 2743 m
Halecium robustum	(s)	B	110 to 787 m
Halecium sessile	(s)	B+	22 to 80 m
Halecium tenellum	(s)	BV	0 to 439 m, on *Sargassum*
Halecium macrocephalum	(s)	C, 39°	0 to 274 m, on *Sargassum*
Family Sertularidae			
Sertularia amplectens	(s)	C+	On *Sargassum*
Sertularia cornicina	(s)	C, 44°	On *Sargassum*, *Zostera*, *etc.*
Sertularia gracilis	(s)	C+	On *Sargassum*
Sertularia inflata	(s)	C+	On *Sargassum*
Sertularia pumila	(s)	B, 40°	Lit. to 55 m
Sertularia stookeyi	(s)	C+	Surface to 15 m, on *Sargassum*
Abietinaria abietina	(s)	B	7 to 93 m
Abietinaria filicula	(s)	B	37 m
Selaginopsis mirabilis	(s)	BV	15 to 110 m
Hydrallmania falcata	(s)	B+	9 to 292 m
Sertularella gayi	(s)	BV	45 to 1453 m
Sertularella geniculata	(s)	B	110 to 787 m
Sertularella gigantea	(s)	B+	27 to 137 m
Sertularella polyzonias	(s)	B, 41°	27 to 274 m
Sertularella rugosa	(s)	B+	0 to 73 m
Sertularella tenella	(s)	BV	0 to 2287 m, on *Sargassum*
Sertularella tricuspidata	(s)	B, 40°	Lit. to 787 m
Thuiaria argentea	(s)	BV	Lit. to 211 m
Thuiaria cupressina	(s)	B, 39°	Lit. to 273 m
Thuiaria fabricii	(s)	B+	Lit. to 73 m
Thuiaria immersa	(s)	B	15 to 91 m
Thuiaria latiuscula	(s)	B, 37°	Lit. to 102 m
Thuiaria lonchitis	(s)	B+	0 to 91 m
Thuiaria plumulifera	(s)	V	0 to 106 m
Thuiaria robusta	(s)	B+	0 to 93 m
Thuiaria similis	(s)	B+	5 to 132 m
Thuiaria tenera	(s)	B+	73 m
Thuiaria thuja	(s)	B+	45 to 84 m
Diphasia fallax	(s)	BV	7 to 311 m
Diphasia rosacea	(s)	B+	0 to 311 m
Diphasia tamarisca	(s)	B	—
Family Plumularidae			
Schizotricha gracillima	(s)	B	18 to 37 m
Schizotricha tenella	(s)	BV	0 to 55 m
Cladocarpus flexilis	(s)	V	59 to 306 m
Theocarpus myriophyllum	(s)	B	110 to 549 m

Aglaophenia latecarinata	(s)	C+	On *Sargassum*
Aglaophenia perpusilla	(s)	C+	On *Sargassum*
Monostaechas quadridens	(s)	V	3 to 542 m
Antennularia americana	(s)	BV	38 to 435 m
Antennularia antennina	(s)	BV	14 to 485 m
Antennularia rugosa	(s)	V	84
Plumularia spp.	(s)	C+	Several spp. on *Sargassum*

Identification

Prompt preservation of material for study is essential or, preferably, living specimens should be studied. The individual polyps are frequently minute, delicate, and are easily destroyed or damaged by careless handling. Standard seawater formalin is a satisfactory preservative. Narcotization is necessary to obtain expanded polyps but, in general, this treatment is not required for the identification procedures employed here. With a few exceptions, reproductive structures are indicated as supplementary evidence only, though such structures are of primary importance in formal systematics.

A substantial number of species has been reported locally only on *Sargassum* and most of them are excluded from the keys. Most of the species are figured in Fraser (1944) or Nutting (1900–1915). Dimensions are maximum adult heights of zooids or colonies, as indicated.

This is a large and hence somewhat unwieldly assemblage. The species are divided initially into nine groups. These groups are unequal in size and to a considerable extent are artificial in composition. The initial separation may be effected as follows.

---

1. Colony unbranched or very sparsely branched, the polyps rising singly from a stolon, stolon network, or crust; polyps naked, i.e., with perisarc obscure, or not forming a collared termination below the hydranth; tentacles scattered on hydranth, or in a single or double whorl; gonophores arise independently from the base of the colony, or emerge from the hydranth below the tentacles, or gonophores lacking; Fig. 5.3. Group One.

1. Colony branched or not; polyps usually with a perisarc that either terminates below the hydranth as a definite collar or is expanded to form a hydrotheca into which the hydranth may withdraw completely or in part (possibly obscure in Group Two, Fig. 5.4)—2.

2. Perisarc expanded to form a hydrotheca which in some species is closed by an operculum; hydrothecae developed on pedicels, or sessile—6.

2. Perisarc without an obvious hydrothecal expansion—3. Note: In *Halecium* spp. the hydrotheca is reduced to, at most, a terminal flange, and in some species of other genera the tubular hydrotheca is undifferentiated from its pedicel. Such forms are here considered "sessile"; proceed to couplet 6, Fig. 5.7.

3. Hydranth relatively large and "flowerlike," with separate proximal and distal whorls of tentacles or with a basal whorl plus scattered tentacles on the hydranth, (frequently the latter tentacles are of a different type); gonophores emerge from the hydranth above the basal whorl; Fig. 5.4. Group Two.
3. Hydranth with either a single or double contiguous whorl of filiform tentacles or with scattered filiform or capitate tentacles on the hydranth but not with a combination of capitate and filiform tentacles on the same hydranth, or of separate distal and proximal whorls—4.
4. Colonies simple and unbranched, or delicate and sparsely branched; in marine forms, colonies are <20 mm and usually <10 mm in extent; freshwater forms reach 60 mm in height; Fig. 5.5. Group Three.
4. Colonies larger, bushy, or, if sparsely branched, then fascicled in part—5.
5. Hydrothecae reduced, usually with a flared rim as in Fig. 5.11 or, if simple, then colony without annulations *Halecium;* see Table 5.8, Fi.g 5.11, pp. 99, 100.
5. Hydrothecae tubular, without flared rim; colony with either annulations or fascicled in part; Fig. 5.6. Group Four.
6. With a cuplike hydrotheca, or cuplike expansion of the perisarc—7.
6. Hydrothecae either sessile or with a conical or wedge-shaped operculum as in Fig. 5.7—9.
7. Perisarc with a wrinkled, cuplike expansion; Fig. 5.8. Group Six. Note: In *Garveia*, in Group One, the gonophores when fully developed have a wrinkled, cuplike expansion but the nutritive zooids do not.
7. Not so—8.
8. Margin of cuplike hydrotheca toothed; Fig. 5.9. Group Seven.
8. Margin of cuplike hydrotheca not toothed; Fig. 5.10. Group Eight. (See also introductory note to Group Seven for somewhat ambiguous forms.)
9. With a conical or wedge-shaped operculum; Fig. 5.7. Group Five.
9. Hydrotheca sessile; Fig. 5.12. Group Nine.

Group One. This somewhat heterogeneous group, representing four families, includes several common, shallow water or littoral genera. *Rhizogeton* is found in tide pools. *Clava* is common on seaweeds, especially *Fucus*, in the lower littoral. The hydractinids grow on small gastropods; *Stylactis* is usually found on live *Nassarius*, and *Podocoryne* and *Hydractinia* most frequently occur on shells inhabited by hermit crabs.

1. Tentacles lacking; zooid solitary, 5 mm; brackish water (See Pennak, 1953). *Protohydra leuckarti.*
1. Tentacles scattered on hydranth—2.
1. Tentacles in a single or double whorl—3.
2. With no more than 12 tentacles; gonophores independent; zooids 6 mm tall; Fig. 5.3A. *Rhizogeton fusiformis.*
2. With 20–30 tentacles; gonophores on hydranth; zoids 10 mm; Fig. 5.3B. *Clava leptostyla.*
3. Base a stoloniferous network—4.
3. Base a stoloniferous network covered by a crust—5.

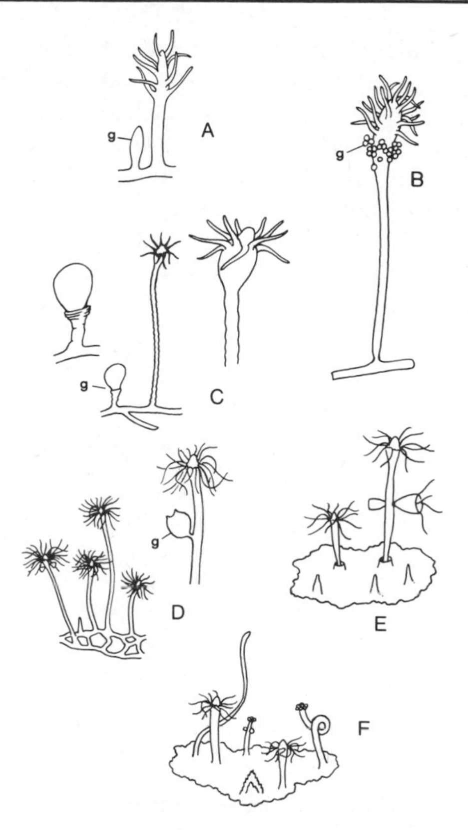

Fig. 5.3. Hydrozoan polyps, Group One. A, *Rhizogeton fusiformis;* B, *Clava leptostyla;* C, *Garveia groenlandica*, with details of hydranth and gonophore; D, *Stylactis* sp., with detail showing gonophore; E, *Podocoryne carnea;* F, *Hydractinia* sp. In this and subsequent figures, gonophores are indicated by "g."

4. With about 10 tentacles; stem wrinkled or somewhat annulated; gonophores independent, when mature consisting of a sporosac emerging from a wrinkled basal cup; zoids 6 mm; Fig. 5.3C. *Garveia groenlandica.*
4. With 12–20 tentacles; stem not wrinkled; colony not conspicuously polymorphic, gonozooids usually smaller than gastrozooids but otherwise similar; Fig. 5.3D. *Stylactis: S. arge* has 12–16 tentacles equally divided in two whorls, zooids 20 mm; *S. hooperi* has about 20 tentacles in a single whorl, zooids 45 mm.
5. Basal spines smooth; colony not conspicuously polymorphic, gonozooids usually smaller than gastrozooids but otherwise similar; zooids 5 mm; Fig. 5.3E. *Podocoryne carnea.*
5. Colony highly polymorphic with nutritive, sensory, defensive, and male and female generative zooids distinctly different; gastrozooids 4.5 mm; Fig. 5.3F. *Hydractinia: H. valens* with 10 tentacles, basal spines smooth; *H. echinata* with up to 30 tentacles, basal spines jagged.

Group Two. This is a natural assemblage of related families, including several peculiar, aberrant forms. *Hypolytus* and *Margelopsis* have pelagic tendencies, and the latter (not figured) has been likened to a pedicel-less hydranth; though pelagic as a polyp, *Margelopsis* also produces free medusae. *Hybocodon prolifer* is a tide pool species, but is better known and more widely distributed in medusoid form. Most of the species require firm substrata, but *Dahlgrenella* and *Acaulis* grow on mud. Several species of *Tubularia*, *Pennaria*, and *Corymorpha* are common in shallow water. Species of *Tubularia* are important fouling organisms on buoys; *T. crocea* is the only species with an extensive Virginian range. The gonophores are borne on the hydranth *above* the basal whorl of tentacles in all of these genera.

1. Zooids with aboral end tapering more or less to a point; not fixed to a firm substratum—2.
1. Zooids with basal attachment a stolon, holdfast, or with rootlets, fixed to a firm substratum—3.
2. Base of hydranth with a ring; basal tentacles 14, distals 10; zooid 15 mm; questionably pelagic, probably a mud dweller; Fig. 5.4B. *Hypolytus peregrinus.*
2. Base of hydranth with small tubercles, basal tentacles 6–16, distals 6–8; zooids 5 mm; benthic on mud; Fig. 5.4C. *Heteractis aurata.* Note that variant *Tubularia crassa* and *T. couthouyi* may also be found with a free, tapered, aboral end. See Table 5.3 for *Tubularia* spp.
3. With two types of tentacles, a basal whorl of filiform tentacles, and scattered capitate tentacles on hydranth, or with more than 2 rows of tentacles on hydranth —4.
3. With filiform tentacles only, in separate basal and distal whorls—6.

Table 5.3. Tubularia Species

Tubularia	Max. Zooid Height, mm	Proximal Tentacles	Distal Tentacles	Stem Annulated	Stem Branched	Gonophores with Apical Process
couthouyi	150	30–40	–50	Yes	No	Yes
crassa	18	32–36	–36	No	No	No
crocea	125	20–24	20–24	No	Few	L*
indivisa	300	–40	–40	No	No/few	No
larynx	50	ca. 20	ca. 20	Yes	Yes	C*
spectabilis	100	ca. 20	ca. 20	No	Few	C*
tenella	20	ca. 18	ca. 18	No	Few	C*

*L = yes, laterally compressed; *C = yes, conical.

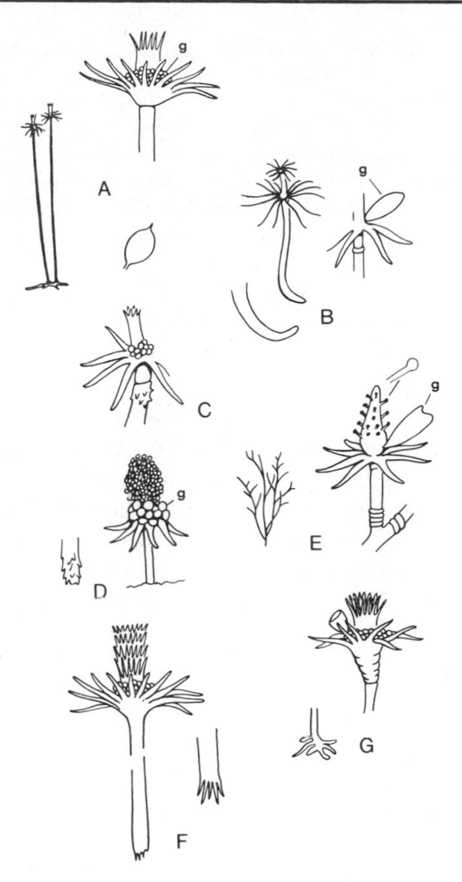

Fig. 5.4. Hydrozoan polyps, Group Two. A, *Tubularia* sp. with detail below showing a gonophore with an apical process; B, *Hypolytus peregrinus*, with details of base and part of lower whorl of tentacles with a gonophore; C, *Heteractis aurata;* D, *Acaulis primarius* with detail of base at left; E, *Pennaria tiarella*, colony form at left; F, *Corymorpha pendula*, with detail of base at right; G, *Hybocodon prolifer*, with detail of holdfast at left.

4. Colonies branching; stems annulated; 10 or 12 filiform, basal tentacles plus 4 or 5 whorls of capitate tentacles on hydranth; base with multiple rootlets; colony 150 mm; Fig. 5.4E. *Pennaria tiarella.*

4. Solitary forms; stem without annulations—5.

5. Stem short; with 8 filiform, basal tentacles plus numerous capitate tentacles on hydranth; zooid 7 mm; Fig. 5.4D. *Acaulis primarius.*

5. Stem long; with about 30 basal tentacles plus numerous filamentary tentacles in whorls on hydranth; zooid 100 mm; Fig. 5.4F. *Corymorphya pendula.*

6. Base of hydranth annulated; distal tentacles in two whorls of about 16 tentacles each; base with a holdfast; zooid 50 mm; Fig. 5.4G. *Hybocodon prolifer.*

6. Not so, base a stolon—7.

7. With fewer distal than basal tentacles; stem annulated at base, with few or no branches. *Ectopleura:* Two questionably distinct species, *dumortieri* with 30 basal tentacles and 24 distals, zooid 12 mm; *prolifica* with 24 basals, 10–12 distals, and zooid 6 mm. (*Ectopleura* produces free medusae, *Tubularia* does not; this is the only real difference between these genera.)

7. With distal tentacles equal to or more numerous than basals; stem annulated or not, branched or not; Fig. 5.4A. *Tubularia* spp. See Table 5.3.

Group Three. This is a heterogeneous assemblage, and some of the same or related genera also appear in Group Four. In addition to several minute or otherwise obscure species, the present group includes two more conspicuous genera. *Cordylophora* is widely distributed in fresh water, and *Eudendrium* has a wide bathymetric and geographical range. The gonophores of *Eudendrium* and of the related *Dicoryne* and *Perigonimus* are usually arranged in clusters. In *Eudendrium* the colonies are sexually dimorphic; male gonophores are often in whorls or fans and are multichambered, Figure 5.6E.

1. Colony a minute, reticular stolen epizoic on other hydroids, worm tubes, etc.; with short, upturned tubular hydrothecae continuous with stolon and elevated; <1mm; Fig. 5.5A. *Fillelum serpens.* Note that young *Lafoea dumosa* are also epizoic and also stolonate; see Group Four.

1. Not so—2.

2. Tentacles capitate, scattered on a somewhat elongate hydranth; stem long or short, annulated or not; Fig. 5.5B,D. Several problematic and in part synonymous forms key out here; at various times, these have been placed in the genera *Sarsia, Zanclea, Syncoryne,* and *Corynitis.* See Fraser (1944), Hargitt (1910).

2. Tentacles filiform, scattered on hydranth or in a single whorl—3.

3. Tentacles scattered on hydranth—4.

3. Tentacles in a whorl—6.

4. Branches and pedicels annulated basally; mature colony well branched; gonophores are sporosacs, on stem; colony 60 mm; fresh to brackish water; Fig. 5.5C. *Cordylophora lacustris.*

4. Colony without annulations; sparsely branched or zooids solitary; gonophores independent, on stolon, or hydranth; marine species—5.

5. Stem unbranched, tapering basally; colonies stolonate, on bivalve mollusks; zooids 5 mm; gonophores arise independently; Fig. 5.5E. *Tubiclava cornucopiae.*

5. Stem sparsely branched, tubular, not tapered; colonies stolonate, on algae, etc., about 5 mm tall; gonophores emerge from base of hydranth; Fig. 5.5G. *Turritopsis nutricola.*

6. With 20 or more tentacles; stem, branches, and pedicels annulated, or with basal annulations only; if without annulations, then tentacles should be 26–32; Fig. 5.5F,H,I. *Eudendrium* spp. See Table 5.4.

6. Tentacles 16 or less; stem, branches, and pedicels annulated, wrinkled, or not—7.

7. Stem, branches, and pedicels annulated basally, wrinkled otherwise; tentacles about 16; stolon reticular with erect stems sparsely branched or with solitary zooids; colonies <5 mm tall; Fig. 5.5J. *Dicoryne conferta.*

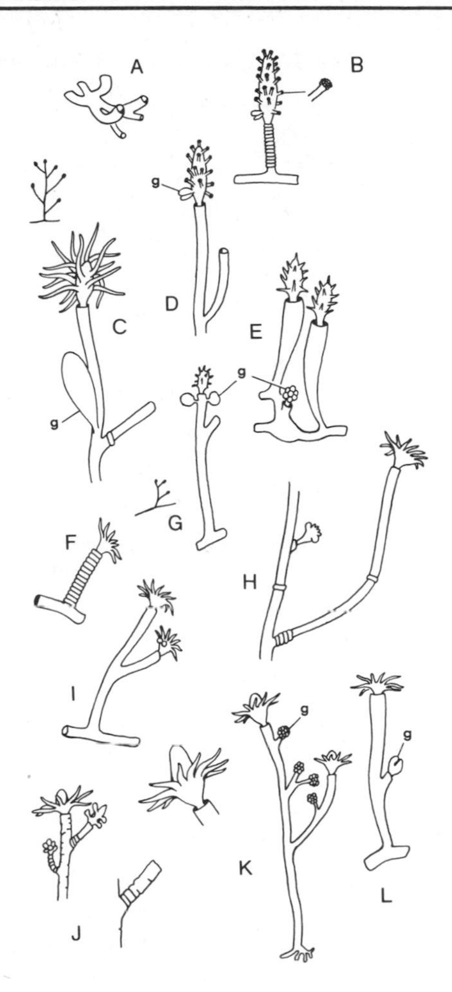

Fig. 5.5. Hydrozoan polyps, Group Three. A, *Fillelum serpens*, hydranths omitted; B, problematic form, family Corynidae; C, *Cordylophora lacustris*, with colony form above; D, problematic form, family Corynidae; E, *Tubiclava cornucopiae;* F, *Eudendrium* sp.; G, *Turritopsis nutricola*, with colony form at left; H, I, *Eudendrium* sp.; J, *Dicoryne conferta;* K, *Dicoryne flexuosa*, with detail of hydranth at left; L, *Perigonimus* sp.

7. Colony without annulations, stems smooth or wrinkled—8.

8. Tentacles about 12; stem thin, flexible, sparsely branched; colony about 7 mm; Fig. 5.5K. *Dicoryne flexuosa*.

8. Stem thicker, stronger; colony about 8 mm; Fig. 5.5L. *Perigonimus* spp. Fraser (1944) lists *P. jonesi* with about 16 tentacles, colonies branching; *P. repens* with about 10 tentacles, colonies unbranched. Note comments in check list.

Table 5.4. *Eudendrium* Species

Eudendrium	Parts Fascicled	Colony Size, mm	Colony Type (1)	Number of Tentacles	See note below	Parts annulated
rameum	sm	40	b, c	24–25	b	p
dispar	s	100	b, c	28	b	p
ramosum	s	150	c	24	a, b	b
carneum	s	120	c	24	a	b
vaginatum	—	40	b	20	b	srp
capillare	—	15	b, c	20–30	a	b
tenue	—	15	b	20	a	b
insigne	—	<6	a	20–25	a, b	srp
tenellum	—	10	a	20	a, b	b
album	—	8	a	26–32	b	—

s, stem; m, main branches; (1) refer to *Halecium* standards, Fig. 5.11A–C; a, gonophores with aborted hydranths; b, gonophores on normal or only slightly aborted hydranths (an "aborted hydranth" is not fully developed); p, pedicels; b, with basal annulations only; r, branches.

Group Four. An extremely heterogeneous lot, superficially similar in having the hydranths emerging from a simple tubular stem or with simple, tubular hydrothecae. *Bougainvillia carolinensis* is a common shallow water species, as are several species of the related genus *Eudendrium* (see also Group Three). For other species of *Bougainvillia*, see Group Six. The remaining species of the present group belong to Thecata, but the hydrothecae are sessile as in Figure 5.6A, or not clearly differentiated from their pedicels. Other species of *Lafoea* are keyed out in Group Eight. The reproductive structures of Lafoeidae are aggregated in masses called *coppinia*.

---

1. Colony without annulations; sparsely but strongly branched, stem and main branches fascicled (in bundles); hydrothecae consisting of relatively short emergent tubules; gonophores in coppinia—2.

1. Colony with annulations; usually profusely branched, fascicled in part or not; pedicels or branches elongate; without coppinia—4.

2. Free part of hydrothecae very short, essentially sessile, curving; colony about 120 mm; Fig. 5.6A. *Grammaria gracilis.*

2. Free part longer—3.

3. At least some of the hydrothecae expanded basally; colony about 50 mm; Fig. 5.6B. *Grammaria abietina.*

3. At least some of the hydrothecae tapered basally; colony about 40 mm; Fig. 5.6C. *Lafoea dumosa.*

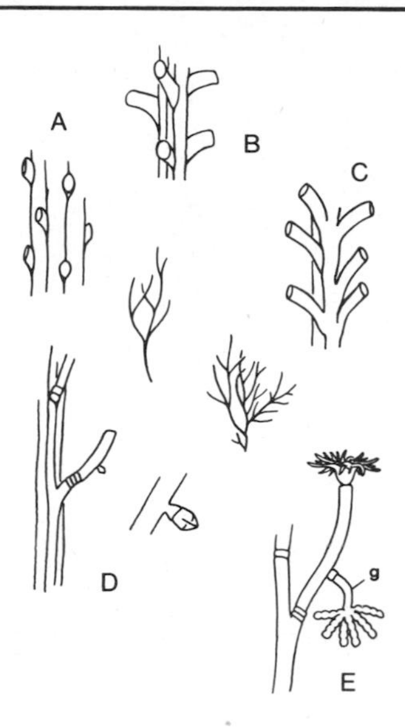

Fig. 5.6. Hydrozoan polyps, Group Four. A, *Grammaria gracilis*, with colony form below, right; B, *Grammaria abietina;* C, *Lafoea dumosa;* D, *Bougainvillia carolinensis*, with detail at right showing gonophore; E, *Eudendrium* sp. with colony form above.

4. Tentacles 10–12; main stem (at least) fascicled; branches annulated basally; colony about 300 mm; Fig. 5.6D. *Bougainvillia carolinensis.*

4. Tentacles 20–25; colony fascicled in part or not; extent of annulation varies; Fig. 5.6E. *Eudendrium* spp. See Table 5.4.

Group Five. This group epitomizes the difficulty in obtaining a unified taxonomy for polyps and medusae. The assembled species were treated by Fraser (1944) as members of a single family.

---

1. Operculum wedge-shaped; stem fascicled; gonophores 5–6 times larger than nutritive hydrothecae; colony 65 mm; Fig. 5.7A. *Stegopoma plicatile.*

1. Operculum cone-shaped—2.

2. Hydrothecae developed on pedicels—3.

2. Hydrothecae sessile, solitary on stolon; opercular segments 10–14; emergent zooids about 0.3 mm; Fig. 5.7E. *Cuspidella: C. costata* about 0.14 mm in diameter; *C. humilus* about 0.075 mm.

3. Opercular segments hinged at base; Fig. 5.7B—4.

3. Opercular segments without a definite hinge line; colonies branching; gonophores produce sporosacs and acrocysts; Fig. 5.7C. *Opercularella*: *O. pumila* has a minute, creeping colony, <4 mm in extent with spindle-shaped sporosacs; *O. lacerata* is more profusely branched, colony 25 mm, and sporosacs egg-shaped.

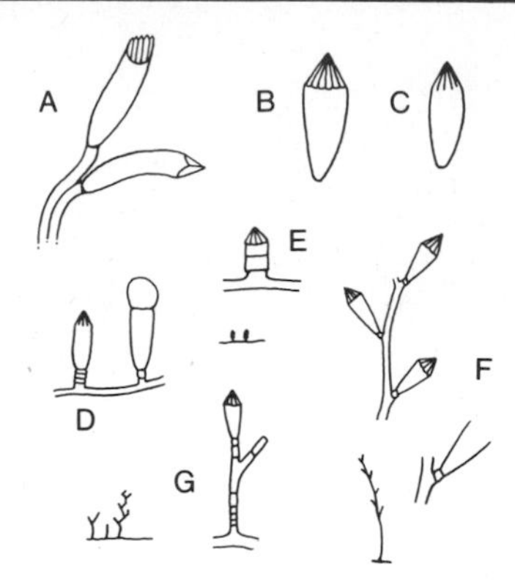

Fig. 5.7. Hydrozoan polyps, Group Five. A, *Stegopoma plicatile;* B,C, hydrothecae indicating opercular segments, with (B) and without (C) a definite hinge line; D, *Calycella syringa*, with sporosac extruding a bubble-like acrocyst; E, *Cuspidella*, sp. with colony form below; F, *Lovenella grandis*, with colony form and detail below; G, *Lovenella gracilis*, colony form at left.

4. Hydrothecae tubular; zooids solitary on short, annulated pedicels; colony stolonate; opercular segments 8 or 9; zooids <1 mm tall; sporosacs extrude acrocysts; Fig. 5.7D. *Calycella syringa.*
4. Hydrothecae top-shaped—5.
5. Pedicels annulated throughout length; very short in *Lov. grand.*, Fig. 5.7F-G–6.
5. Pedicels annulated basally; operculum with at least 12 segments; colony about 3 mm. *Lovenella producta.*
6. Opercular segments 8; hydrothecae about 0.9 mm, colony about 15 mm; zooids solitary or branched; Fig. 5.7G. *Lovenella gracilis.*
6. Opercular segments 10–12; hydrothecae about 1.75 mm, colony about 50 mm; stem unbranched; Fig. 5.7F. *Lovenella grandis.*

Group Six. Additional members of Bougainvilliidae are keyed out in Group Four.

1. Colony a creeping stolon, zooids solitary or sparsely branching, <5 mm tall; tentacles 11 or 12; Fig. 5.8A; gonophores unknown. *Bimeria brevis.*
1. Colony taller, more profusely branched—2.
2. Stem fascicled, smooth or slightly wrinkled, not annulated; colony 75 mm; tentacles 8–10; gonophores small, arising from hydranth pedicels; Fig. 5.8B. *Bougainvillia rugosa.* See also *B. carolinensis*, Group 4 couplet 4.
2. Stem not fascicled, colony with annulations—3.
3. Branches and pedicels annulated; colony much branched, to 50 mm; tentacles 15–20; gonophores as above; Fig. 5.8C. *Bougainvillia superciliaris.*
3. Branches and branchlets annulated basally, perisarc sometimes wrinkled in part elsewhere; colony a main branch plus branchlets, more or less pinnate; 100 mm, tall; tentacles 10–12; gonophores single or in clusters, just below the hydranth cup; Fig. 5.8D. *Calyptospadix cerulea.*

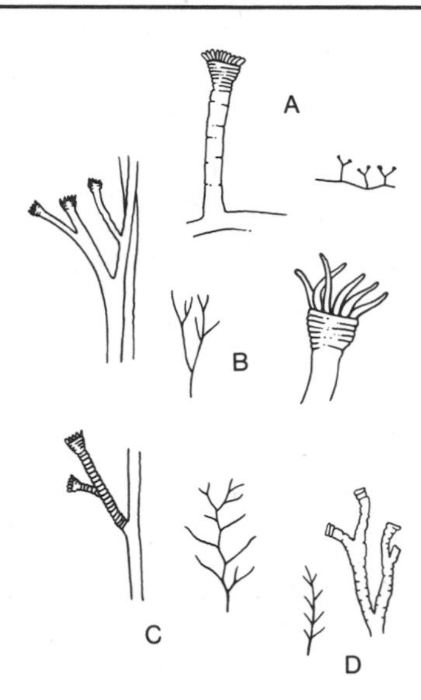

Fig. 5.8. Hydrozoan polyps, Group Six. A, *Bimeria brevis*, colony form at right; B, *Bougainvillia rugosa*, colony form and detail at right; C, *Bougainvillia superciliaris*, colony form at right; D, *Calyptospadix cerulea*, colony form at left.

Group Seven. Campanularidae with toothed, cuplike hydrothecae are keyed here; those with smooth-rimmed, cuplike hydrothecae are keyed in Group Eight with members of the families Lafoeidae and Hebellidae.

The campanularids are an important group with many shallow water species. Some attain a large size and are conspicuous; others are small species growing on seaweeds and other living substrata. The genera can only be differentiated on the basis of generative structures, and the keys and tables given are wholly artificial.

The species are initially divided between (A) small, frequently stolonate, solitary or sparsely branched species, and (B) larger, profusely branched species. The two groups are treated in Table 5.5 p. 96.

Group Eight. The margin of the cup may be wavy rather than perfectly even in *Obelia longissima*, *O. dichotoma*, and *Campanularia volubulis.*

---

1. Stem and/or branches fascicled; gonophores in coppinia—2.

1. No parts fascicled—5.

2. Nematophores present (one small nematophore at base of each branch, otherwise scarce and easily overlooked, Fig. 5.10A); colony pinnate, 60 mm; stems fascicled; pedicels short, with one annulation. *Lictorella crassitheca.*

2. Without nematophores—3.

Table 5.5. Campanularian Hydroids With Toothed Cups

Group	Length of Hydrotheca, mm	Extent of branching	Colony height, mm	Gonophore Type, Fig. 5.9	Tooth Type, Fig. 5.9	Number Teeth	Notes
Group A							
Campanularia volubilis	<1.0	u	<2	F	D	10–12	—
?*Campanularia gigantea*	>1.0	u, f	15	—	E	12–14	—
Campanularia groenlandica	0.9	u	6	F	J	10–12	Cup ribbed, Fig. 5.9C
Campanularia hincksi	1.2	u	<3	B	K	10–12	Cup ribbed, Fig. 5.9C
? *Campanularia pygmaea*	>1.0	u, f	1		L	10–14	—
Gonothyraea gracilis	0.8	f	8	G	G–H	12–14	On *Sargassum*
Clytia coronata	?	u, f	5	G	E–H	7–12	—
Clytia cylindrica	0.7	u, f	<7	G	H	10–12	On *Sargassum*
Clytia edwardsi	1.0	b	25	A	H	12–14	—
Clytia johnstoni	v	u, f	<7	A	E–H	12–16	On *Sargassum*
Clytia inconspicua	0.3	u	<2	G	E–F	7	—
Clytia noliformis	?	u	<2	D	D–E	10–14	On *Sargassum*
Clytia raridentata	0.3	u	<2	G	F	5–10	On *Sargassum*
Group B							
Campanularia gelatinosa	?	b	250	D	L	10	Fascicled
? *Campanularia fasciculata*	0.6	b	20	—	N	16	Fascicled
Campanularia neglecta	?	b	20	D	L	8–12	—
Campanularia verticillata	0.8	b	350	E	E–J	16	Fascicled
Gonothyraea loveni	?	b	30	D	J	10–12	—
Clytia longicyatha	1.0	b	25	D	G	18–24	Fascicled
Clytia minuta	?	f	20	A	H	7–8	—
Obelia articulata	?	b	70	—	E	12–14	Fascicled
Obelia bicuspidata	0.4	b	21	A	L	14–20	Fascicled, cup ribbed, Fig. 5.9C, on *Sargassum*

u, unbranched; f, with few branches; b, profusely branched; v, variable.

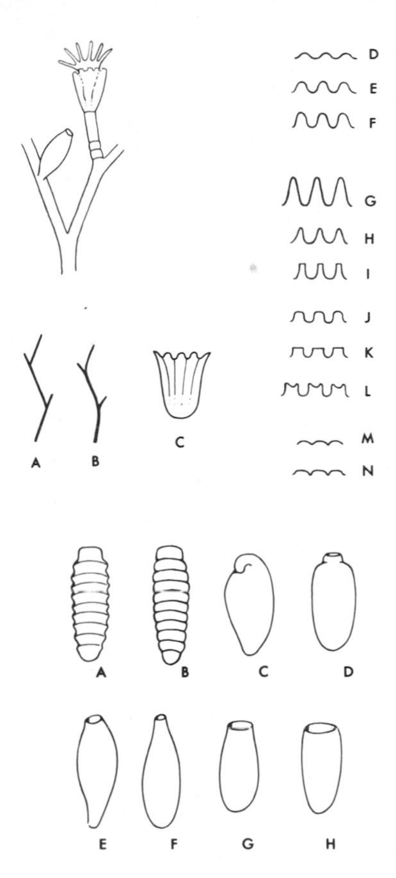

Fig. 5.9. Hydrozoan polyps, Groups Seven and Eight. Standards of tooth, gonophore, and colony growth forms for use with Tables 5.5 and 5.7. See also Fig. 5.10.

3. Branching nearly at right angles; pedicels with 5 or 6 annulations or twists; colony bushy, 35 mm; Fig. 5.10B. *Lafoea symmetrica.*

3. Many of the branches emerging at less than a right angle—4.

4. Branches not pinnate, sometimes all on one side; pedicels with 3 or 4 twists; colony 50 mm; Fig. 5.10C. *Lafoea fruticosa.*

4. Main stem short or indistinct, profusely branched; pedicels with one or more slight twists; colony 35 mm, sometimes prostrate; Fig. 5.10D. *Lafoea gracillima.*

5. Colony creeping or stolonate, no stem; minute species, erect zooids about 0.5 mm. See Table 5.6.

5. Colony erect; larger colonies. See Table 5.7.

Note. *Halecium.* This genus is keyed in the fifth couplet of the introductory key, (p. 86); Table 5.8 and Fig. 5.11 identify the species, pp. 99, 100.

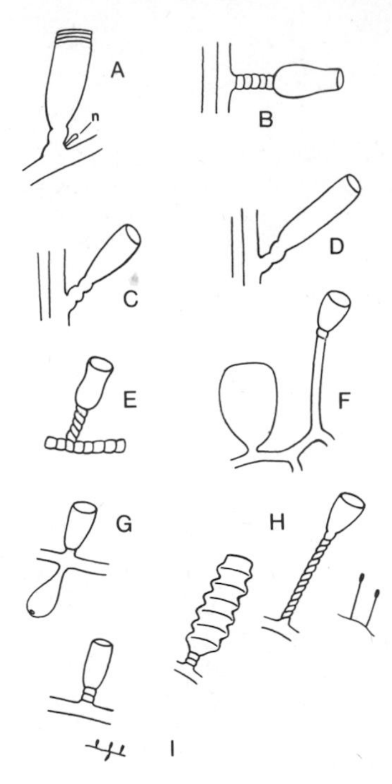

Fig. 5.10. Hydrozoan polyps, Group Eight. A, *Lictorella crassitheca*, "n" indicates the nematophore; B, *Lafoea symmetrica;* C, *Lafoea fruticosa;* D, *Lafoea gracillima;* E, *Hebella pocillum;* F, *Eucopella caliculata*, with sporosac; G, *Hebella calcarata*, with sporosac; H, *Campanularia integer*, with growth form at right and gonophore at left; I, *Lafoea pygmaea*, with growth form below.

Table 5.6. Stolonate Polyps, Group Eight

Species	Stolon	Pedicel		Fig. 5.10
Eucopella caliculata	r	a	s, 1	F
Hebella pocillum	a	a	s	E
Hebella calcarata	—	—	1	G
Lafoea pygmaea	—	a	s	I
Campanularia integra	r, 2	2	g	H

r, reticulated; a, annulated; s, short; g, long; 1, hydrotheca essentially sessile; 2, variable.

Table 5.7. Campanularian Hydroids with Smooth-Rimmed Cups

Species	Extent of Branching	Colony Height, mm	Gonophore Type, Fig. 5.9	Growth Form	Notes
Campanularia angulata	u, f	15	F	g	Stem wrinkled
Campanularia flexuosa	u, f	30	G	x	Stem brown
Campanularia amphora	b	15	F	—	—
Campanularia calceolifera	u, f	30	C	g	Internode 7–8 times longer than wide
Campanularia fragilis	u	3	?	—	Pinnate
Gonothyraea integra	u, f	<8	H	x	On *Sargassum*
Obelia geniculata	u	25	D	g	On *Sargassum*
Obelia commissuralis	b	200	D	x	—
Obelia dichotoma	u, f	25	D	—	On *Sargassum*
Obelia flabellata	b	250	D	x	Stem horn-colored

u, unbranched; f, with few branches; b, profusely branched; g, growth form geniculate, Fig. 5.9A; x, growth form flexuose, Fig. 5.9B. All species have the stem or pedicels all or partly annulated.

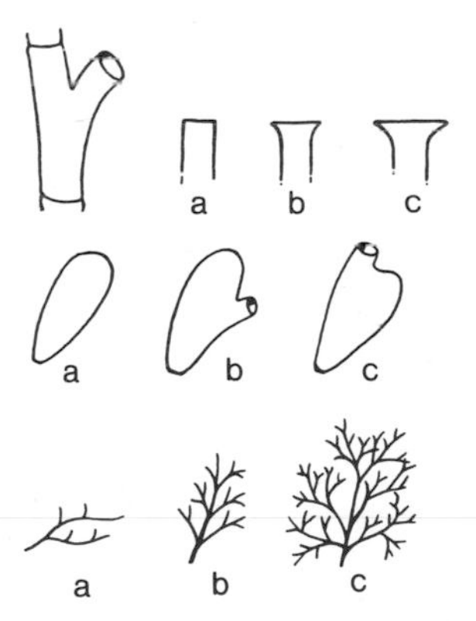

Fig. 5.11. Hydrozoan polyps, *Halecium*. Standards of cup, gonophore (gonotheca), and colony growth forms for use with Table 5.8. See note p. 97.

Table 5.8. *Halecium* Species

Halecium	Fascicled	Colony Height, mm	Hydrotheca, Fig. 5.11	Colony Fig. 5.11	♀ Gonophore, Fig. 5.11	Stem, Branches with Nodes	Annulated, at Least Partly	Notes
curvicaule	—	<3	b	a	c	—	—	—
diminutivum	—	<2	c	a	?	—	/	—
minutum	—	20	b	a	a	/	/	—
sessile	—	?	a	b	b	/	—	Hydrothecae sessile
tenellum	—	15	c	a	a	—	/	Sparsely branched
articulosum	/	50	a	b	a,b	/	—	—
beani	/	40	a, b	c	b	/	—	—
flexile	/	40	a, b	b	?	/	—	—
gracile	/	50	a, b	c	c	/	—	—
halecium	/	70	a, b	b	c	/	—	—
labrosum	/	80	c	b, c	a	/	/	—
muricatum	/	70	b	c	2	/	—	2–spiny

/ equals yes; — equals no.

Group Nine. Containing two large families, Sertulariidae and Plumularidae, with numerous shallow water species.

---

1. Nematophores present; gonophores "naked" except in extralimital forms and some likely to be taken on *Sargassum;* Fig. 5.15. Plumularidae. See separate key, below.

1. Without nematophores. Sertulariidae (gonophore types, Fig. 5.13)—2.

2. Hydrothecae in single row on one side of branch; gonophores of type d, Fig. 5.13; colony bushy, about 300 mm; Fig. 5.12A. *Hydrallmania falcata.*

2. Hydrothecae on all sides of branch; gonophores of type f, Fig. 5.13; main stem to about 70 mm; Fig. 5.12B. *Selaginopsis mirabilis.*

2. Hydrothecae in double row on opposite side of branch—3.

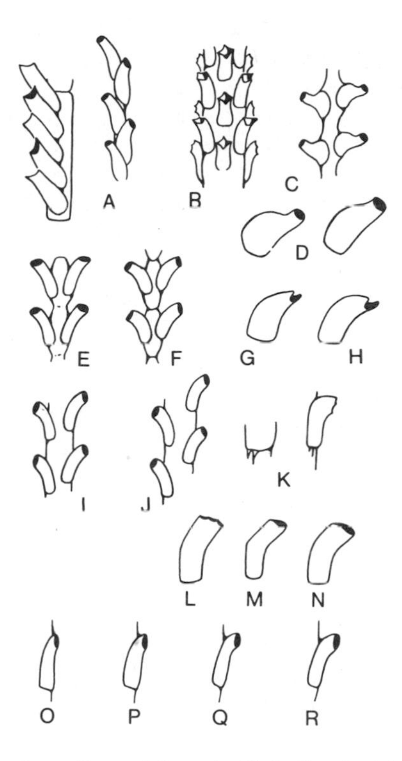

Fig. 5.12. Hydrozoan polyps, Group Nine. A, *Hydrallmania falcata*, lateral view at left, frontal view at right; B, *Selaginopsis mirabilis;* C, *Abietinaria* sp.; D, the same, individual hydrothecae; E, *Diphasia* sp., with hydrothecae opposite and nodes indistinct compared with F, *Sertularia* sp., with nodes distinct; G,H, *Thuiaria similis*, hydrothecae; I, hydrothecae subopposite; J, hydrothecae alternating; K, *Sertularia cornicina*, hydrotheca at right and detail of basal projections at left; L, *Diphasia tamarisca*, hydrotheca; M, *Diphasia rosacea*, hydrotheca; N, *Diphasia fallax*, hydrotheca; O–R, *Thuiaria* sp., hydrothecae showing varying degrees of immersion in branch, for use with Table 5.9.

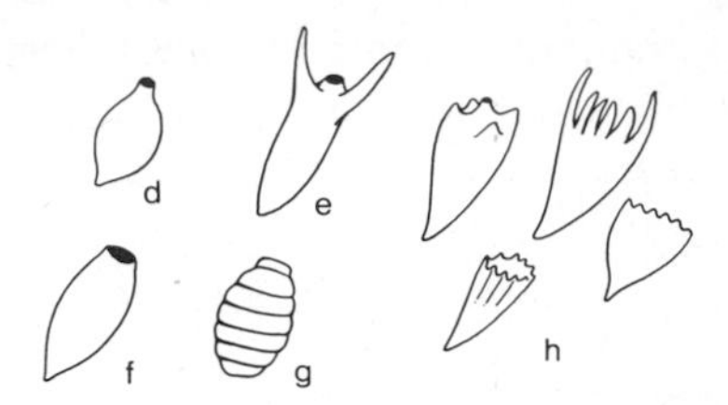

Fig. 5.13. Hydrozoan polyps, Group Nine. Gonophore (gonotheca) types for use with keys and tables for this group.

3. Hydrothecae broadly flask-shaped with a definite neck, alternating or opposite on branch, rim without teeth; Fig. 5.12C,D. *Abietinaria: A. abietina* usually without nodes; colony 300 mm and hydrothecae about 0.85 mm long; gonophores type f, Fig. 5.13; *A. filicula* with a few nodes in branches; colony and hydrothecae about half above sizes; gonophores type d, Fig. 5.13.

3. Hydrothecae more tubular, without neck or with rim distinctly toothed, or hydrothecae more immersed in branch—4.

4. Hydrothecae distinctly alternate with 1 hydrotheca per internode (at least in branches); rim usually with 4 teeth (range 3–6, in different species); gonophores at least partly annulated or transversely ridged; Fig. 5.14. *Sertularella* spp. See separate key below.

4. Hydrothecae distinctly opposite or, if subopposite or alternate, then nodes indistinct or with more than one hydrotheca per internode; rim with zero to three teeth; gonophores not annulated or ridged—5.

5. Hydrothecae subopposite or alternating, Fig. 5.12I,J; rim with one, two, or no teeth, *Thuiaria*. See Table 5.9.

5. Hydrothecae opposite or subopposite (at least in branches), Fig. 5.12EF; rim with two, three, or no teeth. Species of *Diphasia*, *Sertularia*, and *Thuiaria*. See Table 5.10.

Subkey for Sertularella. The common neritic species are all cold water forms.

---

1. Stem fascicled; hydrothecae corrugated on one side only; colony 150 mm; deep water species, usually below 183 m; Fig. 5.14A. *S. gayi.*

1. Stem not fascicled—2.

2. Colonies small, <25 mm, unbranched or sparsely branched; hydrothecae corrugated all around; margin of hydrothecae with 4 teeth; Fig. 5.14B–D—3.

2. Colonies larger, 50–150 mm, usually branched; hydrothecae smooth, not corrugated, margin with 3–6 teeth; Fig. 5.14F,G—5.

3. Internodes long; Fig. 5.14C—4.

3. Internodes short; Fig. 5.14B. *S. rugosa.*

4. Hydrothecae without a smooth collar; Fig. 5.14D. *S. geniculata.*

4. Hydrothecae with a smooth collar; Fig. 5.14C,E. *S. tenella.*

Table 5.9. *Thuiaria* Species

Thuiaria	Branching Type	Hydrothecal Teeth	With Hydrothecae Subopposite	With Secondary Branching	Colony Size, mm	Hydrotheca Type, Fig. 5.12	Stem Flexuose or Geniculate	Gonotheca Type, Fig. 5.13
'enera	o	2	—	—	70	R	x slightly	f
plumulifera	o	2	—	—	250	R	neither	?
latiuscula	o	2	/	/	50	P	neither	?
lonchitus	o	2	/	s	300	P	x slightly	f
cupressina	o	2	/	?	350	O-P	g	e
similis	o	2	/	/	80	Q-R	g	d
thuja	a	0	—	—	300	O	g	d
argentea	a	2	—	—	300	Q-R	g	d
robusta	a	2	—	—	300	O	x	e
fabricii	a	2	/	—	50	P	g	e
immersa	o	0	—	/	40	O	g	g

/ present; — absent; a, alternating; o, opposite; s, some; x, flexuose; g, geniculate. Basal branches are frequently broken off; colonies with alternating branching frequently resemble bottle brushes or are tufted or bushy. Those with opposite branching are frequently flattish or feathery.

Table 5.10. *Sertularia, Diphasia, Thuiaria* in Part

Species	Hydro-thecal Teeth	Colony Size, mm	With Secondary Branching	Gonotheca Type, Fig. 5.13	Notes
Sertularia cornicina	2	15	–	g	(2)
Sertularia pumila	2	50	v	f	—
Diphasia fallax	s	100	/	h	(1) (4); see Fig. 5.12N
Diphasia rosacea	*s*	100	/	h	(1) (4); see Fig. 5.12M
Diphasia tamarisca	3	160	/	h	(1) (4); see Fig. 5.12L
Thuiaria similis	2	80	/	d	(1) (3)

/, present; –, absent; s, rim sinuous; v, variable; (1) nodes indistinct; (2) hydrothecae with basal projections, see Fig. 5.12K; (3) stem much thicker than branches, hydrothecae on stem frequently alternate; (4) gonothecae sexually dimorphic, those of male with short, terminal lobes or bumps, female usually more deeply lobed.

5. Colony usually unbranched; margin with 5–6 teeth; Fig. 5.14 F. *S. gigantea.*
5. Colony usually branched; margin with 3 or 4 teeth—6.
6. Hydrothecae swelling basally; Fig. 5.14G. *S. polyzonias.*
6. Hydrothecae tubular; Fig. 5.14F. *S. tricuspidata.*

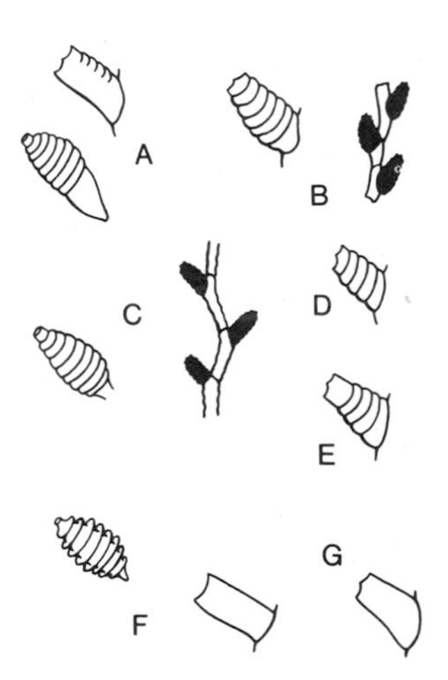

Fig. 5.14. Hydrozoan polyps, *Sertularella.* A, *S. gayi*, hydrotheca above and gonophore below; B, *S. rugosa*, hydrotheca at left, segment of colony at right; C, *S. tenella*, gonophore at left, segment of colony at right; D, *S. geniculata*, hydrotheca; E, *S. tenella*, hydrotheca; F, *S. gigantea*, gonophore at left, hydrotheca at right (*S. tricuspidata* has a similar hydrotheca); G, *S. polyzonias*, hydrotheca. (Armored gonophores are called gonothecae.)

Subkey for Plumularidae. This is the largest single family of hydroids; however, most of its species are extralimital either geographically or bathymetrically. The principal genera, *Plumularia* and *Aglaophenia*, occur here chiefly as strays on *Sargassum*.

Nematophores are a characteristic, though not exclusive, feature of this family. These structures have been regarded as a type of dactylozooid or protective polyp. They vary in form in different species and sometimes, e.g., in *Monostaechas quadridens*, within individual species. In all of the species considered, the nematophores are mounted on pedicels and are somewhat free or movable; in *Theocarpus* and *Cladocarpus* (not keyed), the nematophores have a broad base and are fixed. The gonophores in the species considered are simple hernialike protuberances except in *Schizotrichia* in which they are more or less cuplike.

1. Hydrotheca-bearing branches (hydrocladia) emerging from upper side of main branches; colony 150 mm, with dichotomous branching; Fig. 5.15A. *Monostaechas quadridens.*

1. Not so—2.

2. Hydrocladia arranged in whirls around a single stem; not fascicled; Fig. 5.15C. *Antennularia* spp.—3.

2. Colony a main stem with geniculate branching or stem fascicled; *Schizo Tricha.*—5.

3. Hydrocladial process usually bears a hydrotheca; colony about 200 mm; Fig. 5.15E. *A. americana.*

3. Hydrocladial process separated from the first thecate internode by a node or internode; Fig. 5.15F—4.

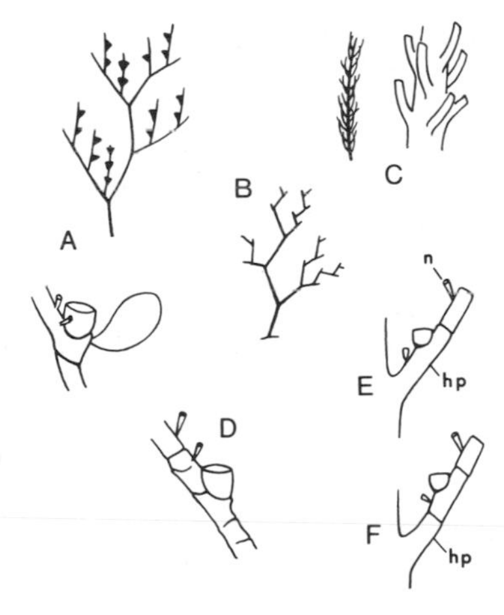

Fig. 5.15. Hydrozoan polyps, Plumularidae. A, *Monostaechas quadridens*, colony form above, detail below with sporosac; B, geniculate branching; C, *Antennularia* sp., colony growth form at left, detail at right; D, *Antennularia rugosa;* E, *Antennularia americana*, "n" indicates a nematophore, "hp" the hydrocladial process; F, *Antennularia antennina.*

4. Internodal septa strongly marked; colony 150 mm; Fig. 5.15D. *A. rugosa.*
4. Internodal septa weak or absent; colony 250 mm; Fig. 5.15F. *A. antennina.*
5. Stem fascicled; gonophore an elongate cup or goblet; colony 65 mm. *S. gracillima.*
5. Stem not fascicled; gonophore curving, cornucopialike; branching geniculate; colony 50 mm; Fig. 5.15B. *S. tenella.*

POSTSCRIPT. *Sargassum* species. Burkenroad (1939) surveyed the principal hydroids on pelagic *Sargassum;* the most important are the sertularian and plumularian genera *Sertularia*, *Aglaophenia*, *Pasya*, and *Plumularia*, Figure 5.16, plus a sprinkling of species from other families, including members of the genera *Clytia*, *Halecium*, *Obelia*, *Zanclea*, and *Sarsia*. Fraser's (1944) account of American hydroids shows that more than four dozen species have been taken on *Sargassum*. Those recorded north of Hatteras are listed but not keyed unless they have been found in neritic situations.

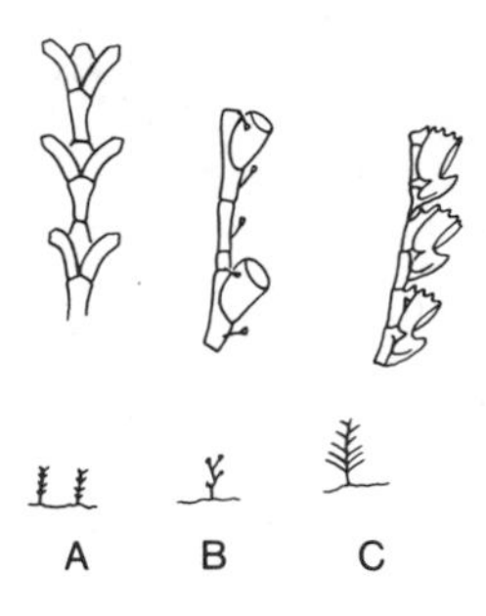

Fig. 5.16. Hydrozoan polyps, *Sargassum* forms. A, *Sertularia stookeyi;* B, *Plumularia* sp.; C, *Aglaophenia* sp. Colony forms indicated below; heights: A,B, 4 mm; C, 7 mm.

References

Burkenroad, D. 1939. Hydroids on pelagic *Sargassum*. *In* Pelagic *Sargassum* vegetation of the North Atlantic. A. E. Parr ed. Bull. Bingham oceanogr. Coll. 6(7):23–25.

Dougherty, E. C., ed. 1963. The Lower Metazoa—Comparative Biology and Phylogeny. Univ. California Press.

Fraser, C. M. 1921. Keys to the hydroids of eastern Canada. Contr. Can. Biol. Fish. 1918–1920: 137–180.

———. 1944. Hydroids of the Atlantic coast of North America. Univ. Toronto Press.

———. 1946. Distribution and relationship in American hydroids. Univ. Toronto Press.

Gudger, E. W. 1928. Association between sessile colonial hydroids and fishes. Ann. Mag. nat. Hist. Ser. 10,1:1–48.

Hargitt, C. W. 1908. Notes on a few coelenterates of Woods Hole. Biol. Bull. mar. biol. Lab., Woods Hole 7(5): 251–253.

———. 1910. The genera *Corynitis*, *Gemmaria* and *Zanclea*. Proc. Seventh Int. Congr. Zool. Boston. 815–818.

Hyman, Libbie H. 1940. The Invertebrates: Protozoa through Ctenophora. McGraw-Hill Book Co., New York.
Kingsley, J. S. 1910. A synopsis of the fixed hydroids of New England. Tufts Coll. Stud. 3(1):13–38.
Kramp, P. L. 1959. The hydromedusae of the Atlantic Ocean and adjacent waters. Dana Rept. 46:283 pp.
———. 1961. Synopsis of the medusae of the world. J. mar. biol. Ass. U.K. 40: 469 pp.
Murbach, L. 1899. Hydroids from Woods Hole, Mass. A. J. micro. Sci. (N.S.) 42(3):341–360.
Nutting, C. C. 1900, 1904, 1915. American hydroids, Parts I, II, III. U.S. natn. Mus. Spec. Bull. 4, Part I: The Plumularidae. ii + 285 pp. Part II: The Sertularidae. ii + 325 pp. Part III: The Campanularidae and Bonneviellidae. iii + 126 pp.
———. 1901. The hydroids of the Woods Hole Region. Bull. U.S. Fish. Commn. (for 1899): 325–386.
Pennak, R. W. 1953. Fresh-water invertebrates of the United States. Ronald Press Co., New York.
Rees, W. J. 1957. Evolutionary trends in the classification of capitate hydroids and medusae. Bull. Br. Mus. nat. hist. Zool. 4: 1–60.
———. 1958. British and Norwegian hydroids. J. mar. biol. Ass. U.K.: 23:1–42.
Rothschild, Lord. 1965. A classification of living animals. Second ed. Longmans.
Russell, F. S. 1953. The medusae of the British Isles. Cambridge Univ. Press.

Class Hydrozoa, Part Two
HYDROMEDUSAE

A preliminary key to medusoid forms is given in the introductory section of this chapter. Check lists for the orders Athecata, Thecata, and Limnomedusae are combined in the preceding section. The free medusoid stages of these orders and of Trachymedusae are keyed in the present section. Distributional data for Trachymedusae and for the remaining hydrozoan orders are given in the sections that follow, pp. 125, 127, and 131.

Deevey (1952, 1956, 1960) reports the presence of hydromedusae in the plankton throughout the year with a peak of abundance in spring and summer. The volumes are usually relatively small, but medusae may sometimes account for as much as 65% of the total plankton sample. The seasonal occurrence of individual species is indicated in the keys.

The orders Athecata and Thecata include numerous species and are divided, for identification, into subgroups. Trachymedusae and Limnomedusae have few species, and for practical purposes are not sharply differentiated from the preceding orders, to quote Russell (1953), "The main divisions of the group are based on the position of the gonads, and on the

Table 5.11. Hydromedusan Orders

Order	Mouth	Radial Canals	Marginal Tentacles	Bell Shape	Statocysts	Gonads
Athecata	Simple tubular Simple tubular + oral tentacles Four-lipped Four-lipped, tentacular	4 Bifurcated in 2 genera Centripetal canals in 1 genus	1 to ca. 90 Usually <30	Deep to hemis-pherical	None	On stomach
Thecata	Four-lipped Rarely six- or eight-lipped or up to about 70 lips	4 to >100	4 to >4000 Usually >12	Saucerlike to hemispherical	Present or none	On radial canals
Trachymedusae	Four- to six-lipped	4 to 8 $\pm$ centri-petal canals	8 to 80	Deep to hemis-pherical	Present	On radial canals

presence or absence of statocysts and on their structure when present." The first criterion requires sexually mature material, and the second involves a determination of the fine structure of elements that are minute to begin with. Another complication arises from fairly substantial ontogenetic changes that are especially important in Thecata; they involve the form of the umbrella, tentacle complements, and other details of systematic value. Immature material, thus, is very likely to cause problems in identification, and no very satisfactory solution can be offered. Indications of ontogenetic change are given as textual comment within the keys but become unwieldy if given strong consideration in the actual construction of the keys.

Table 5.11 should aid in the separation of hydromedusae into more manageable key groups. Limnomedusae have been omitted from the summary. Of the two local genera, *Proboscidactyla* is likely to be interpreted as belonging to Athecata and is keyed with members of this order. The second genus, *Gonionemus*, is keyed with the Thecata, as are the few Trachymedusae in the local fauna.

In the succeeding sections generic keys are based mainly on Russell (1953) and Kramp (1959) with considerable help at the species level from Mayer (1910); these authors are the principal source of illustrations, although Miner (1950) has also figured most of the local species. Mayer's nomenclature frequently differs from that of the more recent references, but Mayer, Russell, and Kramp all give excellent synonymies.

Order Athecata

ANTHOMEDUSAE

Formal criteria identifying the order stress the absence of statocysts and the placement of the gonads on the stomach wall. Note that in *Nemopsis* the gonads extend from the stomach wall onto the radial canals.

The chief variations in the structure of the gonads are defined in terms related to radial symmetry (Fig. 5.17E–G). The gonads may surround the stomach in a *continuous band* (type A); they may be *interradial*, i.e., in four sections lying between the primary radii as delineated by the four radial canals (type B); they may be *adradial* in eight sections lying between the primary radii and a secondary set of interradii (type C).

Most anthomedusae are relatively small, less than 5 mm in diameter or in bell height; the pandeids generally are larger species reaching 20–30 mm in

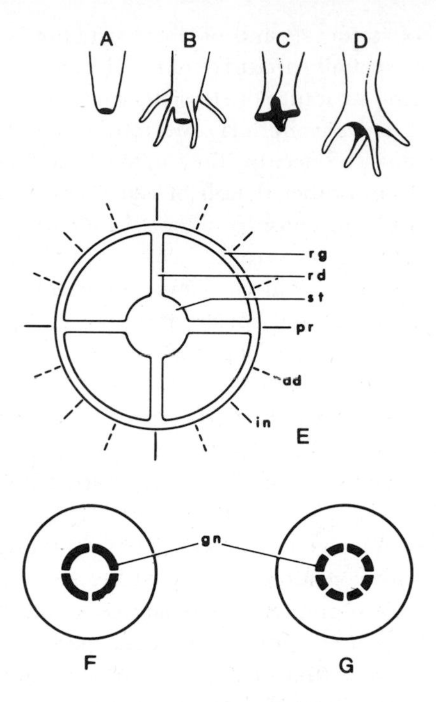

Fig. 5.17. Anthomedusae, key criteria. A-D, mouth types; A, simple tubular; B, with oral tentacles; C, four-lipped; D, with lips drawn out in tentacles; E, diagrammatic section through bell; F, diagrammatic section with interradial gonads; G, the same with adradial gonads. Abbreviations are as follows: rg, ring canal; rd, radial canal; st, stomach; pr, perradius; ad, adradius; in, interradius; gn, gonads.

bell height; the largest of local forms is *Calycopsis typa* with a bell diameter of 40 mm.

For purposes of identification this numerous assemblage is subdivided on the basis of mouth structure, taking note first of two genera of degenerate medusae with rudimentary marginal tentacles. The genera are *Pennaria* and *Stylactis*, the first with a simple, circular mouth (Fig. 5.18J), and the second with the mouth four-lipped and with the lips elongated to form tentacles.

GROUP ONE: Mouth simple, tubular; Figure 5.17A.

GROUP TWO: Mouth simple, tubular, with oral tentacles; Figure 5.17B.

GROUP THREE: Mouth four-lipped, the lips simple, folded, or frilled; Figure 5.17C.

GROUP FOUR: Mouth four-lipped, the lips drawn out in tentacles; Figure 5.17D.

Group One, Mouth Simple, Tubular. The species are deeply bell-shaped, and measurements are maximum bell heights unless otherwise indicated.

---

1. Marginal tentacles 2 or 4, with stalked capsules; no ocelli; gonads type B; 6 mm; summer; Fig. 5.18A. *Zanclea costata.* Mayer (1910) described a second species, *gemmosa*, with 4 very small lips; Kramp (1961) regarded *gemmosa* as a synonym of *costata.*
1. Marginal tentacles without stalked capsules—2.
2. Marginal tentacles solitary—3.
2. With 4 marginal clusters of 5 or 6 tentacles each; no ocelli; gonads type A; slightly taller than a hemisphere, 2.5 mm; late fall-winter; Fig. 5.18B. *Margelopsis gibbesi.*
3. With ocelli, Fig. 5.18C; 4 tentacles—7.
3. Without ocelli; 1–4 tentacles; gonads type A—4.
4. Umbrella with rows of nematocysts, Fig. 5.18D—5.
4. Umbrella without rows of nematocysts—6.
5. Nematocysts in 8 rows; 4 short, coiled tentacles; bell with apical swelling, 3 mm; summer; Fig. 5.18D. *Ectopleura dumortieri.*
5. Nematocysts in 5 rows. *Hybocodon: H. prolifer* with a single cluster of 1–4 tentacles and *H. pendula* with a single long tentacle and 2 vestigial tentacles; both species 5 mm; spring; Fig. 5.18E,F.
6. With a spirelike apical process; 1 long, 1 short, 2 vestigial tentacles; 5 mm; summer; Fig. 5.18I. *Euphysora gracilis.*
6. Without an apical process; one tentacle; 12 mm; summer; Fig. 5.18G. *Euphysa aurata.*
7. With 8 rows of nematocysts; bell somewhat extended apically, miter-shaped; 2.5 mm; gonads type A; summer-fall; Fig. 5.18C. *Linvillia agassizi.*
7. Not so. *Dipurena* and *Sarsia.* See Table 5.12.

Group Two, Mouth Simple, Tubular, with Oral Tentacles. Species deeply bell-shaped. Measurements equal bell height unless otherwise indicated.

---

1. Oral tentacles unbranched; marginal tentacles 8–16; no ocelli; gonads type A; 1 mm; phosphorescent; late summer; Fig. 5.19A. *Lizzia fulgurans.* Allwein (1967) reported a second species, *L. blondina*, from Cape Cod and Beaufort, N.C.
1. Oral tentacles branched; marginal tentacles in four clusters; ocelli present—2.
2. With two types of tentacles in each cluster, i.e., 1 pair short, clublike, and 14–16 simple and more elongate; gonads extend onto radial canals; 11 mm; fall-winter; Fig. 5.19B. *Nemopsis bachei.*
2. Marginal tentacles all alike; gonads type B; Fig. 5.19C–F. *Bougainvillia.* See Table 5.13. Note also *B. rugosa*, a species of questionable generic affinity known only as a juvenile.

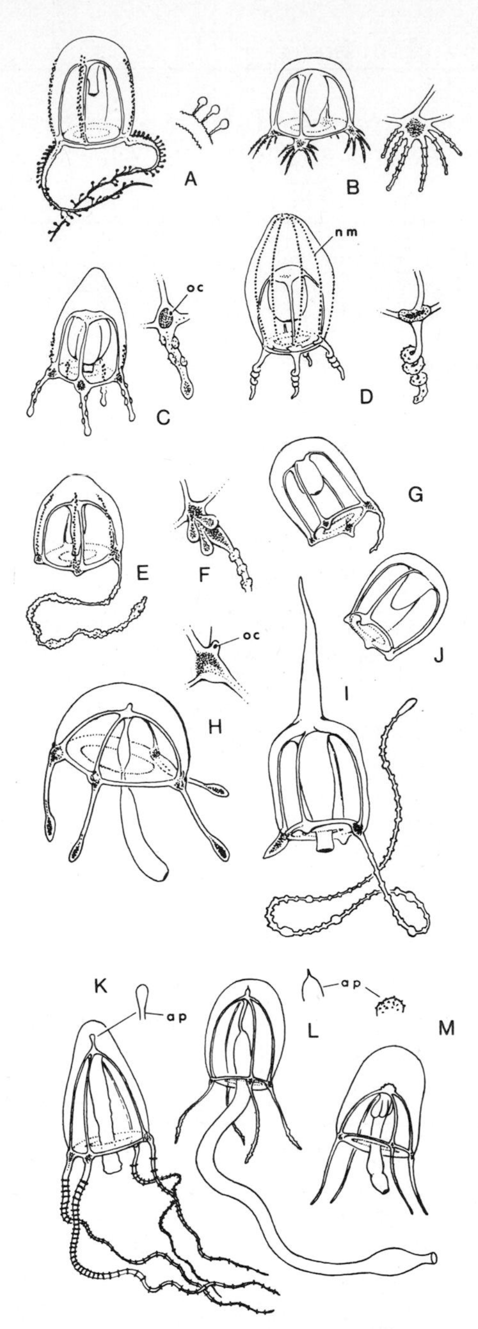

Fig. 5.18. Anthomedusae, Group One. A, *Zanclea costata;* B, *Margelopsis gibbesi;* C, *Linvillia agassizi*, D, *Ectopleura dumortieri;* E, *Hybocodon pendula;* F, *Hybocodon prolifer*, detail of base of tentacle; G, *Euphysa aurata;* H, *Dipurena strangulata;* I, *Euphysora gracilis;* J, *Pennaria* sp.; K, *Sarsia princeps;* L, *Sarsia tubulosa;* M, *Sarsia hargitti*. Abbreviations are: oc, ocellus; nm, nematocyst row; ap, apical chamber.

Table 5.12. *Dipurena* and *Sarsia*

Species	Gonads	Manubrium extends beyond Margin	Apical Chamber	Tentacles	Bell Height, mm	Season of Occurrence	Fig. 5.18
D. strangulata	Two series	Much	Beadlike	Terminal knob	4	Summer-fall	H
S. hargitti	Saclike	Much	Spinous bulb	Simple	1.5	Summer	M
S. tubulosa	Coextensive with manubrium	Much	Pointed bead	Rugose	7	Spring	L
S. princeps	Coextensive with manubrium	Little	Stalked bulb	Rugose	40	Spring?	K
S. eximia	Coextensive with manubrium	Little	None	Rugose, terminal knob	4	Spring	—

Table 5.13. *Bougainvillia*

Species	Stomach	Oral Tentacles Divide	Size, mm	Tentacles per Cluster	Season	Gonads	Fig. 5.19
superciliaris	Conical peduncle	4–7 times	9	11–15	Spring	Interradial	D
ramosa	Long, flasklike	1–3 times	2.5	3–7	Summer-fall	Interradial	F
brittanica	short, broad	4–6 times	12	16–30	Fall	Adradial	E
carolinensis	short, broad	2 times	4	7–9	Summer	Interradial	—

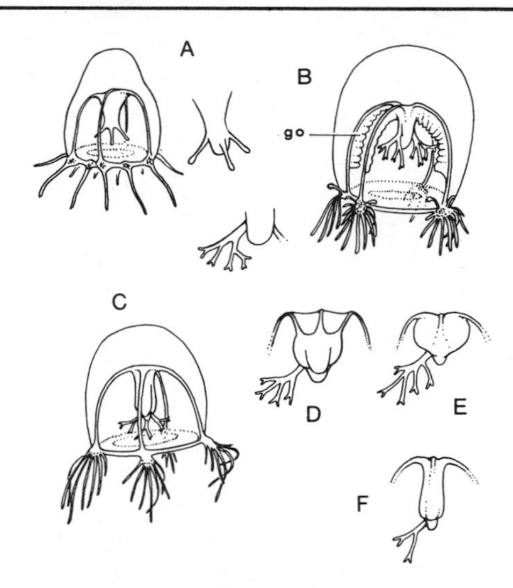

Fig. 5.19. Anthomedusae, Group Two. A, *Lizzia fulgurans;* B, *Nemopsis bachei*, note gonads extending onto radial canals; C–F, *Bougainvillia* spp., C, whole animal, D–F details of peduncle-manubrium of the following species: D, *B. superciliaris*, E, *B. brittanica*, F, *B. ramosa*.

Group Three, Mouth Four-Lipped, the Lips Simple, Folded, or Frilled. Usually deeply bell-shaped; *Niobia* is an exception. Measurements equal bell height unless otherwise indicated.

---

1. With 4 simple radial canals—4.

1. With 4 radial canals plus centripetal canals or with 2 or 4 radial canals branching—2.

2. With 4 radial canals, and 12–16 centripetal canals; 16 primary tentacles plus 16 smaller ones; gonads type C, much convoluted; oceanic. *Calycopsis typa.*

2. With 2 or 4 radial canals branching—3.

3. With 2 radial canals branching; 12 marginal tentacles which may bud medusae; bell flatter than a hemisphere, 4 mm; gonads type B; spring; Fig. 5.20B. *Niobia dendrotentaculata.*

3. With 4 radial canals branching; usually 16–20 tentacles; slightly taller than a hemisphere; 5 mm; gonads type A extending onto stomach lobes; medusae may be budded from stomach or from forks of radial canals; Fig. 5.20A. *Proboscidactyla ornata.* (Order Limnomedusae).

4. With only 2 tentacles but margin with wartlike tentacular bulbs or short, cirrilike tentaculae; bell with well-developed apical process; no ocelli; gonads type C—5.

4. Tentacles numerous, 12 or more in mature medusae; with or without apical process; with ocelli; gonads type B—6.

5. Margin of bell with 60–80 wartlike tentacular bulbs; umbrella subhemispherical with pointed apical process, 30 mm wide; manubrium with well-developed

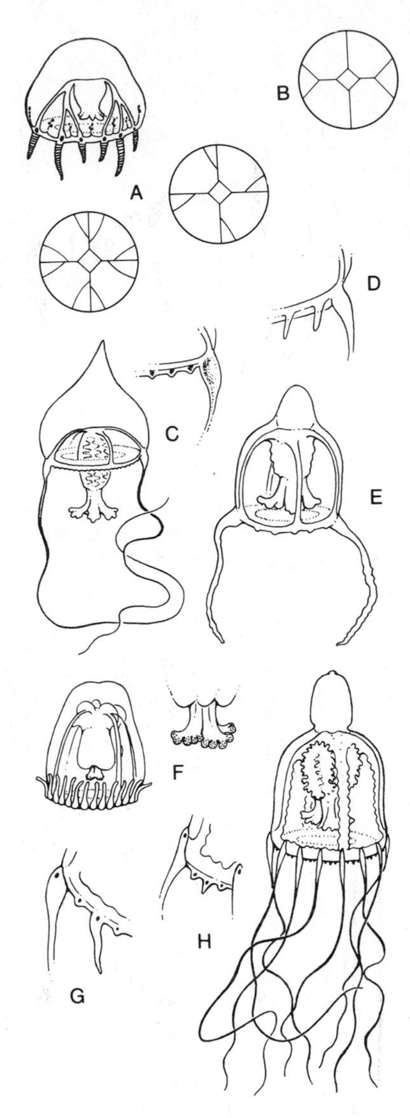

Fig. 5.20. Anthomedusae, Group Three. A, *Proboscidactyla ornata*, whole animal and 2 diagrams of radial canal branching scheme; B, *Niobia dendrotentaculata*, diagram of radial canals; C, *Stomotoca pterophylla;* D, *Amphinema rugosum*, detail of part of bell margin; E, *Amphinema dinema;* F, *Turritopsis nutricola*, with detail of mouth; G, *Leuckartiara octona*, detail of part of bell margin; H, *Catablema vesicarium*.

peduncle, mouth suspended below umbrellar margin; spring-summer; Fig. 5.20C. *Stomotoca pterophylla.*

5. Margin with <30 tentacular bulbs or tentaculae; umbrella taller than wide, 6 mm; no peduncle; summer species. *Amphinema: A. dinema* with about 14–24 wartlike tentacular bulbs, Fig. 5.20E; *A. rugosum* with about 16–24 cirrilike tentaculae, Fig. 5.20D.

6. Lips with a continuous, marginal row of nematocysts; 80–90 marginal tentacles (individuals as small as 3 mm have several dozen tentacles); with ocelli; bell without spical processes; 5 mm; gonads type B; spring-summer; Fig. 5.20F. *Turritopsis nutricola.*

6. Not so; with apical process—7.

7. With 24–32 (rarely 12–48) equally developed tentacles; bell about 25 mm wide; summer; Fig. 5.20H. *Catablema vesicarium.*

7. With 16 (rarely 12–24) primary tentacles plus 16–48 rudimentary ones; 12 mm; late spring-summer; Fig. 5.20G. *Leuckartiara octona.*

Group Four, Mouth Four-Lipped, the Lips Drawn Out in Tentacles. Umbrella deeply bell-shaped; measurements equal height unless otherwise indicated.

1. Marginal tentacles in 8 groups of 3–5 tentacles each; no ocelli; gonads type B, budding medusae when young; 4 mm; spring; Fig. 5.21A. *Rathkea octopunctata.*

1. Marginal tentacles solitary, 16–32 in mature medusae; gonads type B. *Podocoryne*—2.

2. Mouth tentacles unbranched; 3.5 mm; summer; Fig. 5.21C. *P. carnea.*

2. Mouth tentacles branched; 5.0 mm; fall; Fig. 5.21B. *P. borealis.* Kramp (1961) questions the identity of the larger American specimens of *carnea* and suggests that mature medusae seldom have more than 8 marginal tentacles.

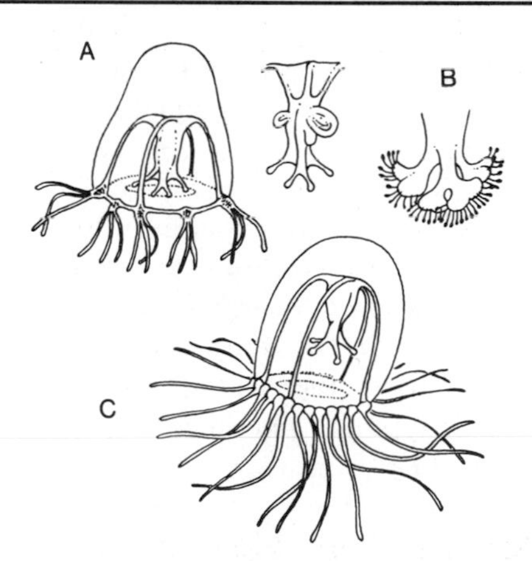

Fig. 5.21. Anthomedusae, Group Four. A, *Rathkea octopunctata* with detail showing budding on stomach wall; B, *Podocoryne borealis*, detail of mouth; C, *Podocoryne carnea.*

Order Thecata (Plus Trachymedusae and Limnomedusae, in Part)

LEPTOMEDUSAE

In all genera except members of the leptomedusan family Melicertidae, marginal sense organs are present; gonads develop on the radial canals except in *Aglaura*, where the gonads develop at the junction of the peduncle and radial canals. The limnomedusan *Gonionemus* also shares these traits. Early references place *Gonionemus* with Trachymedusae.

The present assemblage is less homogeneous than the anthomedusae, with a wider range of adult sizes (less than 5 to greater than 300 mm), greater variation in umbrella shape (deeply bell-shaped to flat and saucerlike), and in the more frequent occurrence of substantial ontogenetic change.

The presence or absence of marginal sense organs and their structure and histological origin when present are major criteria in establishing natural groups. For present purposes, three somewhat arbitrary groupings are utilized:

GROUP ONE: Marginal sense organs absent.

GROUP TWO: Marginal sense organs clublike. This grouping encompasses species with *marginal cordyli* (Leptomedusae), and Trachymedusae with *free marginal sensory clubs*.

GROUP THREE: Marginal sense organs vesicular. The several types, though distinctly different histologically, may prove difficult to distinguish in gross examination. The types include open or closed *marginal vesicles*, both of which are ectodermal in origin and are placed on the velum just inside the margin. In addition there are *enclosed marginal sensory clubs* which are in part endodermal and structurally related to the sensory clubs of Group Two; they are, however, enclosed by ectoderm and may appear to be vesicular. Structures of this type are placed *on* rather than within the margin of the bell.

Group One. Marginal sense organs absent; with or without ocelli; young may have only 4 radial canals, but adults have 5 or more and no centripetal canals. (Note that in *Staurophora* of Group Two the marginal clubs are obscure, and these medusae might be interpreted as belonging in the present group.)

---

1. Umbrella pear-shaped; mouth 8-lipped; usually with 8 radial canals; ocelli absent; tentacles 64–72; 14 mm wide; early summer; Fig. 5.22F. *Melicertum octocostatum.* Very young medusae of *Melicertum* have only 4 radial canals and as few as 4 tentacles; individuals 5 mm in diameter have about half the adult tentacle complement.

1. Umbrella flatter than a hemisphere; mouth 4-lipped; radial canals 5–18, sometimes branching; tentacles of a single type, >100; 15 mm wide. *Dipleurosoma typicum.*

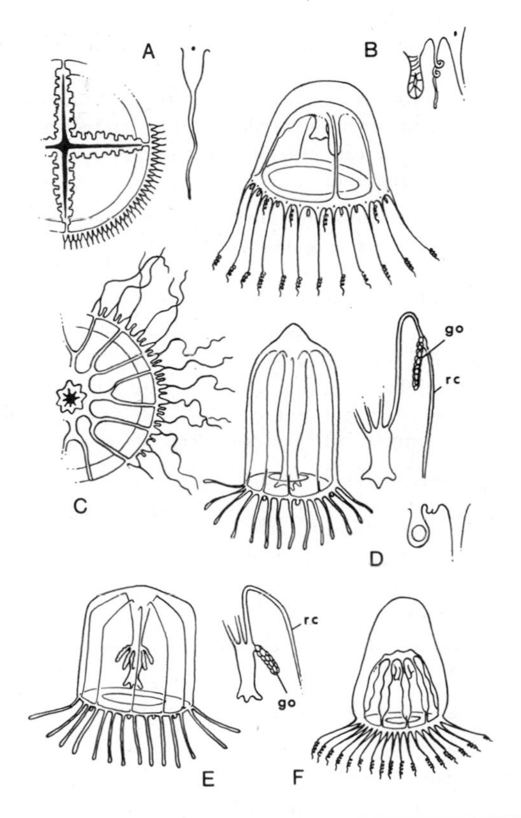

Fig. 5.22. Leptomedusae and Trachymedusae, Groups One and Two. A, *Staurophora mertensi*, part of bell in ventral view; B, *Laodicea undulata*, bell and detail of part of margin showing base of tentacle, cirrus, and club-like sense organ; C, *Toxorchis kellneri*, part of bell in ventral view; D, *Aglantha digitale*, with details showing gonad on radial canal and clublike sense organ; E, *Aglaura hemistoma*, with detail showing gonad on peduncle; F, *Melicertum octocostatum*. Abbreviations are; go, gonads: rc, radial canal.

Group Two. The cordyli or sensory clubs are usually readily discernible except in *Staurophora;* these structures appear early in development but may not be present in the very earliest stages. There is ontogenetic change in umbrella shape, tentacle complement, etc.

---

1. With 4 radial canals—2.
1. With 8 radial canals—4.

2. Ocelli present—3.
2. Ocelli absent; radial canals branching; umbrella about 3 times wider than high, 90 mm diameter; with 300–500 tentacles; bathypelagic. *Ptychogena lactea.*

3. Mouth opening continued as a gutterlike slit along the open undersides of radial canals; umbrella about 6 times wider than high, 300 mm; >4000 tentacles when mature; spring to early summer in south, through summer in north; Fig. 5.22A. *Staurophora mertensi.*
3. Mouth and radial canals not so modified; umbrella subhemispherical, 37 mm; tentacles 400–600, coiled terminally, plus coiled cirri; summer-fall; Fig. 5.22B. *Laodicea undulata.* Young with deep bell; 2 tentacles; cirri present but not cordyli until bell is about 2 mm diameter.

4. Mouth with 8 lips; radial canals usually bifurcated; umbrella subhemispherical, 15 mm wide; about 32 tentacles, no cirri; ocelli present; summer-fall?; Fig. 5.22C. *Toxorchis kellneri.*
4. Mouth 4-lipped; radial canals simple; ocelli absent—5.

5. Deep-belled, 1.5–2 times higher than wide; stomach with peduncle (not apparent in young); tentacles of a single type—6.
5. Hemispherical with slight apical process, 15 mm diameter; stomach without peduncle; with 8 long perradial tentacles and 1–3 smaller tentacles per octant, totaling 32 tentacles; summer-fall?; oceanic. *Rhopalonema velatum* (Trachymedusae).

6. Gonads on radial canals; bell with a small apical process, 30 mm tall; as many as 100 tentacles; spring in the south and spring through summer in north; Fig. 5.22D. *Aglantha digitale* (Trachymedusae).
6. Gonads on peduncle at base of radial canals; bell flat-topped, 6 mm tall; to 85 tentacles; oceanic; Fig. 5.22E. *Aglaura hemistoma* (Trachymedusae). In the last two species the tentacles are easily broken by handling.

Group Three. In the following leptomedusan genera the marginal vesicles are *open: Tiaropsis*, *Mitrocomella*, and *Halopsis*. They are closed in *Tima*, *Eutima*, *Eucheilota*, *Phialidium*, *Blackfordia*, *Obelia*, and *Rhachostoma*.

Because of the size of this assemblage it is convenient to subdivide the species into two sections, and to separate immediately the atypical leptomedusan *Obelia* (Fig. 5.23A). In this genus the velum is virtually absent; the bell is flat, 6 mm diameter; with four simple radial canals; and numerous short, marginal tentacles. Individual species of *Obelia* are regarded as unidentifiable in the medusoid phase.

The remaining species in this group will be dealt with as follows:

PART ONE: With four simple radial canals.

PART TWO: With more than four simple radial canals or with four simple canals plus centripetal canals.

Group Three: Part One.

---

1. With ocelli in the circular canal; bell about twice as wide as high, 30 mm diameter; tentacles of one type, as many as 200; spring; Fig. 5.23B. *Tiaropsis multicirrata.* Young have a deeper bell and a weak apical process. The number of sensory vesicles is constant at 8; tentacles increase in number with age, e.g., individuals 2 mm in diameter have about 24 tentacles.

1. Without ocelli—2.

2. With a well-developed peduncle, manubrium long, clapperlike—3.

2. Without a gastric peduncle—4.

3. With 4 tentacles, and young with cirri; gonads on radial canals and peduncle; umbrella about 1.5 times wider than high, 30 mm; summer; Fig. 5.23D. *Eutima mira*

3. With 32 tentacles in three size groups, 8 long, 8 medium, and 16 short; umbrella about 1.5 times wider than high, 100 mm diameter; gonads on radial canals only; spring in south, autumn-winter north of Cape Cod; Fig. 5.23C. *Tima formosa.* Marginal vesicles may be absent in juveniles; at 35 mm bell diameter there are about 16 tentacles.

4. With 8 or 12 closed marginal vesicles; umbrella more or less hemispherical; with cirri attached to tentacle bulbs plus or minus additional marginal projections; summer-fall. *Eucheilota: E. ventricularis*, 10 mm diameter with 16 tentacles, 8 vesicles; *E. duodecimalis* 2.5 mm with 4 tentacles, 12 vesicles; Fig. 5.23E.

4. With 16 or more vesicles, cirri present or absent—5.

5. With about 300 cirri; 36 tentacles; 16 vesicles; umbrella subhemispherical, 30 mm diameter; spring-summer; rare. *Mitrocomella polydiademata.*

5. Without cirri—6.

6. Tentacles emerge slightly above margin, 60–80 in number in mature medusae, 8 in youngest juvenile, tentacles with a bend near outer end and with a small "sucker" at the bend; vesicles alternate with tentacles; bell subhemispherical, 20 mm; summer; Fig. 5.23H. *Gonionemus vertens* (Limnomedusae).

6. Not so, tentacles at margin—7.

7. Some or all of tentacles with minute, nipplelike projection into bell basally; tentacles 70–80 when mature, alternating with 1–3 vesicles; bell taller than hemisphere, 14 mm tall; Fig. 5.23G. *Blackfordia: B. manhattensis* with 2 or 3 vesicles between tentacles; *B. virginica* with 1, rarely 2, vesicles between tentacles and with dark pigment in the vesicles.

7. Tentacles without basal projections; tentacles about 32 or less; Fig. 5.23F. *Phialidium* spp.—8.

8. With 16 tentacles and 16 vesicles, alternating: *P. folleatum*, hemispherical, 5 mm diameter; gonads on outer ¼ to ⅓ of radial canals; *P. bicophorum* 2–3 times wider

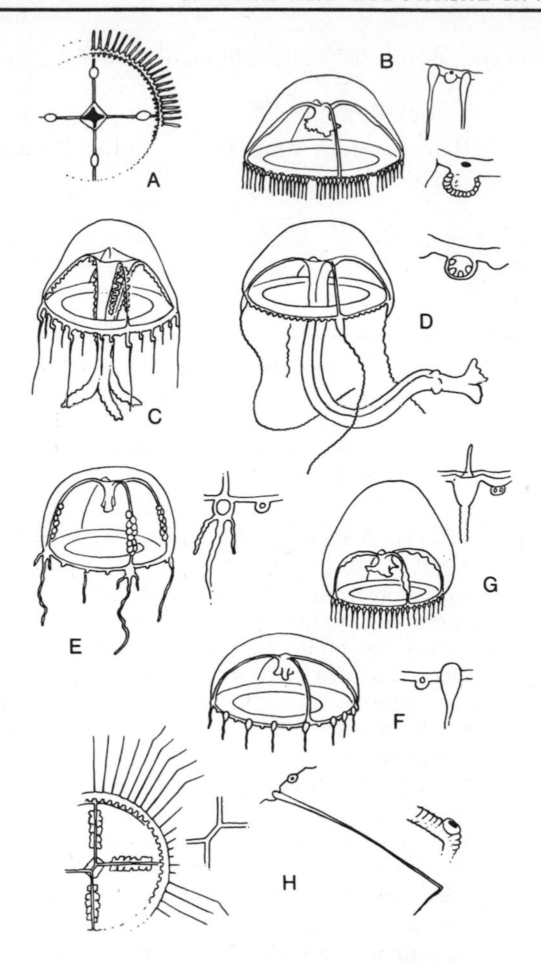

Fig. 5.23. Leptomedusae and Limnomedusae, Group Three, Part One. A, *Obelia*, sp. ventral view of part of bell; B, *Tiaropsis multicirrata*, with detail showing ocellus in circular canal; C, *Tima formosa;* D, *Eutima mira*, with detail of marginal sense organ without ocellus; E, *Eucheilota duodecimalis;* F, *Phialidium* sp.; G, *Blackfordia* sp., with detail showing internal projection of base of tentacle; H, *Gonionemus vertens*, with details of confluence of radial canals, single tentacle, and sucker at bend of tentacle.

than high, 5.5 mm; gonads at mid canal, equal ¼ canal length; both species common in summer.

8. With 16 tentacles plus 16 rudimentary tentacular bulbs and 32 vesicles alternating; gonads within basal half of radial canals; bell a truncated cone with an apical process, 2 mm; summer. *P. singularis.*

8. With 32 tentacles or more and 2 vesicles per interspace; gonads in proximal half of radial canals; bell subhemispherical, 20 mm; common in summer. *P. languidum.*

Group Three: Part Two.

1. With 4 radial canals plus 1–3 short centripetal canals; mouth 4-lipped with long manubrium; bell variable in shape, 1.5 times higher than wide to somewhat wider than high, 30 mm diameter; ocelli absent; 8 enclosed sensory clubs; fall, euryhaline; Fig .5.24A. *Liriope tetraphylla* (Trachymedusae). Note. A related oceanic trachymedusan, *Geryonia proboscidalis,* has 6 radial canals and 7 centripetal canals.

1. With 12 or more radial canals—2.

2. Radial canals 12–16 in 4 clusters; mouth 4-lipped; tentacles to 450 with equal number of cirri, with 3–6 vesicles between pairs of radial canals; no ocelli; bell about 4 times wider than high, 70 mm; summer; Fig. 5.24B. *Halopsis ocellata.*

2. Radial canals more than 16, not in 4 clusters; mouth with variable number of lips—3.

3. Subumbrella with interradial rows of wartlike bumps; radial canals 80–100 with somewhat more numerous tentacles, no cirri, vesicles 8–12 between tentacles; bell saucerlike, about 6 times wider than high, 115 mm; summer-fall, common offshore; Fig. 5.24C. *Rhacostoma atlanticum.* Young *Rhacostoma* ca. 3 mm have 8

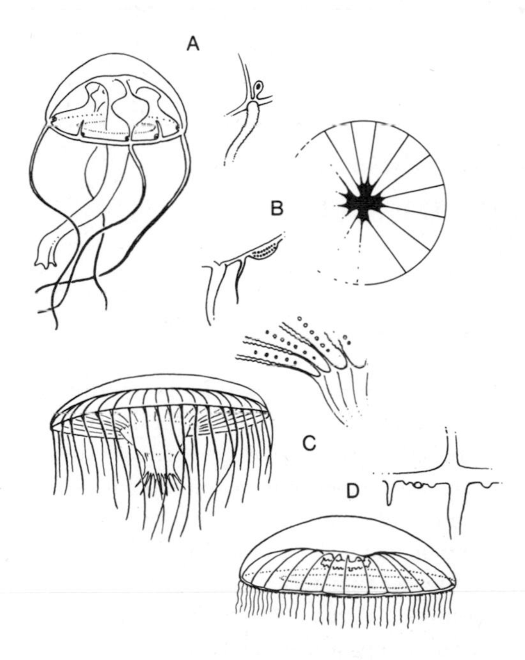

Fig. 5.24. Leptomedusae and Trachymedusae, Group Three Part Two. A, *Liriope tetraphylla;* B, *Halopsis ocellata*, diagram showing clustering of radial canals, and detail showing part of margin; C *Rhacostoma atlanticum*, with detail showing interradial warts; D, *Aequorea* sp.

tentacles and 8 radial canals, 4 of which do not reach the ring canal; the subumbrellar tubercles are present at the 8- canal stage.

3. Subumbrella not so; Fig. 5.24D. *Aequorea*. See Table 5.14.

Table 5.14. Aequorea

Aequorea	Stomach Width Relative to Umbrella Diameter	Radial Canals	Multiply No. of Radial Canals by × to Obtain No. of: Tentacles	Vesicles	Diameter of Adults, mm	Shape
tenuis	1/5	24–32	2–3× + ca. 200 rudimentary	2–3×	100	3–4 times wider than high
albida	1/3	ca. 100	3×	6×	80	Hemispherical
aequorea	1/3	60–80	1/2–2×	5–10×	175	Variable

Order Narcomedusae

NARCOMEDUSAE

Diagnostic criteria are given at the beginning of this chapter in the introductory section on medusae.

These medusae are oceanic and essentially extralimital to the present account in that they are seldom recorded in neritic waters. Lacking a fixed hydroid stage, narcomedusans are independent of the sea bottom. *Cunina octonaria* is somewhat of an exception in that its larvae are parasitic on the anthomedusa *Turritopsis nutricola* which has a sessile polypoid stage confined to shallow water. The larvae of *C. octonaria* and of *C. proboscidea* and *Pegantha clara* also parasitize the oceanic trachymedusan *Geryonia proboscidalis*.

Cunina octonaria (Fig. 5.2G) reaches about 7 mm diameter, with umbrella subhemispherical; 8 relatively stiff tentacles emerging halfway up the sides of the bell; about 24 marginal sense organs, 8 gastric pouches, and, in young medusae, a pendulous manubrium with four-lipped mouth; radial and ring canals absent.

Order Trachymedusae
TRACHYMEDUSAE

Diagnostic criteria are given at the beginning of this chapter in the introductory section on medusae.

The species are primarily oceanic, and some are bathypelagic. Several of those keyed are likely to be encountered only as Gulf Stream waifs. There is no polypoid stage. The one relatively common species, *Aglantha digitale*, is usually reported in winter-spring; additional cold water species will be noted in *Rhopalonema* and *Ptychogastria*.

The formal criteria identifying members of this order have limited value for the general worker seeking to identify unsorted medusae, and the few species to be considered have been keyed out with the leptomedusae in the preceding section.

Systematic List and Distribution.

Family Geryonidae	
Geryonia proboscidalis	CV
Liriope tetraphylla	CV+
Family Rhopalonematidae	
Aglantha digitale	B, 37°
Aglaura hemistoma	CV+
Family Ptychogastriidae	
Ptychogastria polaris	A+

References: Hydromedusae

Allwein, J. 1967. North American Hydromedusae from Beaufort, North Carolina. Vidensk. Meddr. dansk. naturh. Foren. 130:117–136.

Bigelow, H. B. 1915. Exploration of the coast water between Nova Scotia and Chesapeake Bay, July and August ,1913, by the U.S. Fisheries schooner *Grampus*. Oceanography and Plankton. Bull. Mus. comp. Zool. Harv. 59(4):151–359.

———. 1918. Some medusae and Siphonophorae from the Western Atlantic. Bull. Mus. comp. Zool. Harv. 62(8): 365–442.

Deevey, G. B. 1952. Zooplankton of Block Island Sound 1943–1946. Bull. Bingham oceanogr. Coll. 13(3):65–119, 120–164.
———. 1956. Zooplankton. *In* Oceanography of Long Island Sound 1952–1954. Bull. Bingham oceanogr. Coll. 15(5):113–155.
———. 1960. The zooplankton of the surface waters of the Delaware Bay region. Plankton Studies 1. Bull. Bingham oceanogr. Coll. 17(2):5–53.
Fraser, C. M. 1944. Hydroids of the Atlantic Coast of North America. Univ. of Toronto Press.
Halstead, B. W. 1965. Poisonous and venomous marine Animals of the World. Vol. 1. Invertebrates. U.S. Government Printing Office.
Hargitt, C. W. 1904. The medusae of the Woods Hole region. Bull. Bur. Fish., Wash. 24:21–79.
Hyman, Libbie H. 1940. The Invertebrates: Protozoa through Ctenophora. McGraw-Hill Book Co., New York
Kramp, P. L. 1959. The hydromedusae of the Atalntic Ocean and adjacent waters. Dana Rept. 46: 283 pp.
———. 1961. Symposis of the medusae of the world. Jour. mar. biol. Ass. U.K. 40: 469 pp.
Mayer, A. G. 1910. Medusae of the World. Hydromedusae. Vols. I and 2. 498 pp. Carnegie Inst. Wash. Publ. 109.
Miner, R. W. 1950. Field book of seashore life. G. P. Putnam's Sons, New York.
Peterson, Kay W. 1964. Some preliminary results of a taxonomic study of the Hydrozoa. Biol. Bull, mar, biol. lab. Woods Hole 127(2):384–385.
Russell, F. S. 1953. The medusae of the British Isles. Cambridge Univ. Press.

Class Hydrozoa, Part Three

Order Siphonophora

SIPHONOPHORES

A fully developed siphonophore may be regarded as a colonial aggregation of both polypoid and medusoid *persons* derived from a single larval individual. These colonies are complex in structure and not amenable to a simple diagnosis. All of the species are pelagic, and most are oceanic. They appear inshore only as strays and chiefly from the Gulf Stream. The colonies are easily fragmented, and the collector is more likely to encounter parts than intact colonies.

The individual persons arise by budding from a *stem* or from a platelike *coenosarc* developed either directly from the initial larva or by budding from a secondary polypoid. The extension of this budding process produces multipersoned aggregations called *cormidia.* Individual colonies include a number of such cormidia frequently arranged in linear series. These may attain remarkable lengths—colonies of *Apolemia uvaria* are said to grow to 20 m in length.

Within each cormidium, functional specialization is well developed. Such specialization may modify the individual person to such an extent that its resemblance to more typical polypoid or medusoid forms is lost. *Gastrozooids* or "siphons" are the functional equivalent of hydroid feeding polyps. Each has a single tentacle, usually with branches or *tentilla* that are heavily armed with subterminal nematocysts. *Dactylozooids* or "palpons" resemble underdeveloped gastrozooids; each has an unbranched tentacle, and in *Physalia* these are the main fishing organs. *Bracts* were regarded by Hyman (1940) as medusoid persons and by Totton and Bargmann (1965) as polypoids; in their usual form they do not closely resemble either. They are thickly gelatinous, prismatic, leaflike, or helmet-shaped, and in addition to giving buoyancy to the colony are said to have a protective function. *Nectophores*, though often highly modified from the usual medusoid form, are basically swimming bells and are the chief propulsive members. The colony may also be provided with a float or *pneumatophore*. Pneumatophores, including the single large float of *Physalia*, are regarded as specialized individual medusoids—or polypoids depending on the authority consulted. Table 5.15 outlines the structural organization of colonies in the three suborders of siphonophores.

Most siphonophores are hermaphroditic. Individual cormidia may contain both male and female gonophores. The gonophores range in development from recognizably medusoid forms to vestigial, saclike structures. They are frequently developed in specialized aggregations on stalks, called *gonodendrons*. Sexual medusae may be released as free individuals, or not, in different species. In the suborder Calycophorae, whole cormidia may be released as free agents called *eudoxids*, with one of the member gonophores commonly elaborated as a nonsexual, propulsive medusoid. In subtropical waters at least, there is evidence of an "alternation of generations" of eudoxid and unfragmented (*polygastric*) phases; the cycle may be repeated four or five times each year. In sexual reproduction the zygote develops as a planula from which a new colony subsequently develops. For an introductory discussion of larval development in the three suborders see Totton and Bargmann (1965).

Distribution

Siphonophores are scarce in coastal waters, and it seems pointless to list the 20 to 30 or even more species which *might* drift in from oceanic waters. According to Grice and Hart (1962), siphonophores in neritic waters make up less than 1% of the plankton by volume. *Physalia physalia* does occur with some regularity along the coast south of Cape Cod; Sumner *et al.* (1913) state that this species is "taken nearly every summer in Vineyard Sound, some-

Table 5.15. Siphonophore Colony Organization

Calycophorae	Physonectae*	Cystonectae (*Physalia*)
Pneumatophore absent	With a small apical pneumatophore	Pneumatophore a large float
2–4 (1 or >12 in a few genera), relatively large, usually polymorphic nectophores with an elongate appended stem with a linear series of simple cormidia consisting of: 1 bract; 1 gastrozooid; 1 or more gonophores; palpons absent	About 8–25 more or less similar nectophores in 2 rows or in a spiral forming a *nectosome* with an appended *siphosome* with a linear series of complex cormidia consisting of: 1 or more bracts; 1 gastrozooid; gonophophores in clusters or with gonodendra; 1 or more palpons	Nectophores absent, cormidia budded from a platelike ventral coenosarc and consisting of: large and small palpons; gastrozooids without tentacles; dactylozooids with elongate tentacles; complex gonodendra; and miscellaneous polypoid forms
Mature cormidia released as free-living eudoxids with 1 or more gonophores modified as a propulsive medusoid	Without eudoxids	Without eudoxids

* Excluding Rhodaliidae and Athorybiidae.

times in considerable numbers." *Physalia*, with its large pneumatophore, is a surface form that drifts with the wind and currents. Other siphonophores are pelagic, and some swim as vigorously as fish. According to Bigelow (1918), "only one siphonophore, *Stephanomia* (=*Nanomia*) *cara* is anything but accidental . . . between Chesapeake Bay and Nova Scotia." Neither the bathymetric nor the geographical ranges of most species are well known. Cosmopolitanism is common. Most of those listed seem to be essentially warm water forms that occur in our latitude in the Gulf Stream. Moore (1953) has indicated that in the Florida Current most of the species that were common enough to exhibit any trend at all were more common in the cooler seasons. *Diphyes dispar* is an exception, and is one of the more likely species to be encountered locally, since its distribution is boreal as well as temperate-tropical. *Chelophyes appendiculata* is another possible visitor, along with *Diphyes bojani*. See Bigelow (1918), Grice and Hart (1962), and Moore (1953) for distributional data on siphonophores in the western North Atlantic. For revisions in nomenclature see Totten and Bargmann (1965).

Another point of interest with regard to distribution concerns the possible connection between siphonophores and the "deep scattering layer." Barham (1963) presented evidence suggesting that siphonophores may be responsible, in part at least, for this poorly understood phenomenon, since bathyscape observations showed a concentration of 300 to 1000 colonies of *Stephanomia bijuga* per square meter in a deep scattering layer at 260 to 440 m in the Pacific. Bigelow and Sears (1937) presented data on the vertical distribution of siphonophores in the Mediterranean; some of the same species occur in the Western Atlantic.

Identification

The identification of siphonophores presents an appalling problem to the nonspecialist. As previously mentioned, the colonies are easily fragmented. The student, therefore, must be prepared to deal with pieces rather than whole colonies conforming to textbook illustrations, except for *Physalia* which will be found floating or ashore more or less intact. Identification aids for members of the Calycophore would have to include separate keys for superior and inferior nectophores and eudoxids, not to mention bracts and other disconnected fragments. Under these circumstances, we can only refer interested students to Table 5.15, which indicates the structural plan for each suborder, and to Figures 5.25 and 5.26, which show typical representatives of the three suborders. The float of *Physalia physalia* (Cystonectae), Figure 5.26A, may be more than 350 mm long with extended tentacles reaching 15 m. In *Stephanomia cara* (Physonectae) the pneumatophore is about 2 mm in greatest dimension; the large nectophore of *Diphyes* is about 18–20 mm long.

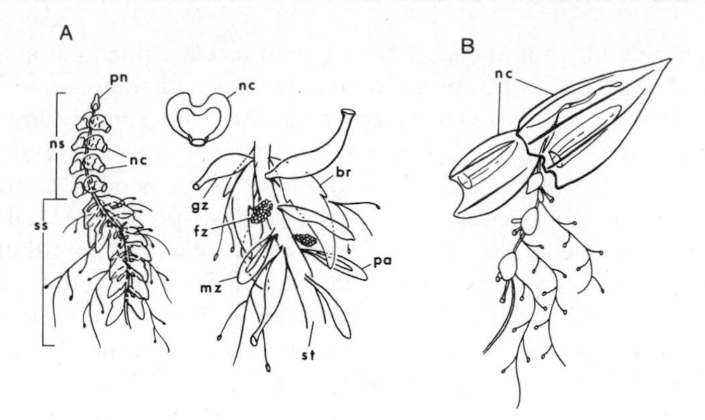

Fig. 5.25. Siphonophora: A, *Stephanomia* sp., suborder Physonectae, part of colony with details; B, *Diphyes* sp., suborder Calycophorae, part of colony. Abbreviations are: ns, nectosome; ss, siphosome; pn, pneumatophore; gz, gastrozooid; br, bract; pa, palpon; mz, male gonophore; fz, female gonophore; st, stalk; nc, nectophores.

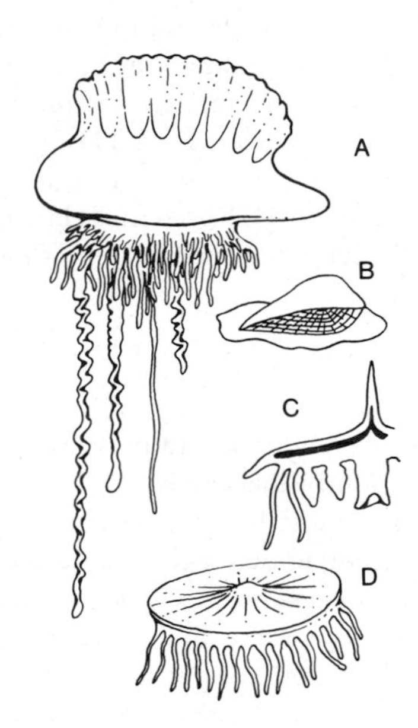

Fig. 5.26. Siphonophora and Chondrophora. A, *Physalia;* B, *Velella*, whole animal; C, same, diagrammatic cross-section of part of animal; D, *Porpita*.

Order Chondrophora
VELELLA AND *PORPITA*

Totton (1954), supported by Rees (1957), elevated the chondrophores from their former position as a suborder of Siphonophora.

Chondrophores have been variously regarded as multipersoned colonies and as highly modified individual polypoids—a theoretical controversy which we are not obliged to explore. In basic plan the mature chondrophore consists of a raftlike disc, round or somewhat lobed in outline, incorporating a horny, chambered float or *pneumatophore* from which are suspended: a central feeding *gastrozooid;* marginal tentaclelike *dactylozooids;* and, in the intervening zone, *gonozooids* bearing medusoid gonophores. (The terms in italics are defined in the siphonophore account, and apply if chondrophores are viewed as colonial.)

Members of the group are hermaphroditic, and male and female medusoids are released as free-living, sexually reproducing individuals. The initial larva is, as usual, a planula. For a brief discussion of larval development, which involves *conaria* and *rataria* stages, see Hyman (1940).

Chondrophores are epipelagic and oceanic in distribution. *Porpita porpita* and *Velella velella* (both genera are now considered monotypic) are warm water forms that may occur inshore occasionally as Gulf Stream waifs. The species may be identified from the illustration (Fig. 5.26); *Velella* has a sail, *Porpita* does not. *Velella* attains a disc diameter of about 100 mm, *Porpita*, about 25 mm.

References: Siphonophora and Chondrophora

Barham, E. G. 1963. Siphonophores and the deep-scattering layer. Science, New York, 140:826–828.

Bigelow, H. B. 1911. The Siphonophora. Reports on the scientific results of the expedition to the eastern Pacific . . . Mem. Mus. comp. Zool. Harv. 38(2):174–401.

———. 1918. Some medusae and siphonophores from the western Atlantic. Bull. Mus. comp. Zool. Harv. 62(8):365–442.

———, and Mary Sears. 1937. Siphonophorae. Rep. Danish oceanogr. exped. Mediterranean. II. Biology. H.2:1–144.

Grice, G. D., and A. D. Hart 1962. The abundance, seasonal occurrence and distribution of the epizooplankton between New York and Bermuda. Ecol. Monogr. 32(4):287–309.

Hyman, Libbie H. 1940. The Invertebrates: Protozoa through Ctenophora. McGraw-Hill Book Co., New York.
Moore, H. B. 1953. Plankton of the Florida Current. II. Siphonophora. Bull. mar. Sci. Gulf Caribb. 2(4):559–573.
Rees, W. J. 1957. Evolutionary trends in the classification of capitate hydroids and medusae. Bull. Br. Mus. nat. Hist. Zool. 4:1–60.
Southcott, R. V. 1967. The Portuguese man-of-war or bluebottle. Aust. nat. Hist. 15(11):337–342.
Sumner et al. 1911 (1913). A biological survey of the waters of Woods Hole and vicinity. Part 1, Sect. 1. Physical and zoological. Bull. U.S. Bur. Fish. 31:1-442.
Totton, A. K. 1954. Siphonophora of the Indian Ocean. Discovery Rep. 27:1–162.
———, and H. E. Bargmann. 1965. A synopsis of the Siphonophora. British Museum.
———, and G. O. Mackie. 1960. Studies on *Physalia physalia* (1): natural history and morphology. Part 2: Behavior and histology. Discovery Rep. 30:301–408.

Class Scyphozoa
SCYPHOZOAN MEDUSAE

The "typical jellyfish," though fewer in number of species than hydromedusae, are larger and more conspicuous. For a general account of medusoid structures see the introductory section to this chapter, where a key to the four typical orders of this class is presented; the fifth order, Stauromedusae, is atypical in being sessile and superficially polyplike (Fig. 5.1G; Fig. 5.27). The order Semaeostomeae contains the principal species of interest to the present account. The species are figured by Mayer (1910); for *Thaumatoscyphus* see Berrill (1962). Checklists are from Kramp (1961).

Order Stauromedusae
SESSILE MEDUSAE

Diagnosis. *Sessile, trumpet-shaped medusae with an aborally adherent peduncle, a calyxlike oral end with eight lobes armed with clublike tentacles, and an abbreviated larval development without scyphistomae.*

The stauromedusae are atypical medusoid forms and are either permanently adherent or detachable at will. The latter, when moving about, maintain position with the *anchors* between the marginal lobes while transferring the peduncular disc to a new hold. The anchors are regarded as modified tentacles; they are cushionlike and usually oval or horseshoe-shaped. They anchor by adhesion. In cross-section the gastrovascular cavity is seen to be partly divided in 4 pockets by thick septa, and in the family Cleistocarpidae each pocket is further divided by a membrane called a *claustrum* (Fig. 5.27B).

The stauromedusan planula metamorphoses through a somewhat polypoidlike phase to the adult medusa. Stauromedusae reach maturity and die in a single year, the adults disappearing after the summer breeding season. These are neritic forms that live attached chiefly to algae such as *Fucus* or especially to laminarians or, in the case of *Haliclystus auricula*, to eelgrass. More than a half century ago Mayer (1910) reported stauromedusae as becoming scarce owing to the pollution of coastal waters. The *Zostera*-attached species, which all but disappeared with the eelgrass epidemic that began in the early 1930s, have apparently not returned to parts of their former ranges.

Systematic List and Distribution

Family Eleutherocarpidae		
Lucernaria quadricornis	B	Lit. to shallow
Haliclystus salpinx	B	Lit. to shallow
Haliclystus auricula	B+	Lit. to shallow
Family Cleistocarpidae		
Craterolophus convolvulus	B+	Lit. to shallow
Thaumatoscyphus atlanticus	B	Lit. to shallow
Halmocyathus lagena	B	Lit. to shallow
Halmocyathus platypus	B	Lit. to shallow

Identification

The species of *Halmocyathus* are omitted from the key; they have anchors, four horseshoe-shaped gonads, a four-chambered peduncle, and claustra.

Measurements are heights of expanded adult medusae.

The axes of symmetry of Stauromedusae are indicated diagramatically in Figure 5.27B.

---

1. Without marginal anchors—2.
1. With marginal anchors; Fig. 5.27C—3.

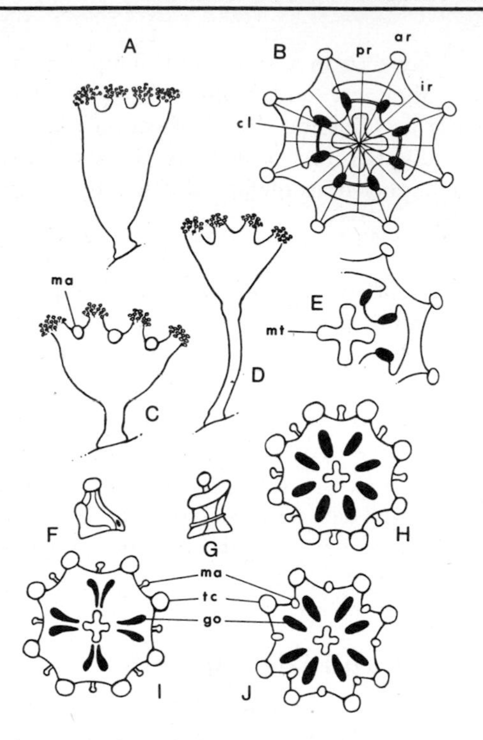

Fig. 5.27. Stauromedusae. A, *Craterolophus convolvulus*, lateral view; B, same, diagram of oral end; C, *Haliclystus auricula*, lateral view; D, *Lucernaria quadricornis*, lateral view; E, same, diagram of part of oral end; F, *Thaumatoscyphus atlanticus*, detail of anchor; G, *Haliclystus salpinx*, detail of anchor; H, same, diagram of oral end; I, *Thaumatoscyphus atlanticus*, diagram of oral end; J. *Haliclystus auricula*, diagram of oral end. Abbreviations are as follows: cl, claustrum; pr, perradius; ar, adradius; ir, interradius; mt, mouth; ma, marginal anchor; tc, tentacle cluster; go, gonad.

2. Stalk much shorter than umbrella; gastric pockets divided by a claustrum; peduncle 4-chambered; 25 mm; Fig. 5.27A,B. *Craterolophus convolvulus.*

2. Stalk somewhat longer than umbrella; without claustrum; peduncle with a single chamber; 70 mm; Fig. 5. 27D,E. *Lucernaria quadricornis.*

3. Marginal lobes paired, adradii unequal in length—4.

3. Marginal lobes equally spaced, adradii about equal; no claustrum; paired condition of the 8 gonads is apparent only where they meet deep in the calyx; 9 mm; 15–30 tentacles per cluster; Fig. 5.27G,H. *Haliclystus salpinx.* The measurement and tentacle complement above are from Berrill (1962); Mayer (1910) gives the maximum size as about 20 mm and the tentacle number as 60–70 per cluster.

4. The 8 gonads in 4 closely aligned pairs; claustrum present; peduncle 4-chambered above, single below; 25 mm; 30 tentacles per cluster; Fig. 5.27F,I. (Berrill, 1962). *Thaumatoscyphus atlanticus.*

4. The 8 gonads in 4 widely aligned pairs; no claustrum; peduncle 4-chambered; 30 mm; 100–120 tentacles per cluster; Fig. 5.27C,J. *Haliclystus auricula.*

Order Cubomedusae
SEA WASPS

Diagnosis. *Deeply bell-shaped with a veillike diaphragm, the velarium, partly closing the subumbrellar cavity; four tentacles, each with a basal pedalium and a distal filamentary part (see Fig. 5.2B,C); margin of umbrella slightly lobed but not deeply divided in lappets.*

Cubomedusae are highly specialized scyphozoans that somewhat superficially resemble hydromedusae. The distinctive form of the tentacles provides a sure identification, however. The umbrella is divided by septa that suggest the divisions of the radial canals of hydromedusae, and the gonads develop on these septa approximating the placement of the gonads in leptomedusae on the radial canals.

Members of the group are notorious for their stinging power. The abilities of our species in this respect have not been well studied. Adults are reported to prey on fish. Cubomedusae are active swimmers, and the bell is capable of rapid pulsations; the pedalia are used for steering

Cubomedusae are described as being bathypelagic until mature, when they may appear in the epiplankton. They are warm water forms and occur uncommonly within the range of this text.

Systematic List and Distribution

Family Carybdeidae		
Tamoya haplonema	CV	Fall
Family Chirodropodae		
Chiropsalmus quadrumanus	C+	Summer-fall

Identification

In *Tamoya* the pedalia are simple, i.e., bladelike with a single, terminal filament (Fig. 5.2B); in *Chiropsalmus* the pedalia are branched or handlike with filamentous tentacles emerging from the "fingers" (Fig. 5.2C). *Tamoya haplonema* attains a bell height of about 90 mm and a diameter of 55 mm; the corresponding measurements for *C. quadrumanus* are 100 by 140 mm.

Order Coronatae
CORONATE MEDUSAE

Diagnosis. *Umbrella with a coronal, i.e., horizontal, groove and margin deeply scalloped with lappets (Fig. 5.2H).*

These are small to medium-sized medusae, about 15–200 mm diameter, readily identified by the criteria indicated. Two of the three species listed as occurring off this coast are bathypelagic and are rare in surface waters. Their appearance inshore is exceptional and is most likely to occur in conjunction with prolonged onshore winds.

Systematic List and Distribution

Family Periphyllidae		
Periphylla periphylla	BV	Bathypelagic, cold water
Family Nausithoidae		
Nausithoe punctata	CV	Epipelagic
Family Atollidae		
Atolla wyvillei	BV	Bathypelagic, Gulf Stream, 600–3600 m

Ontogenetic changes have not been fully described but presumably involve changes in the number of tentacles and other details. The scyphistoma of *Nausithoe* is described as hydroidlike—a branching treelike growth with chitinous perisarc, living as a commensal within sponges such as *Suberites*, *Myxilla*, *Reneira*. The ephyrae are produced by strobilization from this peculiar scyphistoma.

Identification

1. With 8 tentacles, 16 lappets, 8 gonads; umbrella flatter than a hemisphere, 15 mm diameter; summer; Fig. 5.2H. *Nausithoe punctata.*

1. With 12–24 tentacles—2.

2. With 12 tentacles, 16 lappets, gonads not visible owing to the dense, purple-brown pigmentation of the bell; umbrella usually not quite twice as tall as broad, but variable, cone-shaped to hemispherical, 42 mm diameter. *Periphylla periphylla.*

2. With 16–24 tentacles, 36–48 lappets, 8 gonads; bell saucerlike, 150 mm diameter. *Atolla wyvillei.*

Order Semaeostomeae
TYPICAL JELLYFISH

A "diagnosis" might best characterize this group as lacking the features that distinguish the other scyphozoan orders (see key to medusoid forms in the introductory section of this chapter).

This order contains the larger, neritic jellyfish. The common species are familiar enough to have acquired colloquial names, e.g., *Chrysaora* = sea nettle, *Cyanea* = lion's mane, *Aurelia* = moon jelly. The stinging abilities of the sea nettle are well known; the lion's mane's capabilities have been somewhat exaggerated. As is commonly observed in cnidarian-human contacts, susceptibility to being stung varies somewhat in different people and on different parts of the body. The normally calloused human hand may not register the discharge of nematocysts, while more delicate skin surfaces feel such stings as a sharp, burning sensation. Stings may be neutralized by washing the affected parts with alcohol or even 10% formalin (see Halstead, 1965; Barnes, 1966).

The larger jellyfish are predatory and quite capable of subduing fish and other small animals. *Aurelia*, however, is a plankton feeder, the mucous coating of the umbrella serving as a sticky collector of both phyto- and zooplankters which are passed over the exumbrellar surface to the margins of the bell and ultimately to the mouth with the aid of the frilled mouth-arms.

Except for *Pelagia*, local medusae of this order develop by strobilization from scyphistomae. *Pelagia*, in keeping with its "deep sea" habits, has medusae developing directly from a planula larva. The scyphistomae of the remaining species are small and inconspicuous (Fig. 5.1H). Those of *Chrysaora* develop 4 to 16 tentacles, *Cyanea*, 4 to 20, and *Aurelia*, 8 to 24. When budding ephyrae, the scyphistomae are 5 to 15 mm tall. The polypoid form is solitary or weakly colonial with secondary budding from the body of the scyphistoma or from a stolon. The scyphistomae are found attached to algae, eelgrass, or firm substrata in shallow water.

The ephyrae bear little resemblance to the adult medusae. They are radially symmetrical with 8 deeply bifurcated arms (Fig. 3.3M), and measure about 0.4–5.0 mm in diameter. Young medusae frequently have bathypelagic or even benthic tendencies. The adult form is acquired by a gradual metamorphosis in which there is a phylogenetic recapitulation—more or less. Thus *Chrysaora* medusae have only 8 tentacles initially, as do adult *Pelagia*, and *Cyanea* passes through 24 and 40 tentacle "*Chrysaora*" stages. Adult

Cyanea and *Chrysaora* are notably polymorphic, this variation being both ecologically and geographically based.

The common species tolerate a wide range of salinities. *Cyanea* and *Chrysaora* are frequently noted in swarms; sea nettles, for example, are frequently common enough in the Chesapeake in summer to make the waters uninhabitable for swimmers.

Systematic List and Distribution

Family Pelagidae		
Pelagia noctiluca	V+	Oceanic, Gulf Stream
Chrysaora quinquecirrha	V	eury.
Family Cyanidae		
Cyanea capillata	BV	eury.
Family Ulmaridae		
Phacellophora camtschatica	B	—
Aurelia aurita	BV	—

Identification

Measurements are approximate maximum adult diameters.

1. Tentacles marginal, short, very numerous, about 30 or more per octant; bell subhemispherical with 8 shallow, marginal lobes; with a complex system of radiating canals; the 4-lobed stomach conspicuously outlined in mature individuals by 4 horseshoe-shaped gonads; 250 mm; spring-summer in north, winter-spring in south; Fig. 5.28A. *Aurelia aurita.*

1. Tentacles emerge from floor of subumbrella in clusters or from clefts between the marginal lappets—2.

2. Tentacles subumbrellar in clusters; Fig. 5.28B—3.

2. Tentacles emerge from clefts between marginal lappets; Fig. 5.28C—4.

3. Tentacles in 8 clusters, with 3 tentacles per cluster when bell is about 8 mm diameter, about 15 tentacles per cluster when 20 mm, and up to 60 or more when mature; bell lens-shaped with 8 primary, adradial clefts and shallower secondary clefts producing a total of about 32 lappets; south of Cape Cod, bell 200 mm diameter, yellowish brown; north of Cape Cod, 2300 mm in Arctic but rarely more than 800 mm southward, more deeply tinted; winter-spring in south, summer-fall in north; Fig. 5.28 B. *Cyanea capillata.*

3. Tentacles in 16 clusters; bell flat, with 16–32 lappets; 16 rhopalia; 350 mm; rare. *Phacellophora camtschatica.*

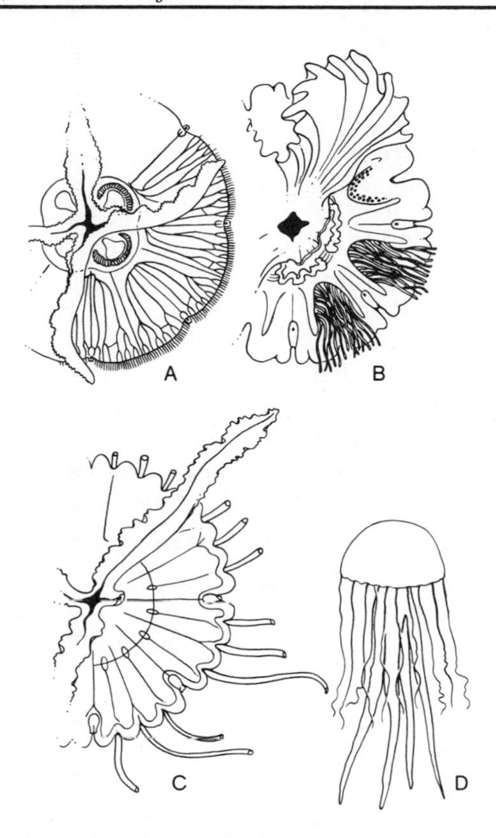

Fig. 5.28. Semaeostomeae. A, *Aurelia aurita*, ventral view of part of adult; B, *Cyanea capillata*, ventral view of part of adult with mouth-arms removed from two octants and tentacles cut basally in the upper right octant; C, *Chrysaora quinquecirrha*, part of medusa with one mouth-arm cut basally and tentacles cut subbasally; D, same, diagrammatic lateral view.

4. With no more than 8 tentacles; bell subhemispherical, 50 mm; 16 lappets; oceanic; summer-fall. *Pelagia noctiluca.*

4. With 24–40 tentacles; bell subhemispherical, with 32–48 lappets; polymorphic. The milky-white form is smaller (about 100 mm diameter), and sexually mature with 24 tentacles and 32 lappets; it ranges into low salinity waters (ca. 3‰) and is particularly common in Chesapeake Bay; the bay mouth form is larger (190 mm), develops a maximum of 40 tentacles and 48 lappets, and is more strongly colored, with 16 radiating reddish speckled bars on the umbrella; summer; Fig. 5.28C,D. *Chrysaora quinquecirrha.* Juveniles of this species pass through a *Pelagia*-like stage.

Order Rhizostomeae

Diagnosis. *Scyphozoan medusae without marginal or subumbrellar tentacles; adults with multiple mouth openings.*

These medium to large medusae have a thick, rigid umbrella shaped like a mushroom cap. They are southern forms that stray northward; *Stomolophus* avoids estuarine situations, but *Rhopilema* is reported from coastal lagoons, sounds, and bays. The tentacle-less condition is present in very young medusae—as small as 3 mm bell diameter in *Stomolophus*.

Systematic List and Distribution

Family Rhizostomatidae		
Rhopilema verrilli	V	Summer-fall
Family Stomolophidae		
Stomolophus meleagris	C, 37°	Summer-fall?

Identification

1. Umbrella with 16 lappets per octant, those adjacent to the marginal sense organs longer than the others; mouth-arms coalesced along their entire length; 180 mm diameter, Fig. 5.2D. *Stomolophus meleagris.*
1. Umbrella with 8 lappets per octant, those adjacent to the marginal sense organs smaller than the others; mouth-arms coalesced basally only; umbrella to 340 mm; Fig. 5.2E,F. *Rhopilema verrilli.*

References

Barnes, J. H. 1966. Studies on three venomous cubomedusae. *In* Cnidaria and their Evolution. W. J. Rees, ed. Symp. Zool. Soc. of London. No. 16. Academic Press, New York, pp. 307–331.

Berrill, M. 1962. The biology of three New England Stauromedusae, with a description of a new species. Can. J. Zool. 40:1249–1262.

Halstead, B. W. 1965. Poisonous and Venomous Marine Animals of the world. Vol. I, Invertebrates. U.S. Government Printing Office.

Hargitt, C. W. 1904. The medusae of the Woods Hole region. Bull. Bur. Fish. Wash. 24:21–79.
Kramp, P. L. 1961. Synopsis of the medusae of the world. J. mar. biol. Ass. U.K. 40:469 pp.
Mayer, A. G. 1910. Medusae of the world. Carnegie Inst. Wash. Publ. 109, Vol. 3: 499–735.

Class Anthozoa

Members of this class are exclusively polypoid. In Actiniaria and Ceriantharia the polyps are solitary, and asexual reproduction by budding is uncommon. The remaining orders are chiefly colonial, and budding is the usual method of colony formation.

Anthozoan polyps are more complex internally than those of Hydrozoa. In brief, the gastrovascular cavity in Anthozoa is partly divided by partitions or mesenteries; in Hydrozoa it is not.

Special larval forms occur in Ceriantharia and Zoanthidea; more or less generalized planulae occur in the other groups but, in Actiniaria, Octocorallia, and Scleractinia, metamorphosis commences while the larvae are still pelagic. Development is known in but a few species in most orders. "Marsupialism" occurs in some anemones, the young developing within the body of the parent or in external brood pouches in a few extralimital species. Anemones reproduce asexually by pedal laceration as well as by transverse and longitudinal fission and budding, but the offspring separate from the parent. Individual anemones often aggregate but do not maintain colonial affinities in a structural sense.

Criteria for the initial separation of ordinal groups were given in the general introduction to the phylum.

Subclass Octocorallia
SEA WHIPS, SEA PENS, AND RELATED FORMS

The only exclusive feature of members of this subclass is their possession of eight pinnate tentacles, but for practical purposes they can be characterized as colonial anthozoans with a skeleton of calcareous spicules.

The colony may be whiplike, arborescent, lobate, or massive, or individual polyps may arise from an encrusting stolon. The spicules are microscopic and, though basically of two types, are diverse in form and size. They are much used in systematics. In many groups the polyps are dimorphic; *autozooids* are typical polyps with eight tentacles; *siphonozooids* are usually smaller, with one tentacle or none and a strongly developed *siphonoglyph* whereby they regulate the hydrostatic pressure in the colony. Typical colonies are flexible but highly cohesive, and are rubbery or fleshy to the touch. Except in pennatulaceans, which have their "pens" seated in unconsolidated sediment, colonies are usually firmly attached to hard substrata.

The sexes are usually separate, but *protogynous hermaphroditism* also occurs, i.e., female gametes are matured before male elements in the same individual. There is no sexual dimorphism between colonies. The autozooids are the sexual members in most groups, but in Alcyanacea and Gorgonacea the siphonozooids are sexual forms. The larvae are planulae.

Octocorals and particularly the gorgonians are perhaps best known as tropical reef animals. The local octocoral fauna is made up of warm water species that barely enter the range of this text and of northern species confined to relatively deep water. The gorgonians *Leptogorgia setacea* and *L. virgulata*, however, do occur in shallow water north, at least to the Chesapeake where *setacea* ranges in mesohaline waters to the mouth of the Potomac.

Systematic List and Distribution

The list is based mainly on Deichmann (1936), Miner (1950), and Bayer (1961).

Order Stolonifera

Family Clavulariidae

Clavularia modesta	B	37 to 1267 m

Deichmann reported a second *Clavularia* from off Salem, Mass.; this is an apparently undescribed species, not *C. modesta*.

Order Telestacea

Family Telestidae

Telesto fruticulosa	C+	27 to 91 m
Telesto nelleae	C+	27 to 293 m

Order Alcyonacea

Family Alcyoniidae

Alcyonium digitatum	B, 39°	15 to 146 m

Family Nephtheidae (Nephthyidae)

Drifa glomerata	B	97 to 1098 m
Gersemia rubiformis	A+	37 to 91 m

Order Gorgonacea		
Family Anthothelidae		
Paragorgia arborea	B	"Fishing banks"
Anthothela grandiflora	B	150 to 240 m
Titandeum frauenfeldii	C+	27 to 124 m
Family Gorgoniidae		
Leptogorgia setacea	C, 39°	eury., shallow
Leptogorgia virgulata	C, 39°	Shallow
Family Plexauridae		
Muricea pendula	C+	13 to 27 m
Family Primnoidae		
Primnoa resedaeformis	B	91 to 273 m
Order Pennatulacea		
Family Renillidae		
Renilla reniformis	C+	Lit. to 5 m
Family Pennatulacea		
Pennatula aculeata	BV	110 to 500 m

Identification

The chief systematic criteria are colony form and spicule complements. Material dry or preserved in alcohol is suitable for study; formalin should never be used, as it dissolves the spicules.

The spicules may be examined by placing a small piece on a microscope slide and dissolving the soft material with Clorox until the spicules are dissociated. Restore evaporated liquid with a drop or two of water, add a cover slip, and examine microscopically at about 100×. The usual range of spicule lengths is from about 0.1 mm to slightly more than 1.0 mm. Different spicule types may be found in different parts of the colony and in different structural layers. See Bayer (1961) for further details on technique. Note that a few local sponges also have calcareous spicules, but they are quite unlike octocoral spicules. The spicules of most sponges are siliceous. Calcareous and siliceous spicules or dried whole specimens containing such may be differentiated by applying a drop or so of dilute hydrochloric acid; the calcareous material is effervescent.

Measurements in the following key indicate approximate maximum colony dimensions. The species are figured by Deichmann (1936); see also Bayer (1961).

1. Colony not attached, but with a fleshy stalk inserted in soft bottom sediment. Pennatulacea—2.

1. Colony fixed, firmly attached to a hard substratum—3.

2. Colony broadly lobate, lily padlike; rose or purple, occasionally white or yellow with purple stalk; ca. 60 mm in diameter; Fig. 5.29A. *Renilla reniformis.*

2. Colony plumelike, with fanlike clusters of polyps (autozoids) in rows along two sides of median stalk; branches red or purplish red, main stalk yellow to pale orange; 100 mm total length; Fig. 5.29B. *Pennatula aculeata.*

3. Arborescent or whiplike; stems or branches in cross-section with differentiated cortical and medullary layers. Gorgonacea—4.

3. Colonies of various types; if arborescent, then cross-sections without distinct cortex and medulla, sometimes hollow—6.

4. Central axis (medulla) spicular; with a ring of canals separating the outer rind or cortex from the central medulla, the two layers with different types of spicules (Fig. 5.29C); colony sparsely branched, yellow-orange to pinkish red; a Carolinian species. *Titandeum frauenfeldii.* Note that two additional species in this family may be taken on the New England fishing banks. For a brief description of *Paragorgia arborea* and *Anthothelia grandiflora* see Miner (1950) or, for a more complete account, Deichmann (1936). Another northern species, *Primnoa re-*

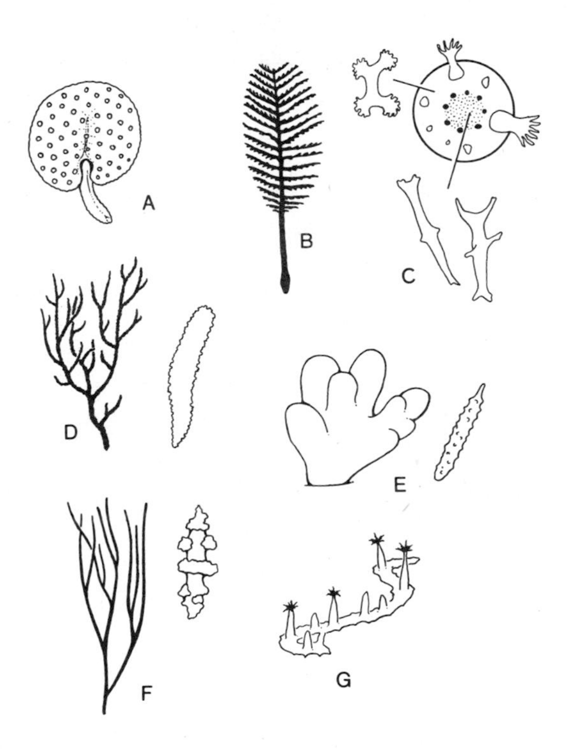

Fig. 5.29. Octocorallia. A, *Renilla reniformis*, whole colony; B, *Pennatula aculeata*, silhouette of whole colony; C, *Titandeum* sp., diagrammatic section of colony with outlines of spicules from medula (below) and cortex (above); D, *Muricea pendula*, colony form and typical spicule from outer rind; E, *Alcyonium* sp., colony form and zooid spicule; F, *Leptogorgia virgulata*, colony form and typical spicule from outer rind; G, *Clavularia*, colony form.

sedaeformis (Primnoidae), also figured in Miner, has the central axis calcified and solid; the colonies are profusely branched, whereas those of the preceding species are more massive or sparsely branched.

4. Central axis without spicules, horny, flexible, core cross-chambered—5.

5. Colonies branching, pinnate, the branches relatively short; outer rind with large (ca. 0.8–1.3 mm), prickly, amber-colored, spindle-shaped spicules; colony brownish, tinted with orange or yellow; twiglets about 20–50 mm long, the colony to ca. 600 mm tall; Fig. 5.29D. *Muricea pendula.*

5. Colonies whiplike, unbranched (*L. setacea*) or with long, slender branches (*L. virgulata*); outer rind with very warty, spindle-shaped spicules <0.2 mm long. *Leptogorgia:* Both species variable in color, yellow to purple; *setacea* grows to 2 m long but colonies 40 mm long are only 1 mm thick; *virgulata* to ca. 960 mm tall; Fig. 5.29F.

6. Colonies developing from a stolon—7.

6. Colonies massive, lobate, to branching or somewhat arborescent. Alcyonacea. Published accounts suggest that these species may be difficult to identify. Appearance of the colony is much affected by the state of expansion of the polyps. The colonies are white, grayish, yellowish, or pink, 40–100 mm in greatest dimension. The polyps are of a single type in nephtheids but dimorphic in *Alcyonium.* The spicules of the common substance of the colony and of the individual zooids are of different types; the zooid spicules of *Alcyonium digitatum* and *Gersemia rubiformis* are chiefly rod-shaped (Fig. 5.29E), those of *Drifa glomerata*, club-shaped. The spicules are small, mostly <0.25 mm. *A. digitatum* (usually cited as *A. carnea*) is the more frequently reported shallow water species; Fig. 5.29E.

7. Soft, tubular polyps arising singly from a creeping stolon, 17 mm tall; Fig. 5.29G. *Clavularia* spp.

7. Primary polyps arising from a creeping stolon and budding secondary polyps, the colony resembling "a branch of spruce" and frequently overgrown by sponge, axial polyp to 100 mm; color orange-yellow. *Telesto* spp. For the differentiation of species see Bayer (1961).

References

Bayer, F. M. 1956. Octocorallia. *In* Treatise on Invertebrate Paleontology. R. C. Moore, ed. Geol. Soc. Amer. and Univ. Kansas Press.

———. 1961. The shallow-water Octocorallia of the West Indian region. Martinus Nijhoff, The Hague.

Deichmann, Elisabeth. 1936. The Alcyonaria of the western part of the Atlantic Ocean. Mem. Mus. comp. Zool. Harv. 53:1–317.

Miner, R. W. 1950. Field Book of Seashore Life. G. P. Putnam's Sons.

Ultinomi, H. 1961. A revision of the nomenclature of the family Nephtheidae (Octocorallia: Alcyonarlia). II The Boreal genera *Gersemia*, *Duva*, *Drifa*, and *Pseudodrifa* (n.g.). Publ. Seto mir. biol. Lab. 9(1):229–246.

Subclass Zoantharia
Order Zoanthidea
ZOANTHID ANEMONES

Diagnosis. *Colonial, anemonelike anthozoans united basally; without a skeleton but usually encrusted with sand or other extraneous matter, Figure 5.1B.*

The only definitely recorded local species is *Epizoanthus incrustatus*, whose range includes the whole coast. This species is usually associated with hermit crabs, but may occur on dead gastropod shells, on the coral *Astrangia*, or on rock. The association with hermit crabs begins when the zoanthid attaches itself to the gastropod housing the pagurid; as the colony develops, the basal mass, or *coenenchyme*, dissolves the mollusk shell, and the coenenchyme itself then partly encloses the crab. The resultant enclosure is termed a *carcinoecium*.

Colonies consist of up to about 18 polyps which produce a mass about 25 by 13 mm with individual polyps about 11 mm tall. Each polyp has a marginal circlet of tentacles. The internal structure is distinctive, for which see Hyman (1940). Zoanthids produce a distinct type of planktonic larvae, but complete life histories are not known. The sexes are separate (or hermaphroditic?) but are not dimorphic.

E. incrustatus is found in 33 to $>$900 m; a form described by Verrill from about 183 m may be a species of *Parazoanthus*, and additional species may occur in very deep water. In the tropics, zoanthids of the genera *Zoanthus* and *Palythoa* are conspicuous in littoral waters, where they form extensive encrusting colonies on rock, dead coral, and even on sandy substrata.

References

Carlgren, O. 1913. Zoantharia. Danish Ingolf-Expedition. 5(4):1–64.

Hyman, Libbie. 1940. The Invertebrates: Protozoa through Ctenophora. McGraw-Hill Book Co., New York.

Order Actiniaria
SEA ANEMONES

Diagnosis. *Solitary anthozoans with neither an internal skeleton of calcareous spicules nor a solid external skeleton; aboral end either an adherent pedal disc or, in some burrowing forms, tapered and rounded terminally; burrowing species with a single whorl of tentacles.*

The oral disc in all local genera is armed with unbranched tentacles; these may be in a single marginal ring, or in multiple rings or *cycles* more or less covering the disc, or in radiating rows. The disposition of the tentacles is related to the number and distribution of the mesenteries. Tentacle complements are of systematic importance but are subject to much variation and present difficulties of interpretation for the novice.

The column may or may not be divided into regions differing in surface ornamentation, relative thickness of body wall, or other details. The principal part of the column is the *scapus* which may have a *collar*. There may also be an upper, thin-walled region, the *capitulum*, or a thick-walled *scapulus*, or both. Burrowing anemones may also have the basal region differentiated as a bulbous, inflatable "digging organ," the *physa*. Most anemones are more or less contractile both in the column and in the oral disc. The ability to retract the tentacles reflects the extent of development of longitudinal retractors on the mesenteries; in addition, the whole oral region may be retracted or "covered" in some species through the operation of an *oral sphincter*. Differences in the degree of development of this sphincter and its histological construction are of major systematic importance.

The walls of the column may be ornamented externally with one or more types of defensive nettle or adhesive structures which are usually in the form of papillae or warts. Some of these have special names: *tenaculi*, solid adhesive organs; *verrucae*, vesicular adhesive organs; and *nemathybomes*, rounded, invaginated nettle organs. Additional bumps, warts, ridges, or other ornamentation of unknown function may also be present. The column in some groups is perforated with small pores called *cinclides* which become apparent especially when the threadlike *acontia* are protruded through them during sudden contraction—as when the anemone is severely disturbed.

Details of the internal structure are of major importance in formal systematics and cannot be wholly escaped in the present more informal account.

In general, such data are included in the keys as supplementary criteria, but some species determinations are impossible without making sections or identifying the different types of nematocysts that are present (see Carlgren, 1949).

If an anemone is sectioned longitudinally (Fig. 5.30A) it will be seen that the mouth opens into a *pharynx* or *stomodaeum* which in turn opens into the gastrovascular cavity. The pharynx may have one or more longitudinal grooves called *siphonoglyphs* (Fig. 5.31A,D). These are lined with flagella and contrast histologically with the rest of the pharyngeal epithelium. The gastrovascular cavity is radially partitioned by mesenteries that project inward from the column wall. Cross-sections through the whole anemone reveal further details including the paired arrangement of the mesenteries and the lack of simple radial symmetry (Fig. 5.30B). Note that some mesenteries extend from the column wall to the stomodaeum and are termed *perfect* mesenteries; *imperfect* mesenteries do not reach the stomodaeum. The gonads are borne on some or all of the mesenteries; those with gonads are said to be *fertile*, those without, *infertile*. The perfect mesenteries of elongate burrowing anemones

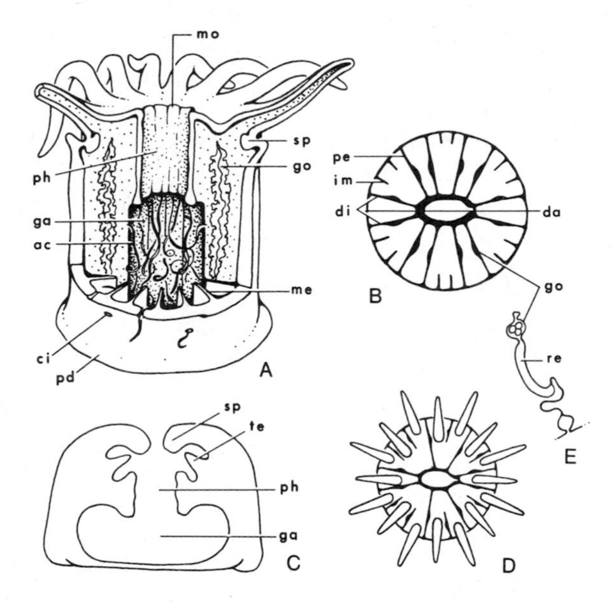

Fig. 5.30. Actiniaria. A, generalized anemone, disected; B, same, section through pharynx; C, *Tealia* sp., diagrammatic longitudinal section showing complete retraction of oral region through operation of oral sphincter; D, *Edwardsia* sp., diagrammatic oral view showing relationship of tentacles to mesenteries. Abbreviations are as follows: ac, acontia; ci, cinclides; da, directive axis; di, directive mesenteries; ga, gastric cavity; go, gonad; im, imperfect mesentery; me, mesentery; mo, mouth; pd, pedal disc; pe, perfect mesentery; ph, pharynx; re, retractor muscle; sp, sphincter; te, tentacle.

(Edwardsiidae) are termed *macronemes* and are more strongly and more complexly developed than the imperfect *micronemes*. In other groups the perfect and imperfect mesenteries may be otherwise similar in development. Mesenteries may extend from the foot to the oral disc—i.e., cross-sections near the top and bottom of the column reveal a similar number of mesenteries—or some of the mesenteries may only extend part way down, or up, in which case the count of mesenteries at top and bottom will be different.

If the anemone is sectioned through the pharynx (Fig. 5.30B), it will be found to have biradial or bilateral symmetry rather than simple radial symmetry. The pharynx is elliptical or lens-shaped in section rather than perfectly round. The long axis of the ellipse is termed the directive axis and is further defined as having "dorsal" and "ventral" poles; the poles may or may not be actually differentiated structurally, e.g., by the presence or absence of siphonoglyphs. The paired mesenteries lying on this axis are called directives; their retractor muscles are on the outer side, as figured. The symmetric relations are bilaterial or biradial depending on the regularity of other internal elements. (See Hyman 1940; Carlgren, 1949; or Stephenson, 1928, for further morphological detail.)

Anemones reproduce asexually, chiefly by longitudinal fission. This and other forms of asexual reproduction frequently result in irregularities in septal patterns and other details of systematic importance. Anemones also reproduce sexually, individual anemones being male or female, or protandric hermaphrodites. There is some recapitulation during ontogeny, the complex species passing through stages whose septal patterns resemble those of less complex genera.

Some anemones, in common with many shallow water octocorallians and most hermatypic corals, enter into commensal relationships with dinoflagellate algae called zooxanthellae. The larvae of some species of *Edwardsia* and *Peachia* are parasitic on ctenophores (*Mnemiopsis*) and on jellyfish (*Cyanea*), respectively; adults of *Calliactis* and *Adamsia* are chiefly associated with hermit crabs, and *Stephanauge spongicola* is found on sponges. Anemones feed on live or dead animal material ranging from planktonic or detrital particles collected by ciliary currents to larger fragments or whole animals which are captured either by mucous secretions or by nematocyst threads on the tentacles. Among local, shallow water species, only *Bolocera tuediae* is capable of producing a very noticeable sting on human contact.

Anemones range from the littoral to great depths. They attach to a wide variety of living and nonliving surfaces, and some species are burrowers in mud or sand, or in cavities in hard substrata. Despite their seemingly sedentary disposition, anemones are motile, moving at a slow pace by gliding on the pedal disc. Some species can "walk" on their tentacles, and a few actually swim; the last, rather unexpected behavior occurs among local genera in *Stomphia* and *Actinostola* according to Robson (1966). Hargitt (1912) describes *Anemonia sargassensis* as being sometimes pelagic, at least in aquaria, by floating upside down with the aid of an inflated pedal disc.

Systematic List and Distribution

The chief source for this list is Carlgren (1949) with additions and modifications from later literature. All of the local species belong to the suborder Nynantheae, which is one of three recognized by Carlgren.

Tribe Athenaria		
Family Edwardsiidae		
Edwardsia elegans	B	Lit. to 117 m
Edwardsia leidyi (larval form)		Parasitic on *Mnemiopsis*
Edwardsia sipunculoides	B	87 to 117 m
Fagesia lineata	V	11 to 22 m
Nematostella vectensis	V	Shallow, eury.
Drillactis pallida	B	Lit.
Family Halcampoididae		
Halcampoides sp.	C+	Shallow
Family Haloclavidae		
Haloclava producta	V	Shallow, eury.
Peachia parasitica, larva	B	Parasitic on Cyanea
Adults not recorded south of Eastport, Maine		
Ilyanthus laevis	B	?
Ilyanthus chloropsis	V	?
Family Halcampidae		
Halcampa duodecimcirrata	B	9 to 164 m
Tribe Thenaria		
Family Actinidae		
Anemone sargassiensis	V	On *Sargassum*
Phymactis cavernata	C+	Lit.
Bolocera tuediae	B	40 to 2023 m
Liponema multicornis	A+	82 to 872 m
Tealia felina	B	Lit. to 73 m
Bunodactis stella	B	Lit. to 102 m

Note that *Bunodactis capitata*, listed by Carlgren (1949) as "Cape Cod to Florida," does not appear in the Virginian lists examined, and Field (1949) cites the species record for North Carolina as probably misidentified and equal to *Diadumene leucolena*.

Family Actinostolidae		
Actinostola callosa	BV	73 to 3660 m
Stomphia coccinea	B	15 m
Paranthus rapiformis	V	Lit.
Antholoba perdix	V	113 to 351 m
Family Isophelliidae		
Telmatactis sp.	C, 39°	—

1 record 70 mi ENE Cape Henry; *see* Wass (1963)

Family Hormathiidae		
Hormathia nodosa	A+	91 to 183 m (Miner, 1950)
Actinauge longicornis	V	73 to 595 m
Actinauge verrillii	B, 37?	91 to 2009 m
Calliactis tricolor	C(+?)	Shallow
Adamsia sociabilis	CV	155 to 549 m
Stephanauge nexilis	B	182 to 547 m
Stephanauge spongicola	B	144 to 578 m
Family Sagartidae		
*Actinothoe modesta**	V	Lit.
*Actinothoe gracillima**	V	Shallow
*Actinothoe eruptaurantia**	C, 37°	Lit.
Sagartiogeton verrilli	A, 39°	91 to 1416 m
Family Metridiidae		
Metridium senile	BV	Lit. to 165 m
Family Aiptasiomorphidae		
Haliplanella luciae	BV	Lit., eury.
Family Diadumenidae		
Diadumene leucolena	V	Lit.

*The integrity of these three names as distinct species is in doubt.

Identification

The identification of individual species in this group is particularly difficult. The systematic arrangement is largely determined by criteria that can only be examined in microscopic sections. These criteria include, among others, various aspects of mesentery development, the histological structure of the oral sphincter and basal musculature, and the types of nematocysts in the collective *cnidome* (the nematocyst complement of the species). There is ontogenetic variation in some traits, and further obfuscation is introduced by irregularities resulting from asexual reproduction. The easily observed details of color, column ornamentation, and growth form are subject to substantial individual variation and may be further affected by preservation and particularly by the highly contractile behavior of most species. Narcotization is recommended as a preliminary to fixing but requires some patience (see Chapter 3). Finally one should note that more than a few species are very imperfectly known from old descriptions that did not include all the criteria now considered to be necessary for a full species description.

In view of the difficulties just outlined, the user of the following keys is advised to proceed with caution. Specimens should initially be identified to tribe.

TRIBE ATHENARIA. Pedal disc lacking; aboral end flattened and adherent in *Fagesia* but in other genera either rounded, tapered, or with a more or less clearly differentiated digging organ or physa.

TRIBE THENARIA. With a clearly differentiated pedal disc, adherent to a solid substratum; burrowing forms are attached basally to stones, shells, or deeper water species may be found clutching a ball of sand or mud.

TRIBE ATHENARIA

These are elongate or wormlike species usually found embedded in sand or mud with only the oral disc exposed. Some also occur in crevices or among stones to which they adhere; in some genera adhesive organs collect a coating of sand or other fine debris. Note that burrowing or sand-dwelling species also occur in Thenaria, e.g., *Paranthus rapiformis* and *Actinothoe* spp. Thenarian anemones are usually attached basally to stones or shells as indicated, and adults usually have more numerous tentacles (40–144) than do most athenarians. *Phymactis cavernata* occurs on sand-covered rocks but is not, strictly speaking, a burrowing form. *Actinauge* and *Actinostola* species are deep water anemones that clasp a ball of mud with the oral disc.

Consider Table 5.16 as well as the following key. Measurements are approximate maximum size of expanded individuals.

---

1. Upper part of column with 20 rows of papillae; tentacles blunt and swollen terminally; 70 mm tall; Fig. 5.31A. *Haloclava producta.*

1. Not so—2.

2. Column divided in two or more regions not including the physa which may be present or not—3.

2. Physa may be present but body regions otherwise not differentiated. For the separation of the four species keying out at this point see Table 5.16. Note that the free-living adults of *Peachia parasitica* (Fig. 5.31D) are unknown south of Eastport, Maine; in addition to the criteria tabulated, this species is distinguished by having cinclides. The imperfectly known *Drillactis pallida* is also boreal. *Nematostella vectensis* (Fig. 5.31B) is Carolinian and Virginian. *Halcampoides* sp. (Fig. 5.31C) is Carolinian and probably extralimital; the species described by Field (1949) may be mistaken for the holothurian *Leptosynapta;* while individuals of this species may attain a length of 55 mm, such a specimen would be only 2–3 mm in diameter.

3. Aboral end flattened or rounded, without a physa; scapus much thickened but without nettle organs; tentacles in cycles 6 + 6 + 12 to about 40, the inner ones longer than the outer; length 35 mm. *Fagesia lineata.*

3. Not so; physa developed—4.

Table 5.16. Burrowing Anemones

Genus	Body Region-ated	Physa Present	Tentacles	Perfect Mesenteries	Macro- and Micro-nemes Distin-guished	Siphono-glyphs	Sphincter Present	Nettle or Adhesive Structures Present
Edwardsia	/	/	12–36	8	/	1	–	/
Fagesia	/	–	12–40	8	/	1	–	/
Nematostella	w	w	12–16	8	/	1	–	–
Drillactis	–	–	12–24	8	/	?	–	–
Haloclava	–	/	20	20	–	1 d	–	/
Peachia	–	–	12	12	–	1 d	–	/
Halcampa	/	/	8–12	8–12	/	–	w	/
Halcampoides	–	/	16	10–16	–	2 w	w	–
Actinothoe	–	p	48–60	>6–>12	–	v	/	–
Paranthus	–	p	ca 144	12–	–	2	/	–

Abbreviations and symbols are: /, present; –, absent; w, weak; d, deep; p, pedal disc present, weak in *Paranthus*.

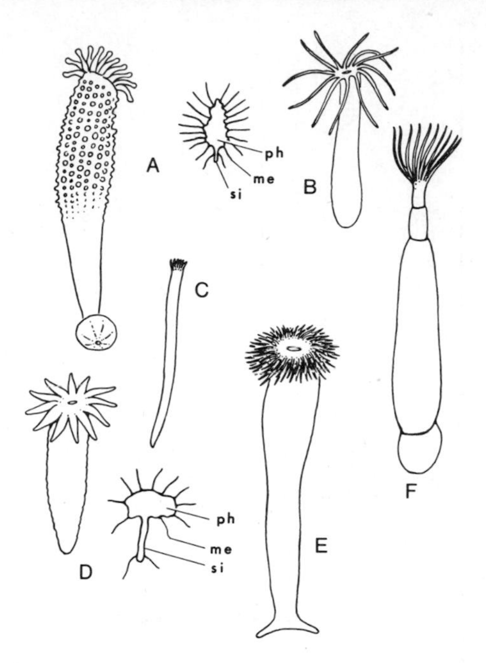

Fig. 5.31. Actiniaria, burrowing anemones. A, *Haloclava producta*, whole animal and diagrammatic section of pharynx; B, *Nematostella vectensis;* C, *Halcampoides* sp.; D, *Peachia parasitica*, whole animal and diagrammatic section of pharynx; E, *Actinothoe* sp.; F, *Edwardsia elegans*. Abbreviations are: ph, pharynx; si, siphonoglyph; me, mesenteries.

4. With nettle organs (nemathybomes) appearing as rounded spots, absent on the physa; tentacles slender, 12–36 in number. *Edwardsia: E. leidyi* is known only as a "larva," parasitic on *Mnemiopsis*, 30 mm long; *E. sipunculoides*, incompletely described, is a deep water form, 100 mm long, with 20–36 tentacles, *E. elegans* is the common shallow water species, 150 mm long, with about 16 tentacles; Fig. 5.31F.
4. With raised adhesive organs (tenaculi); tentacles thick, 8–12; 25 mm long. *Halcampa duodecimcirrata.*

TRIBE THENARIA

This assemblage varies in form and habits, including typical anemones sessile on firm substrata plus burrowing species and deep water forms commensal on hermit crabs, sponges, and octocorals.

The species are considered in two groups, each of which is probably polyphyletic according to Carlgren (1949). Those without acontia, which appear as threadlike or nematodelike structures protruded through cinclides, include Stephenson's subtribes Endomyaria and Mesomyaria as modified by Carlgren; the second group includes the species with acontia, the Acontiaria of Carlgren. The presence of acontia, if they are not visible at once externally, may be confirmed by opening the column. It may also be noted that the mesentery patterns are generally more complex in Thenaria than in Athenaria. They are also individually more variable, particularly in genera that commonly reproduce asexually, e.g., *Metridium* and *Haliplanella*.

Group One, Species without Acontia. Two families are included; they belong to two subtribes which are distinguished by histological differences in the oral sphincter. In Endomyaria (Actinidae) the sphincter (which is present in all local genera) is "endodermal"; it may be strong or weak, diffuse or circumscript (see introduction to Actiniaria). In Mesomyaria (Actinostolidae) the sphincter is "mesogleal" and also varies in development in different genera. The number of perfect mesenteries in the two families is most frequently given as "perfect mesenteries numerous," e.g., in *Stomphia* there are 16–18 pairs, and in *Paranthus* at least 12 pairs. The mesenteries are more numerous basally in *Anemonia*, *Phymactis*, *Bunodactis*, and *Stomphia*, and more numerous at the oral margin in *Paranthus;* in *Bolocera*, *Tealia*, and *Actinostola* they are usually equal in number basally and marginally. A distinctive arrangement of mesenteries occurs in *Actinostola* and *Stomphia* (see Carlgren, 1949, p. 77).

The genera of Actinidae and Actinostolidae are considered in a single key. Omitted from the key are *Anemonia sargassensis* which is found only on *Sargassum*, and two large, deep water species, *Liponema multicornis* and *Antholoba perdix*, both of which attain disc diameters of 100 mm or more.

The nematocyst complements of genera in this group are as follows: in *Phymactis* they are *atrichs*, chiefly in marginal spherules; *spirocysts*, *basitrichs*, and *microbasic-p-mastigophors* are found in all genera; *microbasic-b-mastigophors* occur in *Stomphia* and *Actinostola*. Nematocyst types are defined and figured in Carlgren (1949). The following keys are based, as far as possible, on more easily observed traits, at the risk, however, of diminished infallibility.

---

1. Column with vesicles, verrucae, or other textural ornamentation—2.
1. Column smooth or longitudinally grooved—5.

2. Carolinian, shallow water species; usually found in sand or gravel but attached basally to a firm substratum; column covered with nonadhesive vesicles in about 40 vertical rows, and mature individuals with *marginal spherules;* tentacles sticky, about 96 in 5 cycles; oral sphincter well developed; coloration generally greenish brown, inner tentacles with red stripe; 90 mm tall by 40 mm wide, almost spherical when contracted. *Phymactis cavernata.*
2. Not so, boreal species—3.

3. Sphincter well developed (circumscribed or circumscribed-diffuse), appearing as a ringlike welt around the gastric cavity below the tentacles; tentacles fully contractile—4.

3. Sphincter relatively weak, mesogleal, tentacles short, stout, hardly contractile, numerous and in about four rows, inner ones longer; mouth lobed; column with low, flattish veruccae chiefly on the upper column; coloration nearly unicolorous, pale salmon or orange; 250 mm wide by 180 mm high; generally found below 73 m in mud or sand clasping a ball of sediment with the pedal disc. *Actinostola callosa.*

4. Sphincter strongly developed (circumscribed); verrucae scattered on column; tentacles thick, in 5 cycles (10-10-20-40-80); older mesenteries (10–20 pairs), without gonads; mesenteries equal at top and bottom of column; coloration generally greenish in shallow water, reddish in greater depths, column streaked or blotched, tentacles banded; 120 mm wide by 50 mm high. *Tealia felina.*

4. Sphincter less developed (may be circumscribed-diffuse), verrucae in longitudinal rows; tentacles 48–96 in 4 or 5 cycles; all major mesenteries fertile; mesenteries more numerous basally; coloration pale pinkish or greenish to darker green; 50 mm wide by 35 mm high; in tide pools, sometimes partly buried in sand. *Bunodactis stella.*

5. Elongate, burrowing form; column finely grooved longitudinally; pedal disc small, attached or not; tentacles numerous, to 144 or more; coloration pale yellowish or pinkish; 75 mm high by 25 mm wide; Virginian, littoral. *Paranthus rapiformis.*

5. Boreal, usually below 15 m; coloration whitish, pinkish to red or purple—6.

6. Tentacles with a basal sphincter by which they may be voluntarily pinched off when the anemone is disturbed, in 6 cycles (6-6-12-24-48, etc., to 192); oral sphincter endodermal, diffuse; marginal and basal mesenteries equal in number; 300 mm wide, column much taller than wide. *Bolocera tuediae.*

6. Tentacles without basal sphincter, in 2 or 3 cycles or rows, to 96 or more; oral sphincter mesogleal, strongly developed; basal mesenteries may be more numerous; 37 mm wide by 50 mm high. *Stomphia coccinea.*

Group Two, Species with Acontia. The six families in this group are formally separated in two groups by differences in the histological structure of the oral sphincter; their subsequent differentiation is based mainly on the nematocyst complements of the acontia and tentacles, for which see Carlgren (1949, p. 86). As indicated previously, the mesentery patterns are variable, particularly in species in which pedal laceration is common. There are usually 6 or more pairs of perfect mesenteries and sometimes more than 12 pairs.

This group includes some of the more familiar shallow water forms as well as several deep water species.

---

1. Column smooth or nearly so; common shallow water species; cinclides present—2.

1. Column with tubercles or other ornamentation; cinclides present or not—5.

2. Burrowing species of somewhat doubtful generic affinity; elongate, with pedal disc usually adhering to a stone or shell; pale-colored; tentacles 48–60 or more; height >60 mm; Fig. 5.31E. *Actinothoe:* Carlgren (1949, 1950) listed *modesta* and *gracillima* as doubtful members of this genus; Hand (1964) refers *modesta* to *Sagartia* without discussion. Parker (1900) described *gracillima* as a smaller species, 20 mm tall, with 48 tentacles, and *modesta* 63, mm with 60 tentacles.

2. Nonburrowing species—3.

3. Column dark green to blackish with or without red, orange, or white vertical stripes; sphincter weakly diffuse or absent; up to 50 tentacles in several whorls; 18 mm tall by 6 mm wide. *Haliplanella luciae.* This now common species first appeared in the New Haven, Connecticut, area about 1892 and has since spread through New England and south to North Carolina. It also occurs in Europe but presumably came originally from Japan. It has a "very discouraging synonymy, having been assigned to the genera *Sagartia*, *Chrysoela*, *Diadumene*, *Aiptasiomorpha*, and *Haliplanella*" (Hand, 1964).

3. Pale, translucent whitish to reddish, orange, or brown—4.

4. Tentacles 40–60, sometimes with thick, inner "catch tentacles"; column about 3 or more times taller than wide, 30 mm by 9 mm; translucent whitish or pinkish; sphincter absent. *Diadumene leucolena.*

4. Tentacles to ca. 1000; column less than 2 times higher than wide, 100 mm tall by 75 mm wide; color variable, column white to dark brown, sometimes mottled; sphincter well developed. *Metridium senile.*

5. Littoral, southern species; two color phases, pinkish green or pinkish yellow; upper column with 10 or 11 vertical rows of 2–5 vermilion warts; tentacles 64 or more in 4 cycles; cinclides absent; 25 mm tall by 8 mm wide. *Actinothoe eruptaurantiae.*

5. Deep water (>70 m), boreal (south to 39°) species—6.

6. Pedal disc adherent to pebbles or, frequently, to the tubes of an annelid worm, *Hyalinoecia artifex;* column with a smooth scapulus; scaphus finely papillose, usually coated with sand; upper column with cinclides; tentacles to about 96; coloration pale pinkish to salmon or brown; 40 mm high by 35 mm wide. *Sagartiogeton verrilli.* (For additional details of this and the following species see Carlgren, 1942, and Verrill, 1883.)

6. Pedal disc cuplike, grasping a ball of mud or sand; cinclides absent; tentacles to about 96; "skin" parchmentlike, stiff. *Actinauge* spp.—7.

7. Scapulus smooth, orange-red to chocolate brown; scaphus with scattered tubercles, pearly white to rosy pink; 125 mm tall by 90 mm wide. *Actinauge longicornis.*

7. Column variably tuberculate, the tubercles or warts in longitudinal rows diminishing in size basally; tubercles sometimes so large and closely spaced as to appear coarsely squamose; column pale pinkish to red, scapulus red-orange to brown; 150 mm tall by 100 mm wide. *Actinauge verrilli.* Note also *Hormathia nodosa*, with scattered, compressed tubercles in a transverse row below the oral disc; equals *Actinauge rugosa* of Miner (1950).

References

Carlgren, O. 1942. Actiniaria P. II. The Danish Ingolf-Expedition. 12:92 pp.

———. 1949. A survey of the Ptychodactiaria, Corallimorpharia and Actiniaria. K. Svenska Vetensk-Akad. Handl. 1(1):1–121.

———. 1950. A revision of some Actiniaria described by A. E. Verrill. J. Wash. Acad. Sci. 40(1):22–28.

Crowell, S. 1946. A new sea anemone from Woods Hole, Massachusetts. J. Wash. Acad. Sci. 36:57–60.

Field, L. R. 1949. Sea anemones and corals of Beaufort, North Carolina. Bull. Duke Univ. mar. Stn. 5:1–39.

Hand, C. 1957. The sea anemones of central California, Part III. Wasmann J. Biol. 13:189–251.

———. 1964a. Phylum Cnidaria, Class Anthozoa. *In* Keys to marine invertebrates of the Woods Hole region. R. Smith, ed. Contri. II Systematics-Ecology Program, Mar. Biol. Lab. Woods Hole. pp. 25–28.

———. 1967b. Another sea anemone from California and the types of certain Californian anemones. J. Wash. Acad. Sci. 47(12):411–414. (*Nematostella vectensis*).

Hargitt, C. W. 1914. The Anthozoa of the Woods Hole region. Bull. Bur. Fish. Wash (for 1912) 32:223–254.

Hyman, Libbie H. 1940. The Invertebrates, Protozoa through ctenophora. McGraw-Hill, New York.

Miner, R. W. 1950. Field book of seashore life. G. P. Putnam's Sons, New York.

Parker, G. H. 1900. Snyopses of North American Invertebrates. XIII. the Actiniaria. Amer. Nat. 34:747–758.

Robson, Elaine A. 1966. Swimming in Actiniaria. *In* The Cnidaria and their evolution. Symp. Zool. Soc. London No. 16.

Stephenson, T.A. 1928, 1935. The British Sea Anemones. Ray Society, London, Vol. 1 and Vol. 2.

Verrill, A. E. 1879. Notice of recent additions to the marine fauna of the northeastern coast of America. Proc. U.S. natn. Mus. (for 1878). 2:165.

———. 1882. Notice of the remarkable marine fauna occupying the outer banks of the southern coast of New England. Amer. J. Sci. (20):216–225; (33):309–316.

———. 1883. Reports on the results of dredging by the U.S. Coast Survey steamer "Blake." Bull. Mus. Comp. Zool. Harv. 11(1):1–72.

Wass, M. L. 1963. Check List of the Marine Invertebrates of Virginia. Spec. Sci. Rept. 24 Virginia Inst. mar. Sci.

Order Scleractinia (Madreporaria)
STONY CORALS

Diagnosis. *Colonial or solitary anthozoans with a solid, calcareous, external skeleton.*

The skeleton of the individual polyp of *Astrangia danae*, the only local, shallow water species, consists of a low cuplike or tubular corallite about 5–6 mm in diameter with internal, radiating partitions or *septa* (Fig. 5.1A). The colony or *corallum* may assume a low, branching form but commonly consists of a thin encrustation uniting the individual corallites. Up to about 30 individuals are found in a single colony measuring 100 mm in diameter.

The polyps are anemonelike with the basal part of each individual seated in a corallite; those of *A. danae* are not deep enough to allow for the complete retraction of the polyp. In life the colonies are pale whitish, pinkish, greenish, or brown, the latter two colors being imparted by the presence of symbiotic algae, zooxanthellae. Possession of such algae is typical of *hermatypic* or reef-forming corals. These corals require strong sunlight and are confined to shallow waters; *A. danae*, which frequently lacks zooxanthellae, nevertheless only ranges down to about 40 m. A second species, *Oculina* sp. is said to occur in shallow water off North Carolina. *Dasmosmilia lymani*, reported (as *Parasmilia*) by Verrill (1882) in 104–238 m, is an *ahermatypic* species, and other species are found in bathymetrically extralimital waters off this coast. *Dasmosmilia* is a solitary coral growing in an erect "cow's horn"; the core or *columella* is spongy.

Astrangia danae is Virginian in distribution, with a bathymetric range of the littoral to 40 m; Wass (1963) reported this coral from the York River in the lower Chesapeake—indicating a tolerance for reduced salinity.

References

Bayer, F. M., *et al.* 1956. Coelenterata, Part F. Treatise on Invertebrate Paleontology. R. C. Moore, ed. Geol. Soc. Amer. and Univ. Kansas.

Verrill, A. E. 1882. Notice of the remarkable marine fauna occupying the outer banks of the southern coast of New England. Amer. J. Sci. 33:309–316.

Wass, M. L. 1963. Check List of the Marine Invertebrates of Virginia. Spec. Sci. Rept 24 Virginia Inst. mar. Sci.

Waterman, T. H. 1950. *Astrangia danae. In* Selected Invertebrate Types. F. A. Brown, ed. John Wiley and Sons, New York.

Wells, J. W. 1956. Scleractinia. *In* Treatise on Invertebrate Paleontology. Part F. Coelenterata. R. C. Moore, ed. Geol. Soc. Amer. and Univ. Kansas.

Order Ceriantharia
CERIANTHARIAN ANEMONES

Diagnosis. *Anemonelike anthozoans; elongate, tapering basally, aboral end rounded; tentacles in two distinct whorls, 100 or more in each* (*Fig. 5.1F*).

The resemblance to burrowing anemones is superficial; the arrangement of the mesenteries is unique, and there are substantial histological differences.

For extended accounts of ceriantharian anatomy see Carlgren (1912) and Hyman (1940). Local burrowing anemones (Actiniaria) have their tentacles in a single whorl.

Ceriantharians live in tubes of sand grains and other extraneous material hardened by mucus. The musculature for retracting the disc within the column is undeveloped, but the column itself is strongly retractile. These are large anemones, *Cerianthus borealis* reaching a length of >450 mm with an oral disc ca. 180 mm in diameter; *Ceriantheopsis americanus* is substantially smaller but grows to about 200 mm in length. Ceriantharians are protandrous hermaphrodites. The larvae, many of which have been given distinct generic names, are pelagic and become tentacular before adopting benthic habits.

Systematic List and Distribution

Cerianthus borealis	B+	13 to >400 m
Ceriantheopsis americanus	V	Lit. to 21 m

Identification

Carlgren (1912) separated *Ceriantheopsis* from *Cerianthus* on the basis of internal differences. *Ceriantheopsis americanus* has 100–125 marginal tentacles nearly equal in length; *Cerianthus borealis* has about 150 marginal tentacles which vary in length, the extreme differences being on the order of about 2:1. Both species are usually brownish in color.

References

Carlgren, O. 1912. Ceriantharia. Dan. Ingolf-Exped. 5a(3):1–78.

Field, L. R. 1949. Sea anemones and corals of Beaufort, North Carolina. Duke Univ. mar. Stn. Bull. 5:1–39.

Hyman, Libbie H. 1940. The Invertebrates, Protozoa through Ctenophora. McGraw-Hill, New York.

Kingsley, J. S. 1904. *Cerianthus borealis*. Tufts Coll. Stud. 1.

McMurrich, J. P. 1890. Structure of *Cerianthus americanus*. J. Morphol. 4:131–150.

Parker, G. H. 1900. Synopses of North American Invertebrates. XIII. The Actiniaria. Amer. Nat. 34:747–758.

SIX

Phylum Ctenophora

COMB-JELLIES

Diagnosis. *Gelatinous, transparent, more or less spheroidal, lobed, or sacklike animals with eight rows of comblike ciliary plates; without nematocysts, i.e., stingless.*

Ctenophores have a modified radial organization based on two planes of symmetry. They have a digestive cavity but no developed organ systems and are at the same level of complexity as cnidarians.

The highly modified cestids, of which *Cestum veneris* is the only local representative, are ribbonlike and superficially bilateral; Figure 6.1. The vertical axis is short, but the body extends laterally in two elongate, flattened lobes that may attain a span of 1.5 m. Two rows of combs extend along the upper margin of each lobe; the other four comb-rows are rudimentary. The lower margin has two short rows of tentacles. *Cestum* is primarily an oceanic, Gulf Stream species on this coast.

Though individual ctenophores have great powers of regeneration, asexual reproduction by fission has only been noted in a few, atypical forms. Ctenophores are hermaphroditic, and fertilization is external. There are no distinctive larval forms and no profound metamorphosis; however, young of the more highly evolved genera resemble adults of more primitive taxa. Local species are strictly planktonic and, while exclusively marine, range into nearly freshwater parts of estuaries. Luminescence has been reported from four of the five neritic genera in range.

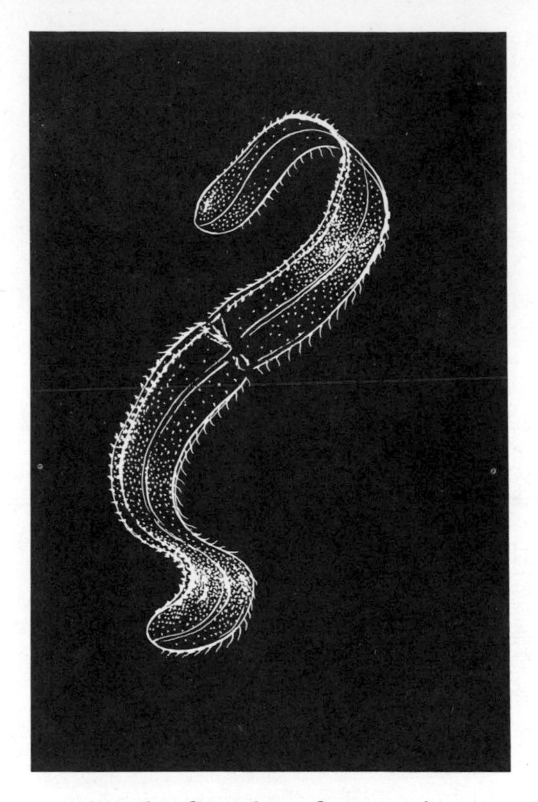

Fig. 6.1. Ctenophora. *Cestum veneris.*

Despite their delicate structure, ctenophores are predaceous and, since they often occur in dense swarms, they have some ecological importance in reducing the number of fish fry, copepods, and other small organisms of the zooplankton.

Systematic List and Distribution

The list is derived from Mayer (1912).

Class Tentaculata		
Order Cydippida		
Family Pleurobrachiidae		
Pleurobrachia pileus (see below)	BV	Inshore winter-spring; offshore summer-fall
Mertensia ovum	A+	Winter

Order Lobata		
Family Bolinopsidae		
Bolinopsis infundibulum	B	Summer
Family Mnemiidae		
Mnemiopsis leidyi (see below)	V	Chiefly summer-fall
Order Cestida		
Family Cestidae		
Cestum veneris		Oceanic, Gulf Stream
Class Nuda		
Order Beroida		
Family Beroidae		
Beroe ovata	V+	Summer-fall
Beroe cucumis	B+	Summer-fall

On the basis of minor differences in size and form, Mayer (1912) described *Pleurobrachia brunnea* as a distinct species from the New Jersey coast, and Agassiz (1860) recognized a brackish water form of *Mnemiopsis*, *M. gardeni*. Neither of these names is in current use. Yet another species, *Lesueuria hyboptera*, collected in Massachusetts Bay more than a century ago by Agassiz, has not been reported since; Mayer suggested that Agassiz' specimens were actually imperfect *Bolinopsis*. Additional species occur in the Gulf Stream; see Mayer (1912) or Miner (1950).

Identification

Because they are so extremely fragile, undamaged ctenophores are more easily taken if individuals are gently dipped or scooped into a container without using a net. In rough weather they avoid the surface. *Pleurobrachia* may be placed directly in 5% formalin, but most ctenophores must be fixed in another reagent. *Mnemiopsis* and *Bolinopsis* in particular are likely to dissolve completely in formalin. The following formulae have been recommended for killing and fixing:

100 cc 1% chromic acid	100 grams mercuris chloride
5 cc glacial acetic acid	3 cc glacial acetic acid
	300 cc water

Specimens are fixed for 15–30 min, washed in fresh water for 15 min, and transferred to 5% formalin or 70% alcohol by stages, starting with 30% alcohol. Some references prohibit the use of formalin entirely.

Cestum is excluded from the following key. Measurements are approximate maximum total length of adults. All of the species are figured by Mayer (1912).

A. With two long tentacles, retractile into sheaths. Cydippida:

 1. Adults globular, more or less egg-shaped; 18 mm; Fig. 6.2A. *Pleurobrachia pileus.*

 1. Adults egg-shaped but compressed; 55 mm. *Mertensia ovum.* Young *Mertensia ovum* resemble *P. pileus* in shape but have orange pigment along the combs.

B. No long tentacles; body oval, somewhat compressed, with two enlarged lobes; combs unequal in length, and near the base of each of the four shorter combs there is a tonguelike *auricle.* Lobata:

 1. Auricles arise from side of body rather than from deep grooves; 150 mm; boreal; Fig. 6.2F. *Bolinopsis infundibulum.*

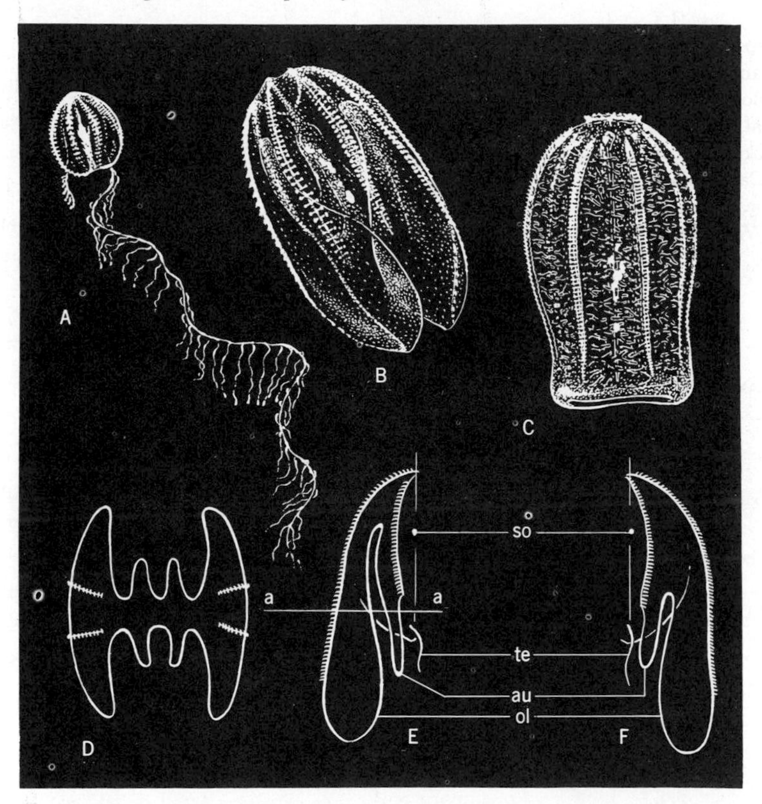

Fig. 6.2. Ctenophora. A, *Pleurobrachia pileus;* B, *Mnemiopsis leidyi;* C, *Beroe* sp.; D, *Mnemiopsis* sp., diagrammatic section on line a–a of the following; E, same, diagrammatic longitudinal section; F, *Bolinopsis infundibulum,* diagrammatic longitudinal section. Abbreviations are as follows: so, sense organ; au, auricles; te, tentacle; o, oral lobe.

1. Auricles arise from deep grooves; 100 mm; Virginian; Fig. 6.2B,D,E. *Mnemiopsis leidyi.* Young of both genera pass through a cydippid-like larval stage with tentacles and without lobes; young *Bolinopsis* are nearly the size of half-grown *Pleurobrachia* before becoming lobate; *Mnemiopsis* loses its tentacles and begins developing lobes by the time it is about 5 mm long, but is then *Bolinopsis*-like until well past 10 mm.

C. With neither tentacles nor lobes; flattened, sacklike; 115 mm. *Beroe:* Beneath each comb is a *meridional canal*, and branchlets from opposing canals fuse in a network in *B. ovata;* in *B. cucumis* they do not. Young differ somewhat from adults in shape, but are sacklike without tentacles or lobes, Fig. 6.2C.

References

Agassiz, L. 1860. Ctenophora. Contrib. Nat. Hist. U.S. 3:269.

Hyman, Libbie H. 1940. The Invertebrates: Protozoa through Ctenophora. McGraw-Hill Book Co., New York.

Komai, T. 1963. A note on the phylogeny of the Ctenophora. *In* The Lower Metazoa—Comparative Biology and Phylogeny, pp. 181–188. E. C. Dougherty, ed. Univ. California Press.

Mayer, A. G. 1912. Ctenophores of the Atlantic coast of North America. Carnegie Inst. Wash. Publ. 162.

Miner, R. W. 1950. Field book of seashore life. G. P. Putnam's Sons, New York.

SEVEN
Phylum Platyhelminthes

FLATWORMS

Free-living flatworms belong to the class Turbellaria. Members of the phylum's other two classes are exclusively parasitic and include tapeworms, class Cestoda, and flukes, class Trematoda. Although a few turbellarians are ecto- or entoparasites, only the nonparasitic turbellarians will be considered here, and of these only the orders Polycladida and Tricladida will receive more than passing attention.

Diagnosis. *Unsegmented, usually flattened, soft-bodied worms without a coelom or appendages; with a ventral mouth, usually either near midbody or near the anterior end; without an anus.*

Local free-living turbellarians range in size from microscopic species in scale with protozoans to others 50 mm or more in length. Body shape ranges from elongate—length exceeding width thirty times or more—to ovoid or almost circular. The body is contractile, soft, and usually flattened, at least ventrally. The head end may be differentiated by a necklike constriction, and is often equipped with sensory bristles, pits, or grooves. None of the local species has more than a single pair of tentacles; they are absent in about as many species as have them. Two main types of tentacles are found. In some planaria-like species the tentacles consist of a pair of lobes or auricles more or less distinct and more or less contractile, placed at the antero-lateral corners of the head. Tentacles of another type occur in some polyclads; these are located well within the body margin where they are easily overlooked in casual inspection. Viewed from the side or with oblique lighting, they appear as small, short, fingerlike projections. Some polyclads also have flaplike tentacles at the anterior margin of the body.

Most turbellarians have eyes in the form of simple ocelli that vary in number and arrangement in different species. Some triclads have but a single pair of round or bean-shaped ocelli, but other triclads and all polyclads have more than two. The number of ocelli in polyclads may increase ontogenetically. The ocelli are occasionally in rows, but more frequently they are in clusters or bands in three main areas. Cerebral ocelli are placed near the brain; tentacular ocelli are on or near the tentacles or in the region where tentacles might be expected in species that do not actually have them.; marginal ocelli extend along the edge of the body and may border the entire body or be confined to the anterior half or third.

Caudal lobes or a taillike filament or cirrus are found in a few turbellarians but not in any of the triclads or polyclads considered here. The mouth in local species is ventral and near midbody. There is no proboscis in any of the more commonly encountered forms. A retractile, proboscislike structure which is not connected with the mouth is found in some turbellarians less than 2 mm long. In some of the smaller forms (order Archoophora) there is no proper gut, and food is digested intercellularly in the interior connective tissue or *mesenchyme*. In higher forms, a pharynx and gut are usually present. In the simplest of these the gut is a saclike structure with or without short diverticula. In triclads the pharynx is directed posteriorly and the gut has three branches, one directed forward and the other two backward; all of the branches have abundant diverticula. In polyclads the pharynx is directed forward or is vertical to the broad axis of the body; the main gut has numerous radiating branches with diverticula. Since the body is more or less translucent, these differences may be detected by examining the worms under transmitted light. A few species are brightly colored, orange or yellow, but most turbellarians are rather drab, whitish, gray, or brownish, frequently with speckles or suffusions of darker or lighter color. The food contents of the digestive tract often influence the coloration of the live animal. The presence of symbiotic algae in smaller, mostly freshwater, species may tinge the whole animal greenish.

Most turbellarians are hermaphroditic. A few triclads are known to be dioecious, and some are suspected of being parthenogenetic. Self-fertilization is uncommon because the male and female reproductive systems are largely, if not completely, separate. Almost all species have copulatory apparatus, and fertilization is usually internal. The eggs are either laid to develop externally, as is usually the case, or they are retained and develop internally. In most species, development is direct without special larval forms. In some polyclads distinctive preadult stages occur which last but a few days; these include Muller's larva, a pelagic stage characterized by the possession of eight ciliated lobes, Figure 3.3G, and Götte's larva with four lobes.

Asexual reproduction, though more characteristic of freshwater and terrestrial forms, also occurs in the local marine species *Microstomum davenporti* (order Archoophora). In this genus and in a few related ones the incompleted fission process results in temporary chains of up to a dozen or more individuals that "become well differentiated before breaking from the chain" according to Hyman (1951). The extraordinary powers of regeneration that have made freshwater planarians such popular experimental animals are not developed to the same high degree by polyclads or, indeed, by most other groups, although regeneration following certain types of injury is possible in most species.

Turbellarians occupy a diversity of habitats. There are freshwater and terrestrial as well as marine and estuarine species, and some of the lower forms tolerate a wide range of salinities. A few species are regularly planktonic or occur on pelagic *Sargassum*. Plehn (1895) lists polyclads taken in plankton tows including *Planocera* (= *Hoploplana?*), *Prosthiostomum*, *Acerosa* (= *Acerotisa*), *Leptoplana* (= *Notoplana?*), and *Stylochoplana* (= *Zygantroplana*). Verrill as reported by Hyman (1939a) found *Zygantroplana angusta* on the bottom of a whaling ship which had presumably imported the worm from the tropics. For flatworms occurring on *Sargassum* see Hyman (1939b). Böhmig (1895) discusses planktonic turbellarians belonging to the lower orders. An important habitat currently being explored is the psammo-littoral—the interstitial water of sands and sandy muds—which apparently contain a multitude of species in the lower orders; see Bush (1968). Several triclads are commensal on *Limulus*, and members of the families Umagillidae and Graffillidae (order Eulecithophora) are commensal on echinoderms, mollusks, and other invertebrates. *Hoploplana inguilina* is found in the mantle cavity of *Busycon* and other gastropods and barnacles, and *Stylochus* species are commensals on mollusks and crustaceans. The shift to outright parasitism has been made by members of the family Fecampidae (order Eulecithophora), but no species of this family has been definitely recorded from this coast. While some of the smaller turbellarians are herbivorous, most species are carnivores or are scavengers on dead animal tissue. Living prey may be incapacitated by mucous and adhesive secretions, and the victim then "sucked dry" or ingested whole. Some turbellarians, e.g., *Microstomum*, are known to eat cnidarians without discharging their nematocysts, later using these sting "cells" as a defensive armament in their own bodies.

Systematic List and Distribution

The classification of Hyman (1951) has been substantially revised in recent years; interested students should compare Hyman's classification with that of Beauchamp (1961), which is followed here. There seems to be little useful purpose in listing the genera previously classified in the orders Acoela, Rhabdocoela, and Alloeocoela. The worms in these orders require special collecting techniques and special preparation for study. The published species lists include but a fraction of those known or currently being described by students of the group. These are mostly microscopic or very small worms and are not likely to engage the attention of the general worker. It seems best therefore to defer their consideration until more comprehensive accounts are available. For figures of more recently discovered psammo-littoral species see Bush (1968). Also see Bush (1964), Hyman (1951), and Miner (1950) where some of the older forms are figured and described; Beauchamp (1961) includes generic synonomies.

Order Trichladida		
Family Dendrocoelidae		
Procotyla fluviatilis	BV	Brackish to fresh water
Family Procerodidae		
Procerodes littoralis	B, 41°	Shallow, eury.
Procerodes warreni	B+	Lit.
Foviella affinis	B+	Shallow
Family Bdellouridae		
Bdelloura candida	V	On *Limulus*
Bdelloura propinqua	V	On *Limulus*
Syncoelidium pellucidum	V	On *Limulus*
Family Uteriporidae (Probursidae)		
Probursa veneris	V	In empty *Mercenaria shells*
Order Polycladida		
Family Discocelidae		
Coronadena mutabilis	V	Shallow, pelagic
Family Plehniidae (Latocestidae)		
Discocelides ellipsoides	B+	7 to 348 m
Family Stylochidae		
Stylochus ellipticus	BV	Shallow, eury.
Stylochus frontalis	C+	Shallow, eury.
Stylochus zebra	V	Shallow, on hermit crabs
Stylochus oculiferus	C+	Shallow, eury.
Family Leptoplanidae		
Euplana gracilis	BV	Lit. to shallow
Zygantroplana angusta	C+	Pelagic?
Notoplana atomata	BV	Lit. to 93 m
Family Hoploplanidae		
Hoploplana inquilina	BV	Commensal on *Busycon* and other gastropods

Hoploplana grubei	C+	On *Sargassum*
Family Planoceridae		
Gnesioceros floridana	V	Shallow, eury.
Gnesioceros sargassicola	C+	On *Sargassum*
Family Euryleptidae		
Prosthecereaus maculosus	V	Shallow
Acerotisa baiae	V	Shallow, eury.
Acerotisa notulata	C+	On *Sargassum*
Family Prosthiostomidae		
Prosthiostomum sp.	BV	Pelagic

Identification

Specimens may be keyed to order by the following summary. Sources of figures are given in parentheses before species names; sizes equal maximum length of adults. Note that, while the number and distribution of ocelli are used as key traits by systematists, the crucial details are not always clearly defined.

- Eyes 2 or lacking, rarely 1 or 4; gut lacking or a simple sac with or without a few diverticula; mostly smaller than 2–5 mm; some 15 mm or even longer but very slender. Lower Orders.
- Eyes 2 or in 2 clusters; gut tripartite, pharynx leading posteriorly, intestine with one forward branch and two posterior branches all with numerous diverticula; adults 3 mm or larger. Tricladida.
- Eyes in 2 or more clusters, bands, or rows and clusters; pharynx leading anteriorly or vertical to broad axis of body, gut with radiating branches with numerous diverticula; adults 1–50 mm. Polycladida.

ORDER TRICLADIDA

1. Auricles absent; commensal on *Limulus* or in empty clam shells—2.

1. Auricles present though sometimes reduced or retracted; not commensal on *Limulus* or confined to empty clam shells—4.

2. Found in empty clam shells (*Mercenaria*); with a copulatory bursa anterior to the penis bulb; brownish black, flecked, and streaked; 3 mm; Fig. 7.1A (Hyman, 1944). *Probursa veneris.*

2. Commensal on *Limulus*—3.

3. With a posterior "sucker"; posterior branches of gut not joined posteriorly; 3 mm; Fig. 7.1B. *Bdelloura* (Hyman, 1951; Bush, 1964): *B. candida*, the common local species, has 60–100 testicular sacs laterally and reaches 15 mm; *B. propinqua* has about 170 testicular sacs and reaches 8 mm.

3. Without a posterior sucker; posterior branches of gut joined posteriorly; 3 mm; Fig. 7.1C (Bush, 1964). *Syncoelidium pellucidum.*

4. With 2 eye clusters, 2 or more eyes in each; anterior margin with an adhesive organ; 20 mm; fresh to brackish water; Fig. 7.1E (Hyman, 1951). *Procotyla fluviatilis.*

4. With 2 eyes; no marginal adhesive organ:
   - *Foviella affinis* has greatly reduced auricles and a vestigial seminal bursa; 12 mm; Hyman (1944) lists this species as ranging from Newfoundland north, but Miner (1950) has it to southern New England. See Bock (1925).
   - *Procerodes warreni* also has reduced auricles; Wilhelmi (1908) lists this species as occurring, with the next one, to southern New England, but this apparently is the less common form. See Hyman (1944).
   - *Procerodes littoralis* (= *wheatlandii*) has well-developed but retractile auricles; 5 mm; a seminal bursa is present in the genus; Fig. 7.1D (Bush, 1964).

ORDER POLYCLADIDA

---

1. With a pair of anterior flaplike tentacles; eyes on tentacles and in 2 rows or clusters in cerebral region; 12 mm; Fig. 7.1F (Bush, 1964). *Prosthecereaus maculosus.*
1. With dorsal entacles—6.
1. Without tentacles—2.

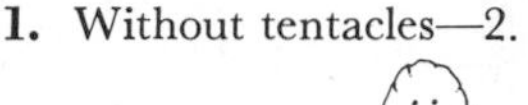

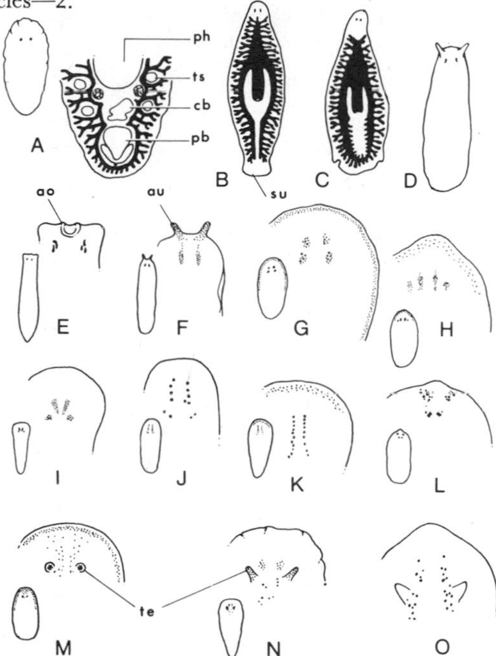

Fig. 7.1. Platyhelminthes. A, *Probursa veneris*, whole animal and detail of posterior end; B, *Bdelloura* sp.; C, *Syncoelidium pellucidum;* D, *Procerodes littoralis;* E–N include figures of whole animals and details of anterior end; E, *Procotyla fluvialtilis;* F, *Prosthecereaus maculosus;* G, *Coronodena mutabilis;* H, *Discocelides ellipsoides;* I, *Notoplana atomata;* J, *Euplana gracilis;* K, *Prosthiostomum* sp.; L, *Acerotisa* sp.; M, *Stylochus* sp.; N, *Gnesioceros* sp.; O, *Hoploplana* sp. Abbreviations are as follows: ao, adhesive organ; au, auricle; cb, copulatory bursa; pb, penis bulb; pg, pharynx; su, sucker; te, tentacles; ts, testes.

2. With eyes in 4 clusters, with or without a marginal band of eyes—3.
2. With eyes partly in 2 rows—5.
3. With 4 eye clusters plus eyes in a marginal band:

- *Coronodena mutabilis* has a band of eyes around half the margin anteriorly; 18 mm; chiefly Carolinian; Fig. 7.1G (Hyman, 1951, p. 116).
- *Discocelides ellipsoides* has a band of eyes around the anterior third of the margin; 25 mm; chiefly Boreal; Fig. 7.1H (Hyman, 1940).

3. Without a marginal band of eyes—4.
4. With 4 eye clusters, 2 marginal and 2 cerebral; Fig. 7.1L. *Acerotisa: A. baiae* is whitish; 3 mm; estuarine. *A. notulata* is 1 mm and is found on *Sargassum* (Hyman, 1939b, 1940).
4. With 4 eye clusters, cerebral and tentacular; 12 mm; a variable species with a complex synonymy; the commonest polyclad on this coast, see Hyman (1939a); Fig. 7.1I (Bush, 1964). *Notoplana atomata.*
5. With 4 or 5 eyes in a row on each side plus 2 more behind; 12 mm; Fig. 7.1J (Bush, 1964). *Euplana gracilis.*
5. With 4 or more eyes in a row on each side plus marginal eyes; 25 mm; Fig. 7.1K (Beauchamp, 1961, p. 67). *Prosthiostomum* sp.
6. With margin and central eyes; Fig. 7.1M. *Stylochus.* (Hyman, 1940.) See Table 7.1.
6. Without marginal eyes—7.
7. Tentacles with eye spots; Fig. 7.1N. *Gnesioceros: G. floridana* is neritic, 8 mm (Bush, 1964). *G. sargassicola* is the common polyclad on *Sargassum*, 10 mm (Hyman, 1939b).
7. Tentacles without eye spots; Fig. 7.10. *Hoploplana: H. inquilina* is white, commensal on gastropods, 6 mm; (Hyman, 1939a). *H. grubei* is brownish with white reticulations; on *Sargassum:* 10 mm.

Table 7.1. *Stylochus*

Stylochus	Marginal eyes	Habits	Size, mm	Color pattern
ellipticus	Anterior ⅓–½	Littoral on barnacles, oysters	25	Mid-dorsal stripe
frontalis	All around	on oysters	50	Mid-dorsal stripe
zebra	All around	In shells with hermit crabs	40	Light–dark crossbars
oculiferus	All around	On oysters	53	Red, pink-spotted

References

Beauchamp, P. de 1961. Classe des Turbellaires. *In* Traité de Zoologie, Tome IV, fasc. 1. P.P. Grassé, ed. Masson et Cie., Paris.

Bock, S. 1925. Oersteds Planaria aiffinis wiederentdeckt. Zool. Anz. 64:149–164.

Böhmig, L. 1895. Die turbellaria Acoela der Plankton-Expedition. Ergebn. Plankton-Exped. 2(H)g:1–48.

Brush, Louise. 1964. Phylum Platyhelminthes. *In* Keys to Marine Invertebrates of the Woods Hole Region. Systematics-Ecology Program, Mar. Biol. Lab. Woods Hole, Mass.

———. 1968. Characteristics of interstitial sand Turbellaria: the significance of body elongation, muscluar development, and adhesive organs. Trans. Amer. Microsc. Soc. 87 (2): 244–251.

Hyman, Libbie H. 1939a. Some polyclads of the New England coast, especially of the Woods Hole region. Biol. Bull. 76:127–152.

———. 1939b. Acoel and polyclad Turbellaria from Bermuda and the Sargassum. Bull. Bingham oceanogr. Coll. 7:1–26.

———. 1940. The polyclad flatworms of the Atlantic coast of the United States and Canada. Proc. U.S. natn. Mus. 89:449–493.

———. 1944. Marine Turbellaria from the Atlantic coast of North America. Amer. Mus. Novitates 1266:1–15.

———. 1951. The Invertebrates: Platyhelminthes and Rhynchocoela. The acoelomate bilateria, Vol. II. McGraw-Hill Book Co., New York.

———. 1952. Further notes on the Turbellarian fauna of the Atlantic coast of the United States. Biol. Bull. 103:195–200.

Miner, R. W. 1950. Field Book of Seashore Life. G. P. Putnam's Sons, New York.

Pearse, A. S. 1938. Polyclads of the east coast of North America. Proc. U.S. natn. Mus. 86:67–98.

Plehn, Marianne. 1895. Die Polycladen der Plankton-Expedition. Ergebn. Plankton-Exped. 2(H)f: 1–12.

Wheeler, W. M. 1894. *Syncoelidium pellucidum*, a new marine triclad. J. Morph. 9:167–194.

Wilhelmi, J. 1908. On the North American triclads. Biol. Bull. Mar. Biol. Lab. Woods Hole 15:1–6.

———. 1909. Tricladen. Fauna and Flora des Golfes von Neapel. Monogr. 32:xi + 401 pp.

EIGHT

Phylum Rhynchocoela (Nemertina, Nemertea, Nemertinea)

NEMERTEAN WORMS

Diagnosis. *Unsegmented, cylindrical or flattened, contractile, soft-bodied worms without a coelom or appendages; with a ventral mouth near the anterior end, an anus at the posterior end, and a completely retractile, tubular proboscis.*

The smallest local nemerteans are only 5 mm long, and the largest are several meters in length. Body shape ranges from elongate and ribbonlike or stringlike to broad and planarialike, or short, thickened, and rounded. Nemerteans are frequently highly contractile, the expanded worm being perhaps ten times its contracted length. There is no external indication of segmentation, and most species are soft, smooth, and slimy. The only appendage in local species is the relatively short, taillike *cirrus* of *Cerebratulus* and related forms. The cirrus is frequently broken off in handling, and many species are easily fragmented generally. The mouth opening may be at the lower front margin or decidedly ventral and back from the fore end; its relation to the brain, either preceding or following it, is of importance in formal systematics. The head frequently is poorly differentiated externally. There may be two or more *ocelli* which are usually black, reddish brown, or, in *Lineus dubius*, white. The ocelli are usually simple, light-sensitive structures, but in some species they are provided with lenses. The head usually has sensory or *cephalic grooves* which may be transverse, longitudinal, or oblique; they may be conspicuous or obscure. The anus is usually not conspicuous.

The nemertean proboscis is a unique feature. This is a tubular structure, typically as long as or several times longer than the body. It can be everted—turned completely inside out—and thrust out through a separate *proboscis pore* located in front of the mouth or through an opening common to both. Like the cirrus, the proboscis may be broken off by rough handling. The proboscis may be simple and relatively little differentiated into regions (except histologically), or it may be regionated, most conspicuously in possessing a bulbous middle portion armed with a *stylet* or stylets. Differences in these structures are of importance in formal systematics.

The proboscis functions both offensively and defensively. In the capture of food it may be literally shot out to impale the prey on the stylet or entrap it in coils with the aid of a sticky mucus. Most nemerteans are carnivorous. In some cases only the body juices are ingested, but the whole prey may be taken in; small-mouthed species necessarily feed on protozoans and other microfauna. Larger-mouthed forms may ingest prey of nearly their own size.

Nemerteans are frequently brightly colored, and color patterns, which can be observed only in fresh material, are important in species identification.

The sexes are separate in most marine species, and may or may not be differentiated by coloration. In some cases sexually active individuals are more brightly colored than nonreproducing specimens. Fertilization is usually external or, when mating involves actual contact, sperm released by the male may enter the female's body. In the latter case, development may proceed internally, the young being born as miniature worms. In forms with external development the eggs are laid in gelatinous strings or masses. Either the development is direct, the initial young being an oval, ciliated, gastrulalike form with an apical tuft of flagella, or there is a metamorphosis with distinct larval forms. In direct development the hatchling gradually transforms into a young worm which at first resembles a rhabdocoel flatworm. In metamorphic species, development proceeds through either a *pilidium* or a Desor larva. The pilidium larva of many species of Lineidae is a free-swimming, more or less helmet-shaped, transparent, ciliated form with an aboral tuft of elongate cilia (Fig. 3.3F). The Desor larva actually remains in the egg; it is oval and ciliated but lacks an aboral tuft.

There are ontogenetic changes in coloration, number of ocelli, and other details used as identification criteria. For example, the young of many highly colored species are whitish.

Nemerteans, as previously mentioned, are much inclined to break apart when handled, and many species possess extraordinary powers of regeneration. Voluntary asexual reproduction also occurs by fragmentation.

The phylum is chiefly marine in distribution. Most nemerteans are benthic, living under rocks or in burrows in soft substrata, or crawling among algae, hydroids, or in bottom debris. They are chiefly littoral or shallow water forms. A few are bathypelagic and some, e.g., *Cerebratulus*, swim during periods of sexual activity. *Carcinonemertes* is questionably parasitic—certainly commensal—on crabs, and *Malacobdella* is "endemic" in the mantle cavity of pelecypod mollusks. In the latter there is an adhesive disc at the rear end; in both genera the proboscis is reduced.

Systematic List and Distribution

This list is based mainly on Coe (1943) with additions from later literature.

Class Anopla		
Order Paleonemertea		
Family Tubulanidae		
Tubulanus pellucidus	V	1 to 20 m
Family Carinomidae		
Carinoma tremaphoros	V	Lit. to Shallow water
Carinomella lactea	C, 37°	>13 m, eury.
Family Cephalothricidae		
Procephalothrix spiralis	BV	Lit. to shallow water
Order Heteronemertea		
Family Lineidae		
Parapolia aurantiacea	V	Lit. to shallow water
Zygeupolia rubens	V	Lit. to shallow water
Lineus arenicola	B, 37°	Lit.
Lineus bicolor	V	2 to 40 m
Lineus dubius	B	Lit.
Lineus pallidus	B	80 m
Lineus ruber	B+	Lit. to shallow, eury.
Lineus socialis	BV	Lit.
Micrura affinis	B+	10 to 300 m
Micrura albida	B	55 to 256 m
Micrura caeca	V	Lit.
Micrura dorsalis	B	Lit.
Micrura leidyi	V+	Lit.
Micrura rubra	B	70 m
Cerebratulus lacteus	BV	Lit.
Cerebratulus luridus	B+	20 to 350 m
Cerebratulus marginatus	B+	Lit. to 100 m

Class Enopla		
Order Hoplonemertea		
Suborder Monostylifera		
Family Emplectonematidae		
Emplectonema giganteum	B+	120 to 1500 m
Family Carcinonemertidae		
Carcinonemertes carcinophila	BV	On portunid crabs
Family Ototyphlonemertidae		
Ototyphlonemertes pellucida	V	Lit. to shallow water
Family Prosorhochmidae		
Oerstedia dorsalis	BV	Lit. to 50 m
Family Amphiporidae		
Zygonemertes virescens	BV	Lit. to 120 m
Amphiporus angulatus	B+	Lit. to 150 m
Amphiporus bioculatus	V	Lit. to 35 m
Amphiporus caecus	B+	35 m
Amphiporus cruentatus	V	Lit.
Amphiporus frontalis	B	Lit.
Amphiporus griseus	V	Lit., eury.
Amphiporus groenlandicus	A+	To 450 m
Amphiporus lactifloreus	B	2 to 200 m
Amphiporus ochraceus	V+	Lit. to 35 m
Amphiporus pulcher	B	Lit. to 200 m
Amphiporus rubropunctus	V	Shallow water
Amphiporus tetrasorus	B	80 m
Proneurotes multioculatus	V	Shallow water
Family Tetrastemmatidae		
Tetrastemma candidum	B+	Lit. to 25 m
Tetrastemma elegans	V	Lit. to 15 m
Tetrastemma vermiculus	BV	Lit. to 60 m
Tetrastemma verrilli	V	40 m
Tetrastemma vittatum	B+	Lit. to 45 m
Tetrastemma wilsoni	V	Shallow water
Suborder Polystylifera		
Family Drepanophoridae		
Drepanophorus lankesteri	B	80 m (off Nova Scotia)
Order Bdellonemertea		
Family Malacobdellidae		
Malacobdella grossa	BV	Commensal in the mantle cavity of pelecypods

Identification

The chief difficulties presented by this group result from ontogenetic changes in color patterns and other details of systematic importance, as well as from the common tendency of individuals to fragment when disturbed. Special preparation of material for study is required for systematic work, since full descriptions require the examination of microscopic sections and/or specially stained and mounted specimens. Color patterns are much used, even in formal systematics, and observation of fresh material is essential for many species determinations.

Standard formalin solutions destroy the stylets of Hoplonemertea, and 80% alcohol is a preferable, general preservative. Bouin's, corrosive sublimate, or other histological fixing agents are used for material to be sectioned—Coe (1943) warns against using agents containing osmic acid. Standard narcotics are used to discourage fragmentation.

The two classes of Rhynchocoela are differentiated as follows:

- Mouth posterior to brain; proboscis not armed with stylets. Anopla.
- Mouth anterior to brain; proboscis armed with stylets. Enopla.

Full keys and descriptions, so far as they are available, are given by Coe (1943). The following largely artificial keys are based in part on those of Coe (1943) and McCaul (1963) and are almost entirely dependent on external characteristics. Dimensions are total lengths of expanded adult individuals; width is also given in most cases.

---

1. Commensal forms associated with pelecypod mollusks or portunid crabs:
- *Carcinonemertes carcinophila* is found on *Ovalipes*, *Carcinus*, *Callinectes;* and other portunids; young worms less than 15 mm total length are found on the gills and are yellowish or rosy in color; adults (♀ to 40 mm and ♂ to 25 mm) live in egg masses of the host and are red or orange; these are slender nemerteans with a minute stylet and a pair of oval or crescentic ocelli. (Note that *Amphiporus bioculatus* also has 2 ocelli.)
- *Malacobdella grossa* lives in the mantle cavity of *Mercenaria*, *Mya*, and *Ostrea;* adults are short, broad, and thick; lack ocelli and stylets; extremely variable in color; with a sucking disc at the rear end; 40 mm by 15 mm Fig. 8.1A.

1. Free-living forms—2.

2. Ocelli present—12.
2. Ocelli absent—3.

species without ocelli

3. Cephalic grooves absent—4.
3. Transverse or longitudinal cephalic grooves present; Fig. 8.1B,C—8.
3. With oblique cephalic grooves; bright orange to vermilion; 250 mm or more by 10 mm. *Parapolia aurantiacea.*

4. Body extremely elongate, threadlike (length about 25–100 times width), contracting in a spiral; whitish tinged with yellowish or rosy:
- *Ototyphlonemertes pellucida*, proboscis slender with needlelike stylet with spiral grooves; mouth opening before the brain; 12 mm by 0.2–0.5 mm.
- *Procephalothrix spiralis*, proboscis without stylets; mouth opening behind the brain; 100 mm by 1 mm.

4. Relative thickness varying but not threadlike; stylets absent—5.

5. Body rounded in section, thicker anteriorly; does not habitually coil in a spiral; white, with or without a median yellow-orange stripe; 25 mm total length, length about 20 to 25 times width; found in parchmentlike tubes attached to bryozoans, shells, etc. *Tubulanus pellucidus.*

5. Body flattened—6.

6. Whitish with dark band in the esophageal region; head long, thin, rounded, no cirrus; 8 mm long by 0.4 mm wide. *Carinomella lactea.*

6. Esophageal region without dark band, pale reddish or yellowish—7.

7. Head rounded anteriorly, notched or not, with a median row of sensory pits; tail broad, flattened, without cirrus; intestinal region yellowish in ripe males, reddish brown in females; head, esophageal region whitish or rosy-tinted; 150 mm long by 5 mm wide. *Carinoma tremaphoros.*

7. Head tapering to a point; tail with a long cirrus; intestinal region variably colored by contents; 80 mm long by 5 mm wide. *Zygeupolia rubens.*

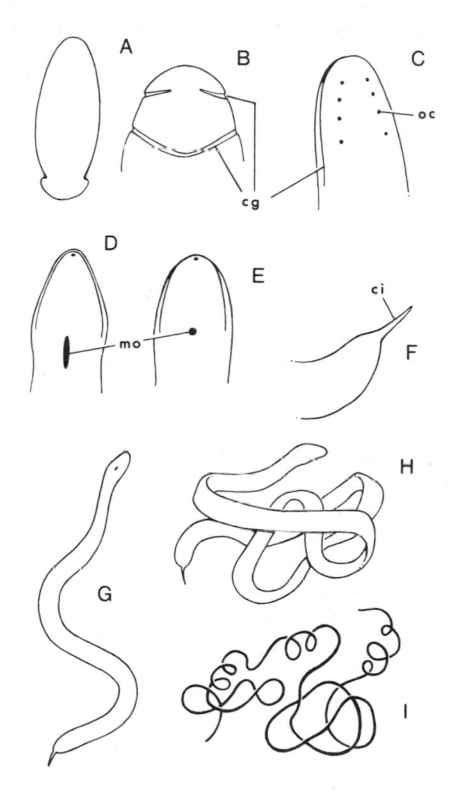

Fig. 8.1. Rhynchocoela. A, *Malacobdella grossa*, whole animal; B, *Amphiporus* sp., dorsal view of head; C, *Lineus* sp., dorsal view of head; D, *Cerebratulus* sp., ventral view of head; E, *Micrura* sp., ventral view of head; F, same, posterior end of body; G, same, whole animal; H, *Cerebratulus* sp., whole animal; I, *Lineus* sp., whole animal. Abbreviations are as follows: cg, cephalic grooves; ci, cirrus; oc, ocellus; mo, mouth.

8. With longitudinal cephalic grooves; Fig. 8.1C—9.
8. With transverse cephalic grooves; Fig. 8.1B—11.
9. Cirrus absent; whitish or pale yellow or reddish anteriorly with an indistinct pale line and two pale spots above (Verrill, 1892, gives this description but figures a green worm); 100 mm long by 0.75 mm wide. *Lineus pallidus.*
9. Cirrus present; Fig. 8.1F—10.
10. Mouth large, slotlike. *Cerebratulus* spp. See subkey below; Fig. 8.1D,H.
10. Mouth small, round. *Micrura* spp. See notes below; Fig. 8.1E,G.
11. Bright orange-red, usually with darker median stripe; 40 mm by 3 mm. *Amphiporus caecus.*
11. Dark brown above, lighter below; 80 mm by 10 mm. *Amphiporus groenlandicus.*

species with ocelli

12. With 2 ocelli; pale yellow to red, paler below, sometimes white to light green; 40 mm long by 3 mm. *Amphiporus bioculatus.* Note the crab commensal, *Carcinonemertes*, and the planarialike larvae of *Procephalothrix* and several other genera, also with 2 ocelli.
12. With 4 ocelli; Fig. 8.2A—13. Note also *Micrura affinis* (Couplet 18) and *Lineus* spp. (subkey).
12. With more than four ocelli—17.
13. Body long, ribbonlike, length about 50 times width, Fig. 8.1I; pale, rose-colored; 100 mm by 2 mm; in sandy-muddy bottom. *Lineus arenicola.*
13. Body relatively short and often nearly cylindrical, length about 7–25 times width, Fig. 8.2A; commonly among algae, hydroids, on rocks, piers, or on shelly bottoms—14.
14. Body longitudinally striped—15.
14. Body not longitudinally striped—16.
15. Length about 7–8 times width; stripes brown; 30 mm by 4 mm. *Tetrastemma vittatum.*
15. Length about 13 times width; stripes green; 20 mm by 1.5 mm. *Tetrastemma elegans.*
16. Variable in color and pattern even in small samples from the same station; head without transverse grooves; 20 mm by 2 mm. *Oerstedia dorsalis.*
16. Head with inconspicuous transverse grooves. *Tetrastemma: T. wilsoni*, translucent whitish, flecked with opaque white; 12 mm by 0.8 mm. *T. candidum*, pale green or yellowish green, head paler; 35 mm by 1.5 mm. *T. vermiculus*, variable in color, light yellowish or reddish spotted with brown, a brown band connecting the ocelli on each side. *T. verrilli*, bright rosy red; 30 mm by 3 mm.
17. With transverse cephalic grooves; Fig. 8.1B—19.
17. With lateral, longitudinal cephalic grooves; Fig. 8.1C—18.
17. With dorsal, longitudinal cephalic grooves and less conspicuous oblique cephalic grooves ventrally; an elongate cluster of 20–30 ocelli anteriorly and a rounded clump of 8–12 ocelli behind; orange-salmon above, paler below; 3.5 *meters* long by 8 mm; a deep water species. *Emplectonema giganteum.*
18. Without a cirrus. *Lineus* spp. See subkey below.
18. With a caudal cirrus, Fig. 8.1F; deep red or brown tinged with red or green, often with indistinct, paler transverse lines; head margined with white; 150 mm by 4 mm; 4–6 ocelli on each side. *Micrura affinis.*

19. Cephalic grooves transverse; Fig. 8.2D,E; *Amphiporus* and *Proneurotes* spp. The genera are only distinguishable histologically. See subkey below.

19. Cephalic grooves oblique, faint; ocelli numerous (there may be more than 80), extending laterally beyond the brain, the number increasing ontogenetically; color highly variable, correlated with the environment; 40 mm by 2 mm; Fig. 8.2C. *Zygonemertes virescens.*

Cerebratulus. While typically found as burrowing worms in soft substrata, sexually active individuals swim to the surface at night in late spring and summer. *C. lacteus* is the common species south of Cape Cod.

---

1. Relatively short and thick, margins less fine; brownish tinged with purple, red, or olive; 240 mm long by 12 mm wide. *C. luridus.*

1. Ribbonlike, margins "feathered," well-adapted for swimming—2.

2. Whitish tinged with yellow or pink, except deep red when breeding; usually to about 1200 mm long by 22 mm wide but reported more than 4 *meters* long. *C. lacteus.*

2. Grayish tinged with brown or olive, paler below; greater than 1 m long by 12 mm wide. *C. marginatus.* Both *marginatus* and *luridus* may have a pair of fine red lines indicating the positions of the lateral nerve cords.

Micura. The species are slender, about 25–50 times longer than wide; *rubra*, *caeca*, and *leidyi* are generally reddish. *M. leidyi* has a pale head; *caeca* is pale generally in comparison to the usually bright red or vermilion of *rubra*. *M. leidyi* reaches about 300 mm, *caeca*, 170 mm; and *rubra*, 75 mm. *M. albida* and *M. dorsalis* are white or yellowish, *dorsalis* being distinguished by dark median stripe. *M. albida* reaches 120 mm, *dorsalis* 160 mm. Note also *M. affinis*, the only local species of the genus with ocelli, couplet 18.

Lineus. Figure 8.2B. Only the species with ocelli are considered in this summary; the others appear in the general key.

---

1. Green with a conspicuous white or pale-yellow median stripe, paler ventrally; with 4–7 ocelli on each side; 50 mm by 1.5 mm. *L. bicolor.*

1. No median pale stripe—2.

2. Ocelli white, about 12 per side; body green; 75 mm. *L. dubius.*

2. Ocelli dark, 2–8 per side—3.

3. Body contracts into a spiral; cephalic grooves long; brown tinged with green or reddish, head pale-margined; body sometimes with pale rings and/or light lateral and mid-dorsal lines; 150 mm by 3 mm. *L. socialis.*

3. Body contracts by thickening, not by coiling; color highly variable, head pale-margined and body sometimes light-ringed; 150 mm by 4 mm. *L. ruber.* Note that the development of ocelli varies ontogenetically, the young having fewer ocelli than adults.

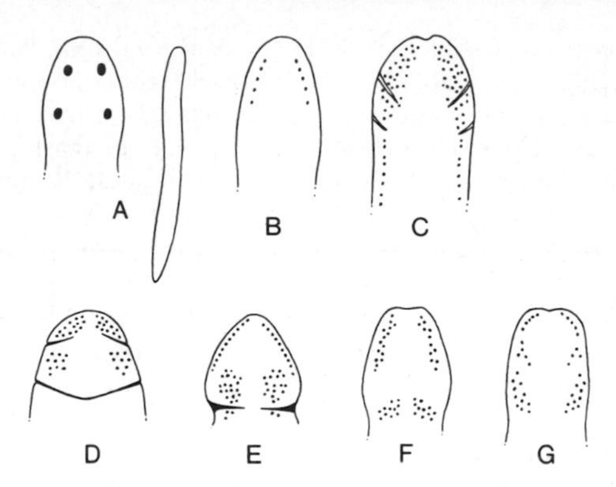

Fig. 8.2. Rhynchocoela. A, *Tetrastemma* sp., dorsal view of head and outline of whole animal; B–G are dorsal views of the head; B, *Lineus* sp.; C, *Zygonemertes* sp.; D, *Amphiporus angulatus;* E, *Amphiporus pulcher;* F, *Amphiporus lactifloreus;* G, *Proneurotes multioculatus.*

Amphiporus, Proneurotes. No attempt is made here to provide a final identification of the numerous species in this group; Figure 8.2D–G.

---

1. Ocelli in one or more rows or scattered but not in two separate groups, less than 12 ocelli per side—2.

1. Ocelli usually in 2 clusters on each side, usually more than 12 ocelli per side; Table 8.1.

2. Ocelli 5–10 in a single row on each side; body pale yellow or rosy with blood vessels visible as bright-red lines; 40 mm by 4 mm. *A. cruentatus.*

2. Ocelli in an irregular double row or scattered; Table 8.2.

Table 8.1.

Species	Color	Size	Distribution	Figure
P. multioculatus	White, pinkish	20 mm/3 mm	Virginian	8.2G
A. angulatus	Brownish–purplish	150 mm/10 mm	Boreal	8.2D
A. pulcher	Orange–rosy red	50 mm/5 mm	Boreal	8.2E
A. lactifloreus	Orange–rosy red	100 mm/5 mm	Boreal	8.2F
A. tetrasorus	Brown	30 mm/2 mm	Boreal	—

Table 8.2.

Species	Color	Size	Distribution
A. griseus	White–pale orange	30 mm/2 mm	Virginian
A. frontalis	pinkish	120 mm/5 mm	Boreal
A. ochraceus	variable	40 mm/3 mm	Virginian
A. rubropunctus	Ochre with minute red spots	20 mm/1 mm	Virginian

References

Coe, W. R. 1943. Biology of the nemerteans of the Atlantic coast of North America. Trans. Conn. Acad. Arts Sci. 35:129–328.

Correa, Diva D. 1961. Nemerteans from Florida and Virgin Islands. Bull. mar. Sci. Gulf and Carribb. 11(1):1–44.

Friedrich, H. 1955. Beitrage zu einer Synopsis der Gattungen der Nemertini monostolifera nebst Bestimmungsschlussel. Z. wiss. Zool. 158(2–3): 133–192.

McCaul, W. E. 1963. Rhynchocoela: nemerteans from marine and estuarine waters of Virginia. J. Elisha Mitchell Sci. Soc. 79(2): 111–124.

Verrill, A. E. 1892. The marine nemerteans of New England and adjacent waters. Trans. Conn. Acad. Arts Sci. 8:328–456.

———. 1895. Supplement to the nemerteans and planarians of New England. Trans. Conn. Acad. Arts. Sci. 9: 523–532.

NINE

Phylum Aschelminthes

The assembling of six related but problematic pseudocoelomate groups in a single phylum was advocated by Hyman (1951) and subsequently won acceptance in many general works. At the same time she acknowledged that "the relationship . . . is looser than is the case with classes of most other phyla," noting that other workers might prefer to regard one or more of the six classes as phyla. While the criteria allying these taxa have little relevance to the immediate problem of species identification, it is convenient to continue with Hyman's arrangement—particularly since the relationships of the taxa remain problematic.

A sufficiently comprehensive diagnosis of Aschelminthes based on external morphology is impractical. The members of the phylum exhibit a substantial range of morphological variation. Criteria for the initial identification of individuals to class were presented in Chapter 3, and it is convenient to proceed directly to the classes, which are: Rotifera, Gastrotricha, Kinorhyncha, Priapulida, Nematoda, and Nematomorpha. Rotifera and Nematoda require special study and are given only cursory treatment in this account. The remaining classes have few local forms, which may be identified without special technique.

Class Rotifera
ROTIFERS OR WHEEL ANIMALCULES

Diagnosis. *Microscopic, with a body more or less regionated, diverse in form; oral end with a ciliated corona that in live individuals of some genera simulates a rotating* *wheel; body wall often stiffened by an intracellular lamella forming an armor case called the lorica, superficially segmented or variously ornamented or divided in plates.*

Most rotifers are smaller than 0.5 mm. They may be globular, or elongate and wormlike, but frequently the body is more or less divisible into a head, trunk, and foot. The head is chiefly distinguished by the corona which in its elemental form consists of a band of cilia encircling the head and expanded in a ventral "field" surrounding the mouth. More commonly the ciliary surfaces are restricted and patterned in lobes, circlets, or tufts reflecting differences in feeding habits or locomotion. In sessile forms especially, the corona may form a funnellike trap. The illusion of a whirling wheel or wheels renders the corona conspicuous in live material, but in preserved specimens or when live individuals are disturbed the head is usually retracted. The mouth opens into a pharynx or *mastax* which contains a complex masticatory apparatus with chitinized *trophi*. The trophi are of major importance in rotifer systematics.

The trunk varies greatly in form and ornamentation, as does the degree of development of the armored lorica. The illustration (Fig. 9.1) indicates the

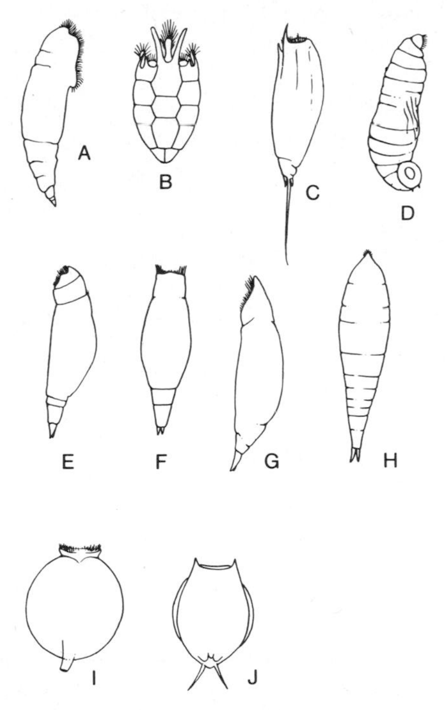

Fig. 9.1. Rotifera, representative marine genera. A, *Lindia*; B, *Keratella*; C, *Trichocerca*; D, *Zelinkiella*; E, *Pleurotrocha*; F, same (different species), dorsal view; G, *Encentrum*; H, *Proales*; I, *Testudinella*; J, *Lecane*.

appearance of these structures in typical marine genera. Pelagic genera tend to have the foot reduced or absent.

The sexes are separate and there is sexual dimorphism in external appearance. Males are smaller than females in most orders, sometimes only slightly so; in other cases, males are no more than a tenth as large as females. In Bdelloidea, males are entirely absent, all members of the order being parthenogenetic so far as known. At hatching, free-swimming rotifers resemble the parental form. In sessile species the larvae differ from adults but are recognizable as rotifers; they are initially free-swimming and they metamorphose on attachment.

Rotifers vary widely in habits and habitat preferences. Most are benthic, but there are pelagic species, chiefly in the genera *Keratella*, *Trichocerca*, and *Synchaeta*. Swedmark (1964) cites *Proales* and *Encentrum* as psammo-littoral genera, but this list will doubtless increase with further study of the interstitial sand fauna. Benthic forms are sessile, or they creep about, inchworm fashion, or swim short distances among plants or in the bottom debris. Food habits also vary. Members of Bdelloidea and of the Philodinidae and Notommatidae draw food into the digestive tract by ciliary currents; this mode of feeding is found expecially in sessile forms. In other groups, e.g., Synchaetidae and Dicranophoridae, the animals are carnivorous, though they also depend to a substantial extent on flagellate algae. Synchaetids feed by ciliary currents, but dicranophorids have a forcepslike mastax that can be extended to seize prey which is then either swallowed whole, chewed, or "sucked dry."

Systematic List and Distribution

While rotifers are not rare in the salt water microfauna, they are not so important or conspicuous in such habitats as in fresh water. Myers (1936) estimated that about 12% of the 1500 species known at that time occur in brackish or marine waters. Many of these are presumed to be essentially freshwater species with a tolerance for increased salinity. Rotifer species are highly cosmopolitan in distribution.

The following abbreviated list does not pretend to completeness but indicates some of the common salt water genera. The classification follows Edmondson (1959) with adjustments to fit the concept of Rotifera as a class of Aschelminthes.

Order Seisonacea
Family Seisonidae
Seison

Order Bdelloidea
Family Philodinidae
Zelinkiella

Order Monogononta
  Suborder Ploima
    Family Brachionidae
      Notholca
      Colurella
      Keratella
    Family Lecanidae
      Lecane
      Monostyla
    Family Notommatidae
      Pleurotrocha
      Cephalodella
    Family Lindiidae
      Lindia
    Family Trichocercidae
      Trichocerca
    Family Dicranophoridae
      Encentrum
  Suborder Flosculariacea
    Family Testudinellidae
      Testudinella
    Family Proalidae
      Proales

Identification

Rotifers require special techniques for species determination. While species with a well-developed lorica may be killed and fixed in 10% formalin, this treatment is not satisfactory for many of the more soft-bodied forms, which must be carefully narcotized. An alternative procedure, and one recommended by Edmondson (1959), is to flood a small sample in a watch glass with an equal amount of boiling water, followed by fixation in formalin. Pennak (1953) recommends 2–5% formalin plus 2% glycerin for preservation. The examination of trophi, a necessity for some determinations, requires painstaking technique; for details of this and other aspects of the preparation of material for study see Pennak (1953).

Freshwater species are important in the estuarine rotifer fauna, and interested students may refer to Pennak and to Edmondson. A selection of typical marine genera is figured (Fig. 9.1).

References

Berzins, B. 1960. Rotatoria I–VI. Fich. Ident. Zoopl. Sheets 84–89.

Edmondson, W. T. 1959. Rotifera. *In* Fresh-Water Biology. W. T. Edmondson, ed. John Wiley and Sons.

Hyman, Libbie H. 1951. The Invertebrates: Acanthocephala, Aschelminthes, and Entoprocta, the pseudocoelomate Bilateria. Vol. III. McGraw-Hill Book Co.

Myers, F. J. 1931. The distribution of Rotifera on Mount Desert Island. Am. Mus. Novit. 494:1–12.

———. 1936. Three new brackish water and one new marine species of Rotatoria. Trans. Am. microsc. Soc. 55(4):428–432.

Pennak, R. 1953. Fresh-water invertebrates of the United States. Ronald Press.

Remane, A. 1929–33. Rotatorien. *In* Bronn's Klassen und Ordnungen des Tierreichs, B. and IV, Abt. II, Buch 1, Akad. Verlags., Leipzig.

Swedmark, B., 1964. The interstitial fauna of marine sand. Biol. Rev. 39:1–42.

Class Gastrotricha
GASTROTRICHS

Diagnosis. *Minute, elongate, ventrally flattened aschelminthines often ornamented with spines, scales, or bristles, and usually with one or more pairs of adhesive tubules; mouth and anus terminal at opposite ends of the body; transparent.*

Adults of local species are less than 2.0 mm, and usually less than 1.0 mm long. The head end may or may not be differentiated by a basal constriction; some species have one or more pairs of red eye spots or ocelli and one or two pairs of tentacles. The rear end may be rounded, bluntly triangular, bilobed, or attenuated with an elongate tail. Cilia are usually present on the head and ventral body surface; frequently the cilia are dispersed in a pattern that is systematically important. The ornamentation of the thin, cuticular body covering is extremely diverse, as are the number and disposition of the adhesive tubules.

Gastrotrichs are hermaphroditic or, as in most chaetonotids, parthenogenetic; protandric hermaphroditism occurs in some genera. Details of development are known for only a few marine species; development is direct in Macrodasyoidea, and the embryology is similar to that of nematodes according to Teuchert (1968).

Gastrotrichs eat bacteria and microscopic algae, feeding primarily by means of a sucking pharynx. They have been reported most often from intertidal and subtidal sand, but also occur in mud and on such substrata as algae and barnacles.

Systematic List and Distribution

The gastrotrich fauna is too poorly known at this time to warrant presenting a species list. At the time of writing (March 1969), only two papers and two abstracts had been published on North American marine species; they list ten named species from the Woods Hole area and more than twice as many species have been found, though not reported in the literature. This probably represents but a fraction of the fauna that awaits further investigation. In the following list, genera that have actually been taken on this coast are marked +.

Order Macrodasyoida
- Family Dactylopodolidae
  - *Dactylopodola* +
  - *Dendrodasys*
  - *Xenodasys*
- Family Lepidodasyidae
  - *Acanthodasys* +
  - *Cephalodasys* +
  - *Lepidodasys* +
  - *Mesodasys* +
  - *Paradasys*
  - *Crasiella*
- Family Macrodasyidae
  - *Macrodasys* +
  - *Pleurodasys*
  - *Urodasys* +
- Family Thaumastodermatidae
  - *Diplodasys* +
  - *Hemidasys*
  - *Platydasys* +
  - *Pseudostomella* +
  - *Ptychostomella*
  - *Tetranchyroderma* +
  - *Thaumastoderma* +
- Family Turbanellidae
  - *Desmodasys*
  - *Dinodasys*
  - *Paraturbanella* +
  - *Turbanella* +
  - *Pseudoturbanella*

Order Chaetonotoida
- Family Chaetonotidae
  - *Aspidiophorus* +
  - *Chaetonotus* +
  - *Heterolepidoderma*
  - *Ichthydium*
  - *Lepidodermella*
  - *Polymerurus*
  - *Musellifer*
- Family Neodasyidae
  - *Neodasys* +
- Family Xenotrichulidae
  - *Heteroxenotricula* +
  - *Xenotrichula* +

Identification

Though gastrotrichs can be fixed and preserved in formalin, better results may be obtained with special treatment. Narcotization is necessary with some material; Brunson (1959) cites the use of crystalline cocaine—a few crystals in a drop of water containing live individuals. Pennak (1953) recommends osmic acid for fixing; specimens in a drop of water on a slide are inverted over an open bottle of 2% osmic acid for 5–10 sec; saturated mercuric chloride or Bouin's may also be used. Hummon (personal communication) recommends the following procedure: relax specimens in 6% magnesium chloride; fix by gradual introduction of 10% formalin; transfer to 1:1 10% formalin:glycerin, allow water to evaporate; mount in glycerin jelly, seal with Murrayite. Hummon also recommends using live narcotized material for taxonomic work. See Pennak (1953) and Edmondson (1959) for further information.

The generic keys offered below are based, with considerable modification, on that of Boaden (1963). The two orders may be separated as follows.

1. Without pharyngeal pores; usually with only 1 or 2 pairs of adhesive tubules and these confined to the posterior foot lobes. Order Chaetonotoida. Note. The pharynx is the initial section of the digestive track extending $\frac{1}{6}$ to $\frac{1}{3}$ of the body length, sometimes with a thin-walled buccal end and with one or more bulbous enlargements.

1. With pharyngeal pores; usually with anterior, lateral, and at least 6 posterior tubules. Order Macrodasyida. The pharyngeal pores, one on each side, extend to the body surface.

ORDER CHAETONOTOIDA

1. Lateral adhesive tubules present; scaleless; posterior feet with 3 or 4 adhesive tubules; Fig. 9.2A. *Neodasys.*

1. No lateral adhesive tubules; scales usually present—2.

2. Ventral cilia form thick tuftlike cirri:

- Dorsal surface with 9 longitudinal rows of large, oblong, stalkless scales; Fig. 9.2B. *Heteroxenotrichula.*
- Dorsal surface with stalked scales, or with tiny scales on the lateral surface only; Fig. 9.2C. *Xenotrichula.*

2. Ventral cilia form two longitudinal rows of normal cilia—3.

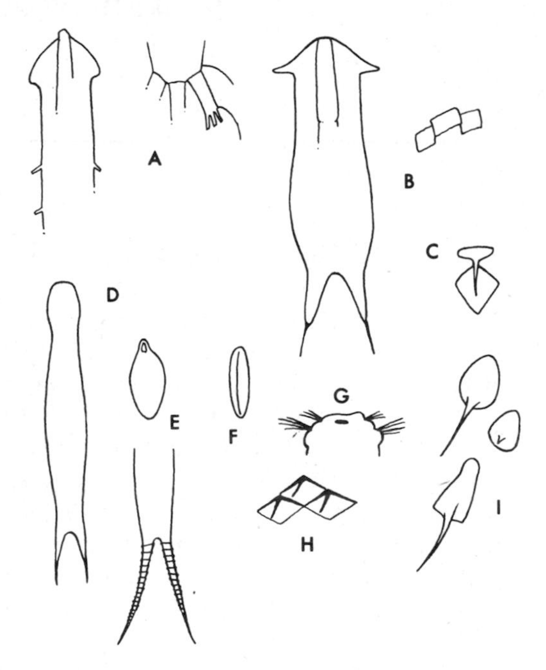

Fig. 9.2. Gastrotricha, Chaetonotoida. Whole animals are shown in outline only. A, *Neodasys*, anterior end and detail showing posterior foot; B, *Heteroxenotrichula*, whole animal and three dorsal scales; C, *Xenotrichula*, dorsal scale (body form similar to preceding); D, *Ichthydium*, whole animal; E, *Polymerurus*, individual scale and detail of posterior end of body; F, *Heterolepidoderma*, individual scale; G, same, head; H, *Aspidiophorus*, three scales; I, *Chaetonotus*, three scales.

3. Body surface naked, without scales or spines; Fig. 9.2D. *Ichthydium.*
3. Body with scales or spines; otherwise somewhat similar in body form to the above —4.
4. With very long, segmented tail forks; Fig. 9.2E. *Polymerurus.* Note. This is a hodgepodge genus; some species have short forks and some are unsegmented; see Hummon (1969) and Remane (1936).
4. Tail forks shorter—5.
5. With simple scales only, no spines or stalked scales:
- Scales with keel; Fig. 9.2F,G. *Heterolepidoderma.*
- Scales without keel. *Lepidodermella.*

5. Scales with stalk or spine:
- Scales stalked, with base plate and rhomboidal, spined end plate (not all species); Fig. 9.2H. *Aspidiophorus.*
- Scales with spine (not all species); Fig. 9.2I. *Chaetonotus.*

ORDER MACRODASYOIDA

1. With scales or modified scales—2.
1. Without scales or modified scales—4.
2. Head with large anterior processes forming a *prebuccal jaw* (or buccal palps); Fig. 9.3A. *Pseudostomella.*
2. Head without prebuccal jaws—3.
3. With modified scales, i.e., with 4 or 5 pronged "hooks" only:
- With at least two pairs of lateral head tentacles; Fig. 9.3B,C. *Thaumastoderma.*
- With 1 pair of lateral head tentacles or none; Fig. 9.3D. *Tetranchyroderma.* Four pronged hooks are also present in *Pseudostomella.*

3. With scalelike cuticular thickenings only; hind end rounded; Fig. 9.3E. *Lepidodasys.*
3. With scales and spines, but no "hooks":
- Hind end bifurcated; Fig. 9.3F. *Acanthodasys.*
- Hind end rounded; Fig. 9.3G. *Diplodasys.*

4. With a long, thin tail, at least as long as rest of body. *Urodasys.*
4. With hind end rounded—5.
4. With hind end bifurcated or truncated—7.
5. Pharyngeal region with a pair of dorso-lateral appendages, each consisting of a stalk with an end knob; Fig. 9.3H. *Pleurodasys.*
5. Without pharyngeal knobs—6.
6. Ovary and testis single (see Hyman, 1951, for descriptions of reproductive anatomy:)
- Male genital pore with cuticular plates. *Hemidasys.*
- Male genital pore without plates; body with numerous papillae; hind end rounded. *Platydasys.*

6. Ovary and testes paired. *Macrodasys.*
7. Ovary and testis both single; Fig. 9.3I. *Ptychostomella.*
7. Testes paired, ovary single; Lepidodasyidae (see subkey).
7. Testes and ovaries paired—8.

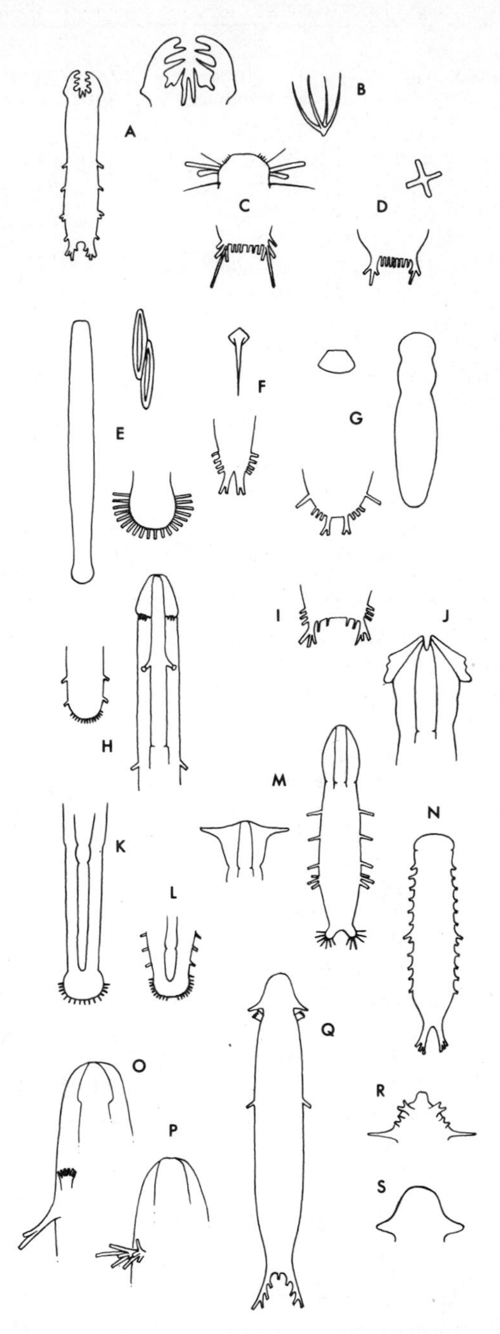

Fig. 9.3. Gastrotricha, Macrodasyida. Whole animals are shown in outline only. A, *Pseudostomella*, whole animal and detail of prebuccal jaw; B, *Thaumastoderma*, modified scale; C, same, head (above) and posterior end (below); D, *Tetranchyroderma*, scale and

8. Head when distinct encloses only anterior part of pharynx: Turbanellidae (see subkey).
8. Head includes all of pharynx: Dactylopodolidae—9.
9. Lateral tubules absent; head with large, blunt side processes; Fig. 9.3J. *Dendrodasys.*
9. With 5 or 6 lateral tubules on each side; head with or without a pair of tentacles; Fig. 9.3M. *Dactylopodola.*
9. With 12 lateral tubules on each side; Fig. 9.3N. *Xenodasys.*

FAMILY LEPIDODASYIDAE

Note that *Acanthodasys* and *Lepidodasys* are scaled genera in this family and have been keyed above.

1. Either with lateral tubules absent and with only two anterior tubules or with lateral tubules present and well-developed anterior "feet" present; bursa not distinct; Fig. 9.30. *Paradasys.*
1. Lateral tubules numerous; distinct anterior feet absent; well-developed bursa present—2.
2. Bursa partly within tail lobe; lateral tubules all posterior to pharyngeal pores; Fig. 9.3L. *Cephalodasys.*
2. Bursa some distance before tail lobe; lateral tubules extend forward to head region; Fig. 9.3K. *Mesodasys.*

FAMILY TURBANELLIDAE

1. With a pair of ventrolateral feet formed by 2 unequal adhesive tubules; with a pair of anterior ventral feet. *Pseudoturbanella.*
1. No ventro-lateral feet; anterior ventral feet present—2.
2. Without lateral tubules; with 1 pair of ventrolateral tubules; Fig. 9.3Q. *Pseudoturbanella.*
2. Lateral tubules present—3.
3. Individual tubules of ventral foot in a tuft; Fig. 9.3P. *Desmodasys.*
3. Individual tubules of ventral foot in a transverse row; Fig. 9.3O—4.

Fig. 9.3. (*Continued*).
detail of posterior end; E. *Lepidodasys*, whole animal, details of scale-like cuticular ornamentation, and posterior end; F, *Acanthodasys*, scale and posterior end; G, *Diplodasys*, whole animal, individual scale, and posterior end; H, *Pleurodasys*, anterior and posterior ends; I, *Ptychostomella*, posterior end; J, *Dendrodasys*, head; K, *Mesodasys*, posterior end; L, *Cephalodasys*, posterior end; M, *Dactylopodola*, whole animal and head; N, *Xenodasys*, whole animal; O, *Paradasys*, anterior end; P, *Desmodasys*, head; Q, *Pseudoturbanella*, whole animal; R, *Dinodasys*, head; S, *Turbanella*, head.

4. Head with tentacles and two pairs of anterior bifurcated lobes; Fig. 9.3R. *Dinodasys*.
4. Head without anterior dorsal tentacles, with elongate or rounded side lobes; Fig. 9.3S. *Turbanella*.

References

Blake, C. H. 1933. Nomenclatural notes on Gastrotricha. Science, New York. 77:606.

Boaden, P. J. S. 1963. Marine Gastrotricha from the interstitial fauna of some North Wales beaches. Proc. zool. Soc. Lond. 140:485–502.

Brunson, R. B. 1959. Gastrotricha. *In* Fresh-water Biology. W. T. Edmondson, ed. John Wiley and Sons. Pp. 406–419.

Clausen, C. 1965. *Desmodasys phocoides* gen. et sp. n., family Turbanellidae (Gastrotricha Macrodasyoidea). Sarsia 21:17–21.

———. 1969. *Crasiella diplura* gen. et sp. n. (Gastrotricha, Macrodasyoidea). Sarsia 33:59–64.

Edmondson, W. T., ed. 1959. Fresh-Water Biology. John Wiley and Sons.

Hondt, J. d'. 1968. Contribution a la connaissance des Gastrotriches intercotidaux du Golfe de Gascogne. Cah. Biol. mar. 9:387–404.

Hummon, W. D. 1966. Morphology, life history, and significance of the marine gastrotrich, *Chaetonotus testiculophorus* N. SP. Trans. Am. microsc. Soc. 85(3):450–457.

———. 1967. Interstitial marine gastrotrichs from Woods Hole, Massachussets. Biol. Bull. mar. biol. Lab. Woods Hole 133(2):452.

———. 1968. Interstitial marine gastrotrichs from Woods Hole, Massachusetts. Part II. Biol. Bull. mar. biol. Lab. Woods Hole 135(2):423–424.

———. 1969. *Musellifer sublittoralis*, a new genus and species of Gastrotricha from the San Juan Archipelago, Washington. Trans. Am. microsc. Soc. 88:282–286 (not seen).

Hyman, Libbie H. 1951. The Invertebrates: Acanthocephala, Aschelminthes, and Entoprocta, the pseudocoelomate Bilateria. Vol. III. McGraw–Hill Book Co.

Packard, C. E. 1936. Observations on the gastrotricha indigenous to New Hampshire. Trans. Am. microsc. Soc. 55:422–427.

Pennak, R. 1953. Fresh-water Invertebrates of the United States. Ronald Press.

Remane, A. 1926. Morphologie und Verwandtschaftbeziehungen der aberranten Gastrotrichen. Z. Morph. Ökol. Tiere. 5:625–754.

———. 1927a. Beitrage zur systematik der Susswassergastrotrichen. Zool. Jb. (Syst.) 53:269–320.

———. 1927b. Neue Gastrotricha Macrodasyoidea. Zool. Jb. (Syst.) 54:203–242.

———. 1934. Die Gastrotrichen des Küstengrundwassers von Schilksee. Schr. Naturw. Ver. Schleswig. Holst. 20:437–478.

———. 1936. Gastrotricha. *In* Bronn's Klassen and Ordnungen des Tierreichs. Band. 4, Abt. 2, Buch 1, Teil 2, Lfrg. 1–2:1–242. Akad. Verlagsges., Leipzig.

———. 1951. *Mesodasys* ein neues Genus der Gastrotricha, Macrodasyoidea aus der Keiler Bucht. Kieler Meeresforsch. 8:102–105.

Swedmark, B. 1957. Nouveaux gastrotriches macrodasyoides de la region de Roscoff. Archs. Zool. exp. gen. (Notes et Revue) 94:43–57.

———. 1967. Trois nouveaux gastrotriches macrodasyoides de la faune interstitielle marine des sables de Roscoff. Cah. Biol. mar. 8:322–330.

Teuchert, G. 1968. Zur Fortpflanzung und Entwicklung der Macrodasyoidea (Gastrotricha). Z. Morph. Ökol. Tiere 63 (4):343–418.

Wilke, U. 1954. Mediterrane gastrotrichen. Zool. Jb. (Syst.) 82:497–550.

Class Kinorhyncha (Echinoderida)

Diagnosis. *Minute, superficially segmented, more or less spinous, wormlike aschelminthes with a retractile head.*

The largest local species are only slightly more than 0.8 mm in total lenth. There are 13 superficial segments or *zonites.* The head constitutes the first zonite; it bears circlets of spines and is wholly retractile into the second zonite or "neck" which is armored with 6, 8, or 16 plates called *placids.* The placids close over the retracted head (*Echinoderes*), or both the head and neck are retractile into the first trunk zonite (*Trachydemus* and *Pycnophyes*). The body is only moderately elongate, length exceeding width by about 3–5 times in different species. In section, the body is flat below with vaulted sides. Conspicuous lateral spines are present on the last trunk segment in four of the six local species. The trunk zonites may be thickly or sparsely sown with minute spines. Mature individuals are usually yellowish brown in color.

The sexes are separate but similar in general configuration, differing only slightly in size and minor external details if at all. Fertilization is internal, but the eggs develop externally. The larvae of *Echinoderes* do not resemble adult kinorhynchs and only gradually assume the adult form through successive moults (Fig. 9.4A). Homalorhagous species have larvae more nearly equal to adults in appearance, but they may not resemble the parent genus initially.

Kinorhynchs have a bathymetric range of 0–5000 m. Serious collecting requires special techniques, including the use of a dredge designed to skim the flocculent surface layer of marine sediment followed by a meticulous sorting procedure. Alcohol (70%) is the preferred preservative. For further details on technique see Higgins (1964a,b). Kinorhynchs are described as feeders on detritus and microscopic algae.

Systematic List and Distribution

Papers by Higgins (1964a, 1964b, and 1965) review the distribution of local species. The group has been little studied on this coast, and further work will doubtless add new forms and clarify sketchily known distribution patterns.

Order Homalorhagida		Order Cyclorhagida	
Family Pycnophyidae		Family Echinoderidae	
Pycnophyes frequens	B+	*Echinoderes remanei*	B+
Pycnophyes beaufortensis	C+	*Echinoderes bookhouti*	C+
Trachydemus mainensis	B+		
Trachydemus langi	C+		

The species above have all been collected from shallow water in coastal bays and sounds. Higgins suggests that the ranges of all of these species probably terminate between Cape Henlopen, Delaware, and Cape Hatteras.

Identification

For full descriptions and additional criteria see Higgins' papers.

1. Second zonite not retractile, with 16 placids, "tail" spines well developed; trunk zonites thickly set with fine spines dorsally; total length about 0.36 mm; Fig. 9.4A,B. *Echinoderes:* In *E. bookhouti* there is a spinelike extension on the rear border of zonite 12; the border of this plate is simple in *remanei*.

1. Second zonite retractile, with 6 or 8 placids; adults reach >0.6 mm; trunk zonites with few spines dorsally—2.

2. Trunk with strongly developed "tail" spines; total length 0.75–0.81 mm; Fig. 9.4C. *Pycnophyes: P. beaufortensis* has 6 placids, *frequens*, 8. Note also that the ventral muscle scars of *P. frequens* are more vertically oriented than in *P. beaufortensis*.

2. Trunk without "tail" spines; total length ca. 0.69 mm; Fig. 9.4D. *Trachydemus: T. mainensis* has strongly developed postlateral spines on the 12th zonite; these are weakly developed in *T. langi*.

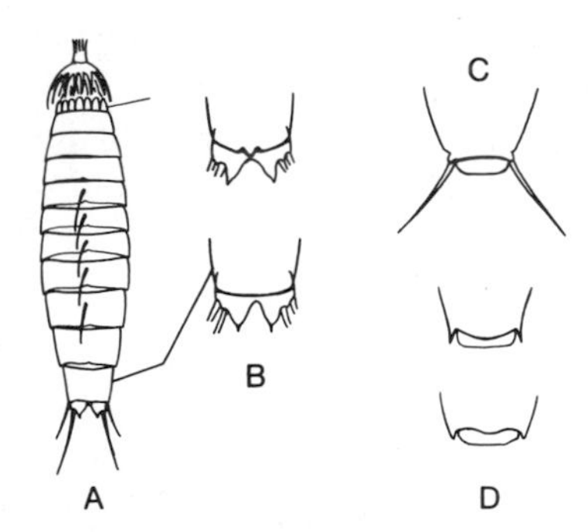

Fig. 9.4. Kinorhyncha. A, *Echinoderes* sp., whole animal with placids and 12th zonite indicated; B, *Echinoderes bookhouti* (above) and *E. remanei* (below), last 2 zonites; C, *Pycnophyes* sp., last 2 zonites; D, *Trachydesmus mainensis* (above) and *T. langi* (below), last 2 zonites.

References

Higgins, R. P. 1964a. Redescription of the kinorhynch *Echinoderes remanei* (Blake 1930) Karling 1954. Trans. Am. microsc. Soc. 83(2):243–247.

———. 1964b. Three new kinorhynchs from the North Carolina coast. Bull. mar. Sci. Gulf Caribb. 14(3): 479–493.

———. 1965. The Homalorhagid Kinorhyncha of northeastern U.S. coastal waters. Trans. Am. Microsc. Soc. 84(1):65–72.

Class Priapulida

Priapulus caudatus is the only local representative of this small class.

Diagnosis. *Stout, contractile, warty, wormlike animals with a two-part body consisting of a shorter, completely retractile presoma and a longer, ringed trunk with a terminal cluster of short, thick appendages, Figure 9.5.*

The presoma consists of a somewhat bulbous *proboscis* with a protrusible, spiny *oral region* and a shorter *collar*, both of these regions being commonly

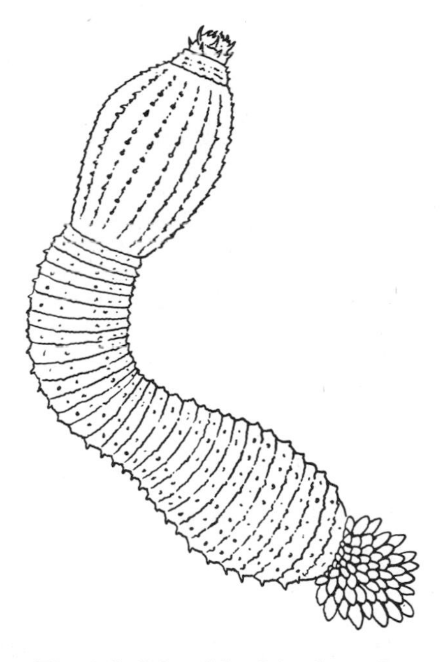

Fig. 9.5. Priapulida. *Priapulus caudatus.*

invaginated within the much larger proboscis. The proboscis is ornamented with 25 longitudinal rows of small, spiny papillae. The trunk is superficially segmented with about 30–40 rings with scattered, wartlike papillae; on the last few rings these are more densely concentrated and form a posterior border. *Priapulus caudatus* attains a length of about 80 mm. The proboscis is whitish, the body flesh-colored, and the caudal appendages yellow according to Verrill (1880); Dawydoff (1959) describes priapulids as generally brownish with a distinct metallic iridescence.

The sexes are separate but not differentiated externally. Fertilization is external. The young hatch in a larval condition with a presoma approximating that of the adult in appearance but with a trunk encased in a *lorica* consisting of 8 large longitudinal plates—3 lateral on each side plus a ventral and dorsal plate—and 2 more, minute dorsal and ventral plates at the anterior margin. They apparently pass 2 years in this condition during which time they occupy the same habitat as the adult.

Priapulids are benthic, lying partly buried in soft mud. They are predaceous, as the well-armed oral apparatus suggests. They are littoral in the Arctic but generally sublittoral at lower latitudes where they range to about 500 m. Verrill (1880) reported this species in 9–11 m in the Bay of Fundy and at 49 m in Massachusetts Bay. Little additional information is available on the distribution of priapulids on this coast.

References

Dawydoff, C. 1959. Classe des Priapuliens. *In* Traité de Zoologie. Anatomie, Systematique, Biologie. Tome V.

Murina, V. V., and J. I. Starobogator. 1961. Classification and zoogeography of Priapuloidea. Akad. Nauk. SSR Inst. Okean. Moskva. 46:179–200.

Verrill, A. E. 1880. Notice of recent additions to the marine invertebrata of the northeastern coast of America with descriptions of new genera and species and critical remarks on others. Proc. U.S. natn. Mus. 2:165–205.

Wesenberg-Lund, Elise. 1930. Priapulidae and Sipunculidae. Dan. Ingolf-Exped. 4(7):1–44.

Class Nematomorpha (Gordiida)
HAIRWORMS

Nectonema agile is the only local representative of this small class, and the genus is the only marine representative of the class.

Diagnosis. *Threadlike, unsegmented worms without externally differentiated regions; with a double row of bristles, Figure 9.6.*

Nectonema is round-bodied, of nearly equal diameter throughout but tapering slightly at the head end and somewhat more toward the tail. Bristles are lacking for a short distance at the tail end and in a longer section at the head. Males attain a length of 50–200 mm, females, 30–60 mm; the thickness is only 0.3–1.0 mm. The body of *Nectonema agile* is pale, translucent whitish or grayish yellow with slate-colored bristles.

The sexes are separate and differ in size, as indicated above. The adults have degenerate mouth parts and do not feed. The young are parasitic in brachyuran and hermit crabs but resemble the adults in external features, except that the very early stages have hooks and stylets for attachment and feeding.

Adults are found swimming with a rapid, undulatory motion at the surface in inshore waters of bays and sounds; they may be attracted to night lights. Ward (1892) reported these worms from July to October with maxima in July and September on moonless nights of a falling tide. Additional species (or the same species?) occur in the Mediterranean and European boreal. *N. agile* has been reported from the Woods Hole area and Narragansett Bay, but its American distribution is otherwise not well known.

Pennak (1953) recommends fixing and preserving nematomorphs in 90% alcohol or, as a preliminary to histological work, in "hot saturated mercuric chloride containing 5–10% of acetic acid." For additional notes on technique see Ward (1892).

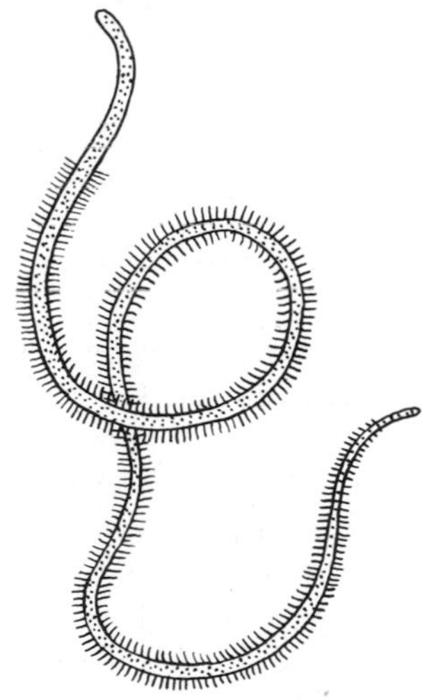

Fig. 9.6. Nematomorpha. *Nectonema agile.*

References

Chitwood, B. G. 1966. Gordiida. *In* Fresh-water biology. W. T. Edmondson, ed. John Wiley and Sons.

Feyel, Thérèse. 1936. Recherches histologiques sur *Nectonema agile* Verr. Étude de la forme parasite. Archs. Anat. microsc. 32:197–232.

Pennak, R. 1953. Fresh-water Invertebrates of the United States. Ronald Press.

Ward, H. B. 1892. On *Nectonema agile*, Verrill. Bull. Mus. comp. Zool. Harvard. 23(3):135–188.

Class Nematoda (Nemata)
NEMATODES or ROUNDWORMS

Diagnosis. *Cylindrical worms without appendages and usually without definite body regions; the body frequently fusiform, resembling an elongate spindle, or filiform, i.e., threadlike; some mature female parasites of plants and animals spherical; body covered with a cuticle, either smooth or ringed, usually white or pale yellowish, often with bristles, warts, ridges, or other ornamentation; parasitic or free-living; mostly* $<$*1 mm, but with some marine species reaching 50 mm.*

The prominence of body rings in some species gives the appearance of segmentation, e.g., in Desmoscolecidae. Definite head and tail regions may be demarcated. The differentiation of head structures is not very advanced, however, and the body segmentation is entirely superficial. Appendages are lacking, but setae are frequently present on the body. The head has circlets of labial and cephalic sensory papillae or setae. There are also lateral sensory receptors on or near the head called *amphids*. The individual amphid takes the form of a pore or narrow transverse slit; it may be circular, spiral, or some derivative of these forms.

The mouth is terminal and varies in form; e.g., it may be tubular or stylet-, funnel-, or goblet-shaped, with or without teeth or jaws. The mouth leads to an internal *buccal cavity*. The esophagous is muscular, with or without one or more bulblike swellings. The gut is usually a straight tube with an anus near the posterior end; most species have a postanal extension or tail which ranges in length from short to very long, and may be rounded, or the terminus may be bluntly or acutely conical, or filamentous.

The sexes are separate in marine species; hermaphroditism and parthenogenesis occur frequently in freshwater and terrestrial species. Males are usually smaller than females, have a curved tail, and may possess accessory copulatory structures in the form of single or paired, heavily cuticularized, sickle- or needle-shaped structures, one on either side of the cloaca, which is the common opening of the male digestive and reproductive systems. Precloacal accessory structures may be present in the form of setae, pores, or papillae. The female genital tract opens via the anus in members of the Lauratonematidae, but a separate opening is present in other families, placed either at about midbody or somewhat posterior or anterior to midbody.

Fertilization is usually internal, and the eggs may be retained through hatching. The young appear as juvenile worms. Hyman (1951) rejected the term "larvae" for young nematodes; while distinctive larval forms are lacking the life history is often complex with both endoparasitic and free-living phases. Life history patterns are extremely diverse and in parasitic forms may involve more than one host species.

Nematodes are ubiquitous, appallingly so! Hyman (1951) estimates that there are at least a half million species, many as yet undescribed. They are often individually abundant as well. A Dutch investigator estimated concentrations of almost four and a half million individuals *per square meter* in bottom muck off Holland. A bottom sample of only 6.7 cc from the Mediterranean yielded 1074 individuals representing 36 species in 20 genera.

Marine nematodes range from the littoral to abyssal depths. They will be encountered in or on all substrata and in all sorts of hosts, both vertebrate and invertebrate.

The study of marine and estuarine nematodes on this coast is still in an exploratory stage. The majority of species are microscopic and require special study not only in laboratory technique but also in collecting and recovery methods. For details see Chitwood and Allen (1959), and for full general accounts of the class see deConinck (1965), and Hyman (1951).

Systematic List

No attempt will be made to present a list of species. The following, abbreviated classification includes the principal higher taxa with free-living marine species. Two subclasses were recognized by deConinck (1965), Adenophorea and Secernentea; the latter, having very few free-living marine species, is omitted. The adenophorean order Dorylaimida is excluded for the same reason along with a few families in other orders. For an introductory account of freshwater nematodes see Chitwood and Allen (1959).

Subclass Adenophorea
- Order Araeolaimoidea, families Araeolaimidae, Diplopeltidae, Haliplectidae, Axonolaimidae, Leptolaimidae, Camacolaimidae, and Tripyloididae
- Order Monhysteroidea, families Linhomoeidae, Sphaerolaimidae, Monhysteridae, and Siphonolaimidae

Order Desmodoroidea, families Spirinidae, Microlaimidae, Desmodoridae, Dasynemellidae, Ceramonematidae, Monoposthiidae, Richtersiidae, Epsilonematidae, and Draconematidae

Order Chromadoroidea, families Comesomatidae, Chromadoridae, Cyatholaimidae, Choanolaimidae, and Selachinematidae

Order Desmoscolecoidea, families Meyliidae, Desmoscolecidae, and Greeffiellidae

Order Enoploidea, families Ironidae, Leptosomatidae, Oxystominidae, Lauratonematidae, Phanodermatidae, Thoracostomopsidae, Enoplidae, Oncholaimidae, and Eurystominidae

Identification

Chitwood and Allen (1959), and deConinck (1965, pp. 396–405) provide useful information on technique. After initial fixation in 5% formalin, nematodes may be transferred to a microscope slide in a mixture of alcohol and glycerin (about 5% glycerin in 50% alcohol), the alcohol evaporated, and the worms examined in pure glycerin or in glycerin jelly under a cover slip.

In the present simplified account, emphasis is placed on the labial and cephalid sense organ formulae, and on the form of the amphids, buccal cavity, and cuticular ornamentation. The keys and tabular data are based on deConinck (1965) but are greatly condensed.

The following formulae are used to describe the anterior sensory structures. The full complement, 6 + 6 + 4, features a circlet of 6 *inner labial papillae*, a circlet of 6 *outer labial papillae* or *bristles*, and a *cephalic circlet* of 4 bristles, Figure 9.7F. In some cases the inner labial circlet is obscure, in which case the formula has been rendered as 0 + 6 + 4. The second and third circlets are sometimes merged, and the total number of elements in the outer cycle is used in such cases, e.g., 6 + 10, Figure 9.7E. Further departures from the primitive or generalized condition are indicated; they result from the occurrence of subcephalic bristles in close proximity to the cephalic series. Because these structures vary greatly in their development, the formulae should be determined with care.

An introductory key separates the principal groups.

1. Amphids in the form of a pocket with a slitlike, transversely elliptical, or reniform, opening; Fig. 9.7B. Order Enoploidea. These are small to medium-sized nematodes. The anterior end, typically, is narrowed, strongly cuticularized, and frequently has a clearly defined cephalic capsule; the anterior sensory formula is usually 6 + 10.

1. Amphids in the form of a single or multiple spiral, circle, or derivative form such as a crook, horseshoe, or loop, or the amphid is vesicular, i.e., hemispherical with a transparent membrane; Fig. 9.7C,D—2.

2. Amphids vesicular or spiral; cuticle usually strongly ringed, with prominent bristles, and some genera with alternate bands naked and encrusted with extraneous debris; Fig. 9.7R. Order Desmoscolecoidea.
2. Not so—3.

3. Cuticle smooth or weakly annulated, never with a cephalic capsule.
- Amphids usually spiral, anterior sensory organs $6 + 6 + 4$, rarely $6 + 10$ or vestigial. Order Araeolaimoidea.
- Amphids usually circular, anterior sensory organs usually $6 + 10$, sometimes $6 + 6 + 4$, the inner circlet sometimes indistinct, or with 8, 10, or 12 bristles in 1 or 2 circlets. Order Monhysteroidea.

3. Cuticle annulated, often strongly so; cephalic capsule often present; formula $6 + 6 + 4$, $6 + 10$, or irregular:
- Cuticle sculptured with raised dots or other ornamentation. Order Chromadoroidea.
- Cuticle without ornamentation. Order Desmodoroidea.

ORDER ENOPLOIDEA

See Filipjev, 1925, and Allgen (1947b, 1951).

---

1. Buccal cavity largely free, only enclosed basally by the esophagus; Fig. 9.7A:
- Esophagus widening posteriorly, often crenellated or with bulbular expansions. Eurystominidae, 16 genera.
- Esophagus cylindrical, never crenellated. Oncholaimidae, 34 genera. See Kreis (1934).

1. Buccal cavity enclosed by the esophagus; Fig. 9.7A—2.

2. Cephalic cuticle simple; esophagus more or less cylindrical. Ironidae, 12 genera.
2. Cephalic cuticle double; esophagus widening posteriorly—3.

3. Buccal cavity with three massive jaws; Fig. 9.7G. Enoplidae, 25 genera.
3. Not so—4.

4. Posterior part of esophagus enlarged and muscular:
- Amphids indistinct; female genital opening in anus, or within three diameters of the anus; finely annulate. Lauratonematidae, 2 genera.
- Amphids distinct; female genital opening well separated from anus: in Leptosomatidae (22 genera) the anterior sense organs are in 2 circlets, $6 + 10$, and the amphids are pouch-like; in Oxystominidae (17 genera) the anterior sensory formula is $6 + 6 + 4$, and the amphids have a large, longitudinally elongate opening.

4. Posterior part of esophagus enlarged, vesicular, and crenellated; in Phanodermatidae (9 genera) the buccal cavity is weakly developed; in Thoracostomopsidae (one genus), the buccal cavity is armed with a sharp spur or stylet.

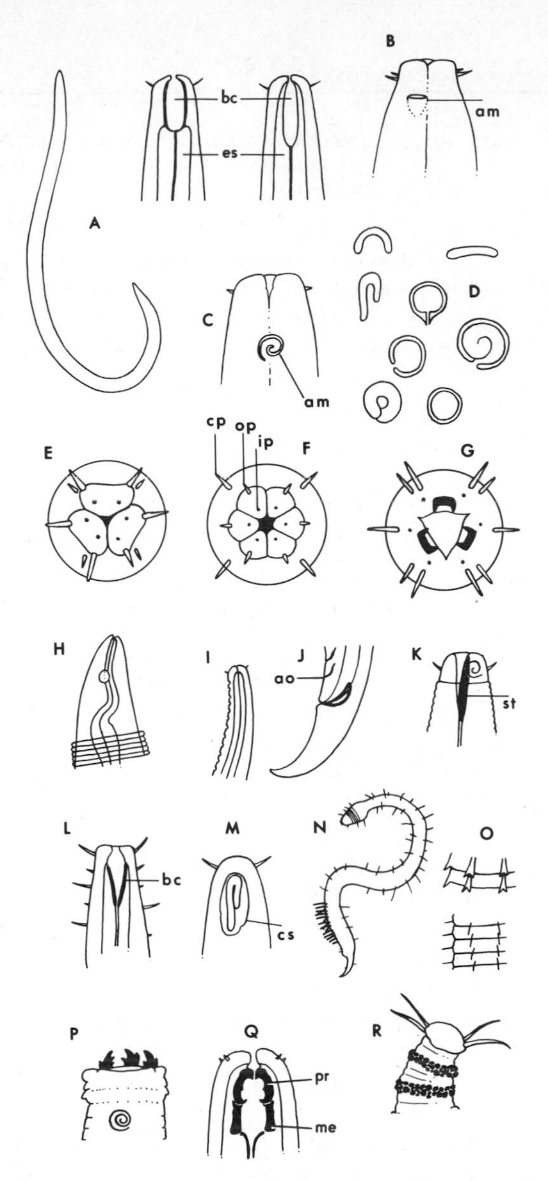

Fig. 9.7. Nematoda, representative structures. A, generalized nematode, whole animal with details at right showing buccal-esophagous relationship, buccal cavity free (left), enclosed (right); B,C, anterior end showing two types of amphids; D, additional variations in amphid form; E–G, variations in anterior end shown schematically as keyed; H, haliplectid, anterior end; I,J, leptolaimid, anterior and posterior ends, respectively; K, camacolaimid, anterior end; L, axonolaimid, anterior end; M, diplopeltid, anterior end; N, draconematid, whole worm; O, monoposthiid (above) and richtersiid (below), details of rings; P, selachinematid, anterior end; Q, choanolaimid, anterior end; R, desmoscolecoid, anterior end. Abbreviations are as follows: am, amphid; ao, accessory male organs; bc, buccal cavity; cp, cephalic circlet; cs, cephalic shield; es, esophagous; ip, inner labial papillae; me, mesostome; op, outer labial papillae; pr, prostome; st, stylet.

ORDER DESMOSCOLECOIDEA

These are short, plump, and rather atypical nematodes with strong annulation. In Greeffielloidea (1 genus) the body is entirely covered with rather long bristles; in the other two families, bristles do not entirely cover the body. Meyliidae (1 genus) has spiral amphids; Desmoscolecidae (6 genera) has vesicular amphids without visible spiral structure, and the body usually has alternating areas of naked cuticle and rings heavily encrusted with foreign debris (Fig. 9.7R). See Chitwood (1936b).

ORDER ARAEOLAIMOIDEA

1. Lips paired, forming 3 double lips; Fig. 9.7E; anterior sensory formula 6 + 10; buccal cavity often distinctly chambered. Tripyloididae, 6 genera.
1. With six lips, Fig. 9.7F—2.
2. Anterior sensory organs vestigial, obscure or apparently absent; Fig. 9.7H. Haliplectidae, 1 genus.
2. Anterior sensory organs clearly present—3.
3. Male with preanal accessory organs (Fig. 9.7J) or with tubular organs in the esophageal region (Fig. 9.7I); cuticle distinctly ringed; buccal cavity usually cylindrical. Leptolaimidae, 14 genera.
3. Not so, male without accessory organs—4.
4. Buccal cavity with a stylet; cuticle distinctly ringed; Fig. 9.7K. Camacolaimidae, 10 genera.
4. Not so; cuticle smooth or feebly annulated—5.
5. Buccal cavity well developed, conical; cuticle smooth; Fig. 9.7L. Axonolaimidae, 9 genera.
5. Buccal cavity weakly developed, or more or less cylincrical—6.
6. Amphids crook-shaped, strongly developed and mounted on a cuticular shield; Fig. 9.7M. Diplopeltidae, 8 genera.
6. Amphids spiral or, if crook-shaped, not mounted on a cuticular shield. Araeolaimidae, 20 genera.

ORDER MONHYSTEROIDEA

1. Esophagus usually enlarged basally, often with a terminal bulb; amphids usually a simple spiral, sometimes circular; anterior sensory organs 6 + 6 + 4, but the inner circlet indistinct, or with 8, 10, or 12 bristles in 1 or 2 circlets. Linhomoeidae, 36 genera.
1. Amphids usually circular—2.
2. Esophagus cylindrical, only slightly enlarged basally, never with a terminal bulb; anterior sensory organs 6 + 10:
- Buccal cavity large, globular, strong-walled and with numerous fine denticles. Sphaerolaimidae, 1 genus.
- Buccal cavity not so, varied in form. Monhysteridae, 30 genera.

2. Esophagus with a terminal bulb; anterior sensory organs include 4–10 bristles. Siphonolaimidae, 4 or 5 genera.

ORDER CHROMODOROIDEA

1. Anterior sensory organs 6 + 6 + 4; amphids spiral, reniform, or indistinct—2.
1. Anterior sensory organs 6 + 10; amphids spiral—3.
2. Rings smooth, punctation internal; amphids a multiple spiral; buccal cavity little developed, sometimes toothed, rarely with a stylet. Comesomatidae, 14 genera.
2. Rings with raised points or other ornamentation; amphids usually indistinct, spiral, or reniform. Chromadoridae, 34 genera.
3. Buccal cavity not very deep, with or without teeth. Cyatholaimidea, 27 genera.
3. Buccal cavity complex—4.
4. Buccal cavity with 2 or 3 jaws, amphids spiral; Fig. 9.7P. Selachinematidae, 4 genera.
4. Buccal cavity without jaws but with well-developed prostome and mesostome; Fig. 9.7Q. Choanolaimidae, 11 genera.

ORDER DESMODOROIDEA

1. With preanal ambulatory bristles in 2–4 tiers; amphids spiral or crooked, horseshoe-shaped; anterior sensory organs in 3 circlets but the inner circlet vestigial, the second circlet often inconspicuous, the outer circlet with 4 bristles; body ringed and bristled:
- With a distinct head and sharp-pointed tail; Fig. 9.7N. Draconematidae, 4 genera.
- Without a distinct head, body sigmoid. Epsilonematidae, 8 genera.

1. Not so; without ambulatory bristles; anterior sensory organs 6 + 6 + 4 or 6 + 10—2.
2. Body with longitudinal ridges—3.
2. Body without longitudinal ridges—4.
3. Cuticle of head reinforced, forming a cephalic carapace; amphids spiral or crooked:
- Rings numerous, >400. Dasynemellidae, 3 genera.
- Rings <300. Ceramonematidae, 4 genera.

3. Cephalic carapace lacking, but head may be clearly delimited; amphids circular or a simple or multiple spiral:
- Rings distinct, ridges with V-shaped spines; Fig. 9.7O. Monoposthiidae, 4 genera.
- Rings less distinct, ridges with short bristles; Fig. 9.7O. Richtersiidae, 2 genera.

4. With a cephalic carapace. Desomodoridae, 17 genera.
4. Cephalic carapace lacking but head may be clearly delimited:
- Anterior sensory organs in 3 circlets but the inner circlet vestigial. Spirinidae, 22 genera.
- Anterior sensory organs 6 + 6 + 4. Microlaimidae, 6 genera.

References

In lieu of a more extensive treatment of the class itself, its literature is presented rather more fully than usual. See also the bibliography of Timm (1954a) below.

Allgen, C. A. 1947a. Die Nematoden-Familie Tripyloididae. Ark. Zool. 39A(15):1–35.

———. 1947b. Papers from Dr. Th. Mortensen's Pacific Expedition 1914–1916 LXXV. West American Marine Nematodes. Vidensk. Meddr. dansk. naturh. Foren. 1947. 110:65–219.

———. 1951. Papers from Dr. Th. Mortensen's Pacific Expedition, 1914–1916. LXXVI. Pacific Freeliving Marine Nematodes. Vidensk. Meddr. dansk. naturh. Foren. 1951. 113:263–411.

Chitwood, B. G. 1936a. Some marine nematodes of the Superfamily Enoploidea. Trans. Am. microsc. Soc. 55:208–213.

———. 1936b. Some marine nematodes from North Carolina. Proc. helminth. Soc. Wash. 3(1):1–16.

———. 1937. A new genus and ten new species of marine nematodes from North Carolina. Proc. helminth. Soc. Wash. 4(2):54–59.

———. 1951. North American marine Nematodes. Tex. J. Sci. 3(4):617–672.

———. 1960. A preliminary contribution on the marine nemas (Adenophorea) of northern California. Trans. Am. microsc. Soc. 79(4):347–384.

———, and M. W. Allen. 1959. Nemata. *In* Freshwater biology. W. T. Edmondson, ed., John Wiley and Sons.

———, and D. G. Murphy. 1964. Observations on two marine monhysterids—their classification, cultivation and behavior. Trans. Am. microsc. Soc. 83(3):311–329.

———, and R. W. Timm, 1954. Free-living nematodes of the Gulf of Mexico. *In* Gulf of Mexico, its Origin, Waters, and Marine Life. Fish, Bull. Fish Wildl. Serv. U.S. 55(84):313–323.

Cobb, N. A. 1917. Nema population of beach sand. Contr. Sci. Nemat. 5.

———. 1920. One hundred new Nemas. Waverly Press, Baltimore, Pp. 217–343.

———. 1929. Initial stratification survey of nemas in the upper 20 mm of marine beach sand, near low tidemark. J. Wash. Acad. Sci. 19:199–200.

deConinck, L. 1965. Nemathelminthes (Nematodes). Traité de Zoologie, Anatomie, Systematique, Biologie. P-P. Grassé, ed.

Ditlevsen, H. 1912. Danish free-living nematodes. Vidensk. Meddr. naturh. Foren.: 63, 70.

Filipjev, I. N. 1925. Les nématodes libres des mers septentrionales appartenant à la famille des Enoplidae. Arch. Naturgesch. 91A, Hefte 6.

Hope, W. D. 1967a. A review of the genus *Pseudocella* Filipjev, 1927 (Nematoda: Leptosomatidae) with a description of *Pseudocella triaulolaimus* n. sp. Proc. helminth. Soc. Wash. 34(1):6–12.

———. 1967b. Free-living marine nematodes of the genera *Pseudocella*, Filipjev, 1927, *Thoracostoma* Marion, 1870, and *Deontostoma* Filipjev, 1916 (Nematoda: Leptosomatidae), from the west coast of North America. Trans. Am. microsc. Soc. 86(3):307–334.

Hopper, B. E. 1961a. Marine Nematodes from the coast line of the Gulf of Mexico. Can. J. Zool. 39:183–199.

———. 1961b. Marine Nematodes from the Coast Line of the Gulf of Mexico, II. Can. J. Zool. 39:359–365.

———. 1962. Free-living marine nematodes of Rhode Island waters. Can. J. Zool. 40:41–52.

———. 1963. Marine nematodes from the coast line of the Gulf of Mexico. III. Additional species from Gulf Shores of Alabama. Can. J. Zool. 41:841–863.
———. 1966. *Theristus polychaetophilus* n. sp. (Nematoda), an external parasite of the spionid polychaete *Scolelepis squamata* (Müller, 1806). Can. J. Zool. 44:787–791.
———. 1967a. Free-living marine nematodes from Biscayne Bay, Florida, I. Comesomatidae: the male of *Laimella longicauda* Cobb 1920, and description of *Actarjania* new genus. Marine Biology 1(2):140–144.
———. 1967b. Free-living marine nematodes from Biscayne Bay, Florida, II. Oncholaimidae: descriptions of five new species and one new genus (*Meyersia*). Marine Biology 1(2):145–151.
———, and S. P. Meyers. 1966. Observations on the Bionomics of the marine nematode. Nature, Lond. 209(5026): 899–900.
———. 1966. Aspects of the life cycle of marine nematodes. Helgoländer wiss. Meersunters. 13:444–449.
———. 1967a. Foliicolous marine nematodes on turtle grass, *Thalassia testudinum* König, in Biscayne Bay, Florida. Bull. mar. sci. Gulf Caribb. 17(2):471–517.
———. 1967b. Population studies on benthic nematodes within a subtropical seagrass community. Marine Biology, 1(2):85–96.
Hyman, Libbie H. 1951. The Invertebrates: Acanthocephala, Aschelminthes, and Entoprocta, the pseudocoelomate Bilateria. Vol. III. McGraw-Hill Book Co.
Kreis, A. 1934. Oncholaiminae Filipjev 1916. Eine monographische Studie. Capita zool. 4(5):1–271.
Murphy, D. G. 1962. Three undescribed nematodes from the coast of Oregon. Limnol. Oceanogr. 7(3): 386–389.
———. 1963a. A new genus and two new species of nematodes from Newport, Oregon. Proc. helminth. Soc. Wash. 30(1):73–78.
———. 1963b. Three new species of marine nematodes from the Pacific near Depot Bay, Oregon. Proc. helminth. Soc. Wash. 30(2):249–256.
———. 1964a. The marine nematode genus *Pseudonchus* Cobb, 1920, with descriptions of *Cheilopseudonchus*, n. g. and *Pseudonchus kosswigi*, n. sp. Mitt. hamb. zool Mus. Inst. S:113–118.
———. 1964b. *Pseudonchus jenseni*, n. sp. and *Comesoma arenae* Gerlach, 1956, marine nematodes from NSW, Australia. Zool. Anz. Bd. 173, Heft 6.
———. 1964c. Free-living marine nematodes, I. *Southerniella youngi*, *Dagda phinneyi*, and *Gammanema smithi*, new species. Proc. helminth. Soc. Wash. 31(2):190–198.
———. 1964d. *Rhynconema subsetosa*, a new species of marine nematode, with a note on the genus *Phylolaimus* Murphy, 1963. Proc. helminth. Soc. Wash., 31(1):26–28.
———. 1965a. The marine nematode genus *Nygmatonchus* Cobb, 1933, rediscovered, with the description of *N. alii*, new species. Veröff. Inst. Meeresforsch. Bremerh. Band IX: S204–209.
———. 1965b. Chilean marine nematodes. Veröff. Inst. Meeresforsch. Bremerh. Band IX. S:173–203.
———. 1965c. *Cynura klunderi* (Leptolaimidae), a new species of marine nematode. Zool. Anz. Bd. 175, Heft 2/3.
———. 1965d. *Praethecacineta oregonensis*, a new species of suctorian associated with a marine nematode. Zool. Anz., Bd. 174, Heft 4/5.
———. 1965e. *Thoracostoma washingtonensis*, n. sp., eine Meeresnematode aus dem pazifischen Küstenbereich vor Washington. Abh. Verh. naturw. Ver. Hamburg, N. F. Bd. IV.
———. 1966. An initial report on a collection of Chilean marine nematodes. Mitt. hamb. zool. Mus. Inst. Band 63. S:29–50.

———, and A. G. Canaris. 1964. *Theristus pratti* n. sp., a marine nematode from Kenya. Proc. helminth. Soc. Wash. 31(2): 203–208.

Murphy, D. G., and H. J. Jensen. 1961. *Lauratonema obtusicaudatum*, n. sp. (Nemata: Enoploidea), a marine nematode from the coast of Oregon. Proc. helminth. Soc. Wash. 28(2): 167–169.

Tietjen, John H. 1967. Observations on the ecology of the marine nematode *Monhystera filicaudata* Allgen, 1929. Trans. Am. microsc. Soc. 86(3):304–306.

Timm, R. W. 1951a. A note on the cell inclusions of *Syringolaimus smarigdus* Cobb, 1928. Proc. helminth. Soc. Wash. 18(2):125–126.

———. 1951b. A new species of marine nematode *Thoracostoma magnificum* with a note on possible "pigment cell" nuclei of the ocelli. J. Wash. Acad. Sci. 41(10):331–333.

———. 1953. Observations on the morphology and histological anatomy of a marine nematode, *Leptosomatum acephalatum* Chitwood, 1963, new combination (Enoplidae: Leptosomatinae). American Midl. Nat. 49(1):229–248.

———. 1954a. A survey of the marine nematodes of Chesapeake Bay, Maryland. Biol. Stud. Cath. Univ. Am. 23:1–70.

———. 1954b. An abnormality of *Oncholaimus marinus* (Nematoda: Oncholaiminae). Proc. helminth. Soc. Wash. 21(1):36.

Wieser, W. 1959a. Free-living nematodes and other small invertebrates of Puget Sound beaches. Univ. Wash. Publ. Biol. 19:1–179.

———. 1959b. The effect of grain size on the distribution of small invertebrates inhabiting the beaches of Puget Sound. Limnol. Oceanogr. 4(2):181–194.

———. 1959c. A note on subterranean nematodes from Chesapeake Bay, Maryland. Limnol. Oceanogr. 4(3):225–227.

———. 1960. Benthic studies in Buzzards Bay II. The Meiofauna. Limnol. Oceanogr. 5(2):121–137.

———. 1967. Marine nematodes of the east coast of North America. I. Florida. Bull. Mus. comp. Zool. Harv. 135(5):239–344.

———. and B. E. Hopper. 1966. The Neotonchinae, new subfamily Cyatholaimidae: Nematoda), with an analysis of its genera, *Neotonchus* Cobb, 1933, and *Gomphionema* new genus. Can. J. Zool. 44:519–532.

TEN

Phylum Entoprocta

ENTOPROCT BRYOZOANS

Arguments for the separation of the entoproct and ectoproct divisions of the long-established phylum Bryozoa or Polyzoa were reviewed by Hyman (1951). The present group sometimes appears as "Endoprocta."

Diagnosis. *Minute, solitary, or colonial stolonate animals with a calyxlike body mounted on a stalk; the calyx with a circlet of tentacles; with a looped digestive tract with mouth and anal openings within the circlet of tentacles.*

The largest local species may be 10 mm tall. Though considerably more complex in internal structure, entoprocts are superficially similar to athecate, hydrozoan polyps. The base is a simple adhesive disc in solitary forms, or it may be cemented to a substratum. In colonial species the individual animals arise from a stolon or by branching from a common stem; the stolon is usually partitioned by septa.

The sexes are apparently separate in local species of *Barentsia*, but members of other genera, e.g., *Pedicellina*, are thought to be hermaphrodites or protandric hermaphrodites. There are no conspicuous differences between the sexes. Fertilization in some species at least is internal, and development proceeds initially within the mother's body and then in a free-living larval stage. The larvae have been regarded as a trochophore type, though this resemblance is disputed by Hyman (1951). In *Pedicellina* and *Barentsia* the larvae have an apical tuft of cilia plus a similar preoral tuft and a girdle of cilia which is ventral rather than equatorial as in typical trochophores. The larvae of loxosomatids superficially resemble rotifers. Entoproct larvae are at first motile and either benthic or pelagic. They subsequently attach by the ventral ciliary rim and metamorphose into the adult form.

Asexual reproduction by budding occurs in the solitary loxosomatids as well as in the colonial genera. There is evidence that the calyxes of some entoprocts—and certainly those of freshwater forms—degenerate under adverse conditions, but the stalks remain viable and regenerate calyxes with the return of a favorable environment.

Adults of most species are sessile, but those of *Loxosomella davenporti* remain motile, moving in the manner of hydras by attaching to the substratum with their tentacles and somersaulting the pedal disc to a new location. Colonial forms are fixed for life. Entoprocts are ciliary feeders subsisting on a variety of plankters and detritus and entrapping larger particles with their tentacles. Being minute animals, they are generally of minor ecological importance except in the case of epizooic species which occasionally are numerous enough to affect the well being of individual hosts.

Systematic List and Distribution

The list is drawn from the references cited. The distribution of these minute and generally inconspicuous animals is not well known.

Family Loxosomatidae		
Loxosomella davenporti	B+	Shallow water
Loxosomella minuta	B	Shallow water
Family Pedicellinidae		
Pedicellina cernua	BV	Shallow water, eury. 15‰
Barentsia major	V	5–24 m
Barentsia discreta	BV	Shallow water
Barentsia laxa	V	Shallow water
Barentsia gracilis	BV	Shallow water

Identification

The following key is modified from Rogick (1964). Dimensions given equal the height of adult zooids. Most of the local species are figured by Osborn (1912) or Rogick (1948, 1964).

1. Individuals solitary, not stolonate; epizooic—2.
1. Colonial and stolonate; epizooic or not—3.

2. Adults cemented to a substratum, fixed for life; 0.5 mm. *Loxosomella minuta.*
2. Adults with pedal disc, motile; 2.0 mm; Fig. 10.1A,D. *Loxosomella davenporti.*

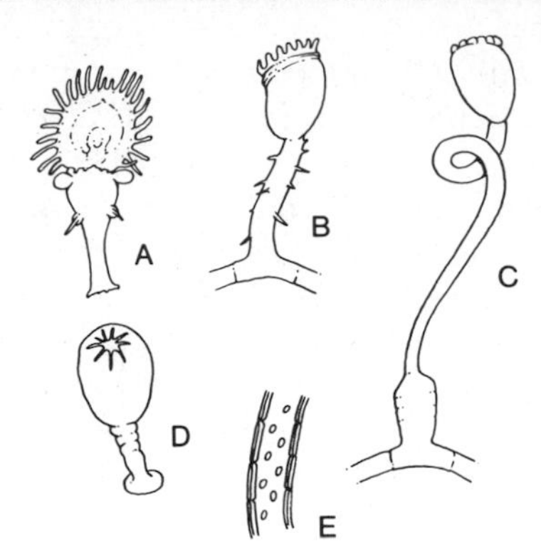

Fig. 10.1. Entoprocta. A, *Loxosomella davenporti*, expanded zooid; B, *Pedicellina cernua*; C, *Barentsia* sp.; D, *Loxosomella davenporti*, contracted zooid; E, *Barentsia gracilis*, detail of stalk showing perforations.

3. Stalks with a distinct basal dilation; spineless; *Barentsia* spp., Fig. 10.1C—4.
3. Stalks not so, tapering; usually with minute spines on calyx and/or stalk; 5.0 mm; Fig. 10.1B. *Pedicellina cernua.*
4. Stalk thick-walled, stiff; about 10.0 mm. *B. major.*
4. Stalk slender—5.
5. Stalk thin-walled, flexible, and motile; calyx small; 6.5 mm. *B. laxa.*
5. Stalk stiff, inflexible—6.
6. Stalk with numerous perforations; calyx small; 2 mm; Fig. 10.1E. *B. gracilis.*
6. Stalk with few or no perforations; calyx large; 3 mm. *B. discreta.*

References

Bobin, Geneviève, and M. Prenant. 1953. La classification des loxosomes selon Mortensen et le *Loxosome singulare* de Keferstein et de Claparède. Bull. Soc. zool. Fr. 8:84–96.

Brien, Paul. 1959. Classe des Endoproctes ou Kamptozoaires. *In* Traité de Zoologie, Anatomie—Systématique Biologie, P-P. Grassé, ed. Masson et Cie.

Hyman, Libbie H. 1951. The Invertebrates: Acanthocephala, Aschelminthes, and Entoprocta, the Pseudocoelomate Bilateria. Vol. 3. McGraw-Hill Book Co.

Nickerson, W. S. 1901. On *Loxosoma davenporti* sp. nov., an endoproct from the New England coast. J. Morph. 17(3):351–380.

Nielson, C. 1966. Some Loxosomatidae (Entoprocta) from the Atlantic Coast of the United States. Ophelia. 3:249–275.

Osburn, R. C. 1912. The Bryozoa of the Woods Hole region. Bull. U.S. Bur. Fish. (for 1910). 30:203–266.

———. 1933. The Bryozoa of the Mt. Desert region. *In* Biol. Survey of the Mt. Desert Region, by W. Proctor Wistar Institute.

———. 1944. A survey of the Bryozoa of Chesapeake Bay. Chesapeake Biol. Lab. Publ., 63:1–59.

Prenant, M. and Geneviève Bobin. 1956. Faune de France. Bryozoaires. 60:1–398. Paul Lechevalier, Paris.

Rogick, Mary. 1948. Studies on marine Bryozoa II, *Barentsia laxa.* Biol. Bull. Mar. Biol. Lab. Woods Hole. 94:128–142.

———. 1964. Phylum Entoprocta. *In* Keys to Marine Invertebrates of the Woods Hole Region. R. Smith, ed. Systematics-Ecology Program, Marine Biological Laboratory, Woods Hole, Mass.

Soule, Dorothy F., and J. D. Soule. 1965. Two new species of *Loxosomella,* Entoprocta, epizoic on Crustacea. Allan Hancock Fdn. Publ. Occasional Pap. 29:1–19.

ELEVEN

Phylum Tardigrada

WATER BEARS

The systematic position of the tardigrades presents some difficulty, and they have at times been allied with either annelids or arthropods. Current opinion, reviewed by Riggin (1962), favors elevation of the group to phylum rank.

Diagnosis. *Minute <1.0 mm; wormlike with head and body only vaguely delimited; segmentation obscure or indicated by dorsal cuticular plates; with four pairs of short, in some cases retractile, legs each with four to eleven claws or pads developed on toes or directly from the ends of the limbs.*

Some species have eye spots, and tactile bristles may be present on the head. The body may also have bristles, spines, thickened cuticular plates, or other ornamentation. Tardigrades may be transparent or opaque, colored or not.

The sexes are separate, but males are often scarce in natural populations. Development of the young is direct without special larval stages. The assumption of a dessicated state with arrested metabolic activity in times of environmental stress is well known in tardigrades inhabiting mosses but not in truly aquatic species; this is termed *cryptobiosis* (or, by some authors, *anabiosis*). The littoral *Echniscoides sigismundi* apparently is capable of somewhat similar though "poorly understood dessications" according to Crowe and Higgins (1967).

Mouth parts are fitted for piercing and sucking, and feeding habits vary in different species from herbivorous to carnivorous. Marine forms reputedly parasitic on echinoderms are known in the European fauna. *Pleocola limnoriae* (not yet recorded from this coast) has been cited as a commensal on the wood-boring isopod *Limnoria lignorum* which does occur here. Similarly, *Echniscoides sigismundi* is found here on littoral algae, particularly *Enteromorpha*, but has been reported in the mantle cavity of the edible mussel, *Mytilus edulus*, and on the littoral barnacles *Balanus balanoides* and *Chthamalus stellatus* in European waters.

Except as indicated above, marine tardigrades are particularly characteristic of the interstitial, i.e., psammo-littoral, fauna. *Batillipes mirus* has been found in concentrations of 300 individuals per 100 cubic centimeters in this habitat.

Systematic List and Distribution

Tardigrades have been neglected in this country. Mathews (1938) listed 32 North American species with only two marine forms. A more recent review by McGinty and Higgins (1968) lists a half-dozen species. Some 28 species in 14 genera constitute the total, worldwide marine fauna. There are strong tendencies to cosmopolitanism, and increasing interest in the psammo-littoral fauna will doubtless add to the American list. The local species are:

Order Heterotardigrada
  Family Batillipedidae
    Batillipes mirus
    Batillipes pennaki
    Batillipes bullacaudatus
  Family Halechiniscidae
    Halechiniscus remanei
  Family Stygarctidae
    Stygarctus bradypus
  Family Echniscidae
    Echniscoides sigismundi

These species have been recorded in the Virginian or adjacent Carolinian provinces; additional species whose occurrence may be anticipated include *Bathyechiniscus tetronyx*, *Batillipes fraufi*, and *Styraconyx sargassi*.

Identification

See Higgins (1960) for special methods of preparing material for study; standard formalin or alcohol preservation is recommended as an initial step. Measurements in the following key are total length of adult individuals.

1. With 4 elongate claws and 2 elongate setae, legs without toes, i.e., claws "sessile"; body armored above with thickened cuticular plates; 0.15 mm; Fig. 11.1B,F. *Stygarctus bradypus*.
1. Not so—2.

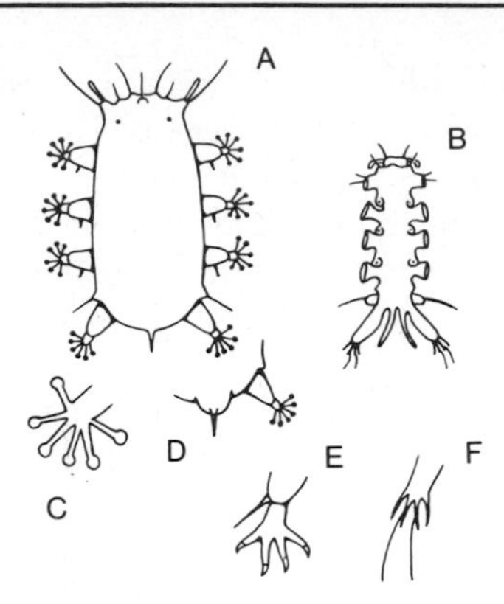

Fig. 11.1. Tardigrada. A, *Batillipes mirus*, whole animal; B, *Stygarctus bradypus*, whole animal with cuticular plates omitted; C, *Batillipes*, detail of toes; D, *Batillipes pennaki*, detail of posterior for comparison with A; E, *Halechiniscus remanei*, detail of foot; F, *Stygarctus bradypus*, detail of foot.

2. Legs with toes with distal pads; *Batillipes: B. bullacaudatus*, 0.13 mm, has the caudal spine ending in a bulb; in *B. mirus* and *B. pennaki* there is no bulb and the species are distinguished as figured; *mirus* reaches 0.72 mm, *pennaki*, 0.20 mm; Fig. 11.1 A,C,D.

2. Not so—3.

3. With 5–11 sessile claws; 0.34 mm *Echniscoides sigismundi*.

3. Legs with 4 toes, a single claw on each; 0.12 mm; Fig. 11.1E. *Halechiniscus remanei*.

References

Crowe, J. H., and R. P. Higgins. 1967. The revival of *Macrobiotus areolatus* Murray (Tardigrada) from the cryptobiotic state. Trans. Am. microsc. Soc. 86(3):286–294.

Higgins, R. P. 1960. Some tardigrades from the Piedmont of North Carolina. J. Elisha Mitchell scient. Soc. 76(1):29–35.

McGinty, M. Maxine, and R. P. Higgins. 1968. Ontogenetic variation of taxonomic characters of two marine tardigrades with the description of *Batillipes bullacaudatus* n. sp. Trans. Am. microsc. Soc. 87 (2): 252–262.

Mathews, G. B. 1938. Tardigrada from North America. Am. Midl. Nat., 19:619–627.

Pennak, R. W. 1953. Fresh-water Invertebrates of the United States. Ronald Press.

Ramazzotti, G. 1962. Il phylum tardigrada. Memorie Ist. ital. Idrobiol. 14:1–595.

———. 1965. Il phylum tardigrada (1° supplement). Memorie Ist. ital. Idrobiol. 19:101–212.

Riggin, G. T. 1962. Tardigrada of southwest Virginia: with the addition of a description of a new marine species from Florida. Tech. Bull. Va. agric, expl. Stn. 152.

Schuster, R. O., and A. A. Grigarick. 1965. Tardigrada from western North America with emphasis on the fauna of California. Univ. Calif. Publ. Zool. 76:1–67.

TWELVE

Phylum Chaetognatha

ARROW WORMS

Diagnosis. *Chaetognaths are transparent, torpedo-shaped animals with one or two pairs of lateral fins and a fanlike tail in the same plane; head with four to fourteen strong, curved spines or hooks on each side, a pair of minute, darkly pigmented eyes, and a retractable hood. With one exception, local species are 20 mm or less in length.*

The fifty or so species of chaetognaths constitute a distinctive phylum with obscure affinities to other coelomate groups.

Chaetognaths are protandric hermaphrodites; the sperm ripen first, initially filling the coelom in the tail and then concentrating in the seminal vesicles, where spermatophores are formed. The spermatophores escape, apparently by rupture of the body wall, become attached to the fins, and release their contents for self-fertilization. The fertilized eggs of the common genus, *Sagitta*, are freely shed in the water where they float near the surface for several days until development to hatching is complete. The individual egg is about 0.2 mm in diameter and the hatchling about 1 mm long. Development is essentially direct from a simplified but clearly chaetognath form.

In local waters, chaetognaths are exclusively planktonic. They are important predators, feeding primarily on copepods and occasionally on fish larvae and other plankters including other chaetognaths. Deevey (1960) recorded densities in excess of 32,000 individuals per ten minute plankton tow in Delaware Bay, and Cronin *et al.* (1962) estimated 12 individuals of *elegans* per cubic meter in the same bay. While *elegans*, *tenuis*, and *enflata* enter estuaries, chaetognaths generally are more common in relatively undiluted seawaters; *elegans* was recorded by Cronin *et al.* (1962) in salinities as low as 11.4‰.

Systematic List and Distribution

The taxonomy of chaetognaths has been confused through lack of knowledge of individual and ontogenetic variation. The majority of species are in a single genus, *Sagitta*, and no suprageneric categories have been defined.

Like many small zooplankters, chaetognaths have seasonal peaks of abundance, and there is evidence of vertical migration in some species. Many of the species are essentially cosmopolitan. Those marked (*) occur in coastal waters as Gulf Stream strays; those marked (†) are not keyed.

Sagitta bipunctata	Epiplankton*	—	—
Sagitta elegans	Epiplankton	Northern	Winter-spring
Sagitta enflata	Epiplankton	Southern	Winter-spring
Sagitta helenae	Epiplankton	Southern	Delaware Bay south
Sagitta hexaptera	Epi-mesoplankton	Oceanic	—
Sagitta minima	Epi-mesoplankton	Southern	Continental Slope
Sagitta serratodentata	Epiplankton*	—	—
Sagitta tenuis	Epiplankton	Southern	To Delaware Bay
Sagitta hispida	Epiplankton	Southern	To Long Island
Sagitta lyra†	Mesoplankton	Oceanic	—
Sagitta maxima†	Mesoplankton	Oceanic	—
Sagitta tasmanica	Epiplankton	Northern	To Virginia
Pterosagitta draco	Lower epiplankton	Southern	Long Island south, fall-winter*
Krohnitta subtilis	Epi-mesoplankton	Oceanic	—
Krohnitta pacifica	Epi-mesoplankton	Oceanic*	—
Eukrohnia hamata	Epiplankton in Arctic, meso-bathyplankton in lower latitudes		

S. elegans and *S. enflata* have been reported in estuarine waters and are common species on northern and southern parts of the Continental Shelf, respectively; *S. tasmanica* is reported to be the prevalent neritic species from Virginia north; *tenuis*, *minima*, and *hispida* may also be common at times. See Hyman (1959) for a general review of chaetognath zoogeography.

Identification

Species of *Sagitta* have 2 pairs of lateral fins; other genera have a single pair of fins and are further distinguished as follows: in *Pterosagitta draco* the lateral skin or *collarette* of the neck region extends along the sides to merge with the lateral fins, and this species has a tuft of sensory bristles at each side (Fig.

12.1A). Collarette and bristles are lacking in *Krohnitta* and *Eukrohnia;* in the first genus the lateral fins are short, in the second, long (Fig. 12.1B,C). *Pterosagitta*, *Krohnitta*, and *Eukrohnia* are figured by Hyman (1959). The species of *Sagitta* are distinguished in the following key. Dimensions are total lengths of adults. The key is based largely on Grant's data from Virginia specimens and incorporates part of his key (Grant, 1963b) with modifications from Grant (personal communication). Names and dates in parentheses in the key are sources of figures.

A. Tail segment equals 16–24% of the total length including tail fin; collarette small or absent; fins only partly rayed in some species; seminal vesicles round or conical:

1. Fins fully rayed; 20 mm; Fig. 12.1D (Fraser 1957) *elegans.*

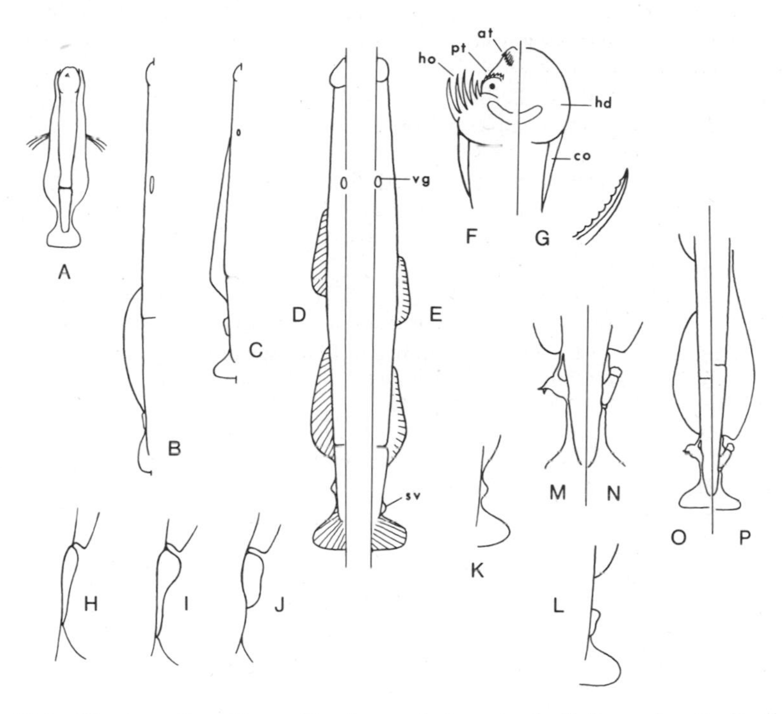

Fig. 12.1. Chaetognatha. *Pterosagitta draco*, whole animal; B–E are longitudinally bisected; B, *Krohnitta* sp.; C, *Eukrohnia* sp.; remaining figures are *Sagitta* spp.; D, *S. elegans*; E, *S. enflata*; F, head with hood retracted at left, extended at right; G, detail of dentate hook; H–N show the outline of the seminal vesicle and adjacent parts on one side of the body; H, *S. tenuis*; I, *S. helenae*; J, *S. hispida*; K, *S. hexaptera*; L, *S. bipunctata*; M, *S. serrodentata*; N, *S. tasmanica*; O, *S. serrodentata*, posterior body left; P, *S. tasmanica*, posterior body right. Abbreviations are as follows: at, anterior teeth; co, collarette; hd, hood; ho, hooks; pt, posterior teeth; sv, seminal vesicle; vg, ventral ganglion.

1. Posterior fins rayed, anterior fins without rays or but partly rayed; 8 mm. *minima.*
1. Fins partly rayed—2.
2. Anterior fins elongate, tapered forward, originating behind rear margin of ventral ganglion a distance equal to 2–4 times the length of the ganglion; seminal vesicles close to tail fin; 20 mm; Fig. 12.1E (Fraser, 1957). *enflata.*
2. Anterior fins short, semicircular, more posteriorly placed; seminal vesicles well separated from tail fins; 71 mm; Fig. 12.1K (Russell, 1939). *hexaptera.*

B. Tail segment equals 24–33% of total length including tail fin; collarette small; fins partly rayed; hooks dentate on inner margin in specimens greater than 5 mm (magnification 100 X); 16 mm; Fig. 12.1G:

1. Mature seminal vesicles urn-shaped with two fleshy papillae; posterior fin wavy in outline and always separated from anterior fin; Fig. 12.1M,O (Fraser, 1957). *serratodentata.*
1. Mature seminal vesicles oblong-ovate to wedge-shaped with numerous epidermal proliferations on tip; posterior fin smoothly convex in outline and often joined to anterior fin in young individuals; Fig. 12.1N,P (Grant, personal communication). *tasmanica.*

C. Tail segment equals 26–33% of total length including tail fin; collarette small or well developed; fins completely rayed; hooks not dentate; seminal vesicles wedge-shaped:

1. Seminal vesicles widely separated from posterior fins; collarette reaches seminal vesicles; 15 mm; Fig. 12.1L. (Fraser, 1957). *bipunctata.*
1. Seminal vesicles close to posterior fins; collarette shorter—2.
2. Collarette extending ⅓ distance from head to ventral ganglion; 11 mm; Fig. 12.1H (Tokioka, 1940). *tenuis.*
2. Collarette extending ⅓ to ½ or more from head to ventral ganglion—3.
3. Anterior teeth broad, overlapping, 8–12 in specimens greater than 8 mm; collarette extending ⅓ to ½ distance from head to ventral ganglion; 14 mm; Fig. 12.1I (Tokioka, 1940). *helenae.*
3. Anterior teeth narrower, 5–8 in specimens greater than 8 mm; collarette extending at least ½ distance from head to ventral ganglion; 15 mm; Fig. 12.1J (Tokioka, 1940). *hispida.*

References

Alvarino, A. 1965. Chaetognaths. Oceanog. Mar. Biol. Ann. Rev. 3:115–194.

Bigelow, H. B. 1915. Explorations of the coast water between Nova Scotia and Chesapeake Bay, July and August, 1913, by the U.S. Fisheries schooner *Grampus*. Oceanography and plankton. Bull. Mus. comp. Zool. Harv. 59(4):151–359.

———, and Mary Sears. 1939. Studies of the waters of the Continental Shelf, Cape Cod to Chesapeake Bay. III. A volumetric study of the zooplankton. Mem. Mus. comp. Zool. 54:183–378.

Cronin, L. E., Joanne C. Daiber, and E. M. Hulbert. 1962. Quantitative seasonal aspects of zooplankton in the Delaware River estuary. Chesapeake Sci. 3(2):63–93.

Deevey, Georgiana B. 1960. The zooplankton of the surface waters of the Delaware Bay region. Bull. Bingham oceanogr. Coll. 17(2):5–53.

Fraser, J. H. 1957. Chaetognatha. *In* Fiches d'ident. du zooplankton. Sheet 1, 6 pp. Conseil International pour l'exploration de la mer.

Grant, G. C. 1963a. Chaetognatha from inshore coastal waters off Delaware, and a northward extension of the known range of *Sagitta tenuis*. Chesapeake Sci. 4(1):38–42.

———. 1963b. Investigations of inner Continental Shelf waters off lower Chesapeake Bay. Part IV. Descriptions of the Chaetognatha and a key to their identification. Chesapeake Sci. 4(3):107–119.

———. 1967. The geographic distribution and taxonomic variation of *Sagitta serratodentata* Krohn 1953 and *Sagitta tasmanica* Thomson 1947 in the North Atlantic Ocean. Unpubl. Ph.D. dissertation, University of Rhode Island, Kingston, R.I. 116 pp.

Hyman, Libbie H. 1959. The Invertebrates: Smaller Coelomate Groups. Vol. 5. McGraw-Hill Book Co.

Pierce, E. L. 1958. The Chaetognatha of the inshore waters of North Carolina. Limnol. Oceanogr. 3:166–170.

Russell, F. S. 1939. Hydrographical and biological conditions in the North Sea as indicated by plankton organisms. J. Cons. Copenhagen. 14:171–192.

Tokioka, T. 1940. A small collection of chaetognaths from the coast of New South Wales. Rec. Aust. Mus. 20(6):367–379.

THIRTEEN

Phylum Bryozoa (Ectoprocta)

ECTOPROCT BRYOZOANS

The majority of recent texts place entoproct and ectoproct bryozoans in separate phyla. Hyman (1951) has reviewed the evidence relating to this question. The majority of recent workers appear to prefer Bryozoa to Ectoprocta as the phylum name.

Diagnosis. *Minute, colonial animals forming encrusting, massive, stolonate, or bushy colonies usually permanently fixed to a substratum; with a circular, tentacle-bearing lophophore; digestive tract more or less U-shaped with mouth and anal openings close together but with only the mouth inside the tentacular circle; individual zooids encased in a chitinous shell with or without an additional calcareous exoskeleton.*

The largest colonies of local species are measured in hundreds of millimeters; individual zooids are essentially microscopic, usually less than 0.5 mm long. The entire colony of some stolonate genera is minute. Nonstolonate, erect forms are anchored by a holdfast or by rhizoidlike "rootlets" that sometimes bear clawlike appendages; such species assume a wide variety of forms ranging from sparsely arborescent to densely bushy, fanlike or lobate colonies. They may be mistaken for algae or for hydroids or other "zoophytes."

Colonies of ectoprocts consist of an aggregation of individual *zooids*, the term including both the live animal and its nonliving skeleton or *zooecium*. The zooids originate by budding and remain fused and joined together either directly or by subsidiary growths. In a few species, e.g., members of the genus *Alcyonidium*, the exoskeleton is a gelatinous crust. The exoskeleton has a primary opening or orifice through which the lophophore is thrust when the animal feeds. In many ectoprocts this opening is provided with a hinged lid or operculum. The position of the primary opening with relation to the symmetry of the enclosed animal varies. In tubular or vaselike forms, e.g., *Tubulipora*, Figure 13.4, the orifice is in what may be regarded as the dorsal position. Many encrusting and some erect genera have exoskeletons that are boxlike or coffin-shaped, Figure 13.3; in these forms the orifice shifts to a ventral or *frontal* position, the shell being adherent by the dorsal surface. (The attempt to relate exoskeletal surfaces to the symmetry of the animal itself is a ready source of confusion, however, and *distal* and *proximal* are currently preferred to dorsal and ventral in descriptive texts.) Though the exoskeleton of most ectoprocts is calcified, the extent of calcification varies among different species and ontogenetically. In some genera, e.g., *Membranipora*, the frontal surface is sealed by a chitinous membrane, while only the base and side walls are calcified. When the softer tissue has been cleaned away, the frontal area is left with a large opening, the *opesium*, Figure 13.7. Frequently the opesium is guarded by calcareous spines. In other genera the frontal wall is calcified, although perforations of various sorts in addition to the main orifice may be present.

Zooids of individual colonies may be essentially identical, but polymorphism is also common. The typical members of the colony, the feeding zooids, are called *autozooids;* specialized polymorphic members include *avicularia* and *vibracula*, Figures 13.7, 13.8, as well as attachment devices such as rhizoids and holdfasts which are regarded as modified zooids. Avicularia and vibracula may replace normal zooids in parts of the colony and may be larger or smaller than the autozooids, or they may occur as appendages to the autozooids. The avicularia may be scarce or abundant, and individual species may have more than one type of avicularium, Figure 13.2F. Their function is believed to be chiefly defensive—to thwart the attachment of fouling organisms. Vibracula are less common structures, with an apparently similar defensive function. They consist of a basal chamber and a smooth or toothed bristle, see Figure 13.8E,F.

Additional departures from the typical pattern of autozooid development include devices for sheltering or brooding the developing eggs. Brooding devices vary considerably. In some species they consist of relatively simple swellings. In others there are external brood chambers called *ovicells* or *ooecia*. Typically these have a hood or helmetlike form and are at the orifice end of the zooid. Ovicells have been classified in about a half-dozen categories (see Bassler, 1953), and differences in their structure are important systematically. Indeed, some species are chiefly characterized by such differences!

From the preceding account it will be apparent that both asexual reproduction by budding and sexual reproduction with brooding of the young are virtually universal in ectoprocts. A very few species shed their eggs freely in the water. The details of fertilization are poorly documented, but most species are thought to be hermaphroditic. The few dioecious species exhibit some sexual dimorphism (see Hyman, 1959, p. 324). Two types of free-living larvae are produced by ectoprocts. A *cyphonautes* larval type is known for *Alcyonidium* and for *Membranipora* and its close relatives. The cyphonautes larva has an apical tuft of bristles, a ciliary girdle, and is bivalved, compressed, and triangular in profile (Fig. 3.3K). The motile period for this type of larva may last for several months. A second (nameless) larval type is oval or rounded, ciliated, and frequently has red "eye spots" and tufts of cilia. The larva in either case is microscopic and after the planktonic period, settles down to metamorphose into a polyplike zooid, the *ancestrula*, from which the colony develops asexually by budding.

The brooded eggs or young of some species are pinkish, and when these are present in sufficient numbers the colony may be temporarily colored. Noncalcareous species are usually "horn-colored," i.e., a yellowish brown Calcareous forms may be glassy or white, or colored yellow, orange, red, or brownish.

Marine ectoprocts are, without exception, sessile and colonial. (A freshwater form, *Cristatella*, is motile.) Adults are planktonic only in the sense of becoming fixed to organisms such as pelagic *Sargassum*. They attach to a wide variety of animate and inanimate substrata. A very few species are almost host specific, occurring, for example, on the stems of hydroids. None are truly parasitic, although *Electra hastingsae* may occur as a commensal in the gill chamber of the crab *Libinia*. The phylum is largely marine, but there are a few freshwater species and some that range into brackish, estuarine situations.

Systematic List and Distribution

The basic systematics of the following list conforms to that of Bassler (1953); in addition, there has been an inordinate number of recent taxonomic changes, and the synonymy problem in this phylum is particularly irksome. *En*toprocts have been elevated to phylum status in agreement with the current practice of most specialists. The list itself is a compilation from sources cited in the references. The bathymetric range of many species is but vaguely known.

Class Gymnolaemata		
Order Ctenostomata		
Family Alcyonidiidae		
Alcyonidum parasiticum	BV	26 to 51 m, eury., 20‰
Alcyonidium mamillatum	A+	Shallow
Alcyonidium polyoum	BV	Lit. to 50 m, >20‰
Alcyonidium gelatinosum	B+	Lit. to 475 m
Alcyonidium verrilli	V	18 m, eury., 13‰
Alcyonidium hirsutum	V	Shallow
Family Flustrellidridae		
Flustrellidra hispida	B+	Lit. to 19 m
Family Victorellidae		
Victorella pavida	V+	Shallow, eury., 1–27‰
Family Nolellidae		
Anguinella palmata	V	Shallow, eury., 13‰
Nolella blakei	V	Shallow, eury.
Nolella stipata	V	To 1 m, eury.
Family Vesicularidae		
Bowerbankia gracilis	BV	Lit. to 15 m, eury., 10‰
Bowerbankia imbricata	V	Lit.
Amathia vidovici	V	Shallow, eury., 11‰
Amathia convoluta	V	To 15 m, eury., 22‰
Family Walkeriidae		
Walkeria uva	BV	Lit. to 15 m
Aeverrillia setigera	V	To 18 m
Aeverrillia armata	V, 44°	To 27 m, eury., 12‰
Family Triticellidae		
Triticella pedicellata	B	Georges Bank
Triticella elongata	V	Shallow
Order Cyclostomata		
Family Crisiidae		
Crisia eburnea	BV	Lit. to 366 m, eury., 18‰
Crisia cribraria	B+	Lit. to 37 m
Crisia denticulata	B	18 m
Family Tubuliporidae		
Idmonea atlantica	B	5 to 2330 m
Tubulipora liliacea	B+	Lit. to 27 m
Tubulipora flabellata	B+	26 to 55 m
?Tubulipora lobulata	B	—
Family Oncousoeciidae		
Oncousoecia canadensis	B	—
Oncousoecia diastoporides	B+	To 100 m
Family Diaperoeciidae		
Diaperoecia harmeri	B	12 to 72 m
Diplosolen obelium	B	4 to 670 m
Family Lichenoporidae		
Lichenopora verrucaria	B+	0 to 700 m
Lichenopora hispida	BV	0 to 360 m

Order Cheilostomata		
Suborder Anasca		
Family Aeteidae		
Aetea anguina	BV	2 to 35 m, eury., 21‰
Aetea recta	V	Shallow
Family Scrupariidae		
Scruparia chelata	V	To 18 m
Scruparia ambigua	V	15 to 18 m
Haplota clavata	B+	11 to 33 m
Eucratea loricata	B+	Lit. to 1359 m
Family Membraniporidae		
Membranipora membranacea	V	Fresh water to 13‰
Membranipora tuberculata	C+	On pelagic *Sargassum*
Membranipora tenuis	V	To 9 m, eury., 6‰
Cupuladria canariensis	C+	9 to 102 m
Conopeum reticulum	BV	Lit.
Conopeum truitti	V	To 1 m, eury, 11‰
Family Electridae		
Electra crustulenta	BV	Lit. to 90 m, eury., 2‰
Electra arctica	A+	0 to 520 m
Electra pilosa	BV	Lit. to shallow, eury., 11‰
Electra monostachys	BV	Lit. to shallow
Family Flustridae		
Flustra foliacea	A+	Lit.
Family Hincksinidae		
Cauloramphus cymbaeformis	B	5 to 228 m
Family Calloporidae		
Callopora dumerili	B	9 to 21 m
Callopora lineata	B+	0 to 378 m
Callopora craticula	B+	0 to 280 m
Callopora aurita	B+	Lit. to 33 m
Amphiblestrum flemingii	B+	13 to 428 m
Amphiblestrum osburni	B	30 to 100 m
Amphiblestrum trifolium	A+	—
Tegella arctica	B+	Lit. to 300 m
Tegella armifera	B+	Lit. to 314 m
Tegella unicornis	B+	Shallow to 100 m
Family Thalamoporellidae		
Thalamoporella falcifera	C+	On pelagic *Sargassum*
Family Bugulidae		
Bugula harmsworthi	B+	26 to 137 m
Bugula simplex	V, 44°	Lit. to 18 m
Bugula turrita	V	0 to 27 m, eury., to 20‰
Bugula fulva	BV	4 to 58 m
Bugula stolonifera	V+	—
Dendrobeania murrayana	B+	0 to 753 m
Kinetoskias smitti	B	174 to 811 m
Family Bicellariellidae		
Bicellariella ciliata	B+	To 176 m

Family Scrupocellariidae		
Scrupocellaria scabra	B	Lit. to 454 m
Tricellaria gracilis	B+	0 to 869 m
Tricellaria ternata	B+	0 to 366 m
Tricellaria peachii	B	7 to 91 m
Caberea ellisii	B+	11 to 110 m
Family Cribrilinidae		
Cribrilina punctata	B+	Shallow to 91 m
Cribrilina annulata	B+	0 to 112 m

Suborder Ascophora (The arrangement of families and genera in this suborder follows Powell, 1968b.)

Family Hincksisporidae		
Hincksipora spinulifera	A+	8 to 186 m
Family Umbonulidae		
Umbonula arctica	AB	1 to 345 m
Family Hippothoidae		
Hippothoa divaricata	B	16 to 33 m
Hippothoa hyalina	BV	Lit. to 2018 m
Family Gigantoporidae		
Cylindroporella tubulosa	B	Shallow
Family Stomachetosellidae		
Stomachetosella sinuosa	B	Lit. to 285 m
Stomachetosella hincksi	A+	23 to 672 m
Posterula sarsi	A+	4 to 1267 m
Family Schizoporellidae		
Schizoporella unicornis	BV	Lit. to 33 m, eury., 18‰
Stephanosella biaperta	BV	5 to 256 m
Schizomavella auriculata	BV	Lit. to 1285 m
Family Microporellidae		
Microporella ciliata	BV	Lit. to 570 m, eury., 20‰
Family Hippoporinidae		
Hippoporina americana	V, 44°	Shallow to 22 m
Hippoporina porosa	BV	5 to 37 m
Hippoporina reticulatopunctata	B	Lit. to 306 m
Hippoporina smitti	A+	6 to 55 m
Hippoporina propinqua	B+	1 to 210 m
Hippoporina verrilli	V	5 to 28 m
Cleidochasma contractum	V	9 to 37 m
Family Celleporinidae		
Turbicellepora canaliculata	B	27 to 93 m
Turbicellepora dichotoma	V+	2 to 25 m
Family Escharellidae		
Escharella immersa	B	Lit. to 1331 m
Escharella ventricosa	B	Lit. to 1262 m
Escharella abyssicola	A+	5 to 500 m
Family Smittinidae		
Smittina rigida	B	14 to 250 m
Smittina majuscula	B+	Lit. to 327 m
Smittina bella	B	Lit. to 100 m

Parasmittina jeffreysi	B+	5 to 400 m
Parasmittina nitida	BV	Lit. to 37 m
Porella reduplicata	B	6 to 18 m
Porella concinna	B	Lit. to 1165 m
Porella minuta	A+	Lit. to 55 m
Porella acutirostris	B+	Lit. to 395 m
Porella smitti	B+	9 to 293 m
Pseudoflustra solida	A+	4 to 1275 m
Rhamphostomella bilaminata	B+	Lit. to 226 m
Rhamphostomella costata	B	Lit. to 360 m
Rhamphostomella scabra	A+	Lit. to 460 m
Rhamphostomella ovata	A+	4 to 180 m
Rhamphostomella radiatula	A+	18 to 29 m
Palmicellaria skenei	A+	102 to 110 m
Family Cheiloporinidae		
Cryptosula pallasiana	BV	Lit. to shallow, eury.
Family Reteporidae		
Rhynchozoon rostratum	V	24 m
Hippoporella hippopus	B	6 to 235 m
Family Myriaporidae		
Myriapora subgracila	B	Lit. to 1267 m
Myriapora coarctata	B	82 m
Myriozoella plana	A+	1 to 206 m

Identification

Ectoprocts present special problems in species identification; these problems result chiefly from inadequate knowledge of many of the species and from the inherent variability of the material. In particular, there are individual variations in growth habit, and even more perplexing variations due to different degrees of calcification. The increase of secondary calcification in the older parts of colonies may produce confusing material, in which case younger zooids in the colony may show key traits more clearly. Much emphasis is placed on brood structures in some groups. While some use has been made of soft parts, e.g., numbers of tentacles, requiring examination of live or well-preserved material, most descriptions in the older literature deal exclusively with hard parts. Ectoprocts found in general collections in formalin, even buffered formalin, should be washed promptly and transferred to 70% alcohol or dried. Soft parts may be dissolved with Clorox. Specimens may be removed from rocks or shells by the technique of "calcining" (see Rogick, 1964); this procedure is not approved by all students.

It is convenient to subdivide this large assemblage initially on the basis of growth habit and calcification as follows.

A. Creeping, stoloniferous, or encrusting forms, mostly minute in size, with uncalcified zooids.

B. Erect, arborescent or bushy forms with uncalcified or very weakly calcified zooids.
C. Creeping, stoloniferous, or encrusting forms with well-calcified zooids.
D. Erect, arborescent, or bushy forms with well-calcified zooids.

Size, in terms of maximum height or width of mature colonies, is given for most species, but it should be borne in mind that the size attained depends on local growth factors and is not of diagnostic importance. Sources for descriptions and illustrations are given in parentheses.

Group A. This artificial group includes all Ctenostomata plus several atypical members of Cheilostomata, i.e., members of the suborder Anasca in the monogeneric family Aeteidae and of the genera *Scruparia* and *Haplota* in the Scruparidae. Most of the stoloniferous species are minute epizoic or epiphytic forms; a few, e.g., *Amathia* spp., may form larger bushy colonies.

---

1. Colony forming a gelatinous or rubbery crust, sometimes elevated in erect lobes—2.
1. Colony developing from a creeping stolon, erect or recumbent, branching or not—3.
1. Colony not stolonate but composed of a creeping series of zooids, more or less monolinear, with erect branches of zooids also in linear series; Fig. 13.1L,M—10.

2. Zooids spiny, except very young individuals; closed orifice slitlike with 2 lips; colony brown or reddish brown; size "extensive"; Fig. 13.1B. (Osburn, 1912a). *Flustrellidra hispida.*
2. Zooids without spines; closed orifice rounded, puckered; colony pale gray to yellow, red or brownish; colonies may reach about 380 mm in diameter; Fig. 13.1A. *Alcyonidium* spp.: The six species are difficult to distinguish; *A. parasiticum only* occurs on animals, whereas several other species may be epizoic or not; *A. verrilli*, *gelatinosum*, and sometimes *hirsutum* develop erect forms (see Osborn, 1912a; Maturo, 1957).

3. Zooids relatively long and slender, tubular; stolon without septa except basally; Fig. 13.1C–E—4.
3. Zooids relatively shorter, typically spindle-shaped; stolon usually divided by septa; Fig. 13.1F–H—6.

4. Colony branching; cuticle impregnated with mud; colony 25 mm tall; gray to brown; Fig. 13:1C (Osburn, 1944). *Anguinella palmata.*
4. Not so—5.

5. Zooids sometimes with an expanded basal part with spinelike projections; zooids 3.8 mm tall; with 8–22 tentacles; on *Zostera* and algae in estuarine situations; Fig. 13.1D (Rogick, 1949; Maturo, 1957). *Nolella* spp.
5. Colonies forming a dense mat 3–6 mm high, occasionally to 15 mm, zooids 4.85 mm; with 8 tentacles; chiefly in mesohaline, i.e., brackish, situations with salinity $<12‰$; Fig. 13.1E (Osburn, 1944). *Victorella pavida.* Osburn (1944) places *Victorella* in Nolellidae with the comment that "the species appear to intergrade with *Nolella*."

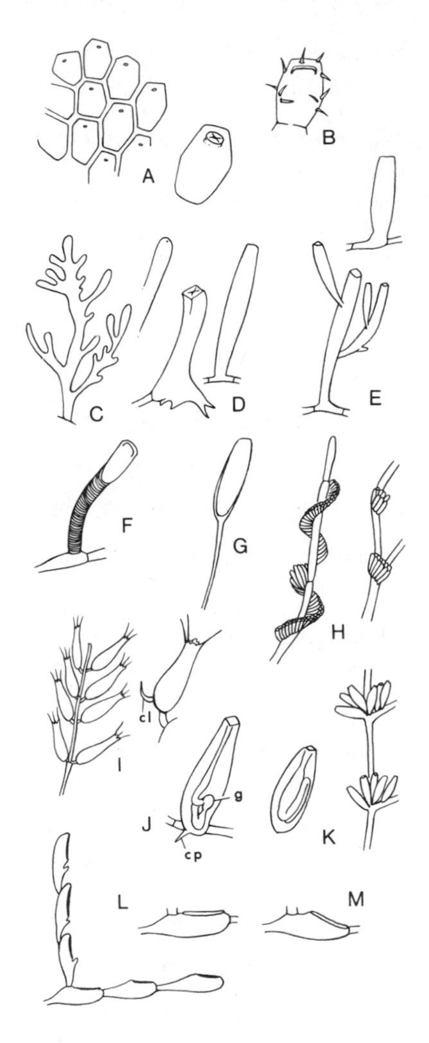

Fig. 13.1. Bryozoa, Group A. A, *Alcyonidium* sp., part of colony and single zooid; B, *Flustrellidra hispida*, zooid; C, *Anguinella palmata*, colony form and single zooid; D, *Nolella* sp., two zooids; E, *Victorella pavida*, part of colony and single zooid; F, *Aetea* sp., zooid; G, *Triticella* sp.; H, *Amathia* sp., part of colony of *A. convoluta* (left) and *A. vidovici* (right); I, *Aeverrillia* sp., part of colony of *A. armata* (left) and single zooid of *A. setigera* (right); J, *Bowerbankia* sp., zooid; K, *Walkeria uva*, zooid (left) and part of colony (right); L, *Scruparia ambigua*, part of colony (left) and single zooid; M, *Scruparia chelata*, zooid. Abbreviations are as follows: cl, clasping process; cp, caudata process; g, gizzard.

6. Zooids with a snakelike basal part; orifice with an operculum; stolon with spindlelike swellings; 0.6 mm tall; sometimes pinkish with contained young; Fig. 13.1F (Rogick, 1964). *Aetea* spp.
6. Zooids with a long, stemlike peduncle rising singly or in pairs from the stolon; sometimes epizoic on crabs; 1.85 mm tall; Fig. 13.1G (Osburn, 1912a). *Triticella* spp.
6. Zooids without peduncle; irregularly dispersed in clusters or paired; Fig. 13.1H—7.
7. Zooids paired, either in a double file winding spirally around the stolon or pinnate; with spines at the border of the orifice; minute or in bushy colonies; 50–150 mm tall—8.
7. Zooids irregularly dispersed or in clumps, or if in pairs not as above; minute species usually epizoic or epiphytic—9.
8. Zooids in a double file winding spirally around the stolon; colony erect, branching; Fig. 13.1H (Osburn, 1944). *Amathia:* In *A. vidovici* about half of each internode is without zooids and colonies reach 50 mm; in *A. convoluta* zooids are more extensive and colonies reach 150 mm.
8. Zooids in pairs more or less pinnately arranged on the stolon; colony erect, branching; Fig. 13.1I. *Aeverrillia: A. setigera* has clasping processes, *armata* does not (Maturo, 1957).
9. With gizzard (examine young individuals under transmitted light); at least some zooids with a caudate process; expanded tentacles forming a complete circle; gray to pinkish with contained young; Fig. 13.1J. *Bowerbankia: B. gracilis* with 8 tentacles, the zooids usually in pairs; *B. imbricata* with 10 tentacles, the zooids usually in clusters (Osburn, 1912a; Rogick, 1964).
9. Without gizzard; the expanded tentacles do not form a perfect circle since two of them are commonly bent away from the others; zooids in clusters; Fig. 13.1K (Osburn, 1912a). *Walkeria uva.*
10. Zooids joined back to back or arising from the back of other zooids (Osburn, 1912a). *Haplota clavata.*
10. Zooids arising from the front of other zooids; Fig. 13.1L,M. *Scruparia: S. ambigua* has the rim of the opesium of nonfertile zooids parallel to the basal wall; Fig. 13.1L; in *S. chelata* the rim dips obliquely; Fig. 13.1M (Hastings, 1941). Note the possibility of confusion here with *Hippothoa* spp. and *Electra arctica* of Group C2, with colonies of branching, uniserial, encrusting, calcareous zooids. *Eucratia loricata* of Group B has bushy colonies of zooids joined back to back, but the zooids are not calcareous.

Group B. Included in this group are some of the larger and more familiar erect, branching species of the phylum. Species traits in *Bugula* overlap and may cause difficulty.

---

1. Colony sparsely branched, fingerlike; orifice of zooids puckered; zooids not in a single or double linear series, without avicularia; Fig. 13.1A. *Alcyonidium: A. hirsutum* has colonies up to 150 mm tall covered with small conical papillae; the following species do not; *A. verrilli* has 10 tentacles and *A. gelatinosum*, 15–17 with colonies reaching 350 mm tall. Note also *Pseudoflustra solida*, a rare arctic

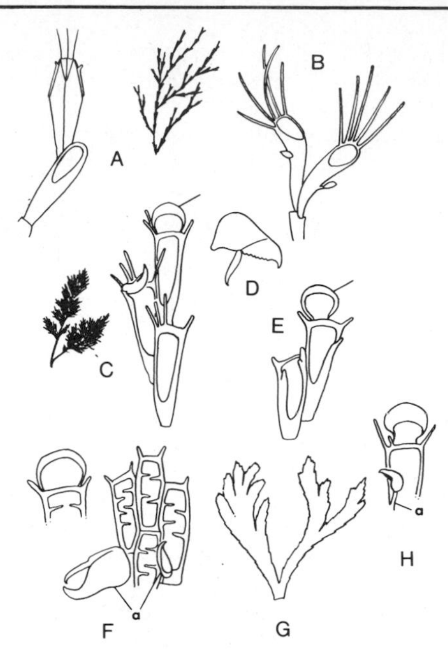

Fig. 13.2. Bryozoa, Group B. A, *Eucratea loricata*, three zooids (left) and colony growth form (right); B, *Bicellariella ciliata*, two zooids; C, *Bugula harmsworthi*, colony growth form (left) and three zooids, one with ooecium indicated (upper right); D, same, single avicularium; E, *Bugula turrita*, two zooids, one with ooecium indicated; F, *Dendrobeania murrayana*, three zooids and detail of ooecium; G, *Bugula simplex*, colony growth form; H, same, detail of part of zooid and ooecium. Avicularia are indicated by letter "a."

form, and *Anguinella palmata*, Fig. 13.1C, in colonies no more than 25 mm tall arising from a stolon. (Osburn, 1912a).

1. Not so—2.

2. Zooids in a double series joined back to back, i.e., with orifices facing in opposite directions; colonies to 250 mm tall; Fig. 13.2A (Osburn, 1921). *Eucratea loricata.*

2. Zooids joined side by side with orifices facing in nearly the same direction—3.

3. Zooids trumpet-shaped, in two rows; orifice oblique, facing upward with 4–8 marginal spines; colony feathery, to 25 mm tall; Fig. 13.2B (Osburn, 1912a). *Bicellariella ciliata.*

3. Zooids nearly tubular in 2 or more rows; orifice frontal—4.

4. Zooids in a double series—5.

4. Zooids in more than 2 rows—6.

5. Avicularium with a serrated beak; upper margin of orifice usually with 4 or 5 spines; Fig. 13.2C,D (Maturo and Schopf, 1968). *Bugula harmsworthi.*

5. Avicularium with simple beak; upper margin of orifice with 2 or 3 spines usually; Fig. 13.2E (Maturo, 1966). *Bugula turrita.* Note that two additional *Bugula* species with biserial zooids have been reported (rarely) from our area; *B. fulva* has 4–7 spines at the orifice; *B. stolonifera* commonly has 3 but differs from *B. turrita* in its manner of branching. See Maturo (1966) and Ryland (1960).

6. Zooids in 4–12 rows; orifice with marginal spines; with two sorts of avicularia; ooecium large, subglobular; colonies to 40 mm tall; Fig. 13.2F (Osburn, 1912a). *Dendrobeania murrayana.*
6. Zooids in 3–6 rows, fronds fanlike; orifice without marginal spines; with only one type of avicularium; ooecium small, hemispherical; colonies to 25 mm tall; Fig. 13.2G,H (Osburn, 1912a). *Bugula simplex.* Note that the zooids are sometimes in more than 2 rows in *Bugula fulva;* this species commonly has 4 spines at the orifice.

Group C. The majority of ectoproct species are encrusting forms with calcified zooids. Three subgroups will be used to permit the construction of less cumbersome keys. These assemblages are more or less natural. Colony size is not always indicated but ranges from a few millimeters to a half-dozen centimeters or more in different species. The subgroups are as follows.

C1. Zooids in the form of simple tubes with a rounded terminal or nearly terminal orifice; avicularia lacking; in some species the zooids are largely immersed in the common crust of the colony with only the orifice apparent; orifice without an operculum and external ooecia lacking. Encrusting Cyclostomata. Note also the cheilostome *Cylindroporella tubulosa* (Fig. 13.4A), with tubular or vase-shaped zooids; opercula and ooecia present, in colonies to 7 mm diameter (Fig. 13.4).

C2. Zooids of various forms, but, if tubular, with a more or less lateral orifice; avicularia present or not; orifice with an operculum; ooecium usually present in fertile zooids; the frontal wall is calcified (in contrast to the next group), but may be pierced by pores or other openings in addition to the primary orifice (Figs. 13.5, 13.6). Encrusting Cheilostomata, exclusive of *Membranipora* and related genera.

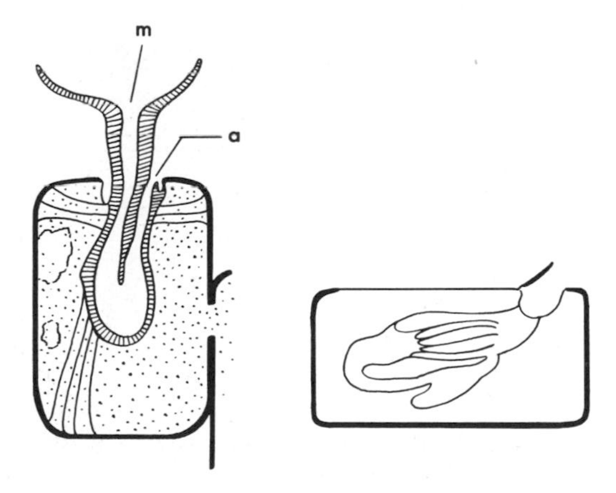

Fig. 13.3. Bryozoa. Relationship of soft anatomy to hard parts in typical bryozoans shown schematically. Mouth and anus are indicated in the figure at left.

C3. Zooids box- or coffin-shaped with membranous frontal wall; the cleaned exoskeleton, after treatment with Clorox, with a large, open opesium (Fig. 13.7). Membranipore Cheilostomata.

Group C1.

1. Colonies disclike or oval with simulated radial symmetry, i.e., with zooids radiating in more or less linear series from a center which may be free of zooids; orifice with toothed margin; colonies minute, <4 mm in diameter—2.

1. Colonies fan-shaped or lobate; zooids frequently in linear series but the colony usually not distinctly symmetrical; margin of orifice without teeth—3.

2. Colony usually circular with a distinct central area free of zooids; zooids usually with a distinct rib extending as a pointed spur or tooth on the margin of the orifice; roof of brood chamber reticulated; Fig. 13.4B (Osburn, 1933). *Lichenopora verrucaria.*

2. Colony circular to oval, the center with slightly raised tubes or jagged projections; colony with a somewhat lichenlike or laminated appearance; roof of brood chamber not reticulated; Fig. 13.4D (Rogick and Croasdale, 1949). *Lichenopora hispida.*

3. Colony a thin crust with zooids largely embedded, i.e., the tubules projecting slightly if at all above the common crust; orifices well separated, the zooids not fused or fascicled; colony to about 20 mm; Fig. 13.4C (Osburn, 1933). *Oncousoecia* spp.

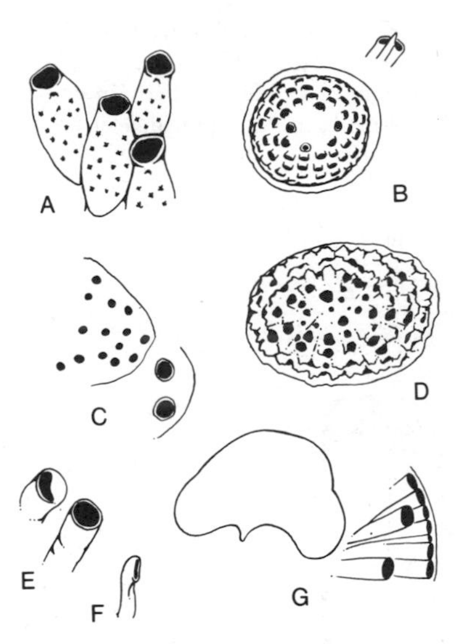

Fig. 13.4. Bryozoa, Group C1. A, *Cylindroporella tubulosa*, part of colony; B, *Lichenopora verrucaria*, colony with details of margin of orifice of single zooid (above right); C, *Oncousoecia*, part of colony with detail of two zooid orifices; D, *Lichenopora hispida*, colony; E, *Tubulipora liliacea*, ooeciostome (left) and autozooid (right); F, *Tubulipora flabellata*, ooeciostome; G, *Tubulipora* sp., young colony form (left) and detail of fascicled zooids.

3. Colony fan-shaped when young, becoming nodular or rounded when older, the tubular zooids projecting considerably from the common crust; at least some of the zooids fused along their whole length and with closely adjoining orifices, i.e., they are fascicled; Fig. 13.4G. *Tubulipora: T. liliacea* and *T. flabellata* are distinguished by the form of the *ooeciostome*—the raised tubular aperture of the brood chamber—the aperture and orifice being of about equal diameter in *liliacea* (Fig. 13.4E), while in *flabellata* the orifice is about twice as large as the ooeciostome (Fig. 13.4F) (Osburn, 1912a). Note also *Diplosolen obelium*, a boreal species with small tubules "interspersed among normal ones" (Osburn, 1933). See Powell (1968a) for all members of this group.

Group C2. This is a large and varied assemblage. Variations in the shape of the orifice provide useful key criteria which must, however, be used with care; difficulties arise in part from individual variations due to wear and to variations in the extent of calcification.

---

1. The orifice with a distinct notch or sinus narrower than the rest of the orifice, i.e., orifice schizostomatous; Fig. 13.5F—2. Note that the form of the sinus ranges from a narrow, sharply defined notch to a broader depression only a little narrower than the orifice proper; *Hippoporina* (Fig. 13.6H) could, for example, be keyed out here. The strongly developed prominence below the orifice of zooids of *Turbicellepora* may obscure the presence of the sinus (Fig. 13.5D).

1. The orifice varying in outline but not schizostomatous (Fig. 13.6)—8. The effects of secondary calcification may be especially confusing in this group and key traits are frequently most clearly expressed in young zooids. Colonies of most of these genera sometimes occur as thickened or nodular, elevated or semierect encrustations.

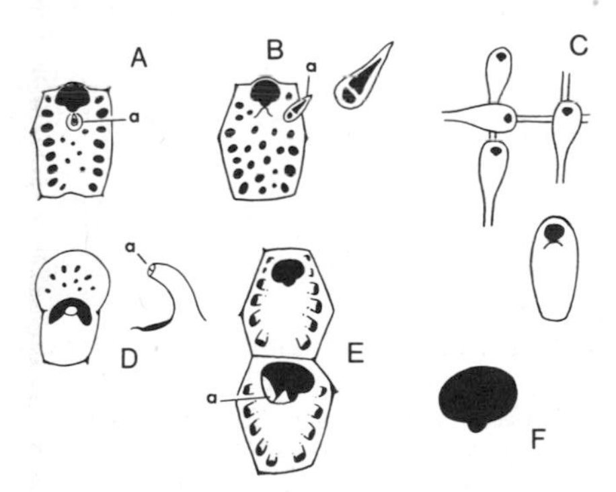

Fig. 13.5. Bryozoa, Group C2. A, *Schizomavella auriculata*, zooid; B, *Schizoporella unicornis*, zooid and detail of avicularium (right); C, *Hippothoa divaricata*, part of colony and individual zooid (below); D, *Turbicellepora canaliculata*, zooid (left) and detail of avicularium (right); E, *Rhynchozoon rostratum*, two zooids, the lower one with submerged avicularium; F, schizostomatous orifice. Avicularia are indicated by the letter "a."

2. Frontal wall perforated all over with numerous coarse pores which may be filled by secondary calcifications in older zooids—3.
2. Frontal wall without coarse pores or with a marginal series of large pores—6.

3. Avicularia absent; colony with heavy secondary calcification; reddish brown. *Stomachetosella* spp. (*S. sinuosa* is the more common species.) (Powell, 1968a,b; Osburn, 1933a).
3. Avicularia present, but note 5 below—4.

4. Avicularium on midline below orifice; colorless to reddish or yellowish; Fig. 13.5A (Osburn, 1912a, 1933a). *Schizomavella auriculata.*
4. Avicularia not on midline below orifice—5.

5. Avicularia sharply pointed, sometimes scarce and lacking over a large part of the colony; pale orange to brick red; Fig. 13.5B (Osburn, 1912a; Rogick and Croasdale, 1949). *Schizoporella unicornis.*
5. Avicularia oval or rounded, rarely pointed; whitish to pink red (Osburn, 1912a, 1933a; Rogick and Croasdale, 1949). *Stephanosella cornuta.*

6. Avicularia absent. *Hippothoa:* In *H. divaricata* (a rare boreal species), the zooids are in a loose, branching series or in branching colonies (Fig. 13.5C); *H. hyalina* forms a crust, to about 25 mm diameter, and is common along the whole coast, especially on algae south of Cape Cod (Osburn, 1933a).
6. Avicularia present—7.
(Note that *Schizomavella auriculata*, with extensive calcification, may key out here; see 4 above).

7. Avicularium mounted on a raised prominence (*umbo* or *rostrum*). *Turbicellepora:* In *T. dichotoma* the avicularium is placed at the side of the umbo; in *T. canaliculata* it is placed at the apex (Fig. 13.5D) (Osburn, 1912a,b).
7. Avicularium ultimately submerged in a large, bulbous chamber; Fig. 13.5E (Maturo, 1957). *Rhynchozoon rostratrum.*

Species With Nonschizostomatous Orifice

8. Orifice horseshoe-shaped, with denticles, *candeles*, or *condyles*, at the lower sides; Fig. 13.6A,B—9.
8. Orifice rounded, semicircular, frequently with a marginal projection, *lyrule* or *umbo*, in the middle, below; sometimes with lateral denticles as well as a lyrule—10.

9. Avicularia absent; ovicells lacking; Fig. 13.6A (Maturo, 1957; Osburn, 1912a). *Cryptosula pallasiana.*
9. Avicularia present, often with several on a single zooid; ovicells present; Fig. 13.6B (Osburn, 1933a). *Hippoporella hippopus.*

10. Frontal wall with transverse or radiating ribs and furrows, the ribbing less evident in *C. punctata;* porous, the pores irregularly dispersed or in rows; colonies white, reddish, or brownish. *Cribrilina: C. punctata* has avicularia, usually 2 per zooid, a large ooecium, and irregularly dispersed pores (Fig. 13.6D); *C. annulata* lacks avicularia, has a small ooecium, and regularly arranged pores (Fig. 13.6C) (Powell, 1967; Rogick and Croasdale, 1949).
10. Not so—11.

11. Orifice usually with a prominent median tooth—15.
11. Orifice usually rounded or semicircular, without tooth—12.

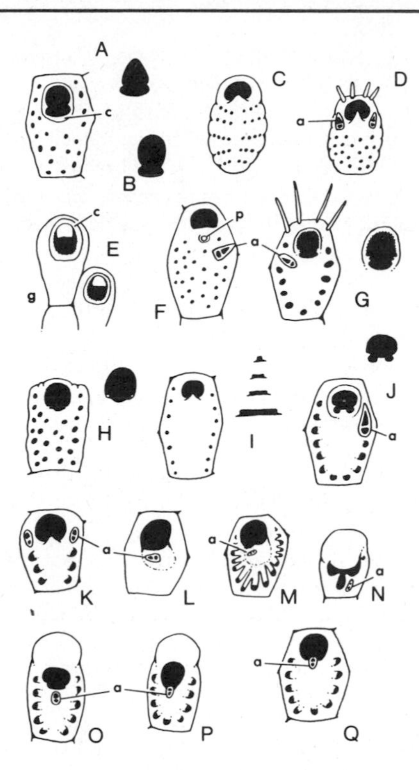

Fig. 13.6. Bryozoa, Group C2. A, *Cryptosula pallasiana*, zooid with variant orifice (right); B, *Hippoporella hippopus*, orifice; C, *Cribrilina annulata*, zooid; D, *Cribrilina punctata*, zooid; E, *Electra arctica*, zooid and parts of adjacent zooids; F, *Microporella ciliata*, zooid; G, *Cleidochasma contractum*, zooid with detail of orifice (right); H, *Hippoporina* sp., zooid and detail of orifice (right); I, *Escharella* sp., zooid and outlines of mucro (right), the species are, from top to bottom, *E. spinulifera*, *E. immersa*, *E. ventricosa*, and *E. abyssicola*; J, *Parasmittina* sp., zooid and detail of orifice (above right); K, *Umbonula arctica*, zooid; L, *Rhamphostomella ovata*, zooid; M, *Rhamphostomella costata*, zooid; N, *Rhamphostomella bilaminata*, zooid; O–Q, *Smittina* and *Porella* species. Abbreviations are as follows: a, avicularium; c, cryptocyst; p, pore.

12. Zooids usually in a single linear series but branching, and the branches sometimes fused to produce a double or triple series; orifice oval without lateral denticles; Fig. 13.6E. *Electra arctica.* Note: This species is actually a "membranipore" with a conspicuous cryptocyst and well-developed gymnocyst (Osburn, 1933a; Powell, 1968a).

12. Zooids not uniserial or, if so, then orifice with lateral denticles—13.

13. Orifice with lateral teeth—14.

13. Orifice semicircular or, if round, without denticles; a median pore below the orifice (often obscure); avicularia usually present; Fig. 13.6F. *Microporella ciliata.*

14. Top of orifice beaded; lateral teeth bicuspid; orifice with or without spines above; colony to 20 mm; Fig. 13.6G. *Cleidochasma contractum.*

14. Top of orifice not beaded; lateral teeth bicuspid or not; Fig. 13.6H. *Hippoporina* spp. The species will very likely prove difficult for the novice to distinguish; see Maturo and Schopf (1968), and Osburn (1933a).

15. Avicularia absent. *Escharella:* The form of the *mucro* or tooth varies in the several local species, as figured; young individuals especially may have spines above the orifice; Fig. 13.6I.

15. Avicularia present—16.

16. Avicularia median and suboral; Fig. 13.6O–Q—18.

16. Avicularia neither median nor suboral—17.

17. Mucro broad, flat-topped; avicularia variously placed; Fig. 13.6J. *Parasmittina: P. jeffreysi* has large pointed avicularia plus smaller oval ones; *P. nitida* has large and small oval avicularia but no large pointed ones; colony white to yellow (Osburn, 1912a).

17. Mucro narrower, blunt or pointed; usually with a small oval avicularium on each side of the orifice; Fig. 13.6K (Osburn, 1933a; Powell, 1968b). *Umbonula arctica.*

18. Avicularium placed on the side of a prominent suboral swelling or umbo; the avicularium oblique or transverse; Fig. 13.6L. *Rhamphostomella:* In *R. costata* and *R. ovata* there is a single umbo; in *bilaminata* there are two liplike projections (Fig. 13.6N); *costata* is strongly ribbed (Fig. 13.6M), *ovata* is not (Fig. 13.6L); *R. radiatula* is a small arctic form. These species occur on various substrata but most frequently as nodular encrustations on hydroid stems (Osburn, 1933a).

18. Avicularia longitudinal; Fig. 13.6O–Q. The differentiation of the dozen or more species in the genera *Smittina* and *Porella* is beyond the scope of this text and is complicated by a wide range of individual variation in the common species. Their synonymies are complex, and additional related genera occur in deep water. The species are boreal or arctic and most are described and figured in Osburn (1933a), Powell (1968a, 1968b), and Rogick and Croasdale (1949). *Porella smitti*, *concinna*, and, *reduplicata* are apparently the most common forms in our range.

Group C3. Membraniporan bryozoans are common forms, familiar as lacy encrustations on a variety of substrata. The differentiation of genera has given bryozoologists considerable difficulty and the classification remains unstable.

1. Avicularia and ovicells absent—2.

1. Avicularia and ovicells usually present, but note the second alternate of 7—7.

2. Frontal wall partly calcified, i.e., with a *gymnocyst* or with opesium partly filled by a calcareous shelf, i.e., with a well-developed *cryptocyst*; Fig. 13.7G—3.

2. Gymnocyst and/or cryptocyst weakly developed or absent, but note couplet 9—5.

3. Cryptocyst distinctly developed, sometimes filling half the opesium, with toothed margin; corners of opesium sometimes with rounded knobs, but border not spiny; Fig. 13.7A (Osburn, 1944). *Membranipora tenuis.*

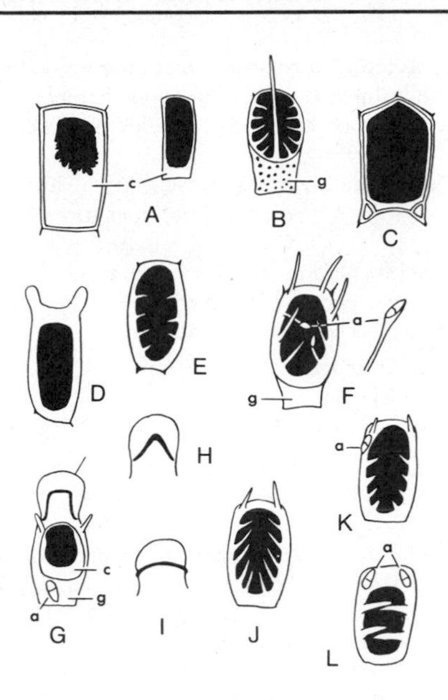

Fig. 13.7. Bryozoa, Group C3. A, *Membranipora tenuis*, two zooids to show variations in cryptocyst development; B, *Electra pilosa*, zooid; C, *Conopeum truitti*, zooid; D, *Membranipora tuberculata*, zooid; E, *Electra crustulenta*, zooid; F, *Cauloramphus cymbaeformis*, zooid and detail of avicularium (right); G, *Amphiblestrum flemingii*, zooid with ooecium indicated; H, *Callopora aurita*, detail of ooecium; I, *Tegella unicornis*, detail of ooecium; J, *Callopora craticula*, zooid; K, *Callopora lineata*, zooid; L, *Tegella arctica*, zooid. Abbreviations are as follows: a, avicularium; c, cryptocyst; g, gymnocyst.

3. Gymnocyst present; opesium border usually spiny including a medial, proximal spine and a pair of semierect spines on either side of opesium—4.

4. Gymnocyst with numerous conspicuous pores; with 10 or 12 lateral spines, the medial spine sometimes greatly elongated; colony to about 300 mm; Fig. 13.7B (Osburn, 1933; Rogick and Croasdale, 1949). *Electra pilosa*.

4. Gymnocyst with few pores; with 6–18 or no lateral spines; the medial spine usually equal to laterals (Rogick and Croasdale, 1949; Powell and Crowell, 1967). *Electra monostachys*.

5. Proximal (posterior) corners of zooids with triangular hollows with erect, cylindrical or conical, chitinous tubes; Fig. 13.7C (Osburn, 1944). *Conopeum truitti*.

5. Not so—6.

6. Opesium usually rectangular, without spines; anterior (distal) corners with a tubercle (especially well-developed in *M. tuberculata*); Fig. 13.7D. *Membranipora: M. tuberculata* is found only on *Sargassum; M. membranacea* is an estuarine species (Osburn, 1944).

6. Opesium usually elliptical; distal tubercles absent but up to 12 delicate spines often present; Fig. 13.7E (Osburn, 1944; Powell and Crowell, 1967). *Electra crustulenta*.

7. Usually with 1 or 2 avicularia mounted on elongate pedicels; gymnocyst present; opesium with 6 or 8 spines; no ovicells; small colonies, almost always attached to the stems of bryozoans or hydroids; Fig. 13.7F (Osburn, 1912a). *Cauloramphus cymbaeformis.*

7. Avicularia not pedicellate, sometimes scarce or absent; gymnocyst present, cryptocyst present or not; ovicells present; opesium with spines in most but not all species—8.
The remaining species belong to three genera, *Tegella*, *Amphiblestrum*, and *Callopora*. All except two rare northern forms are keyed in Osburn (1912), and the remainder of this key is based on this source.

8. Cryptocyst present in older zooids at least; spines <6, sometimes absent; ooecium bears a rib; Fig. 13.7G,H—9.

8. Cryptocyst absent; spines 2–14, rarely absent; ooecium may bear a more or less transverse rib; Fig. 13.7I—10.

9. Ooecial rib encloses a more or less quadrangular area; spines 2–6 or absent; Fig. 13.7G. *Amphiblestrum flemingii.* Note also the less common *A. trifolium*, usually without spines and with a triangular area on the ooecium, and *A. osburni*, a species from deep water (see Powell, 1968a,b, and Osburn, 1933).

9. Ooecial rib encloses a more or less triangular area; spines 2–4, usually only 2 in older zooids and one of these larger than the other; Fig. 13.7H (Osburn, 1933). *Callopora aurita.*

10. Usually with 2 spines which may be unequal in size; if with 4 spines, the anterior pair smaller; Fig. 13.7I. *Tegella unicornis.* Note also the similar, less common *T. armifera* and the even rarer *Callopora dumerili.* See Osburn (1933).

10. Spines usually more numerous, 4 to 14—11.

11. Spines 12–14, long, bent downward over the opesium, flattened in cross-section, directed strongly forward; Fig. 13.7J. (Osburn, 1933). *Callopora craticula.*

11. Spines 4–6 or 8–12, but not strongly directed forward—12.

12. Spines 8–12, nearly erect and pointed, except the most anterior pair which are blunt at the end and directed somewhat forward; Fig. 13.7K (Osburn, 1912a). *Callopora lineata.*

12. Spines 4–6, one or two erect, the others broad and flattened and bent downward over the opesium; Fig. 13.7L (Osburn, 1933). *Tegella arctica.*

Group D.

1. Zooids tubular with orifice more or less terminal—2.

1. Zooids of various forms, if tubular then with orifice more or less lateral—3.

2. Colony articulated with horny joints or nodes; Fig. 13.8A,B. *Crisia:* The species are separated by differences in the ooecium and by the number of zooids per internode as follows: *C. eburnea*, 5–7 zooids per internode; ooeciostome tubular, about 1 ½ times longer than wide; colony to 18 mm; Fig. 13.8A,B. *C. cribraria*, 18–20 zooids per internode; ooeciostome a compressed tube, almost slitlike; colony to 12 mm; Fig. 13.8C. *C. denticulata*, about 11 zooids per internode; ooeciostome not projecting; colony to 25 mm (Osburn, 1912a, 1933).

2. Colony not articulated; with paired branches; at least some of the zooids fused along their whole length, i.e., fascicled; Fig. 13.8D (Osburn, 1912a). *Idmonea atlantica.*

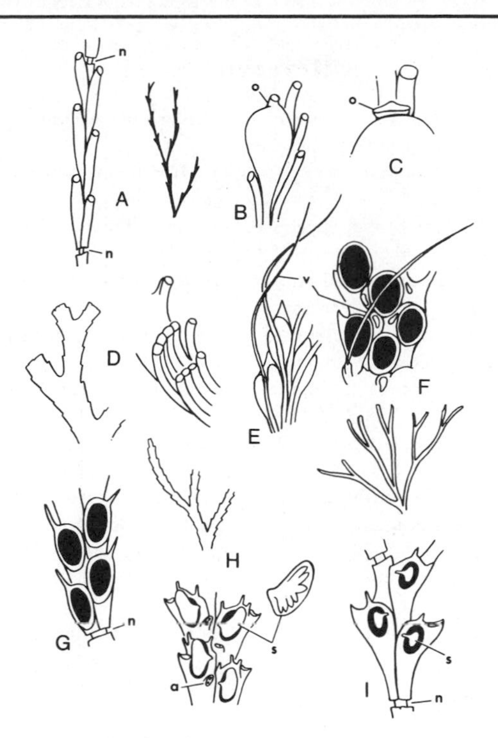

Fig. 13.8. Bryozoa, Group D. A, *Crisia eburnea*, part of colony (left) and colony growth form (right); B, same, detail with ooecium; C, *Crisia cribraria*, detail of ooecium; D, *Idmonea atlantica*, colony form (left) and detail showing fascicled zooids; E, *Caberea ellisii*, part of colony; F, same, different specimen, with colony form (below); G, *Tricellaria peachii* part of colony; H, *Scrupocellaria scabra*, colony form (above), part of colony (below) and detail of shield (right); I, *Tricellaria* sp., part of colony. Abbreviations are as follows: a, avicularium; n, node; o, ooeciostome; s, shield; v, vibracula.

3. Colony jointed—4.

3. Colony not jointed; zooids in 2–4 rows; vibracula present; Fig. 13.8E,F (Osburn, 1933). *Caberea ellisii.*

4. Orifice not protected by a shield, bordered by a single stout spine or with no bordering spine; the terminal zooids in each internode with a median spine; avicularia scarce or usually absent; colonies to about 25 mm tall; Fig. 13.8G (Osburn, 1933). *Tricellaria peachii.*

4. Orifice protected by a shield; avicularia usually present; zooids usually with 2 or 3 spines—5.

5. Orifice protected by a flattened shield frequently nearly as large as the orifice and ornamented with an antlerlike area; with 5–12 zooids per internode; avicularia with short vibracula, i.e., with a short bristle or whip; Fig. 13.8H (Osburn, 1933). *Scrupocellaria scabra.*

5. Shield less developed and lacking antlerlike area; frequently with only 3 zooids per internode but the number ranging to 11; no vibracula; Fig. 13.8I (Osburn, 1933; also see Powell, 1968b). *Tricellaria* spp.

References

Bassler, R. S. 1953. Bryozoa, Part G, Treatise on Invertebrate Paleontology. R. C. Moore, ed. Geol. Soc. Am. and Univ. Kansas.

Borg, F. 1944. The stenolaematous bryozoa. Further Results Swedish Antarct. Exped. 1901–1903. 3(5):1–276.

Hastings, A. B. 1941. The British species of *Scruparia* (Polyzoa). Ann. Mag. nat. Hist. Ser. 11. 7:465–472.

Hutchins, L. W. 1945. An annotated check-list of the salt water Bryozoa of Long Island Sound. Trans. Conn. Acad. Arts Sci. 36:533–551.

Hyman, Libbie H. 1951. The Invertebrates: Acanthocephala, Aschelminthes, and Entoprocta, the Pseudoceolomate Bilateria. Vol. III. McGraw-Hill Book Co.

———. 1959. The Invertebrates: Smaller Coelomate Groups. Vol. V. McGraw-Hill Book Co.

Maturo, F. 1957. Bryozoa of Beaufort, North Carolina. J. Elisha Mitchell scient. Soc. 73(1):11–68.

———. 1966. Bryozoa of the southeast coast of the United States: Bugulidae and Beaniidae (Cheilostomata: Anasca). Bull. mar. Sci. 16(3):556–583.

———, and T. J. M. Schopf. 1968. Ectoproct and entoproct type material: reexamination of species from New England and Bermuda named by A. E. Verrill, J. W. Dawson, and E. Desor. Postilla. 120:1–95.

Osburn, R. C. 1921a. The Bryozoa of the Woods Hole region. Bull. Bur. Fish. Wash. (1910), 30:205–266.

———. 1912b. Bryozoa from Labrador, Newfoundland, and Nova Scotia. Proc. U.S. natn. Mus. 43:275–289.

———. 1933. Bryozoa of the Mount Desert region. Biol. Surv. Mount Desert Reg.: 291–385.

———. 1940. Bryozoa of Porto Rico with a resumé of the West Indian bryozoan fauna. N.Y. Acad. Sci. Surv. Porto Rico and Virgin Islands. 16:321–486.

———. 1944. Bryozoa of Chesapeake Bay. Chesapeake Biol. Lab. Publs. 63:3–55.

Powell, N. A. 1967. Sexual dwarfism in *Cribrilina annulata* (Cribrilinidae-Bryozoa). J. Fish. Res. Bd. Can. 24(9): 1905–1910.

———. 1968a. Studies on Bryozoa (Polyzoa) of the Bay of Fundy region. II. Bryozoa from fifty fathoms, Bay of Fundy. Cah. Biol. mar. 9:247–259.

———. 1968b. Bryozoa (Polyzoa) of Arctic Canada. J. Fish. Res. Bd. Can. 25(11): 2269–2320.

———, and G. D. Crowell. 1967. Studies on Bryozoa (Polyzoa) of the Bay of Funday Region. I. Bryozoa from the intertidal zone of Minas Basin and Bay of Fundy. Cah. Biol. mar. 8:331–347.

Rogick, Mary. 1949. Studies on marine Bryozoa, IV. *Nolella blakei*, n. sp. Biol. Bull. mar. biol. Lab. Woods Hole. 97:158–168.

———. 1964. Ectoprocta. *In* Keys to the Marine Invertebrates of the Woods Hole Region. R. I. Smith, ed. Systematic-Ecology.

———, and Hannah Croasdale. 1949. Studies on marine bryozoa. III. Woods Hole region Bryozoa associated with algae. Biol. Bull. mar. biol. Lab. Woods Hole. 96:32–69.

Ryland, O. 1960. The British species of *Bugula* (Polyzoa). Proc. zool. Soc. London. 134(1): 65–105.

Whiteaves, J. 1901. Catalogue of the marine Invertebrata of eastern Canada. Polyzoa Geol. Surv. Can.: 91–114.

FOURTEEN

Phylum Phoronida

PHORONID WORMS

Phoronis architecta is the only local representative of this small phylum of about 18 species, Figure 14.1.

Diagnosis. *Tube-dwelling, wormlike animals with a horseshoe-shaped lophophore.*

The worm is slender; adults reach 50 mm in length, but individuals 20–25 mm long are only about 1 mm thick. They have been described as faintly annulated, but there is no morphological basis for such a description; this condition, if present, probably results from contractions due to the preservation reagent. The rear end is slightly swollen and apparently serves to anchor the worm in its tube. The anterior two-thirds is flesh color and the remainder in some seasons at least is a dark, yellowish red. The lophophore consists of a horseshoe-shaped row of tentacles with a central mouth.

The tube is cylindrical, straight, and more than twice as long as the worm itself. It is produced as a chitinous secretion and, being initially sticky, becomes covered with sand.

P. architecta has been regarded as a protandric hermaphrodite, but Hyman (1959) questions this view. Fertilization is external, and the eggs hatch as a distinctive *actinotroch* larva that is planktonic, reaching a length of several millimeters, Figure 3.3L. Asexual reproduction by fission is also known in the phylum; in this connection it may be noted that aggregations may occur. Wass (1963) reported densities of 90 individuals per square meter in the

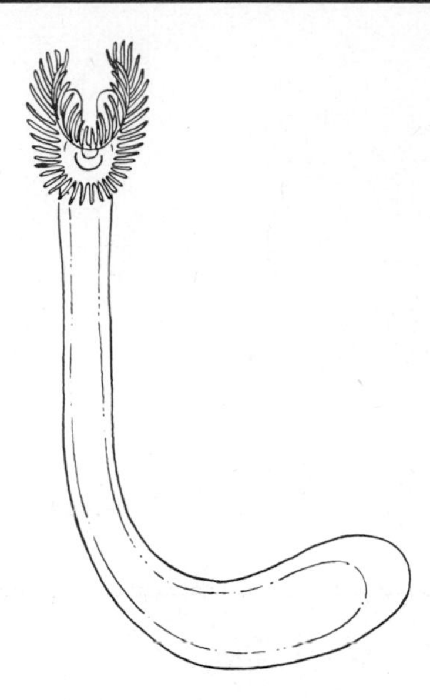

Fig. 14.1. Phoronida. *Phoronis architecta*, whole animal.

Chesapeake. Like other lophophore-equipped animals, phoronids are ciliary-mucus feeders subsisting on plankters or detrital fragments. *P. architecta* is found on sandy substrata in shallow water from the lower littoral to depths of at least 18 m. It is an estuarine species, and Wass (1963) reported it from polyhaline waters with salinities as low as 18‰. The local species is apparently Carolinian and has not been reported north of the Chesapeake. Wilson (1881) reported two species in Chesapeake Bay, but only one local form is recognized. *Phoronis architecta* is figured in Hyman (1959).

References

Andrews, E. A. 1890. On a new American species of the remarkable animal *Phoronis*. Ann. Mag. nat. Hist. 6 Ser. 5(3):445.

Brook, W. K., and R. P. Cowter. 1906. *Phoronis architecta:* its life history, anatomy, and breeding habits. Mem. natn. Acad. Sci. Wash. 10:71–148.

Dawydoff, C. 1959. Class des Phoronidiens. *In* Traité de Zoologie—Anatomie, Systématique, Biologie. Masson et Cie.

Hyman, Libbie H. 1959. The Invertebrates: Smaller Coelomate Groups. Vol. V. McGraw-Hill Book Co.

Wass, M. L. 1963. Check List of the marine invertebrates of Virginia. Va. Inst. mar. Sci. spec. Sci. Rep. 24:1–56.

Wilson, E. B. 1881. The origin and significance of the metamorphosis of Actinotrocha. Q. J. microsc. Sci., New Ser. Mem. 21(82):202

FIFTEEN

Phylum Brachiopoda

LAMP SHELLS

Diagnosis. *Solitary, bivalved animals with a lophophore and with one valve either firmly cemented to the substratum, or fixed by a short pedicel, or burrowing with an elongate, wormlike pedicel distally sheathed in sand.*

This phylum had its heyday in the Paleozoic Era, reaching a peak of diversity in the Devonian with more than 200 genera. Since the beginning of the Mesozoic, brachiopods have been increasingly supplanted as important benthic, marine animals by other forms, notably pelecypod mollusks.

The bivalved condition suggests molluscan affinities, but the soft anatomy is quite different, and the shells of brachiopods are dorsal and ventral rather than lateral as in pelecypods. The valves are *articulate*, i.e., hinged, or held together by muscles only, i.e., *inarticulate;* in the second condition the shells are capable of lateral as well as gaping movements.

The lophophore, the possession of which readily distinguishes brachiopods from mollusks, is a conspicuous structure that occupies much of the space inside the mantle cavity. Its most conspicuous feature is a fringe of tentacles or filaments. In simple form, e.g., in *Argythrotheca*, Figure 15.1N, the lophophore consists of a lobed disc with relatively few, elongate marginal filaments; this is the *schizolophous* condition. The *plectolophous* lophophore of *Terebratulina*, Figure 15.1L, has more extensively developed lateral lobes, and there is a coiled median lobe. The median lobe is lacking in the *spirolophous* lophophore of *Hemithiris*, but the lateral lobes are even more elongate and

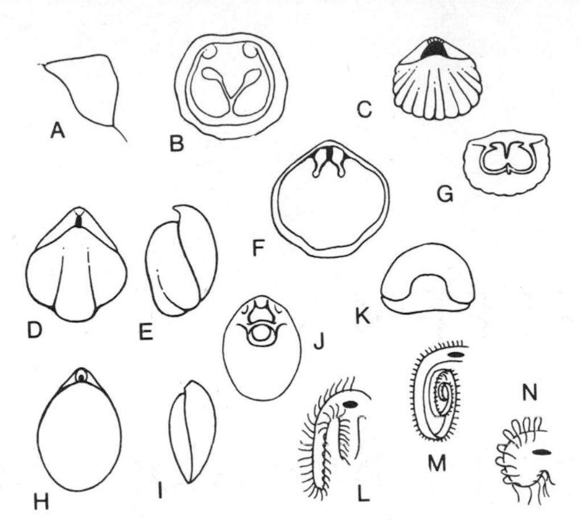

Fig. 15.1. Brachiopoda. A, *Crania anomala*, outline of animal in place, lateral view; B, same, inside brachial valve; C, *Argyrotheca lutea*, dorsal view; D, *Hemithiris psittacea*, dorsal view; E, same, lateral view; F, same, inside brachial valve; G, *Argyrotheca lutea*, inside brachial valve; H, *Terebratulina septentrionalis*, dorsal view; I, same, lateral view; J, same, inside brachial valve; K, *Hemithiris psittacea*, posterior view; L, *Terebratulina septentrionalis*, lophophore, diagrammatic; M, *Hemithiris psittacea*, lophophore, diagrammatic; N, *Argyrotheca lutea*, lophophore, diagrammatic.

are coiled, Figure 15.1M. Lophophore surfaces are ciliated, and both incurrent and excurrent flows are maintained in the mantle cavity. Planktonic and detrital particles are trapped on the filaments and conveyed along a food groove to the mouth, which occupies a basal position on the lophophore. In Articulata the lophophore has a calcareous, skeletal support, and systematic differences in the configuration of this structure provide useful identification criteria. The lophophore is suspended from the dorsal or *brachial* valve. The valves are otherwise dissimilar, the *pedicel* or ventral valve being shaped to allow for the passage of the pedicel; details of this opening and its fine structure are useful in identification.

The sexes are usually separate, but some species of *Argyrotheca*, among others, are hermaphroditic, and protandric hermaphroditism may occur in extralimital species. There is little sexual dimorphism except that ripe male gonads tend to be white or pale yellow, and those of the female, deep yellow to red-brown. In local species the eggs are freely shed and fertilized in the water and develop into a planktonic, trochophore type of larva. Planktonic life is short in terebratulids, ranging from little more than a day to a week or less, so far as known. The young brachiopod then settles down but only gradually develops the diagnostic traits of the adult; there is ontogenetic variation in the fine structure of the lophophore, lophophore skeletal elements, pedicel opening, and other traits.

Though not colonial in a structural sense, brachiopods tend to aggregate. They are exclusively benthic and, except for the more motile lingulids, which are burrowing forms, become permanently fixed. The substratum is usually a "solid" surface of rock or shell, but in *Terebratulina*, at least, the pedicel may be attached to algal stems, worm tubes, and other "live" substrata as well.

Systematic List and Distribution

The list is derived from Dall (1920); the classification follows Williams *et al.* (1965). Note that the nearest geographical representative of the familiar, textbook lingulids is *Glottidia pyramidata* from the Carolinian region.

Class Inarticulata		
Order Acrotretida		
Family Craniidae		
Crania anomala	A+	27 to 183 m
(var. *pourtalesii* reported from deep Carolinian waters)		
Class Articulata		
Order Rhynchonellida		
Family Hemithyrididae		
Hemithiris psittacea	A+	14 to 2200 m
Order Terebratulida		
Family Cancellothyrididae		
Terebratulina septentrionalis	B, 40°	Lit. to 3800 m
Family Megathyrididae		
Argyrotheca lutea	C+	55 to 232 m

Identification

The dimensions given are adult maxima approximately. *Crania*, *Argyrotheca*, and *Terebratulina* have the shell perforated by microscopic canals containing outgrowths of the mantle. Illustration sources are Hyman (1959), Miner (1950), and Williams *et al.* (1965).

1. Pedicel lacking, ventral valve firmly cemented to substratum, (whole specimen in place resembles a limpet); brachial valve conical; margins of valves thickened, granulated, or pustulose internally; reddish brown externally; 20 mm long by 25 mm; Fig. 15.1A,B. *Crania anomala.*

1. Pedicel present, ventral valve not cemented to substratum—2.

2. Shells bluish gray to dark brown, more or less divided in three lobes with surfaces finely striated; lophophore spirolophous; lophophore support consisting of a spurlike *crura*, not forming a loop; pedicel foramen an elongate, triangular slot; length 30 mm; Fig. 15.1D–F,K,M. *Hemithiris psittacea.*

2. Shell pale colored; lophophore support looplike; foramen round or oval—3.

3. Shell with about 12 strongly defined, radiating ribs; light brownish white; lophophore schizolophous; loop curving widely and joined to the median septum of the branchial valves; 8 mm long; Fig. 15.1, C,G,N. *Argyrotheca lutea.*

3. Shell with up to about 240 fine riblets or *striae*, yellowish white; lophophore plectolophous; loop smaller, collarlike; 32 mm long; Fig. 15.1H–J, L. *Terebratulina septentrionalis.* Adults are frequently covered by the sponge *Iophon nigricans.*

References

Dall, W. H. 1920. Annotated list of the recent brachiopods in the collection of the U.S. National Museum. Proc. U.S. Natn. Mus. 57:261–377.

Davidson, T. 1886–1888. A monograph of recent Brachiopoda. Trans. Linn. Soc. Lond. Ser. 2(4):1–248.

Hyman, Libbie H. 1959. The Invertebrates; Smaller Coelomate Groups. Vol. V. McGraw-Hill Book Co.

Miner, R. W. 1950. Field Book of Seashore Life. G. P. Putnam's Sons.

Williams, A., *et al.* 1965. Brachiopoda. Part H, Vols. 1–2. Treatise on Invertebrate Paleontology. R. C. Moore, ed. Geol. Soc. Am. and Univ. of Kansas.

SIXTEEN

Phylum Mollusca

MOLLUSKS

Introduction

This is one of the largest animal phyla, second only to Arthropoda in number of species. As in other such large assemblages, the diversity of adaptation within the phylum thwarts any simple definition of practical value to the identification of individual species.

Mollusks are highly modified coelomates with well-developed organ systems contained in a *visceral mass*. The coelom is substantially restricted in most classes. Except in the extralimital Monoplacophora, there is no true segmentation in the phylum. Most mollusks are bilaterally symmetrical, but in gastropods the body suffers a torsion of 180°, this twisting being reflected in the coiled shell. There is usually a more or less clearly defined *head*, members of Bivalvia being the chief exceptions. The *foot* is a third recognizable unit of the molluscan body, and is a modification of the ventral body wall. It is the usual organ of locomotion, but takes various forms in the different taxa.

Two structures are found in Mollusca and nowhere else in the animal kingdom; these are the *mantle* and the *radula*. The mantle is a fold in the body wall, and in most mollusks it secretes a calcareous shell. The shell is single in most Gastropoda, Scaphopoda, and Cephalopoda, double in most Bivalvia, and consists of eight platelike pieces in Polyplacophora. Shells are absent or vestigial in Aplacophora, a very few aberrant Bivalvia, and many gastropods and cephalopods. The radula is a toothed tongue- or ribbonlike organ contained in the buccal cavity, and generally used as a rasping, food-gathering utensil. The radula is subject to considerable adaptive modification and is important in formal systematics. This structure is absent in all Bivalvia.

The sexes are separate in many mollusks, but hermaphroditism occurs in most classes. The eggs are brooded or are freely shed to hatch externally. The young may develop directly or with a metamorphism involving one or more distinct larval stages. As in many other sedentary forms, the motile, planktonic larval phase is important in species dispersal.

The initial larva of scaphopods, chitons, bivalves, and lower gastropods is a trochophore. The molluscan trochophore is similar in plan to the equivalent stage in annelid worms and other invertebrate phyla, a similarity of great theoretical interest. (See Chapter 3 for a general discussion of invertebrate larval types.) In most gastropods and bivalves the trochophore stage is quickly passed or is found only within the egg. The trochophore usually transforms into a more efficient swimming stage called a *veliger* which may last a few hours, days, or even a month before metamorphosing into the adult form. A shell is present in these early larvae, and many species that are shelless as adults, e.g., nudibranchs, have a larval shell that is lost at metamorphosis. The shell is double in larval as well as adult bivalves, and in larval gastropods the shell becomes torted or asymetrical at an early stage.

The elaboration of the trochophore into a veliger is accomplished by expanding the *protroch*—the main ciliary band—into a *velum* which is bilobed in most gastropods, Figure 3.3E. The primary lobes frequently subdivide into four or more secondary lobes which may become extremely elongate. With the abandonment of a pelagic existence, the veliger metamorphoses into a juvenile with the definitive traits of its class.

Additional larval forms occur in Mollusca, e.g., in deep sea squid. In the gastropod families Lamellariidae and Cypraeidae, the veliger has two shells. This form is called an *echinospira*. It has a normal larval shell plus "an outer thin transparent shell that encloses the larval shell and acts as a float" (Hyman, 1967).

Deevey (1956) has indicated that molluscan larvae are present in the plankton throughout the year and bivalve veligers are the most numerous of larval forms of benthic invertebrates. During the peak season in early summer (in Long Island Sound) bivalve veligers reach densities of 50,000 individuals per cubic meter according to Deevey's study. Thorson (1946) also found bivalve larvae abundant, accounting for about 58% of the larvae of bottom forms in Danish waters. Progress in larval identification has been made, but good data are available for relatively few species. Loosanoff *et al.* (1966) give criteria for identifying the larvae of some of the most common littoral and shallow water bivalves. These veligers reach a maximum length of 0.2–0.3 mm.

Mollusks have evolved to utilize virtually all of the adaptive options open to them, including flight in some squid. Criteria for initial identification to class are given in Chapter 3. Class units within the living Mollusca are generally agreed upon; further systematic refinements are not. The following outline is from Rothschild (1965).

Class Aplacophora	Solenogasters
Class Polyplacophora	Chitons
Class Gastropoda	Snails and slugs
Class Scaphopoda	Tusk shells
Class Bivalvia (Pelecypoda)	Bivalves
Class Cephalopoda	Squids and octopuses

General References

Further citations are given at the conclusion of each class account.

Abbott, R. T. 1954. American Seashells. D. Van Nostrand.

———. 1968. A guide to field identification, seashells of North America. Golden Press, New York.

Binney, W. G. 1863–64. Bibliography of North American Conchology. Previous to the year 1860. Smithson, Misc. Coll. Parts I and II.

Cooke, A. H. 1895. Molluscs and Brachiopods. Vol. 3, Cambridge Natural History. London.

Deevey, Georgiana B. 1956. Zooplankton. *In* Oceanography of Long Island Sound, 1952–1954. Bull. Bingham oceanogr. Coll. 15:113-155.

Gould, A. A. 1870. Report on the Invertebrata of Massachusetts, 2nd ed. W. G. Binney, ed. Wright and Potter, Boston.

Hardy, A. C. 1956. The open sea, its natural history; the world of plankton. Houghton Mifflin Company.

Hyman, Libbie H. 1951. The Invertebrates: Platyhelminthes and Rhynchocoela, the acoelomate Bilateria. Vol. II. McGraw-Hill Book Co.

———. 1967. The Invertebrates. Vol. VI. Mollusca I. McGraw-Hill Book Co.

Jacobson, M. K., and W. K. Emerson. 1961. Shells of the New York City area. Argonaut Books Inc.

Johnson, C. W. 1915. Fauna of New England. 13. List of the Mollusca. Occ. Paps. Boston Soc. Nat. Hist. 231 pp.

———. 1934. List of marine Mollusca of the Atlantic coast from Labrador to Texas. Proc. Boston Soc. Nat. Hist. 40(1):1–203.

Keen, Myra. 1958. Sea shells of Tropical West America. Stanford Univ. Press.

———. 1963. Marine molluscan genera of Western North America. Stanford Univ. Press.

Loosanoff, V. L., H. C. Davis, and P. E. Chanley. 1966. Dimensions and shapes of larvae of some marine bivalve mollusks. Malacologia. 4(2):351–435.

Miner, R. W. 1950. Field Book of Seashore Life. G. P. Putnam's Sons.
Moore, R. C. de. 1960. Treatise on Invertebrate Paleontology. Part I. Mollusca I. Geol. Soc. Am. and Univ. of Kansas.
Morris, P. A. 1947. A field guide to the shells of our Atlantic and Gulf coasts. Houghton Mifflin Co.
Morton, J. E. 1958. Molluscs. Hutchinson Univ. Library.
Rothchild, Lord. 1965. A classification of living animals. 2nd ed. Longmans.
Thorson, G. 1946. Reproduction and larval development of Danish marine bottom inverterbrates, with special reference to the planktonic larvae in the Sound (Øresund). Medd. Komm. Havundersog. Kbh. Ser. (d). Plankton 4:1–523.
Warmke, Germaine L., and T. R. Abbott. 1961. Caribbean Seashells. Livingston Publ. Co.

Class Aplacophora
SOLENOGASTERS

The phyletic position of solenogasters has been disputed, but the consensus places them near the stem of the phylum Mollusca.

Diagnosis. *Bilaterally symmetrical, cylindrical, wormlike mollusks without sharply differentiated body regions; shell lacking, but calcareous spicules present; mouth minute, slitlike, and subterminal; a pair of short, exposed gills at the posterior end. Adults reach 12 to 80 mm in length.*

The only genus reported locally in relatively shallow water is *Crystallophrisson*, more commonly known as *Chaetoderma*, Figure 3.1B. The spindle-shaped "head" is set off by a constriction; the mouth is subterminal and partly surrounded by an area, free of spicules, called the *oral shield.* The body is somewhat contractile and varies in width, being narrow anteriorly and substantially expanded posteriorly. The exposed gills are feathery. The spicules are minute anteriorly but become progressively larger posteriorly and are more or less lance-headed or spikelike in shape. The coloration is pale, buff or drab with a silvery sheen. Two species have been reported locally. *C. nitidulum*, the more common form, reaches 80 mm by about 3 mm but is usually much smaller; its spicules are weakly keeled or lack a keel entirely. *C. vadorum* has been reported to reach 12.5 mm by about 1.1 mm and has spicules with a strong central keel, Figure 3.1B.

Most solenogasters are hermaphroditic, but *Crystallophrisson* has the sexes separate. Sexual behavior is not well known; fertilization is internal in some species, and the young may be brooded in the parent's body. When brooding occurs in this class, a free-swimming larval phase may be omitted, but pelagic "yolk larvae," a modified trochophore type of larva, have been described; see Hyman (1967).

Solenogasters are benthic animals that burrow in soft bottoms, where they feed on a wide variety of microscopic plants and animals.

Systematic List and Distribution

The only report of solenogasters locally is that of Heath (1918). *Crystallophrisson* replaces the preoccupied *Chaetoderma*, as indicated by Hyman (1967, footnote p. 62).

Order Crystallophrissonoidea		
Family Crystallophrissonidae		
Crystallophrisson nitidulum	B, 38°	46 to 540 m
Crystallophrisson vadorum	B	46 to 117 m

References

Heath, H. 1918. Solenogasters from the east coast of North America. Mem. Mus. Comp. Zool. 45:9–179.

Hoffman, H. 1929. Aplacophora. *In* Bronn's Klassen und Ornungen des Tierreichs. 3(1): 1–134.

Hyman, Libbie H. 1967. The Invertebrates: Volume VI. Mollusca I. McGraw-Hill Book Co.

Wiren, A. 1892. Studien uber die Solenogastres 1. Monographie des *Chaetoderma nitidulum* Loven. Svenska vet. akad. Handl. 24(2):1–66.

Class Polyplacophora
CHITONS

Diagnosis. *Body bilaterally symmetrical, more or less flattened; shell in 8 successive pieces bordered by and partly embedded in a fleshy girdle, Figure 16.1.*

Chitons are more or less oval in outline. The body may be quite depressed and flattened, or the back may be raised in a high arch or with a central keel.

The calcareous shell pieces or *valves* may be almost fully embedded and concealed by the mantle or *girdle;* usually, however, they are well exposed and slightly overlapping. The marginal girdle may be naked or ornamented with scales or spines.

The girdle overlaps a broad sole-shaped foot. There is a deep submarginal groove containing the gills, and the head is differentiated but lacks eyes and tentacles. Adults of local species range from about 10 to 50 mm in length.

Details of valve structure are important in species differentiation, Fig. 16.1. The end shells—the head and tail valves—are more or less semicircular in shape; the intermediate valves are somewhat rectangular or winged with their long axes crosswise to the long axis of the body. The exposed surface of each intermediate valve is usually regionated as follows: there is a median keel or narrow triangular area, called the *jugum*, separating side pieces that are divided diagonally into broad, central and lateral triangles. The jugum and *central* and *lateral areas* may be sharply differentiated by sculpturing and ornamentation, or their differentiation may be suppressed and obscure. Each valve has two primary layers. The surface layer is called the *tegmentum* and has a thin, pigmented horny periostracum. The internal, pearly layer is the *articulamentum*. The articulamentum of each intermediate valve has bladelike forward extensions that underlie the next valve in front. There is also an area for mantle attachment lying beneath each lateral area; these surfaces are called *insertion plates*, and they are commonly notched on each side. The tail or both end plates are also notched marginally with about 7–11 teeth in all local genera except *Lepidopleurus* and *Hanleya;* the tegmental border that projects beyond these *insertion teeth* is termed the *eaves*. The inner surface of the valves may be white or colored.

The sexes are separate in almost all chitons, but in most species they are indistinguishable externally. Fertilization is external, and usually the eggs are shed in the water. In some extralimital species the eggs are brooded. The hatchlings are trochophores which are briefly planktonic.

Chitons are chiefly shallow water or littoral forms and are usually confined to hard substrata to which they cling tenaciously. They are notably sedentary. If detached, they tend to curl up in the manner of pill bugs or armadillos. So far as known, they are herbivores.

Systematic List and Distribution

The basic list is that of Johnson (1934) with the taxonomy revised to conform to that of Smith (1960), except that the chitons are raised to class status. Alternatively the name Amphineura may be used with Polyplacophora and Aplacophora as subclasses.

Order Neoloricata		
Family Lepidopleuridae		
Lepidopleurus cancellatus	B	37 to 183 m
Lepidopleurus carinata	B+	22 m
Family Hanleyidae		
Hanleya hanleyi	B	69 m
Hanleya mendicaria	B	46 to 55 m
Family Ischnochitonidae		
Ischnochiton alba	B	Lit. to 617 m
Ischnochiton ruber	B	Lit. to 146 m
Tonicella marmorea	B	Lit. to 91 m
Tonicella blaneyi	B	37 m
Family Chaetopleuridae		
Chaetopleura apiculata	V	5 to 22 m
Family Molpaliidae		
Amicula vestita	B	9 to 55 m

Identification

Dimensions are total length.

1. Girdle largely covering valves; girdle thin, smooth, brown, usually with widely scattered tufts of hairlike spines; exposed valves heartshaped; 50 mm; Fig. 16.1B. *Amicula vestita.*

1. Valves well exposed, not heart-shaped—2.

2. Girdle smooth, leathery, naked, except for microscopic granules; valves apparently smooth but microscopically granular; eaves porous; valves rosy pink inside. *Tonicella: T. blaneyi* has radial riblets in the lateral areas; this sculpturing is absent in *T. marmorea*. *T. marmorea* has buff-colored valves marked with dark red, a banded girdle and is, 40 mm. *T. blaneyi* has rose-colored valves with white lines or reticulations.

2. Girdle scaly or granular—3.

2. Girdle spiny or hairy—5.

3. Girdle "sand-papery" with fine, granular scales; head valve with 8–13 insertion teeth—4.

3. Girdle with overlapping split pea-shaped scales, some scales spinose, especially marginally; head valve without insertion teeth; whitish to orange-gray; 12 mm; Fig. 13.1C. *Lepidopleurus cancellatus.*

4. Head valve with front slope straight or slightly concave; pale buff to whitish; valves white inside; 15 mm. *Ischnochiton alba.*

4. Head valve with front slope convex; pale buff to reddish, marbled red-orange; valves pink inside; 25 mm. *Ischnochiton ruber.*

5. Intermediate valves with central and lateral areas sharply differentiated, the central areas with rows of beads parallel to central keel, lateral areas with scattered

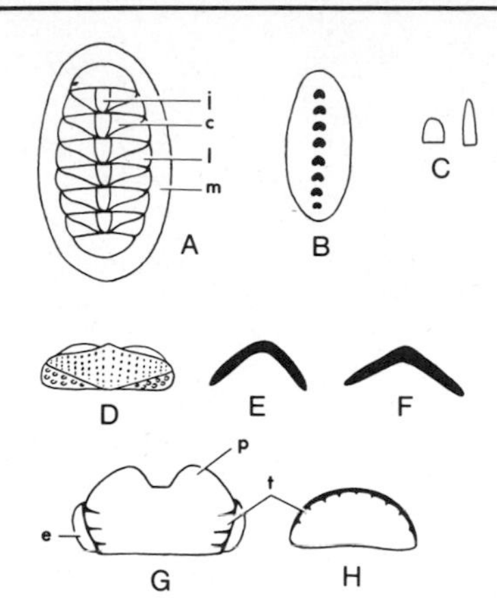

Fig. 16.1. Mollusca, Polyplacophora. A, typical chiton, dorsal view; B, *Amicula vestita*, whole animal with valves shown in solid black; C, *Lepidopleurus cancellatus*, two scales from girdle; D, *Chaetopleura apiculata*, intermediate valve, external view; E, *Lepidopleurus* sp., schematic cross-section; F, *Hanleya* sp., schematic cross-section; G, intermediate valve, internal view; H, head valve, internal view. Abbreviations are as follows: j, jugum; c, central area; l, lateral area; m, mantle; t, insertion teeth; p, insertion plate; e, eaves.

tubercles; girdle with sparsely scattered short hairs; pale buff, unicolored or with dark patches; 17 mm; Fig. 13.1D. *Chaetopleura apiculata.*

5. Intermediate valves with feebly or undifferentiated central and lateral areas; head valve without insertion teeth—6.

6. Valves elevated, crown high, rounded or arched, with divergence about 70–80°, Fig. 16.1E; head valve without a rough-edged insertion plate. *Lepidopleurus:*
- Girdle generally scaly with overlapping split pea-shaped scales but some scales spinose and sometimes numerous enough to present a mossy aspect; valves without a median keel; white to orange-gray; 12 mm. *L. cancellatus.*
- Girdle densely pilose with very fine white spinules; valves with a weak keel; yellow-white; 35 mm. *L. carinata.*

6. Valves more depressed, crown angled, rooflike, with divergence about 100° or more, Fig. 61.1F; head valve with roughened insertion plate. *Hanleya:*
- Girdle with very fine glassy spicules; pale ashy; 27 mm. *H. mendicaria.*
- Girdle with short and long, horn-colored spicules; pale; 10 mm. *H. hanleyi.*

References

Dall, W. H. 1927. Diagnoses of undescribed new species of mollusks in the collection of the United States National Museum. Proc. U.S. Natn. Mus. 70(19):1–11. (*Lepidopleurus carinata.*)

Hoffman, H. 1929–30. Amphineura (Polyplacophora). *In* H. G. Bronn, ed., Klassen und Ordnungen des Their-Reichs 3(1) sup. 2:135–368, sup. 3: 369–382

Hyman, Libbie H. 1967. The Invertebrates: Volume VI. Mollusca I. McGraw-Hill Co.
Johnson, 1934. List o marine Mollusca of the Atlantic coast from Labrador to Texas. Proc. Boston Soc. Nat. Hist. 40(1): 1–203.
Pilsbry, H. A. 1892–1894. Manual of Conchology. Polyplacophora. Acad. nat. Sci. Phila., 14: xxiv + 350 pp., 15:65–180.
Smith, A. G. 1960. Amphineura. *In* Treatise on Invertebrate Paleontology Part I, Mollusca L. R. C. Moore, ed. Pp. 41–176. Geol. Soc. Am. and Univ. of Kansas.

Class Gastropoda
SNAILS AND SLUGS

The largest class of Mollusca includes shell-less as well as univalved shelled species. Gastropod shells exhibit every conceivable variation of form and detail and no concise diagnosis will be attempted. A classification to order provides a frame of reference for the brief summary that follows. The classification is mainly that of Franc (1968). Exclusively shell-less orders are marked "n."

Subclass Prosobranchia
  Order Arachaeogastropoda
  Order Mesogastropoda
  Order Neogastropoda (Stenoglossa)
Subclass Opisthobranchia
  Order Cephalaspidea
n Order Sacoglossa
n Order Notaspidea (Pleurobranchacea)
  Order Thecosomata
n Order Gymnosomata
n Order Nudibranchia
  Order Pyramidellacea
Subclass Pulmonata
  Order Basommatophora

The soft body usually includes the readily defined regions of head, foot, and mantle in addition to the visceral mass. The head usually has 1 pair of tentacles in prosobranches and local marine pulmonates, and 1–3 pairs (most typically 2) in opisthobranchs; the tentacles are reduced or absent in some species of all three classes. A pair of eyes is usually present. In opisthobranches and pulmonates these are commonly placed at the base of the tentacles or embedded in the head. Prosobranchs are much more varied in this respect. In this class the eyes are frequently stalked and placed on or near the tentacles, but the development of the eye stalks or *ommatophores* and their positional relationship to the tentacles varies.

The mouth is terminal or ventral and in prosobranchs is borne on a snout that precedes the tentacles. The snout is greatly developed in such predaceous genera as *Nassarius* into an elongate, retractile proboscis. In pulmonates the mouth is flanked by oral lappets; oral or labial projections sometimes simulating tentacles are also present in some prosobranchs and opisthobranchs. The mouth, either round or slitlike in shape, leads to a buccal cavity or directly to the pharynx. In addition to the radular apparatus there are frequently hardened jaws. Radular and jaw structures are important in systematics.

The foot in typical forms is a flat, creeping sole. It may be produced at the anterior corners in tentaclelike lobes. The foot is simple, without regional divisions in pulmonates and most shell-bearing opisthobranchs and nudibranchs; it is regionated in many prosobranchs and some opisthobranchs. *Parapodia* are an important modification of the basic foot structure; these are lateral lobes that form the swimming organs of the pteropod orders and of a few sea slugs such as *Elysia*. The pelagic heteropods have the foot either greatly reduced or modified anteriorly into a swimming lobe.

A horny or calcareous, platelike *operculum* is developed on the upper surface of the rear of the foot in most prosobranchs but generally not in the other two subclasses. Its purpose is to seal the shell aperture when the animal retreats within. Among prosobranchs the operculum is absent in the Tonnidae, in limpets and limpetlike calyptraeids, in the pelagic *Janthina*, and in genera with a deep, slitlike aperture. Prosobranchs that lack an operculum as adults usually have one when juvenile. Details of operculum structure often provide useful criteria for generic and specific identification.

The mantle, consisting of a fold in the body wall, lines the body whorl in shelled forms. A troughlike, or tubular, respiratory *siphon* is developed as an extension of the mantle on the left side of many prosobranchs, and particularly of Neogastropoda. The shell of such species may be modified with a siphonal canal or notch, Figure 16.2A–D. The mantle in some species is developed in large lobes that extend dorsally to cover the shell, as in the families Marginellidae and Lamellaridae among prosobranchs and in some members of the order Cephalaspidea among opisthobranchs. Conversely, the mantle is reduced or absent in Gymnosomata, most Sacoglossa, and some Nudibranchia.

External gills are found in some limpetlike prosobranchs, but they are lodged in the pallial groove and are inconspicuous. Internal gills are generally present in prosobranchs. Conspicuous external gills occur in such nudibranchs as the Doridacea. They are present or absent in other opisthobranchs where various substitute respiratory structures are provided, especially in nudibranchs other than Doridacea. In pulmonates, gills are absent and the mantle cavity is modified as a pulmonary sac.

The formal division of gastropods into higher systematic categories generally depends on differences in soft anatomy and internal symmetry. Shell shape is an unreliable indicator of phylogenetic relationship. However, for practical purposes, differences in shell structure, form, and ornamentation provide the chief criteria for species identification.

The basic gastropod shell is a spirally wound cone. Shell form varies widely. Most gastropod shells have a *dextral* twist; this means that, when the shell is held with the spire uppermost and with the primary opening or *aperture* facing the observer, the outer lip will be on the right. The individual turns of the shell, called *whorls*, are separated by *sutures;* the last whorl, containing the animal itself, is the *body whorl*, and the remaining whorls make up the *spire*. Terminally in undamaged shells, there may be a series of minute

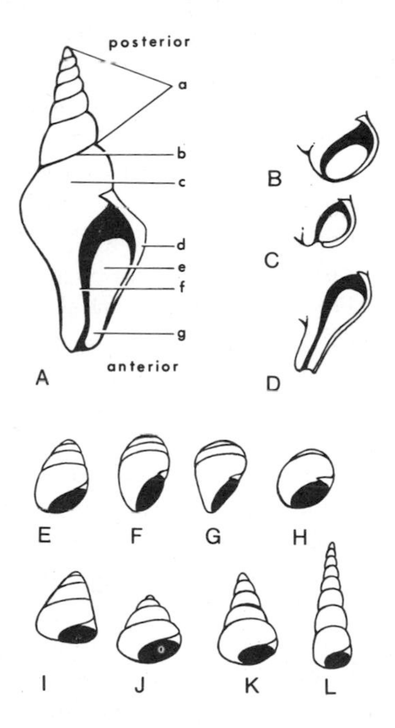

Fig. 16.2. Mollusca, Gastropoda. A, generalized gastropod shell; B–D, variations in aperture; B, without siphonal notch or anterior canal; C, with siphonal notch; D, with anterior canal; E–L are stylized shell forms; E, ovate; F, obovate; G, obconical; H, globular; I, trochoid; J, squat turbinate; K, tall turbinate; L, turriculate. Abbreviations are as follows: a, spire; b, suture; c, body whorl; d, outer lip; e, aperture; f, columella; g, anterior canal.

nuclear whorls, Figure 16.9C, which may contrast with the other whorls in texture or even in symmetry; this is the larval shell retained from the veliger. The sculpturing of the whorls is described as *axial* when it parallels the main growth axis of the shell as defined by the long axis of the *columella* or central column. *Spiral* sculpturing follows the turns of the individual whorls. The columella coils around a central cavity, the *umbilicus*, which may or may not be apparent externally. The presence or absence of an *anterior canal* or notch for the passage of the siphon is also important. In some sytematic accounts the siphonal canal is described as being "obsolete"; shells with this type of aperture are considered here with the truly holostomatous forms, i.e., those technically without a siphonal canal. For a full glossary of shell terms see Cox (1960).

The sexes are separate in most prosobranchs but are seldom conspicuously different externally. Among the exceptions are *Lacuna pallidula*, in which the male is a dwarf attached to the female shell. Most pulmonates and opisthobranchs are simultaneous hermaphrodites. Protandric hermaphroditism occurs among opisthobranchs in pteropods and most aeolidacean nudibranchs but is perhaps best known in the prosobranch *Crepidula* (see Morton, 1960). Similar sex reversals are known in other limpets and in a few genera scattered through the subclass. Self-fertilization is prevented in most hermaphroditic gastropods by one or another of various morphological, physiological, or behavioral mechanisms.

The disposition of the gametes and young also varies in different groups. Thus, in many archaeogastropods, fertilization is external; in other groups, fertilization is internal. Development is external in local marine gastropods generally. The eggs are shed singly in archaeogastropods, but in most others they are laid in gelatinous masses or in more elaborately formed capsules. A relatively few local species are ovoviviparous, i.e., with development within the female's body, *Littorina saxatilis* being one of them.

The initial larva is a trochophore in some archaeogastropods, but in most species with a free, pelagic larval phase the young hatches as a veliger. (See introductory account of Mollusca for a discussion of larval development.) Most neogastropods do not have a free pelagic phase and the young hatch as juvenile snails.

Gastropods occur on land and in fresh and salt water, on all substrata, and at all depths. They are chiefly benthic and some are fixed or sessile. They have also evolved pelagic forms in several unrelated higher taxa. Feeding adaptations include herbivores that feed either by rasping microscopic plant films, or as macrophagous browsers on coarse algae, or as suctorial forms tapping the liquid contents of plant cells. Other gastropods are scavengers on detritus, or they are deposit feeders; still others are carnivorous, and some are mucous-ciliary feeders utilizing plankters. The pyramidellids and a few genera in other groups are ectoparasites, while the most aberrant gastropods are highly modified endoparasites with hardly any gastropod features remaining in the adult state.

Systematic List and Distribution

The primary source for this list is Johnson (1934) with additions and systematic changes from more recent literature. The arrangement is based on Knight *et al.* (1960) and Franc (1968).

Order Archaeogastropoda		
Family Scissurellidae		
Scissurella crispata	B	7 to 1446 m
Family Fissurellidae		
Diodora cayenensis	C, 37°	Lit. to 9 m
Puncturella noachina	B	Lit. to >183 m
Family Acmaeidae		
Acmaea testudinalis	B, 41°	Lit. to shallow
Family Lepetidae		
Lepeta caeca	B	4 to 549 m
Family Cocculinidae		
Cocculina reticulata	C, 38°	128 to 538 m
Cocculina rathbuni	V	183 to 1128 m
Cocculina beani	V, 41°	183 to 1067 m
Family Lepetellidae		
Addisonia paradoxa	V	91 to 1171 m
Family Trochidae		
Calliostoma bairdii	V	119 to 461 m
Calliostoma occidentale	B, 40°	11 to 1674 m
Calliostoma sayanum	C+	119 to 366 m
Calliostoma euglyptum	C+	59 m
Solariella lamellosa	V	27 to 1250 m
Solariella obscura	B	5 to 732 m
Solariella lacunella	C, 37°	33 to 732 m
Margarites helicina	B	4 to 183 m
Margarites olivacea	B	13 to 146 m
Margarites groenlandica	B	Lit. to 274 m
Margarites johnsoni	B	73 m
Margarites multilineata	V	Deep
Margarites minutissima	B	Deep
Margarites ottoi	V	119 to 891 m
Margarites costalis	B	Lit. to 113 m
Seguenzia monocingulata	BV	183 to 3710 m
Moelleria costulata	BV	7 to 538 m
Order Mesogastropoda		
Family Lacunidae		
Lacuna vincta	B, 40°	Lit. to 37 m
Lacuna pallidula	B+	Lit. to 18 m
Lacuna parva	B+	Lit. to 31 m

Taxon	Distribution	Depth
Family Littorinidae		
Littorina littorea	B, 39°	Lit.
Littorina obtusata	B, 40°	Lit.
Littorina irrorata	V (39° southward)	Lit.
Littorina saxatilis	B, 40°	Lit.
Family Hydrobiidae		
Hydrobia minuta	B	Shallow, eury.
Hydrobia laevia	V	Shallow, eury.
Hydrobia salsa	V	Shallow, eury.
Hydrobia stimpsoni	V	Shallow, eury.
Family Rissoidae		
Alvania areolata	B	18 to 238 m
Alvania castanea	BV	4 to 187 m
Alvania exarata	BV	5 to 196 m
Alvania janmayenis	BV	37 to 915 m
Alvania latior	B	Deep
Alvania multilineata	B	9 m
Alvania carinata	B	177 to 366 m
Cingula aculeus	B, 38°	Lit. to shallow
Family Assimineidae		
Assiminea modesta	V	Lit.
Family Skeneopsidae		
Skeneopsis planorbis	BV	Shallow
Family Turritellidae		
Turritellopsis acicula	B	2 to 91 m
Tachyrhynchus erosa	B	13 to 137 m
Family Architectonicidae		
Heliacus bisulcata	V	101 to 805 m
Family Vermetidae		
Vermicularia spirata	V	5 to 320 m
Family Caecidae		
Caecum pulchellum	V+	2 to 95 m
Caecum cooperi	V+	—
Caecum johnsoni	V	15 to 18 m
+ several accidental records		
Family Cerithiidae		
Bittium alternatum	BV+	Lit. to 37 m
Bittium varium	V	Shallow
Bittium virginica	V	Shallow
Cerithiopsis greeni	V	Lit. to 18 m
Cerithiopsis costulata	A+	—
Cerithiopsis subulata	V	2 to 60 m
Seila adamsii	V	Lit. to 73 m
Family Triphoridae		
Triphora perversa	V	Lit. to 55 m
Triphora pyrrha	V	Lit.
Family Epitoniidae		
Epitonium angulatum	V	To 46 m
Epitonium clathratulum	V	90 to 300 m

Epitonium dallianum	V	115 to 351 m
Epitonium humphreysi	V	Lit. to 52 m
Epitonium multistriatum	V	15 to 220 m
Epitonium novangliae	B	Lit. to 457 m
Epitonium pourtalesi	V	90 to 267 m
Epitonium costulatum	B	18 to 73 m
Epitonium greenlandicum	B+	18 to 199 m
Epitonium rupicolum	V	Lit. to 37 m
Epitonium championi	V	Lit. to 40 m
Family Janthinidae		
Janthina janthina	V	Pelagic, oceanic
Family Trichotropidae		
Trichotropis borealis	A+	Lit. to 165 m
Trichotropis conica	B	137 m
Torellia fimbriata	BV	95 to 472 m
Torellia vestita	BV	157 to 274 m
Family Calyptraeidae		
Crucibulum striatum	BV	2 to 344 m
Crepidula fornicata	BV	Lit. to 27 m
Crepidula plana	BV	Shallow
Crepidula convexa	BV	Shallow
Family Capulidae		
Capulus ungaricus	BV	2 to 838 m
Family Aporrhaidae		
Aporrhais occidentalis	BV	91 to 640 m
Family Atlantidae		
Atlanta peroni	V	Pelagic, oceanic
+ additional species		
Oxygyrus keraudreni	V	Pelagic, oceanic
Family Carinariidae		
Carinaria lamarcki	V	Pelagic, oceanic
Family Pterotracheidae		
Firoloida leseurii	V	Pelagic, oceanic
Family Naticidae		
Natica clausa	BV	29 to 2813 m
Natica pusilla	B	4 to 27 m
Polinices nana	A+	40 to 538 m
Polinices groenlandica	A+	—
Polinices duplicata	V+	Lit. to shallow
Polinices heros	BV	Lit. to 435 m
Polinices levicula	B	48 to 183 m
Polinices triseriata	BV	2 to 115 m
Polinices pallida	BV	128 to 146 m
Polinices immaculata	BV	9 to 201 m
Sinum perspectivum	V	Shallow
Amauropsis islandica	B	49 to 146 m
Bulbus smithi	B	91 m
Family Cassididae		
Phalium granulatum	C+	Shallow

Family Lamellariidae		
Marsenina ampla	B	—
Marsenina glabra	B	27 to 838 m
Marsenina prodita	B	—
Velutina laevigata	B	5 to 91 m
Velutina undata	B	37 to 55 m
Order Neogastropoda (Stenoglossa)		
Family Muricidae		
Nucella lapilla	B+	Lit.
Urosalpinx cinereus	BV	Shallow
Eupleura caudata	V	2 to 15 m
Boreotrophon clathratus	B	Deep
Boreotrophon lintoni	B	128 m
Boreotrophon truncata	B	18 to 91 m
Family Columbellidae		
Anachis avara	BV	Lit.
Anachis haliaecti	BV	55 to 1171 m
Anchis translirata	V+	4 to 37 m
Anachis obesa	C, 37°	Shallow water
Mitrella lunata	BV	Lit.
Mitrella zonalis	B	15 to 110 m
Mitrella rosacea	B	5 to 110 m
Mitrella diaphana	V	119 to 891 m
Mitrella pura	V	128 to 891 m
Mitrella multilineata	V	16 to 370 m
Family Buccinidae		
Buccinum undatum	B, 40°	Lit. to 183 m
Buccinum cyaneum	B	128 to 862 m
Buccinum tenue	B	77 to 168 m
Buccinum abyssorum	B	90 to 2624 m
Buccinum tottenii	B	15 to 91 m
Neptunea decemcostata	B	18 to 91 m
Neptunea despecta	B	18 to 862 m
Colus stimpsoni	BV	2 to 862 m
Colus pubescens	BV	33 to 1171 m
Colus ventricosis	B	Deep
Colus sabinii	B	Deep
Colus pygmaeus	BV	2 to 1171 m
Family Melongenidae		
Busycon carica	V	Lit. to 18 m
Busycon canaliculatum	V	2 to 15 m
Busycon contrarium	C, 40°	Lit. to 5 m
Family Nassariidae		
Nassarius vibex	V	Lit. to shallow
Nassarius trivittata	BV	Lit. to 82 m
Ilyanassa obsoleta	BV	Lit.
Family Cancellariidae		
Admete couthouyi	B	18 to 110 m
Family Marginellidae		
Prunum apicinum	V	117 to 183 m

Family Turridae		
Daphnella limacina	V	155 to 1473 m
Mangelia cerina	V	—
Mangelia comatropis	V	91 to 1967 m
Mangelia dalli	V	172 to 2459 m
Pleurotomella packardi	B	155 to 2943 m
Pleurotomella agassizi	V	66 to 2943 m
Pleurotomella blakeana	V	183 to 2943 m
Lora spp. (more than a dozen species)	B or B+	9 to 891 m
Family Terebridae		
Terebra dislocata	C, 37°	Shallow

SUBCLASS OPISTHOBRANCHIA

Shell-less species are preceded by an "n."

Order Cephalaspidea		
Family Acteonidae		
Acteon punctostriatus	V	Lit. to 115 m
Family Ringiculidae		
Ringicula nitida	BV	183 to 935 m
Family Diaphanidae		
Diaphana minuta	B+	11 to 666 m
Family Bullidae		
Bulla gemma	Cape Cod	183 to 210 m
Family Retusidae		
Retusa obesiuscula	V	115 to 307 m
Retusa obtusa	BV	18 to 538 m
Retusa canaliculata	BV	Shallow
Family Scaphandridae		
Scaphander punctorostriatus	BV	37 to 2685 m
Cylichna alba	BV	18 to 1996 m
Cylichna occulta	B	Shallow
Cylichna oryza	B+	4 to 7 m
Cylichna gouldii	B	48 to 62 m
Family Philinidae		
Philine angulata	B	110 to 146 m
Philine finmarchia	B	29 to 165 m
Philine fragilis	B	161 to 168 m
Philine lima	B	—
Philine tineta	B	119 m
Philine quadrata	BV	37 to 732 m
Philine tincta	B	119 m
Family Atyidae		
Haminoea solitaria	V+	Shallow

Order Sacoglossa		
Suborder Polybranchiacea		
Family Hermaeidae (Stiligeridae)		
n *Hermaea bifida*	V	—
n *Stiliger fuscatus*	BV	Lit., eury.
n *Alderia harvardiensis*	B	Lit., eury.
Suborder Elysiacea		
Family Elysiidae		
n *Elysia chlorotica*	BV	Shallow
n *Elysia catula*	BV	Shallow
Family Limapontiidae		
n *Limapontia zonata*	B	Lit.
Order Notaspidea (Pleurobranchacea)		
Family Pleurobranchiidae		
n *Pleurobranchaea tarda*	V	51 to 567 m
Order Thecosomata		
Suborder Euthecosomata		
Family Spiratellidae (Limacinidae)		
Spiratella retroversa	B	Pelagic
Spiratella helicina	B	Pelagic
Spiratella trochiformis	V	Pelagic
Spiratella bulimoides	C, 39°	Pelagic
Spiratella leseuri	V	Pelagic
Spiratella inflata	V	Pelagic
Family Cavoliniidae		
Creseis acicula	BV	Pelagic
Creseis virgula	V	Pelagic
Styliola subula	V	Pelagic
Cavolina tridentata	V	Pelagic
Cavolina inflexa	V	Pelagic
Cavolina longirostris	BV	Pelagic
Cavolina gibbosa	V+	Pelagic
Cavolina uncinata	V	Pelagic
Cavolina quadridentata	V	Pelagic
Cavolina trispinosus	BV	Pelagic
Euclio cuspidata	BV	Pelagic
Euclio pyramidata	BV	Pelagic
Euclio recurva	V	Pelagic
Euclio polita	BV	Pelagic
Hyalocylis striata	V	Pelagic
Suborder Pseudothecosomata		
Family Peraclidae		
Peraclis reticulata	V	Pelagic
Peraclis bispinosa	C, 38°	Pelagic
Family Cymbuliidae		
Cymbulia calceola	V	Pelagic
Order Gymnosomata		
Family Pneumodermatidae		
n *Pneumoderma atlanticum*	BV	Pelagic

Taxon	Region	Habitat
Family Clionidae		
n *Clione limacina*	BV	Pelagic
n *Paedoclione doliiformis*	B	Pelagic
n *Notobranchaea macdonaldi*	V	Pelagic
Order Nudibranchia		
Suborder Doridacea		
Family Corambidae		
n *Corambella depressa*	V	Pelagic, on *Sargassum*
n *Corambe obscura*	V	Pelagic
Family Okeniidae		
n *Okenia pulchella*	B	Pelagic
n *Okenia quadricornis*	V	Pelagic
n *Ancula gibbosa*	B	Lit.
Family Lamellidorididae		
n *Adalaria proxima*	B	Lit.
n *Acanthodoris pilosa*	B+	Lit.
n *Onchidorus fusca*	B+	Lit. to 8 m
n *Onchidorus aspersa*	B+	Lit. to 38 m
n *Onchidorus diaphana*	B	Lit.
n *Onchidorus diademata*	B	Deep
+ additional Boreal species		
Family Polyceridae		
n *Issena lacera*	B	46 to 168 m
n *Issena ramosa*	B	183 to 238 m
n *Polycera emertoni*	B+	Lit. to shallow
n *Polycera davenporti*	B+	5 m
n *Polycera dubia*	B+	Lit. to 37 m
Family Echinochilidae		
n *Echinochila laevis*	B	Lit. to 274 m
Family Geitodorisidae		
n *Geitodoris complanata*	V	155 to 267 m
Suborder Dendronotacea		
Family Scyllaeidae		
n *Scyllaea pelagica*	V	On *Sargassum*
Family Dendronotidae		
n *Dendronotus frondosus*	B+	Lit. to 110 m
n *Dendronotus robustus*	B	35 to 351 m
Family Tethyidae		
n *Tethys fimbria*	V	Lit.
Family Dotonidae		
n *Doto coronata*	BV	Shallow
n *Doto formosa*	B+	18 to 91 m
Family Phylliroidae		
n *Phylliroe bucephala*	BV	Parasitic on *Zanclea*
Suborder Aeolidacea		
Family Coryphellidae		
n *Coryphella verrucosa*	B+	37 to 165 m
n *Coryphella stimpsoni*	B	2 to 93 m
n *Coryphella rutila*	B	Lit.
n *Coryphella nobilis*	B	137 m
+ additional Boreal species		

Family Eubranchidae		
n *Eubranchus exigua*	B	Lit.
n *Eubranchus tricolor*	B	—
Family Cuthonidae		
n *Cuthona pustulata*	B	Lit.
n *Cuthona concinna*	B	—
n *Tergipes despectus*	B	2 to 15 m
n *Embletonia fuscata*	B	Shallow
Family Fionidae		
n *Fiona pinnata*	V	On *Sargassum*
Family Facelinidae		
n *Facelina bostoniensis*	B+	2 to 27 m
Family Cratenidae		
n *Cratena aurantia*	B+	Lit.
n *Cratena gymnota*	B+	Shallow
n *Cratena pilata*	B+	2 to 267 m
n *Cratena veronicae*	B	42 to 57 m
Family Glaucidae		
n *Glaucus atlanticus*	V	Pelagic
Family Aeolidiidae		
n *Aeolidia papillosa*	B	Lit. to 366 m

Families of Uncertain Status. Knight *et al.* (1960) place the following families in Opisthobranchia, but Franc (1968) has them in Mesogastropoda (Prosobranchia) between the Epitonoiidea (families Epitonidae and Janthidae) and the Stromboidea.

Family Stiliferidae (Hermaeidae?)		
Stilifer stimpsoni	B+	64 m
Family Aclididae		
Aclis spp.	BV	183 m or more
Family Pyramidellidae		
Pyramidella (ca. 6 spp.)	BV	Shallow to deep
Turbonilla (ca. 20 spp.)	BV	0.5 to >183 m
Odostomia (ca. 13 spp.)	BV	Lit. to >183 m
Subclass Pulmonata		
Order Basommatophora		
Family Ellobiidae		
Melampus bidentatus	BV	Lit., salt marshes
Melampus floridanus	C, 37°	Lit., salt marshes
Ovatella myosotis	BV	Lit., salt marshes

Identification

The shelled and shell-less gastropods are considered separately.

Shelled Gastropods. The following treatment is completely artificial and is based on shell form. The species are divided in 8 primary groups.

Measurements refer to greatest dimension, height or width, of adult shells. Names in parentheses indicate illustration sources other than Abbott (1954, 1968) or Morris (1951).

---

1. Shell an open coil, uncoiled, tubular or conical, or caplike or, if typically snail-like, then transparent or highly translucent, not bulloid or cylindrical; Figs. 16.3–16.5.
- Shell caplike, uncoiled or feebly coiled: limpets, Fig. 16.3. Group One. (Note Group Five below, Fig. 16.7G–I.)
- Shell tubular or wormlike with an open, often irregular coil, Fig. 16.4 Group Two.
- Shell transparent or highly translucent, typically snaillike, or conical, or highly modified, Fig. 16.5; pelagic species. Group Three.

1. Shell a closed coil or otherwise not as above; for shell terms see Fig. 16.2 in addition to the figures cited—2.

2. Aperture without a siphonal notch or anterior canal:
- Shell bulloid, ovate, obovate, or cylindrical. Fig. 16.6. Group Four.
- Shell planispiral or much broader than wide, or inconspicuously coiled and with a large aperture; Fig. 16.7 Group Five.
- Shell trochoid, squat turbinate, or globular, with height and width nearly equal or less than 1½ times higher than wide; Fig. 16.8. Group Six.
- Shell tall turbinate or turriculate; Fig. 16.9. Group Seven.

2. Aperture with a siphonal notch or anterior canal; shells mostly tall turbinate, turriculate, or fusiform; Figs. 16.10, 16.11. Group Eight.

Group One, Limpets. The limpet mode has been adopted by quite unrelated taxa. In these gastropods the shell is caplike or a depressed cone adapted for close adherence to a firm substratum. Local examples occur especially in littoral and shallow water habitats; a few range into or are confined to deep water. These species are found on a wide variety of firm surfaces including the shells of other mollusks and crustaceans and the fronds of seaweeds and marine grasses. The heteropod, *Carinaria lamarcki* (Group Three), also has a caplike shell; the shell is thin and transparent; Figure 16.3C; see Abbott (1954).

---

1. Concave surface of shell with a deck or other hard structure; Fig. 16.3H,I—6.

1. Concave surface without such structures—2.

2. With an opening or "keyhole" at or near the apex—3.

2. Without an apical hole—4.

3. With a small, slitlike opening in front of the apex; shell delicate, with about 21–26 smooth ribs; white to brownish; 13 mm; Fig. 16.3A. *Puncturella noachina.*

3. With a keyholelike opening at the apex; shell strong, with more than 40 ribs roughened by concentric threads; 50 mm; Fig. 16.3B. *Diodora cayenensis.*

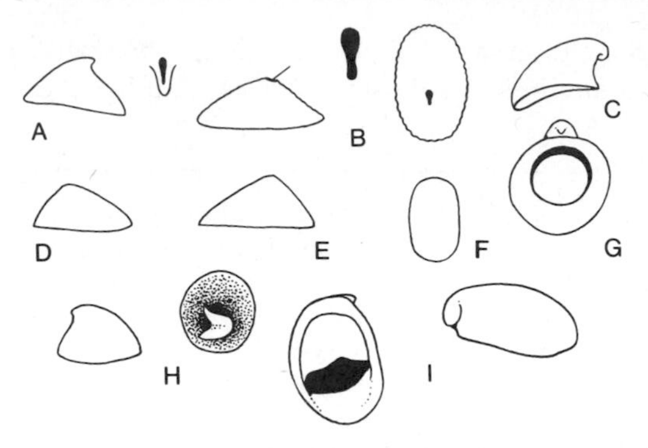

Fig. 16.3. Mollusca, Gastropoda, Group One. A, *Puncturella noachina*, with detail of opening; B, *Diodora cayenensis*, lateral and dorsal views and detail of keyhole; C, *Capulus ungaricus*, lateral view (the figure also serves for *Carinaria lamarcki*); D, *Acmaea testudinalis;* E, *Addisonia paradoxa;* F, *Acmea testudinalis*, form *alveus;* G, *Capulus ungaricus*, aperture view; H, *Crucibulum striatum*, lateral and aperture views; I, *Crepidula fornicata*, aperture and lateral views.

4. Apex curling above the posterior margin; commensal on bivalves including *Pecten*, *Chamys*, *Modiolus*, and *Astarte*, on brachiopods such as *Macandrevia* and possibly on other mollusk or annelid species; 30 mm; Fig. 16.3C,G (Franc, 1968.) *Capulus ungaricus*.

4. Apex near center or only slightly displaced from center—5.

5. Shell moderately high; apex blunt, behind the center; 13 mm; in deep water, usually more than 90 m; Fig. 16.3E (Knight *et al.*, 1960). *Addisonia paradoxa.*

5. Apex at or somewhat in front of center; usually in shallow water or littoral:

- Interior bluish white with brown center and marginal brown bars; shell variable in shape; 38 mm; Fig. 16.3D. *Acmaea testudinalis.* The form *alveus*, with uncertain systematic status, occurs on eelgrass; it is long, thin, and more strongly marked than the typical form; Fig. 16.3F.
- Interior white or pinkish; shell fragile, whitish or brownish; 13 mm. *Lepeta caeca.*

6. Concave surface with a hard, cuplike process; interior yellow-white to orange-brown, the cup white; 25 mm; Fig. 16.3H. *Crucibulum striatum.*

6. Concave surface with a shelf; Fig. 16.3I. *Crepidula:*

- Milky white; thin, flat shell; 32 mm. *C. plana.*
- Shelf occupies about ⅓ of aperture; brown with darker brown spots or stripes; 13 mm. *C. convexa.*
- Shelf occupies about ½ of aperture; whitish marked with spots or stripes of reddish brown; 40 mm; Fig. 16.3I; commonly found attached in a stack or series, the lowest members of which are female, the upper members male, and with transitional stages between. *C. fornicata.*

Group Two. The West Indian *Vermicularia spirata*, Fig. 16.4A, has been found as far north as the Woods Hole area: a second species, *V. knorri*, ranges from North Carolina southward. The mature snail reaches 100 mm by

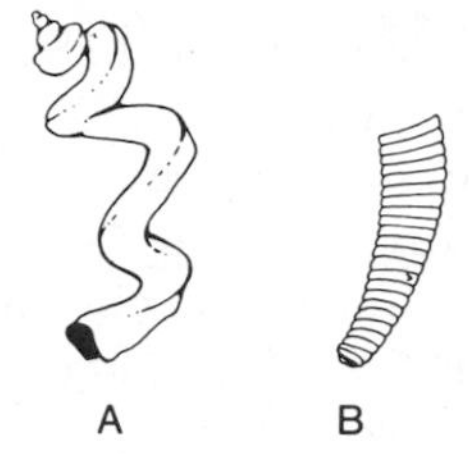

Fig. 16.4. Mollusca, Gastropoda, Group Two. A, *Vermicularia spirata;* B, *Caecum.*

about 6 mm and is unmistakable with its twisted, wormlike shell. The young snail, about 8 mm long, somewhat resembles *Turritellopsis.* The early whorls are dark in *spirata*, white in *knorri.*

Caecum is an equally distinct genus of very small, tubular, or cucumber-shaped snails found in dead sponges or among sand grains (Fig. 16.4B). *C. cooperi* reaches 5 mm in length, but *C. johnsoni* and *C. pulchellum* are only 2.5 and 2.0 mm, respectively. *C. pulchellum* has 20–30 circular ribs; *C. cooperi* has about 15 longitudinal ribs, and *C. johnsoni* is only marked with growth lines. The shells are white to horn-colored. Additional species have been transported accidentally in ballast sands.

Group Three, Pelagic Species. The majority of pelagic species belong to one or the other of two natural groups—the Heteropoda which is variously regarded as a subordinal or superfamilial aggregation of Prosobranchia, and the Thecosomata which contains the shelled pteropods. In life the shell in both groups is transparent, translucent, or only partly calcified; in preserved individuals the shell may become opaque and whitish. Two additional pelagic prosobranch genera are: *Litiopa*, Figure 16.9E, with a fragile, snail-like, brownish shell, found on *Sargassum*, and *Janthina*, Figure 16.7A, also snaillike with a delicate purple shell.

The name Heteropoda is used here for a prosobranch group that is also known as Atlantacea or Atlantoidea. In less modified atlantids the flattened spiral shell is large enough to contain the contracted animal. *Atlanta* has a keel or flange running along almost the entire margin of the shell, Figure 16.7J. *Oxygyrus* is only keeled at one end. The highly modified *Carinaria* has a large, somewhat fishlike body with a finlike foot and a caplike shell that is too small to contain the whole animal, Figure 16.3C. All three genera have conspicuous eyes with a spherical, amber-colored lens backed by black pigment. The shells of *Atlanta* and *Oxygyrus* may reach 13 mm diameter; *Carinaria* reaches about 50 mm shell length, but the whole animal may be 250 mm long. The latter is a predator on fish and jellyfish.

The shell-bearing and shell-less pteropods are now classified as separate orders, Thecosomata and Gymnosomata, respectively, in the subclass

Opisthobranchia; the older polyphyletic name Pteropoda has thus been abandoned. Thecosomes are chiefly oceanic, and some impression of their abundance may be gained by the significant contribution they make to deep ocean sediments. Pteropod oozes cover about 1½% of the Atlantic Ocean bottom at an average depth of more than 2000 m. Some of the Arctic species contribute substantially to the diet of the large baleen whales.

The muscular "wings" or parapodia with which thecosomes swim are responsible for the popular name "sea butterflies." The wings are usually withdrawn in preserved specimens. The shells may be colorless or yellowish brown in life, but they lose some of their transparency in preservative. Many of the species are virtually cosmopolitan. As a group they are most abundant in warm seas. They may occur in swarms and, as with other plankters, their occurrence is often cyclic and distinctly seasonal. For identification refer to Figure 16.5. *Spiratella* may be mistaken for a larval gastropod; the shell is sinistral, however, and most—but by no means all—gastropods are dextral. According to Wickstead (1965) the lack of a prominent eye spot or spots in *Spiratella* distinguishes this genus from larval gastropods; also, the withdrawn wings appear fingerlike through the shell.

The shell, though transparent or nearly so, is actually calcareous in most thecosomes. In *Cymbulia*, however, the shell is internal and cartilaginous, and is called a *pseudoconch*. The shells of the various thecosome species range in size in adults from about 1–30 mm.

For additional information see Franc (1968), Hardy (1956), Simroth (1911), Tesch (1949), and Wickstead (1965), as well as the references cited by Hyman (1967, p. 510). All of the thecosomes are figured by Abbott (1954) except *Cymbula*, for which see Hyman (1967) or Wickstead (1965). *Atlanta* and *Carinaria* are figured by Abbott (1954), and *Oxygyrus* by Keen (1963).

Group Four, Shelled Opisthobranchia and Pulmonata. The shell is internal in some species. Sizes for some species are given in the legend for Fig. 16.6.

---

1. Spire concealed, definitely depressed; aperture as long as shell—2.
1. Spire apparent, at least a little elevated; aperture shorter than total shell length—5.

2. Shell expanded, bubblelike, Fig. 16.6E–L—3.
2. Shell more cylindrical, Fig. 16.6A–D. *Cylichna:* The four species reported from the local fauna differ as figured. See Johnson (1915) for references to the original descriptions.

3. Shell internal, very thin, Fig. 16.6E,F. *Philine:* See Johnson (1915).
3. Shell external, thin—4.

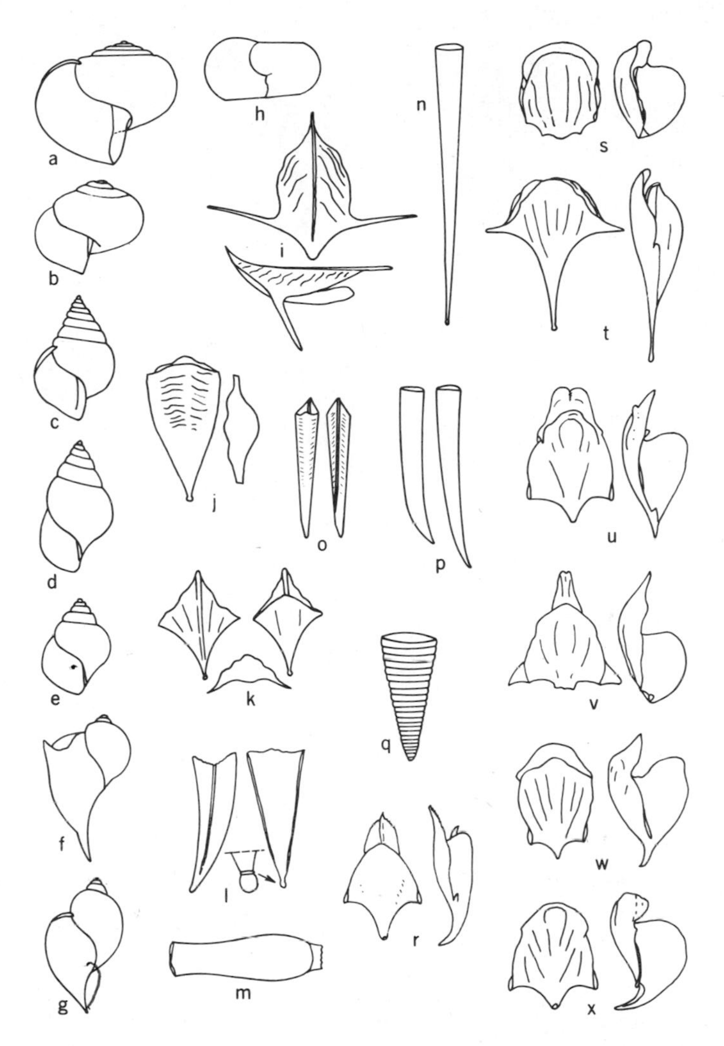

Fig. 16.5. Mollusca, Group Three, Thecosomata. a, *Spiratella helicina;* b, *Spiratella lesueuri;* c, *Spiratella retroversa;* d, *Spiratella bulimoides;* e, *Spiratella trochiformes;* f, *Peraclis bispinosa;* g, *Peraclis reticulata;* h, *Spiratella inflata;* i, *Euclio cuspidata;* j, *Euclio recurva;* k, *Euclio pyramidata;* l, *Euclio polita;* m, *Cuvierina columnella;* n, *Creseis acicula;* o, *Styliola subula;* p, *Creseis virgula;* q, *Hyalocylis striata;* r, *Cavolina inflexa;* s, *Cavolina quadridentata;* t, *Cavolina trispinosa;* u, *Cavolina tridentata;* v, *Cavolina longirostris;* w, *Cavolina gibbosa;* x, *Cavolina uncinata.* Reproduced from Abbott (1954) courtesy of D. Van Nostrand Company.

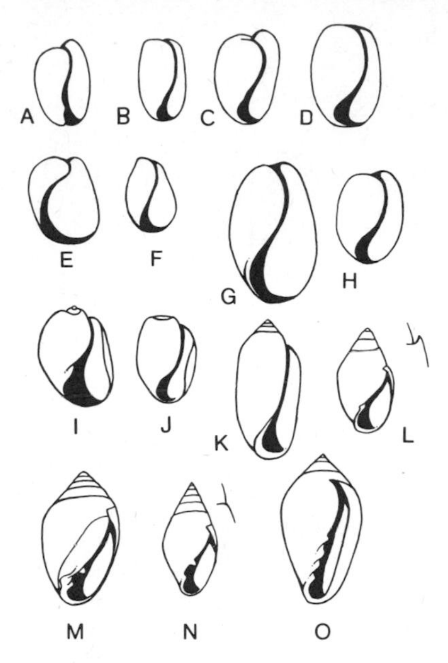

Fig. 16.6. Mollusca, Gastropoda, Group Four. A, *Cylichna oryza*, 3.7 mm; B, *Cylichna alba*, 3.0 mm; C, *Cylichna occulata*, 6.0 mm; D, *Cylichna gouldii*, 10.0 mm; E, *Philine quadrata*, 5.0 mm; F, *Philine lima*, 4.0 mm; G, *Scaphander punctostriatus;* H, *Haminoea solitaria;* I, *Diaphana minuta;* J, *Retusa obtusa;* K, *Retusa canaliculata;* L, *Acteon punctostriatus*, with detail of canaliculate suture; M, *Melampus bidentatus;* N, *Ovatella myosotis*, with detail of adpressed suture; O, *Prunum roscidum.*

4. Shell constricted posteriorly; surface microscopically pitted in numerous spiral rows; white with yellowish periostracum; 40 mm; Fig. 16.6G. *Scaphander punctostriatus.*

4. Shell not markedly constricted posteriorly; surface spirally grooved; white to amber; 13 mm; Fig. 16.6H. *Haminoea solitaria.* Note also *Bulla gemma* from depths greater than 183 mm.

5. Shell ovate or obovate, Fig. 16.6L–O—6.

5. Shell subcylindrical or flaring somewhat anteriorly, Fig. 16.6I–K; columella without folds or teeth:

- Shell thin, with flaring aperture; tan, translucent; 5 mm; Fig. 16.6I. *Diaphana minuta.*
- Shell thin; spire but little elevated; 3 mm; Fig. 16.6J. *Retusa obtusa.*
- Shell heavier; spire elevated; 6 mm; Fig. 16.6K. *Retusa canaliculata.*

6. Shell ovate; columella with one or more folds:

- Sutures canaliculate; body whorl with 10–15 spiral, pitted lines; white; 6 mm; Fig. 16.6L. *Acteon punctostriatus.* This is the only cephalaspid with an operculum.
- Sutures adpressed; white to bluish or brown; 7 mm; Fig. 16.6N. (Jacobson and Emerson, 1961). *Ovatella myosotis.*

6. Shell obovate; columella with 2 or more folds or teeth; with a color pattern of vague to well-defined dark bands:
- Columella with 4 folds; 8 mm; Fig. 16.6O. *Prunum roscidum.*
- Columella with 2 teeth; 10 mm; Fig. 16.6M (Jacobson and Emerson, 1961). *Melampus bidentatus.*

Group Five.

---

1. Shell transparent, glassy; 13 mm; Fig. 16.7J. *Atlanta peroni.*
1. Shell violet, fragile; 25 mm; Fig. 16.7A. *Janthina janthina.* This species is pelagic and oceanic, but may be found in beach drift of Gulf Stream flotsam.
1. Not so—2.

2. Aperture with a notch; 4 mm; Fig. 16.7B (Dall, 1889). *Scissurella crispata.*
2. Aperture without a notch—3.

3. With a distinct umbilicus, Fig. 16.7D,F—4.
3. Umbilicus not apparent—5.

4. Aperture relatively small; without a brown or velvety periostracum:
- Shell smooth; 2.5 mm; Fig. 16.7C (Miner, 1950). *Skeneopsis planorbis.*
- Shell sculptured with beaded whorls; 13 mm; Fig. 16.7D. (Warmke and Abbot, 1961). *Heliacus bisulcata.*
- Shell sculptured with numerous fine, axial ribs; 2.5 mm; Fig. 16.7E (Miner, 1950). *Moelleria costulata.*

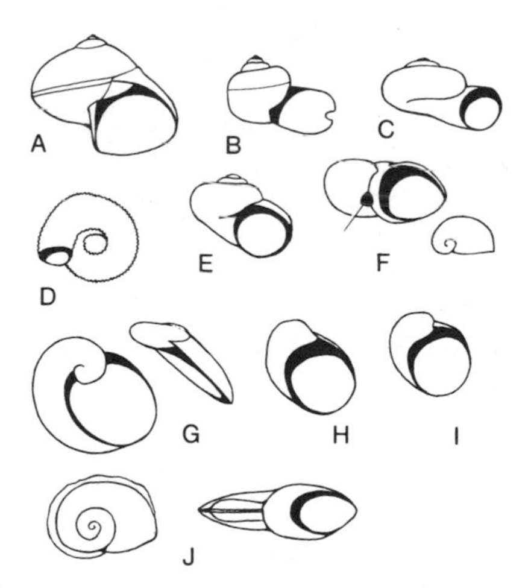

Fig. 16.7. Mollusca, Gastropoda, Group Five. A, *Janthina janthina;* B, *Scissurella crispata;* C, *Skeneopsis planorbis;* D, *Heliacus bisulcata;* E, *Moelleria costulata;* F, *Torellia* sp., two views with umbilicus indicated; G, *Sinum perspectinum,* two views; H, *Velutina* sp.; I, *Marsenina* sp.; J, *Atlanta peroni,* two views.

4. Aperture relatively large; shells light, fragile with a brown or velvety periostracum:
   - Boreal, deep water forms (in 90 m or more) ranging south of Cape Cod but at even greater depths; Fig. 16.7F. *Torellia* spp. See Johnson (1915) for citations.
   - Carolinian, shallow water species; 50 mm; Fig. 16.7G. *Sinum perspectinum.* In life the animal may envelop the shell.
5. Operculum present; aperture squarish; shell planorbid with 1 or 2 beaded cords above and below; 100 mm. *Pseudomalaxis nobilis.* See Abbott (1954),
5. Without operculum; aperture round; shell thin with a velvety periostracum. Lamellaridae. These are small gastropods, the largest species about 18 mm. They are sea sluglike in life with an internal shell. *Velutina* is shown in Fig. 16.7H; see Abbott (1954). *Marsenina* is shown in Fig. 16.7I; for all three local species see Verrill (1882). For literature citations see Johnson (1915).

Group Six.

1. Inside pearly—2.
1. Inside not pearly—5.
2. With umbilicus—3.
2. Umbilicus not apparent. *Calliostoma:*
   - Trochoid in shape, the sutures flush but with strongly beaded cords; brownish yellow, faintly blotched with reddish brown; 32 mm; Fig. 16.8A *C. bairdii.*
   - Whorls clearly differentiated; cords well developed, only vaguely beaded; yellow-ivory or opalescent; 17 mm; Fig. 16.8B. *C. occidentale.* The genus has been monographed by Clench and Turner (1960).
3. Umbilicus set off by a beaded rib or by a change in ribbing pattern; Fig. 16.8D. *Solariella*—4.
3. Ribbing of umbilical area not differentiated. *Margarites: See Table 16.1.*
4. Umbilicus bordered by an angular rim; 6 mm; Fig. 16.8C. *Solariella obscura.*
4. Umbilicus bordered by spiral row or rows of beads; Fig. 16.8D:
   - Whorls with 6 cords; 10 mm. *Solariella lacunella.*
   - Whorls with a single strong cord; 4 mm; Fig. 16.8E. *Solariella lamellosa.*

Table 16.1. *Margarites* Species

Species	Size, mm	Sculpture on Whorls	Color
ottoi	18	Beaded	White
helicina	12	Smooth	Yellow-green-pinkish brown
groenlandica	8	Ribbed	Reddish brown
olivacea	6	Smooth	Olive-brown
costalis	12	Ribbed	Rosy to white or greenish-tinged

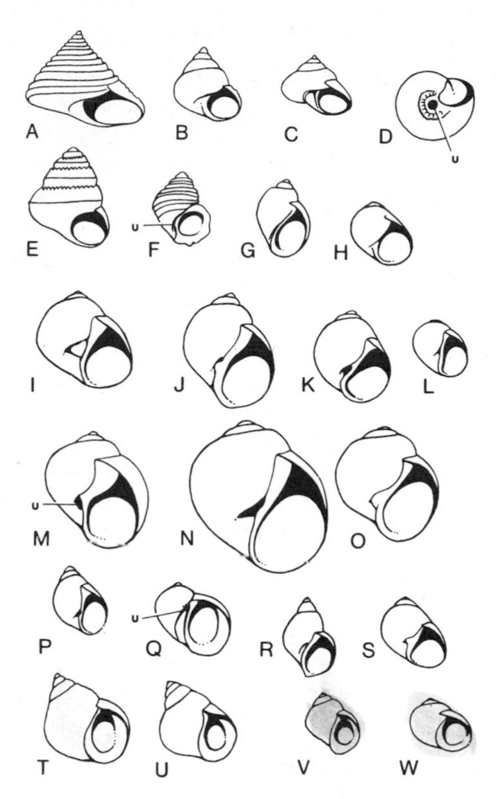

Fig. 16.8. Mollusca, Gastropoda, Group Six, sculpturing omitted from most figures. A, *Calliostoma bairdii;* B, *Calliostoma occidentale;* C, *Solariella obscura;* D, *Solariella lacunella;* E, *Solariella lamellosa;* F, *Fossarus elegans;* G, *Amauropsis islandica;* H, *Bulbus smithii;* I, *Polinices duplicata;* J, *Polinices groenlandica;* K, *Polinices triseriata;* L, *Polinices nana;* M, *Polinices levicula;* *N*, *Polinices heros;* O, *Natica clausa;* P, *Polinices immaculata;* Q, *Lacuna pallidula;* R, *Lacuna vincta;* S, *Natica pusilla;* T, *Littorina littorea;* U, *Littorina irrorata;* V, *Littorina saxatilis;* W, *Littorina obtusata.* The umbilicus is indicated for some species and marked "u."

5. With a round umbilicus, open, or partly, to wholly covered by a callus; Fig. 16.8I–P,S. Naticidae. See Table 16.2.

5. Umbilicus slitlike:
- Sculptured with cords and riblets; 3 mm; Fig. 16.8F. *Fossarus elegans.*
- Shell smooth; 38 mm; Fig. 16.8G. *Amauropsis islandica.*

5. With umbilicus absent or groovelike; Fig. 16.8Q—6.

6. With a conspicuous, greenish yellow periostracum; 20 mm; Fig. 16.8H (Morris, 1951). *Bulbus smithii.*

6. Periostracum thin, brownish, inconspicuous or absent—7.

7. Inner lip with a groovelike umbilicus. *Lacuna: L. vincta*, Fig. 16.8R, 10 mm, with a relatively high spire, contrasts with *L. pallidula* and *L. parva* with a low spire,

Table 16.2. Species of *Polinices* and *Natica*

Species	Size mm	Color	Umbilicus Open or Callous-Covered	Fig. 16.8	Figure Sources
P. nana	6	White	Open?	L	Verrill (1882)
P. duplicata	65	Grayish	Almost covered	I	Abbott (1954)
P. heros	115	Grayish	Open	N	Abbott (1954)
P. levicula	40	White	Open	M	Dall (1889)
P. triseriata	22	Spotted or banded	Open	K	Abbott (1954)
P. groenlandica	25	White	Almost covered	J	Abbott (1954)
P. immaculata	10	White	Open	P	Dall (1889)
N. clausa	32	Spotted	Closed	O	Abbott (1954)
N. pusilla	8	Spotted	Closed	S	Abbott (1954)

Fig. 16.8Q. *L. parva* is frequently banded, *L. pallidula* never is. Abbott (1954) figures *L. vincta;* for *L. pallidula* see Sars (1878), and for *L. parva* see Forbes and Hanley (1853).

7. Inner lip without a groovelike umbilicus; Fig. 16.8T–W. *Littorina: L. littorea* and *L. irrorata* reach 25 mm and are thick shelled, Fig. 16.8T,U; *L. littorea* is relatively smooth, while *L. irrorata* has spiral grooves and an orange-brown columella. *L. irrorata* is a marsh species, extinct north of New Jersey; *L. littorea* is a European Boreal species that extended its range during the 19th century and is now found as far south on this coast as Maryland. *L. saxatilis,* Fig. 16.8V, is a northern species resembling *L. littorea* but with spiral cords. *L. obtusata* is a smooth species, highly variable in color and pattern with a very low spire, Fig. 16.8W. *L. obtusata* and *L. saxatilis* reach 8 mm.

Group Seven. This group contains several assemblages of difficult genera of very small gastropods that will not be keyed to species. For a more extended account of the pyramidellids see Bartsch (1909); many of the species are full- or part-time parasites on echinoderms, annelids, or other mollusks. There is confusion in the literature regarding generic allocations in this family and in Rissoidae.

1. Columella with 1 or more folds—2.

1. Columella simple—3.

2. Columella with 1 fold; 2–3 mm; Fig. 16.9B. *Odostomia* spp. More than a dozen species are listed by Johnson (1934).

2. Columella with 2 or 3 or more folds; to about 6 mm; Fig. 16.9A. *Pyramidella* spp. A half-dozen species are listed by Johnson (1934).

3. Nuclear whorls heterostrophic, Fig. 16.9C; to about 7 mm. *Turbonilla* spp. About 20 species are listed by Johnson (1934).

3. Nuclear whorls normal—4.

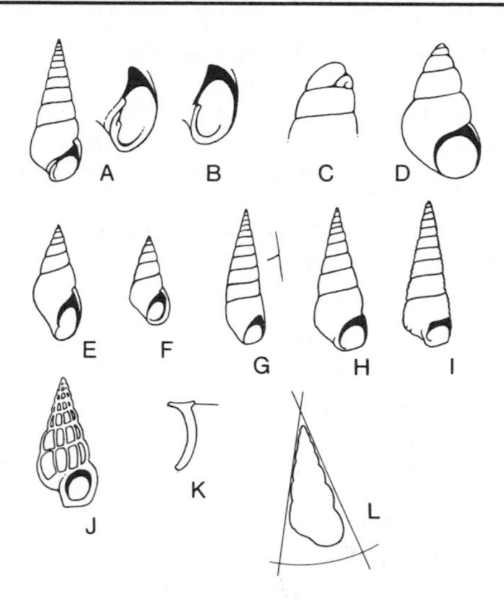

Fig. 16.9. Mollusca, Gastropoda, Group Seven, sculpturing omitted except in J. A, *Pyramidella* sp., with detail of aperture; B, *Odostomia* sp., aperture; C, *Turbonilla* sp., apex with heterostrophic nuclear whorl; D, *Hydrobia minuta;* E, *Litiopa melanostoma;* F, rissoid species; G, *Balcis* sp., with detail of flush suture; H, *Acirsa costulata;* I, *Tachyrhynchus erosa;* J, *Epitonium* sp.; K, costa with hook indicated above; L, *Epitonium* sp., shell outline indicating spiral angle.

4. Whorls 6 or less—5.

4. Whorls 8–12—6. Note also *Bittium* with 6–8 whorls and a weakly developed or absolete anterior notch; see Group 8. Also *Litiopa melanostoma*, a *Sargassum* species, 5.0 mm, Fig. 16.9E.

5. Shell smooth or only finely striated; yellowish to rusty brown; 4 mm; Fig. 16.9D. *Hydrobia minuta*. In brackish waters, salt marshes, on *Ulva* and other plants, as well as on ocean beaches under debris. Two additional species are listed; see Johnson (1915, 1934).

5. Shell sculptured; species keying out here have been assigned by various authors to *Rissoa*, *Cingula*, *Alvania*, *Onoba*, etc.; the sculpturing may consist of either axial or spiral ribbing or striations, or both. The species are smaller than 4 mm; Fig. 16.9F.

6. Smooth; whorls flush; white or tinged with brown; whorls 10–13; 13 mm; Fig. 16.9 G (Warmke and Abbott, 1961). *Balcis* spp. For other genera of Turridae see Group Eight.

6. Shell sculptured—7.

7. Sculpturing consists chiefly of sharply defined axial riblets; whorls 8–12; Fig. 16.9J. *Epitonium*. The genus has been monographed by Clench and Turner (1951, 1952). See paragraph and tabular summary, below.

7. Not so:

- With ill-defined axial riblets and spiral striations; straw yellow; 33 mm; Fig. 16.9H (Clench and Turner, 1950). *Acirsa costulata*.

- With 5 or 6 flat spiral ribs or cords; whitish with brownish periostracum; 25 mm; Fig. 16.9I. *Tachyrhyncha erosum.*
- With 3 strongly defined spiral ribs plus fine axial striations; whitish to brown; 13 mm (Miner, 1950). *Turritellopsis acicula.*

Epitonium. The axial riblets or *costae* may be sharp and finlike or they may be broad and thickened. There is both ontogenetic and individual variation within species and, in some cases, intergradation between species. The shells are usually white, but *E. novangliae* is often banded with brown, as is *E. rupicolium;* the latter may be almost entirely brown.

The species are compared in Table 16.3 in two groups, one with spiral sculpturing and the other without or with sculpturing only visible at a magnification of about 15 times. The spiral angle is a rough measure of the relative broadness of the shell, Figure 16.9L. All of the species are figured by Clench and Turner (1951, 1952). The spiral sculpturing usually consists of fine striae or threads except in *E. championi* and *E. greenlandicum*, which have quite strong spiral cords; *E. greenlandicum* usually has a basal ridge and *E. championi* does not, Figure 16.9L.

Table 16.3. *Epitonium* Species

Species	Costae on Body Whorl	Spiral Angle	Size, mm
Group One			
multistriatum	16–19	22–28°	15
novangliae	9–16	22–28°	14 *
pourtalesi	11–14	35–40°	22 *
championi	8–9	20°	14
greenlandicum	9–12	22°	60
Group Two			
dallianum	20–30	17°	13
angulatum	9–10	28–30°	19 *
humphreysii	8–9	25–27°	18
rupicolum	12–18	35–40°	20

* = costae with hooks or angles above; see Fig. 16.9K. A method for measuring the spiral angle is indicated in Fig. 16.9L.

Group Eight.

1. Aperture equal to about ⅓ or less of total shell length; generally turriform shells with about 6–12 whorls—2.
1. Aperture equal to about ⅖ or more of total shell length—6.

2. Anterior canal short, weakly developed or obsolete; Fig. 16.10A–C. *Bittium: B. virginicum* and *B. varium* are about 3 mm long and have a *former varix*, Fig. 16.10B; *B. alternatum* does not have the varix and reaches about 6 mm; *virginicum* and *varium* differ in proportions as figured.

2. Anterior canal well developed—3.

3. Columella with 2 folds; gray to pinkish or orange; 50 mm; Fig. 16.10D. *Terebra dislocata.*

3. Columella simple—4.

4. Coiling sinistral (i.e., aperture on left facing); whorls 10–12. *Triphora:*

- Body whorl with 4 rows of tubercles, spire whorls with 2; 3 mm (Henderson and Bartsch, 1914). *T. pyrrha.*
- Body whorl with 4 rows of tubercles, spire whorls with 3; 6 mm. *T. perversa nigrocincta.*

4. Coiling dextral—5.

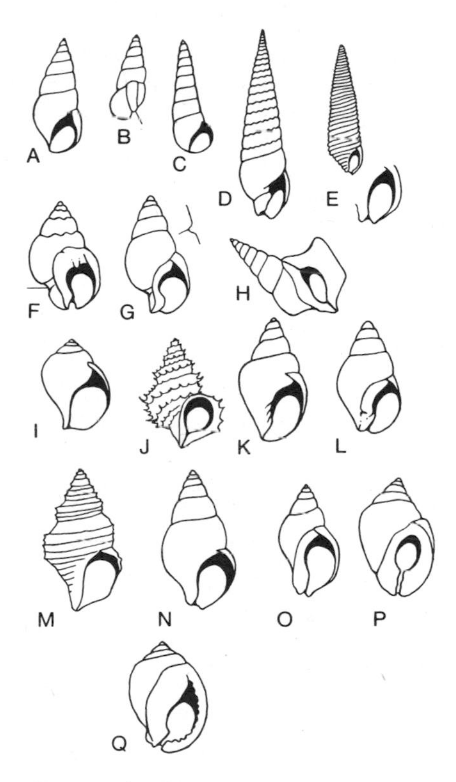

Fig. 16.10. Mollusca, Gastropoda, Group Eight, couplets 1–12, sculpturing indicated in E, J, and M only. A, *Bittium varium;* B, same with varix indicated; C, *Bittium virginicum;* D, *Terebra dislocata;* E, *Seila adamsii;* F, *Nassarius vibex,* with groove indicated; G, *Nassarius trivittatus,* with detail of channeled suture; H, *Apporhais occidentalis,* adult; I, *Eudolium crosseanum;* J, *Trichotropis borealis;* K, *Admete couthouyi;* L, *Ilyanassa obsoleta;* M, *Buccinum abyssorum;* N, *Buccinum tottenii;* O, P, *Nucella lapilla,* two variants; Q, *Phalium granulatum.*

5. Sculpture spiral with or without fine axial threads; whorls 12; 13 mm; Fig. 16.10E. *Seila adamsii.*

5. Sculpture spiral and axial. *Cerithiopsis:*
- *C. greeni* with 9 whorls; 3 mm.
- *C. subulata* with 14 whorls; 19 mm. These two species are Virginian; an arctic species, *C. costulata*, ranges south to Eastport, Maine.

6. Anterior canal and body whorl separated by a groove; Fig. 16.10F. *Nassarius:*
- Outer lip thin; sutures channeled; 18 mm; Fig. 16.10G. *N. trivittatus.*
- Outer lip thick, toothed within; aperture relatively larger; 13 mm; Fig. 16.10F. *N. vibex.*

6. Anterior canal less sharply set off—7.

7. Outer lip flaring in a winglike expansion; 50 mm; Fig. 16.10H. *Apporhais occidentalis* (adult). (See 14 below for young.)

7. Not so—8.

8. Outer profile of aperture curving in a deep arc; whorls about 4–8; Fig. 16.10I–Q. —9.

8. Outer profile of aperture curving in a shallow arc or sinous; Fig. 16.11—13.

9. Spire short; aperture about ⅔ shell length; 90 mm; Fig. 16.10I. *Eudolium crosseanum.* Note also *Phalium granulatum*, 75 mm; Fig. 16.10Q.

9. Spire longer; aperture about ½ shell length or less—10.

10. Columella a simple arc, without folds or teeth; anterior canal scoop-shaped; yellowish brown periostracum with bristly or fibrous extensions; 18 mm; 4 whorls; Fig. 16.10J. *Trichotropis borealis.*

10. Columella angled posteriorly; anterior canal a groove or slot—11.

11. Columella with 2–5 folds; 19 mm; Fig. 16.10K. *Admete couthouyi.*

11. Columella with one or no folds—12.

12. Aperture brown or black; columella with fold above anterior canal; 25 mm; Fig. 16.10L. *Ilyanassa obsoleta.*

12. Aperture light or, if dark-colored, without a fold on columella above anterior canal; color and sculpture highly variable; 30 mm; Fig. 16.10O,P. *Nucella lapillus.*

12. Aperture light. *Buccinum:*
- *B. undatum* with axial and spiral sculpture; 100 mm.
- *B. tenue* with axial sculpture only; 63 mm.
- *B. abyssorum* with spiral sculpture only; 50 mm., Fig. 16.10M.
- *B. tottenii* without sculpture; 50 mm, Fig. 16.10N.

13. Spire short; Fig. 16.11A–C. *Busycon:*
- *B. contrarium* is sinistral; a deep water species; 400 mm; Fig. 16.11A.
- *B. carica* and *B. canaliculatum* are dextral (the latter rarely sinistral); *carica* has knobbed whorls while *canaliculatum* does not; 200 mm, Fig. 16.11B,C.

13. Spire long—14.

14. With about 8–10 whorls; aperture a narrow oval; with strong axial ribbing; 25 mm; Fig. 16.11D. *Apporhais occidentalis* (young).

14. Not so—15.

15. Outer lip with notch at posterior edge, the notch effecting growth lines on the spire; Fig. 16.11F. Turridae. This is another difficult group of mostly small gastropods. The genera have not been well defined, and there are numerous species, more than 20 according to Johnson (1934). For literature citations see

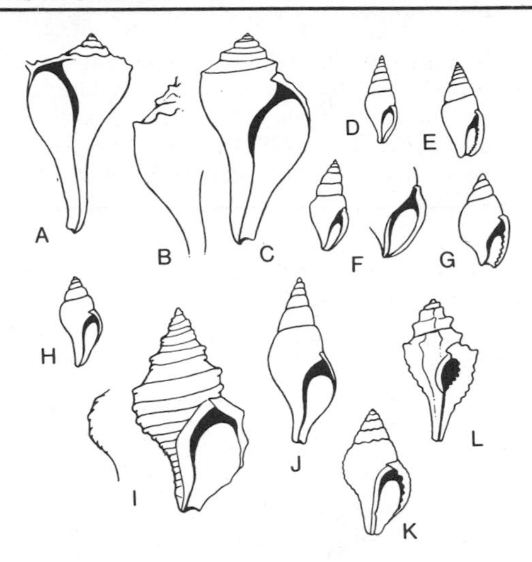

Fig. 16.11. Mollusca, Gastropoda, Group Eight, couplets 13–18, sculpturing omitted from most figures. A, *Busycon contrarium;* B, *Busycon carica*, outline of part of shell; C, *Busycon canaliculatum;* D, *Apporhais occidentalis*, young; E, columbellid species; F, turrid species with detail of aperture; G, columbellid species; H, *Boreotrophon clathratus;* I, *Neptunea decemcostata*, whole shell and outline of part of shell of *N. despecta;* J, *Colus* sp.; K, *Urosalpinx cinereus;* L, *Eupleura caudata.*

the latter. See also Keen (1958). Several of the common species of *Lora* are figured by Abbott (1968); Jacobson and Emerson (1961) figure *Mangelia cerina* and *M. plicosa*, and Morris (1951) figures *Pleurotomella packardi.*

15. Outer lip without notch at posterior edge—16.

16. Outer lip usually toothed within, feebly so in some species—18.

16. Outer lip smooth within—17.

17. With spiral and/or axial sculpture:
- With both axial and spiral sculpture; 13 mm; Fig. 16.11H. *Boreotrophon clathratus.*
- With axial ribs only. *Neptunea:* The ribbing is much stronger in *N. decemcostata* than in *N. despecta;* Fig. 16.11I; 100 mm. Note also *Ptychatractus ligatus* from deeper water; see Mighels and Adams (1842).

17. With spiral sculpture; Fig. 16.11J. *Colus:*
- Olive-gray, small species, 25 mm. *C. pygmaeus.*
- Brown: *C. stimpsoni*, 125 mm with aperture less than half shell length, and *C. pubescens*, 63 mm, with aperture more than ½ shell length.

18. Anterior canal well developed; outer lip of aperture more or less sinuous:
- Anterior canal open; axial ribs rounded; 18 mm; Fig. 16.11K. *Urosalpinx cinereus.*
- Anterior canal constricted, axial ribs sharp; 18 mm; Fig. 16.11L. *Eupleura caudata.*

18. Anterior canal shorter; outer lip of aperture a shallow arc; Fig. 16.11E,G. Columbellidae. These are small shells, generally less than 13 mm long. Ten species in two genera are listed by Johnson (1934), but there has been difficulty in defining the genera. Keen (1958) gives *Anachis* as having axial ribs and *Mitrella* without. Species of *Anachis* are figured by Abbott (1954), Clench and Turner (1950), and Verrill (1882). Abbott (1954) figures *Mitrella lunata*. For additional species, mostly from depths of 15 m or more, see Johnson (1934).

Shell-less Gastropods. The tendency toward shell reduction is chiefly developed in the subclass Opisthobranchia and reaches its logical conclusion in the complete elimination of a shell in the orders Sacoglossa, Notaspidea, Gymnosomata, and Nudibranchia. In species with shell-less adults a juvenile or larval shell may be present. Microscopic calcareous spicules are present in the skin of most doridacean nudibranchs. These vary considerably, and may be spindle-shaped, tri- or tetraradiate or irregular in form. The range of body form is indicated in the figures. Members of the order Gymnosomata are distinctly modified for a pelagic existence. In addition to other structural alterations, the foot is greatly reduced, and locomotion is accomplished with a pair of natatory structures called *parapodia*. Other naked opisthobranchs are generally more or less sluglike or resemble flatworms in general appearance. The dorsal surface may be smooth but is more commonly ornamented with papillae, tubercles, gills, or club-shaped respiratory structures called *cerata* (singular *ceras*). The mantle is entirely lacking in nudibranchs, Sacoglossa, and Gymnosomata. Many of the species are small, 10 mm or less in length, but a few attain lengths of several centimeters.

The head region is developed, but in simple forms is not sharply delimited. There is usually a single pair of eyes. Tentacles are usually present and are often differentiated as a frontal or cephalic pair and a posterior pair which are called *rhinophores*. The rhinophores vary considerably in structure in different groups. They are especially highly elaborated in nudibranchs, where they are often retractile in sheaths.

Naked opisthobranchs, like most other members of the subclass, are hermaphroditic; they may be simultaneously male and female or protandric, sometimes with the cycle repeated. Cross-fertilization is the usual practice with copulation rather than external fertilization. The young either pass through a veliger stage, usually with a spiral larval shell, or the larval stage is suppressed.

The gymnosomes and a few species in other orders are pelagic. Most nudibranchs are benthic; however, Fish (1925) reported *Facelina bostoniensis* occurring regularly in the spring plankton, and others may be found swimming or even crawling beneath the surface film. Some species range into or are confined to brackish waters, e.g., species of *Alderia*, *Embletonia*, *Limapontia*, *Corambe*, and *Tergipes*. Gymnosomes are carnivorous, and their principal prey are the pelagic shelled pteropods, Thecosomata. Nudibranchs are also predatory, and in many cases are quite "host-specific." Thus doridaceans feed mainly on sponges, tunicates, or bryozoans, and dendronotaceans and aeolidaceans prey on hydroids. In some cases, only a single species is preyed on by a particular kind of nudibranch. The pelagic nudibranchs *Glaucus* (Fig. 16.12E) and *Fiona* feed on the colonial hydrozoans, *Velella* and *Porpita*. It is well known that some nudibranchs are capable of incorporating hydroid nematocysts into their own defensive armament. This habit is chiefly known in the Aeolidacea but occurs as well in some dendronotaceans. The nematocysts are concentrated in the cerata in aeolidaceans. Saccoglossans are herbivorous.

Systematic List and Distribution

The naked forms are indicated as such in the general opisthobranch list, p. 265 ff. The classification used is that of Franc (1968). I have followed Pruvot-Fol (1954) in reducing some of the American species to synonymy with European ones, but quite a few species require restudy. Synonymies are especially complex in many groups.

Identification

Naked gastropods require special preparation for study. They are contractile and, unless narcoticized, may become badly distorted and difficult if not impossible to identify. Color is often a useful criterion, and the appearance of live or fresh material should be noted. For species determination in some genera it is necessary to examine the radular apparatus microscopically; no attempt is made here to carry such material beyond the genus, but the appropriate literature is cited. The keys which follow are highly artificial. The naked gastropods are considered in five subgroups.

---

1. Pelagic species highly modified for swimming, most conspicuously by the possession of a pair of winglike parapodia; foot greatly reduced; Fig. 16.12A–D. Group One. Gymnosomata.

1. Crawling or gliding species, generally sluglike or flatwormlike with a well-developed foot; a few species pelagic but not as above—2.

2. With a circlet of gills surrounding the anus, Fig. 16.12G–K. Group Two. Typical Doridacea.

2. Not so—3.

3. Cerata absent; body more or less flattened, fusiform, or sluglike; some species with 1 or 2 pairs of dorsal lobelike folds or processes; Fig. 16.13A–E. Group Three. Miscellaneous genera.

3. Cerata present—4.

4. Rhinophores retractile in sheaths; Fig. 16.13F–I. Group Four. Dendronotacea.
4. Rhinophores not retractile in sheaths; Fig. 16.14. Group Five. Most Aeolidacea and Saccoglossa.

Group One, Order Gymnosomata. The body is divided into a "head" bearing the cephalic structures and winglike parapodia, with a larger trunk section following. Sizes for individual species are not given in the key; gymnosomes rarely exceed 20 mm in length. *Paedoclione doliiformis*, Figure 16.12B, is sexually mature at 1.5 mm and was initially mistaken for a larva. All but *Paedoclione* are figured by Pruvot-Fol (1954); for that genus see Danforth (1907).

---

1. Mouth region with 2 or 3 arms bearing suckers; Fig. 16.12A. *Pneumoderma atlanticum*.
1. Mouth region with cones—2.
2. Minute, larvalike; trunk with two bands of cilia; Fig. 16.12B. *Paedoclione doliiformis*.
2. Not so—3.
3. Foot without posterior lobe; with 3 pairs of cones; Fig. 16.12C. *Clione limacina*.
3. Foot with posterior lobe; with 2 pairs of cones covered with papillae; penis very long, coiled; Fig. 16.12D. *Notobranchaea macdonaldi*.

Group Two, Order Nudibranchia, Doridacea. While the characteristic gill ring encircling the dorsal anus readily identifies members of this order, the differentiation of species or in some cases of genera may prove difficult. (The atypical corambids are considered in Group Three.) For technical keys see Odhner (1939). Dimensions are adult body length.

---

1. Gills and rhinophores retractile in depressions; mantle developed all around; back tuberculate; white, translucent with lateral rows of opaque white or sulfur yellow spots; 25 mm; Fig. 16.12J (Pruvot-Fol, 1954). *Echinochila laevis*.
1. Gills retractile but not in depressions; rhinophores retractile or not—2.
2. Rhinophores retractile in sheaths; back with a distinct margin; white with yellow gills and rhinophores; 25 mm; Fig. 16.12G. (Pruvot-Fol, 1954). *Issena lacera*.
2. Rhinophores without sheaths—3.
3. Mantle distinct from foot, Fig. 16.12L; body form as in Fig. 16.12J—5.
3. Mantle reduced to a low rim or absent, Fig. 16.12H,K—4.
4. Mantle absent; back with few clublike processes in gill area; whitish; 12 mm; Fig. 16.12H (Abbott, 1954). *Ancula gibbosa*.
4. Mantle reduced to a low rim; Fig. 16.12K. *Polycera* and *Okenia*. Five species are recorded, 3 in the first genus, 2 in the second, but no attempt will be made here to distinguish them. For generic diagnoses see Odhner (1939) and Pruvot-Fol (1954), and for citations of original descriptions see Johnson (1915, 1934). The generic synonymies are complex.

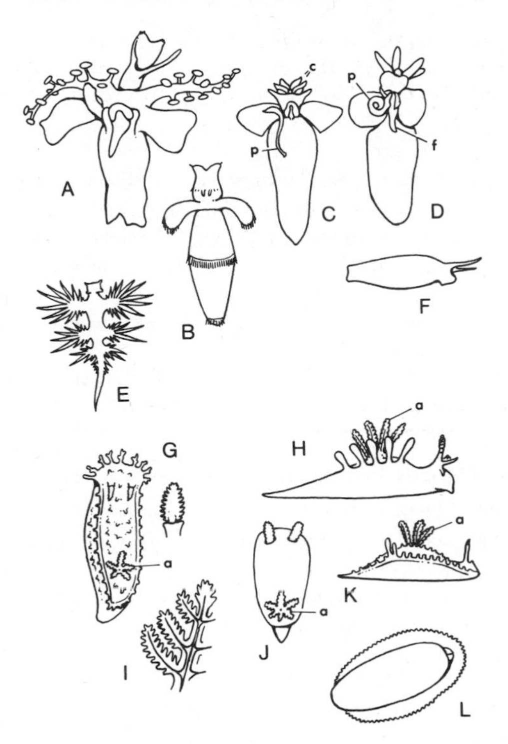

Fig. 16.12. Mollusca, naked Gastropoda, Groups One and Two. A, *Pneumoderma atlanticum;* B, *Paedoclione doliiformis;* C, *Clione limacina,* with penis (p) and cones (c) indicated; D, *Notobranchaea macdonaldi* with penis (p) and posterior foot lobe (f) indicated; E, *Glaucus atlanticus;* F, *Phylliroe bucephala;* G, *Issena lacera,* with detail of rhinophore; H, *Ancula gibbosa;* I, *Acanthodoris pilosa,* detail of gill; J, doridacean nudibranch, body form in *Acanthodoris, Adalaria, Echinochila,* and *Onchidorus;* K, *Polycera* sp; L, doridacean nudibranch, ventral view. Gills are indicated in G, H, J, K, by "a."

5. Gills complexly pinnate, Fig. 16.12I; color variable, white to brown; 23 mm (Abbott, 1968; Pruvot-Fol, 1954). *Acanthodoris pilosa.*

5. Gills simply pinnate. *Adalaria* and *Onchidorus.* The genera are similar, but *Adalaria* has relatively larger tubercles on the back. Most of the species of *Onchidorus* are figured by Miner (1950), and Abbott (1954) figures *Adalaria proxima;* see also Moore (1964) for Woods Hole species.

Group Three, Miscellaneous Genera.

---

1. Back without folds, lobes, or processes—2.

1. Back with 2 pairs of straplike processes; 36 mm; Fig. 16.13A; on *Sargassum* (Moore, 1964; Pruvot-Fol, 1954) *Scyllaea pelagica.*

1. Back with elongate lateral folds; Fig. 16.13B. *Elysia:*
- Folds extending about ⅔ the body length; 10 mm. *E. catula.*
- Folds extending to tip of foot; 30 mm; Fig. 16.13B. *E. chlorotica.* (Both species figured by Miner, 1950).

2. With tentacles—3.
2. Without tentacles; flatwormlike; 2 mm (Hyman, 1967). *Limapontıa zonata.*
3. With a pair of gills between mantle and foot posteriorly; on *Sargassum;* 5 mm:
- With a notch at rear of mantle; Fig. 16.13D (Verrill, 1870). *Corambe obscura.*
- Without notch; Fig. 16.13C (Moore, 1964). *Corambella depressa.*

3. Not so—4.
4. With 2 pairs of tentacles; dorsal gill present; 25 mm; Fig. 16.13E (Moore, 1964). *Pleurobranchaea tarda.*
4. With 1 pair of tentacles; no dorsal gills; 26 mm; Fig. 16:12F: parasitic on medusa of *Zanclea costata.* (Pruvot-Fol, 1954). *Phylliroe bucephala.*

Group Four, Dendronotacea. Two genera (*Scyllaea* and *Phylliroe*) from this family are keyed above in Group Three. The formal classification of the order relies heavily on differences in the anatomy of the hepatic system, see Odhner (1936).

---

1. Cerata arborescent, Fig. 16.13F. *Dendronotus:*
- Whitish with reddish brown markings; 80 mm (Abbott, 1954). *Dendronotus frondosus.*
- Bright red with white spots; 90 mm (Verrill, 1870). *Dendronotus robustus.*

1. Cerata not arborescent—2.
2. With 6–9 cerata with 4–7 circlets of tubercles; Fig. 16.13G. *Doto:*
- Tubercles with a dark apical spot or ring; whitish or yellowish with red spots; 12 mm (Miner, 1950). *D. coronata.*
- Tubercles with pale spots; uniform white; 13 mm (Verrill, 1875). *D. formosa.*

2. Cerata lobed; 20 mm; Fig. 16.13H,I (Pruvot-Fol, 1954). *Fimbria fimbria.*

Group Five, Aeolidacea and Sacoglossa in Part. *Elysia* and *Limapontia* from the latter order are keyed in Group Three above. This group may prove difficult. The systematics of the aeolidaceans has a complex history with numerous problems not only in the differentiation of species but of genera and families as well. This key will doubtless function imperfectly. For common Woods Hole forms see Moore (1964).

---

1. With 2 pairs of tentacles—4.
1. With 1 pair of tentacles, or head lobes, or with the oral tentacles replaced by a velum—2.
2. Oral tentacles replaced by a velum, Fig. 16.14A; 6 mm. *Embletonia fuscata.* A second species, assigned to *remigata* and described by Gould on the page following the description of *fuscata,* was regarded as doubtfully distinct by the author himself; see Gould and Binney (1870)
2. With a single pair of tentacles or lateral head lobes—3.

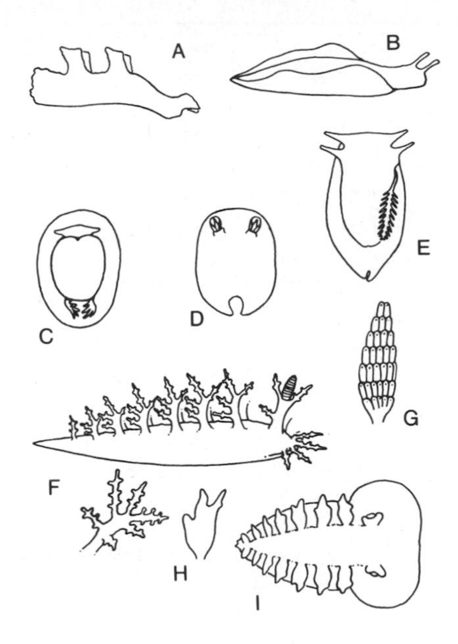

Fig. 16.13. Mollusca, naked Gastropoda, Groups Three and Four. A, *Scyllaea pelagica;* B, *Elysia chlorotica;* C, *Corambella depressa*, ventral view; D, *Corambe obscura*, dorsal view; E, *Pleurobranchaea tarda;* F, *Dendronotus* sp., with detail of ceras (below); G, *Doto* sp., ceras; H, *Fimbria fimbria*, ceras; I, same whole animal.

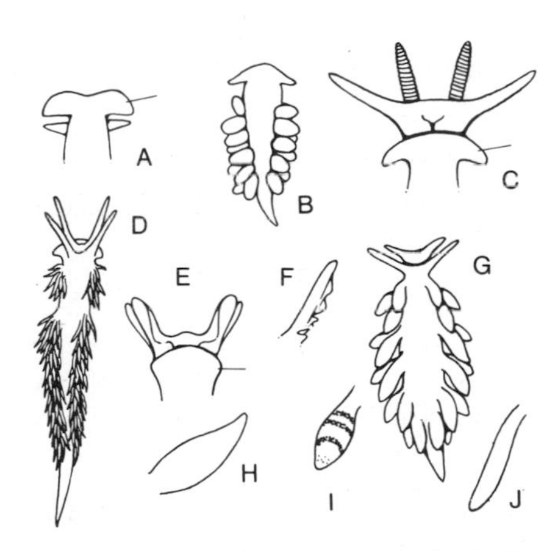

Fig. 16.14. Mollusca, naked Gastropoda, Group Five. A, *Embletonia fuscata*, detail of head with velum indicated; B, *Stiliger fuscatus;* C, aeolidacean nudibranch, detail of head, ventral view, with foot process indicated; D, *Coryphella* sp.; E, *Hermaea bifida*, detail of head, ventral view, with foot process indicated; F, *Fiona pinnata*, ceras; G, *Capellina exigua;* H, *Hermaea bifida*, ceras; I, *Capellina exigua*, ceras; J, *Cratena aurantia*, ceras.

3. Cerata in a single row on each side; 3 mm; Fig. 16.14B (Marcus, 1958). *Stiliger fuscatus.*
3. Cerata in clusters in 2 lateral bands; 13 mm (Gould, 1870). *Alderia harvardiensis.*

4. Anterior corners of foot produced and elongated; Fig. 16.14C—5.
4. Anterior corners of foot may be lobed or not, but not as above—8.

5. Dorsal tentacles with raised rings; cerata in 2 rows of oblique clusters; 25 mm (Moore, 1964). *Facelina bostoniensis.*
5. Dorsal tentacles smooth or slightly wrinkled—6.

6. Cerata very numerous (to about 400 on each side), in oblique rows; body ovate; 100 mm; color variable (Abbott, 1954). *Aeolidia papillosa.*
6. No more than 100 cerata on each side; body slender—7.

7. Cerata in clusters, the first group more or less separated from the rest, Fig. 16.14D; about 20–25 mm. *Coryphella* spp. (in part). Color in these species varies with nutrition, and species differentiation may require examination of the radula; see Pruvot-Fol (1954). For an introduction to the literature of the genus see Johnson (1915).
7. First group of cerata not separated from the rest; cerata in 10–14 rows; anterior part of body striped; 30 mm (Moore, 1964). *Cratena pilata.* The generic status of species variously listed as *Cuthona*, *Cratena*, *Trinchesia*, etc., appears to be in a somewhat muddled state; see Pruvot-Fol (1954, p. 380). Note also that in some species of *Coryphella* the cerata are not clustered, and these may erroneously key out here.

8. Anterior corners of foot lobed, Fig. 16.14E—9.
8. Anterior corners of foot rounded—10.

9. Rhinophores bifurcated; cerata leaflike; transparent; 20 mm; Fig. 16.14E,H (Pruvot-Fol, 1954). *Hermaea bifida.*
9. Not so. *Cuthona* (see notes above under *Cratena pilata*).
- Cerata with a yellow-orange core and with white granules on surface; 6 mm; body transparent, jaws showing through skin (Alder and Handcock, 1855). *C. pustulata.*
- Cerata without surface granules; 12 mm (Moore, 1964). *C. concinna.*

10. Cerata with a membraneous lateral edge; cerata irregularly dispersed; 25 mm; Fig. 16.14F (Moore, 1964). *Fiona pinnata.* On *Sargassum* and drift material.
10. Cerata without a lateral membrane, in single or multiple rows—11.

11. Cerata few, 10 or less, club-shaped:
- Tentacles and cerata with dark bands, cerata 5–10 per side; 6 mm; Fig. 16.14G,I (Alder and Hancock, 1848; Moore 1964). *Capellina exigua.*
- Tentacles and cerata not banded; cerata 4 or 5 per side; 7 mm (Moore, 1964). *Tergipes despectus.*

11. Cerata numerous, 30 or more:
- Dorsal and oral tentacles subequal; cerata slender; 15 mm; Fig. 16.14J (Moore, 1964). *Cratena aurantia.* See notes above under *Cratena pilata.* Two additional species are listed, *C. gymnota* and *C. veronicae;* see Johnson (1915).
- Oral tentacles about half as long as dorsal tentacles; cerata thicker; 11 mm (Alder and Hancock, 1847). *Eubranchus tricolor.*

References

Abbott, R. T. 1954. American Seashells. D. Van Nostrand Co.

———. 1968. A Guide to Field Identification, Sea Shells of North America. Golden Press, New York.

Alder, J., and A. Hancock. 1845–1855. Monograph of the British Nudibranchiate Mollusca. Ray Society, Parts 1–VII.

Balch, F. N. 1899. List of Marine Mollusca of Coldspring Harbor, Long Island, with descriptions of one new genus and two new species of Nudibranchs. Proc. Boston Soc. nat. Hist. 29:133–162.

Bartsch, P. 1909. Pyramidellidae of New England and the adjacent region. Proc. Boston Soc. nat. Hist. 34:67–113.

Bergh, R. 1884. Beiträge zu einer Monographie der Polyceraden III. Verh. Zool. Bot. Ges. Wein. 33:135–180.

Clench, W. J. 1951. The genus *Epitonium* in the Western Atlantic, Part I. Johnsonia 2(30): 249–288.

———. 1952. The genera *Epitonium* (Part II), *Depressiscala*, *Cylindriscala*, *Nystiella* and *Solutiscala* in the western Atlantic. Johnsonia. 2(31):284–356.

———. 1960. The genus *Calliostoma* in the Western Atlantic. Johnsonia. 4(40):1–80.

Clench, W. J., and Ruth D. Turner. 1950. The Western Atlantic marine mollusks described by C. B. Adams. Mus. Comp. Zool. Occ. Pap. on Mollusks. 1(15): 233–403.

Cox, L. R. 1960. Gastropoda, general characteristics of Gastropoda. *In* Treatise on Invertebrate Paleontology, Part I. Mollusca I. R. C. Moore, ed. Geol. Soc. Amer. and Univ. of Kansas.

Dall, W. H. 1899. Scientific results of explorations by the U.S. Fish Commission Steamer *Albatross*. VIII. Preliminary report on the collection of Mollusca and Brachiopoda obtained in 1887–88. Proc. U.S. natn. Mus. 12:219–362.

Danforth, C. H. 1907. A new pteropod from New England. Proc. Boston Soc. nat. Hist. 34:1–19.

Fish, C. J. 1925. Seasonal distribution of the plankton of the Woods Hole region. Bull. U.S. Bur. Fish. 41: 91–179

Forbes, E., and S. Hanley. 1853. A history of British Mollusca and their shells. Four volumes. John van Voost. London.

Franc, A. 1968. Sous-class des Opisthobranches. *In* Traité de Zoologie, Anatomie, Systematique, Biologie. Tome V, Fasc. III. P. P. Grassé, ed. Masson et Cie.

Gould, A. A., and W. G. Binney. 1870. Report on the Invertebrata of Massachusetts. Second ed. Wright and Potter, Boston.

Grusov, E. N. 1965. The endoparasitic mollusk *Asterophila japonica* Randall et Heath (Prosob. Melanellidae) and its relations to the parasitic gastropods. Malacologia 3(1):111–181.

Hardy, A. C. 1956. The Open Sea, Its Natural History: The World of Plankton. Houghton Mifflin Co.

Henderson, J. B., and P. Bartsch. 1914. Littoral marine mollusks of Chincoteague Island, Virginia. Proc. U.S. natn. Mus. 47:411–421.

Hyman, Libbie H. 1967. The Invertebrates. Volume VI. Mollusca I. McGraw-Hill Book Co.

Jacobson, M. K., and W. K. Emerson. 1961. Shells of the New York City Area. Argonaut Books Inc.

Johnson, C. W. 1915. Fauna of New England 13, List of the Mollusca. Occ. Pap. Boston Soc. nat. Hist. 7:1–231.

———. 1934. List of marine molluscs of the Atlantic coast from Labrador to Texas. Proc. Boston Soc. nat. Hist. 40(1): 1–203.

Keen, Myra. 1958. Sea shells of Tropical West America. Stanford Univ. Press.

———. 1963. Marine Molluscan Genera of Western North America. Stanford Univ. Press.

Knight, J. B., *et al.* 1960. Systematic Descriptions. *In* Treatise on Invertebrate Paleontology Part I. Mollusca I. R. C. Moore, ed. Geol. Soc. Amer. and Univ. of Kansas.

Lemche, H. 1956. Anatomy and histology of Cyclichna. Spolia Zool. Musei. Nauniensis 16 Skrifter udg. Un. Zool. Mus. Copenh. 1–278.

Marcus, E. 1958. On Western Atlantic Opisthobranchiate Gastropods. Am. Mus. nat. Hist. Novitates 1906: 1–82.

Mighels, J. W., and C. B. Adams. 1842. Descriptions of twenty-four species of the shells of New England. Boston J. nat. Hist. 4:37–53.

Miner, R. W. 1950. Field Book of Seashore Life. G. P. Putnam's Sons.

Moore, G. W. 1964. Phylum Mollusca, shell-less Opisthobranchia. *In* Keys to Marine Invertebrates of the Woods Hole Region. Systematic-Ecology Program, Marine Biological Laboratory, Wooes Hole.

Morris, P. A. 1951. A field guide to the shells of our Atlantic and Gulf coasts. Houghton Mifflin Co.

Morton, J. E. 1960. Mollusca. Harper and Row.

Odhner, N. 1936. Nudibranchia Dendronotacea. Mem. Mus. Royal. Hist. natur. Belgique, 2(3):1057–1128.

———. 1939. Opisthobranchiate mollusca from the coasts of Norway. Kongelige Norske Videnskaps. Selskabs. Skrifter 1:1–93.

Pruvot-Fol, Alice. 1954. Mollusques Opisthobranches. Faune de France. 58:1–460.

Sars, G. O. 1878. Bidrag Til Kundskaben om Norges Arktiske Fauna. Fauna Reg. Arct. Norvegiae.

Simroth, H. 1911. Die gastropoden des nordischen plankton. Nord. Plankt. 5:1–35.

Tesch, J. J. 1949. Heteropoda. Carlsberg Foundation's Oceanog. Exped. round the world, 1928–1930. Dana Report. 34:1–53.

Totten, J. G. 1835. Descriptions of some shells belonging to the coast of New England. Am. J. Sci. 28(2):347–353.

Verrill, A. E. 1870. Descriptions of some New England nudibranchs. Am. J. Sci. Ser. 2(50):405–408.

———. 1875. Brief contributions to zoology from the Museum of Yale College. No. 33—Results of dredging expeditions off the New England coast. Am. J. Sci. Ser. 3. 10(10):36–43.

———. 1882. Catalog of marine mollusca added to the fauna of the New England region, during the past ten years. Trans. Conn. Acad. Arts Sci. 5:447–587.

———, and Katharine J. Bush. 1897. Revision of the genera of Ledidae and Nuclidae of the Atlantic coast of the United States. Am. J. Sci. Ser. 4. 3(5):51–63.

———, and Katharine J. Bush. 1898. Revision of the deep-water mollusca of the Atlantic coast of North America, with descriptions of new genera and species. Proc. U.S. natn. Mus. 20:775–901.

Warmke, Germaine L., and R. T. Abbott. 1961. Caribbean Seashells. Livingston Publishing Co.

Wickstead, J. H. 1965. An introduction to the study of Tropical Plankton. Hutchinson Tropical Monographs.

Class Scaphopoda
TUSK SHELLS

Diagnosis. *Shell straight or curving in a simple arc, not coiled, tapering to resemble a miniature elephant's tusk, open at both ends.*

Adults of local species range in shell length from less than 5 mm to about 90 mm. The shell is usually whitish but may be reddish, yellowish, or green in part. It may be smooth or sculptured with longitudinal ribs or striae; the sculpturing may change with age and may be obscured by erosion. Similarly the diagnostically important features of the apical rim may be affected by wear. The animal has a simple cylindrical form without a strongly developed head. Eyes are lacking. The foot is conical or capable of being expanded terminally into a disc and bears numerous ciliated, prehensile filaments called *captacula*.

The sexes are separate, but there is no conspicuous sexual dimorphism in shell form. Fertilization is presumably external, the young passing through a trochophore type of larval stage.

Scaphopods are benthic, burrowing forms that lie with the narrow apical end exposed above a sand or mud bottom. They are reported to feed mainly on foraminiferans which are drawn to the mouth by threadlike captaculae. As the species list indicates, the majority of tusk shells are confined to the outer Continental Shelf or Slope; the inshore species are mostly southern and have only been reported as far north as the vicinity of Cape Hatteras. *Dentalium entale* is the most likely species to be encountered in shallow water north of Cape Cod; southward this species ranges in deep water at least as far as the offing of Chesapeake Bay.

Systematic List and Distribution

The species list is from Henderson (1920), the classification from Emerson (1962). There is no ordinal breakdown.

Family Dentaliidae		
Dentalium laqueatum	C+	90 to 227 m
Dentalium entale	B+	15 to 2297 m
Dentalium agile	B	104 to 245 m
Dentalium occidentale	B+	110 to 3085 m

Dentalium taphrium	C+	40 m
Dentalium eboreum	C+	13 to 159 m
Family Siphonodentaliidae		
Siphonodentalium lobatum	BV	110 to 3318 m
Siphonodentalium occidentale	V	157 to 1801 m
Cadulus verrilli	BV	183 to 942 m
Cadulus agassizii	V	115 to 227 m
Cadulus pandionis	V	123 to 706 m
Cadulus quadridentatus	C+	27 to 40 m
Cadulus carolinensis	C+	13 to 115 m
Cadulus minusculus	C+	115 m

Identification

Shell outlines show typical form, sculpturing omitted. Measurements are shell length.

1. Shell widest at aperture—2.
1. Shell contracted toward aperture; Fig. 16.15E–H—6.
2. Shell smooth; apex either with 6 lobes or with neither lobes nor notches. *Siphonodentalium:*
- Apex cut in six lobes; 19 mm; Fig. 16.15A. *S. lobatum.*
- Apex simple with neither lobes nor notches; 3 mm. *S. occidentale.*

2. Shell ribbed, striated, or wrinkled posteriorly; apex with a notch, Fig. 16.15B. *Dentalium*—3.
3. Apex polygonal; with 9–12 strong ribs; 62 mm; Fig. 16.15C. *D. laqueatum.*
3. Apex circular or subcircular—4.

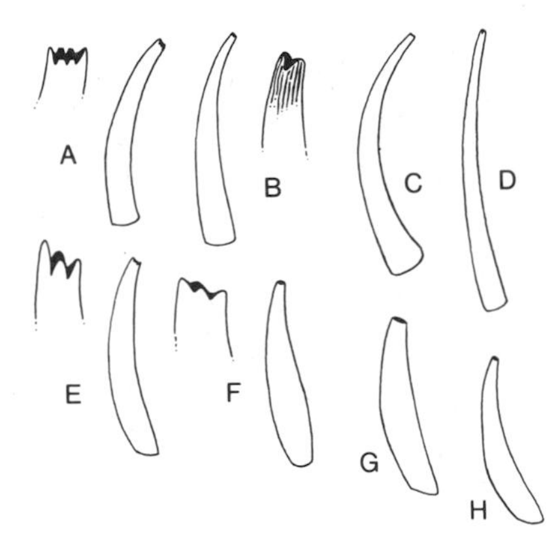

Fig. 16.15. Mollusca, Scaphoda, sculpturing omitted. A, *Siphonodentalium lobatum*, with detail of apex; B, *Dentalium entale*, with detail of apex; C, *Dentalium laqueatum;* D, *Dentalium eboreum;* E, *Cadulus quadridentatus*, with detail of apex; F, *Cadulus carolinensis*, with detail of apex; G, *Cadulus agassizii;* H, *Cadulus pandionis.*

4. Sculpturing feebly developed, with a few longitudinal wrinkles posteriorly; tips of shell commonly eroded; 53 mm; Fig. 16.15B. *D. entale.*
4. Shell striated with 20 or more fine, longitudinal scratches; salmon pink, yellowish, fading to white; 65 mm; Fig. 16.15D. *D. eboreum.*
4. Shell ribbed—5.
5. Shell green, at least anteriorly; with 9–14 (usually 14) primary ribs; 25 mm. *D. taphrium.*
5. Shell whitish; with 16 or 18 ribs;
   - With 16 primary ribs; interspaces coarsely sculptured; 34 mm. *D. occidentale.*
   - With 18 primary ribs; interspaces smooth; 88 mm. *Cadulus* spp.
6. Apex simple, without slits or notches:
   - Apical orifice about ½ diameter of larger aperture; 2.3 mm. *C. minisculus.*
   - Apical orifice about ¾ diameter of larger aperture; 2.8 mm. *C. verrilli.*
6. Apex with 2 or 4 slits or notches—7.
7. Apex with 4 deep slits:
   - Slits relatively shallower; shell more inflated; 10 mm; Fig. 16.15F. *C. carolinensis.*
   - With relatively deeper slits; shell less inflated; 11 mm; Fig. 16.15E. *C. quadridentatus.*
7. Apex with 2 or 4 very shallow, broad notches: *C. pandionis*, Fig. 16.15H, has a bent, "dogtooth-shaped" shell and narrow apical aperture in comparison to *C. agassizii*, Fig. 16.15G.

References

Emerson, W. K. 1962. A classification of the scaphopod mollusks. J. Paleont. 36(3):461–482.

Henderson, J. B. 1920. A monograph of the east American scaphopod mollusks. Bull. U.S. natn. Mus. 111: vi + 177 pp.

Ludbrook, N. H. 1960. Scaphopoda. *In* Treatise on Invertebrate Paleontology. R. C. Moore, ed. Part I, Mollusca. 1:37–40. Geol. Soc. Am. and Univ. of Kansas.

Turner, Ruth. 1955. Scaphopods of the *Atlantis* dredgings in the Western Atlantic. *In* Marine Biology. Pergamon Press, Ltd. London. Pp. 309–320.

Class Bivalvia (.. Lamellibranchiata, Pelecypoda)
BIVALVE MOLLUSKS

Diagnosis. *Shell with two valves lateral in relation to soft body, usually more or less symmetrical, hinged dorsally; head not differentiated, buccal apparatus, radula, and cephalic sense organs lacking.*

The soft body consists of a visceral mass including an anteroventral *foot* and, bilaterally, a pair of *gills* outside which are a pair of mantle lobes adherent to the inner surface of the shells. With the shells agape the mantle cavity is largely open in some forms, but in most genera the mantle lobes are partly fused below. A gape for the protrusion of the foot usually remains anteriorly, and posteriorly the mantle lobes are fused and modified to form a pair of tubes or *siphons*, one incurrent, the other excurrent. The siphons vary greatly in their development. They may be short or long, retractile or not, fused or independent. They may also be quite unequally developed, and in some genera (e.g., *Lucina*) only one siphon is present. Siphons are lacking entirely in some bivalves including scallops, and members of the families Nuculidae, Astartidae, Carditiidae, Arctidae, as well as in oysters and some mussels.

Water currents drawn through the mantle cavity serve in respiration, for food gathering and waste removal, and for other exchanges with the environment. The mantle may also be supplied with sensory structures including eyes in scallops, ark shells, and Carditiidae. In most higher bivalves the siphons are more or less retractile, and the development of the retractor apparatus is reflected in the development of a sinus in the *pallial line* marking the attachment of the mantle musculature, Fig. 16.16. The presence of a *pallial sinus* and its configuration are of systematic importance. Differences in gill structure are also important in taxonomy and have been a major consideration in the determination of higher categories. The four general patterns of gill structure are figured diagrammatically and described briefly by Abbott (1968).

The development of the foot also varies in different taxa. It is substantially reduced in such sedentary forms as oysters, and reaches its greatest extension in more active, burrowing species such as the Tellinidae and razor clams.

The keys in this account depend almost entirely on differences in shell form and structure, including the general shape of the valves, their external sculpture, hinge structure, and patterns of muscle scars. The valves are usually approximately equal in size and development, but they are distinctly

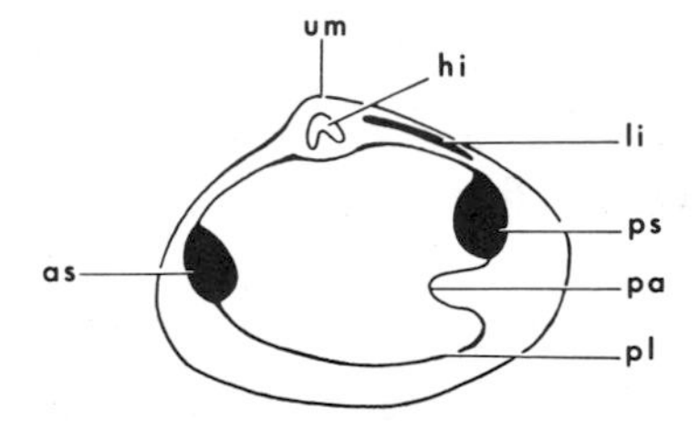

Fig. 16.16. Mollusca, Bivalvia, Internal view of right shell with structural details indicated. Abbreviations are as follows: as, anterior adductor muscle scar; hi, hinge; li, ligament; pl, pallial line; pa, pallial sinus; ps, posterior adductor muscle scar; um, umbo or beak.

unequal or asymmetrical in *Macoma*, Thraciidae, *Anomia*, and in oysters and scallops. In these mollusks one valve is flatter, or larger, or otherwise different in shape. The oyster *Crassostrea* frequently is cemented to a substratum by its left valve, and *Anomia* is attached by a byssal anchor extending through a hole in the left valve. Byssal attachment without conspicuous modification of the shell other than a slight notching occurs in mussels and in ark shells and related families. The anchorage of byssus threads can be broken or disengaged, allowing the animal to move to a new location. In most scallops, only juveniles are attached.

Bivalve shells are hinged dorsally. The shell is closed by one or two *adductor muscles* whose scars mark the inside of the valves. Opposed to these muscles is the *ligament* which may be external to the hinge or internal to it or, somewhat ambiguously, in between. The ligament is really bipartite, but in most genera the two parts are united; in others, e.g., *Mya* and related genera, the parts are separated and one part is internal and is called the *resilium* or "cartilage." It is elastic and resides in a spoonlike *chondrophore*. The hinge itself may be simple, but it is commonly toothed, the teeth of opposing valves interlocking and providing a more secure union of the valves. A complex nomenclature has evolved to deal with diverse tooth arrangements. For our purpose, only the central or *cardinal teeth* and the supplementary *lateral teeth* need be mentioned. These teeth are present or absent and vary in configuration in different taxa.

Viewed laterally, the valves of most species are asymmetrical or *inequilateral*. In the great majority of species the beak or *umbo* points anteriorly and the ligament is posterior to the beak. The pallial sinus, when present, is also posterior, reflecting the posterior position of the siphons in the living animal. The question of orientation is of some importance, as the hinge structure in the left and right valves may be quite different. In the figures accompanying the keys the valves are usually oriented with the anterior end to the left; the muscle scars of the right valve are shown diagrammatically, and the right hinge may be shown enlarged. Departures from this convention are indicated.

The sexes are normally separate in about 96% of bivalve species, but the individuals are usually not distinctly dimorphic in shell structure. Simultaneous hermaphroditism occurs in some bivalves, e.g., in most scallops and some genera and species of freshwater bivalves. The hard-shell clam or quahog *Mercenaria mercenaria* plus some members of Teredinidae are protandric hermaphrodites. Oysters of the genus *Ostrea* are *ambisexual*—the sex changes rhythmically throughout the reproductive life of the individual oyster. In *Crassostrea* the situation is more complex and varies according to environmental conditions and the age of the oyster (see Galtsoff, 1964).

Freshwater bivalves incubate their eggs and so do many marine forms; fertilization in such forms occurs within the mantle cavity. In nearly all bivalves, whether or not the earliest stages are passed within the parental shell, the young hatch as pelagic larvae. They are briefly trochophores, and then become veligers. The veliger stage may last several months and is an active dispersive phase. (See introductory paragraphs on the phylum Mollusca.) In freshwater families, development is direct in Sphaeriidae; in Unionidae a distinctive larval form, the *glochidium*, is parasitic on fish or, in one species, on the neotenic salamander *Necturus*. Young sphaeriids develop in the mantle cavity of the adult and emerge as miniature bivalves.

With few exceptions, bivalves are sedentary animals, but scallops are renowned as active if ungraceful swimmers. Burrowing habits vary considerably; see Morton (1958, p. 51). Boring species are present in several families and are of substantial economic importance. Members of the Pholadidae bore in peat, clay, and solid rock; *Hiatella* and *Petricola* are found in rock and clay or peaty clay, respectively. The teredinids or shipworms bore in wood.

Most bivalve species are either suspension or deposit feeders. Planktonic and minute detrital food items are carried into the body on ciliary currents, collected and partly sorted on the gill and labial apparatus, and further sorted in the stomach. In protobranchs, food is collected by the labial palps which are developed as appendages that can be extended to "graze" on the substratum. Species that feed on coarse particles usually have the stomach modified as a grinding mill. The septibranchs are suctorial scavengers on decaying material and detritus in soft bottom deposits. A few adult bivalves are commensal, e.g., members of Montaculidae. *Entovalva* is an endoparasite in sea cucumbers.

Systematic List and Distribution

The general classification of Keen (1963) has been adopted for this text, but see Morton (1958), Franc (1960), Rothschild (1965), and Abbott (1968) for alternative schemes. The species list is from Johnson (1934) with additions and changes from later literature.

Subclass Prionodesmata		
Order Protobranchia		
Family Solemyacidae		
Solemya borealis	B	Shallow
Solemya velum	BV	Shallow
Solemya grandis	V	55 to 2928 m
Family Nuculidae		
Nucula proxima	V	4 to 55 m
Nucula delphinodonta	B, 40°	—
Nucula tenuis	BV	7 to 183 m

Family Nuculanidae		
Nuculana acuta	V	13 to 412 m
Nuculana caudata	BV	>183 m
Nuculana pernula	B	18 to 395 m
Nuculana tenuisulcata	B+	To 274 m
Nuculana messanensis	BV	59 to 4795 m
Yoldia limatula	BV	2 to 22 m
Yoldia myalis	B	13 to 146 m
Yoldia sapotilla	BV	7 to 183 m
Yoldia cascoensis	B	Deep
Yoldia thraciaeformis	B	18 to 366 m
Yoldia regularis	B	179 to 639 m
Yoldia fraterna	BV	165 to 2928 m
Yoldia frigida	B	161 to 571 m
Yoldia inconspicua	BV	183 to 1290 m
Yoldia iris	BV	37 to 2928 m
Yoldia lucida	BV	40 to 944 m
Yoldia subangulata	B	40 to 944 m
Subclass Pteriomorphia		
Order Prionodontida		
Family Arcidae		
Anadara ovalis	V	2 to 31 m
Anadara transversa	V	Shallow
Bathyarca anomala	B	37 to 49 m
Bathyarca pectunculoides	B	49 to 926 m
Noetia ponderosa	V	Shallow
Family Limopsidae		
Limopsis cristata	V	155 to 2004 m
Limopsis sulcata	V	117 to 639 m
Limopsis minuta	BV	55 to 4064 m
Order Pteroconchida		
Family Pteriidae		
Pteria hirundo	C, 41°	119 to 351 m
Family Pinnidae		
Pinna carnea	C+	Shallow
Family Mytilidae		
Mytilus edulis	BV	Lit. to shallow, eury.
Musculus corrugatus	BV	4 to 183 m
Musculus niger	BV	2 to 110 m
Musculus discors	V	9 to 183 m
Modiolus modiolus	BV	Shallow
Modiolus demissus	BV	Lit., eury.
Crenella decussata	BV	9 to 227 m
Crenella glandula	BV	5 to 110 m
Brachidontes recurvus	V	Lit.
Amygdalum papyria	C, 38°	Shallow, eury.
Crenella faba	B	9 to 549 m
Crenella fragilis	V	128 m
Dacrydium vitreum	BV	110 to 2846 m

Family Ostreidae
Crassostrea virginica BV Lit. to Shallow
Crassostrea gigas (Japan) introduced in Barnstable Bay, Mass.
Ostrea edulis (Europe) introduced in Boothbay Harbor, Maine.
Ostrea equestris C, 37° 110 m
Family Pectinidae
?*Pecten reticulus* C+ 150 to 227 m
?*Pecten thalassinus* V 40 to 580 m
Chlamys islandicus B 18 to 327 m
Chlamys liocymatus C+ 62 m
Aequipecten glyptus V 126 to 285 m
Aequipecten irradians BV Shallow
Aequipecten phrygium V 91 to 1449 m
Placopecten magellanicus BV 18 to 183 m
Family Limidae
Limea subovata V 183 to 915 m
Limatula regularis V 128 m
Limatula subauriculata BV 27 to 91 m
Family Anomiidae
Anomia aculeata BV 2 to 146 m
Anomia simplex BV 2 to 183 m
Subclass Teleodesmata
Order Heterodontida
Family Astartidae
Astarte castanea BV 9 to 119 m
Astarte undata B, 37° 9 to 190 m
Astarte subaequilatera BV 40 to 783 m
Astarte nana C+ 13 to 415 m
Astarte borealis B 27 to 183 m
Astarte elliptica B 15 to 165 m
Astarte quadrans B+ 11 to 73 m
Astarte striata B 18 to 155 m
Astarte laurentiana Pleistocene fossil
Astarte portlandica B 18 m
Family Crassatellidae
Crassinella mactracea V 5 to 183 m
Family Carditidae
Cardita borealis BV 4 to 457 m
Family Corbiculiidae
Polymesoda caroliniana C, 37° Shallow, eury.
Family Arcticidae
Arctica islandica BV 11 to 165 m
Family Leptonidae
Kellia suborbicularis B Lit. to 13 m
Family Montacutidae
Montacuta percompressa V Shallow
Montacuta elevata V 15 to 26 m, commensal on *Clymenella*
Mysella planulata BV Lit. to 88 m
Entovalva perrieri V Parasitic in *Leptosynapta inhaerens*

Family Turtoniidae		
Turtonia minuta	B	Shallow
Family Lucinidae		
Lucina blakeana	V	33 to 849 m
Lucina crenella	C, 37°	4 to 227 m
Myrtaea lens	V	91 to 849 m
Divaricella quadrisulcata	V	18 to 55 m
Lucinoma filosa	BV	11 to 966 m
Family Thyasiridae		
Thyasira insignis	B	119 to 852 m
Thyasira gouldii	B+	9 to 732 m
Thyasira trisinuata	BV	27 to 351 m
Thyasira plana	B	15 to 183 m
Thyasira inaequalis	B	26 to 88 m
Thyasira equalis	V	172 to 1613 m
Thyasira croulinensis	B	14 to 133 m
Thyasira brevis	V	183 to 2140 m
Thyasira ferruginosa	BV	37 to 1591 m
Thyasira succisa	B	168 to 2500 m
Leptaxinus minutus	V	183 m
Axinopsis orbiculatus	B	18 to 55 m
Axinopsis cordata	V	79 to 340 m
Family Dreissenidae		
Congeria leucopheata	V	Shallow, eury.
Family Ungulinidae		
Diplodonta verrilli	V	27 to 126 m
Family Cardiidae		
Dinocardium robustum	C, 37°	1 to 31 m
Cerastoderma pinnulatum	BV	6 to 183 m
Clinocardium ciliatum	V	6 to 183 m
Serripes groenlandicus	B	4 to 110 m
Microcardium peramabile	V	33 to 300 m
Laevicardium mortoni	BV	Lit. to 4 m
Family Veneridae		
Mercenaria mercenaria	BV	Lit. to 12 m
Mercenaria campechiensis	V	1 to 15 m
Gemma gemma	BV	Lit. to 6 m
Pitar morrhuana	BV	4 to 33 m
Dosinia discus	C, 39°	2 to 12 m
Family Petricolidae		
Petricola pholadiformis	BV	7 to 11 m
Family Mactridae		
Rangia cuneata	C, 38°	Fresh-brackish
Anatina anatina	C, 40°	2 to 12 m
Anatina plicatella	C, 40°	Shallow
Spisula solidissima	BV	Lit. to 31 m
Spisula polynyma	B	—
Mulinia lateralis	BV	Shallow, eury.
Family Mesodesmatidae		
Mesodesma arctatum	B, 40°	Lit. to 91 m

Family Tellinidae		
Tellina tenera	V	7 to 18 m
Tellina versicolor	V	2 to 91 m
Tellina agilis	BV	1 to 45 m
Tellina squamifera	C+	115 m
Tellina alternata	C+	18 to 128 m
Tellina nitens	C+	20 to 90 m
Tellina iris	C+	Shallow
Tellina mirabilis	C+	1 to 7 m
Macoma phenax	V	Shallow
Macoma balthica	BV	1 to 18 m
Macoma calcarea	B+	9 to 73 m
Macoma inflata	B, 40°	104 to 377 m
Macoma tenta	BV	4 to 18 m
Macoma brevifrons	V	—
Family Donacidae		
Donax fossor	V	Lit.
Donax variabilis	C, 38°	Lit.
Family Semelidae		
Abra aequalis	C+	—
Arba longicallis	BV	91 to 2685 m
Abra lioica	V	11 to 1574 m
Semele proficua	C, 37°	1 to 9 m
Cumingia tellinoides	BV	5 to 22 m
Family Solecurtidae		
Tagelus plebeius	V	Lit. to 8 m
Tagelus divisus	V	Lit
Family Solenidae		
Solen viridis	V	Lit.
Ensis directus	BV	Lit.
Siliqua costata	BV	Shallow
Siliqua squama	B	New England banks
Family Myidae		
Mya arenaria	BV	Lit. to 9 m, eury.
Mya truncata	B	4 to 31 m
Family Corbulidae		
Corbula contracta	V	5 to 115 m
Family Hiatellidae		
Hiatella arctica	BV	Lit. to 183 m
Hiatella striata	BV	Lit. to 183 m
Cyrtodaria siliqua	B	9 to 91 m
Panomya arctica	B	46 to 210 m
Family Pholadidae		
Pholas campechiensis	C+	Shallow
Barnea truncata	BV	Lit.
Martesia cuneiformis	C+	Lit. 22 m
Martesia fragilis	C+	Floating driftwood
Martesia striata	C+	Floating driftwood
Zirfaea crispata	BV	Lit. to 75 m

Cyrtopleura costata	V	Lit.
Diplothyra smithii	V	In oyster shells
Xylophaga dorsalis	B	183 to 352 m
Family Teredinidae		
Bankia gouldi	V	Shallow
Teredo norvagica	BV	Shallow, rare
Teredora malleolus	V	Shallow
Teredo navalis	BV	Shallow
Psiloteredo megotara	BV	Shallow
Subclass Anomalodesmata		
Order Eudesmodontida		
Family Pandoridae		
Pandora gouldiana	BV	Lit. to 183 m
Pandora glacialis	C, 37°	2 to 44 m
Pandora inornata	B+	18 to 64 m
Pandora arenosa	C+	7 to 37 m
Pandora inflata	C, 41°	48 to 165 m
Family Lyonsiidae		
Lyonsia hyalina	BV	Lit. to 31 m
Lyonsia arenosa	B	24 to 110 m
Family Periplomatidae		
Periploma papyratium	B+	7 to 53 m
Periploma leanum	BV	5 to 29 m
Family Pholadomyidae		
Panacea arata	V	110 to 238 m
Family Thraciidae		
Thracia conradi	BV	5 to 29 m
Thracia myopsis	B	18 to 91 m
Thracia septentrionalis	B+	18 to 53 m
Order Septibranchida		
Family Poromyidae		
Poromya granulata	B	27 to 2653 m
Family Cuspidariidae		
Cuspidaria glacialis	BV	117 to 2685 m
Cuspidaria obesa	BV	37 to 2361 m
Cuspidaria pellucida	B	73 to 174 m
Cuspidaria rostrata	BV	119 to 2999 m
Cuspidaria media	V	115 to 284 m
Cuspidaria perrostrata	V	106 to 761 m
Cuspidaria striata	BV	155 to 2653 m
Cuspidaria multicostata	V	155 to 289 m
Cuspidaria gemma	C+	29 to 31 m
Cuspidaria glypta	C+	88 m
Family Verticordiidae		
Lyonsiella insculpta	V	135 to 891 m
Verticordia ornata	V	183 m
Halicardia flexuosa	New England Banks	137 to 639 m

Identification

The species are dealt with in nine groups that are for the most part highly artificial. Measurements refer to greatest dimension, either height or width of the adult shell. Unless otherwise indicated, the species are figured in one or another of the following: Abbott (1954, 1968), or Morris (1947).

---

1. Muscle scars coalesced; shell winged, or cemented to substratum or fixed by stout byssal cord; Fig. 16.17. Group One.
1. Muscle scars distinctly unequal; shell relatively thin, frequently pearly or iridescent within, more or less elongate with beaks terminal or subterminal at one end of long axis; Fig. 16.18. Group Two.
1. Muscle scars about equal in area if not in shape—2.
2. Dentition taxodont, with teeth in a more or less regular series along a straight or curving hinge line, tooth row interrupted or not at beaks; Fig. 16.19. Group Three.
2. Dentition otherwise—3.
3. Boring species, shells frequently elongate, with strong, scaly, radial ribbing, or shell ornamentation distinctly different anteriorly and posteriorly, or shell divided by radial groove, or animal elongate and worm-like with small tripartite valves; Fig. 16.21. Group Four.
3. Not so—4.
4. Shells marked externally with clearly defined radial ribbing; Fig. 16.22. Group Five.
4. Shells smooth, or concentrically ribbed; radial sculpture, if present, fine, obscure, or confined to ends of shells—5.
5. Shells distinctly inequilateral, i.e., one side spout-shaped, or more or less squared or blunted vertically, or valves distinctly unequal with one valve flatter or otherwise different in shape; Fig. 16.23. Group Six.
5. Not so—6.
6. Shells elongate or subrectangular, usually about twice or more longer than wide; Fig. 16.24. Group Seven.
6. Shells distinctly longer than wide, but more or less ovoid; Fig. 16.25. Group Eight.
6. Shells about as wide as long, or wider than long, more or less circular or triangular in outline; Fig. 16.26. Group Nine.

Group One. These species belong to a series of related families and with those in Group Two constitute the Order Pteroconchida. The economically important oysters and scallops belong here.

---

1. Hinge with distinct "ears" or "wings"; Fig. 16.17D–H—3.
1. Not so—2.
2. Valves translucent; central scar with three smaller scars within; lower valve with slotlike hole; Fig. 16.17A. *Anomia: A. simplex* has smooth, yellow to orange or black valves, 50 mm; *A. aculeata* has the upper valve rough or prickly, brownish or gray, 20 mm.

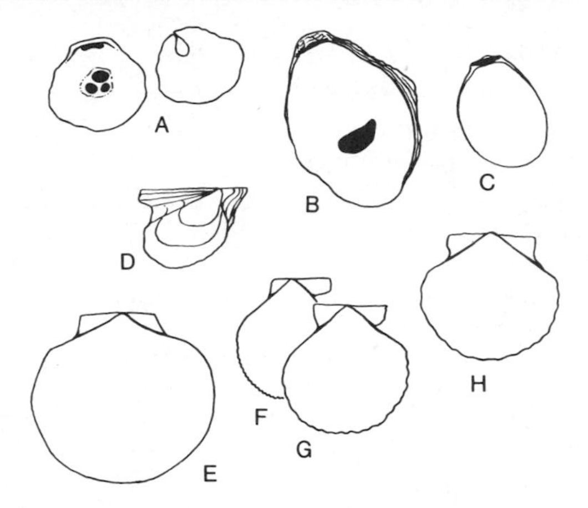

Fig. 16.17. Mollusca, Bivalvia, Group One, sculpturing omitted from all but D. D–H are external views. A, *Anomia* sp., upper and lower valves; B, *Crassostrea virginica;* C, *Limatula subauriculata*, muscle scars omitted; D, *Pteria colymbus;* E, *Placopecten magellanicus;* F, *Chlamys islandicus;* G, *Aequipecten phrygium;* H, *Aequipecten irradians.*

2. Valves thick and opaque, variable in shape; central scar simple; lower valve frequently cemented to a substratum; Fig. 16.17B. *Crassostrea virginica.* This is the common commercial oyster. *Ostrea equestris* is found in deep water as far north as Virginia; the lower valve has a raised, crenulated margin; the upper valve is toothed marginally and the interior is greenish.Japanesse and European oysters have been introduced in Massachusetts and Maine. See Galtsoff (1964).

2. Shell thin, relatively narrow, with numerous fine ribs; inside with two ribs near middle; 13 mm; Fig. 16.17C. *Limatula subauriculata.* Genus figured by Morris; see also Warmke and Abbot (1961). A second species, *L. regularis*, reported in deep water off Delaware Bay, is figured by Verrill and Bush (1898).

3. Valves distinctly inequilateral or asymetrical; 65 mm; Fig. 16.17D. *Pteria colymbus.* Note also *P. hirundo* reported in deep water; see Verrill (1882).

3. Valves nearly equilateral, hinge ears equal or not—4.

4. With numerous fine, threadlike ribs; valves relatively flat; 200 mm; Fig. 16.17E; the commercial sea scallop. *Placopecten magellanicus.*

4. Not so, ribs 17 to 50—5.

5. With about 50 moderately coarse ribs; hinge ears distinctly unequal; 100 mm; Fig. 16.17F. *Chlamys islandicus.* Note also *C. liocymatus* in southern waters.

5. With 17–21 ribs—6.

6. Hinge ears unequal; with 17 ribs, each consisting of 3 rows of fused scales; 25 mm; Fig. 16.17G. *Aequipecten phrygium.*

6. Hinge ears nearly equal; with 17–21 ribs; *Aequipecten glyptus* has one valve white, the other white with rose ribs, 65 mm, in deep water; *Aequipecten irradians* usually has valves gray-brown with darker markings, 75 mm, in shallow water; Fig. 16.17H. The latter is the common bay scallop.

Group Two, Mussels and Related Forms.

1. Beak terminal, Fig. 16.18A–D—2.
1. Beak near anterior end, Fig. 16.18E–L—4.
2. Shell thin, fragile, translucent, with about 10 radial ridges, with or without spines; 150 mm; Fig. 16.18A. *Pinna carnea.*
2. Shell strong, opaque—3.
3. Shell strongly curved, with numerous radial ribs; 25 mm; Fig. 16.18B. *Brachidontes recurvus.*
3. Shell straight or only slightly curved, not ribbed: *Mytilus edulis*, the edible or blue mussel, is bluish black, 75 mm, hinge toothed, in marine to somewhat brackish estuarine waters Fig. 16.18C; *Congeria conradi* is brownish; 13 mm; inside of beak with a platform, fresh to somewhat brackish estuarine waters, Fig. 16.18D.
4. Shell ribbed, at least in part—5.
4. Shell smooth: *Dacrydium vitreum* has a white, vitreous shell, only 3 mm, Fig. 16.18E; *Modiolus modiolus* has a brown, partly hairy periostracum, 155 mm, Fig. 16.18F. Note also the estuarine *Amygdalum papyria* with a smooth shell without hairy periostracum; the other two species are marine.
5. Ribbing confined to end of shell. *Musculus:*
- Shell reddish brown, purple, or black; ribs well developed; 25 mm; Fig. 16.18K. *M. niger.*

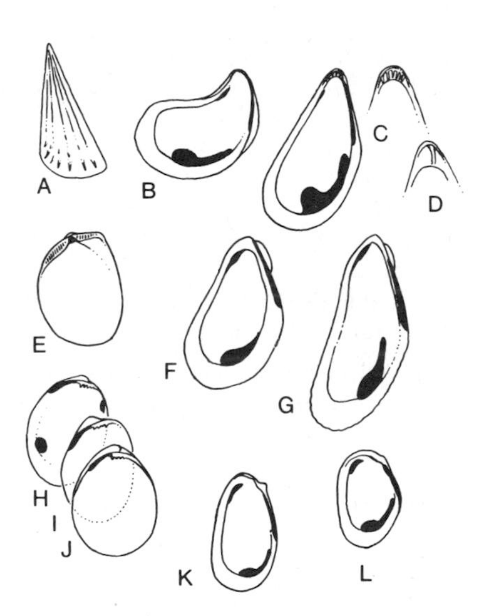

Fig. 16.18. Mollusca, Bivalvia, Group Two. All views internal except A; muscle scars omitted from E,I,J. A, *Pinna carnea;* B, *Brachidontes recurvus;* C, *Mytilus edulis*, with detail of hinge; D, *Congeria conradi*, detail of hinge; E, *Dacrydium vitreum;* F, *Modiolus modiolus;* G, *Modiolus demissus;* H, *Crenella glandula;* I, *Crenella faba;* J, *Crenella decussata;* K, *Musculus niger;* L, *Musculus discors.*

- Shell olive or reddish brown; ribs weakly developed with about 8 ribs anteriorly; 25 mm; Fig. 16.18L. *M. discors.*
- Shell yellowish green; ribs well developed with about 16 ribs anteriorly; 11 mm (Miner, 1947). *M. corrugatus.*

5. Ribbing not confined to ends of shells—6.

6. Shell elongate, strongly ribbed; 100 mm; Fig. 16.18G; a salt marsh species. *Modiolus demissus.*

6. Shell broadly oval. *Crenella:*

- Yellowish gray; 3 mm; Fig. 16.18J. *C. decussata.*
- Olive-brown; 13 mm; Fig. 16.18H. *C. glandula.*
- Yellowish brown; 13 mm; Fig. 16.18I. *C. faba.* A fourth species, *fragilis,* is confined to deeper waters.

Group Three, Ark, Nut Shells, and Related Forms. Sizes for some species given in legend for Figure 16.20.

---

1. Hinge line interrupted by a chondrophore or central pit; periostracum yellowish—2.

1. Hinge line continuous or, if interrupted, without a chondrophore or central pit—6.

2. Shell elongate—4.

2. Shell not much or not at all longer than broad—3.

3. Hinge line strongly arched, Fig. 16.19D. *Nucula:* The ventral edge is smooth in *tenuis* and *delphinodonta;* it is minutely crenulate in *proxima,* 6 mm, Fig. 16.19A. The beaks project slightly in *delphinodonta,* 3 mm, Fig. 16.19B; they do not in *tenuis,* 9 mm, Fig. 16.19C.

3. Hinge line straight; Fig. 16.19F. *Limopsis: L. minuta* has beaded sculpturing, 13 mm; *L. sulcata* and *L. cristata* do not. *L. cristata,* 6 mm, has the inner margin of the valves with a series of small teeth; *L. sulcata,* 13 mm, does not.

4. Shell gapes posteriorly when the valves are closed; Fig. 16.19L. *Yoldia* in part. The species differ in shape as figured, Fig. 16.20B–E. (*Y. cascoensis* is figured by Mighels and Adams, 1842.)

4. Shells close tightly—5.

5. Shell ovate or somewhat quadrangular, posterior end not tapered; Fig. 16.20A–H. *Yoldia,* subgenus *Yoldiella:* The species are somewhat similar in shape but differ in tooth formulae: *Y. lucida* and *Y. inflata* have 9–10 anterior and 11 posterior teeth; *Y. iris* has 12–13 teeth in both series; *Y. subangulata* has 17 anterior, 18 posterior teeth. *Y. inflata* is short and broad in comparison with *Y. lucida.* All are described and figured by Verrill and Bush (1898).

5. Shell with posterior end tapered; Fig. 16.20I,J. *Nuculana: N. messanensis* is small, 6 mm, and nearly smooth; the other species have concentric growth ridges or fine ribs. The posterior end is pointed in *N. acuta,* Fig. 16.20I, 10 mm; it is blunt in *N. pernula* and *N. tenuisulcata,* Fig. 16.20J. *N. tenuisulcata* is found just below low water, *N. pernula* in deeper water; both species reach 20–25 m.

6. Shell length and depth about equal; Fig. 16.19E; about 9 mm. *Bathyarca: B. pectunculoides* has the hinge teeth well developed; in *B. anomala* the teeth are degenerate. See Verrill and Bush (1898).

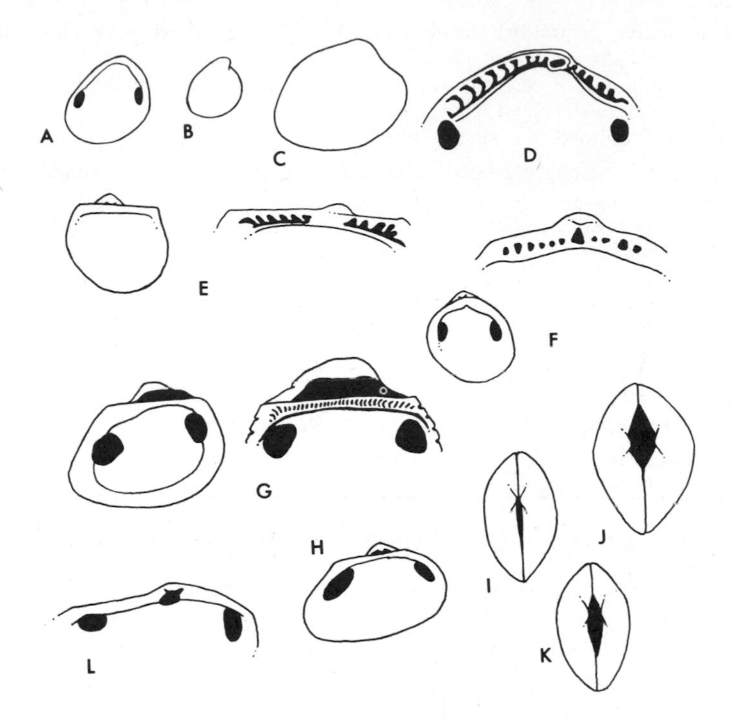

Fig. 16.19. Mollusca, Bivalvia, Group Three. A, *Nucula* sp., pallial line omitted; B, *Nucula delphinodonta*, outline; C, *Nucula tenuis*, outline; D, *Nucula* sp., hinge; E, *Bathyarca* sp., outline of shell and hinge; F, *Limopsis* sp., whole shell, pallial line omitted, and detail of hinge; G, *Noetia ponderosa*, shell and hinge; H, *Anadara transversa*, pallial line omitted; I–K are end views showing ligament; I, *Anadara ovalis;* J, *Noetia ponderosa;* K, *Anadara transversa;* L, *Yoldia* sp., hinge line, teeth omitted.

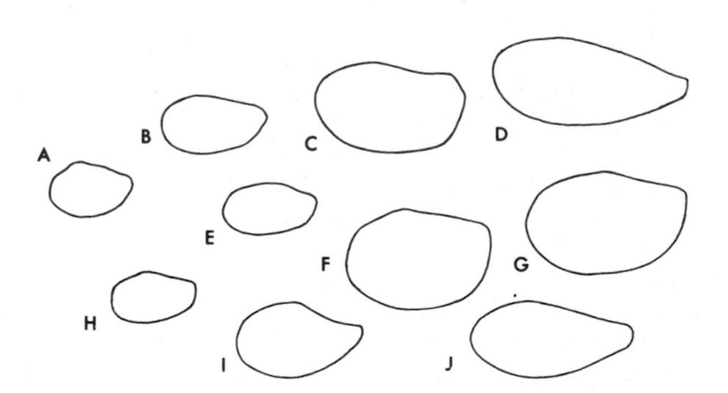

Fig. 16.20. Mollusca, Bivalvia, Group Three. Dimensions are shell lengths. A, *Yoldia subangulata*, <8mm; B, *Yoldia sapotilla*, 38 mm; C, *Yoldia thraciaeformis*, 50 mm; D, *Yoldia limatula*, 63 mm; E, *Yoldia myalis*, 25 mm; F, *Yoldia inflata*, <8mm; G, *Yoldia lucida*, <8mm; H, *Yoldia iris*, <8mm; I, *Nuculana acuta*, 10 mm; J, *Nuculana tenuisulcata*, 25 mm.

6. Shell longer than deep; Fig. 16.19G–K. *Noetia ponderosa* and *Anadara ovalis* reach 60 mm, *Anadara transversa*, 38 mm. The species differ particularly in the relative widths of the ligament, as figured.

Group Four. These species are borers in clay, peat, rock, wood, or shell. Note also that *Hiatella arctica*, a species keyed in Group Seven, may be found in rock crevices.

---

1. Wormlike, elongate, not enclosed by the small, tripartite shells; Fig. 16.21F; siphons with *pallets* that close end of burrow. *Teredo* and related forms have shells usually less than 10 mm long or wide, but the animal may extend to 200 or 300 mm; wood borers—2.

1. Typically clamlike bivalves—3.

2. Pallets with segmented blades; Fig. 16.21A; shell 8.5 mm, pallets 14 mm. *Bankia gouldi*.

2. Pallets with broadly oval, unsegmented blades; Fig. 16.21B–E: *Teredo*, *Psiloteredo*, and *Teredora* are apparently monotypic on this coast; the occurrence of *Teredo norvagica* in the western North Atlantic requires confirmation. The genera differ as figured. *T. navalis* has shells and pallets about 5 mm long; *Psiloteredo* and

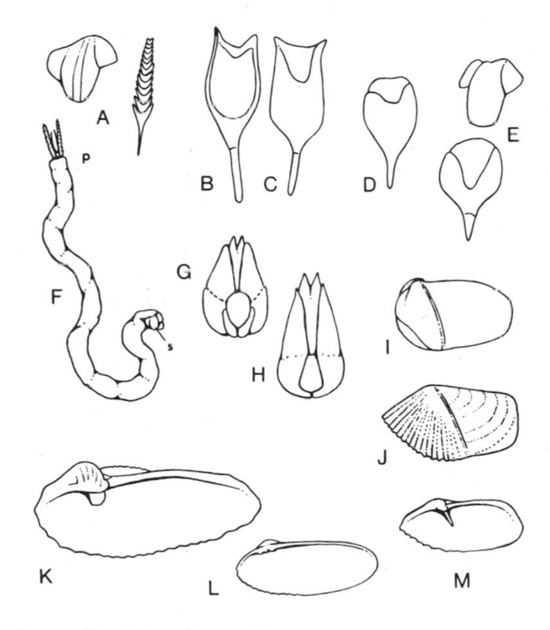

Fig. 16.21. Mollusca, Bivalvia, Group Four. A, *Bankia gouldi*, valve and pallet; B,C, *Teredo navalis*, pallet; D, *Psiloteredo megotara*, pallet; E, *Teredora malleolus*, valve (above) and pallet; F, *Bankia* sp., whole animal; G, *Diplothyra smithii*, dorsal view; H, *Martesia cuneiformis*, dorsal view; I, same, external view, sculpturing omitted; J, *Zirfaea crispata*, external view; K–M are internal views; K, *Cyrtopleura costata;* L, *Petricola pholadiformis;* M, *Barnea truncata*.

Teredora are almost twice as large. The valves of *Psiloteredo* and *Teredo* resemble those of *Bankia* in general form. See Turner (1966). Additional species of normally extralimital wood borers may be carried ashore in driftwood or in wooden ship hulls.

3. Shells divided in 2 areas by groove; Fig. 16.21I,J. Piddocks:
- With coarse, radial scalelike ornamentation anteriorly; 50 mm; Fig. 16.21J: burrows in peat, clay, and less frequently in soft rock or wood. *Zirfaea crispata.*
- With concentric, scalelike ornamentation anteriorly: *Martesia cuneiformis* and *Diplothyra smithii* differ as figured, Fig. 16.21G,H; *M. cuneiformis* reaches 21 mm and is a wood borer; *D. smithii* reaches 15 mm and usually bores in oyster shells; young shells of these two species are pointed anteriorly as in *Zirfaea*. See Turner (1954).

3. Shell not so divided—4.

4. Posterior end of valves truncated and relatively smooth, ribbing increasingly strong anteriorly; Fig. 16.21M; burrows in mud, clay, peat. *Barnea truncata.*

4. Posterior end not truncated—5.

5. Ribbing strongly developed from end to end; 130 mm; Fig. 16.21K; burrows in mud. *Cyrtopleura costata.*

5. Ribbing most strongly developed posteriorly; 50 mm; Fig. 16.21L; burrows in clay or peat. *Petricola pholadiformis.*

Group Five, Cockles and Others. Note additional forms with radial sculpture. *Crenella* spp., keyed in Group Two, and *Serripes groenlandicus* keyed in Group Eight. The latter species is very weakly ribbed at the ends of the shell only, and the ribbing may even be overlooked.

---

1. Anterior and posterior slopes of valves separated by sharp line, sculpture more prominent on posterior third; white; 20 mm; Fig. 16.22A. *Microcardium peramabile.*

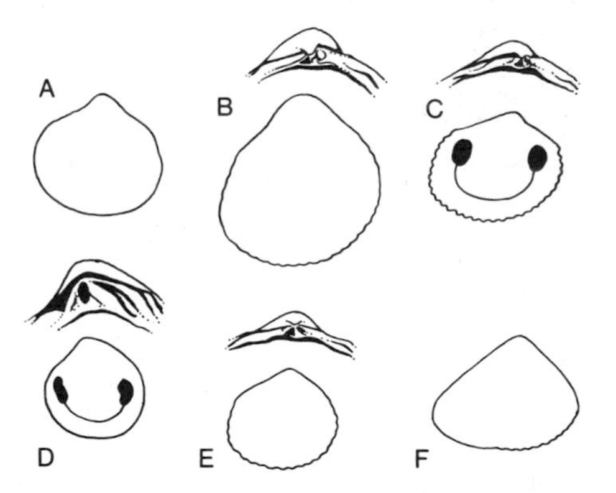

Fig. 16.22. Mollusca, Bivalvia, Group Five. A,B,E,F are shell outlines only; A, *Microcardium peramabile;* B, *Dinocardium robustum*, with detail of hinge; C, *Clinocardium ciliatum*, with detail of hinge; D, *Cardita borealis*, with detail of hinge; E, *Cerastoderma pinnulatum*, with detail of hinge; F, *Panacea arata*.

1. Not so—2.

2. Ribs 32 or more—3.
2. Ribs usually fewer—4.

3. Shell heavy, slightly deeper than long; radial ribs rounded; 100 mm; Carolinian; Fig. 16.22B. *Dinocardium robustum.*
3. Shell thin, somewhat longer than deep; radial ribs ridged; 50 mm; Boreal; Fig. 16.22C. *Clinocardium ciliatum.*

4. Shell strong, about as deep as long; ribs about 20, weakly beaded; 38 mm; Fig. 16.22D. *Cardita borealis.*
4. Shell thin, subcircular or slightly longer than deep; ribs 22–28, delicately scaled; 13 mm; Fig. 16.22E. *Cerastoderma pinnulatum.*
4. Shell thin, triangular, longer than deep; ribs about 30; 50 mm; Fig. 16.22F. *Panacea arata.*

Group Six, Including Pandora, Dipper and Spoon Clams. *Halicardia flexuosa* is figured here, Figure 16.23M, but is omitted from the key; see Morris (1947). *Mya truncata* is keyed with its relatives in Group Eight but is figured here, Figure 16.23E.

---

1. Beak of right valve perforated by beak of left valve; valves unequal; ossicle minute or absent; Fig. 16.23A–C. *Thracia:* The three local species differ in shape as figured. *T. conradi*, the more common, shallow water form, is 75 mm, *T. myopsis*, 25 mm; and *T. septentrionalis*, about 18 mm. (All figured by Gould, 1870.)
1. Right beak not pierced by left—2.

2. Hinge toothless, with small ossicle under beaks; valves unequal, commonly with sand grains "glued" to outside; Fig. 16.23H,I. *Lyonsia:* The more elongate *L. hyalina* is 15 mm and *L. arenosa*, 18 mm. (figures, Gould, 1870).
2. Hinge with well-developed chondrophore, or with 1 or more teeth or toothlike processes—3.

3. Hinge with distinct chondrophore; valves unequal; with or without ossicle; Fig. 16.23D,F. *Periploma: P. leanum*, 38 mm, without an ossicle, and *P. papyratium*, 25 mm, with an ossicle, differ in shape as figured. A third species, *P. fragilis*, resembles *papyratium;* see Abbott (1954, p. 473) (figures, Gould, 1870.)
3. Hinge with 1 or more teeth—4.

4. Hinge with small tooth; valves swollen, about equal, with well-developed "handle"; Fig. 16.23G. *Cuspidaria* spp. Ten species are listed, mostly in depths of about 70 m or more. See Johnson (1915) for references. Most of the species are figured by Verrill and Bush (1898). Note also *Cardiomya glypta*, a Carolinian species of similar form. Keen (1963) separated *Cuspidaria* and *Cardiomya* (on Pacific coast) as follows: *Cardiomya* with radial ribs, *Cuspidaria* without.
4. Hinge with 2 or 3 toothlike processes; valves unequal, flattish; Fig. 16.23J–L,N,O. *Pandora:* The species differ in shape as figured; *P. inornata* and *P. inflata* are about 18 mm; *P. gouldiana* and *P. glacialis* are about 30–35 mm. See Boss and Merrill (1965).

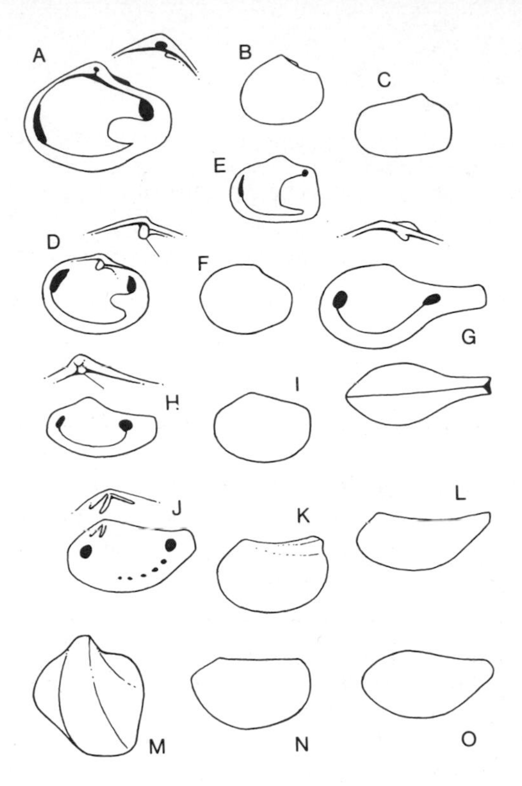

Fig. 16.23. Mollusca, Bivalvia, Group Six. B,C,F,I,K,L–O are shell outlines only. A, *Thracia conradi*, with detail of right beak; B, *Thracia myopsis;* C, *Thracia septentrionalis;* D, *Periploma leanum*, with chondrophore indicated; E, *Mya truncata* (see Group Eight); F, *Periploma papyratium;* G, *Cuspidaria* sp., lateral and dorsal views, with detail of hinge; H, *Lyonsia hyalina*, with ossicle indicated; I, *Lyonsia arenosa;* J, *Pandora inornata*, with detail of hinge; K, *Pandora gouldiana;* L, *Pandora glacialis;* M, *Halicardia flexuosa;* N, *Pandora inflata;* O, *Pandora arenosa.*

Group Seven, Razor Clams and Related Forms. *Panomya arctica* figured, Figure 16.24D, but not keyed, reaches 75 mm; see Morris (1947) or Gould (1870).

---

1. Beak at or near anterior end; shells relatively narrow. Razor and jackknife clams:
- With two cardinal teeth in left valve; length about 6 times width; 150 mm; Fig. 16.24A. *Ensis directus.*
- With 1 cardinal tooth in each valve; length about 4 times width; 50 mm; Fig. 16.24B. *Solen viridis.*

1. Beak near middle of shell—2.

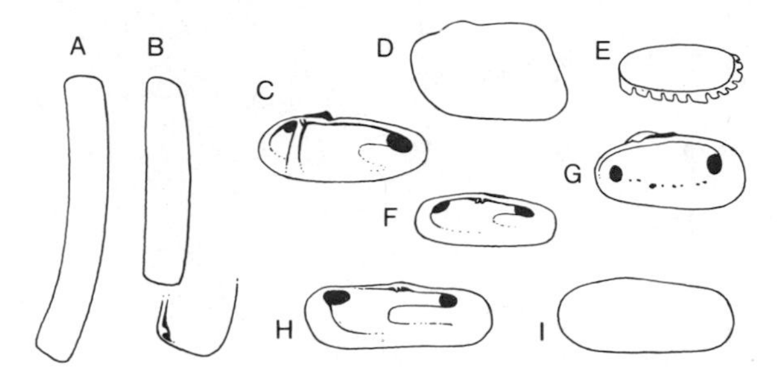

Fig. 16.24. Mollusca, Bivalvia, Group Seven. A, *Ensis directus;* B, *Solen viridis*, with detail of beak; C, *Siliqua costata;* D, *Panomya arctica*, outline only; E, *Solemya* sp.; F, *Tagelus divisus;* G, *Hiatella* sp.; H, *Tagelus plebeius;* I, *Cyrtodaria siliqua*, outline only.

2. With glossy brown periostracum that overlaps valves and is scalloped along edges; Fig. 16.24E. *Solemya:* The siphonal opening has tentaclelike appendages above and below; these number about 40 in *S. borealis* and about 16 in *S. velum.* See Abbott (1968). *S. velum* is about 25 mm; *S. borealis*, 75 mm. A third, deep water species, *S. grandis*, is reported; see Verrill and Bush (1898).

2. Not so—3.

3. Inside of shell with strong, off-center rib; 50 mm; Fig. 16.24C. *Siliqua costata.* Note also *S. squama* reported from cod stomachs from New England fishing banks.

3. Inside of shell without rib or with weak, central rib—4.

4. Hinge with 2 teeth, with or without a weak, often obscure radial rib inside. *Tagelus: T. divisus* has a fragile shell with a thin, smooth periostracum, 38 mm; Fig. 16.24F; *T. plebeius* has a strong, somewhat rectangular shell and thick periostracum; 90 mm, Fig. 16.24H.

4. Hinge without teeth—5.

5. Periostracum thick, glossy black (in dead shells), easily flaked; beaks suppressed; 90 mm; Fig. 16.24I. *Cyrtodaria siliqua.*

5. Periostracum thin, gray; beaks well developed; valves variable in shape; 25 mm (rarely to 75 mm); in hard clay or rock crevices; Fig. 16.24G. *Hiatella:* Adults of *H. striata* and *H. arctica* are almost indistinguishable; young *H. striata* have 2 radial rows of spines while young *H. arctica* do not; see Abbot (1954).

Group Eight, Soft-Shell, Freshwater Clams, and Others.

1. Hinge with large projecting tooth in one valve and socket in other; Fig. 16.25A. *Mya:* The common species is *M. arenaria.* An arctic species, *M. truncata*, Fig. 16.23E, ranging south to Maine, has the posterior margin distinctly truncate, and the pallial sinus a broad U rather than a narrow U as in *M. arenaria*; the latter reaches 100 mm or more, *M. truncata*, 75 mm.

1. Hinges of both valves more or less similar—2.

2. Pallial line without sinus—3.

2. Pallial line with sinus—4.

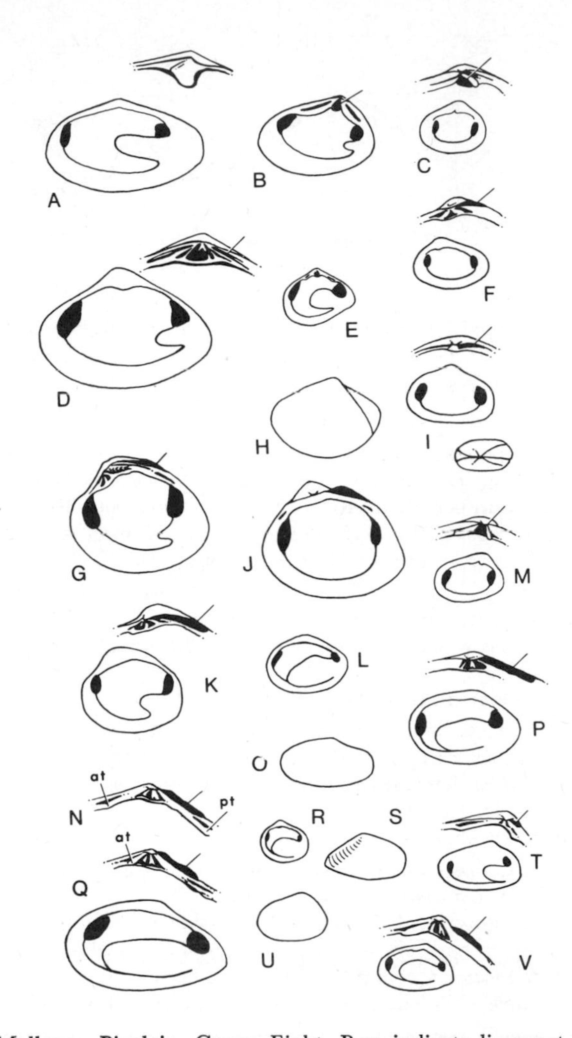

Fig. 16.25. Mollusca, Bivalvia, Group Eight. Bars indicate ligament unless otherwise indicated. A, *Mya arenaria*, with detail of hinge; B, *Mesodesma arctatum;* C, *Montacuta* sp., with detail of hinge; D, *Spisula solidissima*, with detail of hinge; E, *Cumingia tellinoides;* F, *Turtonia* sp., with detail of hinge; G, *Mercenaria mercenaria;* H, *Anatina anatina;* I, *Corbula contracta*, with detail of hinge; J, *Serripes groenlandicus;* K, *Pitar morrhuana*, with detail of hinge; L, *Macoma balthica;* M, *Mysella* sp., with detail of hinge; N, *Tellina* sp., hinge with anterior (at) and posterior (pt) teeth indicated; O, *Macoma tenta;* P, *Macoma calcarea*, with detail of hinge; Q, *Tellina alternata*, with detail of hinge, anterior tooth indicated by "at"; R, *Tellina mirabilis;* S, *Tellina versicolor*, with sculpturing indicated; T, *Donax variabilis*, with detail of hinge; U, *Tellina iris;* V, *Tellina versicolor*, with detail of hinge.

3. Shell somewhat triangular with weakly defined ribs at ends; 75 mm; Fig. 16.25J. *Serripes groenlandicus.*

3. Shell ovoid, proportionately very thick; right valve slightly larger than left; 10 mm; Fig. 16.25I. *Corbula contracta.*

3. Shell ovoid, not especially thick; minute species smaller than 5 mm; commensal on other invertebrates. Several genera, not individually keyed; note hinge patterns and see Keen (1963); Fig. 16.25C,F,M. *Montacuta, Turtonia, Kellia, Mysella.*

4. Ligament completely or largely internal, with prominent chondrophore—5.

4. Ligament external—8.

5. Shell with posterior rib or flat area:

- With diagonal or radial rib posteriorly, shell flaring behind; 20 mm; Fig. 16.25H. *Anatina anatina.*
- With posterior flat area; lower anterior lateral teeth without small denticles (vs young *Spisula*, which have them); 13 mm. *Mulinia lateralis.* (The species is in Abbott, 1954, p. 449, though omitted from the index.)

5. Not so—6.

6. Beaks anterior; shell thick; 50 mm; brackish marsh species. *Rangia cuneata.*

6. Beaks somewhat posterior or directed backward:

- *Anatina plicatella*, shell thin with concentric ribs; white; 64 mm; chiefly Carolinian.
- *Mesodesma arctatum*, shell moderately thick; tan; 38 mm; chiefly Boreal; Fig. 16.25B.

6. Beaks central—7.

7. Pallial sinus shallow; shell strong; 150 mm; Fig. 16.25D. *Spisula solidissima.*

7. Pallial sinus deep; shell thin, posterior end somewhat pointed: 12 mm; Fig. 16.25E. *Cumingia tellinoides.*

8. With 3 cardinal teeth—9.

8. With less than 3 cardinal teeth—10.

9. Interior with purple stain; 100 mm; Fig. 16.25G. *Mercenaria mercenaria.*

9. Interior white; 50 mm; Fig. 16.25K. *Pitar morrhuana.* Note also *Mercenaria campechiensis*, which resembles *M. mercenaria* but lacks the purple stain and is larger and stouter; compare hinges, *Pitar* and *Mercenaria.*

10. Inner margin minutely toothed; color extremely variable; 13 mm; Fig. 16.25T. *Donax variabilis. D. variabilis* and *D. fossor* are doubtfully distinct.

10. Inner margin smooth—11.

11. Hinge with lateral teeth; Fig. 16.25N. *Tellina*—12.

11. Hinge without lateral teeth. *Macoma*—14.

12. With oblique cross-cross sculpturing: *Tellina mirabilis*, 8 mm, Fig. 16.25R, and *T. iris*, 15 mm, Fig. 16.25U, differ in shape as figured.

12. Without oblique cross-cross sculpturing—13.

13. Right postlateral tooth well developed; with orange or yellowish brown periostracum:

- Anterior lateral tooth well separated from cardinal teeth; 39 mm Fig. 16.25N. *Tellina nitens.*
- Anterior lateral tooth close to cardinal teeth; 72 mm; Fig. 16.25Q. *Tellina alternata.*

13. Right postlateral tooth weak or absent; white, tinged or not with pink or yellow: *Tellina versicolor* has the posterior slope of the right valve differentiated by strong sculpturing, Fig. 16.25S,V, 17 mm; *T. agilis* and *T. tenella* lack the above sculpturing. *T. tenella* is marked with raised ridges, while *T. agilis* is marked with fine scratches; both are 10 mm. (See Boss, 1966, 1968.)

14. Shell relatively wide: *Macoma balthica*, 25 mm, and *Macoma calcarea* differ as figured; Fig. 16.25L,P.

14. Shell elongate; 13 mm; Fig. 16.25O. *Macoma tenta*. Note also *M. brevifrons;* see Morris (1947).

Group Nine.

1. Shell more circular than triangular—2.

1. Shell more triangular than circular—10.

2. Pallial line with sinus—3.

2. Pallial line without sinus—4.

3. Beak strongly turned to one side; 75 mm; Fig. 16.26A. *Dosinia discus.*

3. Beak weakly turned to one side; 38 mm; Fig. 16.26B. *Semele proficua.* Note also *Polymesoda caroliniana* with inflated valves and scaly brown periostracum in comparison to the above, which are white or yellowish externally; *P. caroliniana* is 30 mm long, a brackish water species.

4. No hinge teeth; valves with exterior groove posteriorly. Thyasiridae. These are mostly small species, less than 5 mm, from relatively deep water. There are two local genera and about a dozen species. See Johnson (1915) for references. Verrill and Bush (1898) figure most of the species. These are Boreal or Arctic bivalves with a few ranging south of Cape Cod in deep water. *Thyasira insignis,* Fig. 16.26C, is 18 mm.

4. Hinge teeth well developed—5.

5. With thick brown or black periostracum; three cardinal teeth; 100 mm; Fig. 16.26D. *Arctica islandica.*

5. Without black periostracum—6.

6. Anterior muscle scar long and narrow; Fig. 16.26G—7.

6. Anterior muscle scar round—9.

7. Shell sculptured with chevronlike grooves crossing the concentric growth lines obliquely; 25 mm; Fig. 16.26E. *Divaricella quadrisulcata.*

7. Not so—8.

8. With 2 cardinal teeth one or both of which are bifid, no lateral teeth; Fig. 16.26F. *Diplodonta:* Several species are listed as reaching Cape Hatteras and one, *D. verrilli*, is recorded off Cape Cod in more than 100 m. See Johnson (1934). These are small species, less than 20 mm.

8. With 2 cardinal teeth in right valve, 1 in left; 1 cardinal tooth bifid; surface of valves with sharp concentric ridges; 50 mm; Fig. 16.26G. *Lucinoma filosa.* South of Cape Cod this species is confined to deep water. Additional lucinid species may be found in shallow water southward. Note also *Myrtaea lens*, 13 mm, without sharp concentric ridges. See Keen (1958) for remarks on systematics of *Lucina* and for generic and subgeneric keys.

9. Interior of fresh shells bright yellow, fading in dead specimens; 25 mm; Fig. 16.26H. *Laevicardium mortoni.*

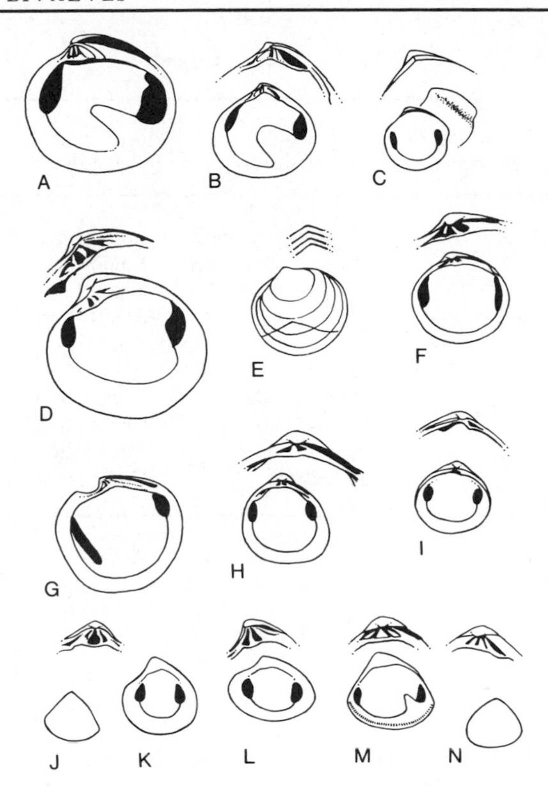

Fig. 16.26. Mollusca, Bivalvia, Group Nine. A, *Dosinia discus;* B, *Semele proficua*, with detail of hinge; C, *Thyasira insignis*, with details of hinge and exterior groove; D, *Arctica islandica;* E, *Divaricella quadrisulcata*, with orientation of chevrons indicated by a single chevron and detail of pattern; F, *Diplodonta* sp., with detail of hinge; G, *Lucinoma filosa;* H, *Laevicardium mortoni*, with detail of hinge; I, *Kellia suborbicularis*, with detail of hinge; J, *Crasinella mactracea*, with detail of hinge; K, *Astarte castanea;* L, *Astarte borealis;* M, *Gemma gemma*, with detail of hinge; N, *Abra* sp., with detail of hinge.

9. Interior white; 10 mm; Fig. 16.26I. *Kellia suborbicularis*. Shell variable in form to ovate-triangular (Gould, 1870).

10. Pallial line without sinus; dark brown; weakly ribbed—11.

10. Pallial line with sinus; pale; smooth—12.

11. Ligament internal; 5 mm; Fig. 16.26J. *Crasinella mactracea*.

11. Ligament external; Fig. 16.26K,L. *Astarte*. See summary in Table 16.4.

12. White to gray-tinted purple; pallial sinus points toward beak; lower inner margin crenulated; 3 mm; Fig. 16.26M. *Gemma gemma*.

12. Right valve with two cardinal teeth; 6 mm; Fig. 16.26N. *Abra:* In *A. aequalis* the anterior margin of the right valve is grooved; in *A. lioica* it is not and the shell is more elongate. A third species, *A. longicallis*, is found in deep water; see Verrill and Bush (1898).

Table 16.4. *Astarte*

Species	Maximum Size, mm	Sculpture	Lower Inner Margin
borealis	45	Ridged near beak	Smooth
elliptica	25	Ridged near beak	Smooth
striata	12	30–40 ridges	Smooth
quadrans	12	Smooth	Smoth with anterior lateral tooth
portlandica	10	Smooth	Smooth
castanea	25	Weakly ridged	Crenulated
undata	25	10–20 ridges	Crenulated
subaequilatera	38	30–40 ridges	Crenulated
nana	6	25 ridges	Crenulated, southern

References

(See also references following the general introduction to Mollusca.)

Abbott, R. T. 1954. American Seashells. D. Van Nostrand.

———. 1968. A guide to field identification, seashells of North America. Golden Press, New York.

Boss, K. J. 1966, 1968. The subfamily Tellininae in the Western Atlantic. The genus Tellina (Part I). Johnsonia 4(45): 217–272; 4(46): 273–344.

———, and A. D. Merrill. 1965. The family Pandoridae in the Western Atlantic. Johnsonia. 4(44): 181–216.

Clench, W. J., and Ruth D. Turner. 1946. The genus *Bankia* in the Western Atlantic. Johnsonia. 2(19):1–28.

Franc, A. 1960. Classe des Bivalves. *In* Traité de Zoologie, Tome V. P.-P. Grassé, ed. Masson et Cie., Paris.

Galtsoff, P. S. 1964. The American oyster *Crassostrea virginica* Gmelin. Fishery Bull. Fish Wildf. Serv. 64: iii + 480 pp.

Gould, A. A. 1870. Report on the Invertebrata of Massachusetts, 2nd ed. W. G. Binney, ed. Wright and Potter, Boston.

Johnson, C. W. 1915. Fauna of New England. 13. List of the Mollusca. Occ. Paps. Boston Soc. Nat. Hist. 231 pp.

———. 1934. List of marine Mollusca of the Atlantic coast from Labrador to Texas. Proc. Boston Soc. Nat. Hist. 40(1):1–203.

Keen, Myra. 1963. Marine molluscan genera of Western North America. Stanford Univ. Press.

Mighels, J. W., and C. B. Adams. 1842. Descriptions of twenty-four species of the shells of New England. Boston J. nat. Hist. 4:37–53.

Miner, R. W. 1950. Field Book of Seashore Life. G. P. Putnam's Sons.

Morris, P. A. 1947. A field guide to the shells of our Atlantic and Gulf coasts. Houghton Mifflin Co.
Morton, J. E. 1958. Molluscs. Hutchinson Univ. Library.
Rothschild, Lord. 1965. A classification of living animals. Longmans.
Turner, Ruth D. 1954. The family Pholadidae in the Western Atlantic and the eastern Pacific. Part II—Martesiinae, Jovannetiinae, and Xylophaginae. Johnsonia 3(34):65–160.
———. 1966. A survey and illustrated catalogue of the Teredinidae. (Mollusca: Bivalvia). Mus. Comp. Zool. Harvard. vii + 265 pp.
Verrill, A. A. 1882. Catalog of marine Mollusca added to the fauna of the New England region, during the past ten years. Trans. Conn. Acad. Arts Sci. 5:447–548.
———, and Katharine J. Bush. 1897. Revision of the genera of Ledidae and Nuculidae of the Atlantic Coast of the United States. Am. J. Sci. Ser. 4. 3(5):51–63.
———. and Katharine J. Bush. 1898. Revision of the deep-water mollusca of the Atlantic coast of North America, with descriptions of new genera and species. Proc. U.S. natn. Mus. 20:775–901.
Warmke, Germaine L., and T. R. Abbott. 1961. Caribbean Seashells. Livingston Publ. Co.

Class Cephalopoda
SQUIDS AND OCTOPUSES

Diagnosis. *Bilaterally symmetrical, the body (mantle) and head differentiated, with the body saclike and without fins in octopods, or more or less elongate and with fins in squid; the head with complex eyes; central mouth armed with a pair of beaklike jaws and encircled by a ring of eight arms, or eight arms and two tentacles armed with suckers and/or hooks; with a siphon opening near the head; the primary shell internal or lacking; a secondary, fragile, calcareous spiral shell developed by the female of Argonauta as an incubatory chamber.*

Squid and octopuses are familiar invertebrate types though quite unlike the usual shelled mollusks. In local squid the shell is reduced to a chitinous "pen" which in fresh material is transparent or translucent, and pliable. In octopuses the shell is absent or reduced to a pair of internal, cartilaginous stylets. An internal calcareous shell in the form of a flat, chambered spiral occurs in *Spirula;* the animal itself is squidlike, but it is usually the shell alone that is found. Similarly, most records of *Argonauta* are based on discoveries of the paper-thin brood shell of the female.

The body contains a mantle cavity as well as the viscera and gills. Water drawn into the mantle cavity can be expelled forcefully through the siphon to provide rapid locomotion through the water or, in some species, even out of the water in the manner of flying fishes. Squid are among the few invertebrates that are sufficiently strong swimmers to qualify as nekton. Both *Loligo* and *Illex*, the common shallow water squid genera of this coast, are seasonal in occurrence, appearing inshore in warm weather and disappearing in the fall.

The skin coloring is carried partly in chromatophores that are especially large and conspicuous in the common squid; these species are capable of rapid color changes. Octopuses are even better known for this ability. Luminescence is common in deep sea cephalopods, including species of *Taonius*. In *Spirula* there is a light organ between the fins consisting of a pore with a luminous pocket beneath containing luciferin. All of the local genera except *Bathypolypus* have ink sacs, usually with a duct expelling the ink into the siphon stream when the animals are disturbed. The ink released by octopods diffuses in the water as a "smoke screen," but in squid it is generally emitted as a "pseudomorph" that approximates the squid in size and shape.

The head is made conspicuous by the eyes which rival those of vertebrates in complexity and efficiency. The jaws are strongly developed, rather parrot-like, and even small individuals can deliver a substantial bite. The salivary secretions of some octopuses are poisonous. Bites of local species, though painful, are not usually dangerous to man according to Halstead (1959); some extralimital forms are deadly. The arms vary in length proportionately in different species and may be relatively equal or not. In squids two of the arms are commonly elongated and modified with terminal clubs to serve as raptorial "tentacles." There is also sexual dimorphism in arm structure.

The sexes are separate. In the male, one or more of the arms is modified as a *hectocotylus* organ that is used to transfer elaborately formed spermatophores to the female. Commonly the terminal suckers of the hectocotylus are reduced in size and the supporting stalks or pedicels are enlarged. In *Bathypolypus*, as in most octopuses, the tip of the hectocotylus forms an expanded spoonlike structure, which is especially large in this genus, Figure 16.28B. The hectocotylus is most elaborately modified in *Argonauta* and its relative *Ocythoe;* in these two genera the hectocotylus is "autotomous," detaching from the male after penetrating the mantle cavity of the female. There is additional sexual dimorphism in *Argonauta*. The first arms of the female of this genus have a brood membrane that secretes the incubatory shell; also the female is about 20 times larger than the male. Less spectacular proportional differences may be found in other species. A ritualized courtship precedes mating in some species.

While fertilization may be internal in some species, the development of the young takes place outside the female's body. Females of *Loligo* may lay their eggs in large communal clusters, but the females do not take further interest in their offspring. Octopuses, on the contrary, may brood their egg clutches.

The eggs of *Argonauta* are brooded in the shell. The young in general are miniature cephalopods and do not have the extreme metamorphosis found in most mollusk life histories. The larvae of many squid, however, are divergent in various details; see Naef (1923) and Pfeffer (1912).

Squid are pelagic and gregarious, ranging from shallow waters to abyssal depths. While the more typical octopuses are solitary, benthic animals, some of the species of Octopodida are pelagic. *Argonauta* and *Ocythoe* are oceanic and occur primarily in surface waters. *Argonauta* in particular is epipelagic. While *Bathypolypus arcticus* has been taken in comparatively shallow water, Robson (1932) regarded this species as typical of the outer edge of the Continental Shelf and upper Continental Slope at an average depth of about 366 m.

Cephalopods are predatory, the principal prey being fish, crustaceans, and smaller squid.

Systematic List and Distribution

The classification to order follows Sweet (1964); alternative ordinal terms preferred by some students are indicated in parentheses. The species list is largely from Verrill (1881) and Berry (1934), and it reflects the usual difficulty in compiling a realistically useful list from a group of planktonic organisms with numerous deep water species that *may* occur inshore but will rarely be encountered by the general worker. See Berry for a more complete account. A notable omission from the present list is *Architeuthis*, the genus of which the notorious giant squids are members; though within the range of this text latitudinally (along the whole coast), they are bathymetrically extralimital under normal conditions.

Subclass Coleoidae		
Order Teuthidida (Teuthoidea)		
Family Ommastrephidae		
Illex illecebrosus	BV	Surface to 481 m
Ommastrephes pteropus	BV	Oceanic, Gulf Stream
Family Gonatidae		
Gonatus fabricii	B+	Surface to 4000 m, young chiefly in surface waters
Family Cranchiidae		
Cranchia scabra	BV	Oceanic
Taonius pavo	BV	Oceanic
Family Loliginidae		
Loligo pealei	V+	Surface to 91 m
Lolliguncula brevis	C, 39°	Shallow eury., to 17‰
Family Onychoteutihidae		
Onykia caribaea	BV	Oceanic, Gulf Stream
Onychoteuthis banksi	BV	Oceanic, Gulf Stream

Order Sepiida (Sepoidea)		
Family Spirulidae		
Spirula spirula	V	200 to 1500 m, shells sometimes cast ashore
Family Sepiolidae		
Stoloteuthus leucoptera	B	172 to 1153 m
Rossia hyatti	B	13 to 580 m
Rossia sublevis	BV	82 to 1171 m
Rossia tenera	BV	33 to 580 m
Order Octopodida (Octopoda)		
Family Argonautidae		
Argonauta argo	C, 42°	Epipelagic, oceanic
Family Ocythoidae		
Ocythoe tuberculata	CV	Shallow to 199 m
Family Octopodidae		
Octopus vulgaris	C+	Lit. and shallow
Normally extralimital but reported by Pickford (1945) from Long Island Sound, two records, regarded as an abnormal occurrence.		
Bathypolypus arcticus	BV	51 to 1543 m

Identification

The species are considered in ordinal groups. The teuthids have fins, 8 arms, and 2 tentacles; the octopods are without fins and have 8 arms. *Spirula* is omitted from the keys, since live individuals are usually extralimital; the whole animal is shown in Figure 16.27Q and the shell in Figure 16.28G. Dimensions are approximate maximum length of adults exclusive of arms, i.e., from tip of mantle to base of arms, unless indicated otherwise.

ORDER TEUTHIDIDA

1. Body short, bluntly rounded terminally (depending partly on state of contraction of preserved material); fins broadly rounded and not united posteriorly; Fig. 16.27A—2.

1. Body elongate, torpedo-shaped; fins usually more or less triangular and united posteriorly—4.

2. Ventral surface with a brown, shield-shaped area bordered except in front by a light band; 31 mm. *Stoloteuthis leucoptera.*

2. Not marked in this way—3.

3. Lower marginal suckers on tentacular clubs enlarged; male with left dorsal arm hectocotylized, suckers smaller and more numerous; in both sexes and young the suckers of lateral arms are larger than others, being about twice as large in females and even larger in males; 48 mm; Fig. 16.27A–C. Note pen shape. *Rossia tenera.*

3. Lower marginal suckers on tentacular clubs not enlarged; males with middle suckers of lateral arms enlarged, females not so; Fig. 16.27D. *Rossia*:

- Fins well forward; sessile arms nearly equal, with suckers in 2 rows; upper surface of body smooth or with few papillae; 46 mm less arms; Fig. 16.27E. Note pen shape. *R. sublevis.*
- Fins farther back; sessile arms unequal, with suckers in 2–4 rows; upper surface covered with conical papillae; 40 mm less arms; Fig. 16.27G. *R. hyatti.*

4. Without eyelids; eyes with a pore in front; Fig. 16.27J—5.

4. With free eyelids; eyes with or without a sinus in front, Fig. 16.27K—6.

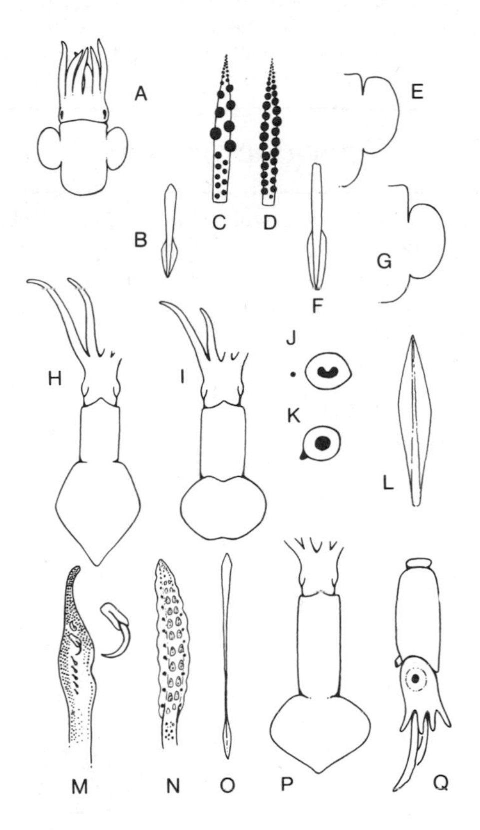

Fig. 16.27. Mollusca, Cephalopoda. A, *Rossia tenera*, whole animal; B, same, pen; C, same, detail of part of tentacular club; D, *Rossia* sp., part of tentacular club; E, *Rossia sublevis*, outline of right side of mantle, dorsal view; F, *Rossia* sp., pen; G, *Rossia hyatti*, outline of right side of mantle, dorsal view; H, *Loligo peali;* I, *Lolliguncula brevis;* J, eye with pore anteriorly; K, eye with sinus anteriorly; L, *Loligo* sp., pen; M, *Gonatus fabricii*, tentacular club and detail of claw; N, *Onykia caribaea*, tentacular club; O, *Illex illecebrosus*, pen; P, same, whole animal with arms cut; Q, *Spirula spirula*, whole animal.

5. Fins triangular, long; body about 6–8 times longer than broad; 1 m total length; Fig. 16.27H,L. *Loligo peali.*

5. Fins rounded, shorter; body relatively shorter, only about 4 times longer than broad; 230 mm total length; Fig. 16.27I. *Lolliguncula brevis.*

6. Tentacular clubs with hooks only; 160 mm, mantle length. *Onychoteuthis banksi.* The general form of the tentacular club is close to that of *Onykia caribaea,* for which see below.

6. Tentacular clubs with hooks and suckers: *Gonatus fabricii* reaches 263 mm total length and *Onykia caribaea,* 32 mm mantle length; the species may be distinguished by differences in the tentacular clubs as figured, Fig. 16.27M,N.

6. Tentacular clubs with suckers only; 460 mm total length; Fig. 16.27O,P; the common inshore Boreal squid. *Illex illecebrosus.* Note that the oceanic *Ommastrephes* is near *Illex,* but has 4 rows of small suckers at the end of the tentacular clubs instead of the eight rows found in *Illex.*

ORDER OCTOPODIDA

1. Sexes generally similar except for hectocotylus, third right or left arm of male—2.

1. Sexes highly dimorphic, both male and female modified—3.

2. First arms longest; dorsal surface warty with a hornlike *cirrus* over each eye; male with an enlarged spoonlike structure on third right arm; 60 mm mantle length; Fig. 16.28A,B. *Bathypolypus arcticus.*

2. First arms shortest or next to shortest; dorsal surface smooth or warty, with or without a cirrus over each eye; male with hectocotylized arm shortened and with tip only slightly modified; 185 mm mantle length; arms webbed for about ¼ to ½ their length; Fig. 16.28C. *Octopus vulgaris.*

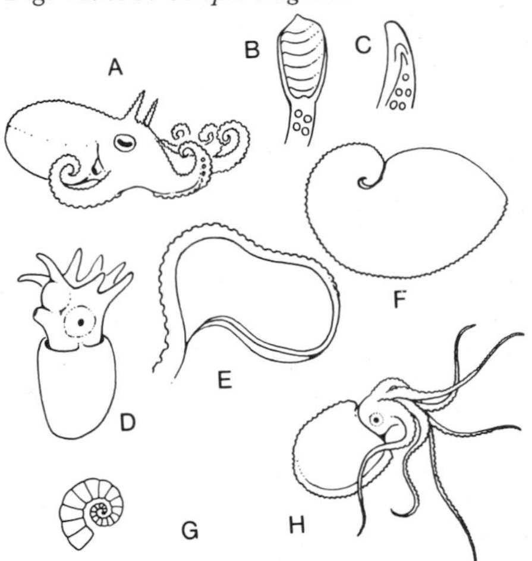

Fig. 16.28. Mollusca, Cephalopoda. A, *Bathypolypus arcticus:* B, same, part of hectocotylus; C, *Octopus vulgaris,* part of hectocotylus; D, *Argonauta argo,* male; E, same, first arm of female; F, same, outline of incubatory shell; G, *Spirula spirula,* shell; H, *Argonauta argo,* female, with first arms omitted.

3. First arms of female with a broad membrane equal to about to about 9 times the arm width (depending on preservation); ventral surface of body thickened and somewhat scalloped; 330 mm mantle length; male much smaller, 15 mm total length; third left arm hectocotylized, up to about 10 times total length of animal when released; before release the hectocotylus is contained in a sac-like swelling; Fig. 16.28D–F,H. *Argonauta argo*. The incubatory shell with brown or black keel, and as long as 350 mm.

3. First arms of female with a membrane about twice the arm width; ventral surface with tubercles connected by ridges; 300 mm less arms; male smaller, 80 mm, and frequently found in empty salp tunics; hectocotylus about 5 times total length; arms not webbed. *Ocythoe tuberculata*.

References

Akimuskin, I. I. 1963. Cephalopods of the seas of the U.S.S.R. Izdatel'stro Akademii SSSR. In English translation. Israel Program for Scientific Translation Ltd. 1384:

Berry, S. S. 1934. Class Cephalopoda. *In* List of marine Mollusca of the Atlantic Coast from Labrador to Texas (C. W. Johnson) Proc. Boston Soc. nat. Hist. 40(1):1–204.

Bruun, A. 1945. Cephalopoda. Zoology of Iceland 4 (64): 1–15.

Halstead, B. W. 1959. Dangerous Marine Animals. Cornell Maritime Press.

Lane, F. W. 1960. Kingdom of the Octopus. xx + 287 pp. Jarrolds, London. Also in paperback without the excellent bibliography.

Naef, A. 1921–28. Die Cephalopoden. *In* Fauna und Flora Neapel. Monogr. 35: ix + 357 p.

Pfeffer, G. 1912. Die cephalopoden der Plankton-Expedition. Ergebn. Plankton-exped. 2: xxi + 815 pp.

Pickford, G. E. 1945. Le poulpe americain: a study of the littoral Octopoda of the Western Atlantic. Trans. Conn. Acad. Arts Sci. 36:701–782.

Robson, G. C. 1929. A monograph of the recent cephalopoda. Part I. Octopodinae. British Museum.

———. 1932. A monograph of the recent cephalopoda, Part II. The Octopoda (excluding the Octopodinae). British Museum.

Sweet, W. C. 1964. Cephalopoda—General Features. *In* Treatise on Invertebrate Paleontology. Part K, Mollusca 3. R. C. Moore, ed. Geol. Soc. Amer. and Univ. of Kansas.

Voss, G. L. 1956a. A review of the cephalopods of the Gulf of Mexico. Bull. mar. Sci. Gulf Caribb, 6:85–178.

———. 1956b. A checklist of the cephalopods of Florida. Quart. J. Fla. Acad. Sci. 19(4): 274–282.

Verrill, A. E. 1880. The cephalopods of the northeastern coast of America. Part I. The gigantic squids (*Architeuthis*) and their allies; with observations on similar large species from foreign localities. Trans. Conn. Acad. Arts Sci. 5:177–257.

———. 1880–81. The cephalopods of the northeastern coast of America. Part II. The smaller cephalopods, including the squids and the Octopi with other allied forms. Trans. Conn. Acad. Arts Sci. 5:259–446.

SEVENTEEN

Phylum Annelida

SEGMENTED WORMS

Introduction

Features of internal anatomy fairly clearly separate annelids from other invertebrate phyla, but external appearance varies so much that no simple diagnosis will suffice to cover the full range of morphological diversity in the phylum. Criteria for the separation of individual classes are presented in Chapter 3; this separation presents little difficulty except for the problematic archiannelids, which are variously regarded as members of a separate class or as no more than neotenic or aberrant polychaetes. All of the annelid classes are represented in marine waters:

Class Polychaeta	Bristle worms
Class Oligochaeta	Aquatic and terrestrial "earthworms"
Class Hirudinea	Leeches
Class Myzostomaria	Myzostomarians (parasitic)

The myzostomarians are also a problematic group now considered to belong to Annelida, and according to Dales (1963) they are "undoubtedly of polychaete derivation." They are disclike, circular, or oval in outline with five pairs of reduced parapodia with hook-shaped bristles. There is evidence of segmentation in the nervous system and in the disposition of the parapodia. They also have a trochophore larva. The trochophore metamorphoses into a creeping form that has been described as tardigradelike, and this in turn changes into the definitive adult form. Adult myzostomids are protandric hermaphrodites, and most are small, a few millimeters to 10 mm usually. An example is figured in Chapter 3, Figure 3.1C. Members of Myzostomaria are parasitic or commensal on echinoderms. They are chiefly known as parasites on crinoids. The host genus of *Myzostomus* is *Hathrometra*, which occurs within the range of this text, and *Protomyzostomus* is an endoparasite in *Gorgonocephalus*. Extralimital forms parasitic on sea stars and brittle stars are also known. For further information see Prennant (1959).

Oligochaetes and leeches are hermaphroditic, and their young develop directly without distinct, free-living larval stages. Development is also direct in some archiannelids, but others pass through a trochophore stage or are essentially neotenic. Myzostomarians and polychaetes have a free, highly metamorphic larval stage and, since the pelagic larvae of polychaetes in particular are likely to be found in plankton samples, this subject is given further attention in the following section and in Chapter 3.

General References

For the literature of individual classes see subsequent sections.

Dales, R. P. 1963. Annelids. Hutchinson University Library. London.

Prennant, M. 1959. Classe des Myzostomides. *In* Traité de Zoologie. Tome V. Annelides, Myzostomides, Sipunculiens, Echiuriens, Priapuliens, Endoproctes, Phoronidiens 5(1)., P.-P. Grassé, ed. Masson et Cie. Paris.

Class Polychaeta
BRISTLE WORMS

Order Archiannelida: *Archiannelids*

The name Archiannelida has been given to a heterogeneous assemblage of small worms that have been variously considered independent derivatives of several polychaete families or specialized relicts of the ancestral polychaete stock. Because of its problematic status the group will be considered apart from the other orders. For a recent evaluation of the status of this assemblage see Hermans (1969).

In lieu of a formal diagnosis the local forms may be characterized in key form; this summary also indicates distributional data and may serve as a species check list. The local families and genera are Dinophilidae (*Dinophilus*), Polygordiidae (*Polygordius*, *Chaetogordius*, *Protodrilus*, and *Nerilla*). See Remane (1932) and Hartman (1959) for additional species and genera and for more recent references. The group has been little studied on this coast, and there is scant information on the ecology of local forms. Dimensions given are maximum lengths for adults.

---

1. Head and body without appendages; Fig. 17.1A. *Dinophilus*. Segmentation indistinct externally but indicated by metamerism of internal parts; segments 5 or 6; with external tufts or bands of cilia; colorless or orange-brown; head with 2 eyes and sensory hairs; last segment with a conical pygidium; 1 mm. Note that published descriptions of individual forms in this genus are inconsistent in some details, and the species limits are uncertain. A fourth species, *D. pygmaeus*, is generally regarded as indeterminate; see Jones and Ferguson (1957). The species are found in upper littoral pools and ditches, salt marshes, etc.—2.

1. With appendages—3.

2. With two broken bands of cilia on head and both rings and scattered cilia on body; orange-brown or reddish; no external sexual dimorphism. Massachusets. *Dinophilus gardineri*.

2. With two broken bands of cilia on head and rings of cilia on body; sexually dimorphic, the males underdeveloped, Fig. 17.1B; colorless or whitish. New Jersey, North Carolina. *Dinophilus gyrociliatus*.

2. With scattered tufts of cilia on head and body but no rings anywhere; male unknown; color not recorded. Virginia, North Carolina *Dinophilus jagersteni*.

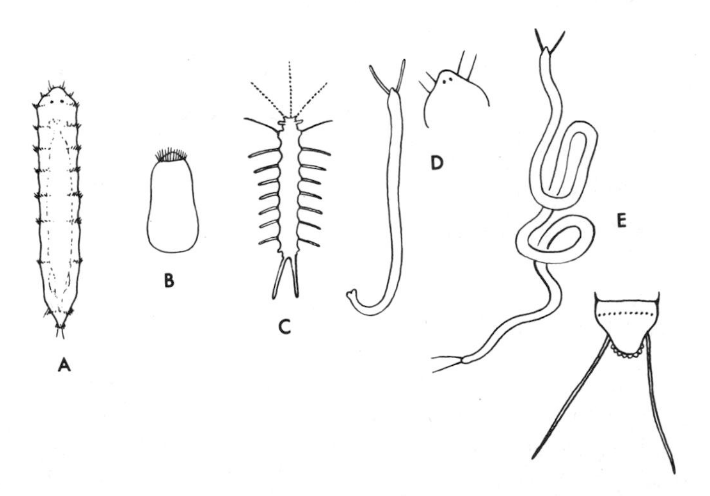

Fig. 17.1. Annelida, Archiannelida. A, *Dinophilus* sp.; B, *Dinophilus gyrociliatus*, male; C, *Nerilla antennata;* D, *Protodrilus* sp., E, *Polygordius appendiculatus*.

3. Head with 3 annulate antennae, 2 club-shaped palps, and 1 pair of tentacular cirri; 4 eyes; 9 segments; with 8 pairs of parapodia, each with dorsal and ventral bundles of capillary setae and the first 7 pairs with cirri; 2 anal cirri; colorless, transparent; 2 mm; Fig. 17.1C. Massachusetts; may occur in marine aquaria (Fauvel, 1927). *Nerilla antennata.*

3. Head with 2 cylindrical tentacles; setae very scarce; segmentation indistinct externally at least anteriorly—4.

4. With a ciliated groove ventrally; *pygidium* (terminal "segment") with 2 or 3 adhesive lobes; setae absent; with or without eyes; color varies in different species, whitish, yellowish, reddish; transparent or opaque; largest species 15 mm; Fig. 17.1D; no local published records but the genus probably occurs at least in the Virginian Province. *Protodrilus.* Wieser (1957) listed 21 species.

4. No ciliated groove ventrally; pygidium not bifurcated:

- With 1 or 2 setae on last 10 or 12 segments; pygidium not enlarged, cirri lacking; 30 mm. Massachusetts. Described from fragments found with marine oligochaetes, and presumably littoral, a questionable species. See Moore (1904), Hermans (1969). *Chaetogordius canaliculatus.*
- Without setae; pygidium swollen basally and with 2 long cirri; salmon-red; 20 mm (to 110 mm, *vide* Beauchamp 1959); Fig. 17.1E (Fauvel, 1927). *Polygordius appendiculatus.* Found "along the Atlantic coast" according to Miner (1950); reported in sand with Amphioxus in Europe; the local American lancelet, *Branchiostoma virginiae,* is reported from the Chesapeake in shallow water.

Note also *Saccocirrus,* a genus without a ventral groove but with a bifurcated pygidium, and retractile parapodia with a bundle of simple setae in each; see Remane (1932). The presence of this genus in the local fauna requires confirmation.

The sexes are separate in most archiannelids, but external dimorphism is relatively minor in most species, a notable exception being *Dinophilus gyrociliatus,* as indicated in the key and Figure 17.1B. Developmental details vary substantially. *Polygordius* has a trochophore type of larva. The larvae of *Protodrilus* somewhat resemble adult *Dinophilus* which have, indeed, been described as neotenic; see Beauchamp (1959) for illustrations of larval *Protodrilus.* The possibly extralimital *Saccocirrus* also has a pelagic larva, but in *Nerilla* development is direct.

References

Beauchamp, P. 1959. Archiannelides. *In* Annelides, Myzostomides, Sipunculiens, Echiuriens, Priapuliens, Endoproctes, Phoronidiens. Traité de Zoologie 5(1). P.-P. Grassé, ed. Masson et Cie.

Fauvel, P. 1927. Polychaetes sedentaires. Faune de France. 16:1–494.

Hartman, O. 1959. Catalogue of the polychaetous annelids of the world. Parts I and II. Occ. Pap. Allan Hancock Fnd. Vol. 23.

Hermans, C. O. 1969. The systematic position of the Archiannelida. Syst. Zool. 18:85–102.

Jouin, C. 1966. Morphologie et anatomie comparée de *Protodrilus chaetifer* Remane et *Protodrilus symbioticus* Giard; création du nouveau genre *Protodriloides*. Cah. Biol. Mar. 7:139–155.

Jones, E. R., and F. F. Ferguson. 1957. The genus *Dinophilus* (Archiannelida) in the United States. Am. Midland Nat. 57:440–449.

Miner, R. W. 1950. Field Book of Seashore Life. G. P. Putnam's Sons.

Moore, Anne. 1899. *Dinophilus gardineri* (sp. nov.). Biol. Bull. mar. biol. Lab. Woods Hole. 1(1):15–18.

Moore, J. P. 1904. A new specific type of Polygordiidae. Am. Nat. 33:519–520.

Nelson, J. A. 1907. The morphology of *Dinophilus conklini* n. sp. Proc. Acad. Nat. Sci. Phila. 59:82–143.

Remane, A. 1932. Archiannelida. *In* Die Tierwelt der Nord- und Ostsee. G. Grimpe, ed. Lief. 22, Teil 6a. Akad. Verlagsges., Leipzig.

Ruebush, T. K. 1940. Morphology, encapsulation and osmoregulation of *Dinophilus gardineri* Moore. Trans. Am. microscop. Soc. 59:205–223.

Wieser, W. 1957. Archiannelids from the intertidal of Puget Sound. Trans. Am. microsc. Soc. 76:275–285.

Typical Polychaete Orders.

Diagnosis. *Usually distinctly segmented annelid worms with parapodia and cephalic appendages; setae present.*

Segmentation is indistinct in only a few local genera, including *Sphaerodorum*, *Ephesiella*, and *Lysilla*. Local adult polychaetes range in size from less than 5 mm to several hundred millimeters long. Body types are extremely diverse, ranging from threadlike or conventionally wormlike to forms that may be characterized as stout conical or ovoid. The body may be distinctly regionated owing to differences in the shape of the segments or of their appendages. In some groups the body is regionated only during the periods of sexual activity. At a minimum it is usually possible to recognize a *prostomium* and *pygidium* which are cephalic and anal lobes, respectively, at opposite ends of a *metastomium* with serial segments numbering anywhere from two dozen or less to 300 or more. Embryologically the prostomium and pygidium are pre- and postsegmental additions.

The polychaete head shows an extraordinary range of adaptive specialization. The head may be taken to include the *prostomium* and a *buccal segment* to which are fused one or more modified trunk segments, the whole forming the *peristomium.* In many sedentary forms the prostomium proper is much reduced, but the head region remains highly specialized. The head commonly bears appendages whose terminology, unfortunately, is not entirely standardized. *Antennae* or *tentacles* are usually dorsal or dorsolateral on the prostomium. *Palpi* range in shape from filamentous to globular. They are often ventral or ventrolateral and border the mouth anteriorly. Dorsal appendages, especially when distinguished by large size or by being grooved, may also be termed palpi; appendages appearing to be the same or similar may be called *tentacular cirri*. True gills with an internal circulatory plexus may also be found on the head. Such gills may or may not be readily distinguishable externally as such.

The morphology of the head is further complicated by the presence in many families of an eversible *proboscis*. This structure is actually a part of the foregut and is often highly modified with teeth, jaws, or papillae. Variations are of systematic importance, but when the proboscis is retracted its characteristic features will not be apparent.

The trunk or metastomium also varies substantially in structural detail. The segments may be repeated with little variation from prostomium to pygidium, or the body may be more or less divided into regions. Frequently the divisions are quite distinct and abrupt, or the changes may be gradual, or involve less conspicuous differences in the form of the appendages and their setation. It should also be noted that many polychaetes have a disconcerting tendency to fragment when handled.

Parapodia are the characteristic lateral trunk appendages of polychaetes. They are unsegmented but may be complexly structured with lobes and subappendages. They are vestigial in only a few local genera, such as *Travisiopsis*. A simple *uniramous* pararodium consists of little more than a lobular or pimplelike protrusion armed with bristles or setae. Commonly the parapodia are *biramous* with a dorsal *notopodium* and a ventral *neuropodium*. Interpretation of these structures may be complicated by secondary lobulation and by the addition of secondary appendages including *cirri* and *branchiae* (gills). The parapodia are, themselves, often highly vascularized and serve as gills. Parapodia are perhaps most highly developed in active, swimming species, and they are frequently reduced in the more sedentary, tubicolous species. In several families, enlarged, scalelike growths, *elytra*, develop from some of the parapodia and cover or partly cover the back. The most important structures of the parapodia are the *setae* (chaetae) or bristles, which are absent or scarce in very few forms. The setae are commonly developed in bundles with each bundle supported by a heavy *aciculum*, Figure 17.2A.

The setae are extremely diverse in form and are of primary systematic importance, Figure 17.2. Initially one may distinguish between *simple* setae, in which the bristle consists of a single piece, and *composite* setae, in which the individual seta is made of two or more pieces, Figure 17.2K–N. Additional named distinctions involve differences in shape and ornamentation as figured. One should also note that intergradation occurs between primary types and that additional special setae are known. Further details are given in the keys

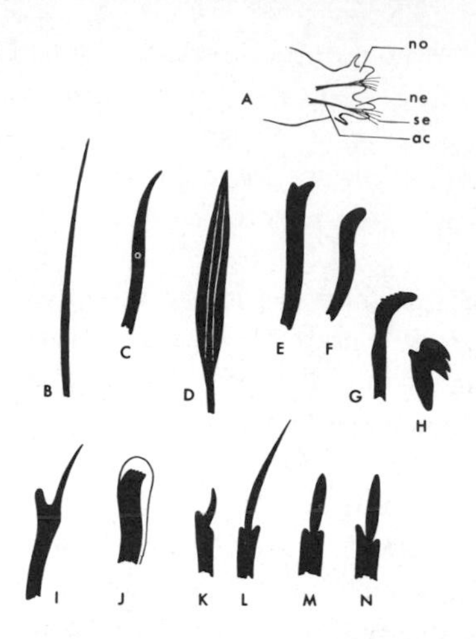

Fig. 17.2. Annelida, Polychaeta, structure of parapodia and types of setae. A, typical biramous parapodium; B–J are *simple* setae; K–N are *composite* setae. Common simple setae include: B, *capillary* seta, long and slender; C, *falcate* seta, curving, sickle-like; D, *limbate* seta, bladed, oar-like; E,F, *crochets* or hooks, broad, with end recurved or bent; G,H, *uncini*, broader, usually toothed; I, *lyrate* seta, bifid, with elongate branches; J, *hooded* seta, with end embedded in a clear matrix. Composite setae include: K, *falcigers*, with outer section sickle-like; L, *spinigers*, with outer section spine-like; M, *homogomphs*, have basal section with equal prongs; N, *heterogomphs*, have basal section with unequal prongs. Additional types and variants are figured as needed in the keys. Abbreviations are as follows: no, notopodium; ne, neuropodium; se, setae; ac, aciculum.

as needed. Setae on the notopodium are *notosetae*, those on the neuropodium are *neurosetae*. Segments with setae are spoken of as *setigers* or *setigerous segments*.

Pelagic polychaetes and some others are colorless and transparent, but many polychaetes are highly colored. The blood may be colorless, yellow, red, or green. Quite a few species are luminescent, including some members of the following families: Aphroditidae, Polynoidae, Alcipidae, Tomopteridae Syllidae, Chaetopteridae, Cirratulidae, and Terrebellidae.

The sexes are separate in most polychaetes. Exceptions include most members of Serpulidae and, less frequently, species in the families Sabellidae, Nereidae, Syllidae, Hesionidae, Cirratulidae, and Eunicidae. Syllids are frequently protandric hermaphrodites, and some serpulids, e.g., *Hydroides*, are protandrous. There is little external sexual dimorphism except that imparted by the color of the gametes showing through the skin. Sperm usually confer a whitish tinge, eggs may be yellow, green, violet, or red. Pronounced external sexual dimorphism occurs in some syllids, alciopids, and nereids, but the differences may only be apparent during periods of sexual activity.

The identification of some polychaete species is greatly complicated by the existence of highly polymorphic reproductive and nonreproductive phases. The phenomenon is called *epitoky*. Characteristically this involves the metamorphosis of a sexually inactive benthic animal into a sexually active pelagic stage. In the simplest cases, e.g., in some members of Nepthyidae, Hesionidae, Scalibregmidae, Euphrosinidae, Phyllodocidae, and Glyceridae, the metamorphosis chiefly involves the appearance of setae adapted for swimming and of modifications in the parapodia for the same purpose. More elaborate alterations are accomplished in "heteronereis," the swarming phase of nereids, in which the body becomes distinctly regionated. In some eunicids, regionation results from a metamorphosis of the posterior, gamete-producing region which subsequently breaks off as a headless, short-lived swarming body—the famed palolos of the Pacific and tropical Atlantic are examples.

Polychaetes generally have considerable regenerative powers, and bizarre individuals undergoing such repair may be encountered. In Syllidae asexual reproduction by serial budding occurs in combinations with sexual metamorphosis.

The gametes are usually shed freely in the water. In a few genera the females have spermathecae (receptacles to receive the sperm which may be packaged in spermatophores). The eggs may be encapsuled when shed, and in some tubicolous species the eggs are "incubated" in the parental tube, or under the elytra in scale worms. Development usually involves a profound metamorphosis based on the trochophore type of larva. The initial larva may, indeed, be a simple trochophore with a single, approximately equatorial girdle of cilia and a small anal circlet of cilia. Often the hatchling is more advanced, but still clearly derived from the basic, top-shaped trochophore; such larvae, with several caudal segments, are called *polytrochulae*. The initial larvae may be more differentiated and wormlike as in the *nectochaeta*, which has simple parapodia, or the *mitraria*, in which the basic equatorial girdle is elaborated in sinuous folds. The latter type of larva is characteristic of the family Owenidae. Polychaete larvae are shown in Figure 3.3.

The earliest larval stages are microscopic, 0.2–0.5 mm. Post-trochophore stages, measuring 0.8 mm to several millimeters, become more wormlike, and the possession of setae usually identifies them as polychaetes. The forms assumed by the larvae of different families vary widely; see Dawydoff (1928), Graveley (1909), Reibisch (1905), Thorson (1950), and Wilson (1932). Polychaete larvae occur in the plankton throughout the year, reaching a peak in the summer months with densities of up to several thousand individuals per cubic meter in local estuaries. With increasing size and decreasing buoyancy, the young sink to the bottom and gradually metamorphose into the adult form.

Polychaetes are chiefly marine or estuarine. As previously indicated, the larvae of most species are pelagic, as are adults of some families, occurring either as permanent plankters or briefly during the breeding season. A few species are especially associated with floating driftwood; among these are *Amphinome rostrata* and *Hipponoe gaudichaudi*. *Harmothoe dearborni* and *Platynereis dumerilii* are found on pelagic *Sargassum*. Benthic forms range from the littoral to abyssal depths, and are found on all types of substrata including living plants and animals. A few species are commensal. *Cirriformia filigera* is found on *Clymenella; Gattyana cirrosa* and *Lepidametria commensalis* are commensal with *Amphitrite*. *Polydora commensalis* has been reported with hermit crabs, and other species of the same genus are especially associated with oysters and sponges. *Polydora* has some importance in oyster ecology, partly because of its activities as a collector of prodigious quantities of mud—which may smother the oysters—and because the worms invade the shells of live oysters. *Polydora* actually bores in the hard structure of mollusk shells, but the mechanism is not well understood. *Dodecaceria coralii* is another boring form, found in colonies of the coral *Astrangia*. Another species of the same genus, *D. concharum*, bores in mollusk shells. Very few polychaetes are actually parasitic. Among local species *Drilonereis caulleryi* is parasitic in another polychaete, *Onuphis conchylega*. Some eunicids are internal parasites in other annelids; *Haematocleptes terebellides* is an endoparasite in *Terebellides*.

Systematic List and Distribution

The systematics of higher categories is not fully agreed upon, but modern students tend to regard the classical division into Polychaeta Errantia and Sedentaria as unnatural. The classification used here is that of Dales (1963); the check list is based on the catalog of Hartman (1959) with changes in nomenclature from more recent literature.

Order Phyllodocida		
Family Phyllodocidae		
Phyllodoce maculata	B+	Lit. to 165 m
Phyllodoce groenlandica	BV	Lit. to 1585 m
Phyllodoce arenae	B, 40°	Lit. to 194 m
Phyllodoce mucosa	BV	Lit. to 37 m, eury., to 20‰
Paranaitis speciosa	BV	Lit. to 183 m
Mystides borealis	A+	7 to 392 m
Eteone lactea	BV	Lit. to 183 m
Eteone trilineata	B	Lit. to 146 m
Eteone heteropoda	BV	Lit. to 18 m
Eteone longa	BV	Lit. to 1716 m
Eteone flava	A+	Lit. to 852 m
Eumida sanguinea	BV	Lit. to 110 m

Nereiphylla fragilis	C, 37°	Lit., eury.
Eulalia viridis	B, 40°	Lit. to 229 m
Eulalia bilineata	BV	Lit. to 2261 m
Notophyllum foliosum	V	128 to 148 m
Notophyllum americanum	V	183 m
Family Alciopidae		
Vanadis longissima	C+	Pelagic
Family Tomopteridae		
Tomopteris helgolandica	BV	Surface to 1830 m, pelagic
Tomopteris septentrionalis	B+	Surface to 1647 m, pelagic
Family Typhloscolecidae		
Travisiopsis lobifera	BV	Surface to 512 m, pelagic
Travisiopsis levinseni	B+	Surface to 1931 m, pelagic
Family Aphroditidae		
Aphrodita hastata	B, 37°	3 to 2024 m
Laetmonice filicornis	BV	35 to 4795 m
Family Polynoidae (not separated from the above by Dales)		
Austrolaenilla mollis	B	37 to 862 m
Antinoella sarsi	B	5 to 2223 m
Antinoella angusta	B	132 to 274 m
Eucranta villosa	B	16 to 1114 m
Lepidametria commensalis	V, 43°	Lit. to 24 m
Lepidonotus squamatus	B, 39°	Lit. to 245 m
Lepidonotus sublevis	V	Lit. to 101 m
Gattyana cirrosa	BV	Lit. to 236 m
Gattyana amondseni	B+	Lit. to 692 m
Gattyana nutti	B	46 to 123 m
Arcteobia anticostiensis	B	5 to 174 m
Hartmania moorei	B	Lit. to 165 m
Enipo gracilis	B+	4 to 197 m
Drieschia pellucida	C+	Pelagic
Harmothoe acanellae	BV	42 to 2251 m
Harmothoe imbricata	B+	Lit. to 3715 m
Harmothoe macginitiei	V	Lit.
Harmothoe fragilis	B	2 to 2948 m
Harmothoe extenuata	B, 37o	Lit. to 1301 m, eury., to 20‰
Harmothoe nodosa	B, 40°	18 to 1263 m
Harmothoe oerstedi	B+	Lit. to 944 m
Harmothoe dearborni	C+	On *Sargassum*
Family Sigalionidae (not separated from Aphroditidae by Dales)		
Sigalion arenicola	V	Lit. to 35 m
Pholoe minuta	B+	Lit. to 2295 m
Sthenelais boa	V	Lit. to 146 m, to eury. to 18‰
Sthenelais limicola	BV	Lit. to 769 m
Leanira tetragona	B, 37°	40 to 2196 m
Leanira hystricis	V	183 to 2641 m
Family Chrysopetalidae		
Dysponetus pygmaeus	A+	Lit. to 51 m

Family Glyceridae		
Glycera capitata	B	Lit. to 3457 m
Glycera americana	V	Lit. to 315 m, eury.
Glycera dibranchiata	V	Lit. to 403 m, eury.
Glycera robusta	B, 37°	Lit. to 377 m
Family Goniadidae		
Goniada maculata	B+	Lit. to 2297 m
Goniada norvegica	B+	30 m to 853 m
Goniada brunnea	BV	68 to 1691 m
Goniadella gracilis	V	Lit. to 86 m
Glycinde solitaria	V	Lit. to 48 m
Ophioglycera gigantea	B+	Lit. to 46 m
Family Sphaerodoridae		
Sphaerodorum gracilis	B	Lit. to 1323 m
Ephesiella minuta	A+	Lit. to 112 m
Family Nephtyidae		
Nephtys bucera	BV	Lit. to 179 m
Nephtys incisa	B, 37°	Lit. to 1746 m
Nephtys paradoxa	B, 39°	5 to 7320 m
Nephtys ciliata	B	Lit. to 915 m
Nephtys discors	A+	Lit. to 1954 m
Nephtys caeca	B+	Lit. to 560 m
Nephtys longosetosa	B	Lit. to 966 m
Nephtys magellanica	V	2 to 27 m, eury. to 18‰
Nephtys squamosa	V	26 to 220 m
Nephtys picta	V	Lit. to 38 m
Aglaophamus verrilli	C, 37°	Lit. to 132 m, eury. to 15‰
Aglaophamus circinata	B+	15 to 787 m, eury.
Family Syllidae		
Autolytus emertoni	B+	Surface (only epitokes known)
Autolytus prismaticus	B	Lit. to 1953 m
Autolytus fasciatus	BV	Lit. to 33 m
Autolytus cornutus	B, 37°	Lit. to 137 m
Autolytus prolifer	BV	Lit. to 55 m
Autolytus alexandri	BV	Lit. to 226 m
Sphaerosyllis erinaceus	BV	Lit. to 138 m
Sphaerosyllis hystrix	BV	Lit. to 29 m
Brania clavata	BV	Lit. to 27 m
Brania wellfleetensis	V	Lit. to 18 m
Parapionosyllis longicirrata	V	Lit. to 18 m
Exogone verugera	B+	Lit. to 159 m
Exogone dispar	BV	Lit. to 128 m
Exogone hebes	BV	Lit. to 143 m
Amblyosyllis finmarchica	B	Lit. to 55 m
Streptosyllis varians	B	Lit.
Streptosyllis arenae	V	Lit.
Syllides longocirrata	B	Lit. to 46 m
Syllides setosa	V	Surface
Odonotosyllis fulgurans	V	Lit. to 27 m
Eusyllis blomstrandi	A+	37 to 220 m

Eusyllis lamelligera	V	7 to 31 m
Syllis spongiphila	V	128 to 580 m
Syllis gracilis	V	Lit. to 232 m
Syllis cornuta	BV	Lit. to 267 m
Family Hesionidae		
Parahesione luteola	V	Lit.
Nereimyra punctata	BV	Lit. to 2361 m
Gyptis vittata	B	Lit. to 55 m
Podarke obscura	V	Lit. to 838 m
Microphthalmus sczelkowii	B	Lit. to 11 m
Microphthalmus aberrans	B	Lit.
Family Pilargidae		
Ancistrosyllis groenlandica	B, 40°	46 to 783 m
Ancistrosyllus hartmanae	V	13 m
Ancistrosyllus jonesi	V	13 m
Sigambra bassi	C, 37°	Lit. to 9 m, eury. to 20‰
Sigambra papillosa	V	9 to 18 m, eury. to 20‰
Sigambra tentaculata	V	Lit. to 5124 m
Sigambra wassi	V	11 to 13 m
Cabira incerta	V	Shallow
Family Nereidae		
Laeonereis culveri	C, 39°	Lit., eury. 5 to 20‰
Ceratonereis irritabilis	V	Lit.
Ceratonereis tridentata	V	Shallow
Ceratonereis scotiae	B	—
Platynereis dumerilii	V	Lit. to 129 m, eury. to 15‰ also on *Sargassum*
Lycastopsis pontica	V	Lit.
Ceratocephale loveni	B, 38°	46 to 2137 m
Nereis arenaceodentata	V	Lit. to 101 m
Nereis succinea	BV	Lit. to 46 m, eury.
Nereis virens	B, 37°	Lit. to 154 m, eury, to 10‰
Nereis diversicolor	B	Lit. to 40 m, eury.
Nereis pelagica	B+	Lit. to 1114 m
Nereis zonata	B	Lit. to 803 m
Nereis grayi	V	Lit. to 18 m
Family Paralacydoniidae		
Paralacydonia paradoxa	V	49 to 1885 m
Order Capitellida		
Family Capitellidae		
Capitella capitata	BV	Lit. to 73 m
Heteromastus filiformis	BV	Lit. to 5 m
Notomastus latericeus	BV	Shallow, eury.
Notomastus luridus	V	Lit.
Dasybranchus lumbricoides	C+	Lit.
Family Arenicolidae		
Arenicola cristata	V	Lit.
Arenicola marina	B(V?)	Lit.
Arenicola brasiliensis	V	Lit.

Family Scalibregmidae		
Scalibregma inflatum	BV	Shallow
Polyphysia crassa	B+	—
Family Maldanidae		
Asychis biceps	B	—
Asychis urceolata	V	—
Nicomache lumbricalis	B	22 to 51 m
Praxillella praetermissa	B	15 to 73 m
Praxillella tricirrata	B+	31 to 37 m
Rhodine loveni	B	15 to 20 m
Rhodine attenuata	B	11 to 53 m
Axiothella mucosa	C	Lit.
Clymenella torquata	BV	Lit. to 110 m, eury. to 15‰
Clymenella zonalis	B+	7 to 37 m, eury.
Maldane sarsi	B	146 m and deeper
Maldanopsis elongata	V	Lit. to 27 m, eury. to 15‰
Leiochone dispar	V	46 to 53 m
Petaloproctus tenuis	A+	—
Family Opheliidae		
Ophelia bicornis	V	Lit. to shallow, eury.
Ophelia denticulata	B	—
Travisia carnea	V	Shallow
Travisia forbesi listed by Miner (1950) "both sides Atlantic"		
Ammotrypane aulogaster	B	Lit. to 18 m
Order Sternaspida		
Family Sternaspidae		
Sternaspis fossor listed from eastern Canada		37 to 73 m
Order Spionida		
Family Spionidae		
Spio filicornis	BV	Lit. to 18 m, eury.
Spio setosa	V	Lit.
Scolecolepides viridis	V	Lit. to 150 m, eury. mostly <15‰
Streblospio benedicti	BV	Lit. 5 to 15‰
Scolelepis squamata	V	Lit.
Pygospio elegans	B+	Lit.
Prionospio heterobranchia	V	Shallow, eury. to 12‰
Prionospio tenuis	V	Lit.
Prionospio steenstrupi	B	18 to 27 m
Prionospio pinnata	V	Shallow
Polydora ligni	V	Lit.
Polydora websteri	V	Shallow
Polydora ciliata	B	Shallow
Polydora gracilis	B	18 to 64 m
Polydora concharum	B	Lit. 55 m
Polydora hamata	V	Shallow
Polydora commensalis	C+	Shallow
Polydora colonia	V	Shallow
Polydora anoculata	V	Shallow
Polydora socialis	C+	Shallow
Laonice cirrata	BV	11 to 15 m

Spiophanes bombyx	V	Lit. to shallow
Spiophanes wigleyi	Georges Bank	148 m
Dispio uncinata	V	—
Family Trochochaetidae		
Trochochaeta multisetosa	B+	37 m
Family Paraonidae		
Paraonis lyra	B	60 to 1940 m
Paraonis gracilis	B	5 to 2035 m
Paraonis fulgens	B+	Lit. to shallow
Aricidea quadrilobata	B	5 to 59 m
Aricidea jeffreysii	B+	2 to 1940 m
Aricidea suecica	A+	5 to 2295 m
Family Apistobranchidae		
Apistobranchus tullbergi	B+	5 to 84 m
Family Chaetopteridae		
Chaetopterus variopedatus	V	To 8 m, eury. to 20‰
Spiochaetopterus oculatus	V	Lit. to shallow
Family Sabellariidae		
Sabellaria vulgaris	V	Lit. to 8m
Order Eunicida		
Family Onuphidae		
Onuphis opalina	B, 37°	26 to 2297 m
Onuphis conchylega	BV	24 to 1963 m
Onuphis eremita	BV	128 to 1479 m
Onuphis quadricuspis	B	—
Diopatra cuprea	V	Lit. to 82 m
Hyalinoecia tubicola	BV	123 to 1768 m
Family Eunicidae		
Eunice norvegica	V	59 to 1098 m
Eunice pennata	BV	4 to 2895 m
Eunice rubra	C, 37°	Lit. to 44 m
Lysidice americana	V	35 m
Marphysa sanguinea	V	Lit. to 91 m
Marphysa belli	V	24 to 115 m
Family Lumbrinereidae		
Lumbrineris coccinea	V	Mostly below 183 m
Lumbrineris latreilli	BV	88 to 2361 m
Lumbrineris acuta	BV	35 to 188 m
Lumbrineris brevipes	V	106 to 3977 m
Lumbrineris fragilis	BV	Lit. to 3446 m
Lumbrineris tenuis	BV	Lit. to 339 m
Lumbrineris impatiens	BV	Lit. to 2361 m
Ninoe nigripes	BV	Lit. to 1171 m
Family Arabellidae		
Arabella iricolor	V	Lit. to 84 m, eury. to 15‰
Notocirrus spiniferus	V	Lit. to 16 m
Drilonereis longa	V	Lit. to 2452 m
Drilonereis magna	B	Lit.
Drilonereis caulleryi	V	Parasitic on *Onuphis conchylega*
Drilonereis filum	V	Lit.

Family Dorvilleidae		
Stauronereis rudolphi	V	Lit. to 263 m
Stauronereis caecus	B+	Lit. to 155 m
Stauronereis sociabilis	V	Shallow
Order Amphinomida		
Family Amphinomidae		
Paramphinome pulchella	BV	37 to 1096 m
Amphinome rostrata	C+	On flotsam
Pseudeurythoe paucibranchiata	V	37 to 192 m, eury., to 15‰
Hipponoe gaudichaudi	C+	On flotsam
Pareurythoe borealis	BV	Lit. to 128 m
Family Euphrosinidae		
Euphrosine cirrata	B, 37°	>183 m locally
Euphrosine borealis	B	9 to 1281 m
Euphrosine armadillo	B, 37°	117 to 529 m
Family Spintheridae		
Spinther citrinus	B	18 to 220 m
Order Magelonida		
Family Magelonidae		
Magelona rosea	V	Shallow
Order Ariciida		
Family Orbiniidae (Ariciidae)		
Orbinia norvegica	B+	62 to 1585 m
Orbinia michaelseni	B	97 m
Orbinia swani	B	46 m
Orbinia ornata	V	Lit. to 33 m
Naineris quadricuspida	B+	Lit. to 2031 m
Naineris laevigata	C	Lit.
Scoloplos riseri	V	Lit.
Scoloplos robustus	BV	Lit. to 57 m
Scoloplos fragilis	BV	Lit. to 102 m
Scoloplos armiger	A+	Lit. to 2013 m
Scoloplos acutus	B, 37°	5 to 179 m
Scoloplos rubra	C, 37°	Lit.
Order Cirratulida		
Family Cirratulidae		
Cirratulus cirratus	B	Lit.
Cirratulus grandis	V	Lit. to 42 m
Chaetozone setosa	B	11 to 20 m
Tharyx setigera	C+	Lit.
Tharyx acutus	B+	11 to 20 m
Tharyx similis	B	37 m
Ledon leidyi	B	15 to 22 m
Dodecaceria coralii	V	In Astrangia
Dodecaceria concharum	B	37 to 55 m
Cossura longocirrata	B	11 to 22 m
Cirriformia filigera	V	Lit., eury. to 20‰
Order Oweniida		
Family Oweniidae		
Owenia fusiformis	BV	46 to 55 m
Myriochele heeri	B	11 to 22 m

Order Terebellida		
Family Pectinariidae		
Pectinaria gouldii	V	Lit. to 30 m, eury. to 5‰
Pectinaria granulata	B	27 to 55 m
Pectinaria hyperborea	B	—
Family Ampharetidae		
Ampharete acutifrons	BV	Lit. to 11 m
Ampharete arctica	A+	—
Samythella elongata	B	—
Samytha sexcirrata	B	ca. 183 m
Asabellides oculata	V	5 to 15 m, eury. to 15‰
Melinna maculata	V	Lit. to shallow, eury.
Melinna cristata	BV	Lit. to 366 m
Anobothrus gracilis	B+	18 to 238 m
Amphicteis gunneri	B	—
Amage auricula	BV	18 m
Family Terebellidae		
Amphitrite ornata	V	Lit. to shallow, eury. to 15‰
Amphitrite cirrata	B	Lit.
Amphitrite affinis	B	—
Amphitrite johnstoni	B	Lit. to shallow
Enoplobranchus sanguineus	V	Lit. to shallow
Terebella lapidaria	V	Shallow
Trichobranchus glacialis	B+	Lit. to 22 m
Terebellides stroemi	B	18 to 55 m
Loimia medusa	V	2 to 24 m, eury. to 10‰
Loimia viridis	V	Lit.
Lysilla alba	V	Lit. to shallow
Nicolea zostericola	B	Shallow
Nicolea venustula	V	Lit.
Pista cristata	V	Shallow, eury.
Pista maculata	V	Shallow, eury.
Pista palmata	V	Lit.
Polycirrus eximius	V	Lit. to 18 m
Polycirrus medusa	B	11 to 22 m
Polycirrus phosphoreus	B	Lit. to 91 m
Artacama proboscidea	B	—
Thelepus cincinnatus	B	Lit. to shallow
Order Flabelligerida		
Family Flabelligeridae		
Flabelligera grubei	B	11 to 18 m
Fabelligera affinis	B+	Lit. to 37 m
Pherusa plumosa	B	15 to 22 m
Pherusa aspera	B	11 to 18 m
Pherusa affinis	V	Lit.
Pherusa arenosa	V	5 m
Brada granosa	B	Lit. to 55 m
Brada villosa	B	—
Brada setosa	B	15 to 55 m
Diplocirrus hirsutus	B	—

Order Sabellida		
Family Sabellidae		
Fabricia sabella	BV	Lit. to 55 m
Euchone elegans	B+	13 to 42 m
Euchone rubrocincta	B+	—
Sabella microphthalma	V	Lit. to 6 m, eury.
Sabella crassicornis	B	Lit. to 55 m
Potamilla reniformis	B+	Lit. to 110 m
Potamilla neglecta	BV	9 to 46 m
Myxicola infundibulum	BV	Lit. to 55 m
Family Serpulidae		
Hydroides dianthus	V	Lit. to 15 m
Hydroides norvegicus	B	Deep
Filograna implexa	B	33 to 55 m
Spirorbis borealis	B+	Lit.
Spirorbis spirillum	B+	Shallow
Spirorbis violaceus	A+	Shallow
Spirorbis granulatus	A+	Shallow
Salmacina dysteri	V	Lit.
Protula tubularia	V?	—

Identification

Since this is one of the larger invertebrate assemblages, the initial problem is one of establishing readily distinguishable subgroups. The following introductory key is based on those of Smith *et al.* (1964) and Fauvel (1923, 1927).

1. Dorsal surface covered with enlarged scales or *elytra* which may be concealed by a dense coating of felt- or furlike setae, or dorsum with bristles in transverse ridges; Figs. 17.3A–D, 17.7 Families Aphroditidae, Polynoidae, Sigalionidae, Euphrosinidae, and Spintheridae. Group One.

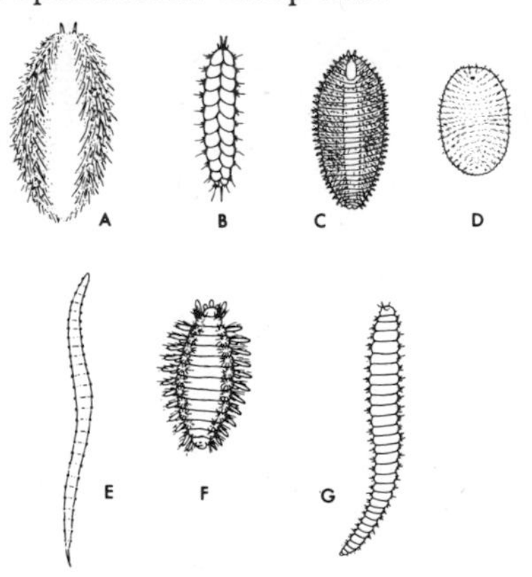

Fig. 17.3. Annelida, Polychaeta, family types in Groups One, Three, Four. A, aphroditid; B, polynoid; C, *Euphrosine;* D, *Spinther;* E, sphaerodorid; F, *Dysponetus;* G, amphinomid.

1. Pelagic forms, transparent; with enlarged leaflike parapodial rami or cirri, or with eyes greatly enlarged, Fig. 17.8. Families Alciopidae, Tomopteridae, and Typhloscolecidae. Group Two.

1. Segmentation indistinct, indicated mainly by spherical capsules and simple, uniramous parapodia; Fig. 17.3E, Fig. 17.9. Family Sphaerodoridae. Group Three.

1. Not as above; segmentation usually distinct—2.

2. Dorsal cirri posterior to fanlike tufts of notosetae; cephalic appendages short, including 3 antennae and 2 palpi; often with a distinct caruncle; Fig. 17.3F,G, Fig. 17.7C–G. Families Amphinomidae and Chrysopetalidae. Group Four.

2. Dorsal cirri lacking or not as above:

- Anterior end with antennae minute or absent or with but a single median antenna; tentacular cirri or palpi short or absent; Fig. 17.4—3. Note that spionids that have lost their palpi might be interpreted as belonging here; see Fig. 17.18.
- Anterior end with well-developed appendages including antennae, tentacles, palpi, or tentacular cirri—6.

3. Prostomium either conical with 4 minute, terminal antennae, or more or less quadrangular with 4 small antennae; Figs. 17.4A,B, 17.10. Families Nephtyidae, Glyceridae, and Goniadidae. Group Five.

3. Not so—4.

4. Body not regionated; body segments not conspicuously longer than wide; prostomium conical, without appendages or with a median antenna; Figs. 17.4C, 17.11. Families Lumbrineridae, Arabellidae, Paraonidae. Group Six.

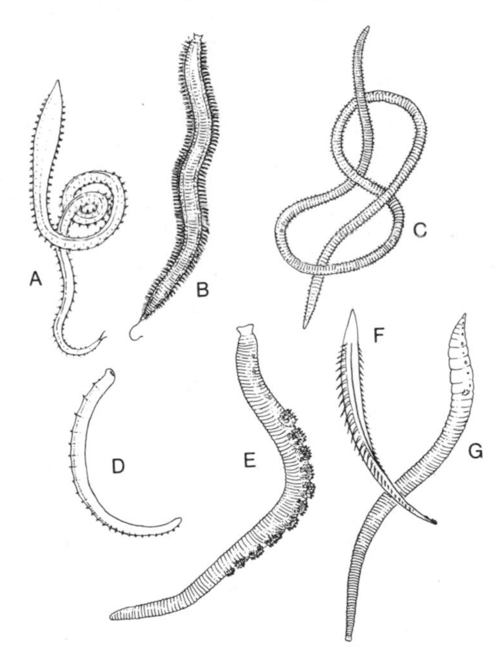

Fig. 17.4. Annelida, Polychaeta, family types in groups Five through Eight. A, *Glycera;* B, nephtyid; C, arabellid; D, oweniid; E, *Arenicola;* F, opheliid; G, *Capitella.*

4. Body usually more or less regionated, or with some body segments much longer than wide; prostomium varies—5.

5. Some body segments much longer than wide; anterior end more or less obliquely flattened or bluntly rounded and hoodlike, without appendages or with a frilled membrane; Figs. 17.4D, 17.12. Families Maldanidae, Oweniidae. Group Seven.

5. Segments shorter than or little longer than wide; anterior end conical or with T-shaped or trilobed prostomium, without appendages or with papillae; Figs. 17.4E–G, 17.13. Families Arenicolidae, Capitellidae, Scalibregmidae, Opheliidae, Orbiniidae. Group Eight.

6. Typically wormlike forms, elongate; with body segments more or less alike throughout, not distinctly regionated (except in epitokes); prostomium usually well developed; parapodia well developed for active crawling or swimming—7.

6. Body relatively short, stout, often regionated; prostomium often reduced or hidden by a concentration of numerous tentacular appendages or setae; parapodia frequently reduced; chiefly burrowing, tubicolous forms; Fig. 17.6—9.

7. Parapodia uniramous, with leaflike or globular dorsal and ventral cirri; 4 or 5 antennae; no palpi; 2–4 tentacular cirri; usually 2 eyes; proboscis jawless but with papillae; Figs. 17.5A, 17.14. Family Phyllodocidae. Group Nine.

7. Parapodia uni- or biramous, cirri when present not as above; antennae 1–7 but not 4 or 5; usually with 2 palpi, which may be globular; with 1–8 pairs of tentacular cirri or none—8.

8. With 2 frontal antennae; a pair of two-jointed palpi; 3 or 4 pairs of tentacular cirri; usually 4 eyes; proboscis with a collarlike "oral ring" and a "maxillary ring" with a pair of sicklelike jaws; Figs. 17.5B, 17.15. Families Nereidae and Paralacydoniidae. Group Ten.

8. With 2 or 3 antennae; a pair of palpi, sometimes two-joined; 1-8 pairs of tentacular cirri; usually 4 eyes; proboscis with a ring of conical papillae and sometimes with a single tooth, or with 6 or 7 teeth; Figs. 17.5C, 17.16. Families Syllidae, Pilargiidae, and Hesionidae. Group Eleven.

8. With 1–7 antennae; 2 globular palpi or none; 2 or 4 eyes or none; 1 or 2 pairs of tentacular cirri or none; proboscis with complex jaw pieces; Fig. 17.17. Families Dorvilleidae, Eunicidae, Onuphidae. Group Twelve.

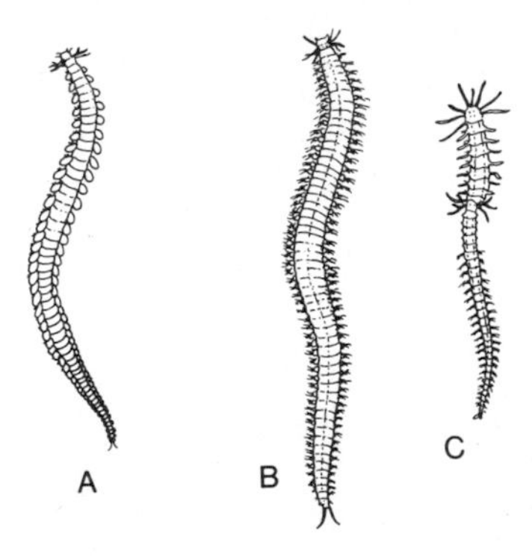

Fig. 17.5. Annelida, Polychaeta, family types in groups Nine through Twelve. A, phyllodocid; B, nereid; C, syllid, epitokous stage.

9. Head with 2 long palpi and sometimes with one or a few pairs of filamentous branchiae but without numerous filamentous appendages; if palpi short, then appendages at midbody modified to form 3 large, circular pallets; Figs. 17.6A, 17.18. Families Spionidae, Magelonidae, Apistobranchidae, Chaetopteridae. Group Thirteen.

9. Head usually with numerous bristles or tentacular appendages—10.

10. Anterior end with tentacular appendages which are pinnate and concentrated in 2 semicircular or spiral lobes; Figs. 17.6B, 17.19. Families Serpulidae and Sabellidae. Group Fourteen.

10. Anterior end with flattened golden setae forming an operculum, or with body covered with papillae and anterior bristles usually directed forward and sometimes forming a cage; Figs. 17.6D,E, 17.20. Families Pectinariidae, Sabellaridae, Flabelligeridae. Group Fifteen.

10. Anterior end with numerous tentacular appendages which may be pinnate or not, but which are not concentrated in two distinct lobes—11.

11. Tentacular appendages usually originating back of the head (note *Ledon* with tentacles at front of prostomium); body not regionated; Figs. 17.6C, 17.21. Family Cirratulidae. Group Sixteen.

11. Tentacular appendages at head end, frequently with a clump of gills behind the tentacles; body regionated—12.

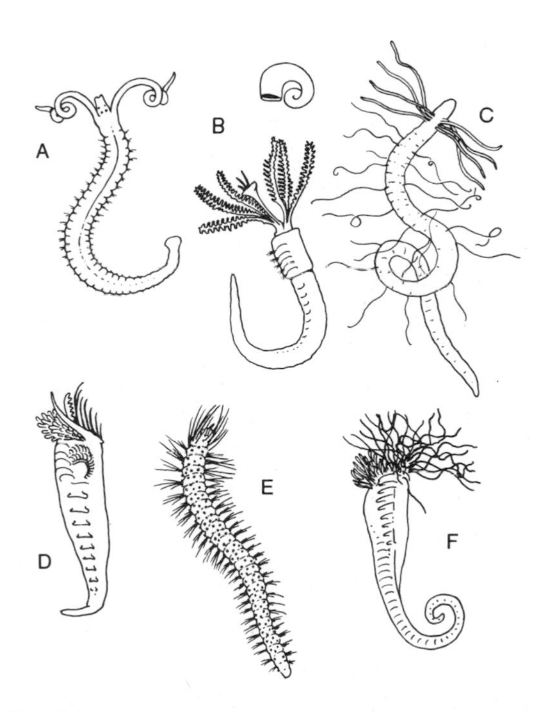

Fig. 17.6. Annelida, Polychaeta, family types in Groups Thirteen through Eighteen. A, *Polydora;* B, serpulid; C, cirratulid; D, pectinariid; E, flabelligerid; F, terebellid.

12. Tentacular appendages not retractile; Figs. 17.6F, 17.22. Family Terebellidae. Group Seventeen.
12. Tentacular appendages retractile; Fig. 17.23. Family Ampharetidea. Group Eighteen.

In the accounts that follow, the chief sources for the development of keys and illustrations are Pettibone (1963a) for Groups One through Six, part of Eight, and Groups Nine through Twelve; for the remaining groups Smith *et al.* (1964) and Fauvel (1927) were most useful. Sources of information for species not included in these references are indicated in parentheses preceding the species name in the keys. Dimensions given are approximate maximum lengths for adults; width is also given in most cases to give some idea of the proportions of the worm.

Group One, Families Aphroditidae, Polynoidae, Sigalionidae, Euphrosinidae, and Spintheridae. Aphroditids, aptly termed "sea mice," dwell on muddy bottoms; sigalionids are burrowers. The larger group of scale worms, Polynoidae, is found on a wide variety of substrata, in crevices, and beneath rocks or about the holdfasts of seaweeds, as fouling organisms, and often as commensals with other benthic invertebrates. *Drieschia* is pelagic, and another scale worm, *Harmothoe dearborni*, is found on floating *Sargassum*. Euphrosinids are small, creeping worms that curl when disturbed, presenting a bristly porcupinelike exterior. *Spinther* is carnivorous or parasitic and is chiefly associated with sponges.

---

1. Dorsum covered by scales which may be exposed to view or concealed by a furry coat of setae—3.
1. Dorsum bristly, with transverse rows of setae—2.
2. Prostomium with a ridgelike caruncle consisting of 3 longitudinal lobes; with a short median antenna; Fig. 17.7A. Euphrosinidae. *Euphrosine:*
- Notosetae and branchiae about equal in length; about a third of mid-dorsum bare, without setae; 10 mm by 4 mm. *E. armadillo.*
- Notosetae much longer than branchiae, giving dorsum a shaggy appearance; dorsum nearly covered with setae; 25 mm by 15 mm. *E. borealis.*

2. Prostomium tiny, nearly covered by a globular, median antenna; no branchiae; 28 mm by 16 mm; Fig. 17.7B. Spintheridae, *Spinther citrinus.*
3. Scales obscured, at least in part, by a feltlike or furry covering. Aphroditidae—4.
3. Scales not concealed:
- With dorsal cirri on segments without elytra Polynoidae—8.
- Without such dorsal cirri, but most genera, except *Pholoe*, with cirriform branchiae on all segments except anterior few Sigalionidae—5.

4. Elytra not completely concealed by loose "fur"; prostomium with ocular peduncles; 90 mm. *Laetmonice filicornis.* A second species, *L. producta*, reported in eastern Canada.
4. Elytra completely concealed by matted "fur"; prostomium without ocular peduncles; 150 mm; Fig. 17.3. *Aphrodite hastata.*

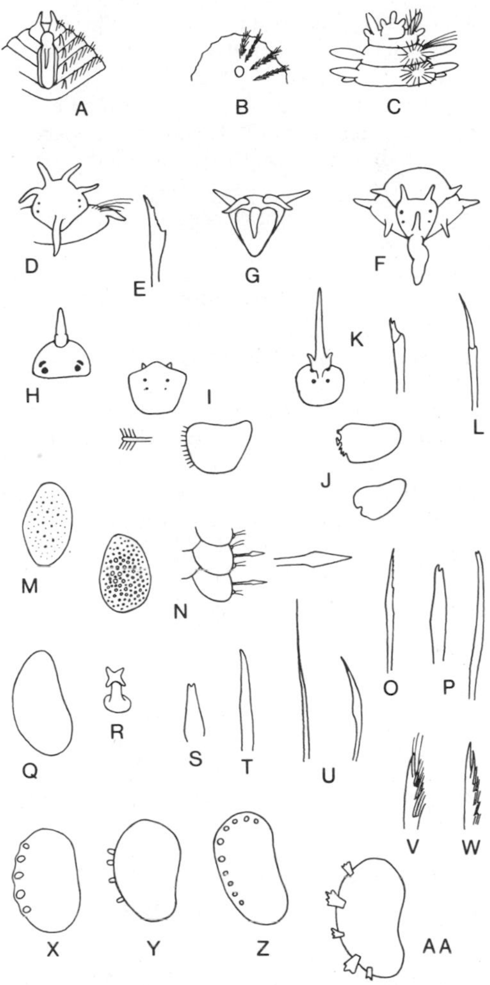

Fig. 17.7. Annelida, Polychaeta, Groups One and Four. A, *Euphrosine borealis*, anterior end, dorsal; B, *Spinther citrinus*, anterior end, dorsal; C, *Dysponetus pygmaeus*, anterior end, dorsal; D, *Hipponoe gaudichaudi*, anterior end, dorsal; E, *Paranamphinome pulchella*, neuroseta; F, *Pareurythoe borealis*, anterior end; G, *Amphinome rostrata*, prostomium; H, *Pholoe minuta*, prostomium; I, *Sigalion arenicola*, prostomium, elytron with detail of pinnate papilla; J, *Sthenelais limicola*, anterior (above) and middle (below) elytra; K, *Sthenelais boa*, prostomium, compound bifid-tipped neuroseta; L, *Leanira tetragona*, compound neuroseta; M, *Lepidonotus sublevis*, elytron; N, *Lepidonotus squamatus*, group of elytra with detail of dorsal cirrus (right) and single elytron (left); O, *Hartmania moorei*, neuroseta; P, *Arcteobia anticostiensis*, upper (right) and lower (left) neurosetae; Q, *Gattyana nutti*, outline of elytron; R, *Gattyana cirrosa*, individual tubercle; S, *Gattyana amondseni*, tubercle; T, *Gattyana cirrosa*, outline of lower neuroseta; U, *Antinoella* sp., neurosetae; V, *Austrolaenilla mollis*, tip of neuroseta; W, *Eucranta villosa*, tip of neuroseta; X, *Harmothoe fragilis*, elytron; Y, *Harmothoe extenuata*, elytron; Z, *Harmothoe nodosa*, elytron; AA, *Harmothoe oerstedi*, elytron.

5. With cirriform branchiae; burrowing species—6.

5. Without cirriform branchiae; 25 mm by 4 mm; a creeping form; Fig. 17.7H. *Pholoe minuta.*

6. Prostomium without median antenna; elytra with an external fringe of pinnate papillae; 300 mm by 8 mm; Fig. 17.7I. *Sigalion arenicola.*

6. Prostomium with median antenna; external border of elytra simple or with simple papillae—7.

7. With several types of neurosetae including some that are compound with bifid tips; Fig. 17.7K. *Sthenelais:*

- All elytra fringed; 200 mm by 5 mm. *S. boa.*
- Anterior elytra fringed, posterior ones notched; 100 mm by 4 mm; Fig. 17.7J. *S. limicola.*

7. Most neurosetae compound with simple pointed tips; 200 mm by 8 mm: Fig. 17.7L. *Leanira tetragona.*

8. With 30–50 pairs of elytra; 100 mm by 9 mm; commensal with terrebellid polychaetes. *Lepidametria commensalis.*

8. Elytral pairs 12 or 13—9.

8. Elytral pairs 15 or 16—10.

9. Elytra with tubercles; benthic species. *Lepidonotus:*

- Tubercles minute, widely spaced; 34 mm by 10 mm; commensal with hermit crabs; Fig. 17.7M. *L. sublevis.*
- Tubercles numerous and crowded, varying in size; 50 mm by 15 mm; a common fouling species; Fig. 17.7N. *L. squamatus.*

9. Elytra smooth; planktonic species, transparent and fragile; 14 mm by 6 mm. *Drieschia pellucida.*

10. Notosetae more slender than neurosetae or at least some with slender capillary tips—11.

10. Notosetae stouter than neurosetae—14.

11. Body without elytra posteriorly; with more than 45 segments; elytra smooth, translucent; 75 mm by 9 mm; notosetae minutely spinous, blunt-tipped; commensal with *Nichomache. Enipo gracilis.*

11. Dorsum covered with elytra; segments about 40—12.

12. Elytra smooth; neurosetae with undivided tips; 15 mm by 5 mm; Fig. 17.7O. *Hartmania moorei.*

12. Elytra tuberculate or, if smooth, then some neurosetae with bifid tips—13.

13. Elytra smooth, with simple border; neurosetae include both bifid and simple types; 26 mm by 8 mm; Fig. 17.7P. *Arcteobia anticostiensis.*

13. Elytra tuberculate with a fringe of papillae; neurosetae simple, slightly hooked; Fig. 17.7T. *Gattyana:*

- With both small and large tubercles; 17 mm by 5 mm; Fig. 17.7Q. *G. nutti.*
- With small tubercles only, one- to four-pronged in *G. cirrosa*, bifid in *G. amondseni;* 47 mm by 12 mm and 30 mm by 12 mm, respectively; Fig. 17.7R, S.

14. At least some neurosetae with capillary tips; elytra with scattered microtubercles; Fig. 17.7U. *Antinoëlla:*

- All neurosetae capillary tipped; 25 mm by 7 mm. *A. angusta.*
- Some neurosetae capillary, some merely slender with curved tips; 68 mm by 27 mm. *A. sarsi.*

14. None of neurosetae capillary tipped—15.

15. Neurosetae with tip hairy; elytra smooth or with few scattered papillae; 50 mm by 23 mm; Fig. 17.7V. *Austrolaenilla mollis.*
15. Neurosetae with slender forked tips; elytra with numerous papillae; 53 mm by 14 mm; Fig. 17:7W. *Eucranta villosa.*
15. Neurosetae simple, hooked, or with or without a small secondary tooth but not forked. *Harmothoe.* A separate key follows.

Harmothoe.

---

1. Segments more than 50; with deep water corals; 90 mm by 15 mm. *H. acanellae.*
1. Segments less than 49—2.
2. Anterior eyes anteroventral, not visible from above unless visible *through* prostomium; 65 mm by 19 mm. *H. imbricata.*
2. Anterior eyes anterolateral, visible from above—3.
3. Elytra with small tubercles only—4.
3. Elytra with both small and large tubercles—5.
4. Neurosetae with bifid tips; on *Sargassum;* 13 mm by 4 mm. *H. dearborni.*
4. Neurosetae with secondary tooth but not bifid; 20 mm by 9 mm. *H. macginitiei.* Note also that *H. extenuata* (in next couplet) may have neurosetae with a secondary tooth and may lack macrotubercles; compare prostomia in figures.
5. Larger tubercles globular or soft:
- Larger tubercles globular, not sharply set off from elytral surface; 26 mm by 12 mm; Fig. 17.7X. *H. fragilis.*
- Larger tubercles either globular or spinelike and sharply set off from elytral surface; 74 mm by 20 mm; Fig. 17.7Y. *H. extenuata.*

5. Larger tubercules nodular or spiny:
- Larger tubercles marginal, roughened; 90 mm by 39 mm; Fig. 17.7Z. *H. nodosa.*
- Larger tubercles branched; 80 mm by 30 mm; Fig. 17.7AA. *H. oerstedi.*

Group Two, Families Alciopidae, Tomopteridae, and Typhloscolecidae. These are planktonic worms, neritic or oceanic but not usually common in estuarine waters. *Tomopteris* is sometimes quantitatively important in the coastal plankton; it is found throughout the year and is a vertical migrant, being more common at the surface at night; *T. helgolandica* is apparently the more common species on this coast.

---

1. With a pair of enormous, globular eyes; parapodia relatively small, uniramous; 200 mm by 3 mm; segments about 200; Fig. 17.8A. Alciopidae. *Vanadis longissima.*
1. Eyes absent or small—2.
2. Parapodia vestigial, uniramous but with large, leafy cirri; Fig. 17.8B. Typhloscolecidae. *Travisiopsis:*
- With nuchal organ extending anteriorly around caruncle; 30 mm by 3 mm; Fig. 17.8B,C. *T. lobifera.*
- With nuchal organ not extending anteriorly around caruncle; 30 mm by 1.5 mm; Fig. 17.8D. *T. levinseni.*

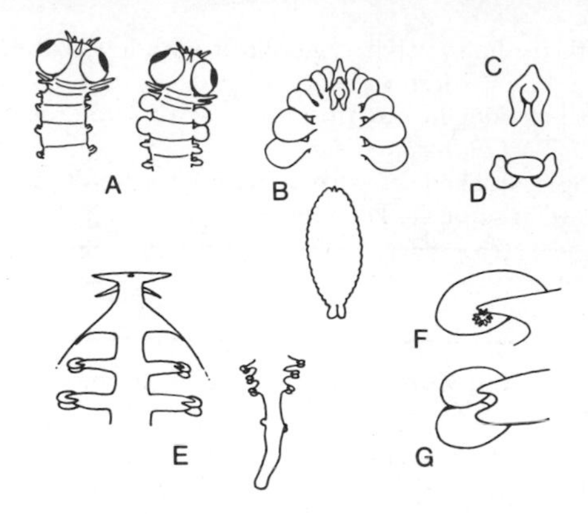

Fig. 17.8. Annelida, Polychaeta, Group Two. A, *Vanadis longissima*, anterior end of male (left), female (right); B, *Travisiopsis lobifera*, anterior end (above), outline of whole animal (below); C, same, nuchal organ; D, *Travisiopsis levinseni*, nuchal organ; E, *Tomopteris* sp., anterior and posterior ends; F, *Tomopteris helgolandica*, part of parapodium with rosette organ; G, *Tomopteris septentrionalis*, parapodium.

2. Parapodia longer than body width, biramous; Fig. 17.8E. Tomopteridae. *Tomopteris:*
- *T. helgolandica* has rosette organs on parapodia and, when adult, has a distinct, cylindrical tail; 87 mm, Fig. 17.8F.
- *T. septentrionalis* has neither; 22 mm; Fig. 17.8G.

Note also the transparent, planktonic scale worm, *Drieschia*, of Group One.

Group Three, Family Sphaerodoridae. These species may be found in the lower littoral algal zones, or on various types of bottom, or pelagically at night. Body short, oval or elongate. Prostomium and tentacular segment indistinct or invaginated. The small spherules are segmentally arranged, as are the parapodia, but segmentation is otherwise indistinct.

---

1. Elongate, 60 mm by 1 mm; segments about 120; setae simple, hooked; Fig. 17.3E, Fig. 17.9A. *Sphaerodorum gracilis.*

1. Short, oval, 6 mm by 1 mm; segments about 30; setae compound; Fig. 17.9B. *Ephesiella minuta.*

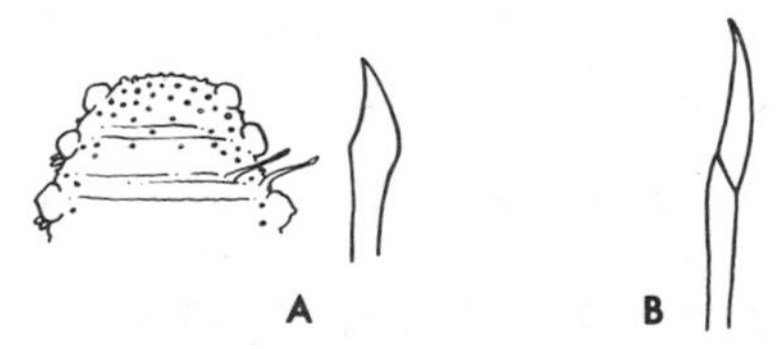

Fig. 17.9. Annelida, Polychaeta, Group Three. A, *Sphaerodorum gracilis*, anterior end and seta; B, *Ephesiella minuta*, compound seta.

Group Four, Families Amphinomidae and Chrysopetalidae. The "fire worms" are so called because their glassy setae, which are sometimes hollow and poison filled, may break off in the skin causing an irritation. They are predaceous worms found in crevices and under rocks, among holdfasts, and on driftwood.

---

1. Without branchiae; neurosetae compound; 3 mm by 1.5 mm; in laminarian holdfasts; Fig. 17.7C. Chrysopetalidae. *Dysponetus pygmaeus.*
1. Branchiae present. Amphinomidae—2.
2. Prostomial caruncle reduced or absent—3.
2. Prostomial caruncle developed, Fig. 17.7F,G—5.
3. Neuropodia ventral; neurosetae short, stout, bifid, hooked; 42 mm by 10 mm; on floating wood with goose barnacles; Fig. 17.7D. *Hipponoe gaudichaudi.*
3. Neuropodia lateral—4.
4. Prostomium without eyes, neurosetae fine, capillary, some short, strongly serrate, and with basal spur; 15 mm by 1.5 mm; Fig. 17.7E. *Paramphinome pulchella.*
4. Prostomium with 4 eyes; neurosetae capillary, some short and bifurcated; about 25 mm. *Pseudoeurythoe: P. paucibranchiata* reported from Virginia by Wass (1963); see Fauvel (1932); a second species, *P. ambigua*, reported from Beaufort, North Carolina, by Hartman (1945).
5. Body subrectangular in section; neurosetae few, stout, hooked; 400 mm by 30 mm; on driftwood with goose barnacles; Fig. 17.7G. *Amphinome rostrata.*
5. Body subcylindrical; neurosetae bifurcated, spinous; 44 mm by 3 mm; benthic; Fig. 17.7F. *Pareurythoe borealis.*

Group Five, Families Nephtyidae, Paralacydoniidae, Glyceridae, and Goniadidae. Members of the first family are common shallow water and littoral burrowers and swimmers; they are active, predatory worms, usually gray with the main circulatory vessels and branchiae reddish. Glycerids and goniadids are also active shore forms; the glycerids, despite their armament, are said to be detritus feeders, the goniadids, predators. From their appearance the glycerids are popularly known as bloodworms. Both of the latter families include epitokous species; changes in *Glycera* include the loss of the proboscis and elaboration of parapodia and setae as more efficient swimming organs. Goniadids retain the proboscis and only the posterior segments become modified. *Glycera capitata* and *G. americana* produce epitokes; *G. dibranchiata* does not. *Glycera* is capable of producing the equivalent of a bee sting and should be handled with care.

---

1. Body cylindrical; parapodia small, without lamellae; prostomium conical with 4 minute terminal antennae, Fig. 17.10E—2.
1. Body flattened; parapodia well developed, with lamellae; prostomium subquadrate with 4 small antennae, Fig. 17.10L,M—5.

2. Body regionated, with a short anterior section with uniramous parapodia and posterior parapodia distinctly biramous; individual segments with one annulation; proboscis eversible, long, cylindrical, covered with papillae and armed with complex accessory structures usually in the form of chevrons, circlets, or bands, Fig. 17.10F. Goniadidae—4.

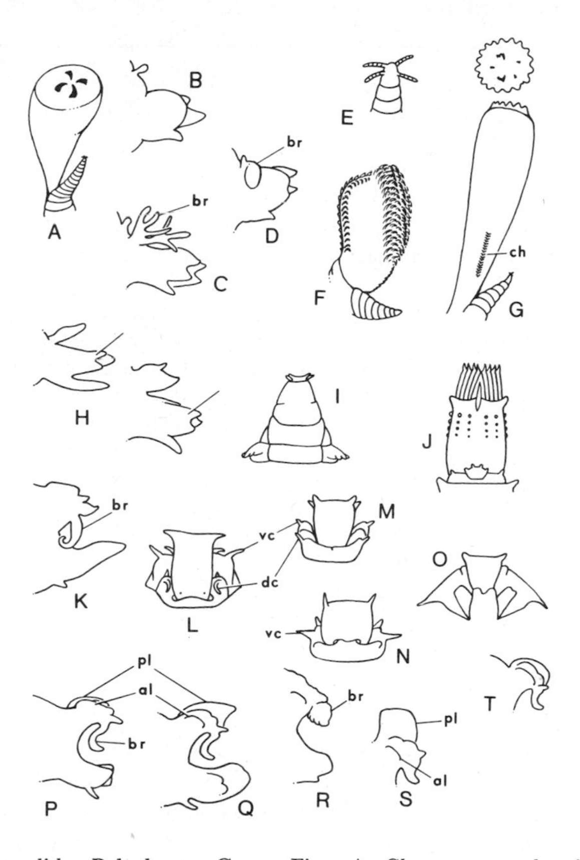

Fig. 17.10. Annelida, Polychaeta, Group Five. A, *Glycera* sp., proboscis and anterior segments; B, *Glycera capitata*, parapodium; C, *Glycera americana*, parapodium; D, *Glycera robusta*, parapodium; E, glycerid prostomium; F, *Glycinde solitaria*, proboscis and anterior segments; G, *Gonaida* sp., proboscis and anterior segments with frontal view (above); H, *Ophioglycera gigantea*, anterior (left) and posterior (right) parapodia with presetal lobe indicated; I, *Paralacydonia paradoxa*, prostomium and anterior segments; J, nephtyid proboscis and anterior segments; K, *Aglaophamus* sp., parapodium; L, *Aglaophamus verrilli*, prostomium; M–O, *Nephtys* species, variations in prostomium; N, *Nephtys picta;* O, *Nephtys bucera;* P, *Nephtys incisa*, parapodium; Q, *Nephtys caeca*, parapodium; R, *Nephtys paradoxa*, parapodium; S, *Nephtys discors*, detail of part of parapodium; T, *Nephtys ciliata*, part of parapodium. Abbreviations are as follows: al, anterior lamella; br, branchia; ch, chevrons; dc, dorsal cirri; pl, posterior lamella; vc, ventral cirrus.

2. Body not regionated, the parapodia essentially similar throughout even in epitokes; individual segments with 2 or 3 annulations; probiscis eversible, with 4 strong jaws; Fig. 17.10A. Glyceridae. *Glycera*—3.

3. Parapodia with 1 postsetal and 2 presetal lobes, no branchiae; 150 mm by 8 mm; Fig. 17.10B. *G. capitata.*

3. Parapodia with 2 postsetal and 2 presetal lobes:

- Branchiae retractile, 370 mm by 13 mm; Fig. 17.10C. *G. americana.* The branchiae in *americana* are fingerlike. In *G. robusta* the branchiae are blister-like; this is a deep burrower, 800 mm by 22 mm; Fig. 17.10D.
- Branchiae not retractile, 370 mm by 11 mm. *G. dibranchiata.*

4. Presetal lobes of neuropodia bilobed, Fig. 17.10H; proboscis with minute papillae, a terminal ring of larger papillae and 3–8 teeth, Fig. 17.10G. (Note that *G. maculata* has about 18 anterior segments with simple presetal lobes and then they become bilobed.):

- Proboscis without chevrons; >60 mm. *Ophioglycera gigantea.*
- Proboscis with chevrons. *Goniada: G. norvegica* has anterior and posterior regions separated by a transitional region, and reaches 300 mm; *G. maculata* and *G. brunnea* lack a transitional body region; the first species has 7–11 chevrons and reaches 100 mm, the second 13–21 chevrons, and reaches 160 mm.

4. Presetal lobes of neuropodia simple, not bilobed; proboscis with patches of minute papillae or conspicuous spinous organs, Fig. 17.10F:

- Proboscis with basal chevrons; neurosetae include compound falcigers and spinigers; 50 mm by 1 mm. *Goniadella gracilis.*
- Proboscis without chevrons but with spinous organs, Fig. 17.10F; neurosetae include compound spinigers only; 35 mm by 1.3 mm. *Glycinde solitaria.*

5. Tentacular head segment with setae and a pair of ventral tentacular cirri; note proboscis, Fig. 17.10J,L–N. Nephtyidae—6.

5. Tentacular segment without setae, without tentacular cirri; proboscis simple, unarmed; 20 mm by 1.5 mm; deep water; Fig. 17.10I. Paralacydoniidae. *Paralacydonia paradoxa.*

6. With branchiae curving inwardly, Fig. 17.10K. *Aglaophamus:*

- With 2 minute eyes; neuropodia with an erect lobe or cirrus above; 44 mm by 3 mm; Fig. 17.10L. *A. verrilli.*
- Without eyes; no neuropodial lobe. *A. circinata.*

A third species, *A. malmgreni*, is reported in deep water.

6. With branchiae curving outwardly; Fig. 17.10P,Q. *Nephtys*—7.

7. With equal dorsal and ventral tentacular cirri; Fig. 17.10M—8.

7. With a ventral tentacular cirrus only, Fig. 17.10N,O: *N. picta* and *N. bucera* differ as figured; both species reach 300 mm by 20 mm.

8. Anterior and posterior parapodial lamellae equally develped; Fig. 1710P; 150 mm by 15 mm. *N. incisa.*

8. Anterior lamella (at least) reduced; Fig. 17.10Q—9.

9. Branchiae wide, foliaceous; 200 mm by 13 mm; Fig. 17.10R. *N. paradoxa.*

9. Branchiae cirriform, inflated basally, rudimentary or absent posteriorly; 300 mm by 12 mm; Fig. 17.10S. *N. discors.*

9. Branchiae cirriform, not inflated basally (as in Fig. 17.10P,Q)—10.

10. Posterior parapodial lamellae short, Fig. 17.10T; 300 mm by 13 mm. *N. ciliata.*
10. Posterior parapodial lamellae longer; both noto- and neuropodial lamellae foliaceous throughout body; 250 mm by 15 mm. *N. caeca. N. longosetosa* has the notopodial lamellae of the middle and posterior body regions reduced; 170 mm by 6 mm. Note also *N. magellanica*, a species near *longosetosa*, reported from Virginia by Wass (1963); see Augener (1912).

Group Six, Families Lumbrineridae, Arabellidae, Paraonidae. These worms are elongate and cylindrical, usually threadlike and easily fragmented; regeneration of lost parts may lead to segmental abnormalities in some genera. The lumbrinerids and arabellids are generally carnivorous, burrowing forms, but some arabellids are endoparasitic in other invertebrates until sexual maturity. The paraonids are described as deposit feeders.

1. Parapodia with projecting setal lobes supported by internal acicula; Fig. 17.2—2.
1. Parapodia without projecting setal lobes and without internal acicula. Paraonidae —8.
2. Neurosetae include hooded crochets, Fig. 17.11A. Lumbrineridae—3.
2. Neurosetae do not include hooded crochets. Arabellidae—6.
3. Some parapodia with palmate branchiae; 100 mm by 4 mm; Fig. 17.11A. *Ninoe nigripes.*
3. Parapodia without branchiae. *Lumbrineris*—4.

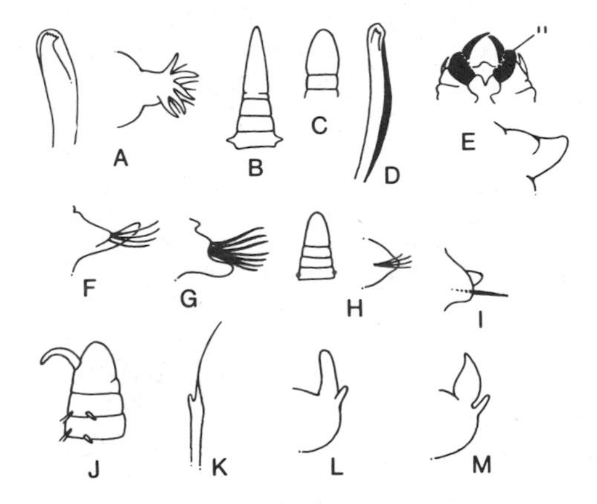

Fig. 17.11. Annelida, Polychaeta, Group Six. A, *Ninoe nigripes*, hooded crochet (left) and palmate branchia; B, *Lumbrineris acuta*, prostomium and anterior segments; C, *Lumbrineris impatiens*, prostomium and adjacent segment; D, *Lumbrineris tenuis*, crochet; E, *Lumbrineris fragilis*, jaws and anterior parapodium; F, *Arabella iricolor*, parapodium; G, *Drilonereis* sp., parapodium; H, *Drilonereis longa*, prostomium and anterior segments (left) and parapodium (right); I, *Drilonereis magna*, parapodium, J, *Aricidea* sp., anterior segments in lateral view; K, *Paraonis lyra*, lyrate seta; L, *Paraonis gracilis*, parapodium; M, *Paraonis fulgens*, parapodium.

4. Anterior parapodia with compound crochets; 300 mm by 5 mm; acicula yellow, *L. latreilli*. Note also *L. coccinea* in deep water.
4. Parapod a without compound crochets—5.
5. Acicula yellow:
- Prostomium 2 or 3 times longer than wide; 40 mm by 1 mm; Fig. 17:11B *L. acuta*.
- Prostomium much shorter, Fig. 17.11C. *L. impatiens* has anterior hooded crochets with long "wings"; in *L. tenuis* the crochets are short-winged, Fig. 17.11D; the species reach 400 mm and 150 mm, respectively.

5. Acicula black:
- Anterior parapodia with longer postsetal lobe; maxillae II with 4 or 5 teeth; littoral to deep water; 380 mm by 12 mm; Fig. 17.11E. *L. fragilis*.
- Anterior parapodia with short postsetal lobe; maxilla II with 3 large teeth; deep water. *L. brevipes*.

6. Parapodia with heavy acicular setae; Fig. 17.11G—7.
6. Parapodia without heavy acicular setae; 600 mm by 5 mm; 4 eyes; Fig. 17.11F. *Arabella iricolor*.
7. With 4 eyes; prostomium conical; 110 mm by 4 mm; young parasitic in *Diopatra*. *Notocirrus spiniferus*.
7. Without eyes; prostomium flattened. *Drilonereis:*
- First setiger with prominent parapodia and similarly developed throughout body; 200 mm by 2.5 mm; Fig. 17.11I. *D. magna*.
- First setiger with inconspicuous parapodia, the parapodia changing form along the body; 710 mm by 1.5 mm; Fig. 17.11H. *D. longa*.
  A third species, *D. caulleryi*, is an endoparasite in *Onuphis conchylega*. Wass (1963) lists yet another species, *D. filum*, as "abundant in intertidal fine sand" in Virginia.

8. With a dorsal, median antenna; Fig. 17.11J. *Aricidea:*
- Antenna extending to about setigers 4–6; 6 mm by 0.6 mm. *A. quadrilobata*.
- Antenna extending to about setiger two at most; *A. jeffreysii* darkly pigmented *A. suecica* whitish; both species about 20 mm by 1.0 to 1.5 mm.

8. Without a median dorsal antenna. *Paraonis:*
- With a few short lyrate setae in addition to capillary setae on notopodia beginning in branchial region; 20 mm by 0.3 mm; Fig. 17.11K. *P. lyra*.
- With capillary setae only on notopodia; *P. gracilis* has straplike branchiae beginning on setigers 6–7; 25 mm by 0.5 mm; *P. fulgens* has foliaceous branchiae beginning on setiger four; Fig. 17.11L,M.

Group Seven, Families Maldanidae and Oweniidae. The elongated segments suggest the name "bamboo worms" sometimes applied to maldanids. These are tubicolous, burrowing worms living in soft tubes encrusted with sand, mud, or shell fragments. The maldanids are typical deposit feeders; the oweniids (*Owenia* at least) feed by sweeping the substratum with the frilly "crown" or use the frill to entrap suspended organisms and particles. Oweniids have a distinctive larva called a mitraria.

---

1. Anal segment with a funnel, flattened plate, or spoonlike; prostomium hoodlike or with a flattened plate. Maldanidae—2.

1. Anal segment simply rounded. Oweniidae:
- Prostomium with deeply lobed branchial membrane; body rigid with 20–30 segments, first 5–7 abdominal segments long, those following progressively shorter; 100 mm by 3 mm; Fig. 17.12B. *Owenia fusiformis.*
- Prostomium rounded, mouth ventral and oblique; with about 27 segments, first abdominal ones very long, shorter posteriorly; 30 mm by 1.5 mm; Fig. 17.12A. *Myriochele heeri.*

2. Prostomium with distinct fringe or rim, more or less tilted posteriorly—4.
2. Prostomium without distinct fringe or rim—3.

3. Some segments with collars; first setigers long; Fig. 17.12C. (Miner, 1950; see also Verrill and Smith, 1873). *Rhodine attenuata.* A second species, *R. loveni*, may also occur, according to Miner.
3. Without collars; first setigers short; pygidium cup- or funnellike:
- Pygidium with 15–25 cirri and central anus; 160 mm by 5 mm; with 22 or 23 setigers; Fig. 17.12D,E. *Nichomache lumbricalis.*
- Pygidium with 1 long ventral cirrus, 2 short lateral ones, and additional papillae or toothlike projections with anus at the summit of a central cone; 50 mm by 2.5 mm; with 18 setigers. *Leiochone dispar.*
  Note also *Petaloproctus tenuis* with pygidium with a concave, lobed plate without cirri.

4. Pygidium oblique, platelike, with anus dorsal; the following may be distinguished by differences in the last posterior segments as figured; *Maldanopsis elongata*, 300

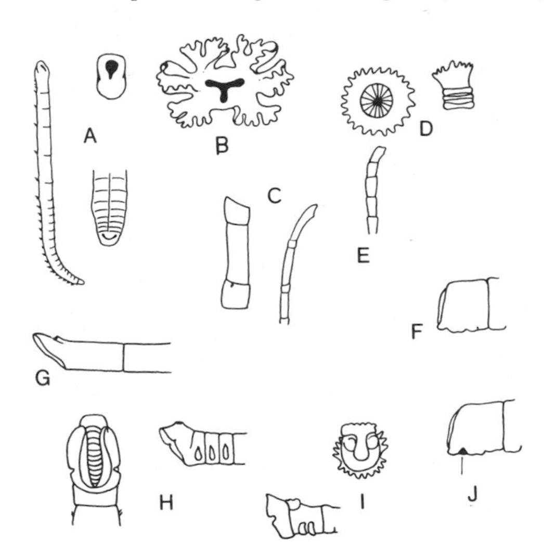

Fig. 17.12. Annelida, Polychaeta, Group Seven. A, *Myriochele heeri*, whole animal, mouth (above right) and posterior end (below right); B, *Owenia fusiformis*, branchial membrane; C, *Rhodine attenuata*, anterior end and detail of collared segment; D, *Nichomache lumbricalis*, pygidium, lateral and end views: E, same, anterior segments; F, *Clymenella torquata*, pygidium; G, *Maldanopsis elongata*, pygidium; H, *Maldane sarsi*, anterior end and pygidium; I, *Asychis biceps*, anterior end and pygidium; J. *Clymenella zonalis*, pygidium with spot indicated.

mm by 5 mm, Fig. 17.12G (Miner 1950); *Maldane sarsi*, 100 mm by 3 mm, Fig. 17.12H; *Asychis biceps*, 180 mm by 5 mm, Fig. 17.12I.

4. Pygidium distinctly funnel-shaped with central anus. *Clymenella: C. zonalis*, 20 mm, has distinct bands in the midregion and a ventral spot on the pygidium as figured, Fig. 17.12J; *C. torquata*, 160 mm, lacks both the spot and bands, Fig. 17.12F. See Mangum (1962).

Group Eight, Families Arenicolidae, Capitellidae, Scalibregmidae, Opheliidae, Aspidobranchidae, and Orbiniidae. Members of these families are burrowers and all are deposit feeders. The burrow of *Arenicola* is well known; it is L- or U-shaped with the head tube diffuse, the tail tube well formed; the worm occupies the bottom loop. Capitellids form mucus-lined galleries. The orbiniids, though burrowers, do not form definite tubes. In gross appearance these worms are quite diverse, ranging from thick and earthwormlike to threadlike; they may be distinctly regionated or not.

---

1. Parapodia with capillary setae or some with rows of crochets on low mounds anteriorly—5.

1. Parapodia with uncini in low rami on some parapodia—2.

2. Body thick, with 11–13 pairs of conspicuous tufted branchiae at about midbody; Fig. 17.13O. Arenicolidae, *Arenicola*—3

2. Body more slender, fragile, with branchiae absent or on posterior body only; Fig. 17.13P. Capitellidae—4.

3. With 12 or 13 pairs of branchiae; green; 200 mm; boreal. *A. marina.*

3. With 11 pairs of branchiae:

- Firm-bodied; dark greenish black; 300 mm; castings formless; egg mass elongate. *A. cristata.*
- Soft-bodied; pale, pinkish tan; 150 mm; castings coiled, cylindrical; egg mass egg-shaped. *A. brasiliensis.*

4. With golden-yellow capillary setae in 4 rows on first 7 setigers; no branchiae; blood red; 100 mm by 2 mm; Fig. 17.13P. *Capitella capitata.*

4. With capillary setae in 4 rows on first 5 setigers; with branchiae beginning on about 80th setiger; thorax dark red, abdomen yellow or greenish red; 100 mm by 1 mm. *Heteromastus filiformis.*

4. With capillary setae on first 11 setigers. *Notomastus:*

- With setae on both noto- and neuropodia of first setiger; blood red anteriorly, yellowish posteriorly; 300 mm by 5 mm. *N. latericeus.*
- With setae on notopodium only of first setiger; dark purplish brown with bluish iridescence anteriorly, with a darker median dorsal line posteriorly; 150 mm by 2 mm (Verrill and Smith 1873). *N. luridus.*

5. Prostomium bilobed or T-shaped; with only 4 pairs of branchiae, limited to a few anterior segments; the species are epitokous. Scalibregmidae:

- With a T-shaped prostomium; 75 mm; Fig. 17.13A. (Miner, 1950). *Scalibregma inflatum.*
- With a bilobed prostomium; 30 mm; Fig. 17.13B. *Polyphysia crassa.*

5. Prostomium conical, rounded, or oval—6.

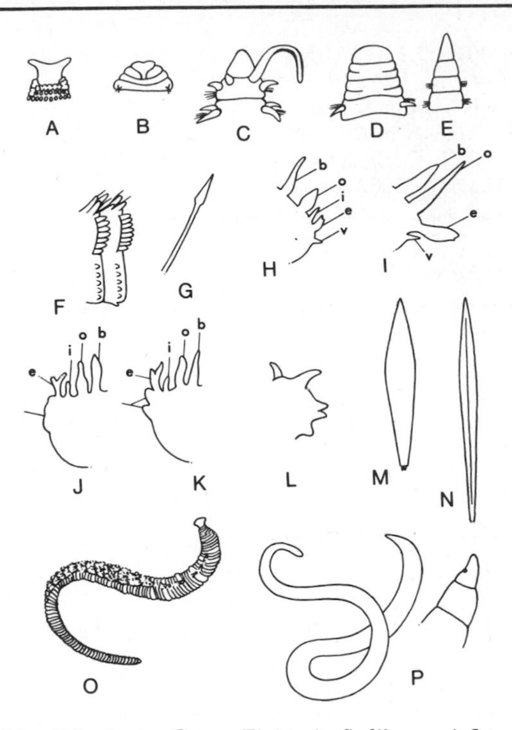

Fig. 17.13. Annelida, Polychaeta, Group Eight. A, *Scalibregma inflatum*, anterior end; B, *Polyphysia crassa*, anterior end; C, *Apistobranchus tullbergi*, anterior end with tentacular palp missing on left; D, *Naineris quadricuspida*, anterior end; E, *Orbinia* sp., anterior end; F, *Orbinia michaelseni*, two thoracic segments, lateral view, showing fringed postsetal lobes (above) and stomach papilla (below); G, same spear-like thoracic spine; H, same, anterior parapodium; I, *Orbinia kupfferi*, anterior parapodium; J, *Scoloplos robustus*, transitional parapodium with subpodal lobe indicated; K, *Scoloplos fragilis*, transitional parapodium with subpodal lobe indicated; L, *Scoloplos armiger*, transitional parapodium; M, *Travisia carnea*, body form, schematic; N, *Ammotrypane aulogaster*, body form, schematic, with ventral groove indicated, O, *Arenicola* sp., whole worm; P, *Capitella capitata*, whole worm and detail of anterior end; Abbreviations are as follows: b, branchiae; o, notopodium; e, neuropodium; i, interramal cirrus; v, ventral cirrus.

6. Parapodia reduced to bundles of setae; branchiae lateral or absent; species sexually dimorphic in color. Opheliidae—13.

6. Parapodia with well-developed and complex parapodia varying in detail on different parts of the body—7.

7. First 1 or 2 segments behind prostomium without appendages or setae. Orbinidae —8.

7. First segment with appendages and setae; no branchiae; prostomium with tentacular palpi which are contractile and easily broken off; 12 mm or more by 2 mm; Fig. 17.13C. Apistobranchidae. *Apistobranchus tullbergi.*

8. Prostomium truncated, oval or rounded; 80 mm by 3 mm; Fig. 17.13D. *Naineris quadricuspida.*

8. Prostomium conical, Fig. 17.13E—9.

9. Thoracic neuropodia with postsetal lobes fringed with papillae; branchiae begin on setigers four to six; Fig. 17.13F. *Orbinia*—10.

9. Thoracic neuropodia without fringed postsetal lobes; branchiae begin on setigers 9–32. *Scoloplos*—11.

10. Some (3–5) thoracic segments with strong, spearlike spines, the upper ones frequently protruding; Fig. 17.13G:

- With "stomach papillae" and interramal cirri on some anterior segments; Fig. 17.13F,H. *O. michaelseni.*
- With stomach papillae but no interramal cirri; Fig. 17.13I. *O. kupfferi.*
- Without stomach papillae. *O. norvegica.*

These species reach lengths of 20–50 mm or more and widths of 2–6 mm.

10. Without spearlike spines:

- With ventral cirri on abdominal parapodia; 50 mm by 3 mm. *O. swani.*
- Without ventral cirri on abdominal parapodia; 250 mm by 7 mm. *O. ornata.*

11. Branchiae begin on setigers five or six; thoracic neuropodia with thick spinelike setae; 65 mm (Webster, 1879; Hartman, 1945). *Scoloplos rubra.*

11. Branchiae begin on setigers 8–32—12.

12. Anterior abdominal segments with interramal cirri:

- With stomach papillae on some anterior segments; 55 by 3 mm. *S. riseri.*
- Without stomach papillae: *S. robustus* and *S. fragilis* differ in the form of the parapodia from the middle or transitional region, as figured; species 375 mm by 10 mm and 150 mm by 3 mm, respectively; Fig. 17.13J,K.

12. Without interramal cirri: *S. armiger* and *S. acutus* differ in their transitional parapodia, *armiger* having 1 or 2 subpodal papillae; 120 mm by 2.5 mm and 40 mm by 1 mm, respectively; Fig. 17.13L.

13. Body slender, with ventral groove—14.

13. Body stouter, without ventral groove; 75 mm; Fig. 17.13M. (Miner, 1950). *Travisia carnea.*

14. Ventral groove extends whole body length; 75 mm; Fig. 17.13N. (Miner, 1950). *Ammotrypane aulogaster.*

14. Ventral groove extends from segments 10 to 12 posteriorly. *Ophelia:*

- With 18 pairs of crenulated branchiae. *O. denticulata.*
- With 11–15 pairs of smooth branchiae; 65 mm. *O. bicornis.*

Group Nine, Family Phyllodocidae. These are elongate, active worms. A few species are particularly associated with other benthic invertebrates, e.g., *Paranaitis speciosa* on *Diopatra*, *Notophyllum foliosum* on large acorn barnacles, *Eulalia viridis* and *Eumida sanguinea* on *Amaroucium.* A few are burrowers, but most are reported as crawling species from mixed sediments, algal holdfasts, or as fouling organisms. The proboscis is eversible, without visible jaws. Some individual species show a substantial range of color; they are frequently banded or spotted. Phyllodocids may be taken at the water surface but are not epitokous.

---

1. With 2 pairs of tentacular cirri; prostomium subtriangular; Fig. 17.14M,N. *Eteone*—6.

1. With 3 pairs of tentacular cirri; prostomium suboval; 16 mm by 0.8 mm; Fig. 17.14A. *Mystides borealis.*

1. With 4 pairs of tentacular cirri; prostomium varying in shape—2.

2. Prostomium with 4 frontal antennae; no median antenna—3.

2. Prostomium with 4 antennae plus a median antenna—4.

3. Prostomium subtriangular; Fig. 17.14E. *Paranaitis:* The species differ in the shape of the dorsal cirri as figured; Fig. 17.14D,E; *P. speciosa*, 18 mm by 3 mm, *P. kosteriensis*, 85 mm by 4 mm.

3. Prostomium heart-shaped, with posterior notch; Fig. 17.14C. *Phyllodoce:*

- Ventral cirri oval; 100 mm by 2 mm; Fig. 17.14F. *P. maculata.*
- Ventral cirri oval with asymmetrical acuminate tip; 450 mm by 9 mm; Fig. 17.14H. *P. groenlandica.*
- Ventral cirri tapered, pointed, Fig. 17.14G: *P. mucosa* has a more or less continuous longitudinal, dark band; *P. arenae* has transverse, dark bands; the species are 150 mm by 3 mm and 100 mm by 2.5 mm, respectively.

3. Prostomium subquadrate or rounded; 80 mm; Fig. 17.14B (Hartman, 1945). *Nereiphylla fragilis.*

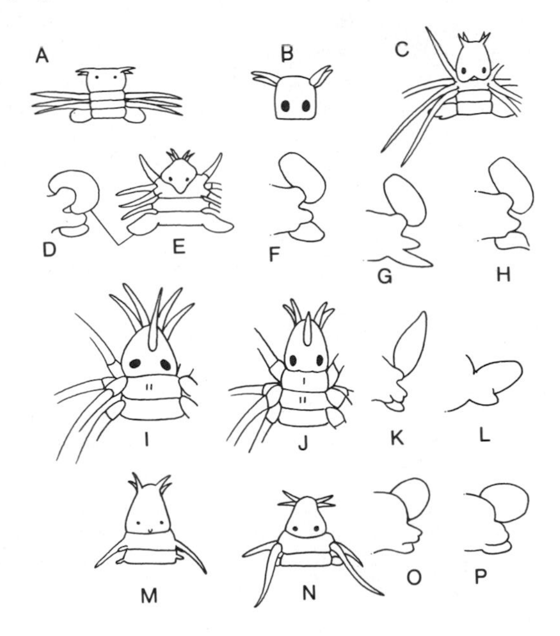

Fig. 17.14. Annelida, Polychaeta, Group Nine. A, *Mystides borealis*, anterior end; B, *Nereiphylla fragilis*, prostomium; C, *Phyllodoce* sp., anterior end; D, *Paranaitis kosteriensis*, parapodium with dorsal cirrus indicated, for comparison with the following; E, *Paranaitis speciosa*, anterior end; F, *Phyllodoce maculata*, parapodium; G, *Phyllodoce mucosa*, parapodium; H, *Phyllodoce groenlandica*, parapodium; I, *Eumida sanguinea*, anterior end with second segment numbered; J, *Eulalia* sp., anterior end with first and second segments numbered; K, *Eulalia viridis*, parapodium; L, *Eulalia bilineata*, parapodium; M, *Eteone lactea*, anterior end; N, *Eteone trilineata*, anterior end; O, *Eteone longa*, parapodium; P, *Eteone flava*, parapodium.

4. Prostomium bordered posteriorly by fingerlike *nuchal epaulettes;* 50 mm by 5 mm. *Notophyllum*, with 2 species from deep water; see Pettibone (1963a).
4. Without nuchal epaulettes—5.

5. With two tentacular segments distinctly visible dorsally, first tentacular segment rudimentary, indistinct or not visible dorsally; 60 mm by 4 mm; Fig. 17.14I; with the tunicate *Amauroecium pellucidum* especially. *Eumida sanguinea.*
5. With 3 tentacular segments distinctly visible dorsally; Fig. 17.14J. *Eulalia:* The species differ in the shape of the dorsal cirri as figured; *E. viridis* is 150 mm by 3 mm, Fig. 17.14K; *E. bilineata*, 100 mm by 2 mm, Fig. 17.14L; the first species greenish, spotted with brown or not; the second, greenish with 2 dark longitudinal bands.

Eteone spp.

6. Ventral pair of tentacular cirri 2 or 3 times longer than dorsal pair; anal cirri tapering; 230 mm by 3 mm; Fig. 17.14M. *E. lactea.*
6. Dorsal pair of tentacular cirri twice as long as ventral pair; anal cirri elongate-oval; 10 mm by 1 mm; Fig. 17.14N. *E. trilineata.*
6. Dorsal and ventral pairs subequal or the ventral pair slightly longer:
   - Anal cirri tapering; parapodia with asymmetrical dorsal cirri; 93 mm by 3 mm. *E. heteropoda.*
   - Anal cirri almost spherical; *E. longa* and *E. flava* differ in the shape and relative size of the parapodial cirri as figured; 160 mm by 5 mm and 120 by 4 mm, respectively; Fig. 17.14O,P.

Group Ten, Family Nereidae. The nereids are found in crevices, under rocks, and as burrowers in soft substrata. They are predators with a pair of strongly developed, darkly colored jaws. Nereids emerge from their burrows and may be found actively swimming. Reproductive behavior varies. *N. arenaceodentata* and *N. diversicolor* are atokous, and pairing takes place in the burrows in the first species. *N. virens* is a swarming species, but the sexually active individuals are only slightly modified. Extreme examples of epitoky, the so-called heteronereids, occur in *N. succinea*, *N. pelagica*, and *Platynereis dumerilii*, the sexually active individuals being highly modified surface swarmers that die after the gametes have been shed. *Lycastopsis*, in contrast, is hermaphroditic.

---

1. Parapodia essentially uniramous; with 3 pairs of tentacular cirri; 57 mm by 1 mm; Fig. 17.15A. *Lycastopsis pontica.*
1. Parapodia biramous; with 4 pairs of tentacular cirri—2.

2. Parapodia, except on first 3 setigers, with double ventral cirri; no paragnaths (dental armament on oral and maxillary rings); >25 mm by 7 mm; Fig. 17.15B. *Ceratocephale loveni.*
2. Ventral cirri single; Fig. 17.15G,H—3.

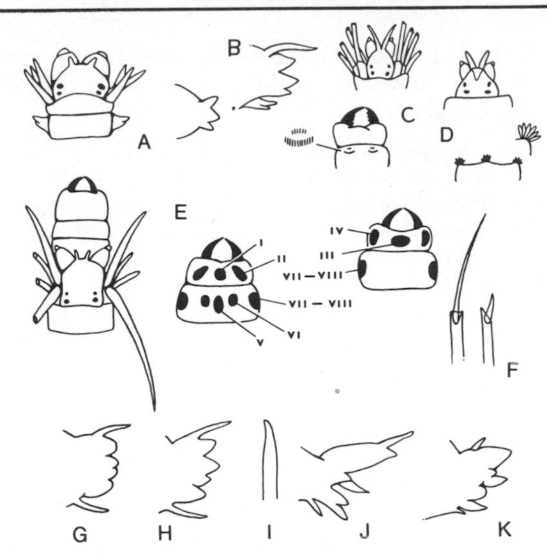

Fig. 17.15. Annelida, Polychaeta, Group Ten. A, *Lycastopsis pontica*, anterior end and parapodium; B, *Ceratocephale loveni*, parapodium; C, *Platynereis dumerilli;* anterior end and proboscis with comb-like paragnaths indicated (below); D, *Laeonereis culveri*, prostomium and detail showing tufts of papillae on proboscis (below); E, nereid, anterior end showing proboscis-prostomium relationship (left) and (right) oral and maxillary rings of proboscis with areas indicated, dorsal (left) and ventral (right); F, *Nereis grayi*, spiniger (left) and falciger (right); G, *Nereis pelagica*, parapodium; H, *Nereis zonata*, parapodium; I, *Nereis diversicolor*, falciger; J, *Nereis succinea*, parapodium; K, *Nereis virens*, parapodium.

3. Prostomium suboval; proboscis with minute, comblike paragnaths (which may be difficult to detect); 75 mm by 6 mm; epitokous; found especially on pelagic seaweeds; Fig. 17.15C. *Platynereis dumerilii.*

3. Prostomium pear-shaped; proboscis with horny, conical paragnaths or none—4.

3. Prostomium pear-shaped; proboscis with soft papillae in tufts; 60 mm by 4 mm; Fig. 17.15D. See Hartman (1945) and Webster (1886). *Laeonereis culveri.*

4. Paragnaths few or absent, lacking on areas VII–VIII; Fig. 17.15E—5.

4. Paragnaths more numerous, present on areas VII-VIII. *Nereis*—6.

5. Notopodia with spinigers only (Fig. 17.15F). *Ceratonereis:*
- Paragnaths entirely lacking or with 3 in area III; 50 mm (Hartman, 1945; Webster, 1886). *C. tridentata.*
- With paragnaths in areas II, III, IV; 80 mm (Hartman, 1945, and Webster, 1879). *C. irritabilis.*

5. Notopodia of middle and posterior regions with spinigers and falcigers; Fig. 17.15F; commensal with *Maldanopsis elongata. Nereis grayi.*

6. Notopodia with spinigers only—7.

6. Notopodia of middle and posterior regions with spinigers and compound falcigers:
- With thick, evenly rounded parapodial ligules; 155 mm; littoral and shallow water; Fig. 17.15G. *N. pelagica.*
- With triangular to conical parapodial ligules; 125 mm; in 18–137 m; Fig. 17.15H. *N. zonata.*

7. Neuropodia of posterior region with 1–3 falcigers with end pieces wholly or partly fused to shaft; Fig. 17.15I; almost exclusively littoral and showing a preference for areas with reduced salinity; 200 mm. *N. diversicolor.*

7. Posterior neuropodia without fused falcigers in upper setal bundles—8.

8. Paragnaths of oral ring minute, forming a continuous band; upper ligule of posterior parapodia triangular with basal cirrus; at low water and below; 70 mm. *N. arenaceodentata.*

8. Paragnaths in groups; upper ligule of posterior parapodia leaflike with basal cirrus; chiefly littoral and a nocturnal swimmer, the most common New England species; Fig. 17.15K; 900 mm. *N. virens.*

8. Paragnaths in groups; upper ligule of posterior parapodia straplike with terminal cirrus; chiefly littoral, the most common southern species, i.e., in Chesapeake Bay and the Carolinas; 190 mm; Fig. 17.15J. *N. succinea.*

Group Eleven, Families Syllidae, Hesionidae, and Pilargidae. Most hesionids are carnivorous; *Parahesione luteola* is commensal with the mud shrimp, *Upogebia*. The species are often brightly colored, and they are active and easily fragmented. Syllids are small carnivorous worms found on sponges, hydroids, ascidians, etc. Their reproductive arrangements are complex. Some, e.g., *Eusyllis*, *Odontosyllis*, *Syllides*, *Streptosyllis*, and *Amblyosyllis*, are epitokous and have pelagic larvae; *Exogene*, *Brania*, *Sphaerosyllis*, and *Parapionosyllis* are also epitokous, but the fertilized eggs are carried by the female until development has passed the trochophore stages. *Syllis* and *Autolytus* may reproduce asexually by serial budding. The budded individuals are sexual and are released as surface swarmers. *Autolytus* is commonly reported in plankton tows, chiefly in the warmer months. Pilargids range in body form from flattened and ribbonlike to cylindrical; they are burrowers on sand or mud flats but also occur in rocky crevices. They are generally small and little known worms.

This is one of the more difficult groups to key effectively.

---

1. Neurosetae compound (or secondarily fused in some Syllidae)—2.

1. Neurosetae simple. Pilargidae (see Pettibone, 1966)—3.

2. Parapodia biramous, the notopodia represented at least by internal acicula, Fig. 17.2. Hesionidae—5.

2. Parapodia uniramous (except in epitokes). Syllidae—7.

3. Body subcylindrical; parapodia poorly developed; dorsal and ventral cirri small or absent; 18 mm by 1.5 mm; Fig. 17.16R. *Cabira incerta.*

3. Body depressed; parapodia deeply cut; dorsal and ventral cirri distinct—4.

4. Epidermis with numerous papillae. *Ancistrosyllis:* In *A. groenlandica* the ventral cirri begin on setiger 1; in *A. hartmanae* and *A. jonesi* they begin on setigers 3–4; *A. hartmanae* has a pair of eyes and notopodial hooked setae beginning on setiger 3; *A. jonesi* lacks eyes and these setae begin on setiger 6; the species are 40 by 1, 25 by 0.5 and 16 by 1.5 mm, respectively; Fig. 17.16S.

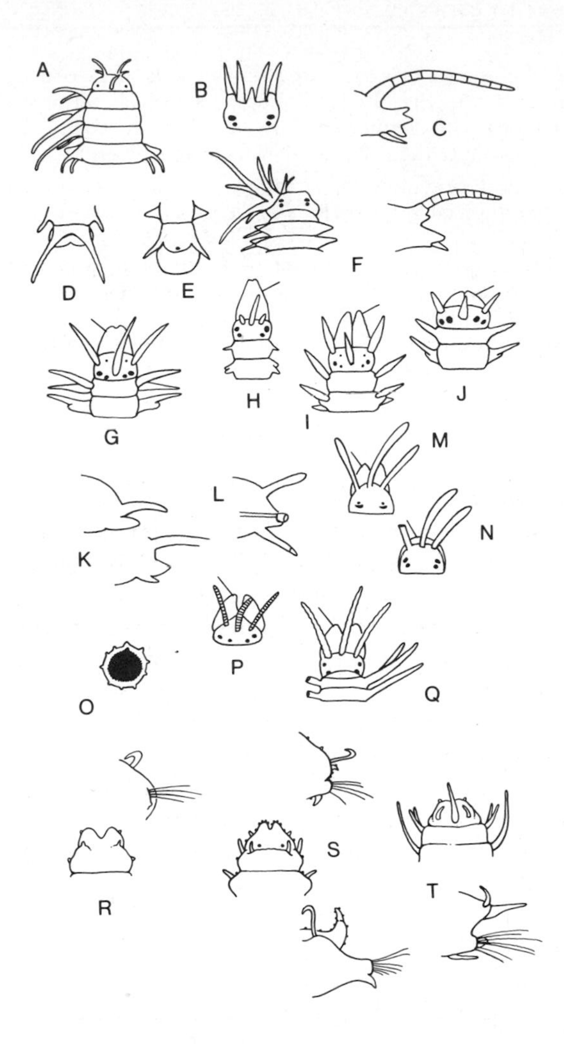

Fig. 17.16. Annelida, Polychaeta, Group Eleven. Bars indicate palpi unless stated otherwise. A, *Microphthalmus aberrans*, anterior end; B, *Podarke obscura*, prostomium; C, *Nereimyra punctata*, parapodium; D, *Microphthalmus aberrans*, posterior end; E, *Microphthalmus scelkowii*, posterior end; F, *Parahesione luteola*, anterior end and parapodium; G, *Brania clavata*, anterior end; H, *Exogone dispar*, anterior end; I, *Parapionosyllis longocirrata*, anterior end; J. *Sphaerosyllis* sp., anterior end; K, parapodia without (above) and with (below) ventral cirrus; *L*, *Streptosyllis* sp., M, *Syllides longocirrata*, prostomium; N, *Syllides setosa*, prostomium; O, *Eusyllis blomstrandi* sp., frontal view of proboscis with tooth at 12 o'clock position; P, *Syllis* sp., prostomium; Q, *Eusyllis bloomstrandi*, anterior end; R, *Cabira incerta*, prostomium and posterior parapodium; S, *Ancistrosyllis hartmanae*, prostomium and parapodium from midbody (above) and midbody parapodium of *A. jonesi* (below); T, *Sigambra tentaculata*, prostomium and midbody parapodium.

4. Epidermis smooth. *Sigambra:* Stout, hooked notosetae begin on setiger 4 in *S. tentaculata*, on 11–15 in *S. bassi*, and on 23–30 in *S. wassi;* the species are 15 by 2, 40 by 2, and 45 by 5 mm, respectively; Fig. 17.16T.

5. With short anal cirri and pygidium with a flattened disclike or bilobed plate; with 2 rudimentary eyes. *Microphthalmus:*
- Anal plate simple, not lobed; 6 mm by 0.5 mm; Fig. 17.16E. *M. sczelkowii.*
- Anal plate bilobed; 9 mm by 1.0 mm; Fig. 17.16A,D. *M. aberrans.*

5. Without an anal plate; prostomium with 4 eyes—6.

6. Prostomium with short, median antenna:
- With 6 pairs of tentacular cirri; 40 mm by 3 mm; Fig. 17.16B. *Podarke obscura.*
- With 8 pairs of tentacular cirri; 6 mm by 2 mm. *Gyptis vittata.*

6. Prostomium without median antenna:
- Neuropodia trilobed; 25 mm by 4 mm; Fig. 17.16C. *Nereimyra punctata.*
- Neuropodia without lobes; 15 mm by 4 mm; Fig. 17.16F; commensal with *Upogebia affinis. Parahesione luteola.*

7. Body threadlike, i.e., worms less than 10 mm long and less than 0.5 mm wide—8.

7. Body not threadlike, thicker—11.

8. With 2 pairs of tentacular cirri; epitokous. *Brania:*
- With a pair of ocular spots in addition to 4 eyes; 4 mm by 0.3 mm; Fig. 17.16G. *Brania clavata.*
- Without ocular spots; 7 mm by 0.4 mm. *B. wellfleetensis.*

8. With 1 pair of tentacular cirri, sometimes rudimentary—9.

9. Tentacular cirri rudimentary, Fig. 17.16H; epitokous. *Exogone:*
- Antennae subequal; 8 mm by 0.3 mm. *E. verugera.*
- Median antenna longer: *E. dispar* has anterior parapodia with simple curved setae, compound spinigers, and compound falcigers, and is 8 mm by 0.5 mm, Fig. 17.16H; *E. hebes* has compound setae only, and is 10 mm by 0.4 mm.

9. Tentacular cirri more strongly developed; Fig. 17.16I,J—10.

10. Palpi fused on basal third; body smooth; 5 mm by 0.3 mm; Fig. 17.16I; epitokous. *Parapionosyllis longicirrata.*

10. Palpi fused for almost whole length; body with adhesive papillae, often mud-encrusted; Fig. 17.16J; epitokous. *Sphaerosyllis:*
- With 6 eyes; 4.5 mm by 0.5 mm. *S. erinaceus.*
- With 4 eyes; 5 mm by 0.3 mm. *S. hystrix.*

11. Parapodia without ventral cirri, Fig. 17.16K; reproduction by sexually dimorphic "stolons" produced singly or in chains, frequently planktonic. *Autolytus;* see subkey following.

11. Parapodia with ventral cirri—12.

12. Body short, flattened; segments about 15; epitokous; 10 mm. (Note figure of whole animal in Miner, 1950). *Amblyosyllis finmarchica.*

12. Body cylindrical; segments more than 50—13.

13. Ventral cirri extending beyond parapodial lobes—14.

13. Ventral cirri shorter—15.

14. Some anterior segments with knobbed acicula, Fig. 17.16L; 8 mm by 0.5 mm; epitokous. *Streptosyllis:*
- Setigers 2–20 or more with knobbed acicula. *S. varians.*
- Setigers 2–5 with knobbed acicula. *S. arenae.*

14. Without knobbed acicula; epitokous. *Syllides:*
- With palpi about as long as prostomium; 7 mm by 1 mm; Fig. 17.16M. *S. longocirrata.*
- With short palpi; 3 mm by 0.3 mm; Fig. 17.16N. *S. setosa.*

15. Proboscis with circlet of strong teeth; 22 mm by 1 mm; epitokous. *Odontosyllis fulgurans.*

15. Proboscis with a single large tooth; Fig. 17.16O—16.

16. Antennae and dorsal parapodial cirri annulated; Fig. 17.16P; reproduces by stolons. *Syllis*:
- With some heavy, bifurcated, simple (not compound) setae at midbody; dorsal parapodial cirri with about 10 annulae (7–16); 50 mm by 1 mm. *S. gracilis.*
- With compound setae only; dorsal cirri with about 22 annulae (11–40); 45 mm by 1.2 mm. *S. cornuta.*

  A third species, *spongiphila*, has simple setae and 40–70 annulae in the dorsal cirri.

16. Antennae and dorsal cirri smooth or indistinctly annulated; Fig. 17.16Q; epitokous. *Eusyllis:*
- With distinct nuchal fold; ventral cirri of first setiger not enlarged; 32 mm by 1.2 mm; *E. blomstrandi.*
- Without nuchal fold; ventral cirri of first setiger much larger than those following; 15 mm by 0.5 mm. *E. lamelligera.*

Autolytus.

---

1. Neurosetae simple, with bulbous tips; known only from sexual stolons. *A. emertoni.*

1. Neurosetae compound—2.

2. Dorsal parapodial cirri equal about half body width:
- With conspicuous dark, longitudinal bands at least in anterior quarter of body; 26 by 1 mm. *A. prismaticus.*
- Colorless or with faint longitudinal bands; 18 mm by 0.7 mm. *A. cornutus.*
- With dark transverse bands, one per segment regularly or at intervals; 30 mm by 1 mm. *A. fasciatus.*

2. Dorsal parapodial cirri more than half body width; colorless or with transverse bands, sometimes 2 per segment:
- Forms chains of 2–8 sexual individuals, head of first after setiger 32; sexual stolons with 2–4 prenatatory setigers; 20 mm. *A. prolifer.*
- Sexual individuals budded singly (presumably), head of the first at about 25–26; sexual stolons with 14 prenatatory setigers; 18 mm. *A. alexandri.*

Group Twelve, Families Dorvilleidae, Eunicidae, Onuphidae. The dorvilleids are small, wandering carnivores that burrow or form temporary tubes under rocks. Onuphids and eunicids are also tubicolous, but the tubes are permanent. Those of Onuphidae may be portable or buried in a soft substratum and are characteristically encrusted with foreign matter; the tubes are parchmentlike. Members of this family are carnivorous but have comparatively weak jaws. The eunicids are carnivorous and herbivorous. The body is not regionated in these worms. Eunicids are epitokous.

---

1. First apparent segment without parapodia or setae, Fig. 17.17A. Onuphidae—2.
1. First 2 segments without parapodia or setae—4.
2. First apodous segment without tentacular cirri; usually without eyes except in young; tube portable, quill-like, tapered, transparent, clean or fouled with sessile organisms; worm 215 mm by 8 mm; Fig. 17.17A. *Hyalinoecia tubciola.*
2. First apodous segment with tentacular cirri—3.
3. Branchiae strongly spiraled and bearing setae, Fig. 17.17B, changing form posteriorly; prostomium with pair of sensory organs which may be mistaken for eyes; tube fixed and mostly embedded, up to a meter in length; worm 300 mm by 10 mm. *Diopatra cuprea.*
3. Branchiae a single filament anteriorly (on segments 10–22 about), then branching with 2–7 comblike filaments, becoming single again posteriorly, Fig. 17.17C; tube thin, membraneous, embedded, encrusted with fine particles; 120 mm by 2 mm. *Onuphis eremita.*
3. Branchiae are simple tendrils, Fig. 17.17D:
- With branchiae on first setiger; tube embedded, with thin, inner, parchment-like lining, and coating of mud; worm 125 mm by 4 mm; no eyes. *Onuphis opalina.*
- Branchiae begin on setigers 9–13; parapodia of first setiger enlarged and bearing conspicuous amber-colored setae; tube free, flattened, scabbardlike, covered with coarse particles; with 2 large and 2 small eyes. *Onuphis conchylega.*

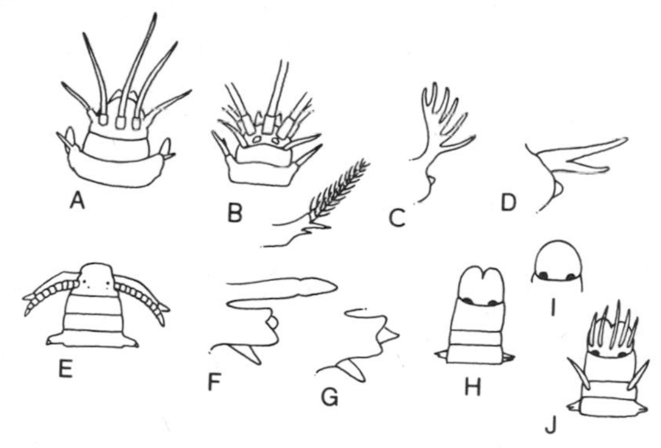

Fig. 17.17. Annelida, Polychaeta, Group Twelve. A, *Hyalinoecia tubicola*, anterior end; B, *Diopatra cuprea*, anterior end and detail of branchia; C, *Onuphis eremita*, branchia with comb-like filaments; D, *Onuphis opalina*, branchia; E, *Stauronereis* sp., anterior end; F, *Stauronereis rudolphi*, parapodium; G, *Stauronereis sociabilis*, parapodium; H, *Marphysa sanguinea*, anterior end, occipital antennae omitted; I, *Marphysa belli*, prostomium with occipital antennae omitted; J, *Eunice* sp., anterior end, occipital antennae shown.

4. Prostomium with pair of articulated antennae and a pair of ventral palpi, Fig. 17.17E; free-living burrowers or in temporary tubes; pale pinkish or white. Dorvilleidae. *Stauronereis: S. caecus* lacks eyes and is 8 mm by 0.6 mm; *S. rudolphi* (Fig. 17.17F) and *S. sociabilis* (Fig. 17.17G) have eyes but differ in the form of the parapodial lobes as figured; the species reach 50 mm and 15 mm, respectively. (For *S. sociabilis* see Hartman, 1945, and Webster, 1879).

4. Prostomium with 1–5 occipital antennae, Fig. 17.17J, and a pair of globular, ventral palpi; species burrow and some form parchment like tubes. Eunicidae—5.

5. Without tentacular cirri. *Marphysa:*
- Prostomium and palpi fused and notched anteriorly; 600 mm by 11 mm; Fig. 17.17H. *M. sanguinea.*
- Prostomium and palpi fused and rounded anteriorly; 200 mm by 3 mm; Fig. 17.17I. *M. belli.*
  Note also *Lysidice americana* with three antennae compared to 5 for the above species; see Verrill and Smith (1873).

5. With pair of short tentacular cirri, Fig. 17.17J; branchiae comblike when fully developed (as in Fig. 17.17C). *Eunice*—6.

6. Branchiae present on segments to near posterior end:
- With branchiae beginning on fifth setiger; Carolinian; does not produce a permanent tube; see Hartman (1944). *E. rubra.*
- With branchiae beginning on seventh or tenth setiger (usually on 8); boreal; 200 mm by 12 mm. *E. norvegica.*

6. Branchiae begin on setigers 3–5, continue to about setiger 40, leaving middle and posterior segments without branchiae; 150 mm by 8 mm. *E. pennata.*

Group Thirteen, Families Spionidae, Magelonidae, Apistobranchidae, and Chaetopteridae. These are sedentary worms chiefly characterized by a pair of elongate palpi that are used to sweep the surface of the substratum for microscopic food items. *Chaetopterus* is a notable exception in that it is a filter feeder, the enlarged pallets being used to produce a complex food-bearing current. Except for *Magelona* these are tubicolous forms. The tubes of spionids are membraneous and mud covered. The U-shaped tube of *Chaetopterus* is tough and parchmentlike with protruding chimneys. Some species of *Polydora* bore in mollusk shells.

---

1. Body with 2 or 3 distinct regions—2.

1. Body not sharply divided in regions, with a gradual transition in segment form. Spionidae—4.

2. Slender worms with two body regions, the shorter anterior region separated from the posterior region by a segment different from the others; tentacles with a fringe of papillae or branchlets on one side; 40 mm by 1 mm; Fig. 17.18A. Magelonidae. *Magelona rosea.*

2. Stout worms with three distinct body regions, the appendages of the middle region modified to form 3 large, circular pallets; 250 mm by 25 mm; Fig. 17.18B; tube U-shaped, parchmentlike. Chaetopteridae. *Chaetopterus variopedatus.*

2. Not as above—3.

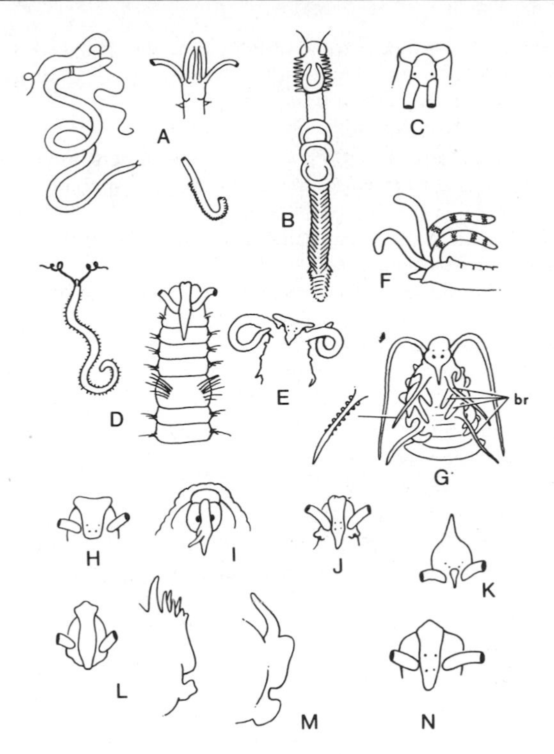

Fig. 17.18. Annelida, Polychaeta, Group Thirteen. A, *Magelona rosea*, whole worm, head, and part of tentacle detailed; B, *Chaetopterus variopedatus*, whole worm; C, *Spiochaetopterus oculatus*, head; D, *Polydora* sp., whole worm and detail of anterior end; E, *Spiophanes bombyx*, anterior end; F, *Streblospio benedicti*, anterior end, lateral view; G, *Prionospio* sp., anterior end with branchiae indicated and detail of individual branchia; H, *Scolecolepides viridis*, head; I, *Laonice cirrata*, head; J, *Pygospio elegans*, head; K, *Scolelepis squamata;* L, *Dispio uncinata*, head, and parapodium; M, *Spio* sp., parapodium; N, same, head.

3. Head rounded; pygidium lobed, without cirri; 50 mm by 1 mm; tube cylindrical, annulated; Fig. 17.18C (Webster, 1879). *Spiochaetopterus oculatus.*
3. Head conical; pygidium with cirri; >12 mm by 2 mm. Apistobranchidae. *Apistobranchus tullbergi.*

4. Fifth setiger highly modified, Fig. 17.18D. *Polydora,* see subkey below.
4. Fifth setiger not modified—5.

5. Without branchiae or with but single pair next to palpi:
- Without branchiae. *Spiophanes: S. bombyx*, Fig. 17.18E, has a prostomium with frontal horns, 60 mm by 1.5 mm; *S. wigleyi* lacks frontal horns, 15 mm by 1 mm. (For *S. wigleyi* see Pettibone, 1962).
- With pair of branchiae marked with cross-bars next to pair of unmarked palpi similar in size; 7 mm by 0.7 mm; Fig. 17.18F; common on *Mytilus* beds (Hartman, 1945). *Streblospio benedicti.*

5. With 4 or 5 pairs of branchiae, some at least pinnate; Fig. 17.18G. *Prionospio.* No attempt will be made here to separate the half dozen local forms; the branchiae vary in number and may all be pinnate or not.
5. With 7 or more pairs of branchiae—6.
6. Branchiae begin on first to third segment and continue to rear of body—7.
6. Branchiae begin on first to third segments but absent on about last half or more of body:
- Branchiae begin on first segment; 100 mm by 3 mm; Fig. 17.18H (Smith *et al.*, 1964). *Scolecolepides viridis.*
- Branchiae begin on second or third segment; 120 mm by 5 mm; Fig. 17.18I. *Laonice cirrata.*

6. Branchiae do not begin until eleventh to twentieth segments and number only 7–9 pairs in females, up to 78 pairs in males; 15 mm by 0.7 mm; Fig. 17.18J (Webster and Benedict, 1884). *Pygospio elegans.*
7. Branchiae begin on setiger two; 60 mm by 2 mm; Fig. 17.18K. (Pettibone, 1963b). *Scolecolepis squamata.*
7. Branchiae begin on setiger 1—8.
8. Two pairs of eyes lateral, concealed by base of palpi; first few notopodial lobes serrate, Fig. 17.18L; 32 mm by 2.5 mm (Hartman, 1951). *Dispio uncinata.*
8. With 2 pairs of eyes forming a square, visible dorsally; first few notopodial lobes not serrate; Fig. 17.18M,N. *Spio:*
- Ventral lamellae greatly reduced; neuropodia with about 16 crochets; 50 mm (Hartman, 1945). *S. setosa.*
- Ventral lamellae little reduced; neuropodia with about 6 crochets; 30 mm by 2 mm. *S. filicornis.*

Polydora (Mainly from Hartman, 1945).

---

1. Pygidium with papillae; commensal with hermit crabs, especially *Pagurus longicarpus. P. commensalis.*
1. Pygidium with disclike flange; not commensal with hermit crabs—2.
2. Posterior segments with falcate hooks; commensal with sponges. *P. colonia.*
2. Without falcate hooks, not commensal with sponges—3.
3. With median antenna; 25 mm; in wood crevices. *P. ligni.*
3. Without median antenna—4.
4. Branchiae present from about tenth segment; no eye spots. *P. anoculata.*
4. Branchiae usually present from segments 6 to 8—5.
5. Setae of fifth setiger falcate or hooded, without an accessory tooth:
- Pygidium a broad disc with a dorsal incision. *P. socialis.*
- Pygidium 4-lobed; 140 mm; forms colonies in shells. *P. concharum.*
- Pygidium bilobed. *P. hamata.*

5. Setae of fifth setiger not falcate:
- Spines of setiger 5 modified with an accessory sheath; 20 mm. *P. websteri.*
- Spines of setiger 5 modified, with an accessory tooth; 20 mm. *P. ciliata.*

(Questionably present in the western Atlantic.)

Group Fourteen, Families Serpulidae and Sabellidae. These are tubicolous worms with pinnate tentacular appendages that are used as a filtration device to entrap microscopic food. In serpulids the tube is calcareous. Sabellids form tubes made of mucus, or the tubes are membraneous or horny and may be covered with mud, or less often, with sand, gravel, or shell fragments. The tubes may be buried in a soft substratum or attached to rocks, or in crevices; the smallest forms are found on algae, hydroids, or the tubes of other annelids.

---

1. Tubes white, calcareous; usually with a pluglike operculum (actually a highly modified tentacle). Serpulidae—2.

1. Tubes flexible, not calcareous, but mucoid, leathery, parchmentlike, or rarely undeveloped. Sabellidae—5.

2. Tube a flat coil, small (2–3 mm diameter); Fig. 17.19A,B. *Spirorbis*—3.

2. Tubes not regularly coiled; worm symmetrical, with at least 15 setigers—4.

3. Tubes coiled dextrally, Fig. 17.19A:
- Tube smooth, white; with 3 thoracic setigers; diameter of tube 2 mm. *S. spirillum.*
- Tube with 3 longitudinal ridges, these projecting as teeth at the opening; with 4 thoracic setigers; diameter of tube 3 mm; Fig. 17.19A. *S. violaceus.*

3. Tubes coiled sinistrally; highly variable in sculpture, almost smooth to rough, or ridged; Fig. 17.19B.
- Operculum simple, without incubatory pouch; tube often rough, sometimes with 2–4 bumps around the umbilicus; diameter of tube 2 mm. *S. borealis.* This is evidently the common littoral species.
- Operculum with incubatory pouch, Fig. 17.19C; tube sometimes with 2 or 3 ridges with opening quadrangular or toothed, or tube smooth to rough without definite ridges; diameter of tube 3 mm. *S. granulatus.*

4. Tubes solitary or in irregular groups, not fascicled; large worms (50 mm to 75 mm by 3 mm to 8 mm):
- Operculum present, 2-tiered, with smooth peduncle; tube attached for most of its length, solitary or in masses 100 mm or more in diameter; tubes irregular and often contorted, with growth lines and sometimes ridged; worm 75 mm by 3 mm; Fig. 17.19D (Smith *et al.*, 1964). *Hydroides dianthus.*
- No operculum; tube solitary, attached basally but with an erect, free part, smooth; worm 50 mm by 8 mm; Fig. 17.19E. *Protula tubularia.*

4. Tubes fascicled, Fig. 17.19F; minute forms (6 mm by 0.4 mm):
- Operculum present, spoon-shaped; body of worm grayish; 5 mm by 0.2 mm; Fig. 17.19F. *Filograna implexa.*
- No operculum; body of worm red or orange; 7 mm by 0.4 mm. *Salmacina dysteri.*

Sabellidae.

5. With 10–12 setigers; minute species (3 mm by 0.25 mm); 2 eyes on first and last segments; with soft tube embedded in mud; Fig. 17.19G. *Fabricia sabella.*

5. With more than 12 setigers; larger species, adults 25 mm or more in length—6.

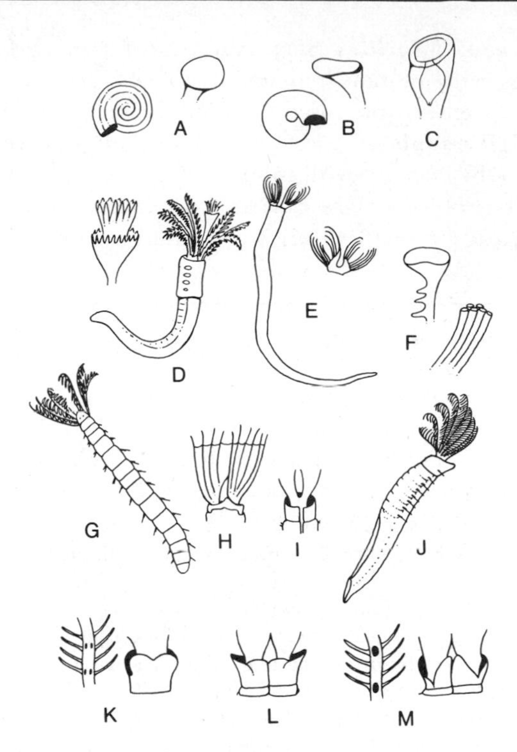

Fig. 17.19. Annelida, Polychaeta, Group Fourteen. A, *Spirorbis violaceus*, tube and operculum; B, *Spirorbis* sp., sinistrally coiled tube, and operculum; C, *Spirorbis granulatus*, operculum with incubatory pouch; D, *Hydroides dianthus*, whole worm and operculum (left); E, *Protula tubularia*, whole worm and tentacular appendages; F, *Filograna implexa*, operculum and fascicled tubes; G, *Fabricia sabella*, whole worm; H, *Myxicola infundibulum*, branchial filaments and branchial collar; I, *Euchone rubrocincta*, branchial collar and base of branchial filaments; J, same, whole worm; K, *Sabella crassicornis*, branchial collar and part of branchial filament; L, *Sabella micropnthalma*, branchial collar; M, *Potamilla reniformis*, branchial collar and part of branchial filament.

6. Branchial filaments united by membrane for half or more of their length, Fig. 17.19H—7.

6. Branchial filaments only united basally—8.

7. Collar weakly developed, Fig. 17.19H; with thick, mucous tube buried in sand; worm 200 mm by 30 mm. *Myxicola infundibulum.*

7. Collar well developed, Fig. 17.19I; tube covered with fine sand or mud:

- With gutter on underside posteriorly; 25 mm by 5 mm; Fig. 17.19I,J. *Euchone rubrocincta.* (Verrill and Smith, 1873, described and figured *E. elegans* with a collar, without a ventral slit.)
- Without gutter on underside posteriorly; 120 mm by 6 mm. *Chone infundibuliformis.*

8. Branchial filaments without eyes; collar lobed; tube leathery, covered with sand or mud; 60 mm by 2 to 3 mm. *Potamilla:* Webster and Benedict (1884) describe *P. neglecta* as pure white; according to Fauvel (1927) *P. neglecta* is reddish or brown.

8. Branchial filaments with eyes; collars as figured, Fig. 17.19K–M:

- Eyes paired, 2–6 per filament; collar 4-lobed; tube transparent and flexible above, rigid and corneous below; 50 mm by 4 mm; Fig. 17.19K. *Sabella crassicornis.*
- Eyes numerous, in 2 irregular rows on each filament; 30 mm by 3 mm; Fig. 17.19L (Smith *et al.*, 1964). *Sabella microphthalma.*
- Eyes 1–8, in single row on each filament; tube leathery, sand covered; 100 mm by 2 mm; Fig. 17.19M. *Potamilla reniformis.*

Group Fifteen, Families Pectinariidae, Sabellaridae, Flabelligeridae. The flabelligerids are mud dwellers characterized in part by the possession of mucus-secreting papillae to which sand or mud particles adhere. The other two families include highly specialized, short-bodied worms with a truncated head and a specialized tentacle-feeding apparatus. Both families are tubicolous. The pectinariids are solitary, although sometimes found locally in concentrations of large numbers of individuals; the pectinariid tube is an elongate cone of closely cemented sand grains. Sabellarids also produce tubes made of sand grains, but they are commonly found in aggregated, colonial masses and may form reeflike structures.

---

1. Body covered with papillae, Fig. 17.20D; anterior end usually with setae directed forward and frequently forming a sort of cage; head not conspicuously truncated; not tubicolous. Flabelligeridae—3.

1. Not as above; body short; anterior end decidedly truncated with an oblique operculum formed of flattened golden setae; forming tube of cemented sand—2.

2. Operculum consisting of 2 comblike series of setae; the sand tube conical, solitary, open at both ends, free; Fig. 17.20A; 40 mm by 7 mm. Pectinariidae. *Pectinaria gouldii. P. gouldii* is Virginian. Two additional species are reported from deeper waters north of Cape Cod.

2. Operculum consisting of 2 semicircular peduncles bearing concentric circlets of setae; sand tubes cemented to a substratum or in masses; 25 mm; Fig. 17.20B. Sabellariidae. *Sabellaria vulgaris.*

3. With thick mucous mantle or collar enclosing two types of pedunculate papillae; neurosetae are compound crochets; 60 mm by 10 mm; Fig. 17.20C. *Flabelligera affinis.*

3. Papillae simple, not pedunculate; neurosetae simple—4.

4. With pair of large, projecting nephridial papillae on ventral surface of fourth or fifth segment; anterior setae do not form cage; 40 mm by 5 mm; Fig. 17.20D. *Brada.* Several species are recorded locally; see Webster and Benedict (1887) for *granosa*, Fauvel (1927) for *villosa*, and Verrill and Smith (1873) for *setosa.*

4. Without nephridial papillae—5.

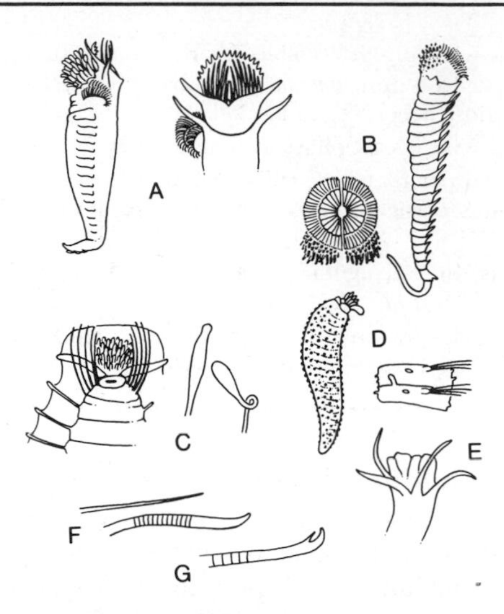

Fig. 17.20. Annelida, Polychaeta, Group Fifteen. A, *Pectinaria gouldii*, whole worm and detail of head; B, *Sabellaria vulgaris*, whole worm and operculum; C, *Flabelligera affinis*, anterior end and two pedunculate papillae; D, *Brada* sp., whole worm and two anterior segments, one with nephridial papilla; E, *Diplocirrus hirsutus*, branchia; F, *Pherusa plumosa*, neurosetae; G, *Pherusa arenosa*, neuroseta.

5. Branchiae all alike. *Pherusa*—6.
5. Branchiae of two kinds, one filiform, the other wider and flatter, bordering anterior end; neuro- and notosetae capillary; 25 mm by 2 mm; Fig. 17.20E. *Diplocirrus hirsutus.*
6. Neurosetae simple—7.
6. Neurosetae bifid; 60 mm by 5 mm; Fig. 17.20G. *Pherusa arenosa.*
7. With hooked neurosetae beginning on fourth segment; 60 mm by 5 mm; Fig. 17.20F. *Pherusa plumosa.*
7. With hooked setae beginning on fifth segment; 60 mm by 3.5 mm (Verrill and Smith, 1873). *Pherusa affinis.* A fourth species, *P. aspera*, inadequately described originally, was identified with reservations by Webster and Benedict (1887).

Group Sixteen, Family Cirratulidae. The members of this family are mostly mud dwellers and tentacle feeders. They do not form permanent tubes.

---

1. With one elongate, median tentacle; no branchiae; 6 mm by 0.8 mm; Fig. 17.21A. (Webster and Benedict, 1887). *Cossura longocirrata.*
1. With 2 elongate, lateral tentacles as well as additional tentacular appendages, Fig. 17.21D,E—3.
1. Without large tentacles; with filamentous tentacles in 2 groups or united on dorsal surface of 1 or 2 anterior setigers; with pair of long branchiae on most segments, Fig. 17.21B—2.

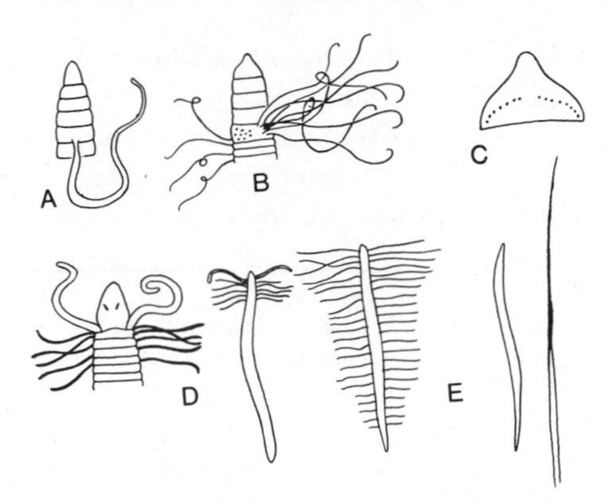

Fig. 17.21. Annelida, Polychaeta, Group Sixteen. A, *Cossura longocirrata*, anterior end; B, *Cirratulus* sp., anterior end; C, *Cirratulus cirratus*, prostomium; D, *Dodecaceria* sp., whole worm and detail of anterior end; E, *Chaetozone setosa*, whole worm and two setae.

2. With 2–9 eyes on each side of prostomium; 120 mm by 3 mm; Fig. 17.21C. *Cirratulus cirratus.*

2. Without eyes:

- Filamentous tentacles on fourth to fifth or fifth to sixth setigers; 250 mm by 6 mm. *Cirriformia filigera.*
- Filamentous tentacles on first and second setigers; 150 mm by 7 mm (Verrill and Smith, 1873; Miner, 1950). *Cirratulus grandis.*

3. With no more than 8 pairs of branchiae—4.

3. With branchiae along entire body—5.

4. Tentacles at front of prostomium; with 5 or 6 pairs of branchiae; 1 pair of cirri on second segment; 6 eyes; 8 mm by 0.5 mm (Webster and Benedict, 1887). *Ledon leidyi.*

4. Tentacles at back of buccal segment; with 4–6 pairs of branchiae; no eyes or 2; dark green to black; in pelecypod shells or in colonies of *Astrangia;* 60 mm by 3 mm; Fig. 17.21D. *Dodecaceria: D. concharum*, as described by Fauvel (1927), is polymorphic with both pelagic and sedentary epitokous forms (rare) in addition to an atokous, sedentary form. The pelagic epitoke has reduced tentacles and cirri; both epitokous forms have 2 eyes. A second species, *D. coralii*, was described by Leidy (1855).

5. Setae all of capillary type; 15 mm by 7 mm (Webster and Benedict, 1887). *Tharyx acutus.* Webster and Benedict described a second boreal species, *T. similis* on the same page as the above, and differing in minor details. Hartman (1945) described *T. setigera*, a Carolinian species that ranges at least as far north as Virginia. *T. setigera* has setae on the first 3 segments bearing tentacular cirri; the boreal species lack setae on these segments.

5. Setae include capillary types and crochets, the crochets almost encircling the body posteriorly; Fig. 17.21E. *Chaetozone setosa.*

Group Seventeen, Family Terebellidae. Terebellids construct membraneous tubes coated with mud, sand, or shell fragments buried in the substratum or fixed to stones, algae, or hydroids. They are tentacle feeders. In some cases the tentacles are extraordinarily extensile and are used to sweep the surface of the substratum for food.

---

1. Setae absent or scarce; segmentation not sharply defined, the body generally covered with transverse lines of papillae; somewhat resembling the holothurian *Leptosynapta;* 400 mm. *Lysilla alba.*

1. Setae present including uncini—2.

1. Setae present but no uncini; parapodia of midbody branching and gill like; 350 mm by 7 mm; Fig. 17.22A. (Smith *et al.*, 1964). *Enoplobranchus sanguineus.*

2. Buccal segment with proboscis covered with minute spicules or papillae; 80 mm by 5 mm (Malmgren, 1865); Fig. 17.22B. *Artacama proboscidea.*

2. Without a proboscis as above—3.

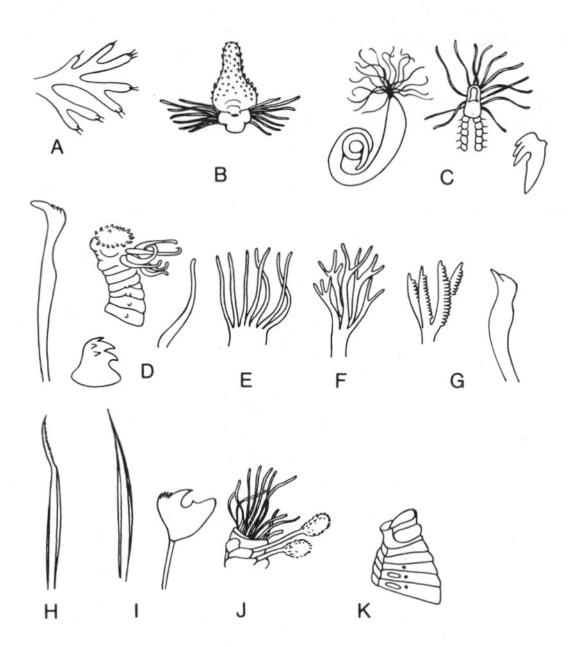

Fig. 17.22. Annelida, Polychaeta, Group Seventeen. A, *Enoplobranchus sanguineus*, parapodium from midbody; B, *Artacama proboscidea*, anterior end; C, *Polycirrus* sp., whole worm, anterior end and uncinus; D, *Trichobranchus glacialis*, anterior end, single gill (right), and two uncini (left); E, *Amphitrite cirrata*, gill; F, branching, tree-like gill; G, *Terebellides stroemi*, gill, and uncinus; H, *Amphitrite affinis*, notoseta; I, *Pista* sp., notoseta and uncinus; J, *Pista cristata*, anterior end; K, *Nicolea* sp., anterior end, gills and tentacles omitted.

3. Without gills; tentacles numerous and extensile; uncini in simple series; Fig. 17.22C. *Polycirrus:*
- Bright red; setae on segments 18–25; 25 mm (Miner, 1950). *P. eximius.*
- Red, orange, to whitish; setae on segments 11–13; 70 mm. *P. medusa.*
- Lemon-yellow (red according to Verrill); setae on segments 21–34; 80 mm; phosphorescent. *P. phosphoreus.*

3. Gills present, at least one of them branching, treelike, Fig. 17.22F—5.
3. Gills present, simple, Fig. 17.22D, or compound, Fig. 17.22E, but not treelike—4.

4. With 3 pairs of simple filamentous gills; 30 mm by 3 mm; Fig. 17.22D. *Trichobranchus glacialis.*
4. With 2 pairs of compound filamentous gills; with numerous eye spots; 200 mm by 10 mm. *Thelepus cincinnatus.*
4. With 3 pairs of compound filamentous gills; no eye spots; Fig. 17.22E:
- Notosetae begin on first gill-bearing segment and continue over most of body; uncini begin on fourth setiger. *Streblosoma spiralis.*
- Notosetae begin in third gill-bearing segment and continue on 17 segments; uncini begin on second setiger; 300 mm by 20 mm; Fig. 17.22E. *Amphitrite cirrata.*

5. With a single large gill with 4 pectinate lobes; thoracic uncini acicular, abdominal ones not; 60 mm by 8 mm; Fig. 17.22G. *Terebellides stroemi.*
5. With 1 pair of gills treelike, the others filiform or filamentous; with 16 setigers; 150 mm by 5 mm. *Pista maculata.*
5. With 2 or 3 pairs of gills treelike—6.

6. Setae on 20 or more segments:
- Setae on 40–50 segments; 380 mm; Virginian (Miner, 1950). *Amphitrite ornata.*
- Setae on 23–45 segments; 250 mm by 12 mm; boreal. *Amphitrite johnstoni.*
- Setae to end of body; 50 mm by 3 mm (Hartman, 1945). *Terebella* sp.

6. Setae on 15–17 segments—7.

7. Capillary notosetae toothed at tip, Fig. 17.22H; gills subequal; 110 mm by 7 mm. *Amphitrite affinis.*
7. Capillary notosetae smooth-tipped—8.

8. Uncini of first segments with basal prolongation, Fig. 17.22I; anterior segments lobed. *Pista:*
- Branchiae contracting to stalked red pompon; 90 mm by 6 mm; Fig. 17.22J. *P. cristata.*
- Branchiae with palmate branches; 70 mm by 3 mm (Webster and Benedict, 1884). *P. palmata.*

8. Uncini of first segments not so; anterior segments lobed or not—9.

9. With 2 pairs of gills; no lateral lobes on anterior segments; Fig. 17.22K. *Nicolea:*
- Coloration uniform rose or pale brown; 20 mm by 2 mm. *N. zostericola.*
- Spotted with white; 60 mm by 5 mm. *N. venustula.*

These species may be indistinguishable in preservative; the first species deposits eggs in a cocoon, the second sheds them freely in the water; see Fauvel (1927).

9. With 3 pairs of gills; buccal segments lobed; 30 mm by 3.5 mm. *Loimia: L. medusa* and *L. viridis* differ in tube construction and minor structural details as indicated by Hartman (1945).

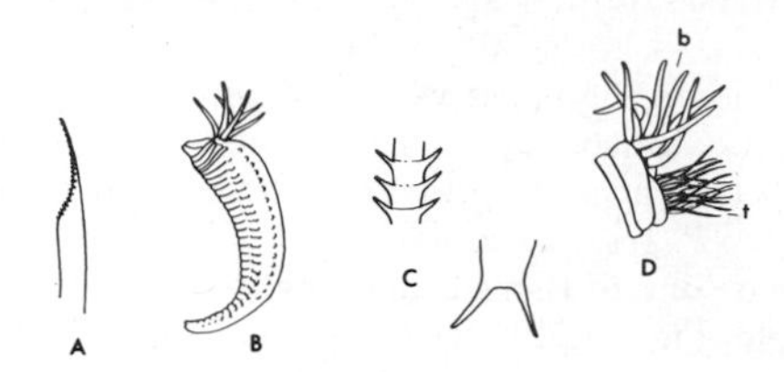

Fig. 17.23. Annelida, Polychaeta, Group Eighteen. A, *Anobothrus gracilis*, seta from eleventh setiger; B, *Hypaniola grayi*, whole worm; C, pinnate tentacle and pygidium with cirri; D, ampharetid prostomium in lateral view. Abbreviations are as follows: b, branchiae; t, tentacles.

Group Eighteen, Family Ampharetidae. Like the terebellids, the ampharetids are tentacle feeders. They differ in being able to withdraw their tentacles into the mouth. They are tubicolous, with a membraneous tube covered with mud or with foreign material. Fig. 17.23D.

---

1. Pygidium without cirri, with or without minute papillae; eleventh setiger with special setae, Fig. 17.23A; tentacles simple or with lateral papillae suggesting a pinnate condition; 13 abdominal segments with uncini; 35 mm by 3 mm. *Anobothrus gracilis.*

1. Pygidium without cirri; tentacles simple; 20 or more abdominal segments with uncini; 15 mm by 1.5 mm; Fig. 17.23B (Pettibone, 1953). *Hypaniola grayi.*

1. Pygidium with well-developed cirri—2.

2. Tentacles pinnate; Fig. 17.23C—3.

2. Tentacles simple—4.

3. With 2 anal cirri; thoracic uncini with 3–5 teeth in a single row: *Sabellides octocirrata* has 3 or 4 teeth and is 10 mm by 1 mm; *Asabellides oculata* has uncini with 5 teeth and is 20 mm (Webster, 1880).

3. With about 10–14 anal cirri; thoracic uncini with 8–10 teeth; with well-developed golden-yellow thoracic setae; males greenish white, females rose; 24 mm by 4 mm, larger in the Arctic. *Ampharete acutifrons.*

4. Nine abdominal segments with uncini; uncini with 5 or 6 teeth; without golden-yellow thoracic setae; 19 mm by 4 mm (Miner, 1950). *Amage auricula.*

4. Fifteen abdominal segments with uncini; uncini with 5–7 teeth; with 8–20 pairs of well-developed golden-yellow thoracic setae; 40 mm by 5 mm. *Amphicteis gunneri.*

4. Fifty abdominal segments with uncini; uncini with 4 teeth; 50 mm by 3 mm. *Melinna cristata. M. maculata* is a closely related form; see Webster (1879).

References

Augener, H. 1912. Beitrag zur Kenntnis verschiedener Anneliden und Bemerkungen über die nordischen *Nephthys*-Arten und deren epitoke Formen. Arch. Natg. Berlin. 78 Abt. A. 162–212.

Bousfield, E. L., and A. H. Leim. 1959. The fauna of Minas Basin and Minas Channel. Bull. nat. Mus. Can. 166:1–30.

Dales, R. P. 1963. Annelids. Hutchinson University Library.

Dawydoff, C. 1928. Traité d'embryologie comparée des invertébrés. Masson et Cie.

Fauvel, P. 1923. Polychaetes errantes. Faune de France. 5:1–488.

———. 1927. Polychaetes sédentaires. Faune de France. 16:1–494.

———. 1932. Annelida Polychaeta of the Indian Museum, Calcutta. Mem. Indian Mus. 12:1–262.

———. 1959. Classe des annélides polychètes. *In* Traité de Zoologie. Tome V. P.-P. Grassé, ed. Masson et Cie.

Graveley, F. H. 1909. Polychaet larvae. Liverpool mar. biol. Comm. Mem. 19:1–79.

Hartman, Olga. 1944. Polychaetous annelids. Part 5. Eunicea. Allan Hancock Pacific Exped. 10:1–237.

———. 1945. The marine annelids of North Carolina. Duke Univ. mar. Stn. Bull. 2:1–53.

———. 1951. The littoral marine annelids of the Gulf of Mexico. Publ. Inst. mar. Sci. Univ. Texas. 2:7–124.

———. 1959. Catalogue of the polychaetous annelids of the world. Parts 1 and 2. Allan Hancock Fnd. Publ. Occ. Pap. 23:1–353, 355–628.

Jones, M. L. 1963. Four new species of *Magelona* (Annelida, Polychaeta) and a redescription of *Magelona longicornis* Johnson. Amer. Mus. nat. Hist. Novitates. 2164:1–31.

Leidy, J. 1855. Contributions toward a knowledge of the marine invertebrates of the coasts of Rhode Island and New Jersey. J. Acad. Nat. Sci. Philad. 3:135–158.

McIntosh, W. C. 1879. On the Annelida obtained during the cruise of HMS "Valorous" to Davis Strait in 1875. Trans. Linn. Soc. N.S. 1:499–511.

Malmgren, A. J. 1865. Nordiska Hafs-Annulater. Öfvers K. vet. Akad. Forh. 22(5):355–410.

Mangum, C. P. 1962. Studies on speciation in maldanid polychaetes of the North American Atlantic coast. Postila Yale Peabody Mus. 65:1–12.

Miner, R. W. 1950. Field Book of Seashore Life. G. P. Putnam's Sons.

Monro, C. C. A. 1933. The Polychaeta Errantia collected by Dr. Crossland at Colon in the Panama region and the Galapagos Islands during the expedition of the S.Y. St. George. Proc. zool. Soc. Lond. 1:1–96.

Pettibone, Marian H. 1953. A new species of polychaete worm of the family Ampharetidae from Massachusetts. J. Wash. Acad. Sci. 43:384–386.

———. 1957a. Endoparasitic polychaetous annelids of the family Arabellidae with descriptions of new species. Biol. Bull. mar. biol. Lab. Woods Hole. 113:170–187.

———. 1957b. North American genera of the family Orbinidae (Annelida: Polychaeta) with descriptions of new species. J. Wash. Acad. Sci. 47:159–167.

———. 1961. New species of polychaete worms from the Atlantic Ocean, with a revision of the Dorvilleidae. Proc. biol. Soc. Wash. 74:167–185.

———. 1962. New species of polychaete worms (Spionidae: *Spiophanes*) from the East and West Coast of North America. Proc. biol. Soc. Wash. 75:77–88.

———. 1963a. Marine Polychaete worms of the New England region. Bull. U.S. natn. Mus. 227(1):1–356.

———. 1963b. Revision of some genera of Polychaete worms of the family Spionidae, including the description of a new species of *Scolelepis*. Proc. biol. Soc. Wash. 76:89–104.

———. 1966. Revision of the Pilargidae (Annelida: Polychaeta) including descriptions of new species, and redescription of the pelagic *Podarmus ploa* Chamberlin (Polynoidae). Proc. U.S. natn. Mus. 118:155–208.

Reibisch, J. 1905. Anneliden. Nord. Plankt. 7:1–10.
Smith, R. I., *et al.* 1964. Keys to marine invertebrates of the Woods Hole region. Systematics-Ecology Program, Marine Biological Laboratory, Contr. 11:x+208.
Thorson, G. 1950. Reproductive and larval ecology of marine bottom invertebrates. Biol. Rev. 25(1):1–45.
Treadwell, A. L. 1931. Three new species of polychaetous annelids from Chesapeake Bay. Proc. U.S. natn. Mus. 79(1):1–5.
———. 1941. Polychaetous annelids from the New England region, Porto Rico, and Brazil. Am. Mus. nat. Hist. Novitats. 1138:1–4.
Verrill, A. E. 1879. Notice of recent additions to the marine invertebrata of the northeastern coast of America, with descriptions of new genera and species and critical remarks on others. Proc. U.S. natn. Mus. 2:165–205.
———, and S. I. Smith. 1873. Report upon the invertebrate animals of Vineyard Sound and adjacent waters. Rept. U.S. Fish Comm. 1871–1872:295–778.
Wass, M. L. 1963. Check list of the marine invertebrates of Virginia. Spec. sci. Rept. 24. Va. Inst. mar. Sci.:1–56.
Webster, H. E. 1879. Annelida Chaetopoda of the Virginian coast. Trans. Albany Inst. 9:1–69.
———. 1880, 1886. The annelida Chaetopoda of New Jersey. Rept. N.Y. State Mus. 32:101–128; 39:128–159.
———, and J. E. Benedict. 1884. Annelida Chaetopoda from Provincetown and Wellfleet, Massachusetts. Rept. Comm. Fish Fisheries (1881): 699–747.
———, and J. E. Benedict. 1887. The Annelida Chaetopoda from Eastport, Maine. Rept. U.S. Fish Comm. (1885):707–755.
Wells, H. W., and I. E. Gray. 1964. Polychaetous annelids of the Cape Hatteras area. J. Elisha Mitchell sci. Soc. 80:70–78.
Wilson, D. P. 1932. On the Mitraria larva of *Owenia fusiformis* Delle Chiaje. Phil. Trans. B.: 221, 231–234.

Class Oligochaeta
AQUATIC EARTHWORMS

Diagnosis. *Distinctly segmented annelid worms with neither parapodia nor cephalic appendages; setae sometimes single but usually in bundles of two or more; setae absent on prostomium and mouth segment or peristomium.*

The common earthworm is perhaps the most familiar representative of the class. The aquatic and semiterrestrial species are small, elongate, cylindrical worms. Most are extremely slender; individuals 40–65 mm long are only 1–1.5 mm in diameter. Oligochaetes may be transparent in life but are commonly tinted white to pink, red, brown, or purple. The segments are essentially similar externally, and the body is not marked in regions. Even the head is little differentiated; the segment bearing the mouth appears little different from the succeeding segments in dorsal aspect except that it lacks setae.

The setae are usually symmetrically placed in two dorso-lateral and two ventro-lateral bundles in each segment; Fig. 17.24B. Ventral setae may be absent on genital segments, and special copulatory setae may occur. In some species the setae are single in part of the body, but bundles usually contain two or more. However, the number and form of the setae may vary not only on different parts of the body but also in dorsal and ventral bundles. Seta on segments before the clitellum are considered "anterior," those behind, "posterior." Quite commonly, the number per bundle diminishes posteriorly.

Oligochaetes are hermaphroditic. Self-fertilization is possible in some genera and has occurred under experimental conditions in *Limnodrilus hoffmeisteri*. The more common technique is probably for two individuals to cross-fertilize each other simultaneously, as has been well demonstrated in earthworms. At the time of copulation the sperm or spermatophores are received in special storage chambers called *spermathecae* for use when the eggs are laid. The eggs are deposited in a cocoon produced by a specialized skin region called the *clitellum*. The clitellum is frequently inconspicuous in aquatic oligochaetes except in breeding season. The cocoon is produced as an elastic belt that slips off and continues an independent existence while the eggs are incubated. Development is essentially direct, and there are no distinct larval forms.

The arrangement and structural details of the sexual apparatus are of primary systematic importance including the location of external perforations or invaginations in the body wall, i.e., the male and female pores and the spermathecae. Male openings are paired or single. The genital pores may lie in the furrow between segments, and they may be so obscure that they can only be detected in microscopic sections.

The testes and genital ducts occupy different segments. Sperm are freely released in the segment in which the testes lie. A funnel, which is the terminus of the genital duct, opens into this segment, collects the sperm, and the gametes are then conveyed through the duct to the exterior pore, Figure 17.24. Parts of the male duct are variously differentiated as *vas deferens*, *atrium*, and, in some species, *ejaculatory duct;* there is usually a *prostate gland* which appears either as a distinct structure appended to the atrium or as a cellular covering on the atrium—in which case the prostate is said to be diffused. In some genera a true intromittent organ, or *penis*, is present with or without a chitinous sheath. The ovaries lie in a different segment from the testes and have relatively little accessory equipment; the female genital ducts are paired but are very inconspicuous.

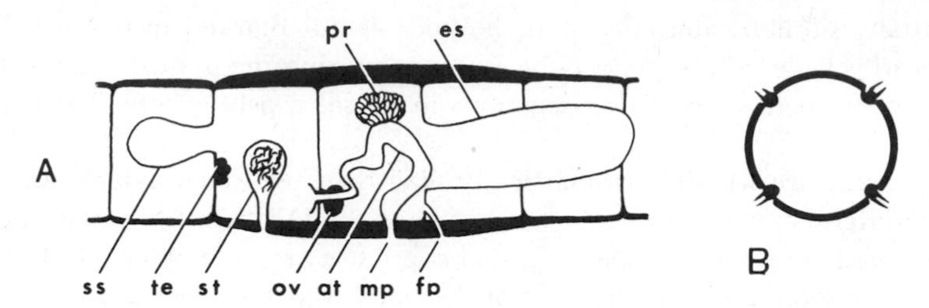

Fig. 17.24. Annelida, Oligochaeta. A, genital segments in schematic longitudinal section. B, schematic cross-section showing typical location of bundles of setae. Abbreviations are as follows: ss, anterior sperm sac; te, testes; st, spermatheca; ov, ovary; at, atrium of male duct; pr, prostate gland; mp, male pore; fp, female pore; es, egg sac.

It is not possible at this time to present either a comprehensive list of local marine and brackish water oligochaetes or a useful summary for the identification of the known species. Work in progress or actually in press would shortly render any such attempt obsolete. Details necessary for species identification can sometimes be determined in transparent or cleared whole mounts or even in life, but frequently more involved techniques are necessary, including special preparation of material and microscopic sectioning. In lieu of a more extended account this text will consider local distributions in general terms. Notes on technique are appended for students interested in taking up the rather special study of these worms.

Local marine oligochaetes are perhaps best known as littoral or semi-terrestrial animals; most are euryhaline, and some occur in fresh water as well as salt. They are particularly common in or under eelgrass or seaweeds stranded in the upper intertidal zone or under stones or other debris. *Enchytraeus albidus* has been reported in terrestrial situations apparently as a result of the use of eelgrass as a fertilizer. Marine oligochaetes also occur below tide level. Cook (1969) lists seven species from depths of 7–46 m in Cape Cod Bay, and *Clitellio arenarius*, a species that may be found in spring seepages along the shore, has been reported in 130 m.

The species that concern us belong in the Order Plesiopora; three families are represented, Naididae, Tubificidae, and Enchytraeidae. For a recent account of the Naididae see Brinkhurst (1964). Brinkhurst (1965, 1966) and Cook (1969) deal with the tubificids with about a dozen and a half species in nine genera locally. At the moment (March 1970) *Enchytraeus albidus* and *Lumbricillus lineatus* are the only enchytraeids reported locally in the literature (Welch, 1917), but additional species are known from work in progress. A comprehensive account of aquatic Oligochaetes of the world with contributions by Brinkhurst and Cook is also in press, and the same authors are preparing keys for New England species under the sponsorship of the

Systematics-Ecology Program at the Marine Biological Laboratory, Woods Hole.

For systematic work the worms may be examined initially in glycerin. Brinkhurst (1963a,b), working with tubificids, recommends fixing in 5% formalin or 70% alcohol. For study, worms are transferred from 70% alcohol to 30% alcohol, to water, and then to Amman's lactophenol which has the following formula:

Carbolic acid	400 gm
Lactic acid	400 gm
Glycerol	800 gm
Water	400 gm

After several hours in this solution the worms are put on a microscope slide and flattened with a cover slip. Pennak (1953) suggests an even simpler technique in which the live worm is placed on a slide under a cover slip; any water present is drawn off; the worm is flattened, and glycerin admitted under the slip. These techniques are mainly for the demonstration of setae but may satisfactorily reveal details of the genital ducts, depending on the transparency of the animal. Setae may be prepared from larger specimens by allowing them to macerate for a day or so in water. Further study of the genital system requires microscopic sections.

References

Bell, A. W. 1958. The anatomy of the oligochaete *Enchytraeus albidus*, with a key to the species of the genus *Enchytraeus*. Am. Mus. nat. Hist. Novitates. 1902:1–13.

Brinkhurst, R. O. 1963a. Notes on the brackish-water and marine species of Tubificidae (Annelida, Oligochaeta). J. mar. biol. Ass. U.K.:709–715.

———. 1963b. Taxonomical studies on the Tubificidae (Annelida, Oligochaeta). Int. Revue Gesamten Hydrobiologie, Syst. Behfte 2:7–89.

———. 1964. Studies on the North American aquatic Oligochaeta I; Naididae and Opistocystidae. Proc. Acad. nat. Sci. Phila. 116:195–230.

———. 1965. Studies on the North American aquatic Oligochaeta II; Tubificidae. Proc. Acad. nat. Sci. Phila. 117:117–172.

———. 1966. A contribution to the systematics of the marine Tubificidae (Annelida, Oligochaeta). Biol. Bull. mar. biol. Lab. Woods Hole. 130:297–303.

———, and D. G. Cook. 1966. Studies on the North American aquatic Oligochaeta III; Lumbriculidae and additional notes and records of other families. Proc. Acad. nat. Sci. Phila. 118:1–33.

Cook, D. G. 1967. Studies on the Lumbriculidae (Oligochaeta) in Britain. J. Zool. Lond. 153:353–368.

———. 1969. The Tubificidae (Annelida, Oligochaeta) of Cape Cod Bay with a taxonomic revision of the genera *Phallodrilus* Pierantoni, 1902, *Limnodriloides* Pierantoni, 1903, and *Spiridion* Knollner, 1935. Biol. Bull. mar. biol. Lab. Woods Hole. 136(1): 9–27.

Dales, R. P. 1963. Annelids. Hutchinson University Library.
Goodnight, C. F. 1959. Oligochaetes. *In* Fresh-Water Biology. W. T. Edmondson, ed. John Wiley and Sons.
Moore, J. P. 1905. Some oligochaeta of New England. Proc. Acad. nat. Sci. Phila. 57:373–399.
Pennak, R. W. 1953. Fresh-Water Invertebrates of the United States. Ronald Press.
Stephenson, J. 1930. Oligochaeta. Oxford Univ. Press.
Welch, P. S. 1917. Enchytraeidae (Oligochaeta) of the Woods Hole region, Mass. Trans. Am. microsc. Soc. 36:119–138.

Class Hirudinea
LEECHES

Diagnosis. *Annelids with a distinct sucker at each end of the body, the anterior one sometimes small; without setae or parapodia but some species with conspicuous branchiae, tubercles, or lateral vesicles; body cylindrical, often divided into two regions, with segmentation obscure externally but with more or less distinct annulations; coelom largely obliterated by parenchyma.*

The largest marine leeches attain a length of about 200 mm, the smallest about 10 mm when full grown. Variations in external appearance are shown in Figure 17.25. All leeches have 33 segments, but external indications may be obscure or ambiguous. Segmentation is indicated most reliably by the structure of the nervous system. For practical purposes the segments can usually be determined externally by reference to the serial repetition of color pattern elements, or of such structural features as vesicles or branchiae, or sometimes by the presence of whitish dots called *sensillae* that may be apparent in transverse rows on the middle annulus of each segment, Figure 17.25B. The number of annulations per segment ranges from 3 to 14. Eyes may be present or not.

Leeches are hermaphroditic. The male and female pores are single and open on the tenth or eleventh segments. Cross-fertilization is the rule, and spermatophores are attached near the female openings. Egg cocoons are attached to the substratum which may be rock, plant, or the living host. The cocoons are rounded, and the outside is spongy or ornamented in various genera. Larval development takes place within the cocoon, and development after hatching is essentially direct without distinctive free-living phases.

Most leeches leave their host after a meal of blood, but *Callobdella lophii* remains more or less permanently attached to its host, the angler fish *Lophius*.

Marine forms attack a wide variety of teleost and elasmobranch fishes, crustaceans, and pycnogonids. *Ozobranchus* has been reported on the loggerhead turtle, *Caretta caretta*. Prey and predator are indicated in the systematic list. In some cases, leeches appear to be fairly or quite host-specific; others feed on a wide variety of organisms. *Platybdella*, for example, attacks various fishes as well as shrimp and pycnogonids.

Systematic List

Marine leeches belong to a single family, Piscicolidae, within the order Rhynchobdellae. The marine forms on this coast are not well known. The following list from Knight-Jones (1962) has some genera not actually recorded on this coast; they are included on the basis of host lists, i.e., if the recorded host occurs within the range of this text the genus is listed. Additional genera may be encountered, and for a more complete coverage see Knight-Jones.

Genus	Host
Arctobdella	On artic halibut
Branchellion	On flattened elasmobranch fishes, i.e., skates and rays
Calliobdella	On *Lophius*, gadoids, dogfish, etc.
Crangonobdella	On *Sclerocrangon boreas* (a shrimp)
Johanssonia	On wolffish, also on the pycnogonid *Nymphon*
Myzobdella	On mysid shrimp, the blue crab *Callinectes*, etc.
Notostomobdella	On nurse shark, greenland halibut
Oceanobdella	On blennies, cod, lumpfish *Cyclopterus*
Ozobranchus	On the loggerhead turtle *Caretta caretta*
Piscicola	On sea lamprey
Platybdella	On wolffish, labrids, *Macrozoarces*, halibut, and on the crustaceans *Sclerocrangon* and *Hyas*
Pontobdella	On flattened elasmobrachs
Pterobdellina	On flattened elasmobrachs
Trachelobdella	On a variety of teleosts

Identification

Material for study should be narcoticized by gradually (over a period of a half hour) adding 70% alcohol to water containing the leech. When relaxed, the leech should be extended, held flat with a glass plate, and fixed in seawater formalin or, as a preliminary to sectioning, in Bouin's or Flemming's. See Mann (1962) for further notes on technique. In some darkly pigmented leeches the eyes, if any, may be obscured; such material can be bleached in caustic potash to reveal the eyes in greater contrast. The following key is adapted from that of Knight-Jones (1962). Dimensions are maximum adult sizes.

---

1. Body with branchiae, tubercles, or lateral vesicles—2.
1. Body smooth or with lateral fins—3.
2. With branchiae; with or without tubercles:
- Branchiae finger-shaped, branching; Fig. 17.25D. *Ozobranchus* spp.
- Branchiae leaf-shaped, unbranched; Fig. 17.25C. *Branchellion* spp.

2. Without branchiae; with tubercles; annuli 4–6 per segment, 1 annulus bearing large tubercles, the others with small tubercles; body regions ill-defined; 200 mm; Fig. 17.25F. *Pontobdella* spp.
2. Without branchiae; with 11–13 vesicles on each side:
- Body regions ill-defined; posterior sucker wider than body, 2 genera: most species of *Calliobdella* have no eyes and 3 faintly double annuli on each midbody segment, while *Piscicola* has 2 eyes and 7 faintly divided annuli; both genera Boreal or Arctic.
- Body distinctly divided in two regions; warm water leeches; Fig. 17.25A. *Trachelobdella* spp.

3. Body with a lateral fin on each side extending the full length of the body; 40 mm; Fig. 17.25E. Reported from deep water >400 m. *Pterobdellina* spp.
3. Body without lateral fins—4.
4. With mouth at or near center of oral sucker—5.
4. With mouth in anterior border of sucker; 150 mm. *Notostomobdella* spp.
5. Posterior sucker little or no wider than rear end of body; a pair of eyes on the tenth segment; on blue crabs, *Callinectes sapidus*. *Myzobdella* spp.
5. Posterior sucker much wider than anterior one—6.
6. Midbody segments with 12–14 annuli:
- Without eyes. *Johanssonia* spp.
- With 2 or 3 pairs of eyes. *Crangonobdella* spp.

6. Midbody segments with 2–6 annuli—7.

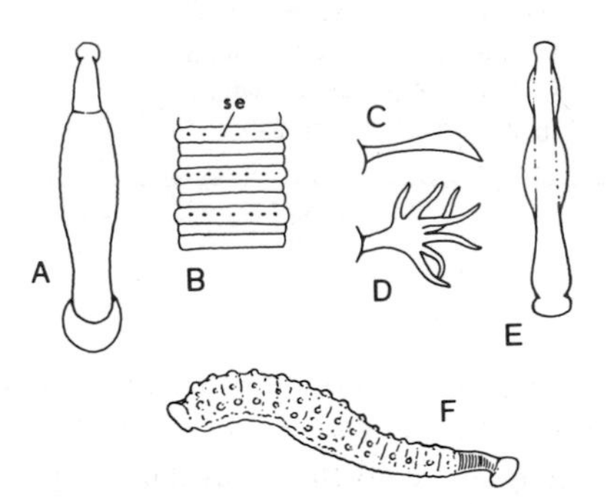

Fig. 17.25. Annelida, Hirudinea. A, *Trachelobdella* sp., whole worm; B, generalized body region with three segments, each with three annulations, one of which has sensillae (se); C, *Branchellion* sp., branchia; D, *Ozobranchus* sp., branchia; E, *Pterobdellina* sp., whole worm; F, *Pontobdella* sp., whole worm.

7. Without eyes; coelom relatively spacious and not divided into small pockets; Arctic. *Arctobdella* spp.
7. Posterior sucker usually with a ring of ocelli; skin transparent; 11–25 mm; with 3 pairs of eyes; Boreal and Arctic. *Oceanobdella* spp.
7. Coelom reduced to small cavities; with 2 or 3 pairs of eyes, or none in one species; Arctic and Boreal species chiefly but one (with 2 eyes) on the oceanpout, *Macrozoarces*. *Platybdella* spp.

References

Knight-Jones, E. W. 1962. The systematics of marine leeches. *In* Leeches. K. H. Mann, ed. Pergamon Press.

Mann, K. H. 1962. Leeches (Hirudinea), Their Structure, Physiology, Ecology and Embryology. Pergamon Press.

Meyer, C. M., and A. A. Barden, Jr. 1955. Leeches symbiotic on Arthropoda, especially decapod Crustacea. Wasmann J. Biol. 13:297–311.

Pinto, C. 1923. Ensaio Monographico dos Hirudineos. Rev. Mus. Paulista. 13:857–1118.

Silva, P. H. D. H. de. 1960. A key to the genera of Piscicolidae. Ceylon J. Sci. (Biol. Sci.). 3:223–233.

EIGHTEEN

Phylum Sipuncula

SIPUNCULAN WORMS

The sipunculans were long allied systematically, as an ordinal group, with annelid worms or in the polyphyletic assemblage Gephyrea with priapulid and echiurid worms. They are now generally accorded phylum rank.

Diagnosis. *Unsegmented, wormlike animals with a two-part body consisting of a slender, retractile introvert followed by a thicker trunk; the introvert with a terminal mouth more or less surrounded by tentacles or lobes; introvert and trunk otherwise without appendages; anus opens anteriorly on the dorsal side of the trunk.*

The highly contractile introvert may be fully withdrawn into the stouter trunk, which is itself contractile; the extent of contraction affects the shape of the posterior end of the trunk, which may be blunt-ended, tapered, or even "tailed" in consequence. The surface of the body may be smooth or ornamented with spines or papillae. Adults range in size (trunk length) from less than 4 mm to more than 300 mm.

Sipunculans are burrowing or otherwise sedentary, benthic, non-parasitic worms; they are exclusively marine. Feeding habits have not been well studied. Sipunculans apparently scavenge organic particles from ingested sand and mud, and there is also evidence that they are mucous-ciliary feeders.

These worms have some power to regenerate lost or damaged parts but do not reproduce asexually. The sexes are separate, *Golfingia minuta* being the only known exception; however the sexes are not conspicuously different externally. Females frequently predominate in collected samples. Fertilization is external, and the offspring of some species develop as planktonic trochophore larvae before assuming benthic habits and metamorphosing into the adult form. This metamorphosis is usually accomplished while the animal is less than 1 mm in length. Planktonic life may last but 1 or 2 days, or several months. Some species have direct development with no planktonic stage; see Åkesson (1958), Rice (1967).

Systematic List and Distribution

The basic list is from Gerould (1913), but the taxonomy has been revised to conform with Fisher (1952) and Stephen (1964). Family divisions have not been established.

Aspidosiphon parvulus	C+	29 m
Golfingia eremita	B	64 to 2009 m
Golfingia improvisa	V	9 to 1482 m
Golfingia verrillii	V	6 to 29 m
Phascolion strombi	BV	1 to 1942 m, eury., to 15‰
Phascolopsis gouldii	B, 39°	Lit. to several meters, eury.
Themiste alutacea	C+	29 m

Additional species of *Golfingia*, e.g., *margaritacea*, *procera*, and *sabellariae* are found in deeper waters offshore; *sabellariae* was recorded in 128 m off Chatham, Cape Cod, but other records are from 587 to 2481 m. The other two species are reported from depths >183 m. *G. eremita* ranges south of Cape Cod in depths >800 m. Several new species, also from deep water, are currently (April 1969) in press under the authorship of E. B. Cutler.

Identification

With due allowance for difficulties resulting from the contractile nature of these animals, local species usually may be differentiated by external criteria. Dimensions given are adult trunk length in extended state.

1. With distinct shield at each end of trunk, the anterior one with short spines, the posterior one divided into rounded or squarish plates; introvert shorter than trunk, emerging from ventral side of anterior shield, covered with minute, slender hooks (visible at 60X magnification); yellowish or grayish brown; 4 mm; Fig. 18.1A. *Aspidosiphon parvulus*. Additional deep water species of this genus are known.

1. Without shields—2.

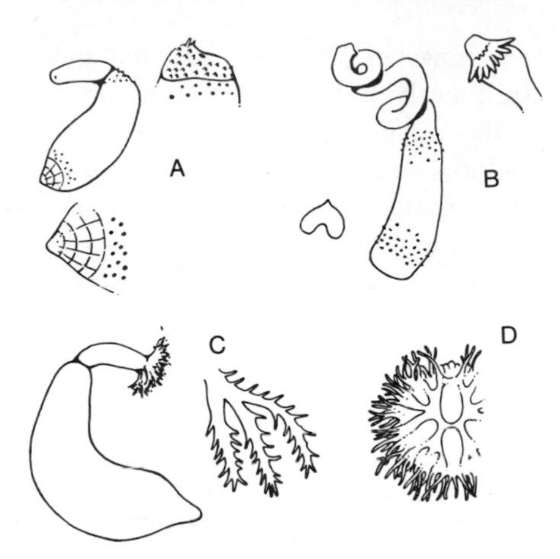

Fig. 18.1. Sipuncula. A, *Aspidosiphon parvulus*, whole animal and details of anterior (right) and posterior (below) ends of the trunk; B, *Phascolion strombi*, whole animal and details of anterior end (right) and enlarged chitinous hook (left); C, *Themiste alutacea*, whole animal and detail of individual tentacle; D, *Phascolopsis gouldii*, frontal view of part of tentacular whorl.

2. With a circle of tentacles around mouth; with zonally arranged bands of papillae or hooks on trunk and introvert as follows: a band of small hooks behind tentacles; a zone of papillae at junction of trunk and introvert; a zone of papillae crowned with triangular, heart-shaped, chitinous hooks or holdfasts on posterior half of trunk, and sometimes with additional papillae at end of trunk; color white to brown tinted with yellow, orange, or gray; 25 mm; Fig. 18.1B. *Phascolion strombi.* The species is variable in form and color with the long, slender introvert often twisted spirally in adaptation to the internal shape of the gastropod shells which this worm often inhabits; less twisted specimens occur in *Dentalium* shells or sand tubes of their own construction.

2. Not so—3.

3. With 4–6 large, palmate tentacles each divided into 2 or 3 main stems and smaller branches; the introvert smooth anteriorly, posteriorly with recurved hooks; trunk pear-shaped, microscopically furrowed transversely, and with a pattern of rectangular blocks at the base of the introvert; brownish to pale pink; 10 mm; Fig. 18.1C. *Themiste alutacea.*

3. Tentacles fingerlike or filiform, sometimes arranged in a double whorl—4.

4. Skin smooth and lustrous; elongate; pale-colored; longitudinal muscles gathered in bands; 300 mm; Fig. 18.1D. *Phascolopsis gouldii.* Note that the deep water *Golfingia margaritacea* is similar but has longitudinal muscles in a continuous sheet.

4. Skin covered entirely or in part with papillae or hooks; longitudinal muscles in a continuous sheet. *Golfingia* (the generic name commemorates a golfing holiday at St. Andrews, Scotland, involving the naturalist Lankester and a Professor MacIntosh): *G. eremita* has a rough skin with minute, fingerlike papillae; body encircled by parallel ridges and furrows; tentacles 22–24; 20 mm. *G. verrillii* has both trunk and introvert covered with prominent papillae; tentacles 34; 15 mm;

dark gray or brownish with darker papillae. *G. improvisa* has distinct papillae on posterior end of trunk and on introvert and a girdle of hooks behind the tentacles; otherwise superficially smooth, but with exceedingly small papillae on both trunk and introvert; 10 mm. Gerould (1913) notes the possibility of confusing contracted, preserved specimens of this species with *Aspidosiphon*.

References

Åkesson, B. 1958. A study of the nervous system of Sipunculoidae with some remarks on the development of the two species *Phascolion strombi* Montagu and *Golfingia minuta* Keferstein. C. W. K. Gleerup. Lund. (Unders, över Oresund, 38).

Andrews, E. A. 1890. Notes on the anatomy of Sipunculus Gouldii Pourtales. Stud. Biol. Lab. Johns Hopkins Univ. 4:384–430.

Cutler, E. B. 1967. Sipuncula of the western North Atlantic, their systematics, distribution, and ecology. Doctoral Dissertation. University of Rhode Island.

Fisher, W. K. 1950. The sipunculid genus *Phascolosoma*. Ann. Mag. nat. Hist. 12(3):547–552.

———. 1952. The sipunculid worms of California and Baja, California. Proc. U.S. natn. Mus. 102:371–450.

Gerould, J. H. 1913. The sipunculids of the eastern coast of North America. Proc. U.S. natn. Mus. 44(1059):373–437.

Hyman, Libbie H. 1959. The Invertebrates: Smaller Coelomate Groups. Vol. V. McGraw-Hill Book Co.

Rice, Mary E. 1967. A comparative study of the development of *Phascolosoma agassizii*, *Golfingia pugettensis*, and *Themiste pyroides* with a discussion of developmental patterns in the Sipuncula. Ophelia 4(2):143–171.

Stephen, A. C. 1964. A revision of the classification of the phylum Sipuncula. Ann. Mag. nat. Hist. 13 ser. 7:457–462.

NINETEEN

Phylum Echiurida

ECHIURID WORMS

This small group of marine worms was formerly allied with sipunculan and priapulid worms in the now abandoned class Gephyrea, phylum Annelida.

Diagnosis. *Unsegmented worms, cylindrical or ovoid in section with a two-part body consisting of a proboscis and trunk; the proboscis nonretractile but mobile and extensible with a gutter or trough ventrally; the trunk relatively short and stout, smooth or ornamented with spines or papillae, superficially ringed or not.*

Local species range from less than 25 mm to more than 300 mm in total length. The mouth lies at the base of the proboscis. The anus is terminal. The sexes are separate, but little different in local forms; in some extralimital genera the males are diminutive, planarialike forms parasitic on or in the body of the female. In these species, fertilization may be internal, but in our species the eggs are shed in the water. The initial larvae—of *Echiurus* and presumably of *Thalassema*—are trochophores. They have been reported to be pelagic for several weeks before undergoing a progressive metamorphosis to the adult form.

Echiurids are exclusively marine. They are benthic, burrowing worms. *Thalassema viridis* was reported from hard nodules of blue clay in the Gulf of Maine. *Thalassema mellita*, however, is "endemic" in the tests of the Carolinian sand dollar *Mellita*. Echiurids are ciliary feeders, sweeping microorganisms along the trough of the proboscis to the mouth.

Systematic List and Distribution

The distribution of echiurids is poorly known on this coast. The classification to family follows Bock (1942); Dawydoff (1959) questions the value of recognizing ordinal categories for this small assemblage.

Family Echiuridae	
Echiurus echiurus	AB

Miner (1950) stated that this species is found (southward?) "from Casco Bay . . . along the entire eastern seaboard," but it does not appear in any of the local faunal lists. Dawydoff (1959) gives the distribution of the genus as "les mers tempérées de la region subarctique."

Family Thalassemidae		
? *Thalassema viridis*	B	141 m
Thalassema mellita	C+	—
Thalassema hartmani	C, 37°	—

Identification

1. With 2 circles of bristles at posterior end; trunk with about 22 spinous rings; color grayish to yellowish or orange; total length about 300 mm. Fig. 19.1A. *Echiurus echiurus.*
1. No posterior bristles; trunk not ringed; Fig. 19.1B. *Thalassema* spp: *T. mellita* is reddish with a yellow preoral lobe; trunk about 25 mm long; *T. viridis* is bright

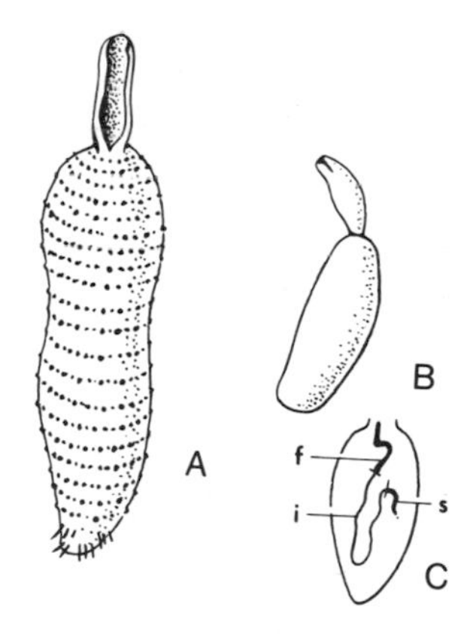

Fig. 19.1. Echiura. A, *Echiurus echiurus*, whole animal; B, *Thalassema*, whole animal; C, *Thalassema hartmani*, schematic view of trunk showing relative length of intestine. Abbreviations are: f, foregut; i, intestine; s, postintestinal segment.

green, about 6 mm in length; *T. hartmani* is also reddish, about 40 mm long but is distinguished by a very long intestine equal to about 5 times the length of the foregut.

References

Bock, S. 1942. On the structure and affinities of *Thalassema lankesteri* and the classification of the group Echiuroids. Göteborgs K. vetenskaps och vitterhetssammhälles Handl., SB2(6):1–94.

Conn, H. W. 1886. Life history of *Thalassema*. Stud. Biol. Lab. Johns Hopkins Univ. 3:351–401.

Dawydoff, C. 1959. Class des Echiuriens. *In* Traité de Zoologie. Tome 5. P.-P. Grasse, ed. Masson et Cie.

Fisher, W. K. 1946. Echiuroid worms of the North Pacific Ocean. Proc. U.S. natn. Mus. 96(3198):215–292.

———. 1947. New genera and species of echiuroid and sipunculoid worms. Proc. U.S. natn. Mus. 97(3218):351–372.

Miner, R. W. 1950. Field Book of Seashore Life. G. P. Putnam's Sons.

Monro, C. G. A. 1927. On the families and genera of the class Echiuroidea. Ann. Mag. nat. Hist. (9)20:615–620.

Spengel, J. W. 1912. Revision der Gattung Echiurus. Zool. Jb. Abt. Syst. 33:173–212.

Verrill, A. E. 1879. Recent additions to the marine invertebrata of northeastern coast of America, with descriptions of new genera and species and critical remarks on others. Proc. U.S. natn. Mus. 2(1):165–205.

Wharton, L. D. 1913. A description of some Philippine Thalassemae with a revision of the genus. Philipp. J. Sci. 8(D):243–270.

TWENTY

Phylum Arthropoda, Part One

JOINTED-LEGGED ANIMALS—NONCRUSTACEAN CLASSES

Introduction

This is the most important marine phylum in terms of numbers of species and, barring protists and perhaps nematodes, Arthropoda leads in abundance of individuals as well. The total number of arthropod species in the present fauna, composed largely of members of the class Crustacea, is approximately double that of Mollusca, the next largest phylum.

Except for a few highly modified forms, arthropods are not likely to be confused with members of other groups; the exceptions are chiefly parasitic, larval, or microscopic. The jointed exoskeleton may be soft and transparent, but its segmented structure in body regions as well as appendages is usually obvious.

The arthropod classes of interest to this account are grouped in three subphyla:

Subphylum Pycnogonida	
Class Pantopoda	Sea spiders

Subphylum Chelicerata	
Class Merostomata	Horseshoe crabs
Class Arachnida	Mites, spiders, pseudoscorpions
Subphylum Mandibulata	
Class Crustacea	Crustaceans
For ordinal classification	see Chapter 21
Class Insecta	Insects
For ordinal classification	see Table 20.3

Of the remaining largely terrestrial classes, Pauropoda and Symphyla are absent from littoral habitats, and Diplopoda very nearly so. Members of the latter class, more popularly known as millipeds, include at least one Italian species that is intertidal, and another from Florida may also be so. A growing number of centipedes (Chilopoda) belonging to the order Geophilomorpha is being discovered in intertidal beach habitats, but none appears to have been reported locally.

The extent to which such essentially terrestrial or limnetic arthropods as insects and spiders inhabit marine and estuarine habitats is impossible to determine fully from the available literature. The data are scattered in systematic works and regional check lists which may or may not consistently take note of such distributions.

Four possible situations must be considered in evaluating the presence of a "terrestrial" arthropod in a salt water habitat:

1. The species is endemic in the habitat.
2. Its ecological tolerances may be sufficiently wide to permit its occurrence there without precluding a wider distribution.
3. Occupancy of the habitat may be temporary and dependent on transiently favorable conditions.
4. The occurrence may be purely fortuitous and essentially irrelevant to the normal ecology of the species.

No attempt will be made to provide full check lists of spiders, mites, or insects. It is desirable, however, to give some indication of their occurrence and to provide the means for distinguishing them from other forms.

Tables 20.1 and 20.2 summarize differential characters that may be used in conjunction with Figure 20.1 to facilitate identification to class. The appendage formulae are subject to numerous qualifications and exceptions, the most serious of which are indicated. The tables should not be interpreted as indicating precise homologies; a discussion of this esoteric problem is more likely to elicit confusion than promote enlightenment and would in any event provide little help in identifying specimens.

Table 20.1. Body Divisions

Pantopoda (1)	Proboscis	Prosoma	Opisthosoma
Merostomata	—	Prosoma	Opisthosoma
Arachnida (2)	Gnathosome	Prosoma	Opisthosoma
Crustacea (3)	Head	Thorax	Abdomen
	Cephalothorax		Abdomen
Insecta	Head	Thorax	Abdomen

(1) Distinction between prosoma and opisthosoma obscure, and segmentation partly vestigial.

(2) Prosoma and opisthosoma fused in Acari; segmentation obscure in prosoma of Pseudoscorpionida and in prosoma and opisthosoma of Araneae.

(3) Body segmentation obscure in some lower Crustacea; carapace covering all or part of cephalothorax and thorax in many.

The most anterior appendages of chelicerates and of pycnogonids when the full complement is present are nominally *chelate*, i.e., "with pinchers." Neither pycnogonids nor chelicerates have jaws or *mandibles*. The *chelicerae* of arachnids are sometimes minute and are usually modified for piercing and sucking; the conspicuous and quite literally pincherlike *chelae* of pseudoscorpions are actually modified pedipalps. The *chelifores* of pycnogonids may

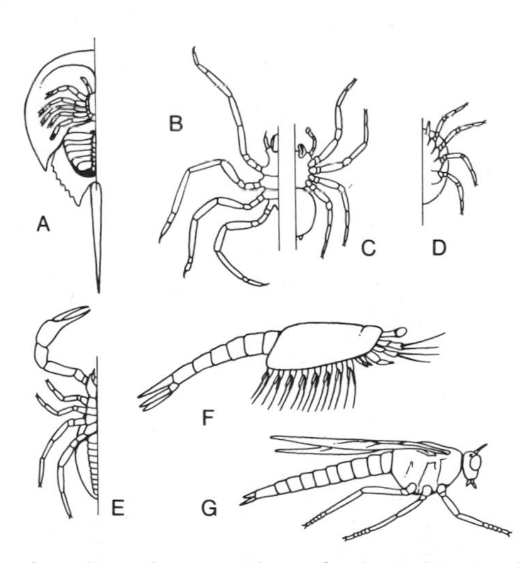

Fig. 20.1. Arthropoda, schematic comparison of principal types. A, *Limulus*; B, pycnogonid; C, spider; D, mite; E, pseudoscorpion; F, crustacean; G, insect.

Table 20.2. Arthropod Appendage Formulae

	Subphylum				
	Pantopoda	Chelicerata		Mandibulata	
	Class				
Appendage	Pycnogonida	Merostomata	Arachnida	Crustacea	Insecta
Cephalothoracic appendages	Chelifores*	—	—	1st antennae*	Antennae
	Palpi*	Chelicerae	Chelicerae	2nd antennae	—
	Ovigers*	Legs I (4)	Pedipalpi	Mandibles	Mandibles
	Legs I	Legs II	Legs I	1st maxillae	Maxillae
	Legs II	Legs III	Legs II	2nd maxillae	Labrum
	Legs III	Legs IV	Legs III	1st maxillipeds	Legs I*
	Legs IV	Legs V	Legs IV*	2nd maxillipeds	Legs II (1)
	(2)	—	—	3rd maxillipeds/ legs I	Legs III (1)
	(2)	—	—	Legs II, etc.	—
Abdominal appendages	—	5 book gills	—	0–5 pleopods	(3)
Posterior end	—	Tail spine or telson	Spinnerets (in spiders)	Caudal rami, telson, 1–3 uropods	Cerci, caudal setae, ovipositors, stings, claspers,etc.

*Sometimes lacking. (1) Plus wings in some. (2) Legs V and VI in extralimital groups. (3) In larvae or primitive groups. (4) Pedipalpi.

be distinctly chelate or not, or they may be absent altogether. The mouth parts of pycnogonids are at the tip of the proboscis and consist of a three-lobed biting "lip" plus an internal filtering apparatus. In *Limulus* (Merostomata) the bases of the appendages serve as a masticatory mill. In all of these groups there is a trend toward simplification of the body plan, including the suppression of body segmentation through fusion of the somites. In Arachnida, for example, there are two main body regions in spiders and essentially only one in mites, while in pycnogonids the body is so reduced that "the animal gives the impression of consisting merely of a bunch of legs." The major part of the alimentary and reproductive organs are in the legs of pycnogonids and consist of diverticula from the virtually nonexistent body.

Among mandibulates the most anterior appendages are antennae, and the body plan is normally trifid with a head, thorax, and abdomen. In many crustaceans the "head" is actually a *cephalothorax* produced by the fusion of the head proper and one or more thoracic segments. In dorsal aspect the crustacean *carapace* may cover head and thorax.

Adult insects almost invariably have three pairs of thoracic legs and lack abdominal appendages except terminally. Larval forms in higher insect orders present exceptions to these generalizations; most of these larvae can be described as more or less wormlike.

For a more detailed account of variations in crustacean appendage formulae see Table 21.2. Note that adult members of a few "entomostracan" subclasses have only four pairs of thoracic appendages. Larval crustaceans of the nauplius type have only three pairs of appendages, but the body plan of this stage is not trifid. The appendages of adult entomostracans and of larval crustaceans are usually at least partly biramous. The exceptions do not present a serious source of confusion between mandibulate, chelicerate, and pycnogonid subphyla.

References

Additional references are given at the conclusion of each major systematic account.

Manton, S. M. 1969a. Introduction to Classification of Arthropoda. *In* Treatise on Invertebrate Paleontology. Part R, Arthropoda 4, Vol. 1. R. C. Moore, ed. Geol. Soc. Amer. and Univ. Kansas.

———. 1969b. Evolution and Affinities of Onychophora, Myriapoda, Hexapoda, and Crustacea. *In* Treatise on Invertebrate Paleontology, Part R, Arthropoda 4, Vol. 1. R. C. Moore, ed. Geol. Soc. Amer. and Univ. of Kansas.

Stormer, L. 1959. Arthropoda—General Features. *In* Treatise on Invertebrate Paleontology, Part O, Arthropoda 1. R. C. Moore, ed. Geol. Soc. Amer. and Univ. of Kansas.

Subphylum Pycnogonida
Class Pantopoda
SEA SPIDERS

Diagnosis. *Small arthropods with a prosoma that usually consists of little more than a base for the 4 pairs of walking legs; legs with 8 segments plus, in most species, a terminal claw; abdomen or opisthosome reduced to a small knob; head with a proboscis and sometimes with 1 or 2 pairs of appendages, the chelifores and palpi; a third pair of appendages, the ovigers, present in males, next in sequence to the chelifores and palpi and less strongly developed than the walking legs.*

The largest shallow water species are about 50 mm in leg span. Species from the Continental Slope may be ten times that size.

The sexes are separate, but sexual dimorphism, except for the smaller size in males, is not conspicuous. Males of all local species have *ovigerous* appendages which are employed in carrying the eggs. Females of some genera also have these appendages. The young commonly have only three pairs of walking legs at hatching.

Pycnogonids are benthic, slow-crawling animals. A few, e.g., *Anoplodactylus* and *Endeis* spp., have become "planktonic" through associaiton with pelagic *Sargassum*. The mouth parts are located on the proboscis and are rather peculiar. They have been described as being adapted for "sucking and sifting." Quite a few species are associated with hydroids or other sessile cnidarians, and in *Phoxichilidium* the larvae are parasitic in the gastric cavity of the hydroids *Tubularia* and *Syncoryne*. Shallow water pycnogonids, though often abundant, are usually small in size and are easily overlooked.

Systematic List and Distribution

Hedgpeth's 1948 review of western North Atlantic pycnogonids is the source of this list. The orders have not been adequately defined.

Family Nymphonidae		
Nymphon hirtipes	A, 43°	46 to 399 m
Nymphon grossipes	B, 41°	22 to 1239 m
Nymphon longitarse	B	29 to 783 m
Nymphon stromi	BV	13 to 959 m
Nymphon macrum	B+	64 to 1543 m

Family Ammotheidae		
Achelia spinosa	B, 41°	Lit.
Achelia scabra	B	Lit. to 82 m
Family Tanystylidae		
Tanystylum orbiculare	V	Lit. to 27 m
Family Phoxichilidiidae		
Phoxichilidium femoratum	B	Lit. to 101 m
Anoplodactylus petiolatus	V	On *Sargassum*
Anoplodactylus parvus	V	22 to 82 m
Anoplodactylus lentus	V	Lit. to 274 m
Family Endeidae		
Endeis spinosa	V	On *Sargassum*
Family Pycnogonidae		
Pycnogonum littorale	B, 41°	Lit. to 1482 m
Family Pallenidae		
Callipallene brevirostris	BV	15 m
Pseudopallene circularis	A, 44°	Lit. to 101 m

Additional species are known in most of these families from the Continental Slope, plus members of the family Colossendeidae, not known from waters shallower than 155 m in the region treated.

Identification

Measurements equal approximate body length of adult females.

A. Chelifores and palpi lacking; males alone with ovigers.

- Body slender; auxiliary claws present; 6.5 mm; Fig. 20.2A. *Endeis spinosa.*
- Body stout; auxiliary claws absent; 5.0 mm; Fig. 20.2B. *Pycnogonum littorale.*

B. Palpi more strongly developed than chelifores; males and females with ovigers.

- Chelifores reduced to inconspicuous stubs; 0.7 mm; Fig. 20.2C. *Tanystylum orbiculare.*
- Chelifores 2-jointed; 1.0 mm; Fig. 20.2D. *Achelia:* In *A. spinosa* the auxiliary claws are at least half as long as the principal claws; in *A. scabra*, they are less than a third as long.

C. Palpi lacking.

- Chelifores stout; males and females with ovigers; *Callipallene* (Fig. 20.2E) has auxiliary claws, *Pseudopallene* does not. *Callipallene* 1.5 mm; *Pseudopallene* 1.8 mm.
- Chelifores slender; males alone with ovigers; *Phoxichilidium* (Fig. 20.2F) has well-developed auxiliary claws; *Anoplodactylus* has only vestigial ones (Fig. 20.2G). *Phoxichilidium* 1.4 mm; *Anoplodactylus* 0.5–4.0 mm; for the differentiation of the 3 local species of the latter genus see Hedgpeth (1948).

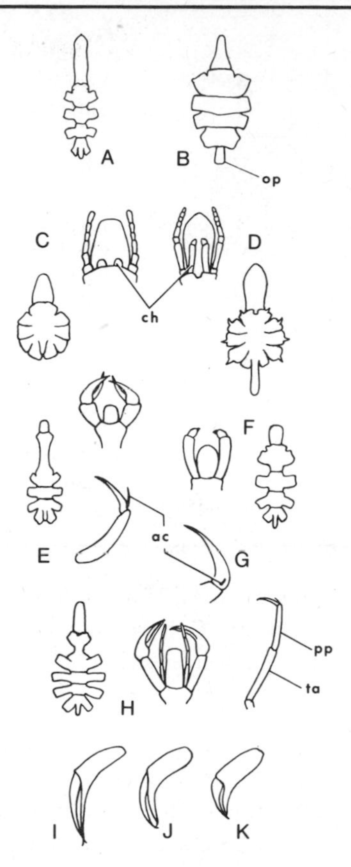

Fig. 20.2. Arthropoda, Pycnogonida. A–F,H show the body outline including prosoma, opisthosoma, and proboscis with appendages omitted; C–F,H also include details of proboscis and appendages, other details as indicated; A, *Endeis spinosa;* B, *Pycnogonum littorale*; C, *Tanystylum orbiculare*; D, *Achelia* sp.; E. *Callipallene* sp., with detail showing auxiliary claw; F, *Phoxichilidium* sp.; G, *Anoplodactylus* sp., detail showing auxiliary claw; H, *Nymphon* sp., with individual leg; I, *Nymphon macrum*, chela; J, *Nymphon stromi*, chela; K, *Nymphon grossipes*, chela. Abbreviations are as follows: ac, auxiliary claws; ch, chelifores; op, opisthosome; pp, propodus; ta, tarsus.

D. Chelifores more strongly developed than palpi; males and females with ovigers; 6–12 mm; Fig. 20.2H. *Nymphon:*

- Tarsus no more than half as long as propodus. *Nymphon hirtipes.*
- Tarsus twice as long as propodus. *Nymphon longitarse.*
- Tarsus about as long as propodus: In *N. grossipes* (Fig. 20.2K) the fingers of the chelae are relatively short and thick; in *N. stromi* (Fig. 20.2J) and *N. macrum* (Fig. 20.2I) they are long and slender. In *stromi* the claws are at least 4 times as long as the auxiliary claws; in *macrum* they are no more than twice as long.

Reference

Hedgpeth, J. W. 1948. The Pycnogonida of the western North Atlantic and the Caribbean. Proc. U.S. natn. Mus. 97:157–342.

Subphylum Chelicerata
Class Merostomata (Xiphosura)
HORSESHOE CRABS

The only local representative of this largely extinct class is *Limulus polyphemus.* Its appearance is unique and confusion with any other organism is unlikely.

Diagnosis. *With a horseshoe-shaped prosoma; hexagonal opisthosoma armed at each side with 6 small spines; a long, spikelike telson; a pair of compound eyes laterally and a pair of inconspicuous median ocelli; 5 pairs of "walking legs" including pedipalps; opisthosoma with 5 pairs of book gills beneath* (*Fig. 20.3*).

The chelicerae are small, and the last pair of walking legs is modified for burrowing. The mouth lies at the center of the cephalothoracic limb complex, and the bases of the pedipalps and walking legs form a spinous chewing mill. The diet consists of worms, mollusks, and other small invertebrates and algae. The tail despite its formidable appearance is harmless, although young individuals half-buried in the sand may erect the spine, presenting a hazard to bare feet. Ordinarily the telson's principal function is to aid in righting capsized individuals.

The sexes are separate; males average substantially smaller than females and, while similar in form, differ in having the first pair of legs modified to form claspers used in clinging to the opisthosoma of the female. Adult females reach a length including telson of more than 600 mm; size varies geographically. The greenish eggs are less than 3 mm in diameter and are laid at a depth of 50–70 mm in sand bars and salt marshes in the upper littoral. The hatchlings lack a telson until the first moult and are rather trilobitelike (Fig. 20.3). Adults and young are benthic, ranging in shallow coastal waters to about 23 m. Both can swim, awkwardly and in an inverted position, by beating the legs and gill plates.

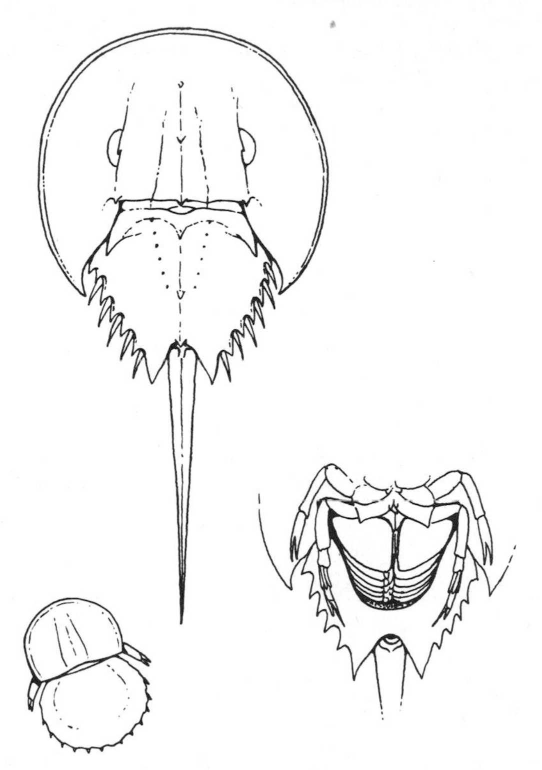

Fig. 20.3. Arthropoda, Merostomata. *Limulus polyphemus*, whole animal in dorsal view, detail of opisthosoma in ventral view (right), and hatchling (lower left).

At one time, horseshoe crabs were collected in large numbers for fertilizer and as food for domestic animals. A few are still taken for lobster bait. A more significant recent use has been in physiological research on vision. Horseshoe crabs are locally abundant on parts of the Middle Atlantic coast.

Living members of the class are contained in a single order, Xiphosurida, and family, Limulidae. The fossil history of *Limulus* itself extends back to the Triassic; its subordinal relatives date back to the Devonian. *Limulus polyphemus* ranges from about 45°N to the Gulf of Mexico.

References

Kingsley, J. S. 1892–3. The embryology of *Limulus*. J. Morph. 7:35; 8:195.
Patten, W., and W. A. Redenbaugh. 1900. Studies on Limulus. J. Morph. 16:1–26.

Savory, T. 1964. Arachnida. Academic Press.
Schuster, Carl N., Jr. 1950. Observations on the natural history of the American horseshoe crab, *Limulus polyphemus*. Third Report, Investigation of Methods of Improving Shellfish Resources, Mass. Pp. 18–22.
———. 1957. Xiphosura. *In* Treatise on Marine Ecology and Paleoecology. J. W. Hedgpeth, ed. Geol. Soc. Amer. Mem. 67(1):1171–1174.

Class Arachnida
Order Pseudoscorpionida (Chelonethida)
PSEUDOSCORPIONS

Diagnosis. *Small arachnids with a prosoma covered dorsally by an unsegmented carapace and with a broadly attached opisthosoma of twelve well-defined segments, the twelfth consisting, however, of a minute circum-anal ring; prosoma with six pairs of appendages as follows, four pairs of walking legs, one pair of powerfully chelate pedipalpi, and one pair of chelate chelicerae of two segments; opisthosoma without appendages; body length generally 1–3 mm.*

The pedipalpi, in addition to their grasping and manipulative function, serve as a venom-injecting apparatus in many species (absent in Chthoniidaea). These animals are not noxious to man. Psuedoscorpions produce silk, spinning nests with the aid of spinning organs on the chelicerae.

The sexes are separate with some dimorphism, particularly in the Cheliferoidea, in the form of the pedipalp, abdomen, and basal segments (coxa) of the fourth legs. There is a suspicion of parthenogenesis in some species; in others there is a courtship "dance" and a transfer of spermatophores followed by internal fertilization. The young are brooded by the female whose ovaries become modified to produce a milklike nutrient. The initial larva is described as "parasitic" on the mother, but there is no distinct, free-living larval stage. The young, called nymphs, resemble the adults and have similar habits.

Pseudoscorpions are terrestrial, air-breathing arachnids. A few species occur in the upper littoral where they may be common, though easily overlooked, under surface debris. Pseudoscorpions are predatory chiefly on other arthropods. The prey is pierced and sucked dry with a complex oral apparatus. For a full account of the biology of the group see Chamberlin (1931).

Systematic List and Distribution

There are few published data on the littoral distribution of pseudoscorpions. Some of the species are extremely wide-ranging and evidently ecologically adaptable. One of these, *Lamprochernes oblongus*, has been recorded from the upper littoral at Woods Hole, Massachusetts. *Dinocheirus tristis* may be more strictly maritime; it was described from Long Island, but no further information has been published about its coastal distribution. Both *Lamprochernes* and *Dinocheirus* belong to the family Chernetidae. Bousfield (1962) lists *Chthonius tetrachelatus* (family Chthoniidae) from the littoral in the Bay of Fundy region.

Species of Garypidae have been reported from littoral situations in the southern United States but not from north of Cape Hatteras. Several additional species found at Beaufort, North Carolina, may occur much farther north. They are *Dinocheirus tumidus*, which may be conspecific with *D. tristus* above, *Serianus carolinensis* reported by Muchmore (1968) in dune areas, and *Parachernes littoralis* (Muchmore, in press) from drift material above high water.

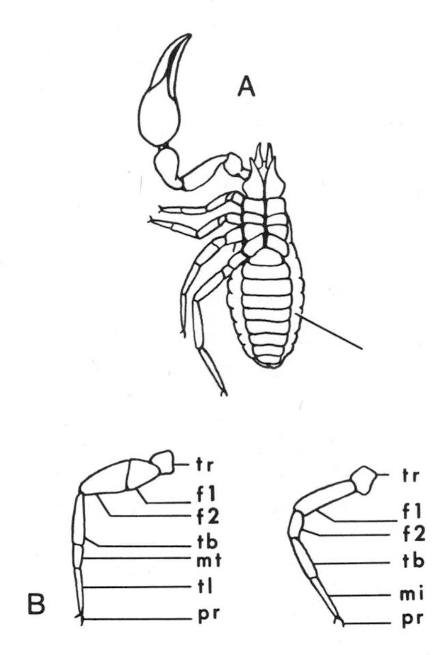

Fig. 20.4. Arthropoda, Pseudoscorpionida. A, *Lamprochernes*, whole animal in ventral view with appendages on right side removed, with pleural membrane indicated; B, *Chthonius* sp., showing segmentation of legs III and IV (left), and legs I and II (right). Abbreviations are as follows: f1 and f2, femur; mi, miotarsus (meta- and telotarsus combined); mt, metatarsus; pr, praetarsus; tb, tibia; tl, telotarsus; tr, trochanter.

Identification

The animals are small and must be studied under magnification. The clarity of fine details may be enhanced by soaking in glycerin. The segmentation of the legs is important in establishing family groups; the femur is usually divided into 2 segments, and the tarsus may or may not be divided. Dimensions are body length, i.e., prosoma plus opisthosoma.

---

1. Segmentation of legs I and II similar to that of legs III and IV, tarsus not divided; all legs with 5 segments not counting the coxae or praetarsus—2.
1. Legs I and II with 5 segments, tarsus not divided; legs III and IV with 6 segments, tarsus divided; 1.5 mm; Fig. 20.4B. *Chthonius tetrachelatus.*
2. Carapace smooth; pleural membrane smoothly striate; tibia of leg IV with strong tactile seta; 2.5 mm; Fig. 20.4A. *Lamprochernes oblongus.*
2. Carapace granular; pleural membrane granularly striate; tibia of leg IV without strong tactile seta; 2.4 mm. *Dinocheirus tristis.*

References

Bousfield, E. L. 1962. Studies on littoral marine arthropods from the Bay of Fundy region. Bull. natn. Mus. Can. 183:42–62.
Chamberlin, J. C. 1931. The arachnid order Chelonethida. Stanford Univ. Publs. biol. Sci. 7(1):1–284.
Hoff, C. C. 1947. The species of the pseudoscorpion genus *Chelanops* described by Banks. Bull. Mus. comp. Zool. Harv. 98(2):473–550.
———. 1949. The pseudoscorpions of Illinois. Bull. Ill. nat. Hist. Surv. 24:413–498.
———. 1958. List of the pseudoscorpions of North America north of Mexico. Am. Mus. Novitates. 1875:1–50 (including a key to genera).
Muchmore, W. B. 1968. A new species of the Pseudoscorpion genus *Serianus* (Arachnida, Chelonethida, Olpiidae) from North Carolina. Ent. News. 79(6):145–150.

Order Acari (Acarina, Acarida)
MITES

For arguments regarding the ordinal-subclass status of Acari, see "Systematic List."

Diagnosis. *Microarachnids with suppressed body segmentation; the body regions combined in a single, globular, or oval unit and with the mouth parts (chelicerae and* *pedipalps) borne on a projecting structure, the gnathosoma ("capitulum" is an equivalent term, now considered obsolete); adults with four pairs of legs. Fig. 20.1 D.*

The marine forms are small; most do not exceed 0.7 mm, and the adults of the majority of local salt water species are between 0.3 and 0.4 mm. Indications of segmentation are, at best, subdued in the body regions but are obvious in the appendages.

The sexes are separate, but dimorphism is confined mainly to the genital apparatus. Development is essentially direct, without profound metamorphosis. The larvae have only 6 legs; nymphs have 8 legs but are sexually immature.

The mouth parts are formed for piercing and sucking. The principal marine family, Halacaridae, includes both predaceous and herbivorous species; very few are parasitic. Species parasitic on insects occur in freshwater groups, and members of the family Astacocrotonidae (Parasitengona) are parasitic on crabs.

The acari are terrestrial as well as aquatic in fresh and salt water. The marine forms range from the littoral to depths of 5200 m.

Systematic List and Distribution

Mites are essentially microscopic and are not treated here in detail. The group has traditionally been treated as an order; see, for example, Petrunkevitch (1955) and Savory (1964). Newell (personal communication) strongly favors elevation of Acari to subclass, but the present author is hardly qualified to make the further adjustments in the disposition of higher arachnid taxa that this change would require. With apologies to Newell, we therefore follow the older nomenclature.

Only two of the taxa next in sequence concern us; they are Trombidiformes and Sarcoptiformes. Sarcoptiformes has only a few genera in marine habitats—*Pontoppidania* and *Hyadesia* on littoral seaweeds and *Ameronothrus* intertidally and in waters of reduced salinity. Within Trombidiformes, *Nanorchestes amphibius* is described as "practically marine" and probably worldwide in distribution, while *Pseudoleptus* and *Dolichotetranychus* are reported from salt grasses "throughout the world." In addition, members of the family Erythraeidae may be expected on sandy shores. The "cohort" Parasitengona with 27 families comprises the principal group of freshwater mites; a very few genera, e.g., *Thyopsis*, *Hydryphantes*, *Pontarachna*, and *Nautarachna*, have estuarine or marine representatives.

The principal family of marine mites is the Halacaridae. Of the forms listed by Newell (1947), 70% are Boreal and occur in both Europe and America. The halacarids occur on a variety of substrata from the intertidal to deep water. Several species are particularly associated with the littoral mussels *Mytilis* and *Modiolus*. *Halixodes* and *Parhalixodes* are parasitic on chitons and nemerteans, respectively. A few halacarids occur in estuarine and fresh waters.

Newell (1947) gives a detailed account of the Halacaridae, and freshwater mites are covered in Newell (1959) and Pennak (1953). In addition, Baker and Wharton (1952) presented a comprehensive survey of all the Acari; this account treats the great majority of aquatic species as a polyphyletic assemblage, the "Hydrachnellae," acknowledging that this is an ecological rather than properly taxonomic concept.

Identification

Mites require special treatment for study, including dissection, in anticipation of which they should be preserved in 65–70% alcohol and not in formalin. Table 20.3 may serve to differentiate the principal groups.

Table 20.3.

	Number of Segments in	
	Legs	Pedipalps
Trombidiformes		
Parasitengona	6 + 2 claws	5
Pleuromerengona	6–7 + 2 or 3 claws	3–4
Sarcoptiformes	4–5 + 3 or 1 claw	2–5

References

Baker, E. W., and G. W. Wharton. 1952. An Introduction to Acarology. Macmillan, New York.

Banks, N. 1904. A treatise on the Acarina or Mites. Proc. U.S. natn. Mus. 28:1–114.

Habeeb, H. 1967. A checklist of North American water-mites. Leafl. Acadian Biol. 43:1–8.

Newell, I. M. 1947. A systematic and ecological study of the Halacaridae of eastern North America. Bull. Bingham oceanogr. Coll. 10(3):1–266.

———. 1959. Acari. *In* Fresh Water Biology. W. T. Edmondson, ed. John Wiley and Sons. Pp. 1080–1116.

Pennak, R. W. 1953. Fresh Water Invertebrates of the United States. Ronald Press.

Petrunkevitch, A. 1955. Arachnida. *In* Treatise on Invertebrate Paleontology. Part P. Arthropoda. Geol. Soc. Amer. and Univ. of Kansas Press.

Savory, T. 1964. Arachnida. Academic Press.

Order Araneae
SPIDERS

Diagnosis. *Arachnids with both carapace (prosoma) and abdomen (opisthosoma) unsegmented and joined by a narrow pedicel or stalk; carapace with four pairs of walking legs, a pair of non-chelate pedipalps which are either leglike (females) or highly specialized for transfer of sperm (males), a pair of chelicerae which are two-segmented with the distal segment fanglike and modified for the injection of venom; abdomen with six (rarely two or four) small, fingerlike spinnerets terminally plus, in most groups, a minute to small structure in front of the spinnerets, called the colulus; abdomen otherwise without appendages; eyes usually consisting of eight simple ocelli. Fig. 20.1 C.*

The sexes are separate and dimorphic in external appearance; males are usually longer-legged and, in some families, much smaller and slimmer than females, with different setation, sometimes differently colored, and with pedipalps highly modified for "sperm induction." Courtship behavior is well developed. The eggs are usually brooded by the female. There are no distinct larval forms, and the young have the ordinal traits.

Spiders are essentially terrestrial. They are air-breathing and, in local forms, there is 1 pair of internal *book lungs* with a pair of lung slits on the ventral surface of the abdomen; there is also a tracheal system with 1 or 2 spiracles also opening on the underside of the abdomen.

Spiders are predatory, mainly on other arthropods. The prey may be snared in webs set for the purpose or run down and attacked directly. The oral mechanism is fitted for crushing and macerating the prey and sucking the liquid fraction.

Maritime Distribution

Roth (1968, personal communication) surveyed the literature on marine littoral spiders, and the following briefly reviews studies pertinent to the present region. For a systematic list see the references cited.

Barnes (1953, 1954) and Arndt (1914) discuss maritime spider distributions in North Carolina and Long Island, respectively. For a systematic account including descriptions and keys covering most of the species mentioned see Kaston (1948); this is the most complete regional study available for any part of the eastern United States. Additional mention of spiders in maritime situations appears in Proctor (1938) and Bousfield (1958). See also Kaston (1952) and the general accounts of Comstock (1948) and Gertsch (1949). Systematic representation in littoral or near-littoral situations includes the major groups of spiders with the Lycosidae, Gnaphosidae, Micryphantidae, Salticidae, and Clubionidae predominating. Web-spinning spiders included in Barnes's studies on beach or marsh forms include the families Pholcidae, Theridiidae, Argiopidae, Linyphidae, Micryphantidae, and Dictynidae.

Though there are no true marine spiders on this coast, there are forms that withstand submersion and some that deliberately seek temporarily submerged situations. The data of Arndt (1914) is specific; *Grammonata trivitatta*, one of the sheetweb spiders, shelters in the base of salt marsh grasses (*Spartina*) where air is trapped between tides. Barnes also mentions this species occurring in marshes "wet at all stages of tide." Experimentally, *G. trivittata* survived submersion for 3 days, and a clubionid, *Clubiona littoralis* remained submerged for 30 hours and survived. Species of the genera *Tetragnatha*, *Epeira*, and *Lycosa*, and of the family Salticidae living at higher levels in salt marshes can withstand up to 12 hours of submersion.

Web-building species usually spin above normal tide levels. The transient spiders that forage in the littoral during periods of emergence retreat as the water rises; of these, *Lycosa*, with several species on beaches, salt marshes, and in the rocky littoral as well, is perhaps the most frequently mentioned genus. Other such genera include *Clubiona*, *Poecilochroa*, *Habronattus*, *Gnaphosa*, *Arctosa*, *Schizocosa*, *Eperigone*, and *Erigone*, all members of the heterogeneous assemblage of essentially terrestrial invertebrates found at the upper limit of the littoral.

For the most part, maritime spiders are not endemic in coastal habitats but occupy similar situations close to or far from the sea. Thus, *Arctosa littoralis*, the principal sand beach species in Barnes's Carolina study also occurs inland in sandy fields. Similarly *Tetragnatha* species are found in grassy meadows and stream borders as well as in salt marshes.

References

Arndt, C. H. 1914. Some insects of the between tides zone. Proc. Indiana Acad. Sci. 323–333.

Barnes, R. D. 1953. The ecological distribution of spiders in non-forest maritime communities at Beaufort, North Carolina. Ecol. Monogr. 23:315–337.

———, and Betty M. Barnes. 1954. The ecology of the spiders of maritime drift lines. Ecology. 35(1):25–35.

Bousfield, E. L. 1958. Distributional ecology of the terrestrial Talitridae (Crustacea: Amphipoda) of Canada. Proc. Tenth Internatn. Congr. Ent. 1(1956):883–898.

Comstock, J. H. (revised by W. J. Gertsch). 1948. The Spider Book. Comstock Publishing Co.
Gertsch, W. J. 1949. American Spiders. D. Van Nostrand Co.
Kaston, B. J. 1948. Spiders of Connecticut. Bull. State Geol. nat. Hist. Surv. (Conn.): 5–874.
———. 1952. How to know the spiders. Wm. C. Brown Co.
Proctor, W. 1938. Biological Survey of the Mount Desert Region. Part VI. The Insect Fauna Wistar Inst.

Class Insecta
INSECTS

Diagnosis. *Arthropods with a body divided into three, usually well marked regions, head, thorax, and abdomen; the head with a single pair of antennae and mouth parts consisting of a pair of mandibles, a pair of maxillae, a labium and labrum, with or without accessory structures and often highly modified for particular feeding habits; thorax with three pairs of legs and, in some orders, with one or two pairs of wings; abdomen without jointed appendages except terminal or subterminal accessory sexual apparatus in the form of claspers, stings, ovipositors, cerci, caudal setae, etc. Fig. 20.1 G.*

Of the insect orders with aquatic species, only Hemiptera, Coleoptera, and Collembola have aquatic adults in the sense of the individual insect spending a substantial part of its time on or beneath the water surface. In other orders aquatic phases are subadult. The abbreviated classification of Table 20.4 may serve as a frame of reference for the discussion that follows; included is a summary of aquatic distributions and of types of life history. Only taxa pertinent to this discussion are included. The classification follows Rothschild (1961).

Life History

Four life history patterns occur in insects; in the first three, the immature stages are sufficiently like the adults to present no difficulty in identifying them as insects according to the diagnosis given. The fourth type is metamorphic. The types are: *ametabolous*, without distinct external changes during development except in size; *paurometabolous*, with a gradual metamorphosis involving changes in size, body proportions, and including the development of wings; *hemimetabolous*, with a gradual metamorphosis as in the preceding

Table 20.4. Summary of Insect Life History Patterns and Aquatic Distribution

Class Insecta	Common Name	Fresh	Salt	Life History
Subclass Apterygota				
Order Collembola	Springtails	+	+	am
Order Thysanura	Bristle tails	Littoral		am
Subclass Pterygota				
Order Plecoptera	Stone flies	+	−	he
Order Odonata	Dragonflies	+	+	he
Order Ephemeroptera	Mayflies	+	+	he
Order Hemiptera	True bugs	+	+	pa
Order Trichoptera	Caddis flies	+	+	ho
Order Megaloptera	Alder flies	+	−	ho
Order Neuroptera	Lacewings	+	−	ho
Order Lepidoptera	Butterflies, moths	+	+	ho
Order Diptera	Flies, mosquitoes	+	+	ho
Order Coleoptera	Beetles	+	+	ho
Order Hymenoptera	Wasps	Parasitic		ho

am = ametabolous, pa = paurometabolous; he = hemimetabolous; ho = holometabolous.

but with accessory gills in the young; *holometabolous*, with three distinctly different post-hatching stages, i.e., larva, pupa, and adult. Immature paurometabolous insects are called *nymphs*, and immature hemimetabolous insects are *naiads*. Holometabolous species present the greatest opportunity for misidentification at the phylum level since, in many cases, the young are wormlike or otherwise distinctly unlike adult members of the class. They do not conform to the diagnosis given; see general introductory keys, Chapter 3.

In some groups the appendages and general body form are highly adapted for swimming. The adaptations of primary interest, however, have to do with adjusting respiration to the marine environment. Insects in general are air breathers, and many aquatic insects in both immature and adult stages retain this habit. Collembolans are essentially surface forms but may be found beneath the surface in trapped air bubbles. Some diving insects, e.g., most aquatic Hemiptera and some Coleoptera, come to the surface periodically for air or carry with them beneath the surface bubbles of air trapped under the wings or in body hairs. In Hemiptera neither the young nor adults have gills. A few aquatic larvae in other orders, notably Diptera, are also air breathers, and are equipped with snorkel devices or with behavioral adaptations for obtaining air from the surface. Many aquatic insects, however, do have structural modifications which enable them to obtain oxygen directly from the water. Oxygen transfer may be effected through the skin either through the cuticle generally or through tracheal or blood gills. Insects have an open circulatory system; blood fills the general body cavity but is kept moving by the pulsations of an open, tubular "heart" aided by accessory structures. In some aquatic insects thin-walled, blood-filled, external projections from the body wall facilitate exchanges between the blood and the environment. These are the so-called blood gills. Tracheal gills are thin-walled projections containing ramifications of the respiratory system. Blood gills are found only in Coleoptera, Diptera, and Trichoptera, and their function may be chiefly, or perhaps exclusively, osmoregulatory rather than respiratory. Edwards (1926) noted that marine chironomids tend to lack blood gills and freshwater species have them. Salt water mosquito larvae tend to have smaller gills than freshwater species. Tracheal gills vary considerably in form and location. They are conspicuous, feathery or plate-like structures in Ephemeroptera and in some Odonata, e.g., damsel flies, while dragonflies have the gills reduced to minute projections lining the walls of a highly modified rectal chamber. In Trichoptera the gills, when present, are relatively simple filamentous or conical projections. They vary considerably in location and degree of conspicuousness in those larval beetles and flies that have them.

Young insects develop through periodic molts, the number varying in different species. With the final moult the insect becomes a mature adult and does not grow any larger. In holometabolous species the pupal stage differs in appearance from the larva as well as from the adult, and it is an inactive phase except in mosquitoes and midges. The pupa of many species resides in a cocoon or puparium. Dipteran pupae of aquatic species are also aquatic, but in most beetles the larva emerges from the water in the last stage, and the individual pupates ashore. With its final metamorphosis the insect's life changes drastically in holometabolous species. Immature insects may be crawlers, swimmers, or burrowers; the adult is usually a flying animal and is often relatively short-lived. Food habits may also change with metamorphosis, as in the case of mosquitoes; the larvae are generally filter feeders subsisting on minute plankters or detritus while adults may not feed at all or subsist on nectar or blood.

Maritime Distribution

Less than 4% of insect species are aquatic according to Pennak (1953), and only a fraction of these inhabit brackish or marine waters; Usinger

(1957) reviewed the literature on salt water insects. Insects are, nevertheless, of substantial importance in estuarine situations—one has but to consider the sometimes overwhelming abundance of noxious dipterans in coastal environments to appreciate this truth. A brief resume of the marine distributions of individual orders follows.

ORDER COLLEMBOLA

Anurida maritima is the best-known springtail found in littoral habitats. Blue-gray, about 3 mm long, and widely distributed, this species is often abundant at the tide line on a wide variety of substrata. It is found in marshes as well as on rocky shores. Dexter (1943) records 100 per square inch as a not unusual abundance. See also Imms (1906) (Fig. 20.5).

ORDER THYSANURA

Bousfield (1962) listed *Petrobius maritimus* as a littoral species in the Bay of Fundy region, and Usinger (1957) mentions a marine bristletail, *Machilis maritima*, around intertidal rocks; the latter genus is generally distributed along the Atlantic coast according to Smith (1910).

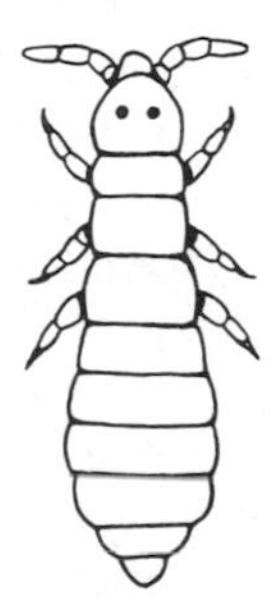

Fig. 20.5. Arthropoda, Insecta. The collembolan, *Anurida maritima.*

ORDER ODONATA

The literature on this group includes Butler and Popham (1958), Berner and Sloan (1954), Garman (1927), Osburn (1906), and Uhler (1879). These authors record larval development in brackish water in about a dozen genera of dragonflies. Of the species listed, only *Micrathyria berenice* appears to be confined to brackish water. Osburn's paper details experimental work indicating normal development in salinities of up to about 7‰ with more limited tolerances in some species to as high as 23‰ depending on the stage of development and period of exposure.

ORDER EPHEMEROPTERA

The only published notices of this group in brackish water (2–10‰) appear to be Berner and Sloan (1954) and Trost and Berner (1963) for *Callibaetis*. The records are for the southeastern United States, but the genus ranges to Canada.

ORDER HEMIPTERA

As indicated previously, this is one of the few orders with aquatic adults. There is even an oceanic species, *Halobates micans*. References include Berner and Sloan (1954), Britton (1923), Esaki (1924), Hungerford (1948), Hutchinson (1931), van Kaakon Lindberg (1936), Osburn and Metcalf (1920), Torre-Bueno (1915). Torre-Bueno listed 49 species in 9 families in stranded material on ocean beaches; the most frequently noted species were flying adults of such terrestrial families as Pentatomidae, Coreidae, and Reduviidae, and their occurrence in beach drift would appear to be fortuitous. A few aquatic hemipterans are found in salt water, including members of the families Corixidae, Saldidae, Veliidae, Gerridae, and Mesoveliidae. Corixids (water boatmen) are often encountered in brackish or even hypersaline pools, and Hutchinson (1931) reported a pond species, *Trichorixa verticalis* in plankton tows in Delaware Bay. The semiaquatic Saldidae occur on sea beaches, salt marshes, and tidal flats; Britton (1923) mentions 2 species of *Pentacora* plus *Saldula pallipes* in such habitats. The water striders are represented by species of *Rhagovelia*, *Mesovelia*, *Irepobates*, *Rheumatobates*, and *Geris* in addition to the oceanic *Halobates*, according to Britton (1923) and Berner and Sloan (1954). A few homopterans have also been reported in littoral habitats. Osburn and Metcalf (1920) cite a littoral distribution for nymphs of the salt marsh cicada, *Tibicen viridifascia*, the adults occurring on sea oats, *Unciola*, on the Carolina dunes. Jamnback and Wall (1959) listed *Coccidella maritima* as a common species on salt marsh sods along with salidids and fulgorids in lesser numbers.

ORDER TRICHOPTERA

Salt water records for caddis flies were reviewed by Siltala (1906). There is little information on American East Coast distributions. Verrill and Smith (1871) mention finding caddis flies (*Molanna*) on pier pilings in Vineyard Sound, and Flint (personal communication) writes that he has found or received from other collectors *Oecetis* (Leptoceridae) from Woods Hole and from Currituck Sound and the lower Pamlico River in North Carolina. Barton-Browne (1964) also reported a species of *Limnephilus* in brackish water.

ORDER DIPTERA

Less than half of the approximately two dozen families of dipterans with aquatic species have representatives in salt water. Three dipteran families are of primary importance in coastal habitats because they are noxious to man and frequently abundant, they are Culicidae (subfamily Culicinae, mosquitoes), Ceratopogonidae (= Heleidae) containing the sand flies or no-seeums, and Tabanidae which includes the deerflies and horseflies, of which the greenhead fly is a notable example. Aside from their nuisance effects, salt water dipterans include potential vectors of equine encephalitis; they are *Aedes cantator* and *Aedes solicitans*. Sand flies, *Culicoides*, have somewhat questionably been implicated as well. The problem is of more than academic interest because of recent outbreaks of this disease in man on parts of the heavily populated coast of New Jersey and Massachusetts.

Of four species of *Culicoides* listed by Jamnback (1965) as salt water forms, one, *C. bermudensis*, apparently does not attack man. This species, along with *C. furens* and *C. hollensis*, are salt marsh species, the larvae developing in upper littoral marsh plant associations as well as in lower *Spartina alterniflora* communities in salinities (for *bermudensis* at least) ranging from 1.2 to 15.0‰. *C. melleus* breeds in intertidal sand on bay beaches. *Culicoides* species range the entire east coast; *melleus* and *hollensis* are the most important pests on the northern coast, while the essentially tropical *furens* is more prevalent from the Carolinas southward. In South America, *furens* is the vector of a type of filariasis in man.

The salt marsh mosquitoes include members of three genera, *Aedes*, *Anopheles*, and *Culex*. Members of the first genus lay their eggs on the mud at midtide levels. The other genera lay their eggs on the surface of the water, often near salt marsh plants. Because of their economic importance, mosquitoes have received considerable attention; see Carpenter and La Casse (1955), Carpenter, Middlekauff, and Chamberlain (1946), and Headlee (1945).

Members of the family Tabanidae, commonly known as deerflies or horseflies are hardly less annoying than mosquitoes as diurnal pests. They can make parts of the barrier beaches and coastal marshes virtually uninhabitable at times. These biting flies belong to the genera *Tabanus* and *Chrysops*. The most infamous species, *Tabanus nigrovittatus*, lays its eggs on vegetation, chiefly at *Spartina alterniflora* and *Spartina patens* levels, and the larvae remain in the mud and peaty soil near the bases of these plants; see Jamnback and Wall (1959). Yet another dipteran, the stable fly *Stomoxys calcitrans* in the family Muscidae, requires mention although it is not aquatic in any stage—nor is it strictly maritime in distribution. The adults are biting flies and the larvae develop in the windrows of eelgrass cast up on bay shores.

The presence of several additional families in the local fauna requires confirmation; in Empididae and Psychodidae, for example, only some species have aquatic larvae. These are, in any event, small families with few species and they are not likely to be of much importance locally. Larval crane flies, Tipulidae, do not tolerate salt water in most cases, but Rogers (1933) reported a Florida species in *Juncus* marshes. In addition Usinger (1957) stated that "crane flies emerge from *Fucus* beds in tremendous numbers." Additional families of minor importance in local estuaries include Ephydridae and Stratiomyiidae whose occurrence in salt marshes is mentioned by Smith (1910), Curran (1934), and others. Larvae of some aquatic dipterans survive in extraordinary environments; the existence of ephydrids in hypersaline coastal lagoons and continental salt lakes is well known. They are also found in hot springs and surface pools of crude petroleum in oil fields. Some larval psychodids are noted by Curran (1934) as being tolerant of "hot water and soap." More pertinent to the present account are the chironomids, members of a large family (Chironomidae, or at times Tendipedidae) with aquatic larvae some of which are well known in salt water. Sumner, Osburn, and Cole (1913), for example, reported *Chironomus halophilus* at depths of 18 m in the Woods Hole region. *Chironomus* as also one of the few insect genera inhabiting the profundal zone of freshwater lakes.

ORDER COLEOPTERA

Walsh (1926) is a general source on the maritime distribution of beetles. According to this author, of the couple of hundred species listed from coastal habitats in Britain only a quarter are actually "halophiles." A similar impression was obtained from a perusal of the New Jersey state list of Smith (1910) in which members of at least twenty families are indicated as occurring in maritime situations. They include a relatively few listed as "strictly maritime"; nearly all belong to terrestrial families, and the information in most cases is not specific enough to determine precise habitat preferences. Such listings as "especially along the shore," "on salt meadows," "frequently in the wash-up," are little more than suggestive. In a few cases the statement is more definite, e.g., larvae of the lampyrid *Pyraciomena eostata* (according to Smith) are found "in marshes among snails." As in the case of the hemipterans discussed by Torre-Bueno (1915), a large number of essentially terrestrial species are mentioned as being found washed ashore in beach drift. Pearse, Humm, and Wharton (1942) mention a species of *Cicindela* (tiger beetles) whose larvae occur in littoral sands. The literature indicates that staphylinids are especially common shore dwellers, and Smith's list indicates at least two dozen genera of Carabidae in maritime situations. The truly aquatic beetle families do not appear to be common in saline habitats; only one genus of Dytiscidae and two of Hydrophilidae are listed by Smith in localities suggesting a tolerance for salt water.

ORDER LEPIDOPTERA

Larvae of a few genera are found on a wide variety of freshwater plants including *Vallisnera*, *Potamogeton*, *Nuphar*, *Typha*, and *Pontederia* as well as on sedges, *Scirpus*, and may thus be expected in salt to fresh water transitional habitats. Jamnback (1959) also mentions lepidopteran larvae on salt marshes.

ORDER HYMENOPTERA

Parasitic hymenopterans deposit their eggs on the aquatic larvae of other insects, including maritime species. Jamnback (1959) mentions ants as occurring on salt marsh soils.

Identification

Keys to adult insects are given in standard entomological texts such as Brues, Melander, and Carpenter (1954); for aquatic insects see Pennak (1953), and Edmondson (1959).

A short key to aquatic immature stages follows; its principal purpose is to indicate the range of larval types for comparison with members of other invertebrate taxa with which they might be confused.

---

1. Worm- or maggotlike, without jointed legs. Larvae of Diptera—2.
1. Jointed legs present—5.

2. Head retractile in prothorax, vestigial or weakly developed; mouth parts consisting of "hooks":
- Respiratory openings at end of long, retractile, caudal tubes; Fig. 20.6A. Ephydridae.
- Respiratory openings not so; Fig. 20.6C. Tabanidae: The larvae of *Tabanus* grow to more than 20 mm in length and have antennae with the terminal segments shortest; *Chrysops* larvae are less than 20 mm long and have the terminal antennal segments longest (see Jamnback and Wall, 1959).

2. Head well developed, not retractile—3.

3. With a pair of prolegs (pseudopods) on thorax and another pair at end of abdomen; Fig. 20.6D; Larvae to 30 mm long. Chironomidae.
3. Without prolegs—4.

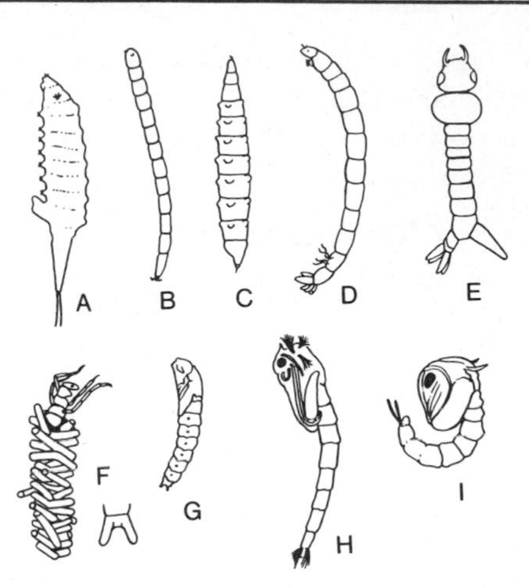

Fig. 20.6. Arthropoda, Insecta, aquatic insect larvae and pupae. A–E are larvae of dipteran families; A, Ephydridae; B, Ceratopogonidae; C, Tabanidae; D, Chironomidae; E, Culicidae; F, Trichoptera, larva, with posterior end of abdomen in dorsal view (below right); G–I are pupae of dipteran families; G, Tabanidae; H, Chironomidae; I, Culicidae.

4. With a distinctly enlarged thorax formed of fused segments; respiratory openings at end of caudal siphon; mature larvae 3–15 mm long; Fig. 20.6E. Culicidae.
4. Wormlike, thoracic segments no broader than those of abdomen; mature larvae 8 mm or less; Fig. 20.6B. Ceratopogonidae.

5. Tracheal gills wells developed, located laterally on abdomen; with 2 or 3 long, segmented, filamentous caudal appendages; mature naiads 3–38 mm long. Naiads of Ephemeroptera.
5. Not so—6.

6. With wing rudiments—7.
6. Without wing rudiments—8.

7. With sucking mouth parts. Nymphs of Hemiptera.
7. With chewing mouth parts, hidden by retracted labrum which is long and scoop-like when extended. Naiads of Odonata.

8. With 5 pairs of abdominal prolegs. Larvae of Lepidoptera.
8. With a pair of terminal abdominal prolegs or none at all—9.

9. With a pair of terminal abdominal prolegs; larva usually in a case which may be be attached to substratum or portable; Fig. 20.6F. Larvae of Trichoptera.
9. Prolegs absent; some species with well-developed tracheal gills on abdomen and some with long caudal appendages. Larvae of Coleoptera.

Some genera of aquatic lepidoptera also construct a portable case or live beneath a fixed canopy (see Pennak, 1953). Representative pupae of holometabolous forms are also figured (Fig. 20.6G–I).

References

The preceding summary is but a cursory review, and a more thorough search of the very large entomological literature would doubtless add many titles to the reference list.

Barton-Browne, L. B. 1964. Water Regulation in Insects. *In* Annual Review of Entomology. R. F. Smith and T. E. Mittler, eds. Annual Reviews, Inc.

Berner, L., and W. C. Sloan. 1954. The occurrence of a mayfly nymph in brackish water. Ecology. 35:98.

Bousfield, E. L. 1962. Studies on littoral marine arthropods from the Bay of Fundy Region. Bull. natn. Mus. Can. 183:42–62.

Britton, W. H. 1923. The Hemiptera or Sucking Insects of Connecticut. Part IV. *In* Guide to the insects of Connecticut. State Geol. nat. Hist. Surv. (Conn.). Bull. 34.

Brues, C. T., A. L. Melander, and F. M. Carpenter. 1954. Classification of Insects: Key to the living and extinct families of insects and to the living families of other terrestrial arthropods. Bull. Mus. comp. Zool. Harv. 108:1–917.

Butler, P. M., and E. J. Popham. 1958. The effects of the floods of 1953 on the aquatic insect fauna of Spurn (Yorkshire). Proc. Roy. ent. Soc. Lond. (A)33(10–12):149–158.

Carpenter, S. J., and W. J. La Casse. 1955. Mosquitoes of North America. Univ. Calif. Press.

Carpenter, S. J., W. W. Middlekauff, and R. W. Chamberlain. 1946. The Mosquitoes of the southern United States east of Oklahoma and Texas. Am. Midl. Nat. Monogr. 3.

Curran, C. H. 1934. The families and genera of North American Diptera. Ballou Press (privately printed).

Edmondson, W. T. 1959. Fresh-water Biology. 2nd ed. John Wiley and Sons.

Edwards, F. W. 1926. On marine Chironomidae (Diptera); with descriptions of a new genus and four new species from Samoa. Proc. zool. Soc. Lond. 2:779–806.

Esaki, T. 1924. On the curious halophilous water strider *Halovelia maritima* Bergroth (Hemiptera: Gerridae). Bull. Brooklyn ent. Soc. 19:29–34.

Dexter, R. W. 1943. *Anurida maritima:* an important sea-shore scavenger. J. econ. Ent. 36(5):797.

Garman, P. 1927. The Odonata or Dragonflies of Connecticut. Part V. Guide to the Insects of Connecticut. State Geol. nat. Hist. Surv. (Conn.). Bull. 39.

Håkon Lindberg, von, R. 1936. *In* Die Tierwelt der Nord und Ost see. G. Grimpe.

Hansens, E. J. 1951. The stable fly and its effects on seashore recreational areas in New Jersey. J. econ. Ent. 44:482–487.

Headlee, T. J. 1945. The mosquitoes of New Jersey and their control. Rutgers Univ. Press.

Hungerford, H. B. 1948. The Corixidae of the Western Hemisphere. Univ. Kansas Sci. Bull. 32:1–827.

Hutchinson, G. E. 1931. On the occurrence of *Trichorixa* Kirkaldy (Corixidae, Hemiptera-Heteroptera) in salt water and its zoogeographical significance. Am. Nat. 65:573–74.

Hynes, H. B.N. 1963. The Biology of Polluted Waters. Liverpool Univ. Press.

Imms, A. D. 1906. Anurida. Liverpool Mar. Biol. Comm. Mem. L3 Proc. Trans. Liverpool biol. Soc. 20:353–451.

Jamnback, H. 1965. The Culicoides of New York State (Diptera: Ceratopogonidae). N.Y. State Mus. Sci. Serv. Bull. 399:1–154.

———, and W. Wall. 1959. The common salt-marsh Tabanidae of Long Island, New York. N.Y. State Mus. Sci. Serv. Bull. 375:1–77.

Oldroyd, E. 1964. Natural History of Flies. The Norton Library.

Osburn, H., and Z. P. Metcalf. 1920. Notes on the life history of the salt marsh cicada (*Tibicen viridifascia* Walk.). Ent. News. 31:248–252.

Osburn, R. C. 1906. Observations and experiments on dragonflies in brackish water. Am. Nat. 40:395–399.

Pearse, A. S., H. J. Humm, and G. W. Wharton. 1942. Ecology of sand beaches at Beaufort, N.C. Ecol. Monogr. 12(2):136–190.

Pennak, R. W. 1953. Fresh-water Invertebrates of the United States. Ronald Press Co.

Proctor, Wm. 1938. Biological Survey of the Mount Desert Region. Part VI. The Insect Fauna. Wistar Inst.

Rogers, J. S. 1933. The ecological distribution of thee crane flies of northern Florida. Ecol. Monogr. 3(1):1–74.

Rothschild, Lord. 1961. A Classification of Living Animals. Longmans.

Siltala, A. S. 1906. Zur Trichopteren fauna des Finnischen Meerbusens. Acta Soc. Fauna Flora fenn. 28(6):1–21.

Smith, J. B. 1910. The insects of New Jersey. Ann. Rept. N.J. State Mus. (for 1909).

Sumner, F. B., R. C. Osburn, and L. J. Cole. 1913. A biological survey of the waters of Woods Hole and vicinity. Bull. Bur. Fish. Wash. 31(2):545–860.

Torre-Bueno, J. R. de la. 1915. Heteroptera in beach drift. Ent. News. 26:274–279.

Trost, L. M. W., and L. Berner. 1963. The biology of *Callibaetis floridanus* Banks (Ephemeroptera; Baetidae). Fla. Ent. 46:216–236.

Uhler, P. R. 1879. List of animals observed at Fort Wool, Va. Chesapeake zool. Lab. Scientific Results of the session of 1878: 17–34.

Usinger, R. L. 1957. Marine Insects. *In* Treatise on Marine Ecology and Paleoecology. J. Hedgpeth, ed. Geol. Soc. Amer. Mem. 67:1177–1182.

Verrill, A. E., and S. I. Smith. 1871. Report upon the invertebrate animals of Vineyard Sound and adjacent waters Rept. U.S. Fish Comm. 1871–1872:295–778.

Walsh, G. B. 1926. The origin and distribution of the coast Coleoptera of the British Isles. Entomologist's mon. Mag. 62:221–31, 257–263.

TWENTY-ONE

Phylum Arthropoda, Part Two

CLASS CRUSTACEA

Introduction

Crustaceans are so diverse an assemblage that, beyond defining them as arthropods with mandibles and 2 pairs of antennae, a useful and yet concise statement of their collective traits is impractical. The phyletic position of such highly modified animals as barnacles and the more aberrant parasitic isopods and copepods will not be immediately obvious. In forms such as ostracods and cladocerans, segmentation is suppressed or obscure, and numerous planktonic larval forms are still another source of obfuscation.

The abbreviated classification in Table 21.1 is intended as a frame of reference for the introductory text and keys that follow. Exclusively fresh-water or otherwise extralimital categories are omitted. In Malacostraca the subdivisions have been carried to progressively lower levels. The classification follows Waterman (1960) but agrees with Green (1961) and Schmitt (1965) in most essentials; for a later classification see Moore and McCormick (1969). The latter text was received after most of the work on this chapter had been completed, but the revised classification of Decapoda has been incorporated (see Glaessner, 1969).

For appendage formulae and other differential traits see Table 21.2.

Table 21.1. Classification of Crustacea

Subclass Cephalocarida
Subclass Branchiopoda
Subclass Ostracoda
Subclass Mystacocarida
Subclass Copepoda
Subclass Branchiura
Subclass Cirripedia
Subclass Malacostraca
  Series Leptostraca
    Superorder Phyllocarida
      Order Nebaliacea
  Series Eumalacostraca
    Superorder Hoplocarida
      Order Stomatopoda
    Superorder Peracarida
      Order Cumacea
      Order Tanaidacea
      Order Isopoda
      Order Amphipoda
      Order Mysidacea
    Superorder Eucarida
      Order Euphausiacea
      Order Decapoda
        Infraorder Penaeidea
        Infraorder Caridea
        Infraorder Astacidea
        Infraorder Anomura
        Infraorder Brachyura

Major Patterns of Crustacean Life History

The identification of crustaceans is complicated by the existence of a diversity of larval forms. A brief review of crustacean development will give some idea of the scope and direction of this problem.

The eggs of a few crustaceans are shed freely in the water. Most euphausiid and penaeid shrimp and many planktonic copepods and ostracods have this habit. The greater number of crustaceans, however, give some protection to their eggs. Some copepods and the cephalocarid *Hutchinsoniella* carry theirs in a sac attached to the abdomen. Marine cladocerans, some ostracods, and members of the superorder Peracarida have brood pouches; in peracarids the pouch is formed by a pair or pairs of plates called *oostegites* suspended from the female thorax—specifically, from the coxae of the pereiopods—as shown in Figure 21.6E; in ostracods and cladocerans the eggs are carried between the carapace and body proper. Stomatopods and *Nebalia* carry their eggs at the base of the thoracic legs, and free-living barnacles brood theirs in the mantle cavity. Decapods, such as the crab "in berry," hold eggs attached to the abdominal appendages.

After hatching, crustaceans develop by discrete steps or *instars*, each terminated by a moult. Increases in size occur immediately after the shedding of the old skin and before the new one has hardened.

The degree to which hatchlings approximate the adult form varies. Morphological differences between larva and adult and between successive larval stages may be so great that one would not guess them to be consecutive manifestations of the same ontogeny. In groups with "direct development" on the other hand, a new stage adds little more than minor changes to the form, setation, segmentation, or appendage formula of its predecessor. Among such groups, young animals resemble the adults at least to the extent that their subclass status is obvious. In still other groups the life history includes both direct and profoundly metamorphic phases. Gurney (1942) attempted to reduce the array of crustacean larvae to a few basic categories, and Table 21.3 summarizes the life histories of different systematic groups with an approximate correlation of the variously named developmental stages. Many of these names were originally applied as generic designations for forms now recognized as developmental phases.

Nauplius. Discounting the more bizarre sorts of nauplii found in oceanic plankton, most nauplii of neritic waters conform to a basic pattern (Fig. 21.1A). The three pairs of appendages characteristic of this stage are equivalent to the first and second pairs of antennae and mandibles; rudiments of additional cephalic appendages appear at a later stage, which is sometimes called a *metanauplius* (Fig. 21.1C).

Most nauplii found in estuaries and inshore neritic waters are the young of copepods or barnacles, though penaeid and euphausiid shrimp also have naupliar young. (Note that penaeids are essentially extralimital as breeding animals, and euphausiids are primarily oceanic in distribution; hence the larvae of these groups are not likely to be common inshore.) For other crustacean groups within our scope, the nauplius is an embryonic stage passed within the egg.

Copepodid-Protozoea. In the transition from nauplius to copepodid or to protozoea the differentiation of a generalized larva into a more specific form is begun. In copepods the free-living larvae now become recognizable as copepods; parasitic forms including branchiurans seek hosts by the second to fourth postnaupliar stage and begin a more or less profound metamorphosis. Barnacle larvae, also, are abruptly transformed into a bivalved,

Table 21.2. Crustacean Appendage Formulae and Other Differential traits

	Cephalic Appendages (1), Pairs					Thoracic Appendages (2), Pairs			Abdominal Appendages (3), Pairs					
	Antennae I (Antennules)	Antennae II (Antennae)	Mandibles	Maxillae I (Maxillules)	Maxillae II (Maxillae)	Total Thoracopods	Differentiated	Undifferentiated (Pereiopods)	Pleopods	Caudal Rami (Furca)	Telson	Uropods, No. of Pairs	Paired Eyes	Carapace-Cephalothorax
Cephalocarida	+	+	+	+	+	2/8	—	2/8	–	+	–	–	–	C
Cladocera	+	+	+	+	–	4–6	—	4–6	–	v	–	–	–	C
Ostracoda	+	+	+	+	–	1–3	—	1–3	–	+	–	–	–	C
Mystacocarida	+	+	+	+	+	5	mpI	4	–	+	+	–	–	c

Copepoda	+	+	+	+	+	6–7	mpI	5–6	–	+	–	–	(s)	c
Branchiura	+	+	+	+	+	5	mpI	4	–	+	–	–	s	c
Cirripedia	+	+	+	+	+	6	—	6	–	–	–	–	–	–
Nebaliacea	+	+	+	+	+	8	—	8	2/4	+	+	–	S	C
Stomatopoda	+	+	+	+	+	8	mpI, rc, mpII–IV	3	5	–	+	$+_1$	S	C
Cumacea	+	+	+	+	+	8	mpI, mpII, (mpIII)	5–6	0–5	–	(+)	$+_1$	(s)	C
Tanaidacea	+	+	+	+	+	8	mpI, gnI	6	0–5	–	–	$+_1$	S	C
Isopoda	+	+	+	+	+	8	mpI, (gnI), (gnII)	5–7	5	–	+	$+_1$	s	–
Amphipoda	+	+	+	+	+	8	mpI, gnI, gnII	5	3	–	+	$+_3$	s	–
Mysidacea	+	+	+	+	+	8	mpI, gnI	6	0–5	–	+	$+_1$	S	C
Euphausiacea	+	+	+	+	+	8	(4)	8	5	–	+	$+_1$	S	C
Decapoda	+	+	+	+	+	8	mpI, mpII, mpIII, (mp IV) (5)	4–5	5	–	(+)	(+)	S	C

(1) The first somite bearing appendages is actually numbered 2.

(2) "Undifferentiated" appendages are essentially similar in form and function. Nominally thoracic appendages may be modified as functional mouth parts; "differentiated" appendages are listed in order from front to back.

(3) The telson or last segment on the abdomen is not a true somite. Uropods are single pairs in most orders, but Amphipoda has 3 pairs.

(4) The last two pairs of pereiopods are vestigial in some euphausids.

(5) One or more pairs of posterior pereiopods are reduced in some crablike decapods.

Abbreviations are as follows: C, carapace; c, cephalothorax; gn, gnathopods; mp, maxillipeds; rc, raptorial claws; S, stalked; s, sessile; v, vestigial. Abbreviations or symbols in parentheses indicate presence or absence in different taxa.

Table 21.3. Crustacean Metamorphosis.

Taxon	Larval Equivalents				
	Nauplius	Copepodid or Protozoea	Zoea or Mysis	Postlarva	Adult
Cephalocarida	|Nauplius (late)	. . . . .	. . . . .	. . . . .	*Hutchinsoniella*
Branchiopoda (Cladocera)				|Postlarva	Water fleas
Ostracoda	|Nauplius	(Bivalved)	. . . . .	. . . . .	Ostracods
Mystacocarida	|Nauplius (late)	. . . . .	. . . . .	. . . . .	*Derocheilocaris*
Copepoda	|Nauplius	Copepodid	(Parasitic forms atypical)		Copepods
Branchiura	|Nauplius	(Varies)	. . . . .	. . . . .	Fish lice
Cirripedia	|Nauplius	Cypris	. . . . .	. . . . .	Barnacles
Malacostraca					
Nebaliacea				|Manca	*Nebalia*
Peracarida (some parasitic forms metamorphic)				|Manca	Various
Eucarida					
Euphausiacea	|Nauplius	Calyptopis	Furcilla	Cyrtopia	Krill Shrimp
Penaeidea	|Nauplius	Protozoea	Mysis	Mastigopus	Shrimp
Caridea		|Protozoea	Mysis	Parva	Prawns
Thalassinidae			|Mysis	Postlarva	Lobster
Astacidea			|Zoea	Megalopa	Hippids
Anomura		|Protozoea	Mysis	Postlarva	Mud Shrimp
				Glaucothoe	Pagurids
Brachyura		|Protozoea (late)	Zoea	Megalopa	Crabs
Hoplocarida		|Antizoea or Pseudozoea (late)	Erichthus or Alima	Stomatopodid	Mantis Shrimp

|Indicates hatching stage. Dotted line indicates more or less direct development.

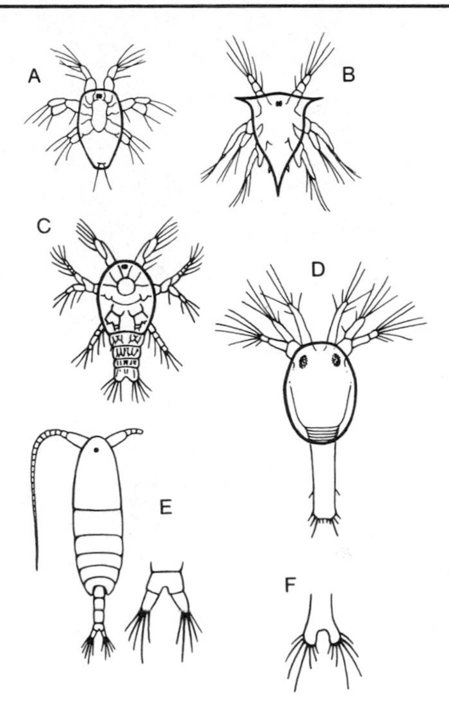

Fig. 21.1. Crustacea, naupli, copepodids. A, generalized nauplius; B, nauplius of *Balanus* sp.; C, metanauplius; D, early euphausiid larva; E, adult calanoid copepod, with detail of caudal rami; F, early penaeid shrimp larva, detail of posterior end.

ostracodlike *cypris* larva (Fig. 21.2A), that shortly selects a permanent attachment and undergoes metamorphosis to the adult form.

The early postnaupliar forms of cephalocarids and of penaeid and euphausiid shrimp (Fig. 21.1D,F), though briefly and superficially copepodid-like, shortly differentiate. The full complement of definitive traits appears gradually. The eyes are paired in the early protozoea of penaeids and euphausiids but do not become stalked until a later stage; in euphausiids luminescent organs appear in the protozoea or *calyptopis* stage. Caridean and thalassinid shrimp, and lobsters hatch at a more advanced stage of development; in these the carapace is fully fused to the thorax at hatching, but this is not true of lower shrimplike forms. Even the higher forms, however, hatch before the uropods have developed.

Zoea-Mysis. The development of shrimplike forms is essentially direct through the mysis stage, though the taxonomically definitive pattern of thoracic and abdominal appendages does not become established until the next, postlarval, stage. The superficial resemblance of decapod zoeal limbs

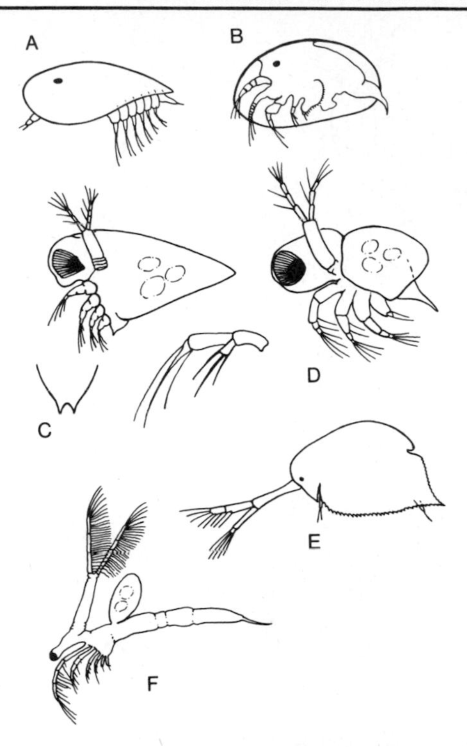

Fig. 21.2. Crustacea, Cladocera and bivalved larvae. A, cypris larva of *Balanus* sp.; B, adult ostracod; C, *Evadne* sp., whole animal and details of abdominal claws (below left) and thoracic limb (right); D, *Podon* sp., whole animal; E, *Penilia* sp., whole animal, thoracic appendages not shown; F, *Leptodora* sp., whole animal.

to those of adult mysids is, in part, the basis for calling these forms "mysis" larvae. (All local mysids, both young and adult, have a *statocyst* in the inner branch of the uropods. See key to macrocrustacea.) Crablike crustaceans, unlike the various shrimplike forms, are distinctly different in zoea and postlarva or *megalopa*, and the zoea in particular bears no resemblance to the adult form.

Preliminary Identification

For this purpose, crustaceans may be divided into two groups—microcrustacea and macrocrustacea—with the dividing point set rather arbitrarily at about 5 mm. The size differences are not absolute, as Table 21.4 indicates, and the subdivisions have no status as taxonomic categories.

Microcrustacea. This division includes developmental forms as well as adults of subclasses above (or below, in the phylogenetic sense) Cirripedia

Table 21.4. Size range of Typical Larval and Adult Small Crustaceans

(Most forms fall within range of solid line)

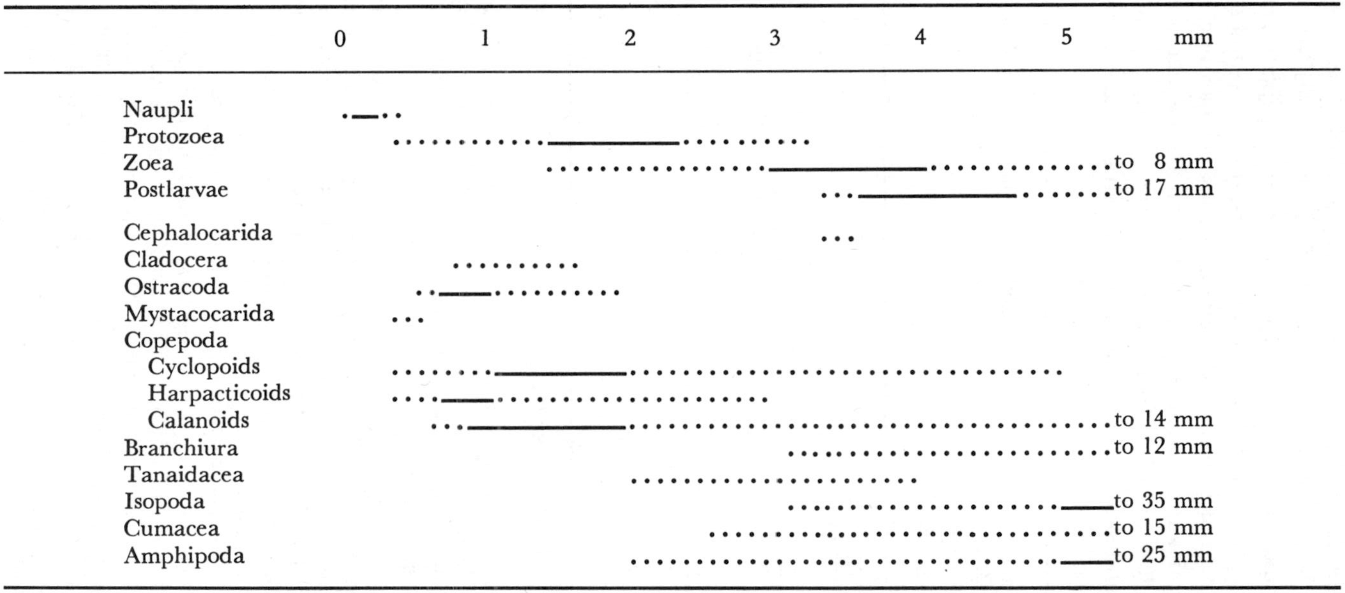

	0 1 2 3 4 5 mm
Naupli	
Protozoea	
Zoea	to 8 mm
Postlarvae	to 17 mm
Cephalocarida	
Cladocera	
Ostracoda	
Mystacocarida	
Copepoda	
Cyclopoids	
Harpacticoids	
Calanoids	to 14 mm
Branchiura	to 12 mm
Tanaidacea	
Isopoda	to 35 mm
Cumacea	to 15 mm
Amphipoda	to 25 mm

in Table 21.1. The larval period is the chief time of dispersal for most benthic and sessile crustaceans; developmental forms, are therefore primarily planktonic, as are marine cladocerans, many copepods, and some ostracods. Adult cephalocarids, mystacocarids, most ostracods, and some copepods are benthic; branchiurans and some copepods are parasitic. Microcrustaceans are important ecologically and may dominate fine net samples.

The following key includes criteria for separating the principal larval forms and adult small crustaceans. Aberrant parasitic and less common forms are omitted.

---

1. Carapace forming a bivalved shell enveloping all or most of the head and body; Fig. 21.2—2.
1. Not so—3.
2. Tabular summary:

	Antennae I	Antennae II	Mandibles	Maxillae I	Maxillae II	Pereiopods	Median eye	Terminal Abdominal Appendages
Cladocera (in part)	v	/	/	/	—	4–6	/	Terminal claws
Ostracoda	/	/	/	/	—	1–3	/	Caudal rami or furca
Cirripedia (cypris)	/	—	/	/	/	6	/	Caudal rami

v = vestigial, / = present, — = absent.

3. Animal globular or teardrop-shaped, body regions undifferentiated; 3 pairs of appendages; median eye; minute in size, 0.40–0.57 mm; Fig. 21.1A,B. Nauplii of Copepoda, Cirripedia, Euphausiacea, and Penaeidea.
3. Not so—4.
4. Cephalothoracic and abdominal regions differentiated but segmentation suppressed except in limbs; with a single, large, compound eye; carapace forming bivalved brood sac; abdomen vestigial, ending in pair of spines; adults to 1.0–1.6 mm; Fig. 21.2C,D. Cladocera in part. Note *Leptodora*, an aberrant oligohaline cladoceran with adults reaching at least 10 mm, Fig. 21.2F.
4. Body regions developed, segmentation apparent; abdomen ending in pair of *caudal rami*, or in claws, or with telson—5. (Note also *Argulus*, Fig. 21.12B.)
5. Abdomen with a pair of unsegmented caudal rami or claws, Fig. 21.1—6.
5. Abdomen with telson; larval forms of higher crustacea, Fig. 21.4, Fig. 21.5—10. (Note that some adult cumaceans lack a telson, Fig. 21.17.)
6. With carapace covering thorax; eyes stalked; Fig. 21.6G. Subadult *Nebalia;* adults reach a length of 10 mm or more, and have 8 pairs of similar thoracic appendages and 6 pairs of abdominal appendages.
6. Not so—7.

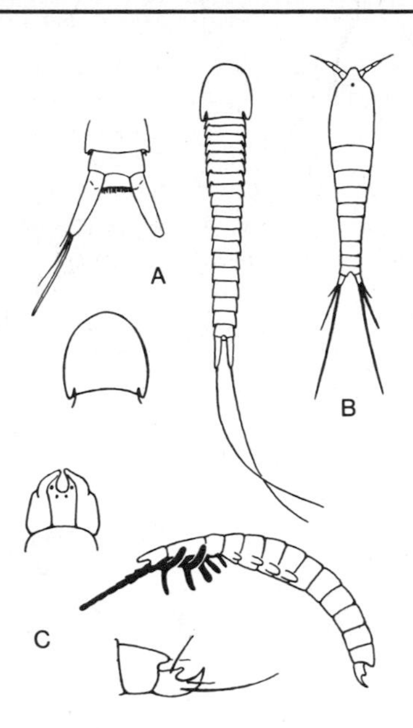

Fig. 21.3. Crustacea, Cephalocarida, Mystacocarida, and Copepoda. A, *Hutchinsoniella macracantha*, dorsal view with appendages omitted, details of posterior (above left) and anterior ends (below left); B, harpacticoid copepod; C, *Derocheilocaris typicus*, lateral view with appendages shown schematically, details of anterior end in dorsal view (above left) and posterior end in lateral view (below).

7. Abdomen ending in a pair of clawlike caudal rami; animal elongate, grublike; adults about 0.46 mm; Fig. 21.3C. Mystacocarida, *Derocheilocaris typicus*.
7. With caudal rami not clawlike—8.
8. Caudal rami with single elongate and 2 or more much shorter branches or setae—9.
8. Caudal rami with fan of 5 or 6 more nearly equal setae; Fig. 21.1E. Copepoda in part, common local forms of Calanoida, Cyclopoida, and Monstrilloida.
9. Adults about 3.5 mm; with 10 pairs of thoracic appendages, the last pair rudimentary; Fig. 21.3A. Cephalocarida, *Hutchinsoniella macracantha.*
9. Adults usually less than 1.0 mm long; with 5 or 6 pairs of thoracic appendages; Fig. 21.3B. Copepoda in part, Harpacticoida.

Developmental Forms of Higher Crustacea, including Mysidacea but Exclusive of Other Peracarida (for the latter see key to macrocrustacea)

10. Second maxillipeds enlarged as raptorial claws; Fig. 21.16. Larvae of Stomatopoda.
10. Not so—11.
11. With 1–3 pairs of legs chelate. Postlarvae and megalopae—17.
11. With no legs chelate. Zoea and protozoeal forms including Euphausiacea and Mysidacea—12. (Note postlarva and adult *Emerita*, also without chelae, Fig. 21.54.)

12. Carapace short and deep; usually with an erect dorsal spine; rostrum down-pointing; telson swallow-tailed or, if trilobed, then dorsal spine missing; length about 0.8–2.5 mm; Fig. 21.4A. Zoea of Brachyura.

12. Carapace elongate, animal shrimplike; rostrum pointing forward; telson varies in form; Fig. 21.5. Protozoea and mysis larvae of shrimplike forms—13.

13. Rostrum greatly exceeding carapace length; Fig. 21.4B; total length 3.0–5.0 mm. Zoea of anomuran "crabs," families Hippidae and Porcellanidae.

13. Rostrum about equal to or shorter than carapace—14.

14. Carapace "free," attached to no more than 3 thoracic segments; telson with inner branch of uropods with a *statocyst;* Fig. 21.40. Subadult Mysidacea.

14. Carapace attached to all thoracic segments—15.

15. Telson essentially rectangular; Fig. 21.1D and Fig. 21.5A,B. Protozoea, mysis, and postlarvae of Euphausiacea and Penaeidea.

15. Telson widening distally—16.

16. Telson with median spine or with next to outermost spines reduced to a fine "hair"; Fig. 21.5F,G. Protozoea and mysis larvae of Thalassinoidea and Paguroidea.

16. Telson without median spine or hairlike spine; Fig. 21.5C–E. Protozoea and mysis larvae of Caridea.

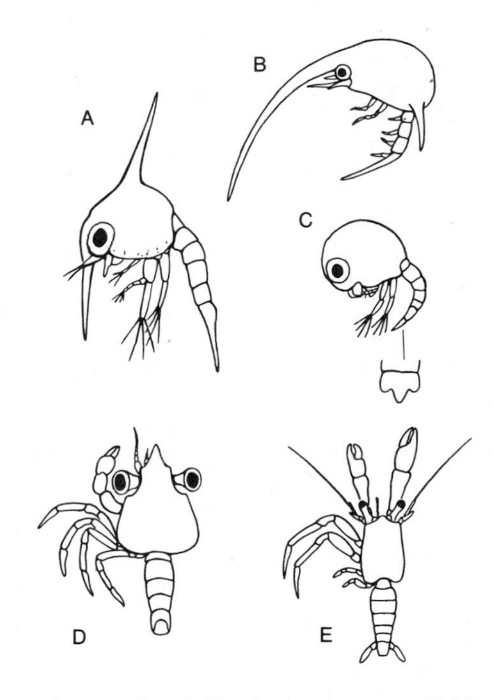

Fig. 21.4. Crustacea, larvae of crab-like forms. A, zoea of typical brachyuran crab; B, zoea of anomuran crab; C, zoea of pinnotherid crab with detail of telson; D, megalopa of brachyuran crab; E, glaucothoe of pagurid crab. A–C are lateral views; D,E are dorsal with all or part of the appendages on the right side omitted.

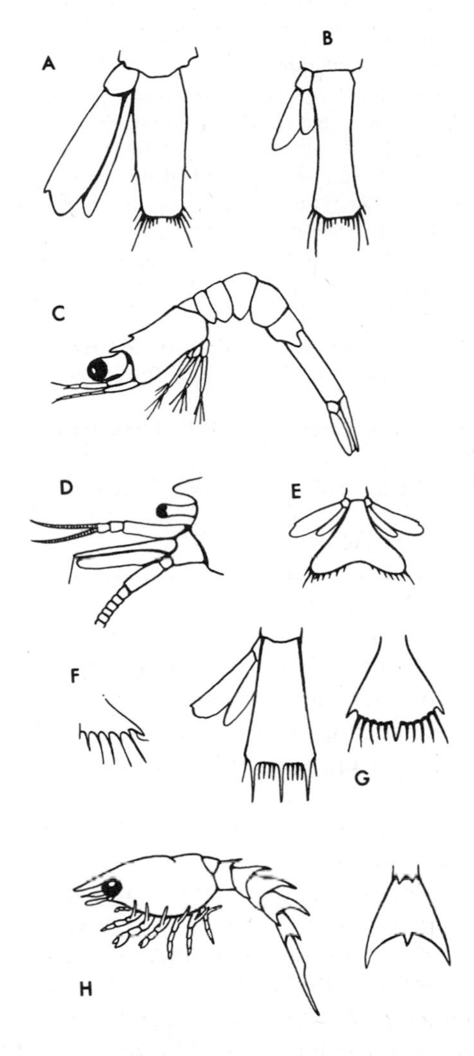

Fig. 21.5. Crustacea, larvae of shrimp-like forms. A, *Penaeus* sp., telson and left uropod; B, *Meganyctiphanes* sp., telson and left uropod; C, caridean shrimp, lateral view of whole larva; D, same, detail of anterior end, with antennal scale indicated; E, same, telson and uropods; F, thalassinid shrimp (*Upogebia*), detail of right half of telson; G, thalassinid shrimp (*Callianasa*), telson and uropod of mysis (left) and telson of protozoea (right); H, *Homarus americanus*, lateral view and detail of telson.

Postlarvae and Megalopae

17. Three pairs of legs chelate—18.
17. One or two pairs of legs chelate—19.
18. Chelae minute, first not larger than others; telson essentially rectangular, not deeply excavated behind; abdominal segments without dorsal spines; Fig. 21.5A. Postlarvae of Penaeidae.
18. First chelae stronger than second and third pairs; telson deeply excavated behind, with median spine; abdominal segments with dorsal spines; Fig. 21.5H. Mysis larvae and post larvae of *Homarus*.
19. Uropods ventral; only first pair of legs chelate; carapace wide and short; Fig. 21.4D. Megalopae of Brachyura.
19. Uropods and telson forming a fan—20.
20. Antennal scale large, well developed; first and second or second pair of legs chelate, sometimes minutely; Fig. 21.5D. Postlarvae of Caridea; see also Fig. 21.42.
20. Antennal scale small or absent; first or first and second pairs of legs chelate or subchelate—21.
21. Second through fifth pairs of legs subequal. Postlarvae of Thalassinoidea.
21. Fifth or fourth and fifth pairs of legs greatly reduced; first pair only chelate; Fig. 21.4E. Glaucothoe of Paguroidea.

Macrocrustacea. Barnacles—subclass Cirripedia—belong here because of their size; anatomically they are allied to the "entomostracan" subclasses above them in Table 21.1; note also the summary of traits in Table 21.2. Barnacles are not included in the key that follows.

The great majority of the more familiar crustaceans belong to a single subclass, Malacostraca. The members of this subclass, except Leptostraca, have 19 body segments or *somites*, which are of interest chiefly in connection with their associated appendages.

The first 5 pairs of malacostracan appendages are regarded as cephalic; the next 8 pairs are counted as thoracic. However, the first thoracic pair—first maxillipeds—usually function with the mouth parts, and in most higher Crustacea the "head" actually includes one or more thoracic segments and is properly termed a *cephalothorax*. The appendages behind the first maxillipeds may be more or less equally adapted for locomotion, in which case the general term *pereiopods* suffices; more often 1 or 2 front pairs are modified for grasping and are chelate or subchelate. In isopods, tanaids, and amphipods such limbs are called *gnathopods*. In most decapods the first 3 pairs of thoracic appendages are functionally maxillipeds. These often figure in formal keys; because of their frequently small size and crowded situation in the mouth area, they may require dissection and careful scrutiny. In most crabs the third maxillipeds cover and conceal the first 2 pairs. Conversely, in some shrimp 1 or more pairs of maxillipeds are quite leglike; to avoid confusion it is convenient to locate specific pairs of legs by counting from the rear forward. In decapods it is the remaining 5 pairs of functionally thoracic appendages behind the maxillipeds that give the decapodous formula.

The abdomen typically has 5 pairs of biramous appendages called *pleopods* (note that thoracic limbs are *pereio*pods) plus a terminal pair of *uropods* functionally allied with the flattened *telson* which is considered an adjunct to the abdomen proper. In hermit crabs the abdomen may be asymetrical and the pleopods are reduced; also, in male brachyuran crabs there are only 2 pairs of pleopods, and they are hidden under the abdomen, which has a telson but lacks uropods. Other departures from the typical pattern are indicated where they influence identification procedures.

The general introductory key to larger crustaceans which follows is intended to give direct access to convenient natural groups, and the end points are variously orders, suborders, or even superfamilies. The sequence in which these groups are considered subsequently follows the formal classification of Table 21.1. Note that barnacles and the more aberrant parasitic forms of Isopoda are excluded from the key; Figure 21.6.

A. Carapace essentially lacking, leaving thoracic segments exposed.

- **1.** With 3 pairs of uropods; body usually compressed from side to side; Fig. 21.6A. Amphipoda. Note the aberrant suborder Caprellidea, Fig. 21.6D.
- **1.** With 1 pair of uropods; body cylindrical or depressed from top to bottom—2.
- **2.** First gnathopods chelate; Fig. 21.6C. Tanaidacea.
- **2.** First gnathopods subchelate or simple; Fig. 21.6B. Isopoda.

B. Carapace short, exposing 4 or 5 thoracic segments.

- **1.** Eyes stalked; second maxillipeds developed as raptorial claws; Fig. 21.6F. Stomatopoda.
- **1.** Eyes sessile or absent; second maxillipeds not conspicuously developed; Fig. 21.6E. Cumacea.

C. Carapace covering all, or nearly all, thoracic segments.

- **1.** Abdomen ending in caudal rami; Fig. 21.6G. Nebaliacea.
- **1.** Abdomen with uropods and telson (may be closely folded under body)—2.
- **2.** Shrimp or lobsterlike forms with well-developed abdomen and caudal fan; Fig. 21.6J—3.
- **2.** Crablike or egg-shaped forms with abdomen and its appendages soft and degenerate or curled under body—8.
- **3.** Inner blade of uropods with beadlike *statocyst;* none of limbs chelate; Fig. 21.6H. Mysidacea.
- **3.** No statocyst on uropods—4.
- **4.** Five to eight pairs of thoracic legs *biramous*, none chelate; Fig. 21.6I. Euphausiacea.
- **4.** At least 1 pair of legs chelate, though the chelae may be extremely minute—5.
- **5.** Thoracic legs I–III chelate or legs IV and V absent—6.
- **5.** One or two pairs of thoracic legs chelate but never the third pair—7.

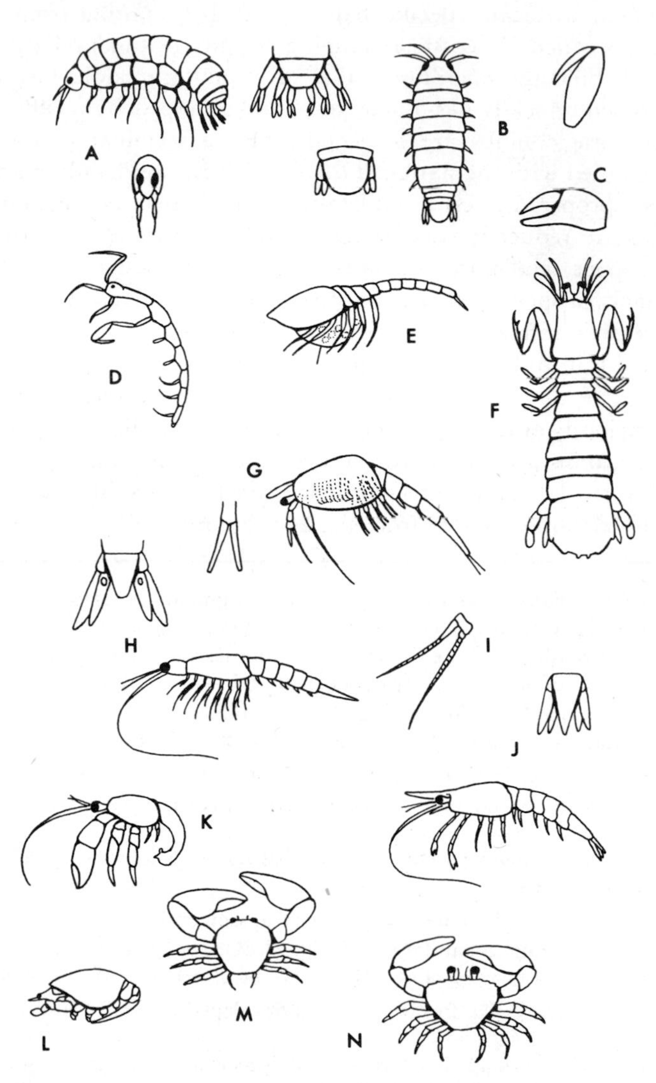

Fig. 21.6. Crustacea, principal types shown schematically. A, amphipod, lateral view of whole animal, frontal view (below), detail of telson and uropods (right); B, isopod, dorsal view of whole animal, detail of telson and uropods (left), and of subchelate gnathopod (above right); C, Tanaidacea, chelated gnathopod; D, caprellid amphipod, whole animal; E, cumacean, whole animal in lateral view with oostegite indicated; F, stomatopod, dorsal view of whole animal; G, *Nebalia* sp., lateral view of whole animal and detail of caudal rami; H, mysid, whole animal and detail of telson and uropods; I, euphausiid, detail of biramous appendates; J, caridean shrimp, whole animal and detail of telson and uropods; K, pagurid, lateral view of whole animal; L, *Emerita* sp., lateral view of whole animal; M, porcellanid crab; N, brachyuran crab.

6. Chelae not strongly developed; legs IV and V missing in some genera; Fig. 21.41. Penaeidea.

6. Chelae of first pair of legs strongly developed; Fig. 21.50. Astacidea.

7. Chelae on first pair of legs strongly developed; antennal scale weak or absent; Fig. 21.51. Thalassinoidea.

7. Chelae on first pair of legs not greatly developed or, if so, then eyes covered by carapace; antennal scale well developed; Figs. 21.5D, 21.6J, 21.45. Caridea.

8. Body egg-shaped; none of the limbs chelate; Fig. 21.6L. Hippoidea. (Note also *Euceramus*, Fig. 21.54B.)

8. No so—9.

9. Crablike in forward part of body, abdomen soft and degenerate, somewhat curled and with reduced appendages; animal "lives" in appropriated gastropod mollusk shell; Fig. 21.6K. Anomura, families Paguridae and Diogenidae.

9. Crablike with abdomen tightly curled against underside of body—10.

10. Last pair of walking legs (fourth pair behind pinchers) greatly reduced or seemingly absent; Fig. 21.6M. Anomura, families Porcellanidae and Lithodidae.

10. Last pair of walking legs proportional to others; Fig. 21.6N. Brachyura.

References

Additional references are given at the conclusion of each major systematic account.

Broad, A. C. 1957. Larval development of *Palaemonetes pugio* Holthuis. Biol. Bull. mar. biol. Lab. Woods Hole 112:144–161.

Glaessner, M. F. 1969. Decapoda. *In* Treatise on Invertebrate Paleontology, Part R, Arthropoda 4, R. C. Moore, ed. Geol. Soc. Amer. and Univ. of Kansas.

Green, J. 1961. A biology of Crustacea. Witherby Ltd.

Gurney, R. 1939. Bibliography of the larvae of decapod Crustacea. Ray Society.

———. 1942. Larvae of decapod Crustacea. Ray Society.

Heegaard, P. 1953. Observations on spawning and larval history of the shrimp *Penaeus setiferus*. Publs. Inst. Mar. Sci. 3:75–105.

Hyman, O. W. 1924. Studies on the larvae of crabs of the family Pinnotheridae. Proc. U.S. natn. Mus. 64(7):1–7.

Lebour, Marie V. 1925. The Euphausiidae in the neighborhood of Plymouth. J. mar. biol. Ass. U.K. 13:810–828.

———. 1931. The larvae of the Plymouth Caridea: I. The larvae of Crangonidae; II. The larvae of Hippolytidae. Proc. zool. Soc. Lond. 1–9.

———. 1940. The larvae of Pandalidae and the larvae of the British species of Spirontocaris. J. mar. biol. Ass. U.K.

Moore, R. C., and L. McCormick. 1969. General features of Crustacea. *In* Treatise on Invertebrate Paleontology. R. C. Moore, ed. Part R, Arthropoda 4. Geol. Soc. Am. and Univ. Kansas (not seen).

Pearson, J. C. 1939. The early life histories of some American Penaeidae, chiefly the commercial shrimp, *Penaeus setiferus*. Bull. Bur. Fish. Wash. 49(30):1–73.

Rathbun, Mary J. 1929. Can. Atlant. Fauna 10: Arthropoda; Decapoda: 1–38. St. Andrews.

Schmitt, W. L. 1965. Crustaceans. Univ. Michigan Press.

Smith, S. I. 1873. The early stages of the American lobster. Trans. Conn. Acad. Arts Sci. 2:351–381.

Snodgrass, R. E. 1956. Crustacean metamorphosis. Smithson. misc. Collns. 131(10):1–78.

Van Engel, W. A. 1958. The blue crab and its fishery in Chesapeake Bay. Comml. Fish. Rev. 20(6):6–17.

Waterman, T. H. ed. 1960, 1961. Physiology of Crustacea. Vols. 1, 2. Academic Press.

Webb, G. E. 1921. The larvae of the Decapoda Macrura and Anomura of Plymouth. J. mar. biol. Ass. U.K. 12:385–417.

Williamson, H. C. 1910. On the larval and later stages of *Portunus*, *Hyas*, *Eupagurus*, *Galathea*, *Crangon*, *Cancer*. Fish. Scotland Sci. Invest. (for 1909):1–20.

———. 1957–1960. Fich. Ident. Zooplancton. Sheet 67. Decapod larvae. Sheet 68 Caridea—Hippolytidae. Sheet 90. Caridea.

———. 1960. The larvae of Spirontocaris and related genera (Decapoda, Hippolytidae). Crustaceana 2:187–208.

Subclass Cephalocarida: *Hutchinsoniella*

Hutchinsoniella macracantha was described in 1955 as the first member of a new subclass.

Diagnosis. *Minute, rather wormlike crustacea with the head extending backward over the first thoracic segment as a short carapace; a thoracic series of eight similar segments followed by a series of eleven somewhat narrower segments, the last bearing caudal rami with three slender branches, the inner branch about half as long as the entire animal; thorax with eight pairs of biramous appendages; the ninth trunk segment with a two-jointed protuberance, the remaining posterior segments without appendages (Fig. 21.3A).*

Adults reach a length of 3.5 mm or less. *Hutchinsoniella* is hermaphroditic, producing probably 2 broods per season. Two egg sacs are carried, attached to the ninth segment; from these the young hatch as metanaupli. There are at least 18 developmental stages, but the ontogeny is essentially direct.

This animal is known from Long Island Sound to Buzzards Bay in polyhaline waters. Adults are benthic, living at depths of about 8–30 m in association with the polychaete *Nephthys* and the bivalve mollusk *Nucula* in sediments with a high silt-clay content.

References

Sanders, H. L. 1955. The Cephalocarida, a new subclass of Crustacea from Long Island Sound. Proc. natn. Acad. Sci. U.S. 41(1):61–66.

———. 1963. The Cephalocarida: functional morphology, larval development, comparative external anatomy. Mem. Conn. Acad. Arts Sci. 15:1–80.

Subclass Branchiopoda: Order Cladocera
WATER FLEAS

Members of this subclass are largely confined to fresh water. The only forms found locally in salt water are representatives of the present order.

Diagnosis. *Semitransparent microcrustacea with body segmentation obscure or absent; carapace either developed as a bivalved brood sac or enclosing most of the thorax and abdomen with only the head exposed; with a median, compound eye; abdomen reduced, without appendages except for a pair of terminal claws; 4 to 6 pairs of thoracic limbs; second antennae strongly developed, biramous; first antenna vestigial; adults reach 1.0 to 1.6 mm in length (Fig. 21.2C–F).*

The more common marine water fleas differ from most freshwater cladocerans in having a reduced carapace. Though a minor group in number of salt water species, the few marine forms are sometimes common in the plankton, reaching concentrations of several thousand individuals per cubic meter.

The sexes are separate, but parthenogenesis has been found in most species that have been studied carefully. In some of these, males are completely unknown. The young are carried in the brood sac and, when released, resemble the adults, except in *Leptodora* whose young hatch as nauplii. The last-mentioned genus is predaceous, while most other water fleas are filter feeders. Most of our forms are euryhaline, but they are often more abundant in coastal waters than in estuaries.

Systematic List and Distribution

The following list has been compiled from the plankton surveys cited in Chapter 2. Among freshwater genera likely to be encountered in the mesohaline parts of estuaries are *Daphnia*, *Bosmina*, and the atypical *Leptodora*.

Freshwater forms are covered by Pennak (1953) and Brooks (1959). All of the species listed below are planktonic. Cladocerans, like other microcrustaceans, usually have seasonal peaks of abundance, and they tend also toward cosmopolitanism in their distributions.

Family Polyphemidae	
Podon intermedius	BV
Podon polyphemoides	BV
Podon leuckarti	BV
Evadne spinifera	V
Evadne tergestina	V
Evadne nordmanni	BV
Family Sididae	
Penilia avirostris	V

Identification

Evadne and *Podon* belong to a cladoceran subgroup with only 4 pairs of thoracic legs and a reduced carapace that serves as a brood sac. *Penilia* has 6 pairs of legs and a fully developed carapace (Fig. 21.2E). *Leptodora* is also figured (Fig. 21.2F). The species of *Podon* and *Evadne* may be identified as follows.

1. Head and carapace divided by a "neck"; Fig. 21.2D. *Podon:* The exopodite of the first leg has 1 bristle (seta) in *P. intermedius*, 2 in *P. polyphemoides*, and 3 in *P. leuckarti;* the species attain lengths of 1.2 mm, 0.6 mm, and 1.0 mm, respectively.
1. Head and carapace produce an uninterrupted curve in profile. *Evadne:* The species range in total length between 1.24 mm and 1.30 mm, and may be distinguished by the number of bristles on the exopodites of the thoracic limbs as follows: *tergestina* with 2, 3, 3, 1 bristles on the first to fourth exopodites, respectively; *spinifera*, 2, 2, 2, 1; and *nordmanni*, 2, 2, 1, 1 (Fig. 21.2C). See Baker (1938).

References

Apstein, C. 1901, 1910. Cladoceren. Nord. Plankt. 4(7):11–20.

Baker, H. M. 1938. Studies on the Cladocera of Monterey Bay. Proc. Calif. Acad. Sci. 23(4):311–365.

Brooks, J. L. 1959. Cladocera. *In* Fresh-Water Biology. 2nd ed. W. T. Edmondson, ed. John Wiley and Sons.

Davis, C. C. 1955. The marine and fresh-water plankton. Michigan State Univ. Press.

Lockhead, J. H. 1954. On the distribution of a marine cladoceran, *Penilia avirostris* Dana (Crustacea, Branchiopoda), with a note on its reported bioluminescence. Biol. Bull. mar. biol. Lab. Woods Hole. 107:92–105.

Pennak, R. W. 1953. Fresh Water Invertebrates of the United States. Ronald Press Co.

Ramner, W. 1939. Cladocera, Sheet 3. Fich. Ident. Zooplancton.

Subclass Ostracoda
OSTRACODS

Diagnosis. *Microcrustacea with the body entirely enclosed in a bivalved carapaec, connected on the dorsal margin by an elastic band and tightly closed by adductor muscles.*

The anatomy of the animal is effectively concealed by the carapace which may be smooth and translucent but is more commonly pigmented and sculptured. There are 3 pairs of thoracic appendages variously adapted for grooming, copulation, locomotion, or other functions, but the chief organs for swimming and walking are usually the antennae. The abdomen ends in caudal rami, also called *furcae* or *furcal rami*, armed with setae or claws. Ostracods are blind or have a single or double eye (Fig. 21.2B).

Adults of most local species are less than 1.0 mm long and many are nearer half that length; a few forms, chiefly in the order Myodocopida, range between 1.0 and 2.0 mm. The sexes are separate and more or less dimorphic in appendage structure and sometimes in the form of the carapace. Parthenogenesis is common, particularly in freshwater forms. The eggs are brooded and, in some families, the young are carried through one or more instars; in others, the eggs are shed freely and hatch into planktonic young which, though nauplii, are bivalved at birth. The members of the family Darwinulidae are viviparous.

Most ostracods are burrowers, or are benthic on plants or among bottom debris. Members of the order Myodocopida, in particular, have planktonic inclinations but have not been reported as important plankters on this coast. Fish (1925) listed species of *Sarsiella* and *Cylindroleberis* (Myodocopida), and the podocopids *Loxoconcha* and *Cythereis* as accidental plankters in the Woods Hole area. Ostracods are presumed to be omnivorous scavengers, and the only known parasitic form, *Entocythere*, is found on crayfish.

These animals occur in fresh, brackish, and marine waters. The individual species evidently have wider tolerances to environmental variables than do most other small crustacea. Some forms, at least, have seasonal cycles of abundance.

Systematic List and Identification

The ostracod fauna of this coast is not well known and will not be treated here in detail. These animals do not generally come to the attention of collectors unless a special effort is made to find them. They require dissection for positive identification and are subject to sexual, ontogenetic, and, in some cases, environmentally induced polymorphism. Such studies as have been made suggest that they are numerous both as individuals and as species. Ten of the 26 species Cushman (1906) found at Woods Hole were described as new, and nearly half of the 13 listed by Tressler and Smith (1948) at Solomons, Maryland, were also new; the two areas had only 3 species in common. Of some thirty-odd forms from eastern Canada only 10 were noted as occurring at Woods Hole.

References

Blake, C. H. 1933. Ostracoda. Biol. Surv. Mt. Desert Region. Part 5. 229–41.

Cushman, J. A. 1906. Marine Ostracoda of Vineyard Sound and Adjacent Waters. Proc. Boston Soc. nat. Hist. 32(10):359–385.

Fish, C. J. 1925. Seasonal distribution of the plankton of the Woods Hole region. Bull. U.S. Bur. Fish. 41:91–179.

Klie, W. 1944. Ostracoda. Sheet 5. Fich. Ident. Zooplancton.

Tressler, W. L., and E. M. Smith. 1948. An ecological study of seasonal distribution of Ostracoda, Solomons Island, Maryland region. Contr. Chesapeake biol. lab. 71:1–58.

Subclass Mystacocarida: *Derocheilocaris*

Derocheilocaris typicus was described in 1943 as the first representative of a new order, since raised to subclass. Additional species are now known; see bibliography in Hessler and Sanders (1966).

Diagnosis. *Elongate, cylindrical microcrustacea with a body consisting of a cephalothorax, a postcephalosome of one segment, a thorax with four segments, and an abdomen with six segments; limbs on cephalothorax and postcephalosome well developed, the four pairs of thoracic appendages rudimentary; abdomen without appendages except clawlike caudal rami; Figure 21.3C.*

Adults are 0.5 mm or less in length. The larvae are metanauplii.

D. typicus is a benthic detritus feeder reported from the interstitial waters of intertidal beaches from Cape Cod to Miami, Florida, where individuals have been found buried 120–160 mm below the surface.

References

Hessler, R. R., and H. L. Sanders. 1966. *Derocheilocaris typicus* Pennak and Zinn (Mystacocarida) revisited. Crustaceana. 11(2):141–155.

Pennak, R. W., and D. J. Zinn. 1943. Mystacocarida, a new order of Crustacea from intertidal beaches in Massachusetts and Connecticut. Smithson. misc. Collns. 103(9):1–11.

Subclass Copepoda
COPEPODS

Diagnosis. *Microcrustacea with a cephalothorax and five or six pairs of biramous thoracic limbs; abdomen with caudal rami but otherwise without appendages; usually with a single median eye but a few, mostly oceanic forms with eyes paired and provided with conspicuous corneal lenses; some oceanic species luminescent.*

The more aberrant parasitic copepods are considered in the concluding paragraphs of this section.

In the common, free-living copepods the complement of cephalic appendages is complete (see Table 21.2). The first antennae are the most conspicuous of these appendages in most planktonic forms, but the smaller, second antennae and the thoracic appendages are the principal organs of locomotion.

The apparent body regions do not conform to standard conceptions of head, thorax, and abdomen. Calanoids and cyclopoids are bottle-shaped, the thickened forward end or *metasome* being made up of the cephalothorax and 3–6 thoracic segments; the narrower tailpiece or *urosome* has 1–5 segments which actually include one or more thoracic segments as well as those of the abdomen (Fig. 21.9). The main difficulty here is that appendages on thoracic urosomal segments may be misinterpreted as being abdominal; see, for example, *Oithona* in Figure 21.7.

Some harpacticoids follow the pattern just outlined, but in many species the body is more or less tapered, isopodlike, and without proportional differences to mark off metasome and urosome (Fig. 21.8).

The sexes are separate, and many species show distinct sexual dimorphism. This dimorphism may be manifested as assymetry of the male antennae, last thoracic appendages, and urosome, in the number and shape of body segments, and in pigmentation. The eggs are shed freely into the water or carried in single or paired, often brilliantly colored, packets appended to the front of the abdomen. The young hatch as nauplii and are planktonic except in Monstrilloida where the nauplii immediately become internal parasites of worms and mollusks, and in Harpacticoida where the nauplii, like most of the adults, are benthic. Adult monstrilloids are planktonic. While there are typically 6 naupliar and 6 copepodid stages, the number is reduced in some parasitic species. Several generations may be produced in a season, and annual cycles of abundance may be more or less bimodal.

Copepods are the most numerous of marine crustaceans both in species and as individuals in many habitats. Because of their extraordinary abundance in the plankton, calanoids are of primary importance in the marine economy. They link phytoplankters and larger predators, including such fish as mackerel and herring and even the largest of baleen whales. Harpacticoids are common in most benthic communities.

Systematic List and Distribution

Copepods usually require special preparation, and only a few of the 373 species listed by Wilson (1932) will be dealt with. They include the most frequently encountered planktonic species, accounting for perhaps 90% or more of the individuals likely to be seen inshore. Representatives of other orders are figured to indicate their general appearance.

ORDER CALANOIDA

Wilson (1932) listed 48 calanoid genera, all of them planktonic. Those in Table 21.5, plus a very few cyclopoids, are the principal species in the neritic and estuarine plankton.

In addition, the following oceanic genera have been reported most frequently inshore: *Eucalanus*, *Euchaeta*, *Rhincalanus*, *Mecynocera*, *Calocalanus*, *Scaphocalanus*, *Scolethrix*, and *Pontella; Gaidius*, *Euchirella*, *Paraeuchaeta*, and *Pleuromamma* are usually taken only in deep tows.

ORDER CYCLOPOIDA

Wilson (1932) listed 26 genera of this order; of these, 7 are parasitic and 8 are restricted to fresh water. *Oithona* is the most common of several genera appearing in the plankton of neritic and estuarine waters. Despite their small size—females less than 1 mm—members of this genus sometimes dominate the coastal plankton. Another, *Oncaea*, has a somewhat more oceanic distribution. *Oithona* and *Oncaea* are shown in Figure 21.7. *Cyclops*, a familiar

Table 21.5. Copepod Distribution

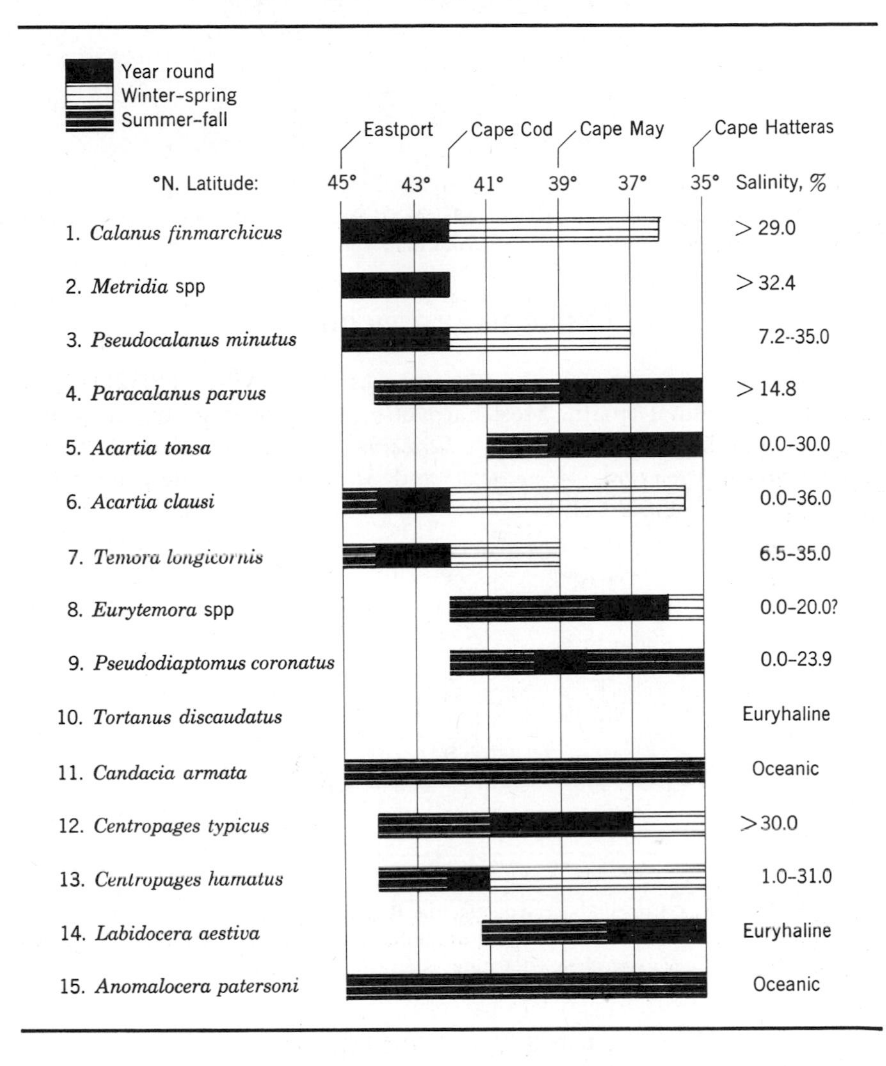

freshwater genus, occurs in the less saline parts of estuaries, while *Halicyclops* and *Microcyclops* are common brackish water genera. Several genera of oceanic cyclopoids appear in local lists chiefly as strays from the Gulf Stream. *Farranula* is a small form, less than 1 mm long, but several genera have species between 1 and 2 mm long, e.g., *Corycaeus*, *Sapphirina*, and species of *Copilia* reach 6 mm. These genera all have eyes with distinct corneal lenses.

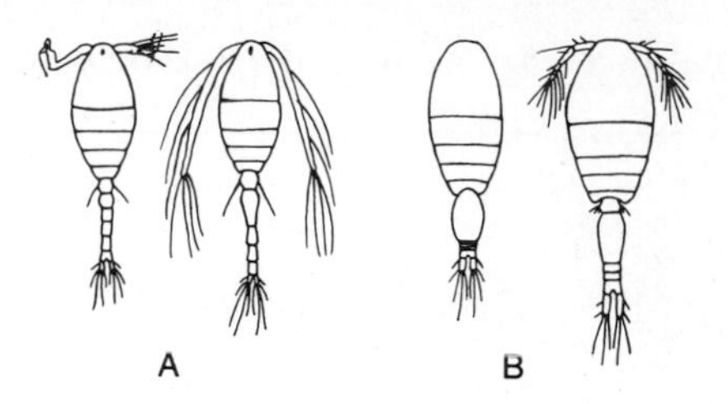

Fig. 21.7. Crustacea, Copepoda, Cyclopoida. A, *Oithona similis;* B, *Oncaea venusta;* males at left in each species; antennae omitted from male *Oncaea*.

ORDER HARPACTICOIDA

Four-fifths of the species in the 44 genera listed by Wilson (1932) do not reach 1 mm in total length. Most harpacticoids are benthic, but they also enter the plankton adventitiously; *Microsetella*, *Harpacticus*, *Tisbe*, *Alteutha*, *Halithalestris*, *Clytemnestra*, *Macrosetella*, and *Metis* are the genera most frequently reported. Figure 21.8.

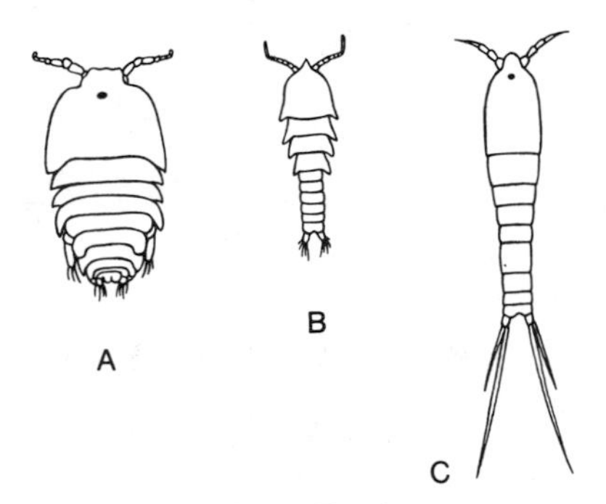

Fig. 21.8. Crustacea, Copepoda, Harpacticoida. Representative genera thoracic appendages omitted; dimensions are total length of adults. A, *Alteutha* sp., 1.4 mm; B, *Clytemestra* sp., 0.8 mm; C, *Harpacticus* sp., 1.4 mm.

ORDER MONSTRILLOIDA

The adults are pelagic and are sometimes common in the plankton. There are only about a half-dozen species, none of which exceeds 4 mm in length.

Identification

Formal keys rely heavily on differences in thoracic appendages. To avoid the painstaking dissection of these parts, Pennak (1953) recommends mounting a half-dozen individuals of what are presumed to be the same species belly up on a microscope slide. The resultant diversity of postures usually suffices to demonstrate the necessary details. Staining with alum cochineal may help clarify details.

The criteria used here are less difficult to observe, but only a relatively few species are covered; all are calanoids. As previously mentioned, these make up the bulk of copepod species in the neritic and estuarine plankton. The common cyclopoids were illustrated in Figure 21.7. Monstrilloids are somewhat harpacticoidlike in being elongate and cylindrical; however, the setae on the caudal rami are short and form a fan as in calanoids; the first antennae are short and strong, but the second antennae and mouth parts are completely lacking. Three species illustrated in Figure 21.8 demonstrate the diversity of appearance found in harpacticoids. For complete keys to all of the orders occurring locally, see Wilson (1932). Developmental stages of a relatively few species have been worked out. The later copepodid stages may differ in only minor details from adults, but these differences may be enough to cause difficulty in identification.

The calanoids considered here fall into two groups according to the shape of the last segment of the metasome (Figure 21.9). There is relatively little sexual variation in this trait except in *Tortanus* and *Eurytemora*, where the sexes

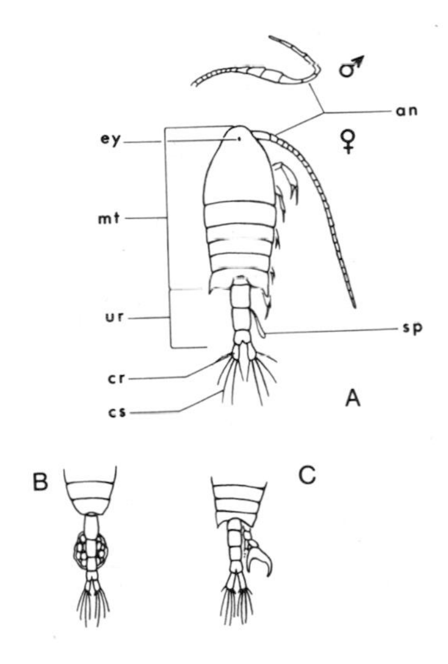

Fig. 21.9. Crustacea, Copepoda, Calanoida. A, typical female in dorsal view with most appendages removed from left side and with detail of modified first antenna of male (above); B, urosome of female showing attached eggs; C, urosome of male including modified thoracic appendage. Abbreviations are as follows: an, first antenna; cr, caudal ramus; cs, caudal seta; ey, eye; mt, metasome; ur, urosome; sp, spermatophore.

Table 21.6. Planktonic Calanoids with Last Metasomal Segment Rounded

	Species	Male Antenna Modified	Females with Ovisacs	Metasomal Segments (Cephalothorax Excluded)	Total Length		
					Females	Males	
A	*Calanus finmarchicus*	–	–	5	2.7–6.5	2.5–4.0	—
D	*Metridia* spp. *	–	–	4	2.5–4.5	2.0–4.0	—
B	*Pseudocalanus minutus*	–	+	3	1.2–1.6	1.0–1.3	Males and Females reddish
B	*Paracalanus parvus*	–	–	3	0.7–1.0	0.9–1.0	Females reddish, males yellowish
E, F	*Acartia tonsa*	s	–	4	1.2–1.6	1.0–1.2	Caudal setae subequal
E, G	*Acartia clausi*	s	–	4	1.1–1.3	1.0–1.1	Outer caudal setae shorter
H	*Temora longicornis*	s	–	4	1.0–1.5	1.0–1.4	Body club-shaped
I	*Eurytemora* spp., males*	s		5		0.7–1.6	Body slender
	Pseudodiaptomus coronatus	+	+	5	1.2–1.5	1.0–1.3	—
C	*Tortanus discaudatus*, males	+		5		1.7–2.0	—

s Means slightly.
* *Metridia:* In *M. lucens* caudal rami are twice as long as wide, in *M. longa*, 3 times.
Eurytemora: 5 local species distinguished mainly by differences in fifth thoracic legs; males of *E. americana* and *E. hirundoides* less than 1.0 mm, *E. affinis* and *E. herdmani* 1.2 mm or longer, and *E. lacustris* between 1.1 and 1.3 mm.

fall in separate groups. Differences in size, body form, segmentation of metasome and urosome, relative length of first antennae, and other details separate individual species. Segmentation of the metasome is the same in both sexes except in *Tortanus;* appraisal of the number of these segments demands care because of the transparency of the body wall in most species. The male urosome commonly has an extra segment, and the form of the individual segments varies sexually. The first antenna on either the right or left side is modified in males of some species for grasping the female, and one of the fifth legs is often formed into a pincher for transferring spermatophores to the female.

In lieu of a key, tentative identification of the common species may be made using Table 21.6 in conjunction with Figure 21.10, and Table 21.7 with Figure 21.11.

Parasitic Copepods. A very few harpacticoids have been described as parasitic, but perhaps a third of the cyclopoid genera recorded on this coast have made the transition from predator to parasite; their hosts include fish, crabs, bivalve mollusks, and other invertebrates. The cyclopoid *Ergasilus* (Fig. 21.12A) is of particular interest since it attacks several genera of

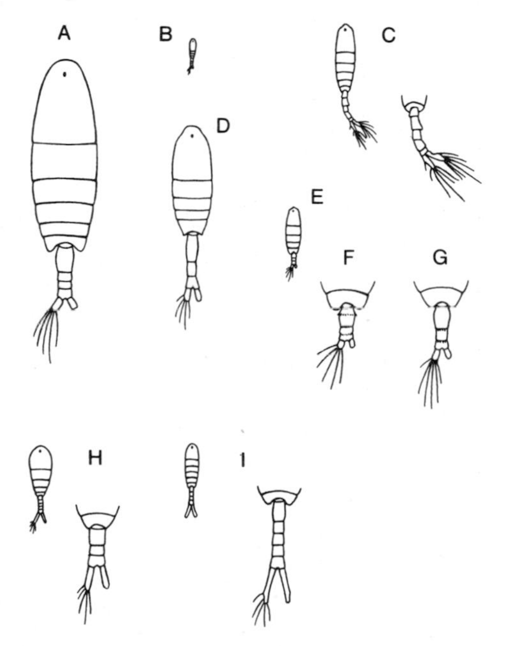

Fig. 21.10. Crustacea, Copepoda, Calanoida. Figures keyed to Table 21.6. Whole figures drawn approximately to scale with enlarged details of female urosome for some species.

Table 21.7. Planktonic Calanoids with Last Metasomal Segment with Pointed "Wings"

	Species	Male Antenna Modified	Females with Ovisacs	Metasomal Segments (Cephalothorax Excluded)	Total Length		
					Females	Males	
E	*Candacia armata*	+	–	4	2.5–2.8	2.3–2.6	—
A, B	*Centropages typicus*	+	–	5	1.2–1.8	1.0–1.6	Male with large pincherlike thoracic leg
A, C	*Centropages hamatus*	+	–	5	1.0–1.4	0.9–1.2	Male with weak pincherlike thoracic leg
H–K	*Eurytemora* spp., females *	–	+	5	1.0–1.9	—	—
F	*Tortanus discaudatus*, females	–	+	4	2.0–2.3	—	—
D	*Labidocera aestiva*	+	–	4	1.7–2.0	1.9–2.2	—
G	*Anomalocera patersoni*	+	–	5	3.0–3.3	2.5–3.0	With four eyes

* *Eurytemora. E. lacustris* is mainly a freshwater species, and both male and female have the last metasomal segment rounded; *E. affinis* resembles *herdmani*, but the genital segment is not swollen.

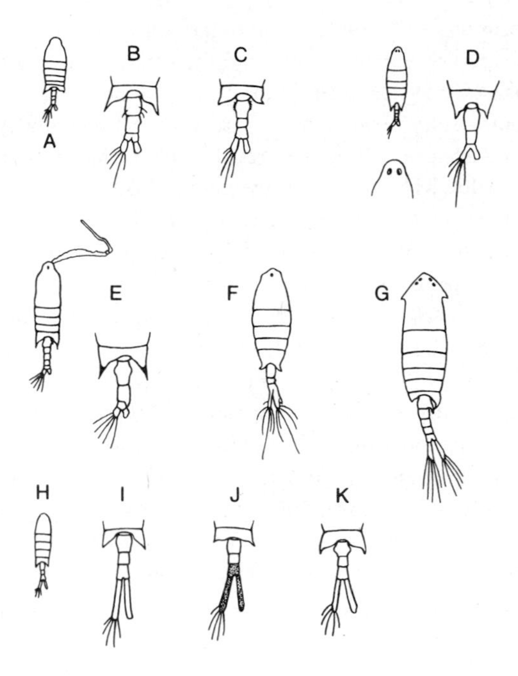

Fig. 21.11. Crustacea, Copepoda, Calanoida. Figures keyed to Table 21.7. Whole figures drawn approximately to scale with enlarged details of female urosome for some species: I is *E. lacustris;* J is *E. americanus*, (*E. affinis*, *E. hirundoides* also have a dorsal covering of spines on the last segment of the urosome and caudal rami); K is *E. herdmani*.

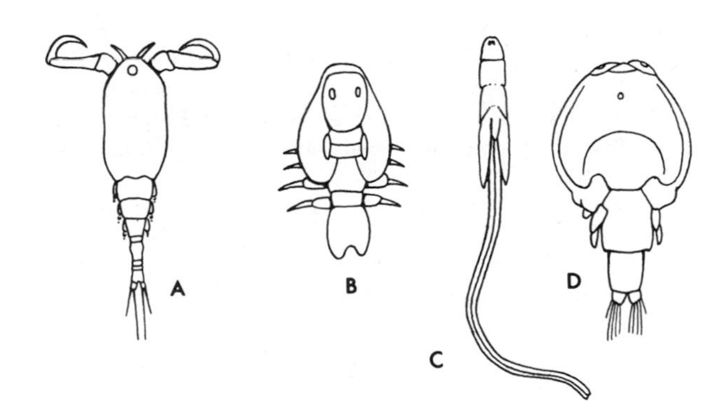

Fig. 21.12. Crustacea, Copepoda, parasitic forms, and Branchiura. A, *Ergasilus* sp.; B, *Argulus* sp.; C, lernaeopod; D, *Caligus* sp.

common estuarine fish. The order Notodelphyoida has been abandoned and its species distributed to various cyclopoid families. Nearly 80% of the species in this assemblage (see Wilson, 1932) are parasitic or commensal on ascidians. Parasitic cyclopoids may be structurally modified but usually not to such an extent that their copepod affinities are obscured. As previously mentioned, members of the order Monstrilloida are pelagic and do not feed when adult; the larval stages are parasitic on worms and mollusks.

There are more than 80 species of Caligoida, most of which attain lengths greater than 5 mm; more than a dozen genera have species ranging between 15 and 50 mm and *Penella* reaches 250 mm. Caligoids are fish parasites. They can disengage from their hosts and, though highly modified, retain some resemblance to "normal" copepods. Fig. 21.12D.

Members of Lernaeopodoida, 18 genera of which are listed by Wilson (1932), are also fish parasites. The grossly modified females become immovably fixed in the host's tissues. Some of the species are large and conspicuous, though hardly recognizable at first sight as copepods or, indeed, as arthropods, since all trace of segmentation is lost in some species (Fig. 21.12C). The much smaller males—many are well under 1.0 mm long—are affixed to the females.

References

See also general plankton references at the end of Chapter 3.

Conover, R. J. 1956. Biology of *Acartia clausi* and *A. tonsa*. *In* Oceanography of Long Island Sound, 1952–1954. Bull. Bingham oceanogr. Coll. 15:156–233.

Davis, C. C. 1944. On four species of Copepoda new to Chesapeake Bay, with a description of a new variety of *Paracalanus crassirostris*. Contr. Chesapeake biol. Lab. 61:1–11.

———. 1955. The marine and fresh-water plankton. Mich. State Univ. Press.

Farran, G. P. 1948. Copepoda. Sheets 11–19. Fich. Ident. Zooplancton.

———, and W. Vervoort. 1951. Copepoda. Sheets 32–49. Fich. Ident. Zooplancton.

Ogilvie, H. S. 1953. Copepod nauplii. Sheet 50. Fich. Ident. Zooplancton.

Pennak, R. W. 1953. Fresh water Invertebrates of the United States. Ronald Press Co.

Wilson, C. B. 1932. The copepods of the Woods Hole Region, Massachusetts. Bull. U.S. natn. Mus. 158:1–635.

Wilson, Mildred S., and H. C. Yeatman. 1959. Free-living Copepoda, Branchiura and Parasitic Copepods. *In* Fresh-water Biology. W. T. Edmondson, ed. John Wiley and Sons.

Yamaguti, S. 1963. Parasitic Copepoda and Branchiura of Fishes. Interscience.

Subclass Branchiura
FISH LICE

Formerly considered a suborder (Arguloida) of Copepoda, this exclusively parasitic group is now generally given subclass rank. Members of a single genus, *Argulus*, are found within the area covered by this text.

Diagnosis. *Microcrustacea with head and first thoracic segment united in a broad, flat, rounded cephalothorax; second to fourth thoracic segments abruptly narrower and lying within the projecting lobes of the cephalothorax; first antennae with claws and modified for seizing; with four pairs of thoracic limbs; fifth and sixth thoracic segments and abdomen fused, unsegmented, without appendages, and bluntly bilobed posteriorly; a pair of large, compound, movable eyes; adults range from 3 to 12 mm in length* (*Fig. 21.12B*).

The mouth parts are fitted for piercing and sucking. Fish lice are exclusively parasitic on fish and attach themselves in the gill chamber or to the body surface. They readily detach, however, and are often found swimming freely. The sexes are separate, and the eggs are attached to a substratum. The degree of metamorphosis varies; in some species there are naupliar stages, but in others development is essentially direct.

Wilson (1932) listed a half-dozen estuarine and neritic species on a considerable variety of fresh and salt water fish.

References

Meehean, O. L. 1940. A review of the parasitic Crustacea of the genus *Argulus* in the collections of the United States National Museum. Proc. U.S. natn. Mus. 88:459–522.

Wilson, C. B. 1932. The copepods of the Woods Hole Region, Massachusetts. Bull. U.S. natn. Mus. 158:1–635.

———. 1944. Parasitic copepods in the United States National Museum. Proc. U.S. Natn. Mus. 94(3177):529–582.

Yamaguti, S. 1963. Parasitic Copepoda and Branchiura of Fishes. Interscience.

Subclass Cirripedia
BARNACLES

Diagnosis. *Sessile crustaceans permanently attached to a substratum; acorn barnacles are enclosed in a squat, conical shell composed of four to six interlocking, calcareous plates with a central opening fitted with four doorlike, opercular valves; goose barnacles are enclosed in a flattened, more or less oval or triangular shell made of five plates and mounted on a fleshy stalk or peduncle.*

Because the animal is concealed by its calcareous shell, structural similarities to other crustaceans are not immediately obvious. Barnacles have been described as shrimplike animals that become fixed head-down by a cementlike secretion of glands on the larval first antennae. The orientation of the animal can be traced through its metamorphosis, but in adults segmentation and the various organ systems are reduced or grossly reorganized. There are 6 pairs of biramous thoracic appendages that are extended, usually in rhythmic sweeps, through the open valves to snare planktonic food. The shell is secreted by a "mantle" which is actually a modification of the crustacean carapace.

Parasitic forms, including some of the goose barnacles that otherwise retain the general form of their subgroup, may have the shells greatly reduced or absent. Even more extreme aberrations completely mask the phylogenetic relationships of other adult parasitic barnacles whose systematic position can only be determined by a reconstruction of their life history.

Most barnacles are hermaphroditic but indulge in cross-fertilization. Eggs are brooded in the mantle cavity, and the young hatch as nauplii that are distinctive in being horned (Fig. 21.1B). Subsequently the nauplius changes into an ostracodlike *cypris* larva (Fig. 21.2A), which abandons the plankton to seek a suitable substratum, attach itself, and metamorphose into a young barnacle. Growth may be rapid with full size attained in a few months. Barnacles are economically important as fouling organisms and are ecologically dominant in some communities.

Systematic List and Distribution

Order Thoracica
  Suborder Lepadomorpha
    Family Lepadidae—goose barnacles
      Lepas anatifera *Lepas pectinata*
      Lepas hillii *Lepas fascicularis*
      Lepas anserifera

Members of this genus are found attached to driftwood and other flotsam including *Sargassum*, and as fouling organisms on ships and offshore buoys. Most of the species are essentially cosmopolitan and have been rafted widely. *L. fascicularis* is truly planktonic in that it can produce its own bubble float in the absence of other support. Several species of *Scalpellum* and *Arcoscalpellum*, all rather heavily "scaled" with small calcareous plates, have been reported from the Continental Slope and offshort fishing banks with *Arcoscalpellum velutinum* in as little as 62 m. Two species of *Conchoderma* may occur locally on whales and oceanic fishes as well as on ship bottoms and buoys; *C. virgatum* is illustrated in Fig. 21.13B. Species of *Octolasmis* and *Poecilasma* occur on crabs, the former primarily on the gills of the blue crab *Callinectes sapidus*, and *Poecilasma inaequilaterale* on *Cancer borealis* on almost any part of the crab's body.

Suborder Balanomorpha—acorn barnacles		
Family Chthamalidae		
Chthamalus fragilis	V	Lit.
Family Balanidae		
*Balanus amphitrite**	V	Lit. to 55 m
Balanus balanoides	BV	Lit.
Balanus balanus	B, 40°	Lit. to 165 m
Balanus crenatus	B, 39°	Lit. to 91 m
Balanus eburneus	V, 42°	Lit. to 1–2 m, eury.
Balanus hameri	B, 37°	18 to 311 m
Balanus improvisus	BV	Lit. to 37 m, eury.

**B. amphitrite amphitrite*, reported north to Cape Cod, is evidently limited as a resident to waters with a winter minimum water temperature above 10°C and, therefore, is extralimital; a subspecies near *B. a. pallidus* has been reported from the Chesapeake. *B. a. niveus* is the common Virginian race.

The balanomorphs listed are the common encrusting barnacles on rocks, wood, and other firm substrata including shelled invertebrates. Additional "parasitic" forms are mentioned in the addendum.

Identification

Full diagnostic accounts of cirriped morphology include descriptions of mouth parts in addition to the criteria used here.

Goose Barnacles. *Scalpellum* and *Arcoscalpellum* have more than the usual 5 plates and have a "scaly" peduncle as well. The parasitic genera usually have reduced plates. Members of the genus *Lepas* are keyed below. Observation of filamentary appendages necessitates the removal of the plates from one side (Fig. 21.13D). Members of this genus attain a length including peduncle of 50–75 mm.

---

1. Shells paper thin; carina in profile with an angular bend and terminating below in a flat, oblong disc; 4–5 filamentary appendages; Fig. 21.13A. *L. fascicularis.*

1. Shells stronger; carina curving in profile and terminating below in a fork—2.

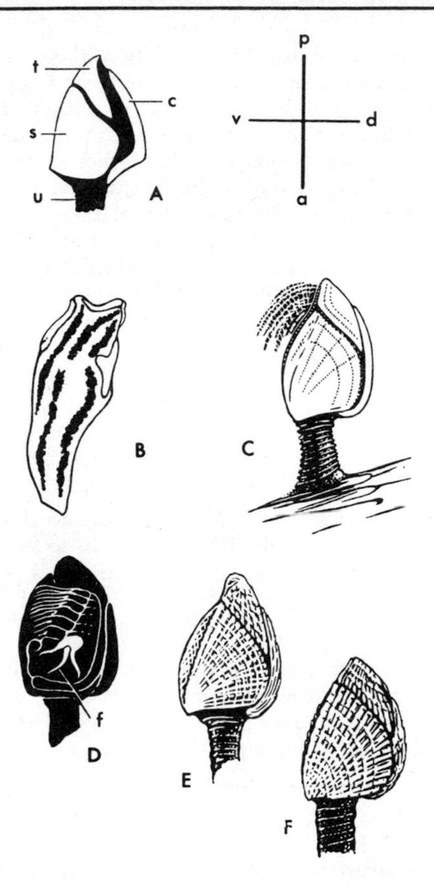

Fig. 21.13. Crustacea, Cirripedia, Lepadomorpha. A, *Lepas fascicularis*, with orientation diagram at right; B, *Conchoderma virgatum*; C, *Lepas anatifera*; D, *Lepas* sp., diagrammatic internal view with plates removed; E, *Lepas anserifera;* F, *Lepas pectinata.* Abbreviations are as follows: a, anterior; c, carina; d, dorsal; f, filamentary appendage; p, posterior; s, scutum; t, tergum; u, peduncle; v, ventral.

2. Plates smooth or finely striated—3.
2. Plates furrowed, strongly striated, sometimes spiny—4.
3. With 3 filamentary appendages. *L. hillii.*
3. With 1 or 2 filamentary appendages; Fig. 21.13C. *L. anatifera.*
4. With 5 or 6 filamentary appendages; Fig. 21.13E. *L. anserifera.*
4. With 0–2 filamentary appendages; Fig. 21.13F. *L. pectinata.*

Acorn Barnacles. Age, conditions of growth such as crowding, exposure, and degree of erosion exaggerate individual variations and complicate the interpretation of criteria. The species are separated into 3 groups according to the condition of the base or *basis* of the shell (Fig. 21.14A–C). Dimensions are maxima for adults.

Group A. Basis membraneous; littoral species; Figure 21.14A.

- Plate walls solid; rostral plate overlapped by adjoining rostrolateral plates; basal diameter 10 mm; color brown or gray; Fig. 21.14E. *Chthamalus fragilis.*
- Plate walls with tubiferous support; rostral plate overlaps adjoining lateral plates; basal diameter 25 mm; color usually whitish; Fig. 21.14 D,F,J. *B. Balanus balanoides.*

Group B. Basis calcareous and solid; distributions mainly boreal; Figure 21.14B,G–I,K. *Balanus* species.

	crenatus	*balanus**	*hameri*	Fig. 21.14
Plate walls	Tubiferous	Tubiferous	Solid	K,L
Radii	Narrow	Wide	Wide	G,H
Scuta grooved longitudinally	No	Yes	Yes	I
Basal diameter	24 mm	50 mm	75 mm	—

*In *B. balanus* the number of supporting ridges in the plate walls exceeds the number of main partitions, Fig. 21.14L.

Group C. Basis calcareous and tubiferous; distributions mainly Virginian; plate walls tubiferous in all; Figure 21.14C. *Balanus* species.

	eburneus	*improvisus*	*amphitrite* (3)	Fig. 21.14
Tergal spur	Long (1)	Long	Varies	M–P
Radii	Wide	Narrow (2)	Wide	Q–S
Scuta grooved longitudinally	Yes	No	No	I
Basal diameter	25 mm	13 mm	10 mm	—
With gray/purple marks	No	No	Yes	—

(1) The tergum is deeply indented at the base of the spur on one side.

(2) Somewhat variable but, even if relatively wide, the radius only meets the parietal border of the adjoining plate at its base.

(3) *B. amphitrite pallidus*, tinged with purple but without stripes or definite pattern; *B. amphitrite amphitrite*, with broad red-purple stripes; *B. a. niveus* with narrow gray or purple stripes; note also differences in the form and position of the tergal spur. South of Hatteras, *B. amphitrite* reaches a larger size than indicated.

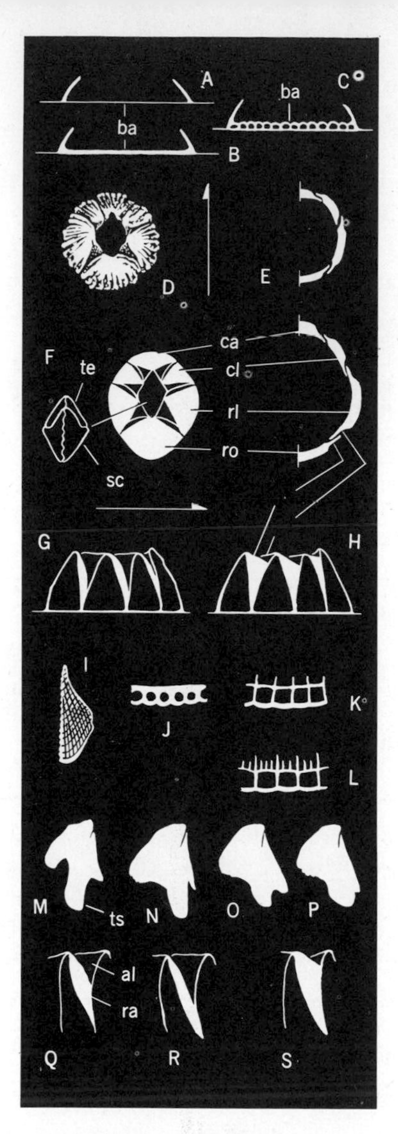

Fig. 21.14. Crustacea, Cirripedia, Balanomorpha. Arrows point anteriorly. A–C are lateral, bisected views indicating basis types schematically; A, basis membraneous; B, basis solid, calcareous; C, basis tubiferous calcareous; D, *Balanus balanoides*, with terga and scuta removed; E, *Chthamalus* sp., schematic to show plate overlap sequence; F–S are *Balanus* spp.; F, *Balanus* sp., schematic view (left center), with detail of terga and scuta (left), and diagram to show plate overlap sequence (right); G, *B. crenatus*, lateral view; H, *B. balanus*, lateral view; I, scuta with longitudinal grooves; J–L are sections of plate walls; J, *B. balanoides*; K, *B. crenatus*; L, *B. balanus*; M–P are outlines of terga; M, *B. eburneus;* N, *B. improvisus;* O, *B. a. amphitrite;* P, *B. a. niveus;* Q–S show relative width of radii (solid white); Q, *B. eburneus;* R, *B. improvisus;* S, *B. amphitrite*. Abbreviations are as follows: al, ala; ba, basis; ca, carina; cl, carinolate ral; ra; ra, radius; rl, rostrolateral; ro, rostrum; sc, scutum; te, tergum; ts, tergal spur. The thickened part of a plate is the *paries*; interlocking parts of lateral plates are *ala* and *radius* (singular). The paired terga and scuta are the opercular valves.

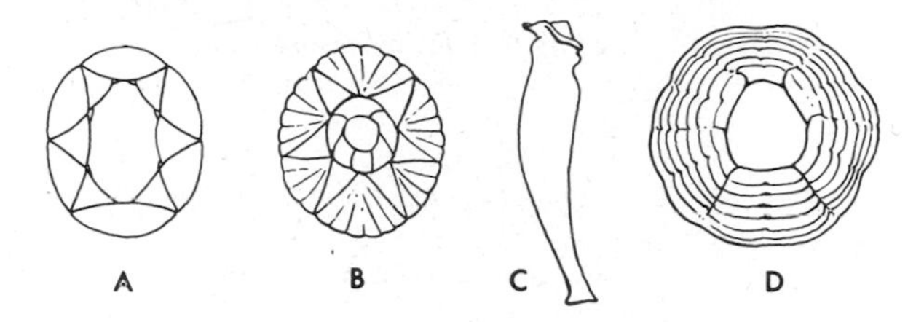

Fig. 21.15. Crustacea, Cirripedia, parasitic balanomorphs with opercular plates omitted. A, *Chelonibia patula*; B, *Coronula reginae*; C, *Xenobalanus globicipitis*; D, *Platylepas hexastylos.*

Parasitic Barnacles. The acorn and goose barnacle suborders include species that attach themselves to living animals, including other arthropods, mollusks, turtles, and whales. Some species are only found on living hosts, and within these groups the gap is bridged between mere encumbrance and outright parasitism.

Lepadomorphs found on living hosts have already been mentioned. Among balanomorphs, *Chelonibia*, *Coronula*, and *Xenobalanus* are found on whales, particularly the humpback whale *Megaptera* and on sea turtles; *Chelonibia patula* is found on the horseshoe crab *Limulus* and on brachyuran crabs, and *Platylepas* settles on sea turtles and crabs, Figure 21.15A. Parasitic members of these two suborders retain the general appearance of their relatives except that the balanomorph *Xenobalanus*, which is parasitic on pilot whales *Globicephala*, approaches the lepadomorph *Conchoderma* in form (Fig. 21.15C).

Members of additional orders are exclusively parasitic; they are grossly atypical as adults but have a nauplius larva. The familiar textbook form *Sacculina*, a member of the order Rhizocephala, has been reported on boreal crabs that occur on both sides of the North Atlantic, but evidently *Sacculina* itself is not known here. An introduced sacculinid, *Loxothylacus panopei*, has been reported on xanthid crabs in the Chesapeake (see Van Engle *et al.*, 1965). Another peculiar form, *Trypetesa* (= *Alcippe*) *lampas*, which belongs to the order Acrothoracica, bores into the shells of moon snails and other mollusks occupied by hermit crabs; its occurrence on this coast may be more widespread than the few records indicate; see Hancock (1849). Members of the other exclusively parasitic order, Ascothoracica, are found on cnidarians and echinoderms, but the extent of local occurrence is poorly known.

References

Darwin, C. 1851. A monograph on the subclass Cirripedia, Lepadidae. Ray Society.
———. 1854. A monograph on the subclass Cirripedia, Balanidae. Ray Society.

Hancock, A. 1849. Notice of the occurrence on the British coast of a burrowing barnacle belonging to a new order of the class Cirripedia. Ann. Mag. nat. Hist. Ser. 2. 4(43): 305–314.

Newman, W. A., V. A. Zullo, and T. H. Withers. 1969. Cirripedia. *In* Treatise on Invertebrate Paleontology. R. C. Moore, ed. Part R, Arthropoda 4. Geol. Soc. Amer. and Univ. of Kansas (not seen).

Pilsbry, H. A. 1907. The barnacles contained in the collection of the U.S. National Museum. Bull. U.S. natn. Mus. 60:1–122.

———. 1916. The sessile barnacles (Cirripedia) contained in the collections of the U.S. National Museum. Bull. U.S. natn. Mus. 93:1–366.

Van Engle, W. A., W. A. Dillon, D. Zwerner, and D. Eldridge. 1965. *Loxothylacus panopei* (Cirripedia, Sacculinidae), an introduced parasite on a xanthid crab in Chesapeake Bay, U.S.A. Crustaceana. 10:111–112.

Zullo, V. A. 1963. A preliminary report on systematics and distribution of barancles (Cirripedia) of the Cape Cod region. Mar. biol. Lab. Woods Hole, Mass. 1–33.

Subclass Malacostraca
HIGHER CRUSTACEA

The Malacostraca, including as it does not only a large number of species but also most of the macroscopic and hence more familiar forms, requires a more detailed treatment than the predominantly microscopic groups just considered.

The general characteristics of the subclass and a key to its principal subdivisions were given in the introduction to this chapter. The breakdown of the subclass into two series (see Table 21.1) is of limited interest here, since the division serves mainly to separate the order Nebaliacea from the higher eumalacostracans.

Eumalacostraca, for our purposes, contains 3 superorders. The allocation of mantis shrimp, Hoplocarida, to one of these recognizes the considerable gap separating these animals from other higher crustacea.

The superorder Peracarida, though diverse in its several orders, is unified not only by internal morphological traits, but also in having the eggs of its members brooded in a "pouch" formed of oostegites appended to the thoracic limbs of the female. The development of the young is accomplished without profound metamorphosis. Five orders of Peracarida will be considered: Cumacea, Tanaidacea, Isopoda, Amphipoda, and Mysidacea.

In the superorder Eucarida the eggs are carried on the abdominal appendages, and larval development with complex metamorphosis is common. There are two orders: Euphausiacea and Decapoda; further comments on the systematics of Eucarida are reserved for the introduction to that assemblage.

Superorder Phyllocarida: Order Nebaliacea
Nebalia

Nebalia bipes is the only local representative of the most primitive living group of the subclass Malacostraca.

Diagnosis. *Small, rather shrimplike Crustacea with well-developed carapace hinged at the midline and covering but not fused with the thoracic segments; rostrum also hinged; abdomen ending in caudal rami; eyes stalked and conspicuous; with eight pairs of short thoracic appendages essentially alike, with major segments flat and paddlelike; abdomen with four pairs of well-developed, biramous, swimming appendages followed by two pairs of smaller uniramous appendages (Fig. 21.6G).*

Adults reach a length of 10–12 mm. The sexes are separate and somewhat dimorphic externally. The eggs are carried by the female, held by setae on the thoracic appendages. Development is direct, and there are no special larval forms.

Nebalia bipes ranges from Maine to Labrador; Packard (1883) reported it in 8–16 m, but in Britain it is known from the littoral where it hides under debris. It is said to favor foul situations and will come to baited lobster pots.

Reference

Packard, A. S. 1883. A monograph of North American Phyllopod Crustacea. 12th Ann. Rept. U.S. geol. and geogr. Surv. Territ. 1:1–433.

Superorder Hoplocarida: Order Stomatopoda
MANTIS SHRIMP

The mantis shrimp are a unique group whose placement in the crustacean hierarchy poses some difficulty. Various authors put them at either the beginning or the end of the eumalocostracan assemblage.

Diagnosis. *Medium to large crustaceans with a small carapace exposing four thoracic segments; abdomen longer than thorax, wide, and often flattened; telson strongly developed, platelike, and flanked with biramous uropods; eyes stalked; second maxillipeds powerfully developed with subchelate raptorial claws reminiscent of mantid insects, Figure 21.16.*

The complement of thoracic and abdominal appendages is complete (see Table 21.2). In addition to the conspicuous second maxillipeds there are 3 pairs of small limbs behind and another pair before the raptorial appendages, all of which are uniramous and function as maxillipeds. The last 3 pairs of thoracic limbs are slender, biramous walking legs. The 5 pairs of pleopods are well developed for swimming.

Local members of the composite genus *Lysiosquilla* reach 40–60 mm, although the (possibly extralimital) adults of *L. scabricauda* grow to 275 mm at least; the more frequently encountered *Squilla empusa* attains a length of 250 mm. *Squilla*, with its spiny body, is prickly to handle, and the mantidlike claws are capable of wounding incautious handlers.

The sexes are separate, but the extent of external sexual dimorphism, if any, varies in different species. The eggs are carried between the thoracic limbs. Young stomatopods hatch at a larval stage that has been equated with the protozoea of other groups. The first stage of the lysiosquillids, so far as known, is termed an *antizoea; Squilla* hatches as a somewhat more advanced *pseudozoea*. The larvae, though not all initially planktonic, shortly become so and may go through as many as nine stages before moulting to the postlarva. The larvae are often common in the plankton. Both adults and larvae are predaceous.

Adults are benthic and are burrowers. While *Squilla* is not rare—it is often taken in otter trawls, for example—the lysiosquillids are practically unknown in the adult state, even to specialists. In an area as heavily frequented by marine biologists as the south shore of Cape Cod, a species of the latter group went undetected until 1958.

Systematic List and Distribution

The present list was compiled from Manning (1963) whose classification of the stomatopods is followed.

Family Gonodactylidae		
Gonodactylus bredini	C	Lit. to 55 m
Gonodactylus torus	C	To 364 m
Family Squillidae		
Meiosquilla quadridens	C	Lit. to 137 m

Squilla deceptrix	C	To 346 m
Squilla empusa	V	Lit. to 154 m, eury.
Squilla neglecta	C	Lit. to 64 m
Family Lysiosquillidae		
Coronis excavatrix	C	Lit.
Heterosquilla armata	V	96 to 220 m
Lysiosquilla scabricauda	C+	See note below
(A tropical-subtropical species, but larvae have been recorded north to New England)		
Nannosquilla grayi	BV	Lit. to 15 m, eury.
Platysquilla enodis	V	31 to 49 m

Identification

The key is from Manning (personal communication, 1967) with modifications.

1. Telson with median longitudinal ridges or carina on dorsal surface—2.

1. Telson without median longitudinal carina—6.

2. Cornea rounded; rostral plate with median spine; movable finger of claw without teeth, Fig. 21.16D. *Gonodactylus:* In *G. torus* the submedian teeth on the telson have movable apices, and in *G. bredini* they do not.

2. Cornea bilobed; rostral plate without apical spine; movable finger with 4 or more teeth—3.

3. Movable finger of claw with 4 teeth; submedian teeth on telson with movable apices. *Meiosquilla quadridens.*

3. Movable finger with 5–6 teeth; submedian teeth on telson with fixed apices; Fig. 21.16A. *Squilla* spp.—4.

4. Median carina of carapace not bifurcated anteriorly; telson with rows of tubercles on dorsal surface. *S. deceptrix.*

4. Median carina of carapace bifurcated anteriorly; telson without rows of tubercles on dorsal surface—5.

5. Movable finger of claw with 5 teeth; lateral process of fifth thoracic somite a spatulate lobe. *S. neglecta.*

5. Movable finger with 6 teeth; lateral process of fifth somite a sharp, curved spine; Fig. 20.16A. *S. empusa.*

6. Posterior portion of body covered with small spinules. *Heterosquilla armata.*

6. Posterior portion of body without spinules—7.

7. Telson eaved, all spines submarginal on ventral side; Fig. 21.16C. *Nannosquilla grayi.*

7. Telson not eaved, at least some spines marginal—8.

8. Telson with 1 pair of fixed marginal teeth. *Coronis excavatrix.*

8. Telson with 4 pairs of fixed marginal teeth; Fig. 21.16D. *Platysquilla enodis.*

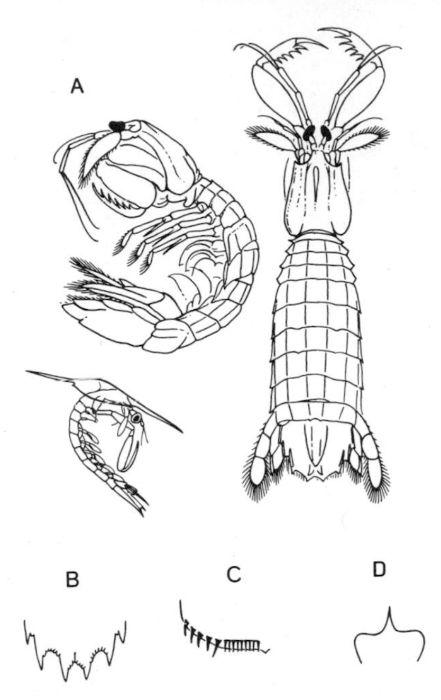

Fig. 21.16. Crustacea, Hoplocarida. A, *Squilla empusa*, dorsal and lateral views with pseudozoea larva at lower left; B, same detail of telson; C, *Nannosquilla grayi*, half telson in ventral view; D, *Gonodactylus* sp., rostral plate.

References

Bigelow, R. P. 1894. Report on the Crustacea of the Order Stomatopoda collected by the steamer *Albatross* between 1885 and 1891, and on other specimens in the U.S. National Museum. Proc. U.S. natn. Mus. 17:484–550.

Chace, F. A., Jr. 1958. A new stomatopod crustacean of the genus *Lysiosquilla* from Cape Cod, Massachusetts. Biol. Bull. mar. biol. Lab. Woods Hole. 114(2):141–145. (*Nannosquilla grayi*.)

Holthuis, L. B., and R. B. Manning. 1969. Stomatopoda. *In* Treatise on Invertebrate Paleontology. R. C. Moore, ed. Part R, Arthropoda. Vol. 2. Geol. Soc. Amer. and Univ. of Kansas (not seen).

Manning, R. B. 1961. Sexual dimorphism in *Lysiosquilla scabricauda* (Lamarck), a stomatopod crustacean. Quart. J. Fla. Acad. Sci. 24(2):101–107.

———. 1961. Stomatopod Crustacea from the Atlantic Coast of Northern South America. Allan Hancock Atlantic Exped. 9:1–46.

———. 1962. Seven new species of stomatopod crustaceans from the northwestern Atlantic. Proc. biol. Soc. Wash. 75:215–222. (*Platysquilla enodis*.)

———. 1963. Preliminary revision of the genera *Pseudosquilla* and *Lysiosquilla* with descriptions of six new genera (Crustacea: Stomatopoda). Bull. mar. Sci. Gulf Caribb. 13(2):308–325.

Superorder Peracarida
Order Cumacea
CUMACEAN SHRIMP

Diagnosis. *Very small peracaridans with an abbreviated carapace that leaves four or five thoracic segments exposed; abdomen with slender biramous uropods, with or without a free telson; females lack abdominal appendages other than uropods, males have up to five pairs of pleopods or none; eyes absent, small and sessile, or fused and median.*

Adults of most species are only 3–8 mm long; a few reach 15 mm, or extralimitally, 35 mm. The fusion of head and thorax and the eyeless condition of most species gives cumaceans a somewhat headless appearance. The swollen carapace is extended in some genera as a *pseudorostrum*. Of the 8 pairs of thoracic appendages, the first 2 or 3 pairs are maxillipeds and the remaining ones are walking legs. The abdomen is slender and cylindrical.

The sexes are separate and show external dimorphism; males of most species not only have abdominal limbs but are also equipped with long, slender second antennae which are rudimentary in females.

Cumaceans are benthic, burrowing forms, and some at least feed by filtering minute food particles from a current maintained by one or more of the anterior thoracic limb pairs. The males of some species are nocturnal plankters and may be attracted to low-intensity light traps. All of the species are free-living. Though primarily a marine group, cumaceans also occur in brackish or nearly fresh waters, and some species are restricted to such parts of estuaries.

Systematic List and Distribution

The classification is that of Calman (1912), with four additions. Distributions are imperfectly known. Of the forms listed, 70% also occur in northern Europe and several more are described as "very near" European species. The apparent dearth of cumaceans in temperate American waters may, however, merely indicate neglect of a group of easily overlooked animals.

Family Bodotriidae		
Cyclaspis varians	V	Surface, eury.
Cyclaspis pustulata	V	Surface, eury.
Leptocuma minor	B	1.8 to 100 m
Mancocuma altera	C, 37°	18 mm, eury.

Family Leuconidae		
Leucon nasicoides	A, 45°	122 m
Leucon americanus	V	Shallow, eury.
Eudorella emarginata	B	1.8 to 100 m
Eudorella truncatula	B, 41°	1.8 to 99 m
Eudorella hispida	B+	1.8 to 130 m
Eudorellopsis deformis	B, 41°	1.8 to 100 m
Family Nannastacidae		
Almyracuma proximoculi	B	Lit., eury.
Campylaspis rubicunda	B+	64 m
Campylaspis affinis	B+	66 m
Family Lampropidae		
Lamprops quadriplicata	B	2 to 150 m
Family Diastylidae		
Diastylis polita	B, 41°	Surface to 33 m
Diastylis sculpta	B, 41°	Lit. to 395 m
Diastylis quadrispinosa	BV	4 to 373 m
Diastylis abbreviata	B	31 to 64 m
Diastylis lucifera	A+	110 to 141 m

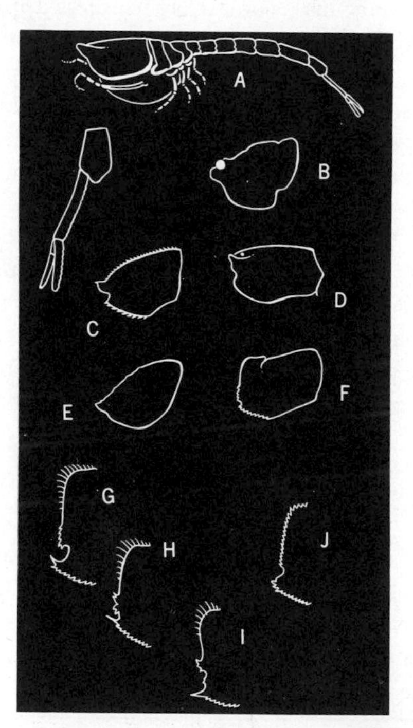

Fig. 21.17. Crustacea, Cumacea. A, *Cyclaspis* sp., whole animal in lateral view and detail of last abdominal segment and left uropod (below left); B–F are carapace outlines viewed from left side; B, *Almyracuma proximoculi*; C, *Leucon* sp.; D, *Leptocuma minor*; E, *Campylaspis* sp.; F, *Eudorella* sp.; G–J are details of left front margin of carapace; G, *Eudorella emarginata*; H, *Eudorella truncatula*; I, *Eudorella hispida*; J, *Eudorellopsis deformis*.

Leptostylis longimana	A+	—
Leptostylis ampullacea	B	95 to 165 m
Oxyurostylis smithi	V, 44°	Surface to 5 m, eury.
Family Pseudocumidae		
Petalosarsia declivis	B+	5 to 55 m

Identification

The species fall readily into two groups: members of the first three families have no free telson, and those of the last three do. In addition to differences noted in the tables, the species are separated by proportional differences in the caudal parts and in the shape and ornamentation of the carapace. In some cases, identification is only carried to genus. Published accounts are wanting in detail for some forms, but most of the species have been figured by either Calman (1912) or Sars (1871, 1900).

A. Species without telson. Table 21.8 and Fig. 21.17.
B. Species with telson. Table 21.9 and Fig. 21.18.

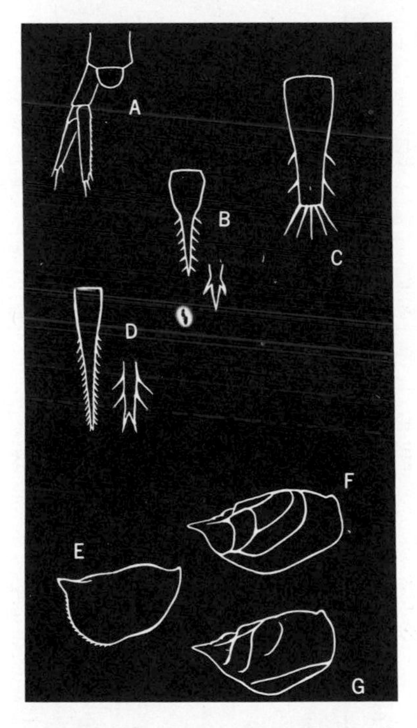

Fig. 21.18. Crustacea, Cumacea. A, *Petalosarsia declivis*, telson and left uropod; B, *Oxyurostylis smithi*, telson and detail of apex; C, *Lamprops quadriplicata*, telson; D, *Diastylis quadrispinosa*, telson; E, *Leptostylis ampullacea*, carapace viewed from left; F, *Diastylis sculpta*, carapace; G, *Diastylis polita*, carapace.

Table 21.8. Differential Traits of Cumaceans without Telson

Species	Segments, Inner Ramus of Uropods	Maximum Length, mm	Male Pleopods	Pseudo-rostrum	Figure Source	Figure 21.17
Mancocuma	2	2.8	2	Short	—	—
Cyclaspis	1	3.4	5	Moderate	Calman	A
Leptocuma	2	7.5	3	—	Calman	D
Leucon	2	8.0	2	Strong	Sars (1871)	C
Eudorella *	2	4.5 8.0	2	—	Sars (1871, 1900)	G–I
Eudorellopsis	1	7	2	—	Sars (1900)	J
Almyracuma	1	4	0	Short	Jones and Burbanck (1959)	B
Campylaspis	1	8.0	0	Strong	Sars (1871)	E

* *Eudorella* with 3 species as follows:

1. Body bristly, lower margin of carapace with 14–18 teeth. *E. hispida.*
1. Body not bristly, lower margin with less than 14 teeth—2.

2. Front margin of carapace notched, deeply in female, slightly in male. *E. emarginata.*
2. Front margin slightly notched in female, serrate or even in male. *E. truncatula.*

Taale 21.9. Differential Traits of Cumaceans with Telson

Species	Segments, Inner Ramus of Uropods	Apical Spines on Telson	Maximum Length, mm	Male Pleopods	Figure Sources	Fig. 21.18
Petalosarsia	1	0	5	2	Sars (1871)	A
Lamprops	3	5	12	0	Sars (1871)	C
Diastylis *	2–3	2	6–14	2	Calman, Sars (1900)	D,F,G
Oxyurostylis	3	0	7	2	Calman	B
Leptostylis	3	2	8	2	Sars (1871)	E

* *Diastylis* with 4 species as follows:

1. Carapace with 4 short, dorsal spines at the front. *D. quadrispinosia.*
1. Carapace without spines as above—2.

2. Carapace smooth laterally. *D. abbreviata.*
2. Carapace grooved laterally—3.

3. Side with 3 grooves. *D. polita.*
3. Side with 4 grooves. *D. sculpta.*

References

Calman, W. T. 1912. The Crustacea of the order Cumacea in the collection of the United States National Museum. Proc. U.S. natn. Mus. 41:603–676.

Jones, N. S., and W. D. Burbanck. 1959. *Almyracuma proximoculi* gen. et sp. nov. (Crustacea, Cumacea) from brackish water of Cape Cod, Massachusetts. Biol. Bull. mar. biol. Lab. Woods Hole. 116:115–124.

Sars, G. O. 1871. Beskrivelse af de paa Fregatten *Josephines* expedition fundne Cumaceer. K. svenska Vetensk. Akad. Handl. 9(13):1–57.

———. 1900. An account of the Crustacea of Norway. 3:i–x, 1–115.

Smith, S. I. 1879. The stalk-eyed crustaceans of the Atlantic coast of North America north of Cape Cod. Trans. Conn. Acad. Arts Sci. 5:28–138.

Zimmer, C. 1943. Über neue und weniger bekante Cumaceen. Zool. Anz. 141:148–167.

Order Tanaidacea
TANAIDS

Long considered a subgroup of Isopoda, the tanaids are now generally given ordinal rank.

Diagnosis. *Very small peracaridans in which the cephalothorax includes the first and second thoracic segments, forming a short carapace; first pair of thoracic appendages modified as maxillipeds; second pair with well-developed, fully chelate gnathopods; third pair adapted for burrowing; the five remaining pairs of walking legs more or less similar; Figure 21.19.*

Adults do not exceed 5 mm in total length except in deep sea Neotanaidae. There is a single pair of uropods with 1 or 2 rami. The last abdominal segment does not have a free telson appended. There are usually 5 pairs of swimming appendages on the abdomen, but there may be as few as 3 or none. The eyes are mounted on immovable stalks; they are usually small but in some species are well developed or, conversely, they may be absent altogether. The sexes are usually separate but hermaphroditism, protandry, and protogyny are known. In some genera the form of the gnathopods is sexually dimorphic, as are the number of pleopods, size of eyes, and other details. The marsupium is formed by a single pair of oostegites in Tanaidae and by 4 pairs in Paratanaidae.

Several tanaids are described as tubicolous, and the tubes of *Tanais robustus* have been found between the plates of the carapace of the loggerhead turtle. Other species are benthic and occur on soft substrata.

Systematic List and Distribution

The primary source of this list is Richardson (1905), who placed all of our species in a single family. *L. filum* is omitted from the key. Dimensions equal total length of adults.

Family Tanaidae		
Tanais cavolini	B, 41°	To 2 m, eury.
Tanais robustus	T	On sea turtle, *Caretta*
Family Paratanaidae		
Leptognatha caeca	B	To 90 m
Heterotanais limicola	B	73 to 110 m
Leptochelia savignyi	T	To 1 m, eury.
Leptochelia rapax	B	To 1 m
Leptochelia filum	A, 44°	15 to 37 m

Identification

1. Uropods uniramous; 3 pairs of pleopods; 5 mm; Fig. 21.19A. *Tanais cavolini.*
1. Uropods biramous; 5 pairs of pleopods or none—2.
2. Eyeless; both rami of uropods with 2 segments (segmental counts are in addition to the peduncle); 2.5 mm. *Leptognatha caeca.*
2. With eyes; longer ramus of uropods with more than 2 segments—3.

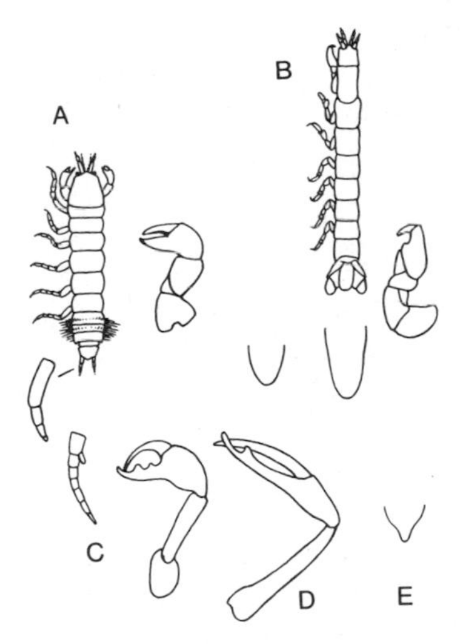

Fig. 21.19. Crustacea, Tanaidacea and Isopoda, Family Anthuridae. A, *Tanais cavolini*, whole animal, dorsal view, with details of gnathopod (similar in make and female) and uropod; B, *Cyathura* sp., female, whole animal and gnathopod (right), and telsons of *C. burbancki* and *C. polita*, below left and right, respectively; C, *Leptochelia savignyi*, male gnathopod and left uropod; D, *Leptochelia rapax*, male gnathopod; E, *Calathura* sp., telson.

3. Shorter ramus of uropods with 2 segments; to 2 mm. *Heterotanais limicola.*
3. Shorter ramus with 1 segment—4.
4. Longer ramus with 6 segments; 2 mm; Fig. 21.19C. *Leptochelia savignyi.*
4. Longer ramus with 5 segments; 4 mm; Fig. 21.19D. *Leptochelia rapax.*

References

Lang, K. 1949. Contribution to the systematics and synonymies of the Tanaidacea. Ark. Zool. 42a. 18:1–14.
Richardson, Harriet. 1905. Isopods of North America. Bull. U.S. natn. Mus. 54:3–54.

Order Isopoda
ISOPODS

Diagnosis. *Small to medium-sized peracaridans without a carapace; body regularly segmented and usually depressed from top to bottom, degree of flattening varies, but lateral thickness usually unmistakably greater than vertical thickness, the chief exceptions being the slender, more or less cylindrical anthurids; first pair of thoracic appendages are maxillipeds, and of the remaining seven pairs of limbs, the first three are sometimes prehensile or subchelate, the rest usually ambulatory (prehensile in Cymothoidae); telson with one pair of usually biramous uropods.*

Adults of local species range between 3 and 35 mm in total length. The first thoracic somite and, rarely, the second are fused with the head. Except in anthurids the telson is fused with the sixth abdominal somite and sometimes with additional ones. The abdominal segmentation also is frequently reduced through fusion, but there are usually 5 pairs of pleopods. The eyes are sessile.

Members of the suborder Epicaridea, parasitic on decapod crustaceans, are aberrant and are dealt with in an addendum.

The sexes are separate, though protandric hermaphroditism occurs in Cymothoidae, individuals being initially free-living and male and later parasitic and female. In this family the genera *Anilocra*, *Ceratothoa*, *Cymothoa*, and *Livoneca* are notably polymorphic. In isopods generally the first and second pleopods, or the second only, are modified in males to assist in sperm transfer. In most parasitic groups the sexes are different in appearance, and frequently distinctive subadult forms also occur. Gnathids lack the usual peracarid brood pouch, and in *Sphaeroma* the brood chamber may be formed by invagination of the ventral body surface rather than by oostegites.

The feeding adaptations of isopods include parasitism and predation, conspicuous examples of the latter being found in cirolanids, which prey on fish and other crustaceans. Most isopods have chewing mouth parts and are general scavengers. Some anthurids have mouth parts fitted for piercing and sucking but are scavengers and detritus feeders as well as predators. *Limnoria* is a destructive wood borer.

Isopods occur on all substrata and at all depths. Shallow water forms are often abundant under rocks, among seaweeds, and in detritus. The freshwater *Asellus* ranges into estuaries, while *Cyathura polita*, *Chiridotea nigrescens* and *C. almyra*, *Idotea balthica*, *Edotea montosa*, and *Erichsonella attenuata* are typically estuarine and may be particularly common on eel grass bottoms. The familiar "pill bugs" of terrestrial situations belong in the suborder Oniscoidea; some species in this group have predominantly maritime distributions, and only these are included in the checklist.

A few isopods are planktonic, e.g., species of *Eurydice*, the larvae of the parasitic suborders, and both adult and juvenile gnathids. In addition some of the larval stages of bopyrids are parasitic on planktonic copepods.

Systematic List and Distribution

The basic list is from Richardson (1909) with amendments and additions from more recent literature. A few groups have had revisionary treatment. The subordinal breakdown follows Waterman (1960).

Suborder Gnathiidea		
Family Gnathiidae		
Gnathia cerina	B+	18 to 891 m
Suborder Anthuridea		
Family Anthuridae		
Cyathura polita	BV	Lit. to 35 m, eury.
Cyathura burbancki	V	17 m
Ptilanthura tenuis	B, 41°	Lit. to 35 m
Calathura branchiata	B+	18 to 457 m
Suborder Flabellifera		
Family Cirolanidae		
Cirolana concharum	BV	Lit. to 82 m
Cirolana polita	B	1 to 587 m
Cirolana borealis	BV	55 to 1479 m
Eurydice sp.	Oceanic	—
Family Aegidae		
Aega psora	B+	55 to 1171 m

Family Cymothoidae		
Nerocila acuminata	C, 37°	—
Nerocila munda	CV	On *Alutera*
Anilocra laticauda	C, 38°	—
Olencira praegustator	CV	On menhaden
Ceratothoa impressa	C, 42°	On flying fishes
Cymothoa excisa	CV	On sparid fish
Cymothoa oestrum	C, 37°	On *Strombus*
Lironeca ovalis	CV	On sparids, bluefish, etc.
Family Limnoridae		
Limnoria lignorum	BV	Shallow to 18 m
Family Sphaeromidae		
Ancinus depressus	V	—
Cassidinisca lunifrons	V	—
Sphaeroma quadridentatum	V	Lit. to 1 m
Ciliacaea caudata	V	4 m
Exosphaeroma papillae	V	Shallow to eury.
Suborder Valvifera		
Family Idoteidae		
Chiridotea coeca	BV	Lit.
Chiridotea tuftsi	B, 41°	Lit. to 82 m
Chiridotea arenicola	B (Georges Bank)	42 to 66 jm
Chiridotea nigrescens	V	Shallow, eury.
Chiridotea almyra	V	Shallow, eury.
Idotea metallica	BV	Lit. to 166 m
Idotea balthica	BV	Lit. to 218 m, 24‰
Idotea phosphorea	B, 41°	Lit. to 55 m
Edotea sp.	B, 41°	Lit. to 46 m, 13–26‰
Erichsonella attenuata	V	Shallow, 6‰
Erichsonella filiformis	B, 41°	Lit. to 13 m
Suborder Asellota		
Family Asellidae		
Asellus sp.	V	Fresh water-oligohaline
Family Janiridae		
Jaera marina	B	Lit.
Janira alta	B, 37°	Lit. to 348 m
Family Munnidae		
Munna fabricii	A, 43°	To 366 m
Suborder Epicaridea		
Family Bopyridae		Host
Hemiarthrus abdominalis	B	*Pandalis*, *Spirontocaris*
Ione thompsoni	M	*Callianassa*
Leidya distorta	V	*Uca*
Pseudione furcata	V	Unknown
Stegophryxus hyptius	B, 43°	*Pagurus*
Probopyrus pandalicola	V, 43°	*Palaemonetes*
Bopyroides hippolytes	B	*Spirontocaris*

Suborder Onoscoidea		
Family Oniscidae		
Philoscia vittata	V+	Maritime terrestrial
Actoniscus ellipticus	V	Maritime terrestrial
Family Ligididae		
Ligia oceanica	B	Lit.
Ligia exotica	C, 38°	Lit. to sublit.
Scyphacella arenicola	V	Lit.

Identification

As a preliminary step the subordinal groups may be distinguished as follows (dimensions are total length of adults):

A. With 5 instead of the usual 7 pairs of pereopods; abdomen abruptly narrower than thorax; sexes and young polymorphic; 4 mm; benthic (adults) or parasitic on fish (juveniles); Fig. 21.21. Gnathiidea, with only one local form, *Gnathia cerina*.

B. Telson and uropods forming a fan, outer branch of uropods arching over telson; elongate, slender, cylindrical species; first pereopod strongly prehensile; Fig. 21.19B. Anthuridea.

C. Telson and uropods forming a fan; Fig. 21.20B. Flabellifera.

D. Uropods folding under the ventral surface; Fig. 21.20A,C. Valvifera.

E. Uropods terminal; Fig. 21.20D.
- Free-living forms. Asellota and Oniscoidea.
- Asymmetrical parasitic forms. Epicaridea.

Group B. The anthurids resemble tanaids in being slender and elongate, but reach a greater size. The best known form, *Cyathura polita*, is characteristic of the ecotone between salt marsh grass, *Spartina*, and freshwater cattail,

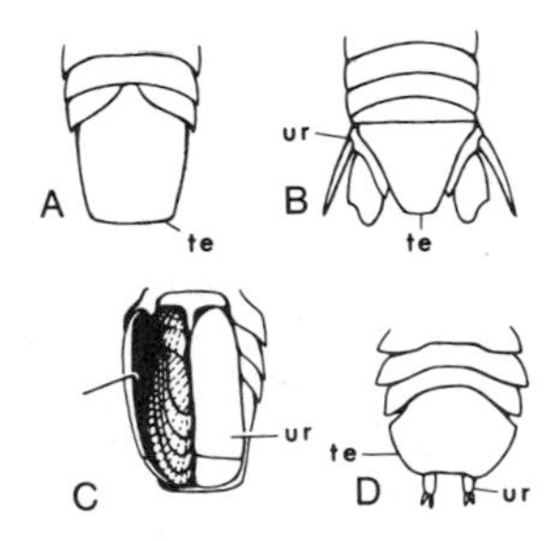

Fig. 21.20. Crustacea, Isopoda, telson-uropod arrangements. A,C, Valvifera, dorsal and ventral views; B, telson and uropods fan-like; D, uropods terminal. Abbreviations are as follows: te, telson; ur, uropods.

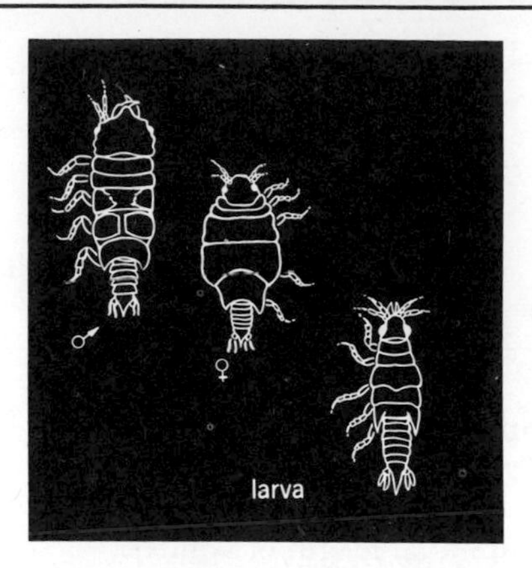

Fig. 21.21. Crustacea, Isopoda, Gnathidea. *Gnathia cerina*, male, female, juvenile (left to right); adults are 4 mm long.

Typha. *Cyathura* is tubicolous, prefers firm substrata, and occurs in concentrations of about 100–4000 per cubic meter. *Cyathura* has been described as a detritus-algal feeder but is also attracted to dead animals and may be predaceous.

The local anthurid species may be separated as follows:

---

1. First 5 or 6 abdominal segments fused into one—2.
1. First 5 abdominal segments distinct, at last laterally—3.

2. Telson shield-shaped or broadly rounded terminally; 27 mm; Fig. 21.19B. *Cyathura:* The telson of *C. polita* and *C. burbancki* differ as figured. See Frankenberg (1965).
2. Telson triangular, attenuated terminally; 26 mm; Fig. 21.19E. *Calathura branchiata*, male

3. Telson triangular. *Calathura branchiata*, female.
3. Telson rounded; 8 mm. *Ptilanthura tenuis*.

Group C. The Flabellifera includes a score of species distributed in 5 families with various habits. The Aegidae and Cymothoidae are parasitic on fish, but they are not grossly modified and may be found free; aegids voluntarily leave their host after feeding, but free cymothoids have either been dislodged or are juveniles. Cymothoidae includes some of the largest local isopods.

Family affinities may be established by Table 21.10. For further identification, examine the illustrations, which are essentially self-explanatory (Figs. 21.22–21.24). The species differ, usually quite unequivocally, in their caudal

Table 21.10. Flabeliferan Isopods

Families	Genera/Species	Ambulatory Legs, Prehensile Legs	Eye	Abdominal Segments	Habits	Maximum Length, mm	Figure
Aegidae	1/1	3p, 4a	Large	6	Parasitic on fish	16	21.22
Cymothoidae	7/10	7p	None–large	6	Parasitic on fish	13–34	21.22, 21.23
Cirolanidae	1/4	3p, 4a	Small	6	Predaceous	12–21	21.24
Limnoridae	1/1	7a	Small	6	Wood borer	4.5	21.24
Sphaeroidae	4/4	7a	Small–medium	2	Scavengers	5–13	21.24

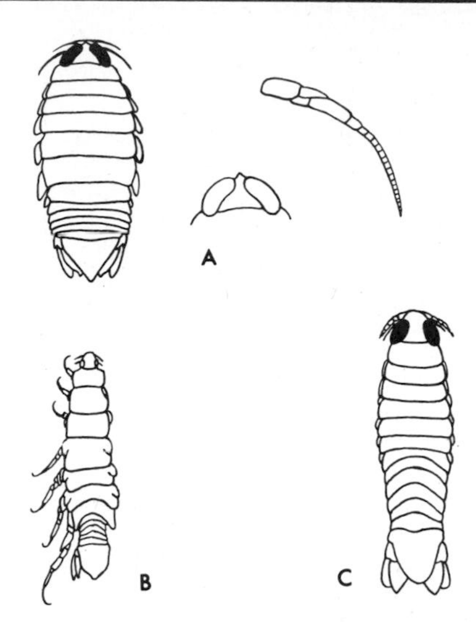

Fig. 21.22. Crustacea, Isopoda, Aegidae and Cymothiodae in part. A, *Aega psora*, whole animal plus details of head and first antenna, 50 mm; B, *Olencira praegustator*, female, 10 mm; C, immature cymothoid.

arrangements and in other details, as shown Sizes are given in the figure captions.

Group D. The single local family of Valvifera has 4 genera and about 11 species, which include some of the more conspicuous littoral and shallow water isopods. The genera are separated in the following key; specific differences are indicated in the illustration, Figure 21.25.

---

1. Eyes dorsal; first 3 pairs of legs behind maxillipeds subchelate; 4 abdominal segments; 6–15 mm; Fig. 21.25A–D. *Chiridotea.*

1. Eyes lateral; 7 pairs of thoracic legs more or less similar; fewer than 4 abdominal segments—2.

2. Second pair of antennae with a multisegmented flagellum; 3 abdominal segments; 18–38 mm; Fig. 21.25E–G. *Idotea.*

2. Second pair of antennae without multisegmented flagellum; 1 abdominal segment—3.

3. Longer antennae more than 3 times longer than shorter antennae; body slender, elongate; 8–15 mm; Fig. 21.25J,K. *Erichsonella.*

3. Longer antennae only about twice as long as shorter antennae; 7–9 mm; Fig. 21.25H,I. *Edotea.*

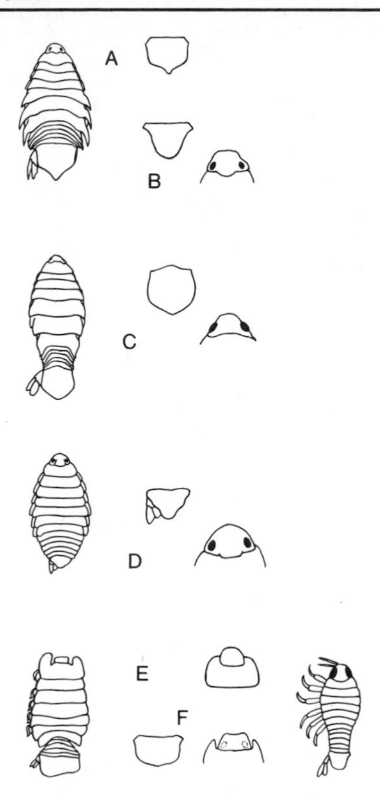

Fig. 21.23. Crustacea, Isopoda, Cymothoidae. A, *Nerocila acuminata*, whole animal plus detail of telson (right), 21 mm; B, *Nerocila munda*, telson and head, 13 mm; C, *Anilocra laticauda*, adult, whole animal and details of telson and head, 34 mm; D. *Lironeca* sp. and details of the telson and head of *L. ovalis*, 21 mm; E, *Cymothoa oestrum*, adult female (left) and head of male (right), 30 mm; F, *Cymothoa excisa*, telson and head of adult male and young whole animal (left), 23 mm.

Group E. The few local species of Asellota and Oniscoidea may be considered together. The freshwater *Asellus* and 3 maritime terrestrial oniscoid genera, *Philoscia*, *Actoniscus*, and *Scyphacella*, are illustrated in Figures 21.26, 21.27, but are not considered in the key; see Richardson (1909). The marine genera are:

---

1. Abdomen with 6 segments; 22 mm; Fig. 21.26D. *Ligia.*

1. Abdomen a single segment—2.

2. Telson narrow; eyes lateral; 3 mm; Fig. 21.27D. *Munna.*

2. Telson broad and deep; eyes dorsal—3.

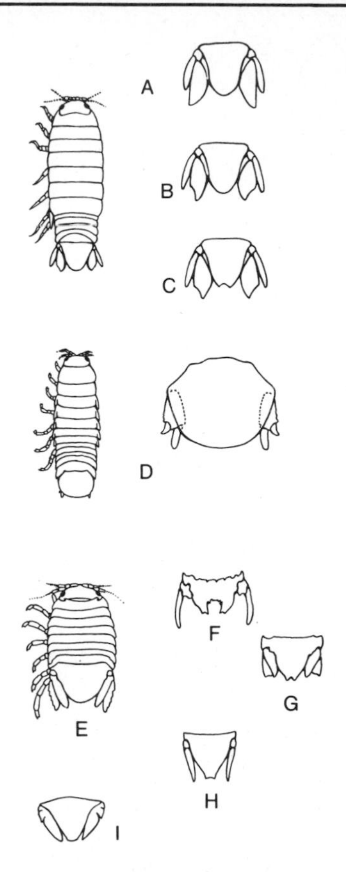

Fig. 21.24. Crustacea, Isopoda, Cirolanidae, Limnoridae, and Sphaeroidae. A, *Cirolana borealis*, telson and uropods, 12 mm; B, *Cirolana polita*, whole animal and detail of telson and uropods, 16 mm; C, *Cirolana concharum*, telson and uropods; D, *Limnoria lignorum*, whole animal and detail of telson and uropods, 3 mm; E, *Sphaeroma quadridentatum*, whole animal, 10 mm; F, *Cilicaea caudata*, telson and uropods, male, 10 mm; G, same, female, 6 mm; H, *Ancinus depressus*, telson and uropods, 13 mm; I, *Cassidisca lunifrons*, telson and uropods, about same size as preceding.

3. Uropods minute; all legs ambulatory; 7 mm; Fig. 21.27C. *Jaera.*
3. Uropods well developed; first pair of legs prehensile; 7 mm; Fig. 21.27B. *Janira.*

Parasitic Isopods. The less aberrant parasitic isopods, which may be taken in a free state and are of conventional form, have already been considered. The Epicaridea, however, are metamorphic and, in the adult female especially, are quite degenerate. Local species belong to the family Bopyridae,

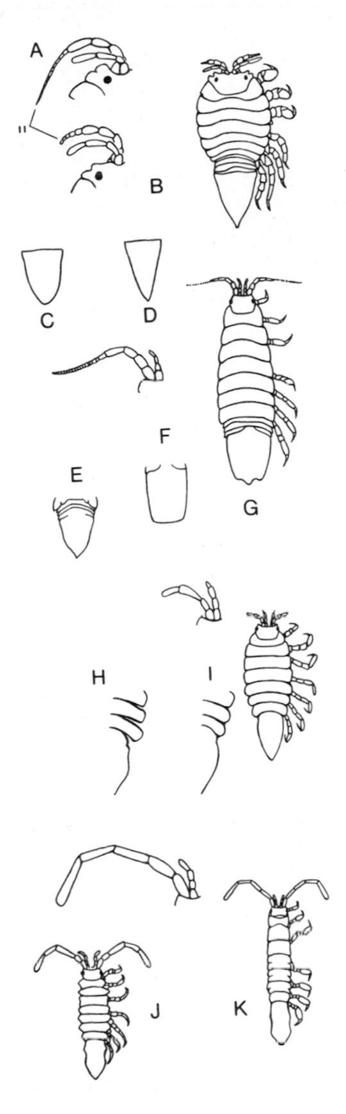

Fig. 21.25. Crustacea, Isopoda, Valvifera. A, *Chiridotea tuftsi*, part of head and left antennae, telson as in D, below; B, *Chiridotea coeca*, whole animal and detail of part of head and left antennae; C, *Chiridotea almyra*, telson (antenna II as in *tuftsi*); D, *Chiridotea arenicola*, telson (antenna II short as in *caeca*); *Chiridotea nigrescens* resembles *C. coeca* but is black; E, *Idotea phosphorea*, telson; F, *Idotea metalica*, telson; G, *Idotea baltica*, whole animal and detail of antenna (left); H, *Edotea triloba*, outline of posterior part of body, right dorsal view; I, *Edotea montosa*, whole animal and antenna (left above) and outline of posterior part of body (left below); J, *Erichsoniella filiformis*, whole animal; K, *Erichsoniella attenuata*, whole animal and detail of antenna (left).

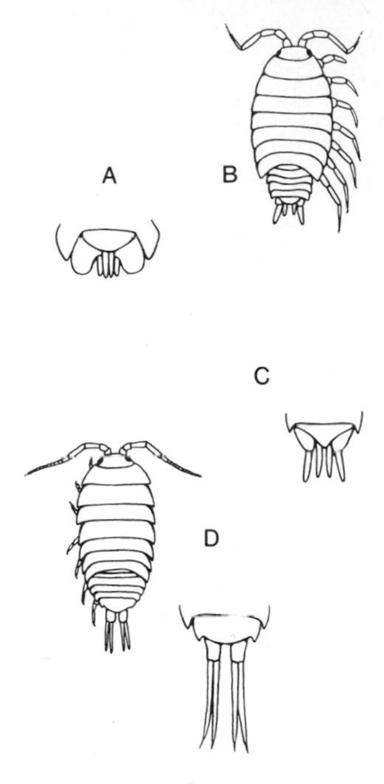

Fig. 21.26. Crustacea, Isopoda, Oniscoidea. A, *Actoniscus ellipticus*, telson and uropods; B, *Philoscia vittata*, whole animal; C, *Scyphacella* sp., telson and uropods; D, *Ligia oceanica*, whole animal (left) and telson and uropods of *L. exotica* (right) showing proportionately longer uropods in the second species.

a representative example of which is shown in Figure 21.28. Of the local species listed by Richardson (1909), *Hemiarthrus* (= *Phryxus*) attaches itself to its host's abdomen; the other genera are found in the gill chamber. All have decapod crustaceans as their final host. In some cases at least the first larval form is free-swimming; the second stage, called a *microniscus*, is parasitic on a copepod; it leaves after several moults as a generalized isopod form, called a *cryptoniscus*, and seeks a final decapod host. The first parasite to arrive becomes a female, but any subsequent individual arriving on an already occupied host becomes a male. Both male and female bopyrids are rather aberrant, the males remaining in a somewhat undifferentiated cryptoniscuslike state, symmetrical with prehensile legs but frequently with a degenerate abdomen. Males are a third or less the size of females, and local forms range in size from about 1–5 mm. The females are asymetrical; the body is segmented, and the full complement of 7 leg-pairs sometimes occurs on one side only. The antennae are rudimentary, the abdomen frequently without uropods. A bopyrid parasite in the branchial chamber usually produces a noticeable

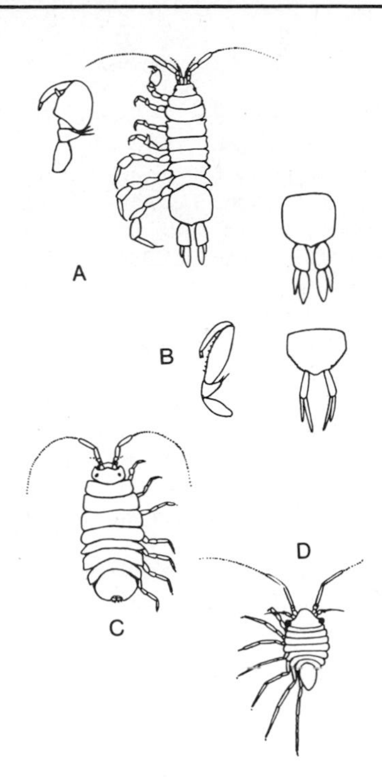

Fig. 21.27. Crustacea, Isopoda, Asellota. A, *Asellus* sp., whole animal and details of gnathopod (left) and posterior end (right); B, *Janira alta*, gnathopod and posterior end; C, *Jaera marina*, whole animal; D, *Munna fabricii*, whole animal. Adult *Asellus* are about 5 mm long.

swelling, since the adult female bopyrid is not minute compared to the size of its host. As in the case of parasitism by sacculinid barnacles, bopyrid infestation may induce a degeneration in the host's reproductive apparatus, and males so castrated resemble females.

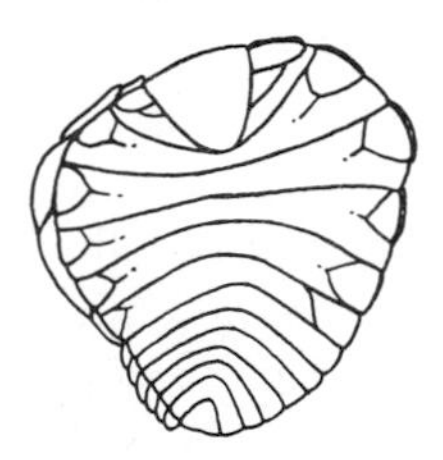

Fig. 21.28. Crustacea, Isopoda, Epicaridea. *Probopyrus pandalicola*, adult female, whole animal.

References

Bayliff, W. H. 1938. A new isopod crustacean (Sphaeromidae) from Cold Spring Harbor, Long Island. Trans. Am. microsc. Soc. 57:213–17.

Bowman, T. E. 1955. The isopod genus *Chiridotea* Harger, with a description of a new species from brackish waters. J. Wash. Acad. Sci. 45:224–29.

Frankenberg, D. 1965. A new species of *Cyathura* (Isopoda, Anthuridae) from coastal waters off Georgia, U.S.A. Crustaceana. 8(2):206–212.

Menzies, R. J. 1957. The marine borer family Limnoridae (Crustacea, Isopoda). Part I. Bull. mar. sci. Gulf Caribb. 7:101–200.

———, and D. Frankenberg. 1966. Handbook on the common marine isopod Crustacea of Georgia. Univ. of Georgia Press.

Miller, M. A., and W. D. Burbanck. 1961. Systematics and distribution of an estuarine isopod crustacean, *Cyathura polita*. Biol. Bull. mar. biol. Lab. Woods Hole. 120:62–84.

Richardson, H. 1909. A monograph of the isopods of North America. Bull. U.S. natn. Mus. 54:1–727.

Tattersall, W. M. 1911. Die nordischen Isopoden. Nord. Plankt. 3(6):179–314 (1915–27). (*Eurydice.*)

Waterman, T. H. ed. 1960, 1961. Physiology of Crustacea. Vols. 1, 2. Academic Press.

Wigley, R. L. 1960. A new species of *Chiridotea* (Crustacea: Isopoda) from New England waters. Biol. Bull. mar. biol. Lab. Woods Hole. 119(1):153–160.

———. 1961. A new isopod, *Chiridotea nigrescens*, from Cape Cod, Massachusetts. Crustaceana. 2:286–292.

Order Amphipoda
BEACH FLEAS, SCUDS, ETC.

Diagnosis. *Small to medium-sized peracaridans without a carapace, body regularly segmented and usually compressed from side to side; the degree of flattening varies, and in some of the tubicolous forms the body may be nearly cylindrical or even somewhat depressed from top to bottom; abdomen ends in a telson and three pairs of uropods.*

The maxillipeds are small and function as mouth parts; most students of amphipods therefore consider them part of the head. There are 7 pairs of functionally thoracic appendages. The first 2 pairs, differentiated as gnathopods, are usually subchelate. Of the remaining thoracic appendages the next 2 pairs are frequently flexed forward and the last 3 backward and outward. The segment joining the legs to the body, including the coxal plate, is often conspicuous, adding considerably to the depth of the compressed body.

Adults of most local species range between 2 and 40 mm in body length; a few, mostly oceanic forms, grow larger. Most amphipods have but 3 pairs of swimming appendages or pleopods on the abdomen. Most have compound, sessile eyes; a few species are blind, and local ampeliscids have 4 eyes, all of which are simple ocelli.

The caprellid amphipods are aberrant, with a slender, more or less cylindrical body, the abdomen reduced to a stub, the first 2 pairs of limbs behind the gnathopods rudimentary, and coxal plates vestigial or lacking (Fig. 21.38).

The sexes are separate in amphipods and frequently externally dimorphic; males are usually larger and may have much more powerfully developed gnathopods and second antennae. There are no special larval forms.

Amphipods occur in virtually all marine habitats. Some genera, including *Ampelisca* and *Haustorius*, are filter feeders, using suspended detritus for food. Others, particularly of the tubicolous families but including some species of *Ampelisca* as well, are described as deposit feeders subsisting as herbivores or on detritus; still others are omnivorous scavengers feeding on dead animals and decaying seaweeds. A few are predacious. In the huge gammaridean assemblage, only a few, such as *Lafystius* among local genera, are fully parasitic. Others, however, are commensal: *Leucothoe* on *Pecten*, sponges, and ascidians; *Dulichia* on the sea scallop; *Cerapus* and *Lembos* on the ascidian *Amaroucium;* and *Listriella* on the annelid *Clymenella* or on the sipunculid *Golfingia*. Several of the hyperiids are associated with jellyfish or salps as parasites and, among caprellids, members of the family Cyamidae are parasitic on whales.

Among characteristic interspecies associations are those of tubicolous forms with the larger hydroids such as *Tubularia;* the pleustids and *Metopa*, *Batea*, *Melita*, and *Gammarellus* are also frequently found with hydroids. *Dexamine*, and *Hyale* are particularly associated with macroscopic algae. Many of the genera just mentioned occur on algae *and* with hydroids. Caprellids are also frequently found in hydroid colonies. The euryhaline species are likely to be especially numerous in eelgrass or *Ruppia* beds; the typical genera of this habitat include *Gammarus*, *Ampithoe*, *Melita*, *Elasmopus*, and *Cymadusa*.

The hyperiids are exclusively planktonic and most are oceanic. A few gammarideans also occur regularly in the plankton and many more appear as tychoplankters. *Calliopius*, *Monoculodes*, *Pontogeneia*, and *Gammarus annulatus* are frequently reported in plankton surveys. A considerable variety has been taken in small nets attached to the wings of otter trawls; some at least of the forty-odd gammarideans reported from Georges Bank by Wheateley (1948) are probably tychoplankters. Many benthic amphipods also swim at times of reproduction, particularly those species in which the male is slimmer, more setose, and has longer antennae than the female. Amphipods are usually but a minor constituent of the plankton, but at times occur in immense numbers.

Amphipods occur on all substrata, at all depths, and in both fresh and fully salt waters.

Systematic List and Distribution

The basic list is that of Holmes (1905) with additions from more recent literature; about 25% of New England amphipod species have been newly described since 1960. A few families or lesser categories have undergone recent revisions, but much of the order awaits such attention. In Gammaridea the nomenclature conforms to Barnard's index (1958) except as indicated.

Suborder Hyperiidea		
Family Oxycephalidae		
Oxycephalus clausi	V	—
Vibilia spp.	CV	Gulf Stream
Family Phronimidae		
Phronima spp.	CV	Gulf Stream (with salps)
Family Phrosinidae		
Anchylomera blossevilli	CV	—
Family Cystisomidae		
Cystisoma spinosum	B	—
Family Hyperiidae		
Hyperoche tauriformis	B	Neritic, 27 to 146 m
Parathemisto compressa	B+	Neritic
Parathemisto oblivia	B	—
Parathemisto gaudichaudi	V	Neritic
Hyperia galba	BV	Neritic, on *Aurelia*
Hyperia medusarum	BV	Neritic, on *Cyanea*
Suborder Gammaridea		

Because of the very large number of species in this order, the largest single assemblage of crustaceans excepting the copepods, the families are listed in alphabetical order.

Family Acanthonotozomatidae		
Acanthonotozoma serratum	A, 44°	9 to 91 m
Family Ampeliscidae		
Ampelisca abdita	V, 44°	Lit. to 27 m, eury.
Ampelisca vadorum	BV	Lit. to 64 m, eury.
Ampelisca agassizi	V, 44°	Lit. to 823 m
Ampelisca macrocephala	B	9 to 183 m
Ampelisca verrilli	V	Lit. to 46 m
Byblis serrata	B, 37°	Lit. to 29 m
Byblis gaimardi	A+	18 to 110 m
Family Ampithoidae		
Ampithoe rubricata	BV	Lit. to 66 m, eury.
Ampithoe longimana	B, 37°	Lit. to 9 m, eury.
Cymadusa compta	V	Lit. to 9 m, eury.
Sunamphithoe pelagica	V	Pelagic
Family Aoridae		
Microdeutopus gryllotalpa	V, 40°	Lit. to 9 m, eury.
Microdeutopus damnoniensis	V, 41°	Lit. to 9 m
Lembos smithi	V	5 to 15 m
Lembos websteri	V	Lit. to 18 m

Family Argissidae		
Argissa hamatipes	B	9 to 183 m
Family Atylidae		
Atylus swammerdami	V	29 m
Family Bateidae		
Batea catharinensis	V	2 to 46 m
Family Calliopidae		
Calliopius laeviusculus	B+	Lit. to 15 m
Halirages fulvocinctus	B	49 to 155 m
Family Cheluridae		
Chelura terebrans	V	2 to 4 m
Family Corophiidae		
Corophium acutum	V	Lit. to 9 m
Corophium acherusicum	V, 43°	Lit. to 9 m, eury.
Corophium crassicorne	B	Lit. to 46 m
Corophium bonelli	B	Lit. to 74 m
Corophium insidiosum	BV	Lit. to 37 m, eury.
Corophium lacustre	V, 46°	Lit. to 9 m, eury.
Corophium tuberculatum	V, 46°	Lit. to 18 m
Corophium volutator	B	Lit. eury.
Cerapus tubularis	V	Lit. to 11 m
Siphonoecetes smithianus	V	37 to 53 m
Erichthonius rubricornis	V, 42°	9 to 91 m
Erichthonius brasiliensis	V	Lit. to 46 m
Erichthonius difformis	B	9 to 91 m
Unciola irrorata	V	Lit. to 284 m
Unciola serrata	V	4 to 18 m
Unciola dissimilis	V, 43°	Lit. to 926 m
Unciola obliqua	B	Lit. to 55 m
Unciola inermis	B	Lit. to 183 m
Unciola spicata	V	37 to 812 m
Unciola laticornis	B, 37°	104 to 3237 m
Family Dexaminidae		
Dexamine thea	B+	Lit. to 9 m
Family Eusiridae		
Eusirus cuspidatus	A+	—
Rhachotropis aculeata	A+	20 to 223 m
Family Gammaridae		
Gammarus oceanicus	B	Lit. to 37 m, eury.
Gammarus setosus	B	Lit. to 37 m, eury.
Gammarus fasciatus	V	Fresh water
Gammarus tigrinus	BV	Lit. to 9 m, low eury.
Gammarus mucronatus	BV	Lit. to 9 m, eury.
Gammarus lawrencianus	B	Lit. to 9 m, eury.
Gammarus annulatus	V	Lit. pelagic
Gammarus daiberi	V	Pelagic
Elasmopus laevis	V	Lit. to 9 m, eury.
Rivulogammarus sp.	BV	Lit. (tide pools), eury.
Marinogammarus finmarchicus	B	Lit. (tide pools), eury.
Marinogammarus obtusatus	B	Lit.

Marinogammarus stoerensis	B	Lit., eury.
Gammarellus angulosus	B+	Lit. to 9 m
Casco bigelowi	B	Lit. to 42 m
Melita dentata	B	Lit. to 292 m
Melita nitida	V, 43°	Lit. to 9 m
Melita parvimana	V	—
Maera danae	B	Lit. to 91 m
Crangonyx pseudogracilis	BV	Fresh water, rivers
Family Haustoriidae		
Amphiporeia lawrenciana	B	Lit. to 45 m
Amphiporeia virginiana	BV	Lit. to 2 m, eury.
Priscillina armata	B	20 to 45 m
Bathyporeia quoddyensis	B	Lit. to 18 m
Protohaustorius deichmannae	V, 44°	Lit. to 2 m
Protohaustorius wigleyi	V, 43°	Lit. to 55 m
Parahaustorius longimerus	V	Lit. to 37 m
Parahaustorius holmesi	V	9 to 55 m
Parahaustorius attenuatus	V	Lit. to 37 m
Acanthohaustorius millsi	V, 44°	Lit. to 55 m
Acanthohaustorius spinosus	B	Lit. to 183 m
Acanthohaustorius shoemakeri	V	16 to 26 m
Acanthohaustorius intermedius	V	Lit. to 37 m
Haustorius canadensis	V, 47°	Lit. to 4 m
Neohaustorius schmitzi	V	Lit., eury.
Neohaustorius biarticulatus	V	Lit. eury.
Lepidactylus dytiscus	V	Lit. to 2 m
Pseudohaustorius caroliniensis	V	Lit. to 2 m
Pseudohaustorius borealis	V	13 to 55 m
Pontoporeia femorata	B	5 m
Family Hyalidae		
Hyale nilssoni	B	Lit.
Hyale plumulosa	V	Lit. to 2 m, eury.
Family Ischyroceridae		
Ischyroceros anguipes	B, 41°	8 to 32 m
Jassa falcata	B, 40°	Lit. to 10 m
Family Lafystiidae		
Lafystius sturionis	V	Lit. to 9 m
Family Leucothoidae		
Leucothoe spinicarpa	B	73 to 110 m
Family Lilljeborgiidae		
Listriella barnardi	V	Lit. to 5 m
Listriella clymenellae	V	Lit. to 5 m
Family Lysianassidae		
Tmetonyx nobilis	B	Lit. to 33 m
Tmetonyx cicada	B+	37 to 1098 m
Hippomedon serratus	V, 37°	13 to 18 m
Anonyx lilljeborgi	B	Lit. to 183 m
Orchomonella pinguis	B+	Lit. to 549 m
Orchomonella minuta	B	Lit. to 183 m
Lysianopsis alba	V	Lit. to 4 m

Family Oedicerotidae		
Monoculodes edwardsi	V	Lit. to 60 m, eury.
Monoculodes tesselatus	B	Lit. to 91 m
Paroediceros lynceus	B	46 to 165 m
Family Paramphithoidae		
Paramphithoe pulchella	A+	73 to 260 m
Paramphithoe hystrix	B	9 to 547 m
Epimeria loricata	A+	64 to 914 m
Family Photidae		
Photis macrocoxa	B	Lit. to 106 m
Photis reinhardi	B	Lit. to 91 m
Photis dentata	BV	18 to 260 m
Microprotopus ranei	V	Lit. to 9 m
Podoceropsis nitida	B+	Lit. to 110 m
Leptocheirus pinguis	B, 37°	Lit. to 274 m
Leptocheirus plumulosus	V	Lit. to 9 m, eury.
Family Phoxocephalidae		
Phoxocephalus holbolli	B+	Lit. to 91 m
Paraphoxus epistomus	V, 43°	Lit. to 9 m
Paraphoxus spinosus	V	Lit. to 9 m
Harpinia propinqua	B	9 to 91 m
Family Pleustidae		
Pleustes panopla	B	5 to 91 m
Stenopleustes gracilis	V	Pelagic
Sympleustes glaber	B	Lit. to 45 m
Neopleustes pulchellus	B	46 to 183 m
Family Podoceridae		
Dulichia porrecta	A+	5 to 37 m
Family Pontogeneiidae		
Pontogeneia inermis	B+	Lit. to 55 m
Family Stegocephalidae		
Stegocephalus inflatus	B+	Lit. to 183 m
Family Stenothoidae		
Stenothoe gallensis	B	Lit. to 9 m
Stenothoe peltata	B	55 to 201 m
Stenothoe minuta	V	Lit. to 9 m
Parametopella cypris	V	Lit. to 9 m
Metopella angusta	B	9 to 106 m
Metopa bruzeli	B	Lit. to 55 m
Family Talitridae		
Orchestia platensis	BV	Lit.
Orchestia grillus	B, 39°	Lit.
Orchestia uhleri	V, 43°	Lit., eury.
Orchestia gammarella	B	Lit.
Talorchestia megalophthalma	BV	Lit.
Talorchestia longicornis	BV	Lit., eury.
Family Tironidae		
Tiron spiniferum	A+	—
Syrrhoe crenulata	A, 42°	22 to 66 m

Identification

The suborders are differentiated as follows:

Suborder Hyperiidea. Typically with the head swollen to accommodate the greatly enlarged eyes; planktonic and primarily oceanic; Fig. 21.29.
Suborder Gammaridea. The eyes varying in size but not filling the head; typical amphipods of varied habits and distribution; Figs. 21.30–21.37.
Suborder Caprellidea. Highly aberrant amphipods with slender, sticklike body and rudimentary abdomen; Fig. 21.38.

A fourth suborder, Ingolfiellidea, has not yet been recorded on this coast, but may be expected in the interstitial fauna; see Mills (1966).

Dimensions in the keys that follow are approximate maximum total body length of adults. For further information see Stebbing (1888, 1906), Chevreux and Fage (1925), Stephenson (1923) and Sars (1895), where most of the species are described and figured.

SUBORDER HYPERIIDEA

Two atypical genera, *Vibilia* and *Oxycephalus*, are figured but are not included in the key.

1. Eyes divided into upper and lower parts; parasitic on salps; 8–10 mm; Fig. 21.29C. *Phronima* sp.
1. Eyes not divided—2.

2. Uropods without rami; 6 mm; Fig. 21.29D. *Anchylomera blossevilli.*
2. Uropods with rami (inner ramus fused to peduncle in *Cystisoma*)—3.

3. Gnathopods pincerlike; parasitic on jellyfish; 6 mm; Fig. 21.29K. *Hyperoche tauriformis.*
3. Gnathopods not pincerlike; Fig. 21.29J—4.

4. Third walking legs much longer than fourth or fifth; 30 mm; Fig. 21.29G. *Parathemisto compressa.*
4. Third, fourth, and fifth legs nearly equal—5.

5. Propodus of third leg stout; 10 mm; Fig. 21.29I. *Parathemisto gaudichaudi.*
5. Propodus of third leg slender; parasitic on jellyfish; 20 mm; Fig. 21.29E,F,H,J. *Hyperia:*
- With very hairy gnathopods; on *Cyanea. H. medusarum.*
- With sparsely haired gnathopods; on *Aurelia. H. galba.*

SUBORDER GAMMARIDEA

This suborder has upwards of 60 genera and well over 100 species in the range of this text and has the reputation of being taxonomically difficult.

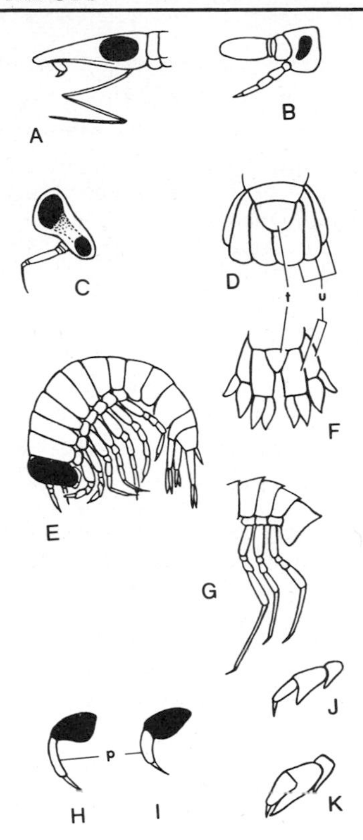

Fig. 21.29. Crustacea, Amphipoda, Hyperiidea. A, *Oxycephalus* sp., head, lateral view; B, *Vibilia* sp., head, lateral view; C, *Phronima* sp., head, lateral view; D, *Anchylomera blossevilli*, telson and uropods; E, *Hyperia* sp., whole animal; F, same, telson and uropods; G, *Parathemisto compressa*, detail of third through fifth walking legs and adjacent thoracic segments; H, *Hyperia medusarum*, third walking leg; I, *Parathemisto gaudichaudi*, third walking leg; J, *Hyperia* sp., gnathopod; K, *Hyperoche tauriformis*, gnathopod. Abbreviations are as follows: p, propodus; t, telson; u, uropods.

The difficulty comes particularly with isolated, immature specimens and from confusion due to sexual dimorphism and individual variation. With care and a little practice, most individuals can be placed with confidence—at least to genus, which is as far as this text attempts to go in some cases. Holmes (1903, 1905) and Kunkel (1918) are generally useful references for local amphipods and, unless other references are given with individual species, should be checked first for descriptions and figures of local forms. Sars (1895), Stephenson (1923–1944), and Shoemaker (1930) are very useful for Boreal species. See also Barnard (1958) for bibliographic citations. Much recent work is in scattered papers as cited.

Gammaridean amphipods are considered in 8 groups for which a preliminary key is given below. The keys are highly artificial in most cases.

---

1. Usually with an uncleft, simply rounded or blunt telson; exceptions with a slightly lobed or notched telson are either littoral "beach fleas," Fig. 21.31, or tubicolous forms, Fig. 21.32A,B, and are not modified for a fossorial existence—2.

1. Usually with a distinctly cleft telson; exceptions with a weakly lobed or simple telson are almost all modified for a fossorial existence with appendages powerfully broadened, spinose, and setose; Fig. 21.36A—3.

2. The four groups placed here may be summarized as follows:

- Group A. Forms readily distinguished by their unusual appearance, or with at least the fourth coxal plate greatly enlarged; Fig. 21.30. Families Stenothoidae, Cheluridae.
- Group B. Littoral and supralittoral "beach fleas" with strongly compressed body and deep coxal plates; uropod III uniramous; Fig. 21.31. Family Talitridae.
- Group C. Tubicolous forms and their relatives, generally less compressed than most amphipods and including cylindrical or even depressed, somewhat isopodlike species; tubes form feltlike or sandy colonial aggregations on piling, buoys, or other hard substrata or among the holdfasts of algae; a few species commensal or free-living, without tube; the telson frequently very small, reduced to a mere stub in some genera; the rostrum not produced to extend down between the antennae; Fig. 21.32. Families Corophiidae, Ischyroceridae, Aoridae, and Photidae.
- Group D. A miscellaneous assemblage of genera, one of which is a fish parasite; another has antennae with calceoli, another is extremely spiny, and the rest have a strongly developed rostrum; Fig. 21.33. Families Lysianassidae, Lafystiidae, Pleustidae, and Oedicerotidae.

3. Antenna I with an accessory flagellum, Fig. 21.32F, which may be multisegmented and conspicuous or no more than a minute, closely appressed scale only detectable by very careful scrutiny—4.

3. Antenna I without an accessory flagellum (note that immature gammarids may have an accessory flagellum that is so minute as to be barely detectable; such individuals may key out here, but will not work convincingly through the group key); Fig. 21.34.

- Group E. Families Ampeliscidae, Hyalidae, Dexaminidae, Pontogeneiidae, Bateidae, Atylidae, and Acanthonotozomatidae.

4. The three groups placed here may be summarized as follows:

- Group F. Telson usually longer than wide and very deeply cleft; antennae I and II about equal in length or with I longer; without a hood or strongly developed rostrum; Fig. 21.35. Family Gammaridae. This family includes some of the most familiar littoral and shallow water species. The accessory flagellum may be extremely small in young individuals (to the point of being undetectable).
- Group G. Telson usually wider than long, sometimes only weakly lobed or simple, in which case the species are usually modified for a fossorial existence with appendages powerfully broadened, spinose, and setose; antennae I and II may be subequal but usually II is longer; without a hood or strongly developed rostrum; Fig. 21.36F. Families Haustoriidae and Lysianassidae.

Members of the first family are common on sandy beaches; lysianassids occur mainly in somewhat deeper water in a variety of habitats.

- Group H. Rostrum well developed as a definite hood, i.e., expanded basally to cover the bases of the antennae, or extending down, somewhat beaklike, between the antennae; chiefly boreal and deep water species; Fig. 21.37. Families Phoxocephalidae, Eusiridae, and Tironidae.

Group A, Families Stenothoidae and Cheluridae. The stenothoids are small-to-minute amphipods with enormously developed coxal plates, giving the smaller ones the appearance of large ostracods at first glance. They are often abundant in hydroid colonies. See standard sources or original descriptions. *Chelura terebrans*, Figure 21.30E, is a wood-boring species, unmistakable in its bizarre appearance. Note that the coxal plates are enlarged in some other families as well, e.g., *Pleustes*, *Tmetonyx*. In Stenothoidae the telson is not notched, and the third uropods are uniramous.

The stenothoids may be identified as follows.

---

1. With only 1 coxal plate greatly enlarged—2.
1. Third, fourth, and fifth coxal plates greatly enlarged—3.
2. Sixth and seventh legs with an expanded basis (basis is second leg joint, counting the coxal plate as the first); 6 mm; Fig. 21.30D. *Stenothoe gallensis*. Note differences between this and the following species in the form of the second gnathopod as figured.
2. Sixth and seventh legs with a narrow basis; 2.0 mm; Fig. 21.30B,C. *Parametopella cypris*.

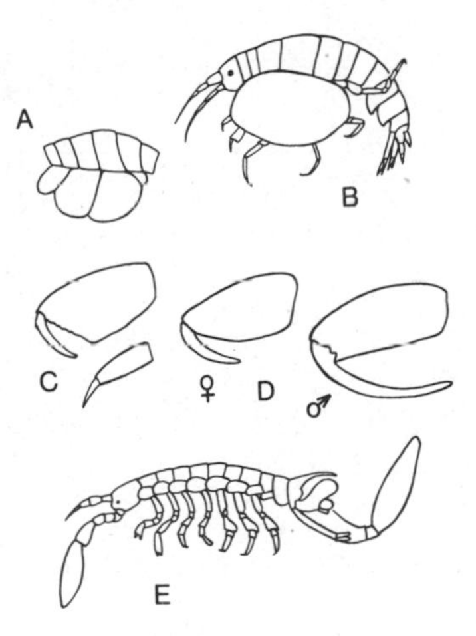

Fig. 21.30. Crustacea, Amphipoda, Gammaridea Group A. A, *Stenothoe minuta*, detail showing coxal plates III–V; B, *Parametopella cypris*, whole animal; C, same, gnathopods I and II, the latter similar in male and female; D, *Stenothoe gallensis*, gnathopod II of female (left) and male (right); E, *Chelura terebrans*, whole animal.

3. Fifth coxal plate much larger than fourth; gnathopod I simple; 4.0 mm. *Metopa bruzelli.*
3. Fifth coxal plate only slightly larger than fourth; gnathopod I subchelate; 2.5 mm; Fig. 21.30A. *Stenothoe minuta. Stenothe minuta* and *S. gallensis* are shallow water, Virginian species; A third species, *S. peltata*, is Boreal and deep water.

Group B, Beach Fleas, Family Talitridae. Species of *Hyale* (Group E) are also strong jumpers. For further details on Talitridae see Bousfield (1958).

---

1. Color very pale; sixth leg longer than seventh; ramus of uropod III longer than peduncle; gnathopod I of female simple; telson notched; 30 mm; Fig. 21.31A–C. *Talorchestia: T. longicornis* and *T. megalopthalma* differ in the proportions of the second antennae and in size of eyes as figured.
1. Color dark; sixth leg shorter than seventh; ramus of uropod III shorter than peduncle; gnathopod I of female subchelate; telson notched or simple; 10–18 mm; Fig. 21.31D. *Orchestia*—2.

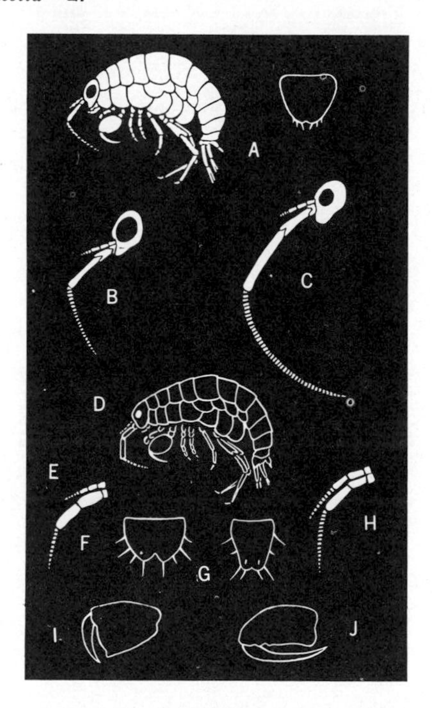

Fig. 21.31. Crustacea, Amphipoda, Gammaridea Group B. A, *Talorchestia* sp., whole animal and detail of telson; B, *Talorchestia megalopthalma*, head; C, *Talorchestia longicornus*, head; D, *Orchestia* sp., whole animal; E, *Orchestia grillus*, antennae; F, *Orchestia gammarella*, telson; G, *Orchestia uhleri*, telson; H, *Orchestia platensis*, antennae; I, *Orchestia gammarella*, gnathopod II, male; J, *Orchestia uhleri*, gnathopod II, male.

2. Telson notched; two species differentiated by form of telson and of male gnathopods as figured; Fig. 21.31F,G,I,J. *O. uhleri*, *O. gammarella*.

2. Telson simple—3.

3. Antenna I longer than next to last segment of peduncle of antenna II; Fig. 21.31E. *O. grillus*.

3. Antenna I shorter; Fig. 21.31H. *O. platensis*.

Group C. As previously noted, the telson is minute in some genera, and the accessory flagellum on the first antennae may be reduced to an appressed scale; careful manipulation of the substage lighting may be required to reveal this easily overlooked flagellum. The genera are not particularly difficult to differentiate, but the student is referred to original sources for species identification. Most of these amphipods are tubicolous. In some cases, individuals are hard to drive from their tubes, but many species have been found free and as tychoplankters. *Siphonoecetes* uses sand in its constructions; *Corophium*, *Jassa*, *Microdeutopus*, and others use finer sediments. *Ampithoe* is a nest builder and uses silk and fine debris. *Lembos* (*smithi*, at least) is a commensal with the ascidian *Amaroucium; Cerapus* makes a portable tube. *Unciola* does not make a shelter but resides in those of other amphipods or with tubicolous worms. *Sunampithoe* is pelagic but is also recorded from littoral algae.

The families are separated in the following key; for the differentiation of the individual genera see Table 21.11 and Figure 21.32, and for species see the references cited in the table.

---

1. Outer ramus of uropod III with terminal hooks on outer edge; Fig. 21.32O:

- Antenna II longer than I; gnathopod II very powerfully developed in male, Fig. 21.32G,H; accessory flagellum on antenna I small but easily detected, Fig. 21.32F. Family Ischyroceridae: genera *Ischyroceros* and *Jassa*.
- Antenna I longer than II; gnathopod II larger than first but not as above; Fig. 21.32M; accessory flagellum absent, or as above. Family Ampithoidae: genera *Ampithoe*, *Sunampithoe*, *Cymadusa*.

1. Outer ramus of uropod III without terminal hooks—2.

2. Gnathopod I more strongly developed than second:

- Uropod III uniramous, Fig. 21.32A–D, but note *Unciola* with a spur on the peduncle simulating a ramus; Fig. 21.32I. Family Corophiidae: genera *Corophium*, *Cerapus*, *Erichthonius*, *Siphonoecetes*, and *Unciola*.
- Uropod III biramous, Fig. 21.32J–L. Family Aoridae: genera *Lembos* and *Microdeutopus*.

2. Gnathopod II more strongly developed than first. Fig. 21.32N,P. Family Photidae: genera *Photis*, *Leptocheirus*, and *Podoceropsis*.

Table 21.11.

Genus	Uropodal Rami (Urop. III)	Accessory Flagellum Present	Telson	Adult Length	Longest Antenna	Gnathopods	Largest Gnathopods	Coxal Plates	Number of Local Species	Reference for Figures Not Included Here
Corophium	1	–	u	4	2	w–m	2	s	8	Shoemaker (1947)
Cerapus	1	–	d	5	e	m	2	s	1	1 local species
Erichthonius	1	–	d	13	e	m–*l*	2	s	3	See original descriptions*
Siphonoecetes	1	–	u	6	2	m	2	s	1	1 local species
Unciola	1	+	u	15	1	m	1	s	7	Shoemaker (1945b)
Microprotopus	1	+	u	3	e	m	2	m–*l*	1	Wigley (1966)
Sunampithoe	2	–	u	9	1	m	2	m	1	Keyed, Mills (1964)
Ampithoe	2	–	u	20	1	m	2	m–*l*	3	Mills (1964)
Cymadusa	2	+	u	12	1	m–*l*	e	*l*	1	1 local species
Ischyrocerus	2	+	u	10	2	*l*	2	m	1	1 local species
Jassa	2	+	u	10	2	*l*	2	m	1	1 local species
Lembos	2	+	u	6	1	*l*	1	s	2	See original description*
Microdeutopus	2	+	u	8	1	*l*	1	s	2	Holmes (1905)
Photis	2	–	u	5	e	w–m	2	*l*	3	Shoemaker (1945a)
Leptocheirus	2	+	u	13	1	w	1	*l*	2	See original descriptions*
Podoceropsis	2	–	u	7	e	m–*l*	2	m–*l*	1	1 local species

* Some species in Kunkel (1918), Holmes (1903, 1905). Abbreviations are: d, lobed; e, equal; *l*, strongly developed or large; m, moderately developed; s, small; u, simple; w, weak.

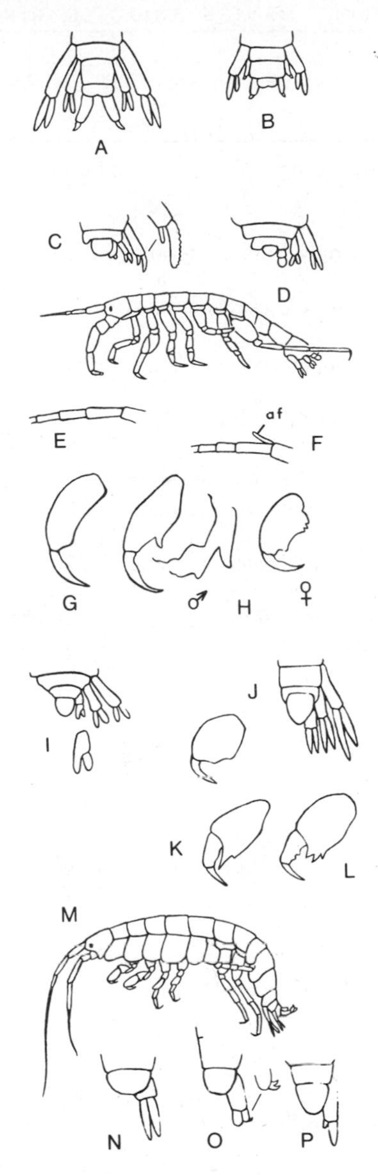

Fig. 21.32. Crustacea, Amphipoda, Gammaridea, Group C. A, *Erichthonius*, telson and uropods; B, *Cerapus*, telson and uropods; C, *Siphonoecetes*, telson and uropods; D, *Corophium*, whole animal and detail of telson and uropods; E, antenna without accessory flagellum; F, antenna with accessory flagellum; G, *Ischyrocerus*, gnathopod II; H, *Jassa*, male gnathopod II with variations in ventral margin (left) and female gnathopod II (right); I, *Unciola*, telson and uropods with detail of uropod III; J, *Lembos*, telson and uropods and gnathopod I: K, *Microdeutopus damnoniensis*, gnathopod I; L, *Microdeutopus gryllotalpa*, gnathopod I; M, *Ampithoe*, whole animal; N, *Leptocheirus* and *Podoceropsis*, telson and uropod III; O, *Ampithoe*, telson and uropod III; P, *Photis*, telson and uropod III.

Group D, Families Lysianassidae, Lafystiidae, Pleustidae, and Oedicerotidae.

1. Antennae with calceoli; 13 mm; planktonic and benthic on algae; Fig. 21.33A. *Calliopius laeviusculus.*
1. Antennae without calceoli—2.
2. Abdominal segments strongly spined laterally and dorsally; 19 mm; Fig. 21.33D. *Paramphithoe hystrix.*
2. Abdominal segments without lateral spines, with or without lateral knobs, and/or dorsal spines—3.
3. Antenna I with well-developed accessory flagellum; 6 mm; Fig. 21.33B. *Lysianopsis alba.*
3. Antenna I without accessory flagellum—4.
4. Abdominal segments spined or knobbed above—5.
4. Abdominal segments without spines or knobs—6.
5. Abdomen with rounded knobs dorsally and laterally; 15 mm; Fig. 21.33E. *Pleustes panopla.*
5. Abdomen with dorsal spines posteriorly, Fig. 21.33G; 5 mm; Virginian. *Stenopleustes gracilis.* Two Boreal species may also key out here; *Haliragesfulvocinctus*, a calliopid, is 17 mm, and has posterior abdominal segment notched as in *Pleustes*, but telson is blunt-ended; *Neopleustes pulchellus* is 17 mm, has a bluntly rounded telson, and lacks the lateral notch in the last abdominal segment.

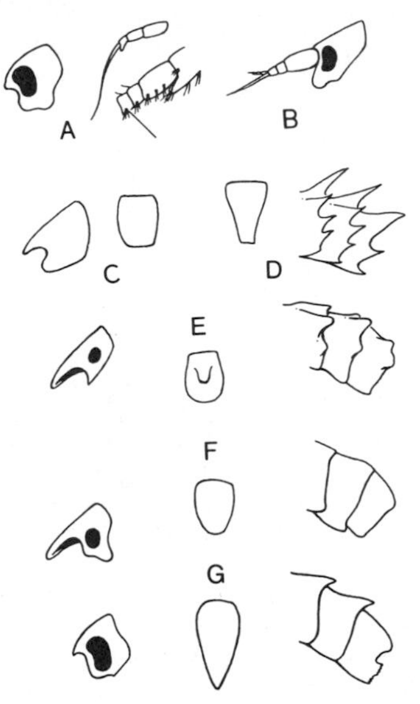

Fig. 21.33. Crustacea, Amphipoda, Gammaridea, Group D. E–G, left column shows head only; middle column, telson; right column, abdominal segments in lateral view; A, *Calliopius laeviusculus*, head, antenna, and detail of base of flagellum with calceoli indicated; B, *Lysianopsis alba*, head and first antenna; C, *Monoculodes edwardsi*, head and telson; D, *Paramphithoe hystrix;* E, *Pleustes panopla;* F, *Lafystius sturionis;* G, *Stenopleustes gracilis.*

6. Antenna II stronger than I; 9 mm; frequently planktonic; Fig. 21.33C. *Monoculodes edwardsi* and *M. tesselatus*, the latter with notched telson.
6. Antennae I and II subequal or I stronger; 6 mm; Fig. 21.33F; parasitic on fish. *Lafystius sturionis*.

Group E, Families Ampeliscidae, Hyalidae, Dexaminidae, Pontogeneiidae, Bateidae, Atylidae, and Acanthonotozomatidae.

1. With 4 small eyes; Fig. 21.34C,D. Ampeliscidae—2.
1. With only 2 eyes—3.

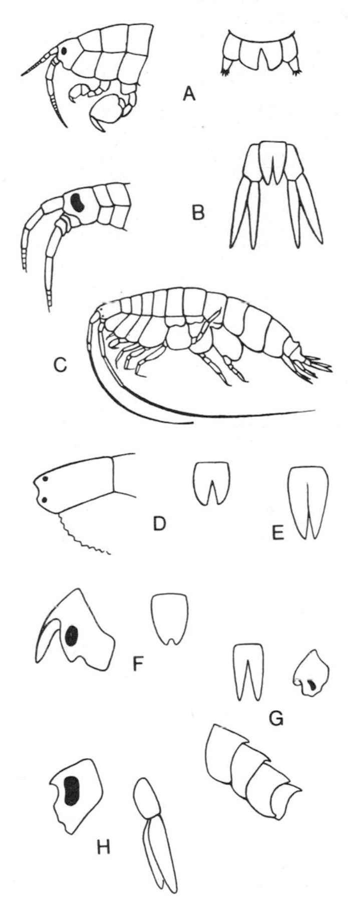

Fig. 21.34. Crustacea, Amphipoda, Gammaridea Group E. A, *Hyale* sp., fore part of body and telson and uropod III; B, *Pontogeneia inermis*, fore part of body and telson and uropod III; C, *Ampelisca* sp., whole animal; D, *Byblis serrata*, head region and telson; E, *Ampelisca* sp., telson; F, *Acanthonotozoma serratum*, head and telson; G, *Dexamine thea*, head, telson, and abdominal segments in lateral view; H, *Atylus swammerdami*, head and uropod III.

2. Telson short; 11 mm; Fig. 21.34D. *Byblis:* In *B. serrata* the lower margin of the first coxal plate is saw-toothed; in *B. gaimardi* it is not.
2. Telson long, Fig. 21.34E; 6–15 mm. *Ampelisca:* See subkey below. (*Ampelisca* constructs a parchmentlike tube with imbedded sand grains.)
3. Uropod III uniramous; 8 mm; Fig. 21.34A. *Hyale:* In *H. plumulosa* the first uropod has a strong inter-ramal spine and the back of the next to last segment is smooth; in *H. nilssoni* this segment has hairs behind and the inter-ramal spine is lacking.
3. Uropod III biramous—4.
4. Antennae with calceoli (see Fig. 21.33A) on peduncle; 11 mm; Fig. 21.34B. *Pontogeneia inermis.*
4. Antennae without calceoli—5.
5. Rostrum produced, Fig. 21.34F,G—6.
5. Rostrum not produced, Fig. 21.34H; 8 mm. *Atylus swammerdami.*
6. Telson with a short notch; Fig. 21.34F. *Acanthonotozoma serratum.*
6. Telson with a deep notch:
- Posterior abdominal segments with a dorsal spine; 3 mm; Fig. 21.34G. *Dexamine thea.*
- Posterior abdominal segments without spine; 5 mm. *Batea catharinensis.*

Subgroup One, Family Ampelisca. (Modified from Mills 1967).

1. Antennae I and II nearly equal. *A. aequicornis.*
1. Antenna I much shorter than antenna II—2.
2. Antenna I longer than the peduncle of antenna II—3.
2. Antenna I equal to or shorter than peduncle of antenna II—4.
3. Outer margin of outer ramus of uropod II with 3–5 spines. *A. vadorum.*
3. Outer margin of outer ramus of uropod II with 1 or 2 spines. *A. abdita.*
4. Antenna I about equal to the peduncle of antenna II. *A. macrocephala.*
4. Antenna I shorter than peduncle of antenna II. *A. verrilli.*

Group F, Family Gammaridae. Adults usually have an easily discerned accessory flagellum on the first antenna, but this accessory may be obscure in juveniles. The following key is largely based on that of Bousfield (1969).

1. Uropod III uniramous; 13 mm. *Crangonyx pseudogracilis.*
1. Uropod III biramous—2.
2. Telson slightly notched; Fig. 21.35A; 10 mm. *Gammarellus angulosus.*
2. Telson deeply cleft, Fig. 21.35B—3.
3. Inner ramus of uropod III rudimentary or very short, Fig. 21.35D,F—4.
3. Inner ramus more than half length of outer ramus—5.
4. Inner ramus scalelike; eye round or oval; 10–16 mm; Fig. 21.35E,F. *Melita:* In *M. dentata* the posterior margins of the abdominal segments are toothed, while in *M. nitida* they are not; *M. parvimana* differs from both in having the "palm" of gnathopod I about as short as broad. See Holmes (1903, 1905).

4. Inner ramus longer, eye bean-shaped; Fig. 21.35C,D. *Marinogammarus*. See Bousfield (1969) for the differentiation of three local species.
5. Last 3 abdominal segments with clusters of spines dorsally—7.
5. Last 3 abdominal segments without spines in clusters—6.

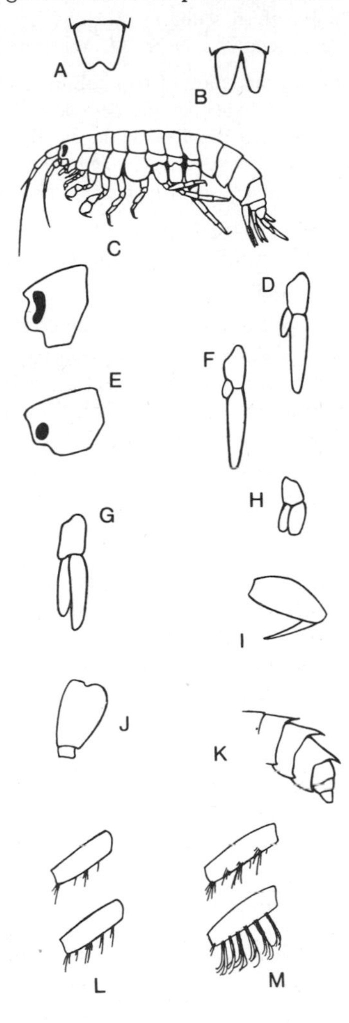

Fig. 21.35. Crustacea, Amphipoda, Gammaridea Group F. A, *Gammarellus angulosus*, telson; B, *Marinogammarus*, telson; C, same, whole animal; and detail of head; D, same, uropod III; E, *Melita*, head; F, same, uropod III; G, *Gammarus*, uropod III; H, *Elasmopus laevis*, uropod III; I, *Casco bigelowi*, gnathopod I; J, *Rivulogammarus* sp., basal segments of next to last pair of legs; K, *Gammarus mucronatus*, abdomen in lateral view; L, *Gammarus fasciatus* (above), *Gammarus tigrinus* (below), second peduncular segment of antenna I to show differences in setation on lower margin; M, same, peduncular segment of antenna II to show differences in setation on lower margin in male.

6. Rami of uropod III short and stout, Fig. 21.35H; 10 mm. *Elasmopus laevis.*

6. Rami of uropod III long:

- Antenna I shorter than antenna II; gnathopod I simple, without a palm (Fig. 21.35I); 25 mm (Shoemaker, 1930). *Casco bigelowi.*
- Antenna I not shorter than antenna II; gnathopod I subchelate (Stebbing, 1906). *Maera danae.* A second species, *M. loveni*, ranges south to Nova Scotia.

7. Second segment of next to last pair of legs lobed; Fig. 21.35J. *Rivulogammarus. R. duebeni* is a boreal, tide pool species; the genus, with two unnamed forms (one near *duebeni*) is recorded from the Chesapeake by Bousfield (1969). Fresh water forms link the genera *Rivulogammarus* and *Gammarus.*

7. Second segment of next to last pair of legs not lobed. *Gammarus*—8.

8. Body with a mid-dorsal keel, first 3 abdominal segments with a backward-projecting tooth or spine, Fig. 21.35K; 6 mm. *Gammarus mucronatus.*

8. Without dorsal keel or spines—9.

9. Abdominal segment IV with dorso-lateral spine clusters—10.

9. Dorso-lateral spine clusters absent on abdominal segment IV:

- Peduncle and base of flagellum of antenna I with long, feathery setae on lower margin; 15 mm (Shoemaker, 1930). *Gammarus annulatus.*
- Peduncle and base of flagellum without long feathery setae. *Gammarus lawrencianus.*

10. First coxal plate with 5–8 long setae at the antero-ventral angle; in *Gammarus fasciatus* the last 3 abdominal segments have a distinct mid-dorsal hump, while *G. daiberi* and *G. tigrinus* have only a small projection at the base of these segments; in *daiberi* the basal segments of the flagellum have posterior setae longer than twice the width of the segment, while in *tigrinus* these setae scarcely exceed the width of the segment; Fig. 21.35L,M.

10. First coxal plate with 1–5 short setae at the antero-ventral angle or with several short setae along the anterior margin—11.

11. Accessory flagellum on antenna I usually with 4 or 5 segments; body banded or mottled in life; 12 mm. *Gammarus palustris.*

11. Accessory flagellum usually with 6–10 segments; body unicolorous; 15–25 mm:

- Telson, last 3 abdominal segments and pereiopods with feathery setae. *Gammarus setosus.*
- These parts with simple setae only. *Gammarus oceanicus.*

Group G, Families Haustoriidae, Lysianassidae, and Liljeborgiidae. The following keys are based in part on Bousfield (1965). As a first step it is convenient to separate this assemblage into three subgroups.

---

1. Head and body segments broadly arched, tumid, or cylindrical; eyes small, weakly pigmented, or lacking; last 3 pairs of thoracic appendages broadly expanded, margins richly spinous and/or plumose; telson broad and short; fossorial species typical of sandy beaches; Fig. 21.36A. Subgroup One: Family Haustoriidae, Subfamily Haustoriinae.

1. Without the above combination of traits—2.

2. Rami of uropod III decidedly unequal. Subgroup Two: Family Haustoriidae, Subfamily Pontoporeiinae.
2. Rami of uropod III about equal. Subgroup Three: Families Lysianassidae and Liljeborgiidae.

Subgroup One, Subfamily Haustoriinae. Most of these amphipods are only keyed to genus; for further differentiation see Bousfield (1965).

1. Telson uncleft or only very slightly so—2.
1. Telson distinctly cleft or notched, Fig. 21.36B–D—3.

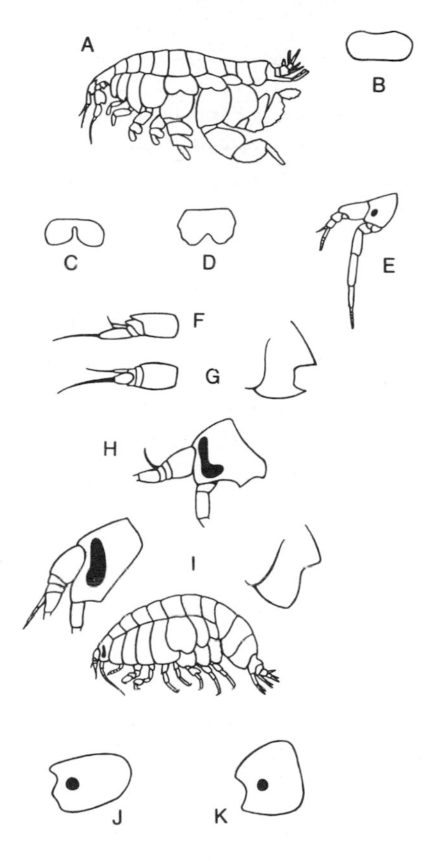

Fig. 21.36. Crustacea, Amphipoda, Gammaridea Group G. A, *Haustorius* sp., whole animal; B, *Pseudohaustorius carolinensis*, telson; C, *Acanthohaustorius* sp., abdomen; D, *Haustorius canadensis*, abdomen; E, *Amphiporeia* sp., head; F, *Hippomedon serratus*, first antenna; G, *Anonyx lilljeborgi*, first antenna and third abdominal segment in lateral view; H, *Tmetonyx cicada*, head; I, *Orchomonella* sp., whole animal, head, and third abdominal segment; J, *Argissa hamatipes*, head; K, *Listriella* sp., head.

2. Uropod I and II with 2 well-developed rami; 6.5 mm. *Pseudohaustorius:* In *P. carolinensis* the telson has a broadly U-shaped cleft; Fig. 21.36B; in *P. borealis* the telson is uncleft.
2. Uropod II uniramous and uropod I either uniramous or with short inner ramus; 5.3 mm. *Neohaustorius.*
3. Posterodorsal border of third abdominal segment strongly overlapping the uropod-bearing segments—4.
3. Posterodorsal border of third abdominal segment not so—5.
4. Lateral lobe of third abdominal segment with a recurving spine, Fig. 21.36C; 10–14 mm. *Acanthohaustorius.*
4. Lateral lobe rounded, Fig. 21.36D; 18 mm. *Haustorius canadensis.*
5. Body relatively slender, head not greatly broadened; 6 or 7 mm. *Protohaustorius.*
5. Body broadly arched; head broad; 9.5–14.0 mm. *Parahaustorius.*

Subgroup Two, Subfamily Pontoporiinae.

1. Antenna I with sharp bend, Fig. 21.36E; littoral, sandy beaches—2.
1. Not so; sublittoral—3.
2. Inner ramus of uropod III longer than peduncle; 7.5 mm; Fig. 21.36E. *Amphiporeia* spp. For the differentiation of the 2 local species see Shoemaker (1933).
2. Inner ramus of uropod III shorter than peduncle; 5.0 mm (Shoemaker, 1949). *Bathyporeia quoddyensis.*
3. Without eyes; antennal flagellum 3–4 segmented; gnathopods I and II similar; 11 mm. *Priscillina armata.*
3. Eyes pigmented; antennal flagellum with 2 segments; gnathopods I and II dissimilar; 14 mm (Sars, 1895). *Pontoporeia femorata.*

Subgroup Three, Families Lysianassidae and Liljeborgiidae.

1. Third abdominal segment deeply notched at the side, Fig. 21.36G; note difference in first antennae, as figured: *Anonyx lilljeborgi*, Fig. 21.36G, 20 mm; *Hippomedon serratus*, Fig. 21.36F, 12 mm.
1. Third abdominal segment without lateral notch; 4 genera differentiated by shape of head, eye, telson, as figured. *Tmetonyx cicada*, Fig. 21.36H, 15 mm; *Orchomonella* spp., Fig. 21.36I, 7 mm; *Argissa hamatipes*, Fig. 21.36J, 6 mm; *Listriella* spp., Fig. 21.36K, 5 mm. For further details on the species of *Anonyx* and *Orchomonella* see Sars (1895); for *Listriella* see Mills (1962) and Wigley (1966).

Group H, Families Phoxocephalidae, Eusiridae, and Tironidae.

1. Eyes absent; 7 mm; Fig. 21.37A. *Harpinia.* See Sars (1895).
1. Eyes present—2.
2. Eyes fused above, Fig. 21.37B; 10 mm. *Syrrhoe crenulata.*
2. Eyes separate—3.

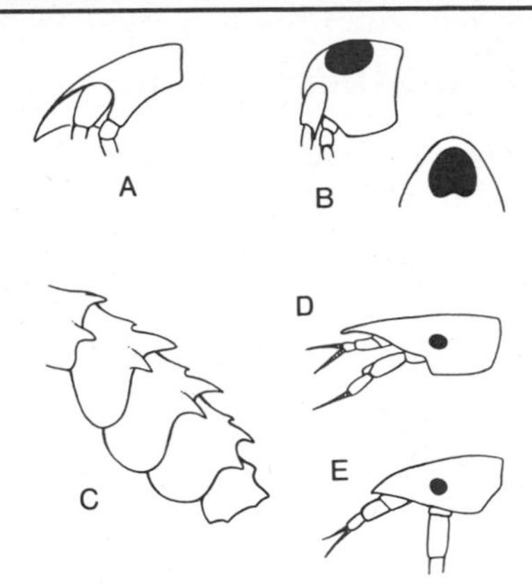

Fig. 21.37. Crustacea, Amphipoda, Gammaridea Group H. A, *Harpinia* sp., head; B, *Syrrhoe crenulata*, head in lateral view and detail of eye region from above; C, *Rhachotropis aculeata*, abdomen; D, *Phoxocephalus holbolli*, head; E, *Paraphoxus* sp., head.

3. Abdomen spiny; 30 mm; Fig. 21.37C; sometimes planktonic, *Rhachotropis aculeata*.
3. Abdomen not spiny—4.
4. Eyes small, Fig. 21.37D; 5 mm; sometimes planktonic. *Phoxocephalus holbolli*.
4. Eyes large, Fig. 21.37E; 5 mm; burrowing form. *Paraphoxus*. See Kunkel (1918) and Shoemaker (1938).

ORDER CAPRELLIDEA

The anatomical relationship between caprellids and more typical amphipods may not be immediately obvious, Figure 21.38. The "head" (or cephalon) and first thoracic segment are, in some cases at least, completely fused and bear the usual complement of cephalic appendages plus the first gnathopods. The second thoracic segment carries the second pair of gnathopods and sometimes a pair of gills. The third and fourth segments carry paddle-shaped gills, oostegites in females, and rudimentary limbs in some local species. The fifth, sixth, and seventh segments usually have 3 pairs of well-developed prehensive legs plus a vestigial abdomen. The largest species in the local fauna reaches 54 mm in body length, but most of the species grow no larger than about 15 mm, and some are mature at 3 or 4 mm.

Caprellids are somewhat "arboreal" in their association with hydroids, bryozoans, algae, and seagrasses. *Caprella unica* is chiefly associated with the sea stars *Asterias forbesi* and *A. vulgaris*. Several species, including *Caprella andreae*, have been reported from the fouling on sea turtles or on spider crabs. *Caprella andrea* is also associated with floating objects. *Hemiaegina minuta*, *Luconacia incerta*, and especially *Phtisica marina* have been taken in plankton tows.

Systematic List and Distribution

The group has recently been monographed by McCain (1968). The species are sublittoral; 0 in the following summary indicates that the species is found a little below low water.

Family Caprellidae		
Aeginella spinosa	B	0 to 1026 m
Aeginina longicornis	BV	0 to 2258 m
Caprella andreae	V	Pelagic
Caprella equilibra	V	0 to 3000 m, eury.
Caprella linearis	B+	0 to 200 m
Caprella penantis	BV	0 to 200 m, eury.
Caprella septentrionalis	B	0 to 1026 m
Caprella unica	B+	0 to 59 m
Hemiaegina minuta	C, 38°	Shallow
Luconacia incerta	V	Shallow
Mayerella limicola	B	9 to 713 m
Paracaprella tenuis	BV	Shallow, eury.
Phtisica marina	C+	Pelagic

Identification

The key is modified from McCain (1968). Measurements are approximate maximum body length of adults. There is substantial variation—individual, ontogenetic, and sexual—in spination, proportion of segments, and other details.

1. Mandible with a 3-segmented palp, Fig. 21.38A—2.
1. Mandible without a palp—6.

2. Third and fourth segments without legs, Fig. 21.38B—3.
2. Third and fourth segments with legs—4.

3. Abdomen with a pair of appendages and a pair of lobes, Fig. 21.38C; body spiny; 20 mm; *Aeginella spinosa*.

3. Abdomen with 2 pairs of appendages and a pair of lobes; Fig. 21.38D; body spiny or not; 54 mm. *Aeginina longicornis*.

4. Legs on segments III and IV two-segmented—5.
4. Legs on segments III and IV six-segmented, Fig. 21.38E; body smooth; 14 mm. *Phtisica marina*.

5. Legs on segment V two- or three-segmented; body smooth; 7 mm. *Mayerella limicola*.

5. Legs on segment V six-segmented:
- Legs in segment V inserted near middle of segment, Fig. 21.38F; body spiny or smooth; 9 mm. *Luconaca incerta*.
- Legs inserted in posterior part of segment, Fig. 21.38G; body usually smooth; 7 mm. *Paracaprella tenuis*.

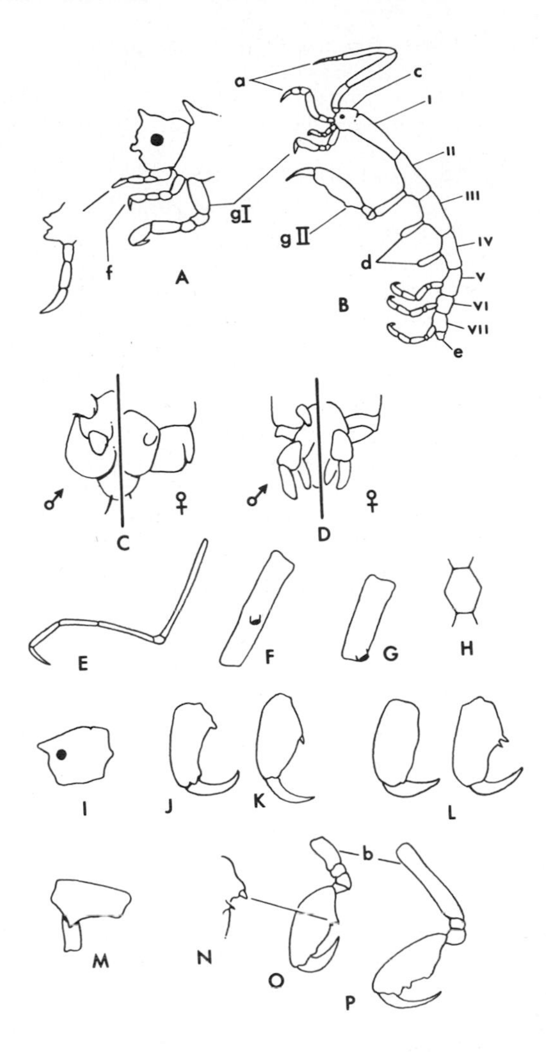

Fig. 21.38. Crustacea, Amphipoda, Caprellidea. A, *Aeginella* sp., head region and detail of mandible and palp; B, *Caprella* sp., whole animal; C, *Aeginella spinosa*, abdomen; D, *Aeginina longicornis*, abdomen; E, *Phtisica marina*, leg from segment III; F, *Luconacia incerta*, segment V; G, *Paracaprella tenuis*, segment V; H, *Hemiaegina minuta*, thoracic segment in dorsal view; I, *Caprella penantis*, cephalon; J, same, part of leg from segment V; K, *Caprella andreae*, part of leg from segment V; L, *Caprella unica* (left) and *Caprella equilibra* (right), part of leg from segment V or VI; M, *Caprella equilibra*, insertion of gnathopod II; N, *Caprella septentrionalis*, detail of gnathopod II; O, same, gnathopod II; P, *Caprella linearis*, gnathopod II. Abbreviations are as follows: a, antennae; b, basis; c, cephalon; gI, gnathopod I; gII, gnathopod II; d, gill; e, abdomen; f, maxilliped; leg-bearing segments (pereonites) numbered I–VII.

6. Legs on segments III and IV with a single segment; thoracic segments hexagonal in dorsal aspect, Fig. 21.38H; body smooth; 4 mm. *Hemiaegina minuta.*
6. Legs on segments III and IV two-segmented; 7 mm. *Paracaprella tenuis.* This species appears twice in the key because the mandibular palp may be present or not.
6. Segments III and IV without legs—7.
7. Cephalon with a "horn," Fig. 21.38I; body smooth:
- Palm of propodus of legs on segments V–VII concave, Fig. 21.38J; 14 mm; the commonest Virginian caprellid. *Caprella penantis.*
- Palm of propodus of legs on segments V–VII convex, Fig. 21.38K; 12 mm; usually on drifting objects. *Caprella andreae.*

7. Cephalon without a "horn"; body smooth or spiny—8.
8. Propodus of legs on segments V–VII without spines, Fig. 21.38L; body smooth or spiny; 18 mm. *Caprella unica.*
8. Propodus of legs on segments V–VII with spines; Fig. 21.38L—9.
9. With or without a ventral spine between the insertions of gnathopods II; propodus of gnathopod II without a small spine on inner surface near poison tooth; body smooth; 22 mm; Fig. 21.38L,M. *Caprella equilibra.*
9. Without a ventral spine between the insertions of gnathopod II; propodus of gnathopod II with a small spine on inner surface near poison tooth, Fig. 21.38N:
- Basis of gnathopod II short (the ratio total body length/length of basis is greater than 13, mean about 15); Fig. 21.38O; body usually spiny or bumpy; 20 mm. *Caprella septentrionalis.*
- Basis of gnathopod II long (ratio less than 13, mean about 10); Fig. 21.38P; body smooth or with only a few spines; 16 mm. *Caprella linearis.*

References

Barnard, J. L. 1957. Revisionary notes on the Phoxocephalidae (Amphipoda), with a key to the genera. Pacif. Sci. 12:146–151.

———. 1958. Index to the families, genera, and species of the gammaridean Amphipoda (Crustacea). Occ. Pap. Allan Hancock Fdn. 19:1–145.

Bate, C. S., and J. O. Westwood. 1868. A history of British sessile-eyed Crustacea. John Van Voorst, London.

Bousfield, E. L. 1958a. Fresh-water Amphipoda of glaciated North America. Can. Fld. Nat. 72:55–113.

———. 1958b. Distributional ecology of the terrestrial Talitridae (Crustacea: Amphipoda) of Canada. Proc. 10th Internatn. Congr. Ent. 1:883–898.

———. 1962. Studies on littoral marine arthropods from the Bay of Fundy Region. Bull. natn. Mus. Can. 183:42–62.

———. 1965. Haustoriidae (Crustacea: Amphipoda) of New England. Proc. U.S. natn. Mus. 117:159–240.

———. 1969. New Records of *Gammarus* from the Middle Atlantic region. Chesapeake Sci. 10(1):1–17.

Chevreux, E., and L. Fage. 1925. Amphipodes. Faune de France 9. Paul Lechevalier, Paris.

Croker, R. A. 1967. Niche diversity in five sympatric species of intertidal amphipods (Crustacea: Haustoriidae). Ecol. Monogr. 37:173–200.

Dexter, D. E. 1967. Distribution and niche diversity of haustoriid amphipods in North Carolina. Chesapeake Sci. 8(3):187–198.

Evans, F. 1968. The subgenera *Parathemisto* and *Euthemisto* of the genus *Parathemisto* (Amphipoda, Hyperiidea). Crustaceana 14:105–106.

Holmes, S. J. 1903. Synopses of North American invertebrates. 18. The Amphipoda. Am. Nat. 37:267–292.

———. 1905. The Amphipoda of southern New England. Bull. U.S. Fish. Commn. 24:457–529.

Jones, N. S. 1947. The ecology of the Amphipoda of the South of the Isle of Man. J. mar. biol. Ass. U.K. 27(2):400–439.

Kunkel, B. W. 1910. Amphipoda of Bermuda. Trans. Conn. Acad. Arts Sci. 16(1):1–115.

———. 1918. The Arthrostraca of Connecticut. St. Geol. Nat. Hist. Surv. Bull. 26:1–181.

McCain, J. C. 1965. The Caprellidae (Crustacea: Amphipoda) of Virginia. Chesapeake Sci. 6(3):190–196.

———. 1968. The Caprellidae (Crustacea: Amphipoda) of the western North Atlantic. Bull. U.S. natn. Mus. 278: vi + 147 pp.

Mills, E. L. 1962. A new species of Liljeboriid amphipod with notes on its biology. Crustaceana. 4(2):158–162.

———. 1963. A new species of Ampelisca (Crustacea: Amphipoda) from eastern North America, with notes on other species of the genus. Can. J. Zool. 41:971–989.

———. 1964. *Ampelisca abdita*, a new amphipod crustacean from eastern North America. Can. J. Zool. 42:559–575.

———. 1966. Deep-sea Amphipoda from the Western North Atlantic Ocean. 1. Ingolfiellidea and an unusual new species in the gammaridean family Pardaliscidae. Can. J. Zool. 45:347–356.

———. 1967. A reexamination of some species of *Ampelisca* (Crustacea: Amphipoda) from the east coast of North America. Can. J. Zool. 45:635–652.

Sars, G. O. 1895. An account of the Crustacea of Norway. Cristiana and Copenhagen.

Shellenberg, A. 1927. Amphipoda des Nordischen Plankton. Nord. Plankt. 6:589–722.

Shoemaker, C. R. 1926. Amphipods of the family Bateidae in the collection of the U.S. National Museum. Proc. U.S. natn. Mus. 68(25):1–26.

———. 1930. The Amphipoda of the Cheticamp Expedition of 1917. Contr. Can. Biol. Fish. 5(10):221–359.

———. 1933. A new amphipod of the genus *Amphiporeia* from Virginia. J. Wash. Acad. Sci. 23:212–216.

———. 1938. Two new species of amphipod crustaceans from the east coast of the United States. J. Wash. Acad. Sci. 28:326–332.

———. 1945a. The Amphipod genus *Photis* on the east coast of North America. Charleston Mus. Leafl. 22:1–17.

———. 1945b. The amphipod genus *Unciola* on the east coast of North America. Am. Midl. Nat. 34(2):446–465.

———. 1947. Further notes on the amphipod genus *Corophium* from the east coast of America. J. Wash. Acad. Sci. 37(2):47–63.

———. 1949. Three new species and one new variety of amphipods from the Bay of Fundy. J. Wash. Acad. Sci. 39(12):389–398.

Stebbing, T. R. R. 1888. Report on the Amphipoda. Challenger Rept. 29.

———. 1906. Das Tierreich, Leif 21, Amphipoda. I. Gammaridea.

Steinberg, Joan, and E. Dougherty. 1957. The skeleton shrimps (Crustacea: Caprellidae) of the Gulf of Mexico. Tulane Stud. Zool. 5(11):267–288.

Stephenson, K. 1923, 1925, 1931, 1944. Crustacea Malacostraca. Amphipoda. Dan.-Ingolf Exped.

Wheateley, G. C., Jr. 1948. Distribution of larger planktonic Crustacea on Georges Bank. Ecol. Monogr. 18(2):233–264.

Wigley, R. L. 1966. Two new marine amphipods from Massachusetts, U.S.A. Crustaceana. 10(3):259–270.

Order Mysidacea
MYSID SHRIMP

Diagnosis. *Small to medium-sized, shrimplike percaridans with carapace nearly covering the thorax but fused to only three segments at most; eyes on stalks; tail with distinctly biramous uropods and telson forming a fan, the inner rami of the uropods with a small, beadlike statocyst; at least five and sometimes all eight pairs of thoracic limbs biramous, the first and second pairs slightly modified as maxillipeds and gnathopods; Figure 21.40A–C.*

The largest local species reach about 30 mm, but the majority do not attain half that length. Mysids lacking the statocyst are extralimital, the nearest geographical form being *Lophogaster americanus* in the Gulf Stream; the statocyst can be detected in even the youngest individuals by the careful manipulation of the substage lighting of the microscope. Swimming appendages (*pleopods*) are rudimentary in all females and variously reduced in males of most species.

The sexes are separate and externally dimorphic. In addition to having rudimentary abdominal appendages, females have 2 or 3 pairs of oostegites forming a brood pouch.

Mysids are benthic, meroplanktonic, and hypoplanktonic; they may also be carried into the plankton accidentally but, of the forms considered, only the oceanic *Siriella* is regularly a surface plankter. Food consists of small plankters or bottom forms as well as detritus filtered from currents set up by the thoracic limbs. American freshwater mysids (two species) are extralimital, but *Neomysis americana* is a common euryhaline form; the other species are marine.

Systematic List and Distribution

Tattersall's (1951) review is the source of this list. All of the species of interest belong in the suborder Mysida and family Mysidae. The planktonic

Siriella thompsoni is virtually cosmopolitan but, of the others, about 5 neritic species occur in Europe as well as eastern North America.

Siriella thompsoni	BV	Oceanic
Erythrops erythropthalma	B	37 to 274 m
Meterythrops robustus	B	60 to 274 m
Mysidopsis bigelowi	V	15 to 196 m
Metamysidopsis munda	C, 38°	Shallow water
Mysis gaspensis	B	Lit. to 3 m, eury.
Mysis mixta	B	Lit. to 200 m
Mysis stenolepis	BV	Lit. to 183 m, eury.
Neomysis americana	BV	Lit. to 214 m, eury.
Heteromysis formosa	V, 43°	Lit. to 248 m

Praunus flexuosus, an intertidal species presumably transported from Europe and initially reported at Barnstable, Massachusetts, now occurs widely from the Gulf of Maine to Nova Scotia.

More than a dozen additional species occupy deeper waters off this coast, some of them ranging down the Continental Slope.

Identification

In *Siriella* the outer ramus of the uropods is divided by a transverse articulation; in the remaining forms the ramus is not divided. Fig. 21.39A.

The remaining species are readily distinguished by differences in the form and spination of the telson and the form of the antennal scale, as indicated in Figure 21.39; sizes are given in the legend to this figure.

Table 21.12 lists additional differential traits of these shrimp.

Table 21.12. Differential Traits of Mysid Shrimp

Genus	Oostegites	Male Pleopods Reduced
Siriella	3	—
Erythrops	2	I
Meterythrops	2	I
Mysidopsis	2	I
Metamysidopsis	2	I
Mysis	2	I, II
Neomysis	2	I, II, III
Heteromysis	2	I–IV

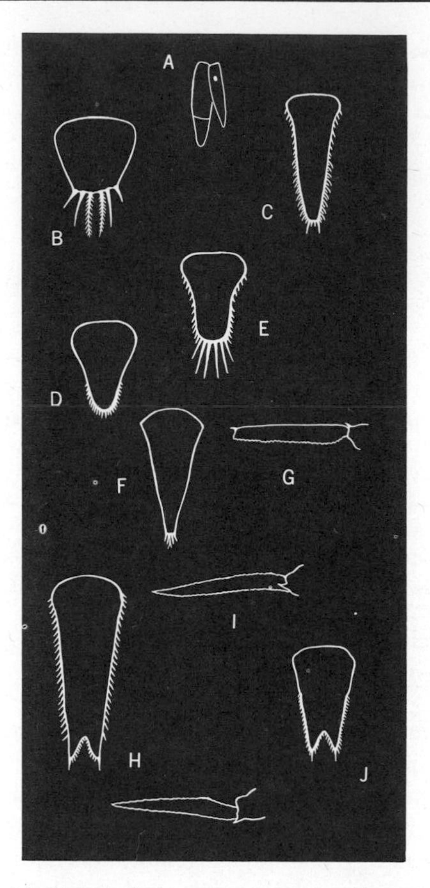

Fig. 21.39. Crustacea, Mysidacea. Sizes are maximum total length of adults less antennae. A, *Siriella thompsoni*, uropods, 12 mm; B, *Erythrops erythropthalma*, telson, 11 mm; C, *Neomysis americana*, telson, 12 mm; D, *Metamysidopsis munda*, telson, 6 mm; E, *Mysidopsis bigelowi*, telson, 8 mm; F, *Meterythrops robustus*, telson, 16 mm; G, *Praunus flexuosus*, antennal scale, telson similar to following but relatively broader, 25 mm; H, *Mysis mixta*, telson and antennal scale, 25 mm; I, *Mysis stenolepis*, antennal scale, telson somewhat similar to preceding but less deeply forked, 30 mm; J, *Heteromysis formosa*, telson, 9 mm.

References

Tattersall, W. M. 1951. A review of the Mysidacea of the United States National Museum. Bull. U.S. natn. Mus. 201:1–292.

———, and Olive S. Tattersall. 1951. The British Mysidacea. Ray Society, London.

Wigley, R. L. 1964. Part I. Order Mysidacea. *In* Keys to marine invertebrates of the Woods Hole region. R. I. Smith, ed. Marine Biological Laboratory, Woods Hole, Mass. Pp. 93–95.

Superorder Eucarida

The classification of higher Crustacea, like that of most large and complex groups, has had a tortured history. The classification used here (see Table 21.1), is that of Glaessner (1969, in general Crustacea references); for a discussion of this and earlier classifications see this reference.

Criteria for differentiating two orders in this group include differences in limb structure. The unspecialized condition of the 8 pairs of thoracic limbs in Euphausiacea contrasts with the situation in Decapoda where the first 3 pairs of "thoracic" limbs are specialized as auxiliary feeding appendages and so are functionally part of the head. The simplified limbs of euphausiids somewhat resemble those of Mysidacea of the last superorder, and they were indeed once allied with them systematically as members of the now defunct taxon "Schizopoda."

Order Euphausiacea
KRILL SHRIMP

Diagnosis. *Small to medium-sized shrimplike crustaceans with carapace covering and fused to all thoracic segments; eyes on stalks; telson with biramous uropods forming a fan; at least six pairs of thoracic limbs biramous but the eighth and sometimes the seventh reduced in size or uniramous; with luminous organs variously placed on eye stalks, base of second and seventh legs, and underside of abdominal segments; full complement of abdominal appendages in males and females. Fig. 21.40F.*

Adults of local species range between 12 and 40 mm in length. The sexes are separate and but slightly dimorphic externally. Males have the first 2 pairs of abdominal limbs modified for copulation. The eggs are usually shed into the water; the young hatch as naupli and pass through protozoeal and zoeal phases known as *calyptopis* and *furcilla* before reaching the post larval or *cyrtopia* stage (see introductory section on larval forms).

Both larval and adult euphausiids are planktonic. They are chiefly oceanic and occur irregularly inshore. In cold seas, euphausiids are sometimes abundant enough to serve as the main food of baleen whales.

Systematic List and Distribution

The check list is derived from Hansen (1915). All of the species are members of the family Euphausiidae.

Meganyctiphanes norvegica	B, 39°	Surface, 48 to 293 m
Euphausia krohni	BV	In water 62–77°F
Euphausia brevis	BV	—
Euphausia americana	C, 40°	—
Thysanoessa inermis	B+	Surface to 196 m
Thysanoessa raschi	B+	—

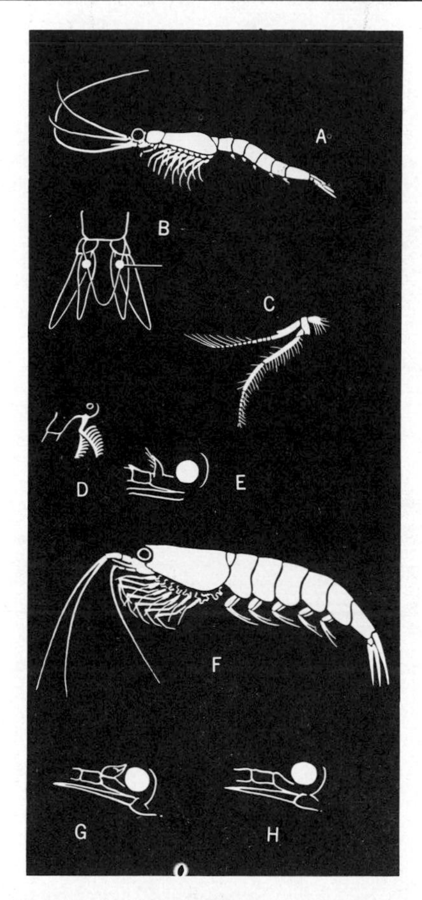

Fig. 21.40. Crustacea, Mysidacea and Euphausiacea. A, typical mysid; B, same, telson and uropods with statocyst indicated; C, biramous limb; D, *Euphausia krohni*, base of seventh thoracic leg; E, same, eye region; F, typical euphausid; G, *Meganyctiphanes norvegica*, eye region; H, *Thysanoessa* sp., eye region.

These species range into neritic waters. *Stylocheiron suhmii* is also a shallow water species that may occur in warmer waters. Additional species of *Euphausia* and *Thysanoessa* plus species of *Nematoscelis* have been reported from oceanic waters in these latitudes, for which see Moore (1952).

Identification

Measurements are of adult total body lengths.

---

1. Exopod of seventh thoracic leg rudimentary; 12–19 mm. *Euphausia:* in *E. americana* a lobe from the first antennular joint has 5 or 6 denticles along the margin; *E. krohni* the lobe has a row of spines plus a thin oblique process at the outer angle with 3 more denticles; Fig. 21.40D,E.

1. Seventh thoracic leg fully biramous; Fig. 21.40C—2.

2. Antennular peduncle with an upstanding leaflet; Fig. 21.40G; 40 mm. *Meganyctiphanes norvegica.*

2. Antennular peduncle without leaflet; Fig. 21.40H. *Thysanoessa:*

- Side of carapace with a tooth; sixth abdominal segment without dorsal spine; 20 mm. *T. raschi.*
- Side of carapace without tooth; sixth abdominal segment with dorsal spine on rear margin; 30 mm. *T. inermis.*

  Note also *Stylocheiron suhmii* with the third pair of legs greatly elongated rather than the second pair, as in *Thysanoessa. S. suhmii* reaches a length of about 8 mm.

References

See also general plankton references, Chapter 2.

Boden, B. P. 1954. The euphausiid crustaceans of South African waters. Trans. Roy. Soc. S. Afr. 34(1):181–243.

Hansen, H. J. 1911. The genera and species of the order Euphausiacea with an account of remarkable variation. Bull. Inst. oceanogr. Monaco. 210:1–54.

———. 1915. The Crustacea Euphausiacea of the United States National Museum. Proc. U.S. natn. Mus. 48:59–114.

Moore, H. B. 1952. Physical factors affecting the distribution of euphausids in the North Atlantic. Bull. mar. Sci. Gulf Caribb. 1(3):278–305.

Order Decapoda

The sequence in which superfamily and higher categories of this order are treated follows the classification of Table 21.1.

Infraorder Penaeidea
PENAEIDEAN SHRIMP

Diagnosis. *Penaeidae: shrimplike animals with small pincers (chelae) on the first three pairs of thoracic legs; pleura of first abdominal segment not overlapped by that of second segment; rostrum and antennal scales well developed.*

Members of a second family, Sergestidae, lack the fourth and fifth pairs of thoracic limbs and have a very short rostrum. In *Acetes* the 3 remaining pairs of thoracic legs have fully formed but minute chelae (Fig. 21.41G); in *Lucifer* only the third legs have chelae and these are rudimentary. *Lucifer* is an elongate, very distinctive form with eyes and antennae at the end of a necklike extension of the thorax (Fig. 21.41F); it is reported to be luminescent, hence the generic name.

Southern commercial shrimp, *Penaeus*, reach a length of about 200 mm. The sergestids are small, 12–15 mm long in adults. In both families the sexes are separate but not strongly dimorphic externally. The eggs are laid in the water, then sink, and the young hatch as nauplii in *Penaeus* and as metanauplii in *Lucifer*. *Penaeus* larvae become shrimplike in the third protozoeal stage when they are about 2.0 mm long; in the postlarval stage (4.0–5.0 mm) the adult penaeid limb pattern is established. In *Penaeus* the eggs are laid at sea, but larval stages are passed as the young move or are carried shoreward, and postlarval to adult growth is accomplished in the bays and sounds.

Acetes and the penaeids are benthic or hypoplanktonic. The commercial forms are fished for with a modified otter trawl, though surface nets are used as well at night. *Lucifer* is planktonic. *Penaeus* breeds on the North Carolina coast where there is a shrimp fishery both in the sounds and offshore. The single Virginia record for *Acetes* is estuarine; *Lucifer* and *Penaeus* are also tolerant of reduced salinity. Adults of *Penaeus* are migratory, both from estuarine to offshore waters and coastwise. *Penaeus* eats a variety of invertebrates, small fish, and plant debris.

Systematic List and Distribution

The list has been compiled from the references cited.

Family Penaeidae		
Penaeus setiferus	C, 41°	Lit. to 91 m, eury.
Penaeus aztecus	C, 39°	Lit. to 91 m, eury.
Penaeus duorarum	C, 37°	Lit. to 91 m, eury.

Trachypeneus constrictus	C, 38°	9 to 27 m
Parapeneus longirostris	C, 42°	"Offshore"
Sicyonia brevirostris	C, 37°	11 to 91 m
Family Sergestidae		
Acetes carolinae	C, 37°	Estuarine
Lucifer faxoni	C, 42°	Oceanic, estuarine in summer

Identification

Criteria for the differentiation of the two families and of the local sergestid genera were given in the diagnosis. Most of the locally occurring forms are figured and described in Verrill (1922) and Williams (1965). The penaeids may be separated as follows. Measurements are body length, adults.

1. Carapace with a dorsal keel and a single tooth near rear margin; 153 mm. *Sicyonia brevirostris*.
1. Carapace with dorsal teeth in forward half only—2.

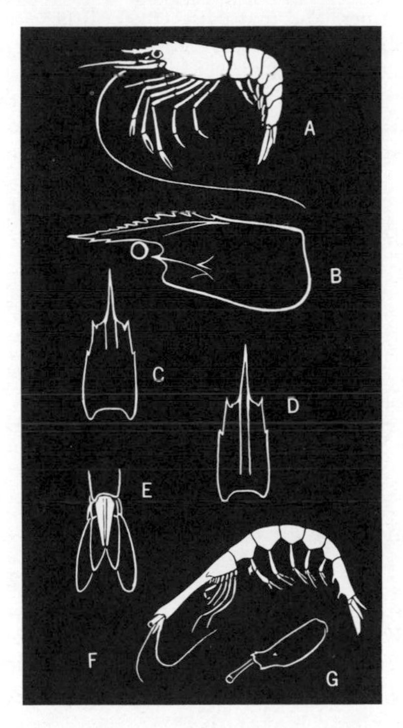

Fig. 21.41. Crustacea, Penaeidea. A, *Penaeus* sp., whole animal; B, same, carapace; C, *Penaeus setiferus*, carapace in dorsal view; D, *Penaeus aztecus*, carapace in dorsal view; E, *Penaeus* sp., telson and uropods; F, *Lucifer faxoni*, whole animal; G, *Acetes carolinae*, carapace.

2. Rostrum toothed above and below; telson without spines; 200 mm. *Penaeus*—3.
2. Rostrum toothed above only—4.
3. Carapace with a pair of dorsal grooves running nearly its whole length; Fig. 21.41D. *P. aztecus* or *P. duorarum*. See Williams (1965).
3. Carapace with grooves running less than half its length; Fig. 21.41C. *P. setiferus*.
4. With 2 teeth on either side of telson, last pair fixed, preceding ones minute and movable; body length 104 mm. *Parapanaeus longirostris*.
4. With 4 movable teeth on each side of telson, 3 penultimate ones minute; 92 mm. *Trachypeneus constrictus*.

References

Broad, C. 1951. The Shrimps of North Carolina. *In* Survey of Marine Fisheries of North Carolina. H. F. Taylor, ed. Univ. North Carolina Press.
Burkenroad, M. D. 1934. The Penaeidae of Louisiana with a discussion of their world relationships. Bull. Amer. Mus. nat. Hist. 68(2):61–143.
———. 1939. Further observations on Penaeidae of the northern Gulf of Mexico. Bull. Bingham oceanogr. Coll. 6:1–62.
Hansen, H. J. 1919. The Sergestidae of the Siboga Expedition. Siboga Exped. 38:1–65.
———. 1933. A North American species of *Acetes*. J. Wash. Acad. Sci. 23(1):30–34.
Verrill, A. E. 1922. Decapod Crustacea of Bermuda. Part 2, Macrura. Trans. Conn. Acad. Arts Sci. 26:1–179.
Williams, A. B. 1965. Marine decapod Crustaceans of the Carolinas. Fishery Bull. Fish Wildl. Serv. U.S. 65(1). xi + 298 pp.

Infraorder Caridea
CARIDEAN SHRIMP

This is a large and ecologically varied assemblage encompassing seven families.

Diagnosis. *Small to large shrimp with third pair of legs never chelate; pleura of first abdominal segment overlapped by those of second segment; antennal scale well developed. Rostrum varying in development in different families.*

The largest species of Pandalidae is said to reach about 175 mm in length. The smaller hippolytids are only about 15 mm long when adult. The sexes are separate and externally dimorphic in varying degrees; protandric hermaphroditism occurs among pandalids, individuals being initially male and later female. Most carideans hatch in a protozoeal stage but, in some species, hatching may be delayed even as late as the postlarval stage.

The majority of these shrimp are benthic or hypoplanktonic; the alpheids burrow or hide in holes in sponges or porous rock. A few species, *Hippolyte coerulescens* and *Leander tenuicornis* among them, are planktonic to the extent that they occur primarily in association with pelagic *Sargassum*. The family Palaemonidae includes freshwater as well as estuarine and marine species, and of the other families only Pandalidae and Pasiphaeidae lack brackish water representatives.

Systematic List and Distribution

The basic list is that of Kingsley (1899) with additions from more recent literature. Generic names follow the nomenclature of Holthuis (1955), with subsequent changes as required.

Family Pasiphaeidae		
Pasiphaea tarda	B	26 to 320 m
Pasiphaea multidentata	B	Deep water
Family Palaemonidae		
Leander tenuicornis	V	Oceanic, on *Sargassum;* also on wharf pilings, Nova Scotia to New Brunswick
Palaemonetes paludosus	C, 39°	Fresh water
Palaemonetes vulgaris	V	Lit. to 9 m, eury.
Palaemonetes intermedius	V	Lit. to 9 m, eury.
Palaemonetes pugio	V	Lit. to 9 m, eury.
Macrobrachium ohione	C, 38°	Fresh water to 14‰
Family Alpheidae		
Alpheus heterochelis	C, 38°	Shallow water
Alpheus normanni	C, 37°	Shallow water
Family Ogyrididae		
Ogyrides alphaerostris	C, 38°	Estuarine
Ogyrides limicola	C, 38°	Estuarine, 9–13‰
Family Hippolytidae		
Hippolyte pleuracantha	V	Shallow water, eury.
Hippolyte zostericola	V	Shallow water, eury.
Hippolyte coerulescens	V	Oceanic, on *Sargassum*
Latreutes fucorum	V	Oceanic, on *Sargassum*
Tozeuma carolinense	V	Estuarine
Hippolysmata wurdemanni	C, 38°	Lit. to 29 m, eury.
Caridion gordoni	B	38 to 400 m
Eualus fabricii	B	Lit. to 400 m
Eualus gaimardii	B	10 to 900 m
Eualus pusiolus	B, 38°	Lit. to 500 m

Spirontocaris spinus	B	6 to 400 m
Spirontocaris phippsii	B	11 to 229 m
Spirontocaris lilljeborgii	B	20 to 1261 m
Lebbeus groenlandicus	B, 41°	2 to 210 m
Lebbeus polaris	B, 38°	Lit. to 930 m
Lebbeus zebra	B	Lit. to 30 m
Family Pandalidae		
Dichelopandalus leptocerus	B, 36°	13 to 787 m
Pandalus borealis	B	37 to 906 m
Pandalus montagui	B, 41°	5 to 787 m
Pandalus propinquus	B, 39°	62 to 1993 m
Family Crangonidae		
Sclerocrangon boreas	B	9 to 66 m
Crangon septemspinosa	BV	Lit. to 128 m, eury.
Sabinea septemcarinata	B	27 to 128 m
Sabinea sarsii	B	152 to 274 m
Pontophilus brevirostris	V	Oceanic, Gulf Stream
Pontophilus norvegicus	V	Oceanic, Gulf Stream

Identification

Family identification may be accomplished with the aid of Table 21.13 and Figure 21.42. Count thoracic legs in reverse order to avoid confusion with the maxillipeds, which are leglike in some species. Note also that the chelae may be extremely minute, and the chelate versus simple condition of small individuals in particular requires close inspection. Also, the multisegmented condition of the carpus of the second pair of legs may be obscure, requiring careful manipulation of substate lighting to detect. Dimensions are total lengths of adults exclusive of antennae.

FAMILIES PASIPHAEIDAE AND OGYRIDIDAE

Pasiphaea and *Ogyrides* are readily distinguished in Figure 21.43. *O. limicola* reaches 16 mm and has a postrostral crest with 8–14 short, fixed spines; *O. alphaerostris* has a single movable spine in this position and is 27 mm. *P. tarda* is figured, and a second North Atlantic species was reported by Sivertsen and Holthuis (1956). *P. tarda* reaches a body length of at least 50 mm.

FAMILY PALAEMONIDAE

The familiar estuarine representatives of this family are members of the genus *Palaemonetes*, widely distributed and abundant in coastal waters; from Virginia south, the river shrimp *Macrobrachium* also enters brackish waters; *Leander* is usually oceanic, on *Sargassum*.

Table 21.13. Differential Traits of the Families of Caridean Shrimp

Family	Legs Chelate, Chela Size	Carpal Segments	Rostrum	Figures
Pasiphaeidae	I II m m	1	Short, erect, toothless	Fig. 21.43A.
Palaemonidae	I II s m	1	Long, toothed	Fig. 21.44
Alpheidae	I II l s	5	Short, spiky	Fig. 21.45.
Ogyrididae	I II s s	4	None	Fig. 21.43B.
Hippolytidae	I II s s	2 to >40	Varies; if short, then toothed	Fig. 21.46, Fig. 21.47
Pandalidae	— II s	5 to >20	Long, toothed	Fig. 21.48
Crangonidae*	I II ml s	1	Short	Fig. 21.49

Abbreviations are: s, small; m, medium; l, large.
* First legs subchelate, second with small chelae except in *Sabinea*.

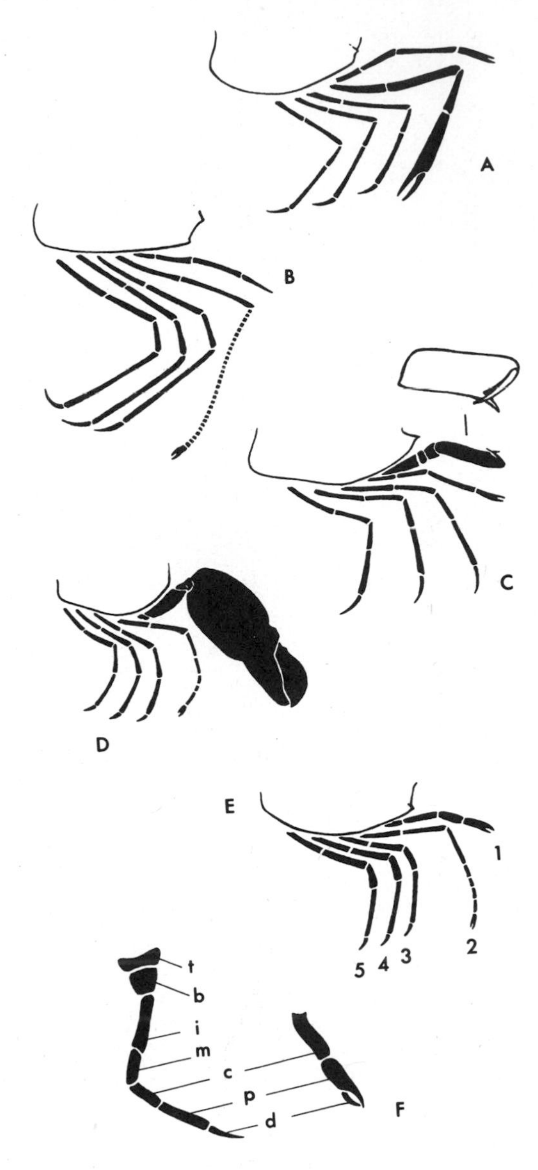

Fig. 21.42. Crustacea, Caridea, family limb patterns shown schematically: A, palaemonid; B, pandalid; C, crangonid with detail of subchelate claw; D, alphaeid; E, hippolytid; F, chelate vs nonchelate thoracic limbs with segments labeled as follows: t, coxa; b, basipodite; i, ischium; m, merus; c, carpus; p, propodus; d, dactylus.

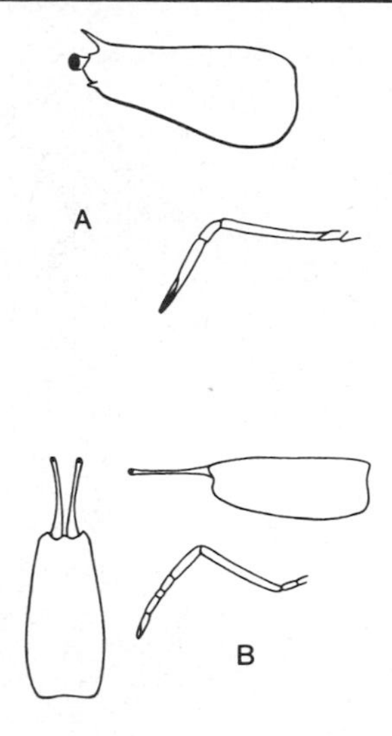

Fig. 21.43. Crustacea, Pasiphaeidae and Ogyrididae. A, *Pasiphaea tarda*, carapace and pereiopod; B, *Ogyrides alphaerostris*, carapace in lateral and dorsal views and second pereiopod.

The local forms are separated by differences in the rostrum as figured and by the following key. Care must be taken not to confuse the nearly marginal hepatic spine of young *Macrobrachium* with the brachiostegal spine of other palaemonids.

---

1. Branchiostegal spine present—2.

1. No branchiostegal spine; chelate second legs enlarged and greatly elongated; fresh to brackish water; 95 mm; Fig. 21.44A. *Macrobrachium ohione.*

2. Greenish, usually oceanic; finger on second pair of claws longer than palm; 47 mm; Fig. 21.44C. *Leander tenuicornis.*

2. Transparent; estuarine; finger on second pair of claws shorter than palm. *Palaemonetes*—3.

3. Carpus of second pair of legs much longer than claw; branched flagellum of antenna I with fused basal element longer than shorter of free branches; 46 mm; Fig. 21.44B; fresh water. *Palaemonetes paludosus.*

3. Carpus of second leg about equal to or shorter than claw, Fig. 21.44D; branched flagellum of antenna I with fused basal element shorter than shorter of 2 branches. Differences between the remaining species of *Palaemonetes* are conveniently presented in tabular form in Table 21.14 and Fig. 21.44E–G.

Note that the hippolytid *Caridion gordoni* has the carpus of the second leg "obsoletely biarticulate"; its limb pattern, therefore, resembles that of the present family. The first pair of legs in *Caridion* is heavier than the second

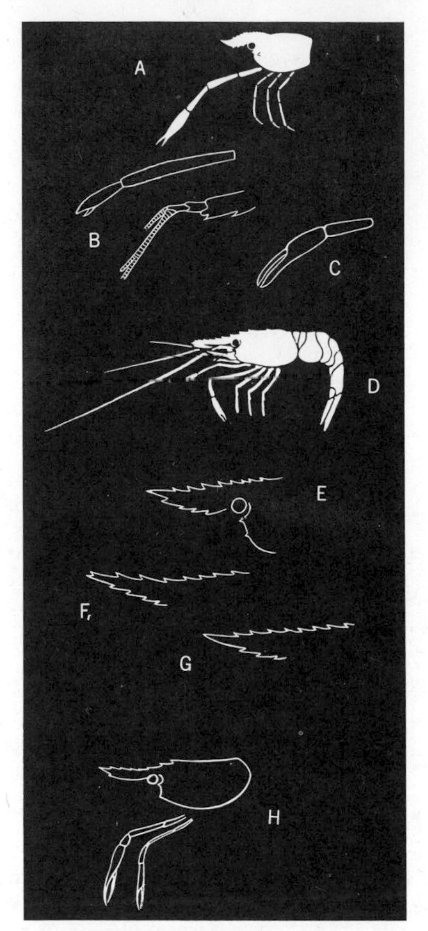

Fig. 21.44. Crustacea, Palaemonidae. A, *Macrobrachium ohione*, limb pattern legs 2–5; B, *Palaemonetes paludosus*, carpus and claw of second leg and base of antenna I; C, *Leander tenuicornis*, second claw; D, *Palaemonetes vulgaris*, whole animal; E, same, detail of front of carapace; F, *Palaemonetes intermedius*, rostrum; G, *Palaemonetes pugio*, rostrum; H, *Caridion gordoni*, carapace and first two legs.

pair, the reverse of the palaemonid condition, and the rostrum has no teeth on the underside except at the tip; Figure 21.44H.

FAMILY ALPHEIDAE

The snapping shrimp are a numerous tropical assemblage but barely enter the range of this text. *Alpheus normanni* and *A. heterochaelis* are 26 mm and 50 mm, respectively, and differ as figured, Figure 21.45.

Table 21.14. Differential Traits of Species of *Palaemonetes*

Trait	Species		
	vulgaris	*intermedius*	*pugio*
Leg II, carpus length/palm length	1.1× or less	1.3× or more	1.3× or more
Dorsal tip of rostrum	Toothed	Toothed	Simple
Ventral teeth on rostrum	3 to 5	4 or 5, rarely 3	2 to 4, usually 3
Dorsal rostral teeth behind eye	2	1	1
Teeth on claws of leg II			
On movable finger	2	1	0
On fixed finger	1	0	0
Total length	42 mm	37 mm	50 mm

FAMILY HIPPOLYTIDAE

This is a large and varied assemblage with a wide geographic and bathymetric range. The following key separates the local forms. For synonomies and literature citations see Holthuis (1947); for illustrations of *Spirontocaris*, *Lebeus*, and *Eualus* spp. see Leim (1921).

1. Carapace with supraorbital spines present—2.
1. Carapace without supraorbital spines—5.
2. Two or more supraorbital spines on each side; Fig. 21.46A,C,E. *Spirontocaris: S. spinus*, 59 mm, *S. phippsii*, 25 mm, and *S. lilljeborgii*, 58 mm, are differentiated by the form of the rostrum as figured.
2. Only one supraorbital spine on each side—3.

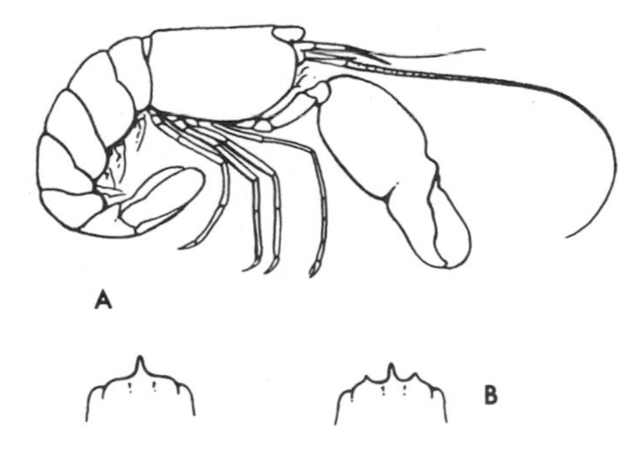

Fig. 21.45. Crustacea, Alpheidae. A, *Alpheus heterochaelis*, whole animal and detail showing frontal margin of carapace in dorsal view; B, *Alpheus normanni*, frontal margin in dorsal view.

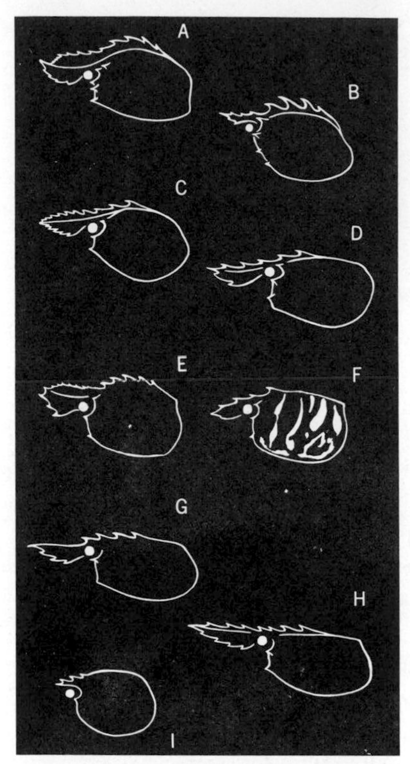

Fig. 21.46. Crustacea, Caridea, northern hippolytids. A, *Spirontocaris lilljeborgii;* B, *Lebeus groenlandicus;* C, *Spirontocaris phippsii;* D, *Lebeus polaris;* E, *Spirontocaris spinus;* F, *Lebeus zebra* (body boldly banded); G, *Eualus fabricii;* H, *Eualus gaimardii;* I, *Eualus pusiolus.*

3. Carpus of second pair of legs with 3 segments—4.

3. Carpus of second pair of legs with 7 segments; Fig. 21.46B,D,F. *Lebeus: L. groenlandicus*, 55 mm*; L. polaris*, 49 mm*;* and *L. zebra*, 49 mm, are differentiated by the form of the rostrum as figured.

4. Rostrum shorter than carapace, with spines above and below; Fig. 21.47D. *Hippolyte: H. coerulescens* lacks the hepatic spine which is present in *H. pleuracantha* and *H. zostericola*, 12–18 mm; the last two species are in need of revision (see Williams, 1965).

4. Rostrum longer than carapace, smooth above, toothed below; 40 mm; Fig. 21.47B. *Tozeuma carolinense.*

5. Carpus of second legs with 2 segments; carapace about 6 mm long; Fig. 21.44H. *Caridion gordoni.* See concluding statement in Family Palaemonidae, above.

5. Carpus of second legs with more than 2 segments—6.

6. Carpus of second legs with 3 segments; 12 mm; Fig. 21.47A. *Latreutes fucorum.*

6. Carpus of second legs with more than 3 segments—7.

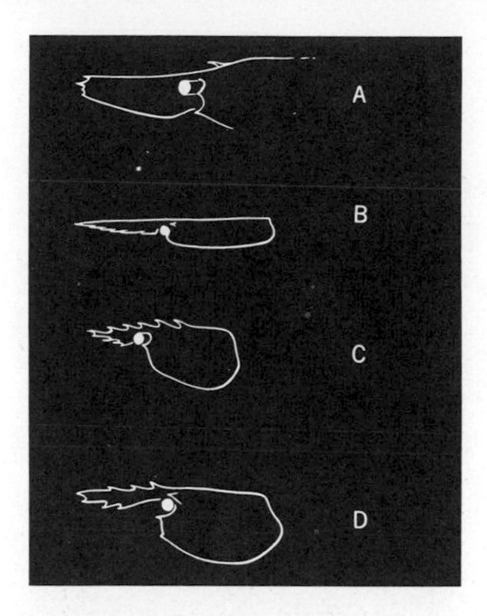

Fig. 21.47. Crustacea, Caridea, southern hippolytids. A, *Latreutes fucorum;* B, *Tozeuma carolinense;* C, *Hippolysmata wurdemanni;* D, *Hippolyte* sp.

7. Carpus of second legs with seven segments; Fig. 21.46G–I. *Eualus: E. fabricii*, 50 mm, *E. gaimardii*, 56 mm, and *E. pusiolus*, 23 mm, are differentiated by form of the rostrum as figured.

7. Carpus of second legs with about 30 segments; 30 mm; Fig. 21.47C. *Hippolysmata wurdemanni*.

FAMILY PANDALIDAE

The commercial fishery for these large, northern, deep water shrimp has been revived in recent years. The presence of a small branch or exopodite on the third maxilliped—a small but easily determined detail—separates *Dichelopandalus* from *Pandalus;* Figure 21.48A. The three local species of *Pandalus* differ as follows.

---

1. Third abdominal segment with a median spine; forward half of rostrum toothed above; 175 mm; Fig. 21.48B. *Pandalus borealis.*

1. Third abdominal segment without spine; top of rostrum without teeth in forward half; Fig. 21.47C—2.

2. Width of antennal scale equal to about ⅓ its length, only slightly tapered; carpus of second leg with about 20 segments; 95 mm; Fig. 21.48C. *Pandalus montagui.*

2. Width of antennal scale equal to about ¼ its length at most, and decidedly tapered; carpus with 5 segments; 110 mm; Fig. 21.48D. *Pandalus propinquus.* The most common pandalid south of Cape Cod is *Dichelopandalus leptocerus*, which reaches a length of about 110 mm.

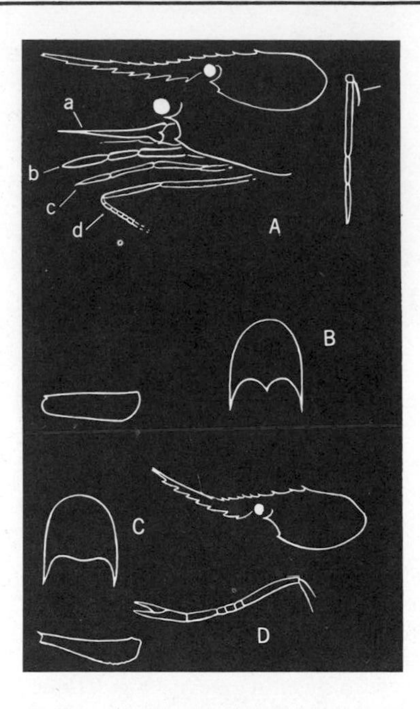

Fig. 21.48. Crustacea, Pandalidae. A, *Dichelopandalus leptocerus*, carapace, details of some anterior appendages, and third maxilliped with exopodite indicated; B, *Pandalus borealis*, third abdominal segment in dorsal view; C, *Pandalus montagui*, third abdominal segment in dorsal view, carapace, and antennal scale; D, *Pandalus propinquus*, antennal scale and second leg. Abbreviations are as follows: a, antennal scale; b, third maxilliped; c, pereiopod I; d, pereiopod II.

FAMILY CRANGONIDAE

The confusion in family and generic names affecting this family and the Alpheidae represents one of the more dismal episodes of crustacean systematics and necessitates constant vigilance in dealing with the older literature.

Crangon septemspinosa, with the palaemonids, is common in shallow water and in estuaries. The other crangonids are less familiar. Generic definitions are not yet stabilized.

---

1. Second pair of legs weakly developed and simple; 60–75 mm; Fig. 21.49D. *Sabinea* spp.

1. Second pair of legs chelate or subchelate, Fig. 21.49A—2.

2. Second pair of legs much shorter than others; oceanic species found in Gulf Stream; 67 mm. *Pontophilus* spp.

2. Second pair of legs not proportionately shorter than others—3.

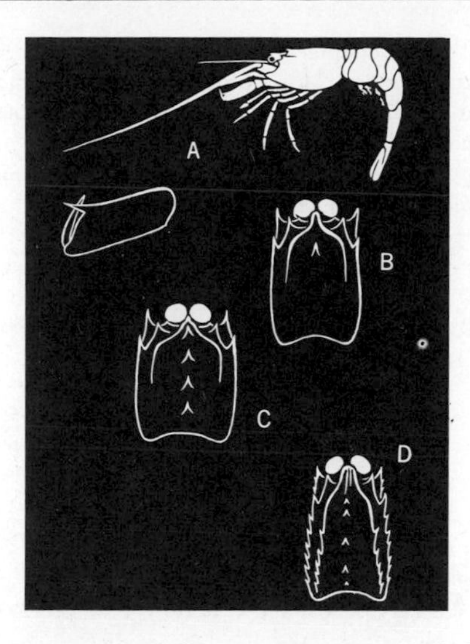

Fig. 21.49. Crustacea, Crangonidae. A, *Crangon septemspinosa*, whole animal and detail of subchelate condition; B–D show the carapace in dorsal view; B, *Crangon septemspinosa;* C, *Sclerocrangon boreas;* D, *Sabinea* sp.

3. A single tooth on midline of carapace behind rostrum; 55 mm; Fig. 21.49B. *Crangon septemspinosa.*

3. Three or four teeth on midline of carapace; 125 mm; Fig. 21.49C. *Sclerocrangon boreas.*

References

Calman, W. T. 1899. On the British Pandalidae. Ann. Mag. nat. Hist. 7th Ser. 3(2):27–39.

Hobbs, H. H., and W. H. Massman. 1952. The river shrimp, *Macrobrachium ohione* (Smith), In Virginia. Va. J. Sci. 3(3):206–207.

Holthuis, L. B. 1947. The Decapoda of the Siboga Expedition—Part IX. The Hippolytidae and Rhynchocinetidae collected by the Siboga and Snellius Expeditions with remarks on other species. Siboga Exped. Monogr. 39 a/8:1–100.

———. 1952. A general revision of the Palaemonidae of the Americas. Occ. Pap. Allan Hancock Fdn. 12:1–396.

———. 1955. The recent genera of Caridean and Stenopodidean shrimps (Class Crustacea, order Decapoda, Supersection Natantia) with keys for their determination. Zool. Verh. Leiden. 26:1–157.

Kingsley, J. S. 1899. Synopses of North American Invertebrates. III. The Caridea of North America. Am. Nat. 33:709–720.

Leim, A. H. 1921. A new species of *Spirontocaris* with notes on other species from the Atlantic coast. Trans. Roy. Canad. Inst. 13(1):133–146.

Man, J. G. de. 1920. The Decapoda of the Siboga Expedition. Part IV. Families Pasiphaeidae, Stylodactylidae, Hoplophoridae, Nematocareinidae, Thalassocaridae, Pandalidae, Psalidopodidae, Gnathophyllidae, Processidae, Glyphocrangonidae, and Crangonidae. Siboga Exped. Monogr. 39. a3:1–318.

Price, K. 1962. Biology of the sand shrimp, *Crangon septemspinosa* in the shore zone of the Delaware Bay region. Chesapeake Sci. 3(4):244–253.

Smith, S. I. 1882. Reports on the results of dredging by the U.S. Coast Survey steamer *Blake*, XVII Report on the Crustacea, Part I. Decapoda. Bull. Mus. comp. Zool. Harvard. 10:1–108.

Wigley, R. L. 1960. Note on the distribution of Pandalidae (Crustacea, Decapoda) in New England waters. Ecology. 41(3):564–570.

Williams, A. B. 1965. Marine Decapod crustaceans of the Carolinas. Fishery Bull. Fish Wildl. Serv. U.S. 65(1): xi + 298 pp.

Infraorder Astacidea: Family Nephropsidae
LOBSTERS

The northern lobster, *Homarus americanus*, which is the only local member of this family, is so familiar that little introduction is necessary, Figure 21.50. Despite the popular association of the lobster with Maine, this species is taken in commercial quantities as far south as Virginia. The related crayfish, Astacidae, are confined to fresh water. The British lobster is congeneric with our species, but the spiny or rock lobsters are warm water forms belonging to the family Palinuridae.

The lobster has the abdomen depressed instead of compressed as in shrimplike forms. The first 3 pairs of legs are chelate, the first pair being the

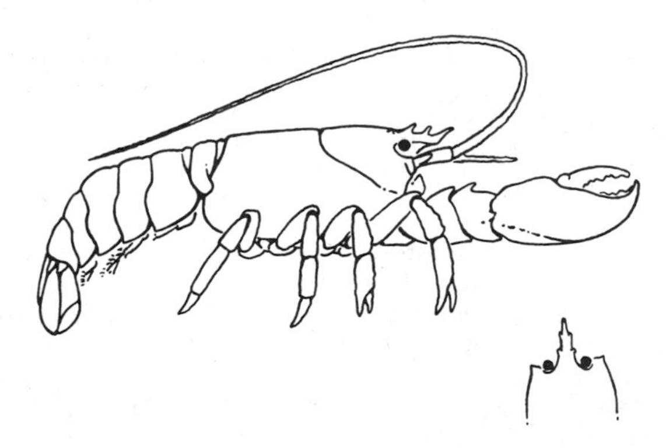

Fig. 21.50. Crustacea, Astacidea. *Homarus americanus*, whole animal and detail of rostrum in dorsal view.

largest. The rostrum is relatively short and triangular, and the antennal scale is moderately developed.

Market-sized lobsters average about 250 mm. in length and 1.5–2.0 pounds in weight, at which time they are about 4 years old. The record length is about 860 mm or 34 inches, and the maximum weight is about 45 pounds. The sexes are separate but not conspicuously dimorphic externally except that males tend to have heavier claws. The eggs are carried for 10–11 months, at the end of which time they hatch in a zoeal or mysis stage and are less than 8 mm long; the thoracic limbs, though biramous, have the rudiments of pinchers on the first 3 pairs of legs behind the leglike maxillipeds in the earliest stage. The abdomen of thc young lobster is spiny, and at hatching the uropods are absent.

After a planktonic larval period, lobsters become benthic. There are local ontogenetic and seasonal movements. Larger and presumably older individuals are found in greater numbers in deeper, offshore waters. As with many other essentially boreal forms, the lobster favors deeper waters in the southern parts of its range though, even in the latitude of New York, trapping has, at times, been commercially feasible in as little as 4 or 5 m. *H. americanus* ranges to the edge of the Continental Shelf.

Reference

Herrick, F. H. 1895. The American Lobster: a study of its habits and development. Bull. U.S. Fish Commn. Wash. 29:1–252.

Infraorder Anomura: Superfamily Thalassinoidea
MUD SHRIMP

Diagnosis. *Small to medium-sized, lobster- or shrimplike decapods with exoskeleton weak and soft; first one or two pairs of legs chelate, with first more heavily developed and usually asymmetrical; rostrum variably developed; antennal scale small or absent.*

Adult mud shrimp range in size from about 35 to 100 mm. The sexes are separate and, in *Callianassa* at least, there is sexual dimorphism in the form of the major claws. The young hatch as zoea, about 2.5–4.0 mm long but without uropods. During the larval period they somewhat resemble the equivalent stages of hermit crabs. The larvae are planktonic until the post-larval stage, when they are about 10 mm long; at this time they assume the benthic, burrowing habits of the adult. Though perhaps not so rare as their scarcity in collections would suggest, mud shrimp are seldom encountered except by special search.

Systematic List and Distribution

Family Axiidae		
Axius serratus	V	37 m
Calocaris templemani	B	Deep water
Family Upogebiidae		
Upogebia affinis	V	Lit. to 18 m
Family Laomediidae		
Naushonia crangonoides	V	Shallow water
Family Callianassidae		
Callianassa atlantica	V	Shallow water

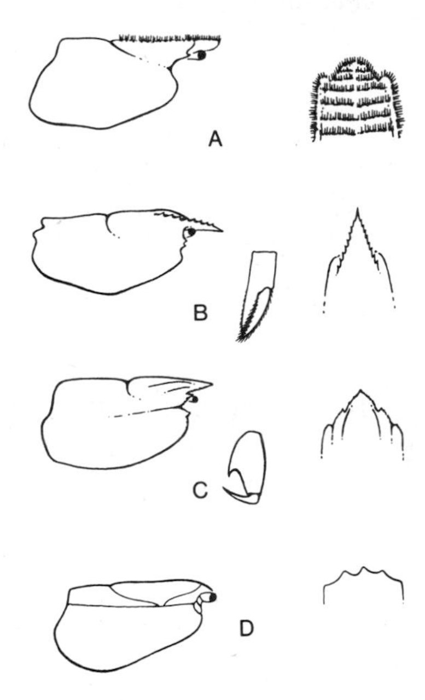

Fig. 21.51. Crustacea, Thalassinoidea. Figures include lateral views of the carapace (left column) and dorsal views of the frontal area of the carapace (right column) plus other details as indicated. A, *Upogebia affinis;* B, *Calocaris* sp., plus principal claw; C, *Naushonia crangonoides;* D, *Callianassa atlantica.*

Identification

1. Principal claw subchelate; rostrum broadly triangular with large central tooth and 2 lateral teeth, margins of rostrum finely denticulate; 35 mm; Fig. 21.51C. *Naushonia crangonoides.*
1. Principal claw chelate—2.
2. Second pair of legs not chelate; claws of first pair subequal; rostrum broad and flat with transverse rows of short bristles and with similar bristles around margin; 100 mm; Fig. 21.51A. *Upogebia affinis.*
2. Second pair of legs chelate—3.
3. With line running lengthwise on either side of carapace (linea thalassinica); rostrum small; antennal scale vestigial; 60 mm; Fig. 21.51D. *Callianassa atlantica.*
3. No linea thalassinica; antennal scale and rostrum well developed—4.
4. Outer blade of uropods with transverse suture; carapace with keel or ridge on midline; eyes pale; 50 mm; Fig. 21.51B. *Calocaris templemani.*
4. No transverse suture; carapace compressed but not keeled; eyes pigmented; to 100 mm. *Axius serratus.*

References

Borradaile, L. A. 1903. On the classification of the Thalassinidae. Ann. Mag. nat. Hist. Ser. 7(12):534–551.

Gurney, R. 1938. Larvae of Decapod Crustacea. Discovery Rept. 17:291–344.

Kingsley, J. S. 1899. Synopsis of Astacoid and Thalassinoid Crustacea. Am. Nat. 33:819–824.

Schmitt, W. C. 1935. Mud shrimps of the Atlantic coast of North America. Smithson. Misc. Collns. 93(2):1–21.

Thompson, M. T. 1904. A rare thalassinid and its larva (*Naushonia crangonoides*). Proc. Boston Soc. nat. Hist. 31(1):1–21.

Williams, A. B. 1965. Marine decapod crustaceans of the Carolinas. Fishery Bull. 65(1): xi + 298 pp.

SUPERFAMILIES PAGUROIDEA, GALATHEOIDEA, HIPPOIDEA

Superficial appearance is a poor indicator of relationship in these anomurans, as the following artificial key indicates.

1. Crablike forms with carapace almost as broad as or broader than long; abdomen broad, folded underneath:
   - Abdomen highly asymmetrical and without uropods, Fig. 21.57C. *Lithodes*, Paguroidea: Lithodidae.
   - Abdomen symmetrical and uropods present, Fig. 21.53. *Porcellana* and *Polyonyx*, Galatheoidea: Porcellanidae.

1. Hermit crabs with carapace about as long as, or longer than, wide; abdomen soft, more or less spirally coiled, and carried in the empty shell of a gastropod mollusk, Fig. 21.6K, Fig. 21.52. All genera, Paguroidea: Paguridae and Diogenidae. (See subkey.)

1. Carapace much longer than wide, egg-shaped or subcylindrical; abdomen elongate, folded underneath:

- With a pair of chelate legs; Fig. 21.54B. *Euceramus*, Galatheoidea: Porcellanidae.
- No legs chelate, Fig. 21.54A. *Emerita*, Hippoidea: Hippidae.

SUPERFAMILY PAGUROIDEA
FAMILIES PAGURIDAE AND DIOGENIDAE

The hermit crabs have well-developed chelae and 2 pairs of walking legs; the fourth and fifth pairs of legs are greatly reduced and modified. The smaller species use *Nassarius*, *Littorina*, or other small gastropod shells, while the largest local forms utilize moon snails (Naticidae) and *Bucycon* shells. Rasps on the telson and uropods are engaged with the shell to lock the animal in place. The crab may retain its position so tenaciously that it cannot be removed without being torn apart.

The sexes are separate but not conspicuously different. The eggs are carried on the abdomen as in most decapods, and the young hatch as shrimp-like zoea. The postlarval or *glaucothoe* young have a distinctly segmented abdomen, Figure 21.4E.

Local species are benthic and are found from the lower littoral to deep water. The association of hermit crabs and sea anemones is found among local species only in fairly deep water, where the anemone *Adamsia* has been reported on shells occupied by *Pagurus pollicaris*. The hydroids *Hydractinia* and *Podocoryne* commonly encrust shells occupied by shallow water hermits.

Systematic List and Distribution

Rathbun (1905), Williams (1965), and Squires (1964) are the principal sources for this list. Most specialists place the local species in two families.

Family Diogenidae		
Paguristes triangulatus	C, 36°	11 to 150 m
Clibanarius vittatus	C, 38°	Lit. to 2 m, eury.
Dardanus insignis	C, 36°	27 to 227 m
Family Paguridae		
Pagurus acadianus	B, 40°	11 to 485 m
Pagurus annulipes	V	Lit. to 42 m
Pagurus impressus	C+	11 to 33 m

Pagurus pubescens	B, 41°	15 to 5490 m
Pagurus longicarpus	V, 44°	Lit. to 52, m, eury.
Pagurus pollicaris	V	Lit. to 46 m, eury.
Pagurus politus	BV	101 to 668 m
Pagurus arcuatus	B, 41°	Lit. to 274 m

Identification

Dimensions are carapace lengths of adults unless otherwise indicated.

1. Claws subequal or left claw larger than right—2.

1. Right claw larger than left—4.

2. Left claw larger than right; claws and larger walking legs with transverse or diagonal ridges or rugosities; color yellowish red, rugosities becoming maroon distally; 17 mm, in shallow water. *Dardanus insignis.*

2. Left and right claws nearly equal in size—3.

3. No appendages on first or second abdominal segments; walking legs dark greenish or brown with light-orange stripes; 32 mm; in littoral and shallow water; Fig. 21.52A. *Clibanarius vittatus.*

3. Abdominal appendages on first and second segments in males and first segment in females; walking legs pinkish; 11 mm; in deeper water. *Paguristes triangulatus.*

4. Principal claw very broad and flat; carapace 31 mm; Fig. 21.52F. *Pagurus pollicaris.*

4. Principal claw strongly sculptured, with ridges or tubercles, and coarse teeth; chiefly boreal species—6.

4. Principal claw smooth or weakly sculptured, teeth weak or absent; chiefly Virginian species—5.

5. Major claw with a few short hairs on inner edges of fingers, otherwise hairless; eyestalk usually less than 3.5 times longer than wide, cornea dilated; 11 mm; Fig. 21.52B. *Pagurus longicarpus.*

5. Major claw hairy; eyestalk usually more than 4 times longer than wide; cornea not dilated; 7 mm; Fig. 21.52C. *Pagurus annulipes.*

6. Minor claw a flattened oval in cross-section; palm rectangular in outline; a broad, red-orange, lengthwise stripe on claws; whole body 75 mm; Fig. 21.52D. *Pagurus acadianus.*

6. Minor claw triangular in cross-section; palm triangular in outline—7.

7. Claws finely pubescent; facets of minor claw separated by a strong ridge of spines, outer face somewhat concave and lightly tuberculate; whole body 75 mm; Fig. 21.52E. *Pagurus pubescens.*

7. Claws coarsely pubescent, hairy enough to obscure shape; facets of minor claw separated by a weak ridge of spines that is double proximally, outer face somewhat convex and coarsely tuberculate; whole body to about 50–75 mm; Fig. 20.52G. *Pagurus arcuatus.*

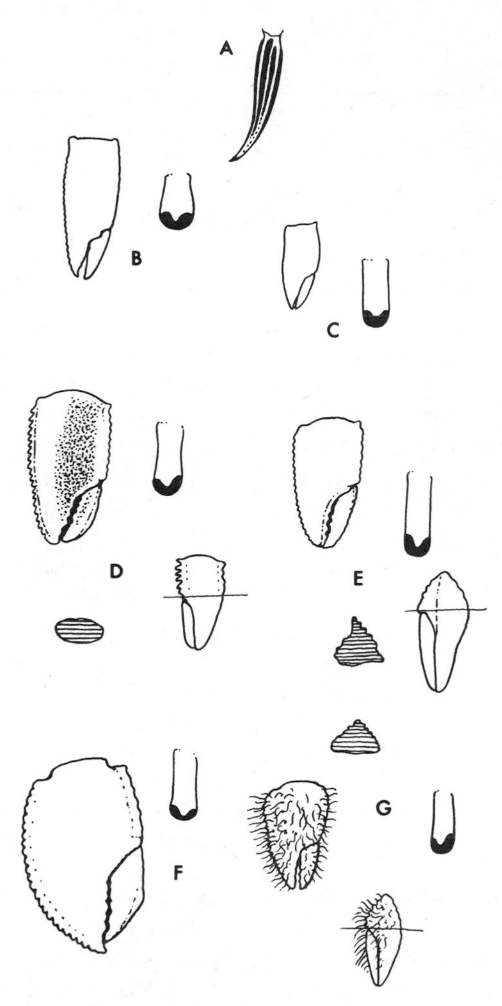

Fig. 21.52. Crustacea, Paguroidea. A, *Clibanarius vittatus*, dactyl of walking leg to show color pattern; B, *Pagurus longicarpus*, outline of major claw and eye; C, *Pagurus annulipes*, outline of major claw and eye; D, *Pagurus acadianus*, major claw with color pattern indicated, eye, and minor claw with schematic cross-section; E, *Pagurus pubescens*, details as in the preceding; F, *Pagurus pollicaris*, major claw and eye; G, *Pagurus arcuatus*, major and minor claws showing pubescence, eye, and schematic cross-section of minor claw (above).

FAMILY LITHODIDAE

The family is monotypic. *Lithodes maia* is a large crablike anomuran with a carapace width reaching 100 mm or more. Despite its close relationship and classification with the hermit crabs it more closely resembles the brachyurans in having a fully calcified abdomen which it carries in a crablike manner; the abdomen is, however, highly asymetrical. Also, the fifth thoracic limbs are hidden in the branchial chamber. This large, spidery, and very spiny crab is essentially boreal and confined to relatively deep waters, 95–532 m, where it is sometimes taken by lobstermen; Figure 21.57C.

The early larvae of *Lithodes* resemble the equivalent stages of hermit crabs.

SUPERFAMILY GALATHEOIDEA, FAMILY PORCELLANIDAE

The porcellanids are primarily southern, but *Polyonyx gibbesi*, Figure 21.53, is reported as a regular commensal in the tubes of the annelid worm *Chaetopterus* which ranges north to Cape Cod. Another typically crablike form, *Porcellana sigsbeiana*, has been reported north to Cape Cod, presumably as a southern waif. *P. sigsbeiana* is mottled reddish and gray, while *P. gibbesi* is nearly unicolorous gray; *P. gibbesi* reaches 16 mm carapace width, *sigsbeiana*, 24 mm.

A quite different form, *Euceramus praelongus*, is found from the lower Chesapeake southward; this species burrows in sandy beaches from just

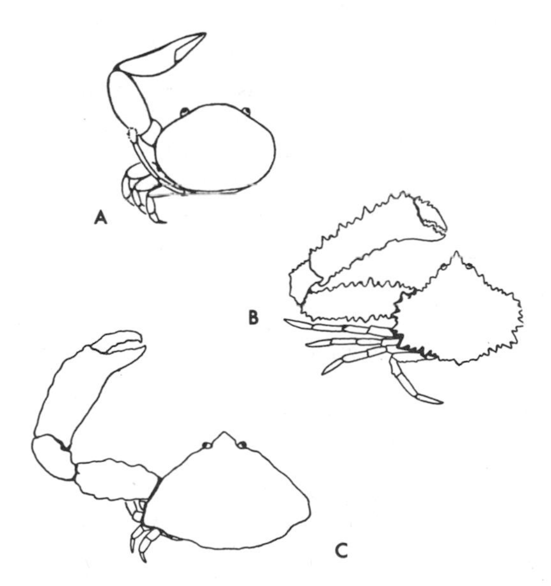

Fig. 21.53. Crustacea, Porcellanidae and Majidae in part. A, *Polyonyx gibbesi*, female; B, *Parthenope pourtalesi;* C, *Heterocrypta granulata.*

below the water line to depths of 38 m. *Euceramus* somewhat resembles the hippids in both habits and general appearance, being elongate and cylindrical in comparison with the typically crablike form of most porcellanids; unlike *Emerita* it has well-developed chelipeds, Figure 21.54B.

Porcellanids have distinctive larvae with a long curving rostrum, Figure 21.4B. Like other southern forms, they may drift northward in the Gulf Stream and are sometimes carried inshore in late summer.

SUPERFAMILY HIPPOIDEA, FAMILY HIPPIDAE.

Emerita talpoida is the only mole crab within the range of this text. It is one of the few inhabitants of the zone of breaking waves on sandy beaches where it is a familiar and often abundant form. Mole crabs move up and down the beaches with the tide, continually burrowing into the sand only to be exhumed by subsequent waves. The species disappears from the beach in winter, reportedly moving into deeper water, down to about 4 m. *Emerita* feeds on microorganisms extracted from the backwash by its antennal net. For a review of the ecological literature see Williams (1965).

Except for a somewhat superficial resemblance to the porcellanid *Euceramus*, *Emerita* should present no difficulty in identification; it has a large egg-shaped carapace, and the abdomen, armed with an elongate, bladelike telson, is folded against the underside of the body; Figure 21.54A. None of the legs is chelate; all are wide, flat, and hairy. Females reach a length of about 30 mm. Males are less than half this length, and during the mating season they are described as being semiparasitic on the females, to which they apparently remain attached for long periods.

The young hatch as zoea with a very long, curving rostrum, Figure 21.4B. Individuals as small as 3 mm in carapace length resemble the adults. The range of *Emerita talpoida* is from Cape Cod south.

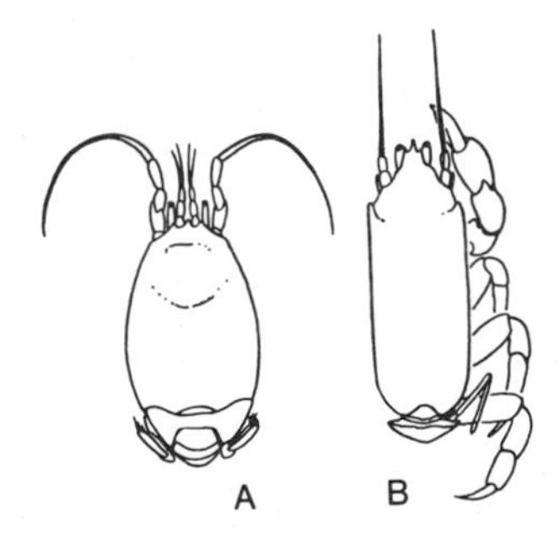

Fig. 21.54. Crustacea, Porcellanidae and Hippidae. A, *Emerita talpoida;* B, *Euceramus praelongus*.

References

Benedict, J. F. 1901. Hermit crabs of the *Pagurus bernhardus* type. Proc. U.S. natn. Mus. 23:451–466.

Haig, Janet. 1956. The Galatheidae (Crustacea, Anomura) of the Allan Hancock Atlantic Expedition with a review of the Porcellanidae of the Western North Atlantic. Allan Hancock Atl. Exped. 8:1–44.

Rathbun, Mary J. 1905. Fauna of New England 5. Crustacea. Occ. Pap. Boston Soc. nat. Hist. 7(2):1–117.

Smith, S. I. 1877. The early stages of *Hippa talpoida*. Trans. Conn. Acad. Arts Sci. 9:311–342.

———. 1882. Occasional occurrence of tropical and subtropical species of decapod Crustacea on the coast of New England. Trans. Conn. Acad. Arts Sci. 4(2):254–267.

Squires, H. J. 1964. *Pagurus pubescens* and a proposed new name for a closely related species in the northwest Atlantic (Decapoda: Anomura). J. Fish. Res. Bd. Can. 21:355–365.

Williams, A. B. 1965. Marine Decapod crustaceans of the Carolinas. Fishery Bull. Fish Wildl. Serv. U.S. 65(1): xi + 298 pp.

Infraorder Brachyura
TRUE CRABS

The crab form is so familiar that confusion with crustacean groups other than crablike anomurans is unlikely.

In brachyurans the fourth pair of walking legs is not markedly reduced in size in comparison with the other pairs; note that Palicidae and Homolidae are exceptions.

Adult males of some pinnotherids have a carapace only 5 mm across while the larger spider crabs, portunids, and rock crabs measure 100–150 mm in carapace width and have a leg span several times larger. The sexes are separate and notably dimorphic in pinnotherids, less so in other groups. However, the sexes can usually be separated by the form of the abdomen as indicated in Figure 21.55. Crabs undergo a complete metamorphosis; the initial larva or *zoea* is planktonic and bears no resemblance to the adult, Figure 21.4A. The postlarva or *megalopa*, though more crablike, has a distinct abdomen carried extended rather than pressed against the underside as in the adult crab, Figure 21.4D.

Most crabs are benthic animals, though the swimming crabs, family Portunidae, have strong pelagic tendencies, and a few species in this family as well as in Grapsidae are planktonic in association with drifting *Sargassum*. Strictly freshwater crabs and land crabs are absent from the local fauna, but

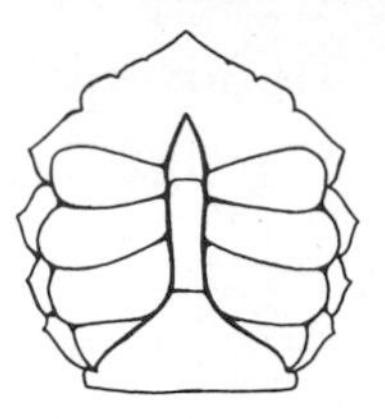

Fig. 21.55. Crustacea, Brachyura. *Callinectes* sp., abdomen of male (above) and female (below).

the principal families have one or more species that enter situations with reduced salinity. A few species are restricted to such conditions. While the great majority of brachyurans are free-living, the diversity of the group extends to parasitism and commensalism in the relationship between pinnotherids and bivalve mollusks, annelid worms, and echinoderms.

Systematic List and Distribution

Glaessner (1969) has reshuffled the arrangements of earlier workers but family composition is nearly unchanged from Rathbun (1918–1938).

Section Dromiacea		
Family Homolidae		
Homola barbata	V	102 to 684 m
Section Oxystomata		
Family Dorippidae		
Ethusa microphthalma	V	123 to 286 m
Family Calappidae		
Calappa flammea	C, 42°	22 to 31 m
Calappa sulcata	C+	29 to 49 m
Acanthocarpus alexandri	V	104 to 284 m
Hepatus epheliticus	C, 38°	20 to 284 m
Family Leucosiidae		
Persephona aquilonaris	C, 40°	4 to 31 m

Section Oxyrhyncha		
Family Majidae		
Euprognatha rastellifera	V	80 to 417 m
Collodes robustus	V	90 to 682 m
Rochinia spp.	V	128 to 481 m
Chionoecetes opilio	A, 44°	Lit. to 640 m
Hyas araneus	B, 41°	9 to 55 m
Hyas coarctatus	BV	9 to 532 m
Pelia mutica	V	Lit. to 24 m
Libinia emarginata	BV	Lit. to 49 m
Libinia dubia	V	Lit. to 46 m
Family Parthenopidae		
Parthenope pourtalesi	C, 40°	90 to 246 m
Heterocrypta granulata	V	6 to 27 m
Section Cancridea		
Family Cancridae		
Cancer irroratus	BV	Lit. to 572 m
Cancer borealis	BV	Lit. to 794 m
Section Brachyrhyncha		
Family Portunidae		
Carcinus maenas	40–45°	Lit. to 9 m, eury.
Ovalipes ocellatus	BV	Lit. to 18 m, eury.
Bathynectes superba	V	128 to 410 m
Portunus sayi	V	Gulf Stream
Portunus gibbesi	V	24 to 88 m
Portunus spinimanus	C, 40°	18 to 49 m
Callinectes sapidus	V	Lit. to 37 m, eury.
Callinectes ornatus	C, 40°	Surface to 26 m
Arenaeus cribrarius	C, 37°	Lit. to 68 m
Family Xanthidae		
Panopeus herbstii	V+	Lit. to 22 m
Neopanope texana sayi	BV	Lit. to 79 m, eury.
Hexapanopeus angustifrons	V	6 to 48 m, eury.
Eurypanopeus depressus	V+	Lit. to 48 m, eury.
Rhithropanopeus harrisii	V, 44°	Lit. to 9 m, eury.
Family Geryonidae		
Geryon quinquedens	BV	40 to 292 m Mostly below 366 m south of Cape Cod
Family Pinnotheridae		
Pinnotheres ostreum	V+	Parasitic on *Crassostrea*
Pinnotheres maculatus	V	Parasitic on *Mytilus* and other bivalves
Dissodactylus mellitae	V	Commensal; see text
Pinnixa retinens	V	37 m
Pinnixa chaetopterana	V	Commensal; see text
Pinnixa sayana	V	Commensal; see text
Pinnixa cylindrica	V	Commensal; see text

Family Grapsidae		
Planes minutus	V	On *Sargassum*
Sesarma reticulatum	V	Lit., salt marshes
Sesarma cinereum	C, 38°	Lit.
Family Ocypodidae		
Ocypode quadrata	C, 41°	Lit.
Uca minax	C, 41+°	Lit.
Uca pugnax	V	Lit.
Uca pugilator	V+	Lit.; eury
Family Palicidae		
Palicus alternatus	C+	7 to 110 m
Palicus spp.	C+, V	90 to 530 m

Identification

The single dromiacid listed above and the oxystomatid *Ethusa* are figured with the spider crabs which they somewhat resemble (Fig. 21.57I,H). The

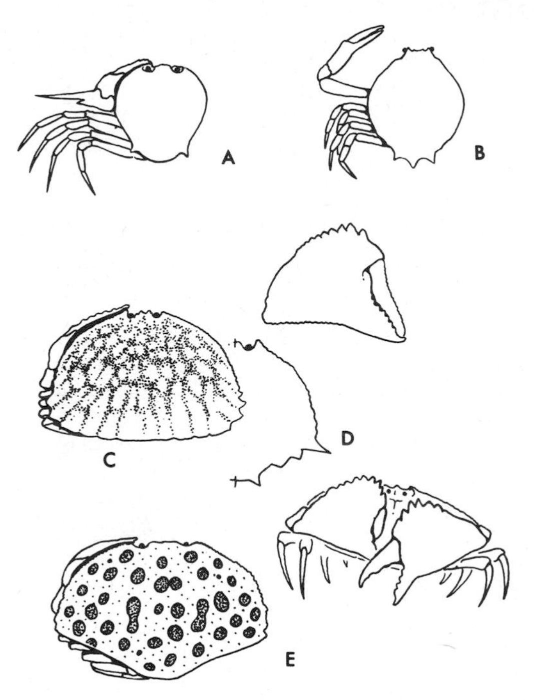

Fig. 21.56. Crustacea, Oxystomata. Dimensions are maximum carapace widths. A, *Acanthocarpus alexandri*, 32 mm; B, *Persephona aquilonaris*, 42 mm; C, *Calappa flammea*, whole animal showing color pattern and principal claw, 135 mm; D, *Calappa sulcata*, outline of right half of carapace, 119 mm; E, *Hepatus epheliticus*, whole animal showing color pattern, 67 mm.

remaining oxystomatids, all of which occur on this coast primarily as southern waifs, are shown in Figure 21.56; *Persephona* has been found clinging to drift nets at sea, while the calappids occur as larvae and metamorphosing young, chiefly in late summer and fall. Dimensions are given in the legend for Figure 21.56.

The remaining species may be identified as follows. Dimensions are carapace width unless otherwise indicated.

---

1. Carapace basically triangular, usually longer than broad with greatest width in posterior half, narrowing forward with a pointed, sometimes bifurcated rostrum extending between eyes. Majidae, Parthenopidae, and similar forms. Fig. 21.57. See subkey. Note also species of Palicidae, Fig. 21.65.

1. Carapace varying in shape, usually wider than long or with greatest width in forward half, or without a projecting rostrum—2.

2. Margin of carapace smooth, without projecting teeth or indentations, or with only 1 or 2 teeth immediately behind or lateral to eyes—3.

2. Carapace margin with 3 or more teeth or indentations—4.

3. Carapace round or oval, without teeth; eyes very small; Fig. 21.62. Pinnotheridae. See subkey.

3. Carapace angular, squarish, or a truncated triangle, sometimes with 1 or more lateral teeth; eyes well developed and on relatively long stalks; Figs. 21.63, 21.64. Grapsidae, Ocypodidae.

4. Last pair of legs with swimming paddles; Fig. 21.59. Portunidae (except *Carcinus*).

4. Last pair of legs with terminal joint pointed—5.

5. Carapace squarish, Fig. 21.65. Palicidae.

5. Carapace rounded or oval, Figs. 21.58, 21.60, 21.61—6.

6. Frontolateral margin of carapace with 7 or more teeth; Fig. 21.58. Cancridae.

6. Frontolateral margin with 5 teeth or less; Fig. 21.61. Xanthidae and similar forms.

FAMILIES MAJIDAE, PARTHENOPIDAE, AND SIMILAR FORMS

The common shallow water spider crabs, Majidae, belong to the genera *Libinia*, *Hyas*, and *Pelia*. Spider crabs are frequently hairy or armed with spines or tubercles, and they are commonly encrusted or all but concealed by debris and living algae that must be cleaned off before the details of the carapace can be examined. The form of the carapace is sufficiently diagnostic in itself, and the illustration is self-explanatory, Figure 21.57. The two species of *Libinia* are similar except that *emarginata* has 9 bumps down the middle of the back and *dubia* only 6. In the family Parthenopidae the genera *Parthenope* and *Heterocrypta* are similar in having a triangular carapace and chelipeds relatively much larger and heavier than the walking legs; in this respect they

Fig. 21.57. Crustacea, spider-like crabs. Carapaces and silhouettes of whole animals are shown. A, *Libinia emarginata;* B, *Chionoecetes opilio;* C, *Lithodes maia;* D, *Rochinia* sp.; E, *Pelia mutica;* F, *Hyas araneus;* G, *Hyas coarctatus*, outline of right half of carapace; H, *Homola barbata;* I, *Ethusa microphthalma;* J, *Euprognatha rastellifera;* K, *Collodes robustus.*

somewhat resemble the porcellanids with which they are figured, Figure 21.53. *Heterocrypta* is found on pebbly substrata where its resemblance to a stone chip provides concealment.

FAMILY CANCRIDAE

The rock crabs, though found in tide pools and under littoral rocks in New England, are chiefly sublittoral south of Cape Cod where *C. borealis* in particular favors deep water. This species is commonly taken in lobster pots on the middle coast, and the claws are marketed. The two species are easily distinguished by the sculpturing on the marginal teeth as indicated in

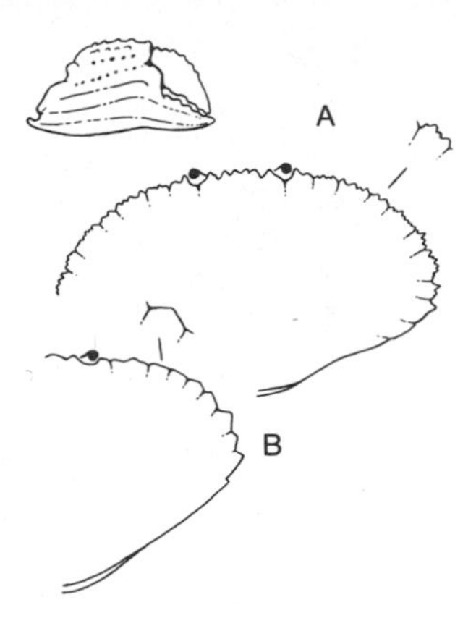

Fig. 21.58. Crustacea, Cancridae. A, *Cancer borealis*, outline of carapace, major claw, and detail of marginal tooth; B, *Cancer irroratus*, outline of right half of carapace and detail of marginal tooth.

Figure 21.58. The predominant colors on both species are dull red and yellowish; carapace width reaches 140–160 mm. These are stout, powerful crabs but not so vicious as the portunids. Compared to their Pacific coast relatives, they are of minor commercial importance.

FAMILY PORTUNIDAE

With the exception of *Carcinus* the portunids are immediately recognizable as such by their paddle-shaped last legs, Figure 21.59F. Adult females of *Carcinus*, Figure 21.60A, belie the common name "green crab" by being orange or brick red; this is a common littoral species under rocks and other debris from northern New Jersey to Maine. The other two common portunids are *Callinectes sapidus* and *Ovalipes ocellatus*. The first, commonly called "blue crab," is the commercially important, edible crab of the Chesapeake. Its distribution is chiefly estuarine extending to nearly fresh water. *Ovalipes* is common over the sandy bottoms of both ocean and bay beaches. Both species are quick and pugnacious. *Portunus sayi* is an oceanic form associated with *Sargassum*. The other portunids listed are essentially southern.

---

1. Carapace only slightly wider than long; Fig. 21.59C,D—2.

1. Carapace much wider than long; Fig. 21.59E,G,J—3.

2. Outermost spines on carapace much longer than other teeth; Fig. 21.59D; in deep water; 50 mm. *Bathynectes superba*.

2. Lateral teeth on carapace subequal, Fig. 21.59C; carapace pale with bright reddish purple spots; in shallow water; 69 mm. *Ovalipes ocellatus*.

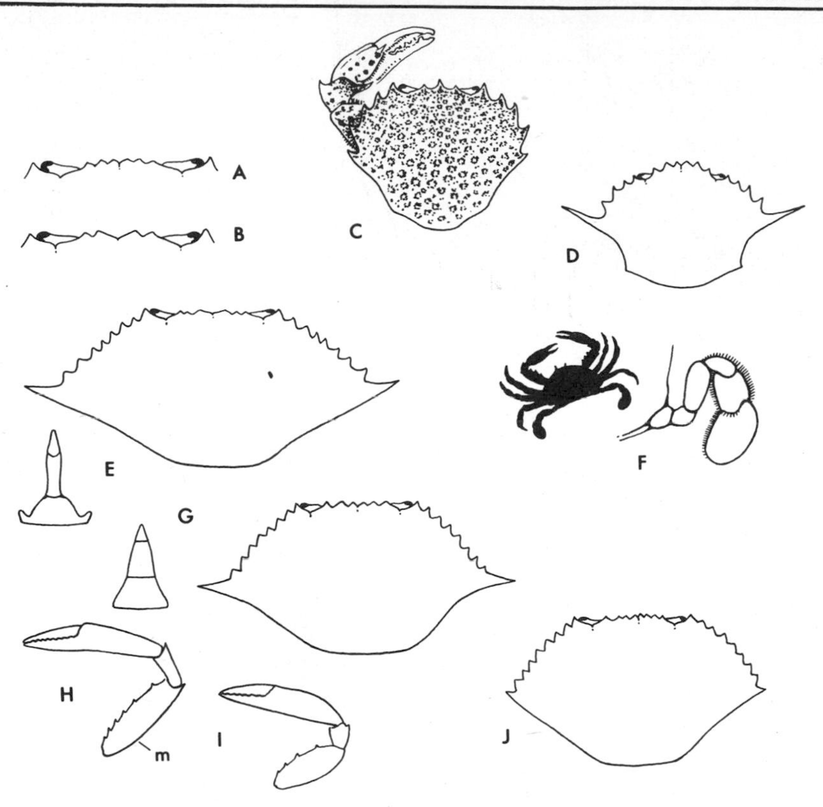

Fig. 21.59. Crustacea, Portunidae. A, *Callinectes ornatus*, frontal margin; B, *Callinectes sapidus*, frontal margin; C, *Ovalipes ocellatus*, carapace and major claw showing color pattern; D, *Bathynectes superba*, outline of carapace; E, *Callinectes* sp., outline of carapace and detail of male abdomen (left); F, swimming leg of typical portunid; G, *Portunus sayi*, outline of carapace and detail of male abdomen (left); H, *Portunus gibbesi*, major cheliped; I, *Portunus sayi*, major cheliped; J, *Portunus spinimanus*, outline of carapace. Merus indicated by the letter "m."

3. Lateral teeth on carapace subequal, Fig. 21.59J; 100 mm. *Portunus spinimanus.*

3. Outermost lateral tooth on each side enlarged as a strong spine; Fig. 21.59E,G—4.

4. Carapace with speckled pattern of light spots on darker ground, pattern persisting in preservative; 150 mm. *Arenaeus cribrarius.*

4. Carapace not speckled—5.

5. Chelipeds very long and slender with 5 or 6 teeth on merus, Fig. 21.59H; 54 mm. *Portunus gibbesi.*

5. Chelipeds moderately developed with 3 teeth on merus; Fig. 21.59I—6.

6. Two teeth on frontal margin between orbits; Fig. 21.59B,E; estuarine; 180 mm. *Callinectes sapidus.*

6. Four teeth on frontal margin; Fig. 21.59A: *Portunus sayi* is oceanic on *Sargassum*, olive or yellow-brown with white marks, 76 mm. *Callinectes ornatus* is a shallow water southern form with habits somewhat like *C. sapidus* and with bright blue on claws, rare north of Hatteras, about 130 mm. Note that mature males of *Portunus* and *Callinectes* differ in the form of the abdomen, Fig. 21.59E,G.

FAMILIES XANTHIDAE AND GERYONIDAE, PLUS *CARCINUS* (PORTUNIDAE)

Mud crabs are typically estuarine, where they are found on a wide variety of substrata and in waters with reduced salinity. They are especially common on oyster bottoms.

1. Frontal area straight or curving, with central notch; Xanthidae—2.
1. Frontal area toothed, Fig. 21.60. *Carcinus maenas*, with 5 regular lateral teeth, is a shallow water or littoral portunid without swimming paddles; 80 mm; see preceding family. *Geryon quinquedens* is a deep water species; 162 mm. The species differ as figured.
2. Fingers of chelipeds dark brown; Fig. 21.61—3.
2. Fingers of chelipeds pale; 19 mm wide; Fig. 21.60B. *Rhithropanopeus harrisii.*
3. Movable finger of major claw with a strong tooth, hidden in *Hexapanopeus* if fingers tightly closed; Fig. 21.61A–B—4.
3. No strong tooth on movable finger; Fig. 21.61C—5.

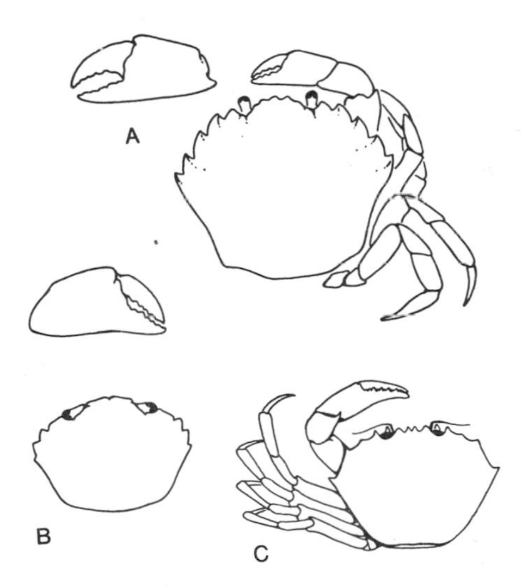

Fig. 21.60. Crustacea, miscellaneous crabs. A, *Carcinus maenas*, whole animal and detail of major claw; B, *Rhithropanopeus harrisii*, carapace and major claw; C, *Geryon quinquedens*, whole animal.

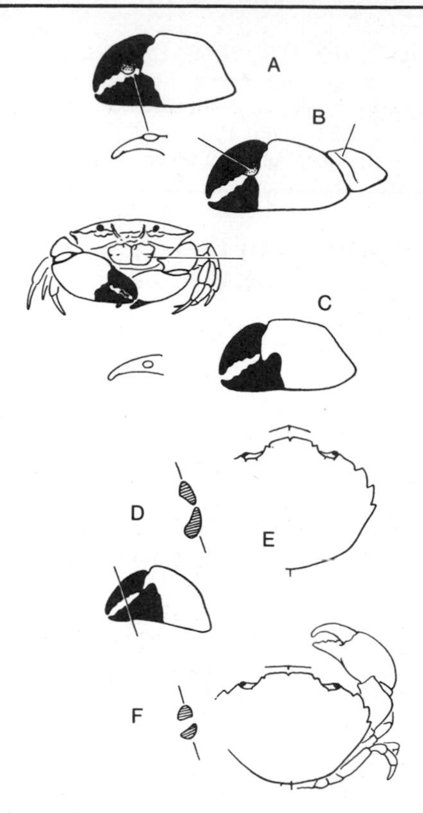

Fig. 21.61. Crustacea, Xanthidae. A, *Panopeus herbstii*, major claw; B, *Hexapanopeus angustifrons*, major claw; C, *Neopanope texana sayi*, frontal view of crab with third maxilliped indicated, and major claw; D, same, minor claw and schematic section showing "bite"; E, *Hexapanopeus angustifrons*, outline of carapace; F, *Eurypanopeus depressus*, outline of carapace, and schematic section of minor claw showing "bite."

4. Tooth on movable finger of claw basal, hidden when claw shut, Fig. 21.61B; wrist with a deep groove; front of carapace advanced as in Fig. 21.61E; 26.5 mm. *Hexapanopeus angustifrons.*

4. Tooth separated from base and visible even when claw shut, Fig. 21.61A; wrist without groove; front continuing lateral curve, not advanced. *Panopeus herbstii.*

5. Fingers of minor claw spoon-shaped at tip; concealed face of maxillipeds with a red spot, Fig. 21.61F; 23 mm. *Eurypanopeus depressus.* Note that *Panopeus herbsti* may also have a red spot on the maxillipeds.

5. Fingers of minor claws meet like a cutting pliers; Fig. 21.61C,D; no red spot on maxillipeds; 27 mm. *Neopanope texana sayi.*

FAMILY GERYONIDAE

Geryon quinquedeus, the only local representative of the family, is figured in Figure 21.60C.

FAMILY PINNOTHERIDAE

The commensal forms occur as follows: *Pinnotheres*, in bivalve mollusks; *Dissodactylus*, on sand dollars, keyhole urchins, and annelid worms; *Pinnixa*, with tubicolous annelid worms. *Pinnotheres ostreum* is the species associated with oysters; because its activities have a debilitating effect on its host, the crab is parasitic rather than commensal.

Pinnotherids are small crabs not exceeding about 16 mm carapace width in the largest species. *Pinnotheres* (at least *P. ostreum*) is sexually dimorphic; mature females are soft, pale, and larger than males, which have a hard exoskeleton and are more strongly pigmented.

---

1. Terminal segments of first 3 pairs of walking legs forked; females 4.5 mm; Fig. 21.62H. *Dissodactylus mellitae.*

1. Terminal segments pointed, not forked—2.

2. Carapace not much wider than long; polymorphic; mature females to 15 mm; Fig. 21.62A. *Pinnotheres: P. ostreum* in oysters; *P. maculatus* in mussels, scallops, *Pinna*, and the annelid *Chaetopterus.*

2. Carapace much wider than long, Fig. 21.62D. *Pinnixa*—3.

3. Propodus of third leg long and slender; 10 mm; host unknown; Fig. 21.62G. *P. sayana.*

3. Propodus of third leg short and thick—4.

4. Length and width of propodus about equal; 15 mm; with *Arenicola;* Fig. 21.62F. *P. cylindrica.*

4. Propodus slightly longer than wide, Fig. 21.62E—5.

5. Immovable finger of claws deflexed, Fig. 21.62C; 14 mm; with *Amphitrite* and *Chaetopterus. P. chaetopterana.*

5. Immovable finger not deflexed, Fig. 21.62B; host unknown; reported from Chesapeake Bay. *P. retinens.*

FAMILY GRAPSIDAE

This family is important in the tropics but is only feebly represented on this coast. The species differ as figured, Figure 21.63. *Planes minutus* is common on drifting *Sargassum*, attains a carapace width of about 19 mm, and is olive or yellowish brown, marked with white. The shore crabs, *Sesarma*, are common in the littoral in salt marshes and on mud banks, ranging north to

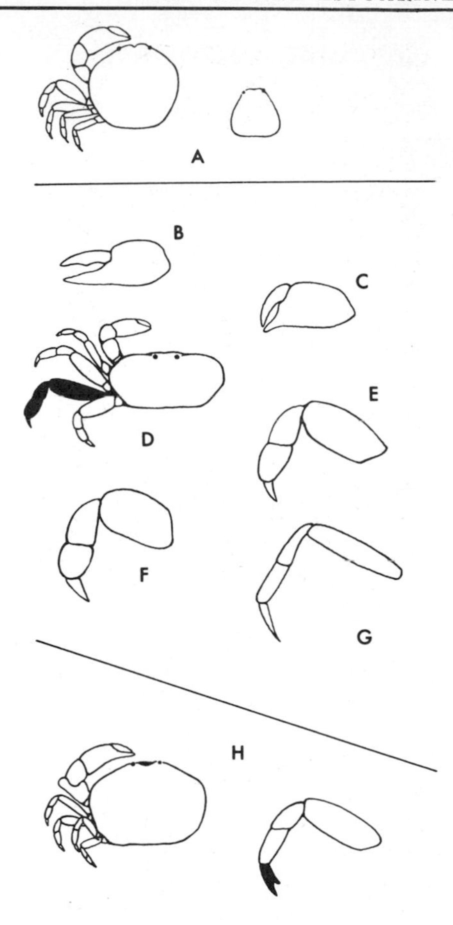

Fig. 21.62. Crustacea, Pinnotheridae. A, *Pinnotheres* sp., female (right), male (left); B, *Pinnixa retinens*, major claw; C, *Pinnixa chaetopterana*, major claw; D, *Pinnixa* sp., whole animal; E, *Pinnixa chaetopterana*, third leg; F, *Pinnixa cylindrica*, third leg; G, *Pinnixa sayana*, third leg; H, *Dissodactylus mellitae*, whole animal and detail of first walking leg.

Cape Cod. They somewhat resemble *Ocypode* in having a squarish carapace but are dark in color and have a broad interorbital front. They reach about 28 mm carapace width; the two local forms are readily distinguished by the presence or absence of a second tooth at the outer, front corner of the carapace.

FAMILY OCYPODIDAE

The ghost crabs and fiddlers are familiar burrowing forms of the littoral and supralittoral. *Ocypode quadrata* is abundant on less disturbed beaches from Maryland south; fiddlers seek more protected beaches, mud flats, and

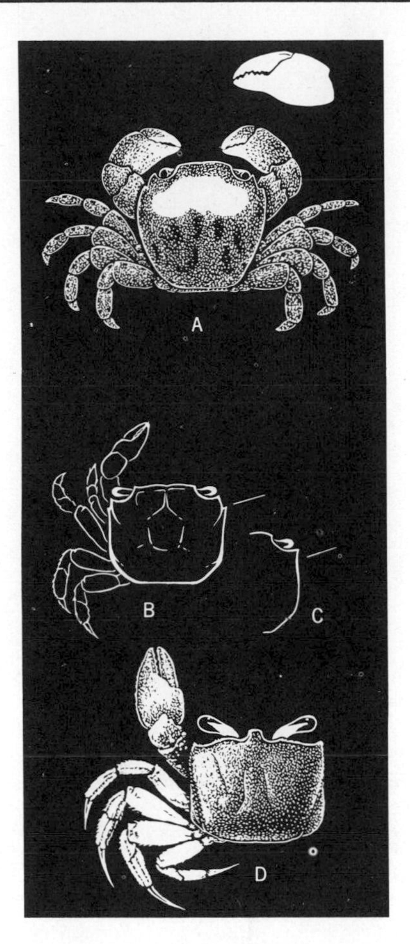

Fig. 21.63. Crustacea, Grapsidae, and Ocypodidae in part. A, *Planes minutus;* B, *Sesarma reticulatum;* C, *Sesarma cinereum*, outline of right half of carapace; D, *Ocypode quadrata.*

marshes. The ghost crab is pale yellowish or sand-colored with whiter claws, reaches about 50 mm in carapace width, and is not likely to be confused with any other local crab; Figure 21.63D.

Sexual differences are conspicuous in fiddlers. In females both claws are all, but the male has one claw enormously developed for use in sexual displays. Fiddlers are colonial, and the substratum in a fiddler colony may be riddled with the holes of these small, diurnally active crabs. The three local species may be distinguished as follows.

---

1. Great claw of male with a conspicuous, roughened, crescentic ridge on inner face of palm, Fig. 21.64B—2.

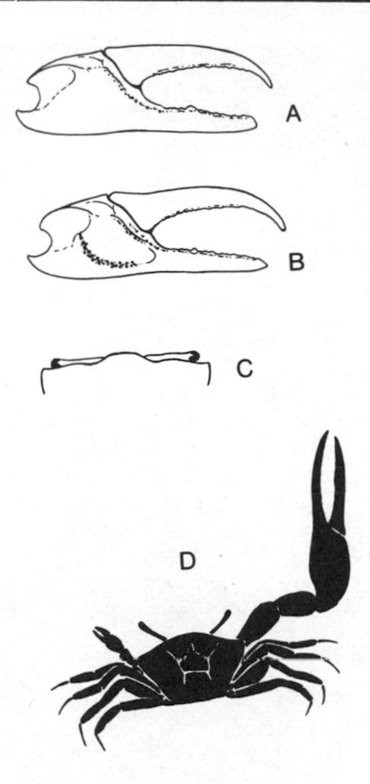

Fig. 21.64. Crustacea, Ocypodidae. A, *Uca pugilator*, major claw of male; B, *Uca pugnax*, major claw of male; C, same, front of carapace; D, *Uca* sp., whole animal in silhouette.

1. Inner surface of palm smooth; back mottled; 26 mm; on sandy substrata; Fig. 21.64A. *Uca pugilator*.

2. Carapace grey with darker central mark; joints on great claw with red spots; 38 mm; in fresh to weakly saline marshes and mud flats. *Uca minax*.

2. Carapace dark brownish or black; 22.5 mm; in salt marshes and mud flats. *Uca pugnax*.

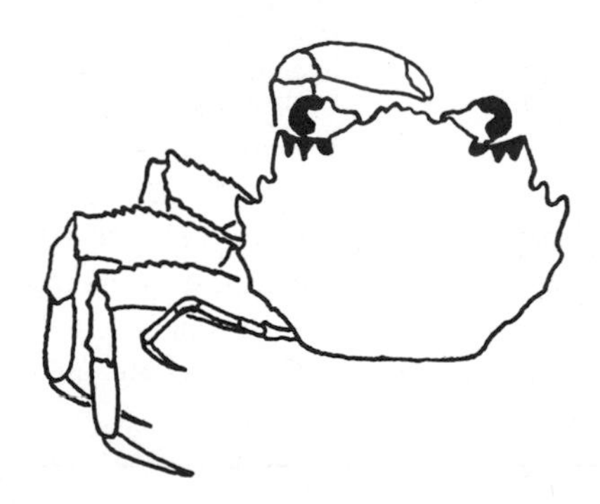

Fig. 21.65. Crustacea, Palicidae. *Palicus alternatus*.

FAMILY PALICIDAE

A representative of the family is illustrated, Figure 21.65; see Williams (1965).

References

Benedict, J. E., and M. J. Rathbun. 1891. The genus *Panopeus*. Proc. U.S. natn. Mus. 14(858):355–385.

Christensen, A. M., and J. J. McDermott. 1958. Life history and biology of the oyster crab, *Pinnotheres ostreum* Say. Biol. Bull. mar. Biol. Lab. Woods Hole. 114(2):146–179.

Cowles, R. P. 1908. Habits, reactions and associations of *Ocypoda arenaria*. Pap. Tortugas Lab. 2(1):1–41.

Crane, J. 1943. Display, breeding, and relationships of fiddler crabs in the northeastern United States. Zoologica, N.Y. 28:217–223.

Glaessner, M. F. 1969. Decapoda. *In* Treatise on Invertebrate Paleontology, Part R, Arthropoda 4, R. C. Moore, ed. Geol. Soc. Amer. and Univ. of Kansas.

Rathbun, M. J. 1918. The grapsoid crabs of America. Bull. U.S. natn. Mus. 97:1–116.

———. 1925. The spider crabs of America. Bull. U.S. natn. Mus. 129:1–613.

———. 1930. The cancroid crabs of America. Bull. U.S. natn. Mus. 152:1–609.

———. 1937. The oxystomatous and allied crabs of America. Bull. U.S. natn. Mus. 166:1–278.

Van Engel, W. A. 1958. The blue crab and its fishery in Chesapeake Bay. Part I. Reproduction, early development, growth and migration. Comm. Fish. Rev. 20(6):6-17.

Verrill, A. E. 1908. Decapod Crustacea of Bermuda; 1—Brachyura and Anomura. Their distribution, variations and habits. Trans. Conn. Acad. Arts Sci. 13:299–474.

Williams, A. B. 1965. Marine decapod crustaceans of the Carolinas. Fishery Bull. Fish Wildl. Serv. U.S. 65(1): xi + 298 pp.

TWENTY-TWO

Phylum Echinodermata

SPINY-SKINNED ANIMALS

Introduction

Diagnosis. *Usually with pentaradial symmetry; body subspherical, ovoid, starlike with simple or branched arms, or cucumber-shaped, or wormlike; free-living or fixed by a stalk; with calcareous plates or ossicles embedded in the skin, and these either few, loosely scattered, or forming a flexible or rigid skeleton; skin usually with spiny warts or projections; with an internal water-vascular system evidenced externally by a system of tube feet or podia, and less conspicuously by an external opening called the hydropore, or multiple openings called the madreporite.*

The principal taxa are as follows.

Class Crinoidea	Sea lilies
Class Holothuroidea	Sea cucumbers
Class Echinoidea	Sea urchins, sand dollars, heart urchins
Class Stelleroidea	
Subclass Asteroidea	Sea stars
Subclass Ophiuroidea	Brittle stars

The common types of echinoderms are such familiar marine forms that there is scarcely any difficulty in placing individuals in their appropriate classes. The chief exceptions are among the holothurians, some of which are distinctly wormlike and lack a dermal skeleton and podia. Some echinoids, popularly known as heart urchins, are secondarily bilaterally symmetrical.

Echinoderms are exclusively marine, and there are relatively few forms that inhabit even moderately brackish waters. They are mostly medium-sized animals—few adult echinoderms are less than 10 mm in greatest dimension. Except for one fixed species of crinoid, all local forms are motile. None is parasitic.

In most echinoderms the sexes are separate. Some brood their young, and a few are viviparous, but in most local species the gametes are shed freely in the water and development is external. Some produce large, yolky eggs, and development is more or less direct, but most species develop with a metamorphosis involving distinctive larval forms. The larvae have received considerable attention in discussions of phylogeny and are the basis for theoretical alliances between echinoderms and chordates. Echinoderms generally are cooperative subjects for embryological study, and the larval development of many species has been worked out. A very superficial resumé of larval types is given below. For further details and for an introduction to the literature see Hyman (1955).

Larval Forms

The initial larva of most species with metamorphic development can be referred to a basic *dipleurula* plan. The actual development of individual larvae proceeds along one or another of three main lines, although there are numerous exceptions. One typical line leads to juvenile sea stars, another to the young holothurian, and the third to either an ophiuroid or an echninoid, the ultimate differentiation in this case being deferred to the later stages. Each of these lines may be viewed in a succession of larval forms differing progressively from the hypothetical dipleurula and from each other.

The following outline, presented in key form and modified from Mortenson (1927), may serve as a frame of reference for the discussion that follows.

---

1. No skeleton or at most with isolated calcareous bodies—2.

1. Skeleton present; body a rounded cone with elongate arms—3.

2. Ciliated band separated into a small preoral band and a larger postoral band; the earliest stages without lobes but subsequently developing lobes or elongate arms: Fig. 22.1D,E; bipinnaria and brachiolaria larvae. Asteroidea.

2. Ciliated band continuous or broken up into rings; calcareous bodies usually present; Fig. 22.1B,C; auricularia and doliolaria larvae. Holothuroidea and Crinoidea.

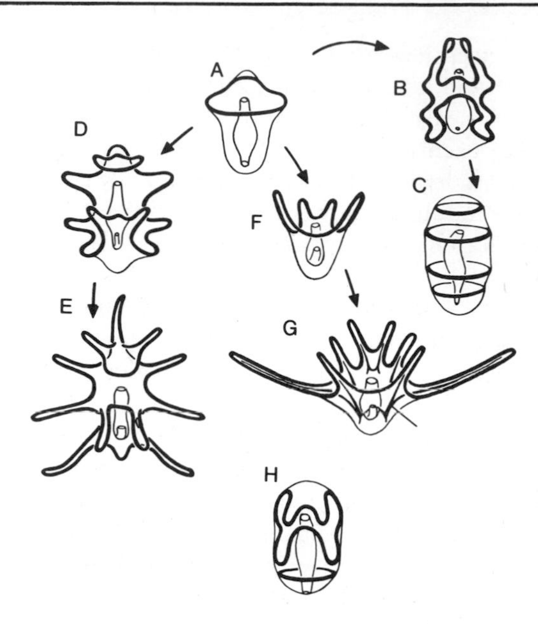

Fig. 22.1. Echinoderm and hemichordate larvae. All highly schematic. A, dipleurula; B, auricularia; C, doliolaria; D, bipinnaria; E, brachiolaria; F, early pluteus; G, later pluteus with skeletal bars indicted; H, tornaria.

3. Skeleton consisting of 2 symmetrical halves; posterolateral arms well developed and directed forward; Fig. 22.1F,G; ophiopluteus larvae. Ophiuroidea.

3. Skeleton consisting of 2 or 4 paired and 1 or 2 unpaired parts; posterolateral arms wanting or directed backward or to the side; echinopluteus larvae. Echinoidea.

The dipleurula plan envisions a short, sausage-shaped larva which in life is essentially microscopic and transparent with a U-shaped gut and an extensive, saddle-shaped ridge or band with cilia (or flagella) that passes in front of both the mouth and the anus (Fig. 22.1A).

The fate of the larval coelomic sacs is crucial to the development of all echinoderm larvae. Initially they are symmetrical with left and right coelomic sacs derived almost always by an outpocketing of the larval gut. Each sac then divides into an anterior and a posterior part, and it is usually the left anterior sac that plays a dominant role in establishing the radial symmetry and general body plan of the adult. It is this sac that is chiefly responsible for the development of the water vascular system and its appendages, the podia.

The initial, free-swimming larva of many holothurians approximates the dipleurula but has the basic pattern of the ciliated band contorted with sinuous diversions (Fig. 22.1B). This larva, called an *auricularia*, is usually <1.0 mm long. It transforms into a *doliolaria* larva (note the very superficial resemblance to the pelagic tunicate *Doliolum* Figure 24.9), in which the ciliated band breaks up into 3–5 pieces that form rings (Fig. 22.1C). This is actually the initial larval stage in some holothurian species. The transformation of the doliolaria into the adult form involves the lobation of the primary coelomic sac to produce, among other things, a set of 5 primary oral tube feet, the precursors of the oral tentacles. By the time the young holothurian, a *pentacula* now, becomes benthic, it has also acquired 2 or 3 more typical podia as well.

The classical descriptions of asteroid development, typified by *Asterias*, include stages termed bipinnaria and brachiolaria, but substantial departures from the typical pattern occur. The *bipinnaria* approximates the dipleurula (Fig. 22.1D). It somewhat resembles the holothurian auricularia and may be compared also with the tornaria larva of the hemichordate *Balanoglossus* (Fig. 22.1H). In the early asteroid the ciliated band is divided into a small preoral and a larger postoral band. Next, the postoral band develops lobes which elongate to become slender arms; the preoral band becomes extended in 3 shorter arms with adhesive tips. A median adhesive sucker also appears. The larva is now a *brachiolaria*, and in *Asterias forbesi* is about 0.75 mm long exclusive of the longer arms (Fig. 22.1E). In metamorphosis the brachiolaria becomes fixed to a substratum, and development, which up to now has required 3 or 4 weeks, is rapidly concluded by the emergence of a juvenile starfish about 1 mm in arm span.

Larval asteroids and holothurians are soft-bodied, although the latter may have ossicles as early as the auricularia stage; skeletal elements do not appear in asteroids until metamorphosis to the juvenile sea star is nearly complete.

The development of ophiuroids and echinoids is similar externally. Both have skeletal elements in early stages. Except in atypical life histories, the initial larva is a pluteus with a body shaped like a rounded cone and initially with 4 arms, later increased to 4, 5, or 6 *pairs* of arms (Fig. 22.1F,G). These elongate considerably, and are supported by long rod-like skeletal elements; the ciliary band follows these extensions, and there may be, in addition, separate patches, rings, or lobes with long cilia. The latter presumably serve as the main locomotor organs.

Ophiuroid and echinoid pluteal larvae differ internally and in the details of metamorphosis. There is no attached stage as in asteroids, the larvae remaining pelagic until the juvenile is formed. As in other groups, the echinoid larva develops asymmetrically with emphasis on the left side. The young sea urchin, though less than 1 mm in diameter, is a rounded form with a few conspicuous spines and, as in other groups, few and disproportionately large juvenile podia. The development of ophiuroids is somewhat atypical in that the asymmetrical form of the adult does not derive from an emphasis on the left side of the larval body; the "larval rudiment" develops, instead, on the ventral side.

The larval period in most echinoderms is quite long, lasting several weeks to a month or more, but final metamorphosis is usually accomplished within a few hours.

A further common trait of members of this phylum is their substantial power to regenerate lost or damaged parts, plus a willingness to sustain damage as a means of escape or in repsonse to molestation. The ability to regenerate lost parts is of practical concern to the problem of identification, in sea stars particularly, because regeneration sometimes results in abnormal morphological details. Asexual reproduction as a typical or voluntary procreative technique is relatively uncommon, though it occurs in asteroids, ophiuroids, and a few holothurians.

References

The following general citations will not be repeated in the more specialized bibliographies following class accounts.

Boolootian, R. A., ed. 1966. Physiology of Edhinodermata. Interscience.

Chadwick, H. C. 1914. Echinoderm larvae. L.M.B.C. Mem. typ. Br. mar. Pl. Anim. 22:1–32.

Clark, H. L. 1904. The echinoderms of the Woods Hole Region. Bull. U.S. Fish Comm. (1902). 22:545–576.

———. 1933. A handbook of the littoral echinoderms of Porto Rico and the other West Indian islands. Sci. Surv. Porto Rico and the Virgin Islands. N.Y. Acad. Sci. 16(1): 1–147.

Coe, W. R. 1912. Echinoderms of Connecticut. Bull. St. Geol. nat. Hist. Surv. Conn. 19:1–152.

Fell, H. B. 1967. Echinoderm Ontogeny. *In* Treatise on Invertebrate Paleontology. R. C. Moore, ed. Part S. Echinodermata 1. Geol. Soc. Amer. and Univ. of Kansas.

Hardy, A. C. 1965. The Open Sea, its Natural History: the world of plankton. Houghton Mifflin Co.

Hyman, Libbie H. 1955. The Invertebrates: Echinodermata, the Coelomate Bilateria. Vol. IV. McGraw-Hill Book Co.

Millot, N. 1967. Echinoderm Biology. Symp. Zool. Soc. Lond. 20.

Mortenson, T. 1901. Die Echinodermen-Larven. Nord. Plankt. 9:1–30.

———. 1921. Studies of the development and larval forms of echinoderms. G. E. C. Gad. Copenhagen.

———. 1927. Handbook of the Echinoderms of the British Isles. Oxford Univ. Press.

Nichols, D. 1962. Echinoderms. Hutchinson University Library.

Ubaghs, G. 1967. General characters of Echinodermata. *In* Treatise on Invertebrate Paleontology. R. C. Moore, ed. Part S. Echinodermata 1. Geol. Soc. Amer. and Univ. of Kansas.

Verrill, A. E. 1882. Notice of the remarkable marine fauna occupying the outer banks of the southern coast of New England. Am. J. Sci. Ser. 3. 23(13):135–142; (16):214.
———. 1895. Distribution of the echinoderms of northeastern America. Am. J. Sci. Ser 3. 49(13):127–141; (19):199–212.

Class Crinoidea
CRINOIDS or FEATHER STARS

Diagnosis. *Pentaradially symmetrical with a crown composed of a cuplike calyx and four to ten slender arms bordered on each side by short, alternating branches or pinnules; oral surface leathery with five branching ambulacral grooves extending into the arms and pinnules and lined with fingerlike podia; fixed throughout life by a long, jointed stalk, or stalked when young but subsequently stalkless and free-moving with a basal circlet of short slender cirri.*

Crinoids had their heyday in the Carboniferous and have been on the decline ever since. At appropriate depths individual species may still be abundant, however; Verrill (1882) reported 10,000 *Hathrometra* in a single dredge haul off Massachusetts, and *Rhizoclinus* appears to be similarly abundant off Florida. Surviving genera are mostly confined to deep water, including the forms reported from this coast. Only a fraction of modern species retains the classic stalked form of the crinoids that fluourished on Paleozoic sea bottoms. Stalkless species such as *Hathrometra tenella* (= *Antedon dentata*) can swim short distances but spend much of their time clinging with their cirri to the sea bottom or to benthic plants and animals. Crinoids feed by trapping plankters on the podia, after which the food is conveyed to the mouth by ciliary currents along the mucus-lined ambulacral grooves.

As in most echinoderms, the sexes are separate. Some species brood their young; in others embryonic and larval development are entirely independent of the adult. Larval development has been fully described for only a few species. The larvae are of the doliolaria type, but the free period is short, lasting a few days at most. The doliolaria then attaches to a substratum by its adhesive disc and metamorphoses into a stalked, juvenile crinoid. Young comatulids may remain as stalked forms for several months before breaking free. The larval stages are figured by Hyman (1955). Crinoids are likely to fragment when handled or severely disturbed, and their regenerative powers are highly developed.

Systematic List and Distribution

The classification of local forms, following Ubaghs (1953), is as follows.

Order Articulata		
Suborder Comatulida		
Family Antedonidae		
Hathrometra tenella	B, 37°	40 to 900 m
Hathrometra sarsi	B, 40°	30 to 1800 m
Suborder Millericrinoida		
Family Bathycrinidae		
Rhizocrinus lofotensis	BV	140 to 3000 m

Published records for these species within the range of this text are scanty. *H. tenella* has been taken in 240 m "off Martha's Vineyard," and *R. lofotensis* was reported by Verrill (1882) in 1100 m off Massachusetts. Also, the Arctic *Heliometra glacialis* has been taken in 10 to 1400 m as far south as Georges Bank. An additional possibility is *Coccometra hageni*, 14–1000 m, from North Carolina to the West Indies.

Identification

1. Adults sessile, stalked; without cirri; 4–7 arms; yellow; height 80 mm, of which 70 mm is stalk; Fig. 22.2A. *Rhizocrinus lofotensis.*
1. Adults free-swimming, without stalk; ten arms—2.

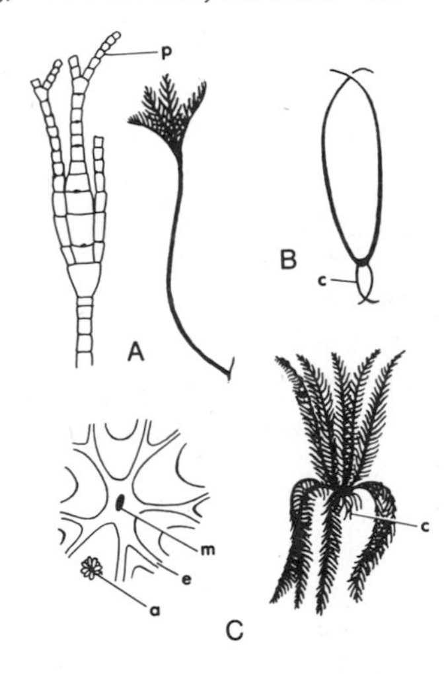

Fig. 22.2. Echinodermata, Crinoidea. A, *Rhizocrinus lofotensis*, growth form (right) and detail of calyx; B, *Heliometra glacialis*, pair of arms and pair of cirri to show proportions; C, *Hathrometra* sp., growth form (right) and detail of oral disc. Abbreviations are as follows: a, anus; e, ambulacral groove; m, mouth; c, cirrus; p, pinnule.

2. First pinnule with 40 or more joints; span 500 mm; yellow, sometimes tinged with purple; Fig. 22.2B. *Heliometra glacialis.* Note that the Carolinian *Coccometra hageni* has the cirri weakly developed, and more or less straight in comparison to the above.
2. First pinnule with 30 or fewer joints; grey, vaguely banded with brown; span 200 mm; Fig. 22.2C. *Hathrometra:*
   - Cirri with 24–33 segments. *H. tenella.*
   - Cirri with 14–24 segments. *H. sarsi.*

References

See also Hyman (1944), Mortensen (1927), Nichols (1962), Verrill (1882, 1895) in the general echinoderm references at the conclusion of the preceding section.

Chadwick, H. C. 1907. *Antedon.* Lpool. Biol. Soc. Proc. 21:371–417.

Clark, A. H. 1915–1950. A monograph of existing crinoids. Vol. 1, parts 1–4c. Bull. U.S. natn. Mus. 82.

———. 1923. Crinoidea. Dan. Ingolf-Exped. 4(5):1–58.

Ubaghs, G. 1953. Classe des Crinoides. *In* Traité de paléontologie. J. Piveteau, ed. III:658–773.

Fell, H. Barraclough. 1966. Ecology of Crinoids. *In* Physiology of Echinodermata. R. A. Boolootian, ed. Interscience.

Class Holothuroidea
SEA CUCUMBERS

Diagnosis. *Body basically cylindrical, elongated, with mouth and anus at opposite ends of the long axis or nearly so, the anterior end sometimes with a collarlike introvert, and the posterior end sometimes with a tail, the mouth surrounded by a ring of retractile tentacles; body more or less differentiated into dorsal and ventral surfaces, tube feet present or not; dermal skeleton absent or consisting of embedded spicules or scalelike surface plaies.*

Adult holothruains of local species range in size from about 20–500 mm. The body form, which is indeed cucumberlike in some species, actually ranges from long, slender, and wormlike to short and thick. The skin may be thin, soft, and translucent or thick, opaque, and leathery; its surface may be smooth but usually is ornamented with warts and papillae. The *podia* may be more or less equally distributed, ranged in rows evidencing a basic pentaradial symmetry, or confined to the sole; in one order they are entirely lacking. The differentiation of functionally dorsal and ventral surfaces—actually, sea cucumbers may be viewed as lying on their sides—also varies, being slight in some genera. In others, e.g., *Psolus*, the lower side is a clearly defined *sole*.

The tentacles may be simple, with few branches, or densely bushy. They are usually in a single circle around the mouth and are more or less similar, although quite commonly 2 tentacles are reduced in size. In *Duasmodactyla commune* the tentacles are paired in inner and outer circles, the inner ones being smaller. The number of tentacles in local genera is either 10, 12 (rarely 13 or 14), 15, or 20. The pattern of branching may be dendritic, pinnate, peltrate, or digitate. *Dendritic* tentacles are bushy with irregular branches (Fig. 22.4B). *Pinnate* tentacles have a series of lateral branchlets (Fig. 22.3B). In the *peltate* form there is a central stalk with a terminal cluster of radial branches (Fig. 22.3E). *Digitate* tentacles are thick and fingerlike, with or without short, terminal branches. The collarlike introvert in species of Dendrochirotida is retractile along with the tentacles.

Dermal skeletal elements, which occur in great variety of form, are not entirely consistent in type even at the family or ordinal level. They are very important in the differentiation of holothurian species, however. In the literature they are described as *deposits*. They are generally microscopic in size but range from about 10 mm or even larger to less than 0.01 mm in maximum dimension. There is often individual, including ontogenetic, variation in the size and form of the deposits. The principal types are rods, rosettes, buttons, tables, plates, baskets, wheels, and, chiefly in Apodida, anchors. *Rods* may be simply linear in form, but they grade into *rosettes* which are branched, often eccentrically, and with exceeding variation in detail. *Buttons* are actually oval in shape with 4, 6, or more holes in rows. *Tables* are perforated discs or plates with an erect, central spire, the latter varying in degree of development. *Baskets* are perforated plates with the rim more or less turned up and toothed. The form of wheels and anchors, the latter associated with perforated plates, is fairly self-evident. Deposits are usually calcareous and whitish or somewhat glassy in cleaned preparations. In *Molpadia* the deposits of the juvenile tend to become transformed into yellow to red *phosphatic deposits* containing iron phosphate.

There are additional calcareous bodies of importance in formal systematics; they form a ring of 10 ossicles around the esophagus and serve for the attachment of muscles (Fig. 22.3G). The *esophageal ossicles* may be simple or complex with long or short posterior projections.

The sexes are usually separate except in *Leptosynapta*, individuals of which, although hermaphroditic, mature male and female gametes at different times. The sexes in other forms are indistinguishable except in extralimital forms where the female broods her offspring. The young may pass through one or more free-swimming larval stages before metamorphosing into juvenile sea cucumbers. Typically the initial stage is an *auricularia*, followed by a *doliolaria*. Among local species the auricularia stage is omitted in *Leptosynapta tenuis*, the initial larva being a doliolaria; in *Cucumaria frondosa* and *Psolus phantapus* neither auriuclaria nor doliolaria occurs and the larva is a ciliated or flagellated oval form. (Developmental forms are considered further in the introduction to this phylum.)

Holothurians, like most other echinoderms, have substantial powers of regeneration. The behavioral response of many species to rough handling—voluntary evisceration—is well known. In some cases, e.g., *Thyone briareus*, the front end is cast off at the same time. These lost parts are regenerated, and most species, other than *Leptosynapta*, have varying abilities to survive and regenerate after transverse fission. However, this is apparently not a normal process of reproduction among local species.

Pelagic holothurians are found extralimitally; these are chiefly bathy-pelagic species reported at depths greater than 800 m off this coast, although *Pelagothuria* is an extralimital epipelagic form. Costello (1946) found young *Leptosynapta* swimming at the surface at night, but local species otherwise are exclusively benthic. *Cucumaria*, *Thyone*, *Molpadia*, and *Caudina* occupy U-shaped burrows. *Leptosynapta* is also a burrower, but only the anterior end remains exposed. Other sea cucumbers such as *Psolus* spp. are creeping animals found on hard substrata, in crevices, or under rocks.

Sea cucumbers of the order Dendrochirotida feed on plankton or detritus or both. Minute plankters and other particles are trapped in the mucous coating of the extended tentacles, which may also be used to sweep the substratum to pick up food items. Other, nondendrochirote burrowers ingest large amounts of bottom material from which nutritive elements are removed in passage.

Systematic List and Distribution

The species list is from Deichman (1930) with the taxonomy revised to conform with Frizzel *et al.* (1966); subclass divisions omitted.

Order Dendrochirotida		
Family Psolidae		
Psolus valvatus	B	ca. 183 m
Psolus phantapus	B	20 to 400 m
Psolus fabricii	B	18 to >183 m

Family Phyllophoridae		
Pentamera calcigera	B	37 to 73 m
Pentamera pulcherrima	V+	Shallow
Havelockia scabra	B, 39°	18 to >183 m
Family Sclerodactylidae		
Sclerodactyla briareus	V	Lit. to 6 m
Family Cucumariidae		
Cucumaria frondosa	B	Shallow to 366 m
Thyonella gemmata	V+	To ca. 50 m
Thyonella pervicax	V+	To ca. 50 m
Stereoderma unisemita	B+	31 to 40 m
Stereoderma parassimilis	Off Cape Cod	146 to 221 m
Duasmodactyla commune	B	Lit. to deep water
Order Apodida		
Family Synaptidae		
Leptosynapta inhaerens	BV	Lit. to 183 m
Leptosynapta roseola	V	Near low water
Labidoplax buskii	C+	18 to 445 m
(A European Boreal species; see Pawson, 1967)		
Family Chiridotidae		
Chiridota laevis	B	Lit. to 82 m
Toxodora ferruginea	V	140 to 280 m
Order Molpadiida		
Family Molpadiidae		
Molpadia oolitica	BV	90 to 720 m
Molpadia musculus	V	101 to 1131 m
Family Caudinidae		
Caudina arenata	B	Lit. to 35 m

Identification

The deposits are easily examined by extracting a small block of the body wall, hardly more than a pinhead-sized piece is needed, and dissolving the soft parts with Clorox on a microscope slide. The remaining hard parts may be examined in a drop of water under a cover slip, or a permanent mount can be prepared with glycerin jelly. The lower powers of a compound microscope provide sufficient magnification. Dimensions are given for deposits and for maximum adult body length.

The ordinal groups are separated by the following key.

1. Tube feet present; tentacles 10–30, richly branched, i.e., dendritic; introvert present. Dendrochirotida.

1. Tube feet absent; tentacles 10–25, simple, digitate, or pinnate; no true introvert present—2.

2. Body cylindrical, wormlike; body wall thin, semitransparent; deposits may include 6-spoked wheels or anchors. Apodida.
2. Body fusiform, often with tapering tail; body wall leathery, opaque; deposits may include anchors, at least in juveniles; wheels lacking. Molpadiida.

ORDER APODIDA

1. Deposits include 6-spoked wheels, no anchors, the wheels about 0.08–0.10 mm in diameter, concentrated in a few large papillae; tentacles peltate; body white to pinkish, 150 mm; Fig. 22.3E,F. *Chiridota laevis.*
1. Deposits include anchors and anchor plates, the anchors about 0.15 mm long; tentacles pinnate; Fig. 22.3A–D—2.
2. Pinkish, with red granules scattered thickly over the body; anchors frequently more delicate than in the next species; 100 mm; in gravel and under rocks. *Leptosynpata roseola.*
2. Whitish; 150 mm; in sand; Fig. 22.3A. *Leptosynapta inhaerens.* In cases of ambiguity the two species of *Leptosynapta* may be distinguished by differences in the esophageal ossicles. In *inhaerens* the radial plates are pierced by a hole; in *roseola,* they are simply notched; Fig. 22.3G.

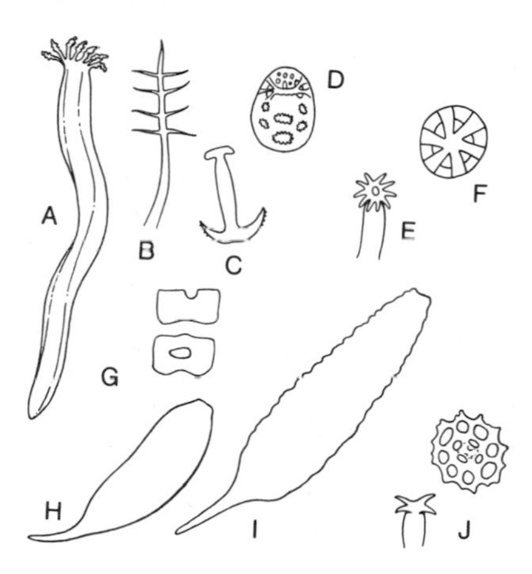

Fig. 22.3. Echinodermata, Holothuroidea. A, *Leptosynapta inhaerens*, whole animal; B, *Leptosynapta* sp., pinnate tentacle; C, same, anchor deposit; D, same, anchor plate; E, *Chiridota laevis*, peltate tentacle; F, same, wheel deposit; G, *Leptosynapta*, esophageal ossicles of *roseola* (above) and *inhaerens* (below); H, *Molpadia musculus*, outline of whole animal; I, *Molpadia oolitica*, outline of whole animal; J, *Caudina arenata*, tentacle and table deposit.

ORDER MOLPADIIDA

1. Without phosphate bodies; deposits include tables with 4 central holes and 8–10 outer holes, the spire with 2 to 4 rods; tentacles with 2 pairs of pointed digits; body pinkish to purplish, 250 mm; shallow water species; Fig. 22.3J. *Caudina arenata.*
1. Phosphate bodies present—2.
2. Deposits present near tail, scarce otherwise, consisting chiefly of irregular tables with up to 10 holes in the disc and a 3-pillared spire; fusiform rods absent, even in the tail; phosphate bodies typically blackish and so numerous as to color the whole animal; 150 mm; Fig. 22.3I. *Molpadia oolitica.*
2. Deposits more numerous, including tables with 6 large holes, fusiform bodies with 3 or 4 holes, and slender rods in the tail; 80 mm; a deep water form usually reported from >183 mm; Fig. 22.3H. *Molpadia musculus.*

ORDER DENDROCHIROTIDA

1. Ventral side flattened, with a distinct sole; dorsal surface with scales; with or without a short subterminal tail; Fig. 22.4A,B. *Psolus*—2.
1. Not so—4.
2. Mouth surrounded by five distinct oral valves; 20 mm. *Psolus valvatus.*
2. Mouth surrounded by numerous indistinct oral valves; 150–200 mm—3.
3. Anus surrounded by numerous scales in numerous rings; tail well developed; scales between oral and anal regions numerous in adults, few in young; large individuals almost black. *Psolus phantapus.*
3. Anus with 1 or 2 rings of 5 or 6 scales each; tail short or absent; adults with about 8–10 scales between mouth and anus; bright red. *Psolus fabricii.*
4. Tentacles 10 (rarely 9 or 11)—5.
4. Tentacles 20, in 5 outer and 5 inner pairs, the inner pairs smaller; deposits scarce, present only on introvert, including tables with several holes and a high spire; tube feet generally distributed or only missing from 8 narrow stripes; color whitish to yellow, the tentacles and oral region violet; 150 mm. *Duasmodactyla commune.*
5. Esophageal ring simple, lacking posterior processes; Fig. 22.4J. Cucumariidae—9
5. Esophageal ring complex, with more or less well-developed posterior processes—6.
6. Posterior processes of esophagael ring long, composed of a mosaic of minute pieces; with a short tail or not; Fig. 22.4K. Phyllophoridae—7.
6. Posterior processes generally short, entire; without a tail; skin smooth, with few or no deposits; brownish or greenish; 120 mm; Fig. 22.4C. *Sclerodactyla briareus.*
7. Tube feet scattered all over body wall, hairlike; posterior end abruptly tapered to form short tail; skin rough with abundant deposits; whitish tinged with brown; 50 mm (rarely to 100 mm); Fig. 22.4D. *Havelockia scabra.*
7. Tube feet restricted to the radii, where they form 5 crowded bands; Fig. 22.4I—8.
8. Table and plates larger, about 0.013–0.016 mm, the plates in a dense layer beneath the tables; 100 mm; white with a dark introvert and tentacles; an Arctic-Boreal species. *Pentamera calcigera.*

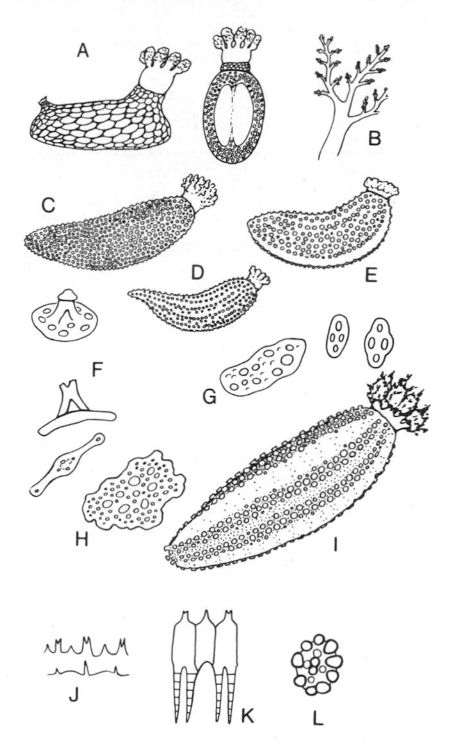

Fig. 22.4. Echinodermata, Holothuroidea. A, *Psolus* sp., whole animal, lateral and ventral views; B, same, tentacle; C, *Sclerodactyla briareus;* D, *Havelockia scabra;* E, *Stereoderma unisemita;* F, three table deposits seen in various aspects; G, deposits, two buttons (right) and a plate (left); H, plate deposit; I, *Pentamera pulcherrima;* J, *Thyonella pervicax*, detail of part of esophageal ring; K, calcareous ring with well-developed posterior processes; L, *Thyonella pervicax*, mulberry-like deposit.

8. Tables and plates smaller, about 0.010 mm; 50 mm (usually about 25 mm); white or pale yellowish or light brownish; a Virginian species; Fig. 22.4I. *Pentamera pulcherima.*

9. Deposits chiefly cups and buttons; no plates. *Thyonella*—10.

9. Deposits buttons or plates; no cups; Fig. 22.4E,G,H—11.

10. Tube feet uniformly spread over body, usually contracted to form low warts; deposits knobbed buttons, cups, and few large mulberry-shaped bodies; 70 mm; Fig. 22.4L. *Thyonella pervicax.*

10. Tube feet in distinct double rows along radii, and scattered in interradii; no large mulberry-shaped deposits; 150 mm. *Thyonella gemmata.*

11. Body wall stiff, filled with numerous deposits; 50 mm. *Stereoderma*—12.

11. Body wall soft, leathery, deposits rare in mature specimens; tube feet completely retractile; brown or (especially in young individuals) paler, yellowish to white; 500 mm. *Cucumaria frondosa.*

12. Deposits four-holed buttons of varying size, with smooth to knobbed surface; Fig. 22.4E. *Stereoderma unisemita.*
12. Deposits roundish or elongated perforated plates with low knobs and few to many holes; Fig. 22.4G. *Stereoderma parassimilis.*

References

See also general echinoderm references following the introductory section of this chapter.

Costello, D. P. 1946. The swimming of *Leptosynapta*. Biol. Bull. mar. biol. Lab. Woods Hole. 90:93–96.

Deichmann, Elizabeth. 1930. The holothurians of the western part of the Atlantic Ocean. Bull. Mus. comp. Zool. Harv. 71(3):43–226.

Frizzell, D. L., Harriet Exline, and D. L. Pawson. 1966. Holothurians. *In* Treatise on Invertebrate Paleontology. Part V. Echinodermata. 3(2):641–572. R. C. Moore, ed. Geol. Soc. Amer. and Univ. of Kansas.

Pawson, D. L. 1967. *Protankyra grayi* new species and *Labidoplax buskii* (McIntosh) from off North Carolina (Holothuroidea: Synaptidae). Proc. biol. Soc. Wash. 80:151–156.

Class Echinoidea
SEA URCHINS, HEART URCHINS, SAND DOLLARS

Diagnosis. *With a rigid globular or disclike test of close-fitting calcareous plates armed with movable spines; mouth ventral, anus dorsal or marginal; with pentamerous radial symmetry, or secondary bilateral symmetry.*

The local members of this class range in size in the adult from <10 mm to about 150 mm in diameter. "Regular" echinoids, such as sea urchins, with unmodified radial symmetry are usually subspherical with the ventral surface more or less flattened and the dorsal surface domed or flattened. The "irregular" echinoids are heart-shaped, ovoid, or flattened to a disclike form. The spines of regular genera are usually well developed but vary in form in different species, on different parts of the same animal, ontogenetically, and individually in the same species. In irregular echinoids the spines are generally smaller, and in sand dollars are reduced to a furlike covering.

The test forms a complete shell perforated only by minute pores for the passage of the tube feet and by anal, oral, and genital openings, plus the madreporite. Keyhole urchins have additional openings, the *lunules*. The fine structure of the test is best examined in cleaned or partly cleaned specimens from which the spines have been removed.

The podia are concentrated in bands which are revealed in the cleaned test as radiating series of fine pores; on close inspection it is apparent that within each of these ambulacral bands the pores are arranged in pairs and the pore pairs in short arcs or zigzags. Differences in these arrangements provide useful criteria for distinguishing species of regular echinoids. The ambulacral areas are, themselves, paired, i.e., the interambulacral areas are usually wider between adjacent pairs than between the individuals of the couplet. In regular echinoids the alternating ambulacral and interambulacral areas simply diverge from the dorsal midpoint of the test. In sand dollars and keyhole urchins the ambulacral pairs from a five-petaled, flowerlike design dorsally (Fig. 22.5L,N). The pattern is further modified in spatangoids, where the ambulacral areas are distinctly unequal (Fig. 22.5K,M).

The mouth, except in spatangoids, has a chewing apparatus which has long been known as *Aristotle's lantern;* this structure has 5 teeth and is surrounded by a soft, more or less flexible area, the *peristome*, which is strengthened by embedded plates. The peristome is reduced to a short crescent filled with scalelike plates in spatangoids. The equivalent aboral area, the *periproct*, is also armored; in *Arbacia* the periproct is actually filled, except for the central anus, by 4 large plates (Fig. 22.5B). In other sea urchins the periproct is strengthened in the adult by numerous small plates irregularly dispersed (Fig. 22.5F). There is ontogenetic variation in this character, and young *Strongylocentrotus*, for example, have a single large plate in the periproct plus several very small ones. In regular echinoids the periproct is surrounded by plates distinguished as forming the *apical system* including 5 large genital plates and 5 smaller plates which may or may not touch the periproct. Each of the genital plates has a single large pore which is the exit aperture for one of the 5 gonads. One of the genital plates is riddled with fine pores and is also the madreporite.

Among irregular echinoids the sand dollars are radially symmetrical in external appearance except that the anus is at the margin of the disc. The spatangoids are more highly modified with a distinct bilateral symmetry; both periproct and peristome are substantially displaced from a central position and establish an anterior-posterior axis, and the test is modified in shape for locomotion in one direction. For a brief explanation of the peculiarities of spatangoid anatomy see Bullough (1960), and for a more extended account see Hyman (1955).

The sexes are separate in echinoids except in occasional aberrant individuals. Males and females are indistinguishable externally in local species. Fertilization is external, and none of the local species is known to brood its young; hence all have a free-living pelagic larva of the pluteus type (see introduction to the phylum). The larval period may last a month or more, by which time the pluteus, with the encumbrance of its developing skeleton, has lost its agility and sunk to the bottom. Metamorphosis is rapid, and the juveniles are generally less than 1 mm in diameter. Echinoids have considerable powers of regeneration of damaged parts of the test as well as of such minor details as individual spines. They do not reproduce asexually.

Echinoids are benthic animals exclusively. Though generally sluggish, they are motile, and sea urchins easily climb vertical surfaces. Urchins are found chiefly on hard substrata. *Strongylocentrotus* has been regarded, somewhat questionably, as a rock borer in the manner of the tropical *Echinometra*. Sea urchins are omnivorous and act to some extent at least as scavengers. A preference for either plant or animal foods is usually evidenced, however, and both *Arbacia* and *Strongylocentrotus* are regarded as primarily herbivorous. Conversely, the digestive tracts of *Genocidaris* and *Echinus* have been reported to contain bryozoans, sponges, and shells.

Irregular echinoids are shallow burrowers in sand or mud. The spatangoids ingest bottom material rich in organic matter, but according to MacGinitie and MacGinitie (1968) the clypeasteroids gather microscopic food particles with the aid of ciliary currents and a fine mucous coating on the dorsal spines, the entrapped food being conveyed eventually to the ventral mouth.

Systematic List and Distribution

The classification is that of Durham *et al.* (1966). The basic list is from Verrill (1895), and Clark (1925).

Order Arbacioida		
Family Arbaciidae		
*Arbacia punctulata**	V	Lit. to 225 m
Order Temnopleuroida		
Family Temnopleuridae		
Genocidaris maculata	C+	12 to 420 m
Family Toxopneustidae		
Lytechinus variegatus	C	Lit. to 55 m
Order Echinoida		
Family Echinidae		
Echinus gracilis	"Off Martha's Vineyard"	120 to 445 m
Family Strongylocentrotidae		
Strongylocentrotus droebachiensis	B+	Lit. to 1150 m
Order Clypeasteroida		
Family Echinarachnidae		
Echinarachnius parma	BV	Lit. to 1600 m

Family Mellitidae		
*Mellita quinquiesperforata**	C, 41°	Shallow
Order Spatangoida		
Family Schizasteridae		
Brisaster fragilis	B	40 to 1300 m
(To Florida, according to Miner, 1950)		
Moira atropos	C, 37°	Lit. to 145 m

**Arbacia punctulata* has a peculiar distribution; it is uncommon in the southern part of the Virginian province, and there is but one Chesapeake record. Nevertheless it occurs locally in the West Indies and Florida. The occurrence of *Mellita* north of the Chesapeake is doubtful.

Identification

Dimensions are maximum test diameters or length. Most of the species are figured by Miner (1950), but see also Mortensen (1928–1951).

1. Test globular, radially symmetrical—2.

1. Test globular or heart-shaped, bilaterally symmetrical—3.

1. Test flat, disclike—4.

2. Color reddish brown, purple, to almost black; spines long, the longest equal to about half the diameter of the test, those in the upper part pointed but the lower spines usually with a spatulate tip or "shoe"; periproct with 4 (rarely 3 or 5) plates; 50 mm; Fig. 22.5A–D. *Arbacia punctulata.*

2. Not so; usually predominantly greenish or greenish and white; spines short, the longest equal to about ⅓ or less of the diameter of the test; periproct with numerous small plates; Fig. 22.5E,F.

- Each ambulacral plate with 5 or 6 (rarely 4–8) pore pairs in a wide arc; color variable, usually predominantly green both in test and spines but sometimes tinged with purplish or reddish; 80 mm; Fig. 22.5G. *Strongylocentrotus droebachiensis.*
- Each ambulacral plate with 3 pore pairs in a shallow arc; test olive, spotted with white, spines olive with 2 or 3 narrow reddish bands, the tips white; 12.7 mm but usually <9 mm; Fig. 22.5H. *Genocidaris maculata.*
- Each ambulacral plate with 3 pore pairs in a deep arc; predominantly green, radially striped with white, the spines white tipped with light brownish; 88 mm; Fig. 22.5I. *Echinus gracilis.*

3. All ambulacral grooves are deep clefts; dried test whitish, grayish, or very pale brown; test usually high, the height equal to 70–80% of the length; 57 mm long; Fig. 22.5K. *Moira atropos.*

3. The longest (anterior) ambulacral groove deep and broad, the others shallow; dried test medium to dark brown; test low and broad; 90 mm long; Fig. 22.5M. *Brisaster fragilis.*

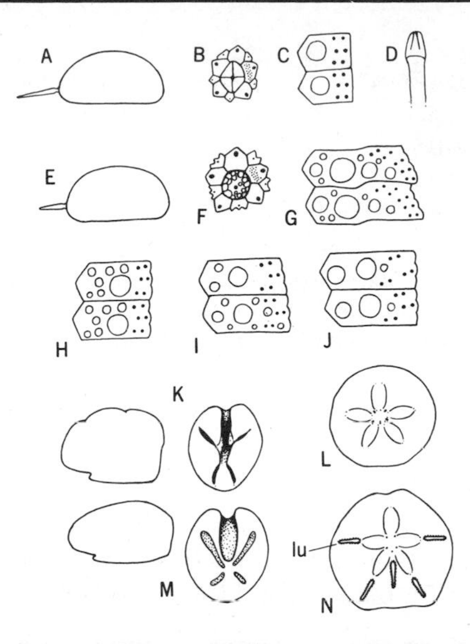

Fig. 22.5. Echinodermata, Echinoidea. A, *Arbacia punctulata*, schematic lateral view showing relative length of spines; B, same, periproct and apical system; C, same, two ambulacral plates; D, same, tip of lower spine; E, *Strongylocentrotus droebachiensis*, schematic lateral view showing relative length of spines; F, same, periproct and apical system; G, same, two ambulacral plates; H, *Genocidaris maculata*, two ambulacral plates; I, *Echinus gracilis*, two ambulacral plates; J, *Lytechinus variegatus*, two ambulacral plates; K, *Moira atropos*, outline of test, lateral view (left) and from above (right); L, *Echinarachnius parma*, test; M, *Brisaster fragilis*, outline of test lateral view (left) and from above (right); N, *Mellita quinquiesperforata*, test. Abbreviation: lu, lunule.

4. Disc with 5 long holes (lunules); brownish yellow to golden ochre tinged with green; 115 mm; Fig. 22.5N. *Mellita quinquiesperforata.*
4. Disc intact, without holes; purplish or reddish brown in life; to about 78 mm; Fig. 22.5L. *Echinarachnius parma.*

References

See also general echinoderm references following the introductory section of this chapter.

Bullough, W. S. 1960. Practical Invertebrate Anatomy. Macmillan and Co.

Clark, H. L. 1925. A catalog of the recent sea urchins (Echinoidea) of the British Museum. British Museum (nat. Hist.).

———. 1940. A revision of the keyhole urchins (Mellita). Proc. U.S. natn. Mus. 89(3099): 435–444.

Durham, J. W., *et al.* 1966. Echinozoans. *In* Treatise on Invertebrate Paleontology. R. C. Moore, ed. Part U, Echinodermata. 3(1):108–366; (2):367–695. Geol. Soc. Amer. and Univ. of Kansas.

MacGinitie, G. E., and Nettie MacGinitie. 1968. Natural History of Marine Animals. 2nd ed. McGraw-Hill Book Co.
Miner, R. W. 1950. Field Book of Seashore Life. G. P. Putnam's Sons.
Mortensen, T. 1907. Echinoidea. Dan. Ingold-Exped. 4(1):1–200.
———. 1928–1951. A Monograph of the Echinoidea. 15 vols. C. A. Reitzel and Oxford Univ. Press.
Verrill, A. E. 1895. Distribution of the echinoderms of Northeastern America. Am. J. Sci. 49:127–141, 199–212.

Class Stelleroidea
Subclass Asteroidea
SEA STARS OR STARFISH

Diagnosis. *Body more or less flattened and radially symmetrical, pentagonal or stellate in form with five or more arms merging insensibly with the central disc; mouth ventral; ventral surface of arms with open ambulacral grooves with two or four rows of podia; madreporite dorsal, conspicuous or not.*

The general form of the typical sea star is so familiar as scarcely to require description. Body shape does vary, however, depending on the number and proportions of the arms in comparison with the central disc. Most sea stars have five arms or *rays*, though some genera, e.g., *Solaster*, *Crossaster*, and *Coronaster*, typically have between 7 and 15 or even more arms. An excess of appendages in normally 5-armed species frequently occurs as a result of the regeneration of damaged individuals.

The endoskeleton consists of individual ossicles bound together by soft tissue. The details of this association are more or less obscured by an external cuticle or dermal covering that is usually ornamented with embedded spines, warts, or other protuberances. The ossicles are most clearly demonstrated externally in orders with conspicuous marginal plates. In Figure 22.6E the arrangement of these skeletal elements is indicated in a schematic section.

Spinous ornamentation on the dorsal surface is all but lacking in the family Poraniidae, in which the dorsal surface is covered with a membranous skin, the principal ornamentation being the *papullae*—papillalike dermal gills present in all sea stars. The hard protuberances found in other genera may be simple spines or warts or may consist of more specialized structures that are embedded in the skin or are extensions of the endodermal ossicles. *Paxillae* are a common form of ornamentation and consist of erect, columnar projections from the skeletal ossicles topped with an expanded,

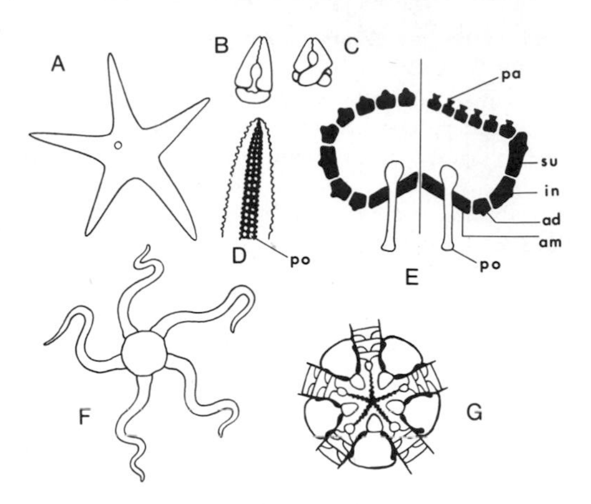

Fig. 22.6. Echinodermata, Stelleroidea. A, typical sea star indicating body form, for comparison with F; B, straight pedicellaria; C, crossed pedicellaria; D, sea star arm, ventral view showing groove with podia; E, sea star, arm, schematic half-sections showing arrangement of ossicles in *Asterias* (left) and *Astropecten* (right); F, typical brittle star; G, brittle star *Ophioderma*, central disc, ventral view. Abbreviations are as follows: ad, adambulacral plate; am, ambulacral plate; in, inframarginal plate; pa, paxilla; po, podium; su, supramarginal plate.

often tabular area with minute granules or spines (Fig. 22.7A,B; Fig. 22.8C,E). Seen from above, the paxillae often produce a pattern of minute rosettes or "flowers."

Pedicellariae are minute pincer-, scissor-, or plierlike appendages of the body wall, important in classification at various levels. They may be pedunculate, with a short, fleshy stalk, or sessile with direct attachment to the endodermal ossicles, or alveolar with an attachment area imbedded in a depression in the skeletal ossicles. Pedunculate pedicellariae may be either straight or crossed; in the straight variety the 2 jaws or valves of the pinchers are simply joined to a common basal piece, while in the crossed type the valves cross basally, scissorlike (Fig. 22.6B,C). Sessile pedicellariae may consist essentially of opposed pairs of straight or curved spines. The alveolar pedicellariae of *Hippasteria* are bivalved, resembling a minute clam.

The sexes, except in atypical specimens, are separate. In some genera, females brood their young, but the sexes are otherwise indistinguishable. The eggs are brooded under the disc in several local genera, e.g., *Henricia* and *Leptasterias*, both of which lack pelagic larvae. In *Pteraster* the dorsal surface is covered with a membrane beneath which in many species is a brood chamber. Direct development is also suspected in *Pedicellaster*, *Ceramaster*, *Hippasteria*, and *Ctenodiscus*, this assumption being based on the occurrence of large, yolky eggs in these genera. Except for the preceding genera, development in most starfish involves a succession of pelagic larvae. In the full sequence the young pass through bipinnaria and brachiolaria stages before changing into the juvenile starfish, as outlined in the introduction to this phylum. The bipinnaria metamorphoses directly in *Astropecten*.

Regeneration of lost or damaged parts is common in asteroids, and a few normally reproduce by fission as well as sexually. Asexual reproduction is found only in *Stephanasterias* among local genera.

Starfish are benthic animals. The common shallow water species of *Asterias* inhabit both soft and hard grounds. Others are more restricted to specific habitats. *Astropecten*, for example, favors sand, and *Ctenodiscus* is a mud dweller. Starfish are predatory animals, feeding on a variety of benthic animals. *Asterias* is best known as a predator on bivalve mollusks, particularly oysters, but will take fish or any other prey it can catch. A few genera, such as *Luidia*, are particularly prone to attack other echinoderms. *Porania* and *Ctenodiscus* are mucus-ciliary feeders subsisting on small particles.

Like other echinoderms, asteroids in general show little tolerance for reduced salinity, although *Asterias forbesi* and *A. vulgaris* can survive in waters of 14 and 18‰ respectively; *Urasterias lincki* is found with a salinity as low as 15‰.

Systematic List and Distribution

The classification adopted is that of Spencer and Wright (1966); the species list is from Verrill (1895) and Sladen (1889), with additions and taxonomic revisions from more recent literature.

Order Platyasterida		
Family Luidiidae		
Luidia clathrata	C, 39°	2 to 175 m
Luidia elegans	C, 42°	97 to 366 m
Order Paxillosida		
Family Astropectinidae		
Astropecten americanus	C, 40°	55 to 542 m
Astropecten articulatus	C+	7 to 165 m
Leptychaster arcticus	B, 38°	91 to 1765 m
Psilaster florae	BV	5 to 1801 m
Tethyaster vestitus	C, 40°	37 to 393 m
Family Goniopectinidae		
Ctenodiscus crispatus	B	9 to 1830 m
Family Benthopectinidae		
Pontaster tenuispina	B+	18 to 2013 m
Order Valvatida		
Family Odontasteridae		
Odontaster hispidus	BV	30 to 875 m
Odontaster setosus	V	102 to 732 m

Family Goniasteridae		
Hippasteria phrygiana	B+	20 to 800 m
Pentagonaster eximius	B	146 to 223 m
Ceramaster granularis	B+	132 to 962 m
Pseudarchaster intermedius	B, 38°	155 to 2943 m
Mediaster bairdii	"Northeast U.S."	90 to 360 m
Peltaster planus	V	180 to 365 m
Order Spinulosida		
Family Solasteridae		
Solaster endeca	B	Lit. to 274 m
Solaster syrtensis	B	55 to 110 m
Solaster papposus	B	Lit. to 327 m
Lophaster furcifer	B	70 to 200 m
Family Pterasteridae		
Pteraster pulvillus	B+	37 to 370 m
Pteraster militaris	B	18 to 366 m
Diplopteraster multipes	BV	123 to 1171 m
Family Asterinidae		
Asterina pygmaea	B	93 to 165 m
Family Poraniidae		
Porania insignis	V	35 to 690 m
Porania grandis	V	121 to 683 m
Poraniomorpha hispida	B, 37°	90 to 1171 m
Family Echinasteridae		
Henricia spp. (see key notes)	BV	Lit. to 2470 m
Order Forcipulatida		
Family Asteriidae (see key notes)		
Asterias forbesii	V+	Lit. to 49 m
Asterias vulgaris	B+	Lit. to 655 m
Asterias tanneri	V	35 to 355 m
Urasterias lincki	A+	Lit. to 549 m
Leptasterias tenera	BV	18 to 150 m
Leptasterias groenlandica	A+	Lit. to 225 m
Leptasterias hispidella	45°14′	91 m
Leptasterias littoralis	A, 44°	Lit. to 42 m
Leptasterias polaris	A+	Lit. to 110 m
Leptasterias austera	B	60 to 64 m
Stephanasterias albula	BV	5 to 314 m
Family Pedicellasteridae		
Pedicellaster typicus	B+	40 to 223 m
Coronaster briareus	C, 37°	37 to 683 m

Identification

The species are considered in three more or less natural groups. Measurements are given in arm radii. Whole animal drawings are only intended to indicate general form. Almost all of the species are figured in Miner (1950); for additional northern species see Grainger (1966).

---

1. With well-developed marginal plates usually forming a distinctly sided margin, separating arms and disc into dorsal and ventral parts; the aboral side usually with paxillae; pedicellariae never of the crossed type; podia in 2 rows; Fig. 22.6E. Group One. Note the somewhat atypical genus *Luidia*, with marginal plates on the lower side only; compare with *Pteraster* in Group Two, below.

1. Marginal plates not conspicuously differentiated externally from the other endoskeletal ossicles—2.

2. Dorsal surface either with paxillaelike groups of spines, or with scalelike, spiny plates, or naked with a thick skin or membrane with no more than a small cluster of spines at the anal pore plus soft papullae elsewhere; pedicellariae rare, never of the crossed type; podia in two rows. Group Two.

2. Dorsal spines usually single, not forming paxillae or paxillaelike groups; with crossed pedicellariae; podia in 2 or 4 rows. Group Three.

Group One, Orders Platyasterida, Paxillosida, and Valvatida. The species are all normally 5-armed.

---

1. Upper marginal plates undeveloped; dorsal paxillae in blocklike groups or clustered in rosettes; Fig. 22.7A,B. *Luidia.*

- Dorsal paxillae in quadrangular groups in rows on the arms; bluish or grayish, frequently with a darker stripe or stripes; 145 mm R; Fig. 22.7A. *L. clathrata.*
- Dorsal paxillae in rosettes; orange; 180 mm R; Fig. 22.7B. *L. elegans.*

1. Upper and lower marginal plates well developed—2.

2. Upper and lower marginal plates naked except for a single, vertical, flattened spine on each; the arms stiff, broad-based, sometimes giving a nearly pentagonal form; yellowish; 50 mm R; Fig. 22.7C. *Ctenodiscus crispatus.*

2. Upper and lower marginals with a granular or spiny ornamentation—3.

3. Podia with a terminal sucker; Fig. 22.7D—6.

3. Podia tapering, without a terminal sucker; Fig. 22.7D—4.

4. Marginal plates with granular spinules or with paxillae, without spines; the upper marginals smaller than the lower; orange or reddish; 30 mm R; Fig. 22.7E. *Leptychaster arcticus.*

4. Lower marginals with spines; upper and lower plates about equal—5.

5. Ambulacral and marginal plates separated from about midarm to base by a series of interradial plates or ossicles; whitish to pale pink; 100 mm R; Fig. 22.7I. *Psilaster florae.* The primary trait may appear ambiguous without comparative material, but note differences in the contours of the marginal borders of Figs. 22.7F and 22.7I.

5. Ambulacral and marginal plates in contact for almost their whole lengths with interradial ossicles only at the base; Fig. 22.7F,G; yellowish brown or orange. *Astropecten:* The differentiation of *Astropecten* species may prove difficult because there is substantial variation in key traits; see Verrill (1915). Verrill regarded

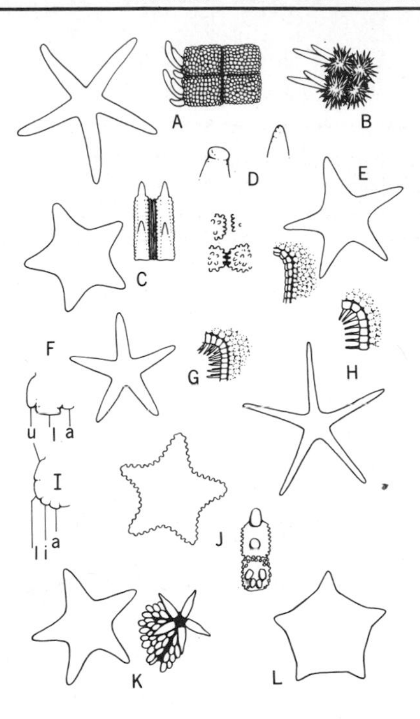

Fig. 22.7. Echinodermata, Asteroidea, Group One. A, *Luidia clathrata*, outline of animal and detail of dorsal margin; B, *Luidia elegans*, detail of dorsal margin; C, *Ctenodiscus crispatus*, outline of animal and lateral view of margin; D, podia with (left) and without (right) terminal sucker; E, *Leptychaster arcticus*, outline of animal with details of dorsal margin (left) and marginal plates enlarged; F, *Astropecten*, outline of animal and outline cross-sectional view of ventral margin; G, same, detail of dorsal margin; H, *Pontaster tenuispinus*, outline of animal and detail of dorsal margin (above right); I, *Psilaster florae*, outline cross-sectional view of ventral margin; J, *Hippasteria phrygiana*, outline of animal and detail of marginal plates in lateral view; K, *Odontaster* sp., outline of animal and detail of mouth area; L, *Ceramaster granularis*, outline of animal. Abbreviations are as follows: a, ambulacral plates; i, interradial plates; l, lower marginal plates; u, upper marginal plates.

A. americanus as the most abundant offshore starfish along this coast, chiefly in depths of 110–274 m and north to about 40°23′. A second species, *A. articulatus*, is found in shallower water at the southern border of our area; *A. comptus*, and presumably *A. vestitus* as well, have been reduced to synonymy with *A. articulatus*.

6. Upper and lower marginal plates alternating; pink, orange, or red above, whitish below; 130 mm R; Fig. 22.7H. *Pontaster tenuispinus*.

6. Upper and lower marginals opposite, arms relatively short, disc large; Fig. 22.7J–L—7.

7. Jaws with large, conspicuous spines; Fig. 22.7K. *Odontaster:* In *O. setosus* the ambulacral plates bear numerous slender spinules while in *O. hispidus* the plates are fewer and relatively sturdy; *O. setosus* 37 mm R, *O. hispidus* 55 mm R.

7. Jaws without spines—8.

8. Marginal plates with thick spines; disc moderately broad; red; to 200 mm R; Fig. 22.7J. *Hippasteria phrygiana.*

8. Marginal plates granular; disc very broad, the body subpentagonal; red; to 40 mm R; Fig. 22.7L. *Ceramaster granularis.* Note also *Pentagonaster eximius* and somewhat similar species of *Pseudarchaster* in depths of about 150 m or more which may key out here.

Group Two, Order Spinulosida.

1. Normally 5-armed—2.

1. Normally with 7–15 arms; note also *Coronaster briareus* in Group Three with 8 or more arms—6.

2. Margin of arms sharp; disc broad, arms wide basally, body form frequently subpentagonal—3.

2. Margin of arms rounded, the arms subcylindrical and slender; dorsum evenly covered with minute paxillalike spinules; bright red, purple, or yellow; 100 mm R; Fig. 22.8A. *Henricia:* The species usually cited is *H. sanguinolenta;* however, there is some dispute as to how many species of this genus should be recognized in the western North Atlantic, and it may be that none of them is *H. sanguinolenta sensu stricto;* see Grainger (1966).

3. Dorsum with scalelike plates with small spines in clumps or paxillae. *Asterina pygmaea.* (Described by Verrill in 1878 from a specimen only 5 mm R, unrecorded since.)

3. Dorsum with a thick skin or membrane—4.

4. Margin with a finlike membrane supported by long, lateral spines; Fig. 22.8B. *Pteraster: P. militaris* has a pair of strong spines on the outer surface of the jaws; *P. pulvillus,* which may be bathymetrically extralimital, does not. *P. militaris* is yellow or reddish above, 7 mm R.

4. Margin with a fringe of small spines; Fig. 22.8D—5.

5. Dorsum without spines except in a small clump at the anal pore, but with indefinite bands of papullae or with scattered papullae. *Porania: P. grandis* has the marginal spines poorly developed and the papullae in petallike bands; *P. insignis* has 3 or 4 spines on each lower marginal plate, and papullae not in definite bands; both species reach 70 mm R; Fig. 22.8D. (See Verrill, 1895.)

5. Dorsum covered with fine spines; yellow to reddish brown; 40 mm R. *Poraniomorpha hispida.*

6. Paxillae with short, close-set spinules in a double series; arms usually 9 or 10 (range from 7 to 13); yellowish red to violet; 200 mm R; Fig. 22.8C. *Solaster endeca.*

6. Paxillae with long, fine spines giving the animal a bristly appearance; arms usually 10–12 (range 8–13); disc purple-red, arms whitish banded; 170 mm R; Fig. 22.8E. *Solaster papposus.*

Group Three, Order Forcipulatida.

1. Tube feet in 2 rows—2.

1. Tube feet in 4 rows—3.

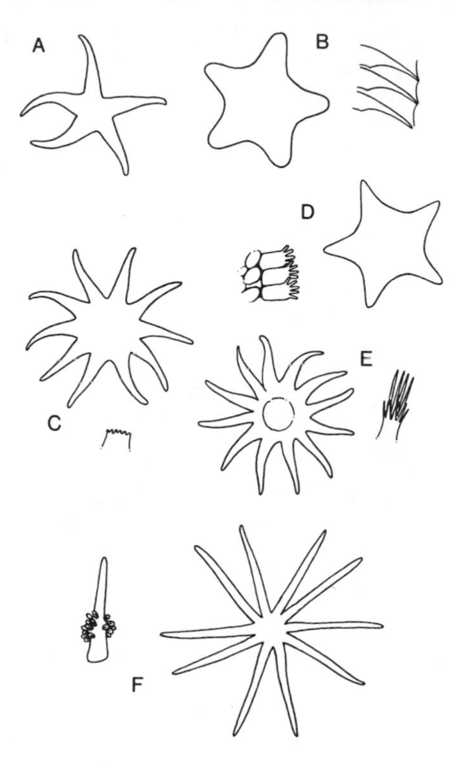

Fig. 22.8. Echinodermata, Asteroidea, Groups Two and Three. A, *Henricia* sp., outline of animal; B, *Pteraster* sp., outline of animal and detail of margin in ventral view; C, *Solaster endeca*, outline of animal and detail of paxilla in lateral view; D, *Porania* sp., outline of animal and detail of ventral margin; E, *Solaster papposus*, outline of animal and detail of paxilla in lateral view; F, *Coronaster briareus*, outline of animal and detail of dorsal spine and pedicellariae.

2. Arms 8 or more; lateral and dorsal spines with clusters of pedicellariae; 75 mm R; Fig. 22.8F. *Coronaster briareus.*

2. Arms normally 5 but note *Stephanasterias* below; lateral and dorsal spines without pedicellariae; 30 mm R. *Pedicellaster typicus.*

3. Adambulacrals and inframarginals without intervening ossicles, the adambulacrals with long spines; 24 mm R. *Stephanasterias albula.* This species reproduces by fission and frequently has more than 5 arms, the maximum being about 9. There is little available information on the relative abundance or general distribution of this species. Determination of the primary criterion may require maceration.

3. With a row of ossicles between the adambulacrals and the inframarginals. *Asterias* and *Leptasterias.* The differentiation of these genera and of the individual species has caused considerable difficulty to students. The most constant differential trait separating the genera has limited value for the general worker. In *Leptasterias* the gonads open ventrally, whereas in *Asterias* they open dorsally. *Asterias* has pelagic larvae, but in *Leptasterias* the young are brooded; this habit suggests a more limited potential for dispersal in the latter genus, which is, indeed, reflected in a tendency toward the formation of local races.

South of Cape Cod the common shallow water species (in <30 m) is *Asterias forbesii*, a relatively sturdy form with a bright red-orange madreporite; the arm radius is about 4.5–5.5 times the radius of the disc. *A. vulgaris* occurs in depths of about 30 m or more south of Cape Cod, has somewhat more tapered arms, and, most conspicuously, a yellowish madreporite; R is also about 5 times r (the disc radius) in this species. Adults of *forbesii* and *vulgaris* reach 130 mm R and 200 mm R, respectively. *Leptasterias tenera* also ranges south of Cape Cod. This species has long, slender arms (R equals 6 or 7 times r), and the madreporite is whitish.

The differentiation of asteriids north of Cape Cod becomes more difficult because of the increase in recorded species, some of doubtful validity. See Fisher (1930) for revisions in taxonomy. The literature does not inspire confidence in the prospects for nonexpert identification of offshore material particularly. In shallow waters *A. vulgaris* replaces *A. forbesii*, and *L. tenera* also ranges into the littoral. However, so, apparently, do several additional species of *Leptasterias*.

References

See also the general echinoderm references following the introductory section of this chapter.

Danielssen, D. C., and J. Koren. 1887. Den Norske Nordhaus-Exped. 1876–1878. Zoologi xi Asteroidea.

Doderlein, L. 1917–1920. Die Asteriden der Siboga Expedition, II. Die Gattung Luidia. Siboga Exped. Monogr. 46A,B.

Fisher, W. K. 1923. A preliminary synopsis of the Asteriidae. Ann. Mag. nat. Hist. Ser. 9. 12:254–258;595–607.

———. 1930. Asteroidea of the North Pacific and adjacent waters. Bull. U.S. natn. Mus. 76:1–356.

Grainger, E. H. 1966. Sea Stars (Echinodermata: Asteroidea) of Arctic North America. Bull. Fish. Res. Bd. Can. 152: vii, 1–70.

Gray, I. E., Maureen E. Downey, and M. J. Cerame-Vivas. 1968. Sea stars of North Carolina. Fishery Bull. Fish Wildl. Serv. U.S. 67(1):127–163.

Miner, R. W. 1950. Field Book of Seashore Life. G. P. Putnam's Sons.

Sanford, S. N. E. 1931. Starfishes of New England. Bull. Boston Soc. nat. Hist. 60:3–9.

Sladen, W. P. 1889. Report on the Asteroidea collected during the voyage of H.M.S. *Challenger* during the years 1873–1876. Challenger Rept. 30:893 pp.

Spencer, W. K., and C. W. Wright. 1966. Asterozoans. *In* Treatise on Invertebrate Paleontology. R. C. Moore, ed. Part U. Echinodermata 3:4–78. Geol. Soc. Amer. and Univ. of Kansas.

Verrill, A. E. 1878. Notice of recent additions to the marine fauna of the eastern coast of North America. Am. J. Sci. 16:207, 371.

———. 1894. Descriptions of new species of starfishes and ophiurans, with a revision of certain species formerly described. Proc. U.S. natn. Mus. 17:245–297.

———. 1895. Distribution of the echinoderms of northeastern America. Am. J. Sci. Ser. 3. 49(13):127–141; (19):199–212.
———. 1899. Revision of certain genera and species of starfish. Trans. Conn. Acad. Arts Sci. 10:145–234.
———. 1915. Report on the starfishes of the West Indies, Florida, and Brazil. Univ. Iowa Monogr. Bull. Lab. nat. Hist. 7:1–232.

Subclass Ophiuroidea
BRITTLE OR SERPENT STARS

Diagnosis. *Body consisting of a flattened, more or less circular or subpentagonal central disc with five sharply differentiated arms symmetrically placed; mouth ventral; podia reduced, papillalike, in small pores on the ventral surface of the arms; madreporite obscure, ventral.*

Adults of local species have a disc diameter ranging from a few millimeters to about thirty-five millimeters with the arms from about 5 to more than 15 times longer. Serpent stars are remarkably constant in general configuration, the chief exceptions being the basket stars, Gorgonocephalidae, with branching arms. As the name brittle star suggests, the arms are easily fragmented, and regenerating specimens are common. There are rarely more than 5 arms. In comparison with the sluggish sea stars, serpent stars are active, briskly moving animals with highly flexible arms.

The deep endoskeleton consists of a series of vertebralike ossicles in the arms and a well-developed mouth frame in the disc. A superficial endoskeleton is usually present, consisting of embedded plates or shields variously ornamented with granules or spines; the most conspicuous of these dermal plates are the *radial shields*. In a few genera the disc is skin-covered, without conspicuous endodermal elements. Arm plates are also absent in some genera; the upper and lower plates may be absent, but laterals are always present. In most genera all of the plates are present and the spination and arrangement of the arm plates are important key criteria. Note that the podia are called "tentacles" in this group, and the pores in which they are set are *tentacle pores*, frequently protected by one or more *tentacle scales* (Fig. 22.11G).

Differences in mouth structure also provide useful key criteria. Hyman (1955) describes the externally visible, star-shaped opening as a "sort of preoral cavity leading to the true mouth." There are 5 wedge-shaped jaws variously armed along the edge with papillae. Each jaw is made of 2 pieces, but this condition may be obscured by superficial ornamentation. In *Ophioderma*, for example, the jaws are covered by a large, central *buccal shield* and smaller granules (Figs. 22.6G and 22.11A). In addition to the single buccal shield there may be a pair of flanking *adoral shields*. One of the buccal shields functions as the madreporite, but this may be scarcely differentiated externally. The half jaws may be visible externally even when the buccal and adoral shields are well developed, or the jaws may barely show. Yet another skeletal element, the *maxiller*, may be apparent at the tip of the jaw. Laterally the jaws may be armed with *oral* or *mouth papillae*, and at the tip there may be a group of *tooth papillae;* one or both of these sets of papillae may be absent. The figures should suffice to explain these arrangements as they are referred to in the keys. Within the mouth, and usually visible only when the mouth is well opened, are heavier *teeth*. Note that the maxiller bears the tooth papillae and the teeth. Dissection of these elements may be accomplished by macerating with Clorox.

The sexes are separate in most species, the hermaphroditic *Amphipholis squamata* being an exception among local species. The deep water *Ophiomitrella clavigera* is a protandric hermaphrodite. The dimorphic pigmentation of the ripe gonads of some species may allow the sexes to be distinguished by external examination, but ordinarily male and female are similar externally. In a few extralimital species the male is dwarfed and permanently attached to the female.

The gametes may be shed in the water, or the young may be brooded internally, e.g., in *Amphipholis squamata* and *Ophiomitrella clavigera*. A pelagic stage is also presumed to be absent in *Asteronyx*.

When development is external, there is usually a pelagic phase with a pluteus larva superficially similar to the larvae of echinoids. The pluteus stage is lacking in *Ophioderma brevispinnum* whose larvae are of the doliolaria type. (See introduction to the phylum for a brief account of echinoderm larval development.) In species with internal development the young emerge as miniature brittle stars, though not necessarily conforming in key traits to the adult. Asexual reproduction by fission occurs but not apparently in any of the local species. Powers of regeneration are well developed in the arms but sometimes less so in the disc.

Color is a useful trait in distinguishing many species, and it may also be noted that some are luminescent, e.g., *Ophiacantha bidentata*, *Amphipholis squamata*, and *Ophioscolex glacialis*.

Ophiuroids occur on all bottoms and from the littoral to abyssal depths. They are secretive and show a strong negative response to light. They are usually found under bottom debris, keeping out of sight during the day. The amphiurids, in particular, burrow superficially in sandy substrata. The food of brittle stars consists of detritus and minute plant and animal organisms that are ingested along with bottom debris, plus larger prey including annelid worms and small crustaceans. The gorgonocephalids have microscopic hooks on the end branches of the arms, and these snare small plankters; the young of *Gorgonocephalus eucnemis* are described as "parasitic" on alcyonarians.

Systematic List and Distribution

The classification follows that of Spencer and Wright (1966). The main sources for the species list are Clark (1915) and Mortensen (1933).

Order Phrynophiurida		
Family Ophiomyxidae		
Ophioscolex glacialis	B, 38°	49 to 1830 m
Ophioscolex purpureus	B	73 to 1391 m
Family Gorgonocephalida		
Gorgonocephalus arcticus	B+	Lit. to 1464 m
Gorgonocephalus eucnemis	B	15 to 1850 m
Family Asteronychidae		
Asteronyx loveni	BV	101 to 1830 m
Order Ophiurida		
Family Ophiuridae		
Ophiura robusta	B	6 to 1001 m
Ophiura sarsi	B+	10 to 2818 m
Ophiura signata	B	115 to 900 m
Ophiocten sericeum	B	9 to 4560 m
Ophiomusium lymani	BV	130 to 3435 m
Family Ophiodermatidae		
Ophioderma brevispinum	V	Lit. to 115 m
Family Ophiacanthidae		
Ophiacantha bidentata	BV	33 to 2470 m
Ophiacantha spectabilis	B	ca. 183 m
Ophiomitrella clavigera	BV	160 to 1500 m
Amphilimna olivacea	V	73 to 353 m
Family Ophiactidae		
Ophiopholis aculeata	B, 40°	Lit. to 1647 m
Family Amphiuridae		
Amphipholis squamata	BV	Lit. to 740 m
Amphioplus abdita	BV	77 to 137 m
Ophiophragmus septus	C+	86 to 95 m
Family Ophiothricidae		
Ophiothrix angulata	C, 38°	Shallow to 1830 m

Identification

Identification to species in this group is not usually difficult but does require close inspection of the jaw structure and other fine details. Measurements are disc diameters.

---

1. Disc with plates or scales dorsally, sometimes concealed by a covering of spines or fine granules—4.
1. Disc naked, skin covered, scales or spines, if present, microscopic (Note that young may have plates.)—2.
2. Arms branched; Fig. 22.9A. *Gorgonocephalus:* The disc is naked except on the radial ribs, which bear rounded grains in *eucnemis* and spines in *arcticus.* Both species are brownish or yellowish brown with disc reaching about 90 mm diameter.
2. Arms not branching—3.
3. Arms distinctly unequal in length; arm shield on the dorsal side with 8 or 9 spines which are hook- or club-shaped; reddish; 35 mm. *Asteronyx loveni.*
3. Arms equal, with 3 or 4 spines on the arm shields. *Ophioscolex: O. glacialis* lacks tentacle scales, has a naked disc, and no apparent dorsal arm plates. *O. purpureus* has a single slender tentacle scale, scattered small spines on the disc (often indistinct), and distinct dorsal arm plates; Fig. 22.9B. Both species are reddish, yellowish, or violet and reach about 25 mm disc diameter.

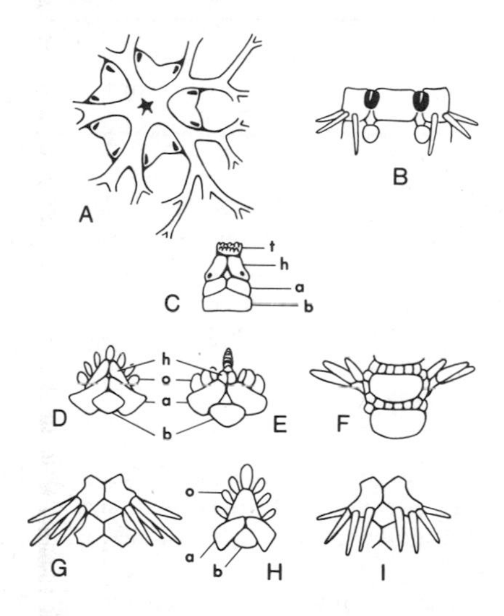

Fig. 22.9. Echinodermata, Ophiuroidea. A, *Gorgonocephalus* sp., central disc and base of arms, ventral view; B, *Ophioscolex purpureus*, arm joint, ventral view; C, *Ophiothrix angulata*, jaw; D, *Ophiacantha bidentata*, jaw; E, *Ophiopholis aculeata*, jaw; F, same, arm joint, dorsal view; G, *Ophiacantha bidentata*, arm joint, dorsal view; H, *Ophiomitrella clavigera*, jaw; I, same, arm joint. Abbreviations are as follows: a, adoral shield; b, buccal shield; h, half jaw; o, oral papillae; t, tooth paillae.

4. Arm spines appressed or rudimentary, usually no longer than the arm plates; Fig. 22.11A. See Subkey One.

4. Arm spines erect, more or less projecting—5.

5. Arm spines short, scarcely or not exceeding the length of the segment; tooth papillae absent. See Subkey Two.

5. Not so—6.

6. Arms spines long and glassy; tooth papillae present but no oral papillae; color extremely variable, brownish, greenish, reddish, etc., but almost invariably with a median, dorsal stripe on each arm, at least at the tip; 12 mm; Fig. 22.9C. *Ophiothrix angulata.*

6. Arm spines not glassy; no tooth papillae but oral papillae present—7.

7. Dorsal arm plates surrounded by small scales; disc with granules and scattered small spines often concealing all but the primary shields, the radial shields usually concealed; reddish, often variegated, the arms often dark-banded; 20 mm; Fig. 22.9E,F. *Ophiopholis aculeata.*

7. Dorsal arm plates not as described above—8.

8. Disc plates not concealed, the radial shields large and distinct; with 5 or 6 arm spines; to 7 mm disc diameter; Fig. 22.9H,I. *Ophiomitrella clavigera.*

8. Disc plates indistinct, the dorsum covered by short, stumpy spines; with 6–8 arm spines; 12 mm; Fig. 22.9D,G. *Ophiacantha bidentata.* Note additional deep water species, reported off this coast and keyed in Mortensen (1927).

Subkey One. These species are mostly mud burrowers.

---

1. Jaw apex with one oral papilla on each side, well separated from the additional papillae at the base of the jaw; tentacle scales single; Fig. 22.10A. *Amphiura.* This is a large and important genus, but the nearest geographical representative is *A. fragilis* from the Continental Slope.

1. Jaws with 2 or more papillae on each side without a gap separating them into an apical and basal series; tentacle scale double—2.

2. Arm spines 6–10; disc scales covered with spinules; olive; 10 mm, the arms very long; Fig. 22.10B. *Amphilimna olivacea.*

2. Arm spines 3 or 4—3.

3. Jaws with 2 oral papillae on each side plus a long basal tooth; grayish with a white spot at the base of each arm (juveniles orange); 5 mm, the arms about 5 times as long; in dead mollusk shells, under stones, in eelgrass, etc.; Fig. 22.10C. *Amphipholis squamata.*

3. Jaws with 3 or more oral papillae; the arms about 10–15 times the disc diameter—4.

4. Margin of disc with thick, blunt spinelets; with 3 oral papillae; disc gray, arms yellowish blotched with greenish plus a dark stripe above; 9 mm; Fig. 22.10E. *Ophiophragmus septus.*

4. Margin of disc without spinelets; 11 mm; gray or brown; Fig. 22.10D. *Amphioplus abdita.*

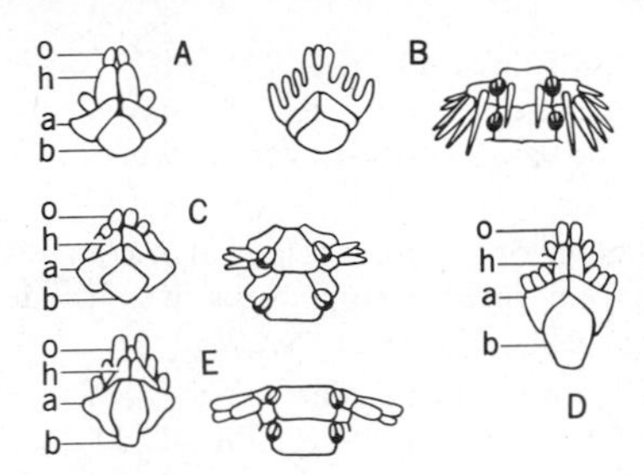

Fig. 22.10. Echinodermata, Ophiuroidea. A, *Amphiura* sp., jaw; B, *Amphilimna olivacea*, jaw and arm joint, ventral view; C, *Amphipholis squamata*, jaw and two arm joints, ventral view, spines omitted from the lower joint; D, *Amphioplus abdita*, jaw; E, *Ophiophragmus septus*, jaw and two arm joints, ventral view. Abbreviations are as follows: a, adoral shield; b buccal shield; h, half jaw; o, oral papillae.

Subkey Two

1. Disc plates concealed by a covering of granules; greenish, brownish, or black, uniformly colored or indistinctly mottled or banded; 15 mm; note jaws; Fig. 22.11A. *Ophioderma brevispinum.*

1. Disc plates distinct—2.

2. Arm combs forming a continuous series of papillae across the base of the arms dorsally; edge of disc sharp; bluish gray to violet or black; 18 mm; Fig. 22.11B. *Ophiocten sericeum.* Note that there are additional deep water species.

2. Arm combs absent or in 2 series; Fig. 22.11D—3.

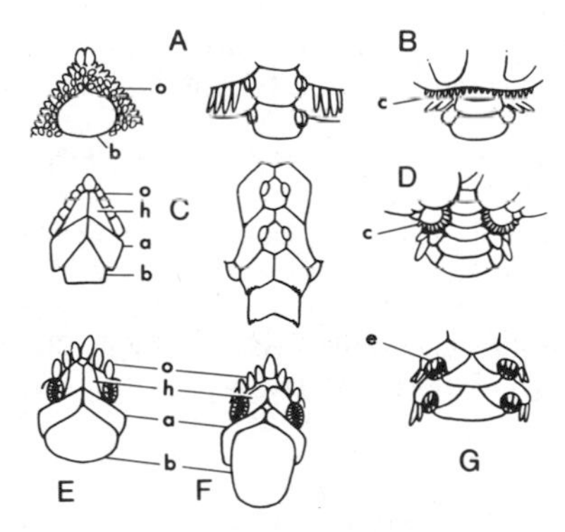

Fig. 22.11. Echinodermata, Ophiuroidea. A, *Ophioderma brevispinum*, jaw and arm joint, ventral view; B, *Ophiocten sericeum*, base of arm and margin of central disc, dorsal view; C, *Ophiomusium lymanni*, jaw and basal joints of arm, ventral view; D, *Ophiura* sp., basal joints of arm; E, *Ophiura robusta*, jaw; F, *Ophiura sarsi*, jaw; G, same, two arm joints, ventral view. Abbreviations are as follows, a, adoral shielf; c, arm combs; b, buccal shield; e, tentacle pore and scales; h, half jaw; o, oral papillae.

3. Arm combs present; arm spines three—4.
3. Arm combs absent; 6–8 rudimentary arm spines; a central series of plates (*ventral plates*) present on the two basal joints only; yellowish brown; 35 mm; Fig. 22.11C. *Ophiomusium lymani.*
4. Buccal shield as long as or longer than wide; arm combs well developed—5.
4. Buccal shield wider than long; arm combs weakly developed; blue-gray or brownish; 10 mm; Fig. 22.11E. *Ophiura robusta.*
5. With 2–4 tentacle scales on the basal plates of the arms (i.e., those within the disc margin); reddish; to 35 mm disc diameter; Fig. 22.11F,G. *Ophiura sarsi.*
5. With 1 or 2 tentacle scales on the basal arm plates; reddish or gray; to 8 mm disc diameter. *Ophiura signata.*

References

See also general echinoderm references following introductory section, this chapter.

Clark, H.L. 1915. Catalogue of recent ophiurans based on the collections of the Museum of Comparative Zoology. Mem. Mus. comp. Zool. Harv. 25:163–376.
Koehler, R. 1907. Revision des ophiures du Muséum d'histoire Naturelle. Bull. Sci. Fr. Belg. 44:279–351.
———. 1914. A contribution to the study of ophiurans of the U.S. National Museum. Bull. U.S. natn. Mus. 84:vii, 1–173.
Lyman, T. 1869. Preliminary report on the Ophiuridae and Astrophytidae dredged in deep water between Cuba and the Florida Reef by L. F. de Pourtales. Bull. Mus. comp. Zool. Harv. 10:309–354.
Mortensen, T. 1927. Handbook of the echinoderms of the British Isles. Oxford Univ. Press.
———. 1933. Ophiuroidea. Dan. Ingolf-Exped. 4(8):5–121.
Spencer, W. K. and C. W. Wright. 1966. Asterozoans. *In* Treatise on Invertebrate Paleontology. Part U, Echinodermata 3:4–78. R. C. Moore, ed. Geol. Soc. Amer. and Univ. Kansas.
Thomas, L. P. 1967. The systematic position of *Amphilimna* (Echinodermata; Ophiuroidea). Proc. biol. Soc. Wash. 80:123–130.
Verrill, A. E. 1894. Descriptions of new species of starfishes and ophiurans, with a revision of certain species formerly described. Proc. U.S. natn. Mus. 17:245–297.
———. 1899a. North American Ophiuroides. Trans. Conn. Acad. Arts Sci. 10(7):301–386.
———. 1899b. Report on the Ophiuroidea collected by the Bahama Expedition of 1893. Bull. Lab. nat. Hist. Univ. Iowa. 7:1–86.

TWENTY-THREE
Phylum Hemichordata

ACORN WORMS

Diagnosis. *Solitary, soft-skinned, wormlike animals; without appendages; with a three-part body consisting of a proboscis, more or less extensible and separated from the much longer trunk by a short, cylindrical collar.*

The proboscis may be elongate or short, conical or bulbous; it is joined to the collar by a narrow stalk. The mouth lies at the base of the proboscis within the collar. The anus is terminal or nearly so on the trunk. The trunk may or may not be subdivided or regionated, and it varies in cross-section. It is basically oval or rounded in section but has mid-dorsal and mid-ventral ridges and, in some genera, much wider, flaplike, or partly folded *genital wings*. The anterior, branchial or *branchiogenital region* of the trunk has longitudinal series of gill openings mounted on a ridge or, conversely, concealed in a groove. In some genera a *hepatic region* is also apparent, differentiated by color or by a longitudinal series of swellings or sacculations. The *caudal region* may be tapered or may continue the subcylindrical contours of the midtrunk to a blunt terminus. Local species attain total lengths of about 150 mm.

Acorn worms are easily fragmented and may be difficult to collect or preserve intact, narcotization being recommended as with annelids.

The sexes are separate but may be indistinguishable externally; in *Saccoglossus kowalevskii* the collar is said to be brighter-colored in males, and the gonads are gray in females, yellow in males. Fertilization is external, but spawning has only been observed in extralimital forms. The eggs are extruded in a mucous cord, and the sperm in a more tenuous cord that is readily dispersed in the water. Development varies as follows: In *S. kowalevskii* hatching is delayed, and the young hatch as miniature, tripartite, wormlike, juvenile acorn worms sufficiently similar to the adult form for their phyletic status to be apparent. They have at this time a conical appendage at the terminus of the trunk. The young are benthic at hatching. In contrast to this direct development other local species are presumed to be profoundly metamorphic, passing through *bipinnaria*like and *tornaria* larval stages similar to those found in echinoderms. The resemblance is of major theoretical importance. The tornaria (usually <1 mm in size) are planktonic for days or weeks and then undergo further metamorphosis to the benthic, wormlike juvenile. Asexual reproduction apparently does not occur in local species and is not common in enteropneusts generally, though they have substantial regenerative powers.

Adult acorn worms are benthic and most are sluggish burrowers living in mucus-lined burrows with 2 openings. The production of piles of stringlike castings 10–20 mm high in *S. kowalevskii* is characteristic. They are mucus-ciliary feeders, but also swallow a considerable amount of sand and presumably obtain nourishment from detritus ingested at the same time. Enteropneusts are exclusively marine but are reported in estuarine waters in salinities at least as low as 19‰. They produce a fetid odor similar to iodoform.

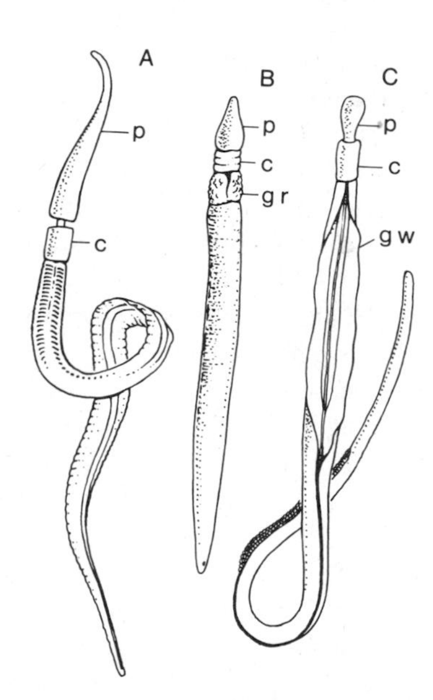

Fig. 23.1. Hemichordata. A, *Saccoglossus kowalevskii;* B, *Stereobalanus canadensis;* C, *Balanoglossus aurantiacus*. Abbreviations are as follows: c, collar; gr, genital ridges; gw, genital wings; p, proboscis.

Systematic List and Distribution

Only the class Enteropneusta occurs locally; there are no ordinal divisions. Bathymetric data are rather scant, but most species are described as littoral-shallow water forms.

Family Harrimaniidae		
Saccoglossus kowalevskii to Cape Ann at least	V+	Lit. to a few meters, eury.

Wass (1963) reported *S. kowalevskii* and "a much larger species . . . at a depth of 30 feet" in Virginia.

Stereobalanus canadensis	B	—
Family Ptychoderidae		
Balanoglossus aurantiacus	C (+?)	—

Identification

Measurements given are adult total lengths. The species are figured by Hyman (1959).

---

1. Proboscis elongate; trunk orange-yellow to brown, collar orange or red-orange, proboscis white to salmon pinkish; 150 mm; Fig. 23.1A. *Saccoglossus kowalevskii.*
1. Proboscis short—2.
2. Trunk regions ill-defined; with short, fluffy dorsal and ventral genital ridges; gill pores fused to a common slit; pale-lemon yellow, the liver region brown; 50 mm; Fig. 23.1B. *Stereobalanus canadensis.*
2. Trunk regions well defined, genital wings, hepatic sacculations, caudal region differentiated; body purplish or greenish with bright golden bands; 150 mm; Fig. 23.1C. *Balanoglossus aurantiacus.* This species, a member of the Ptychoderidae, is further defined by internal criteria requiring sectional preparations; see Hyman (1959, p. 135) and Dawydoff (1948, pp. 450–451).

References

Agassiz, A. 1873. The history of Balanoglossus and Tornaria. Mem. Am. Acad. Arts Sci. 9(2):421–436.
Dawydoff, C. 1948. Embranchement des Stomocordés. *In* Traité de Zoologie XI. P.-P. Grassé, ed. Masson et Cie. Pp. 367–532.
Girard, C. 1853. New nemerteans from the coast of the Carolinas. Proc. Acad. nat. Sci. Phila. 6:365–367.
Hyman, Libbie H. 1959. The Invertebrates: Smaller Coelomate Groups. Vol. V. McGraw-Hill Book Co.
Reinhard, E. 1942. Stereobalanus canadensis. J. Wash. Acad. Sci. 32:309–311.
Wass, M. L. 1963. Check list of the marine invertebrates of Virginia. Va. Inst. mar. Sci. spec. Sci. Rep. 24:1–56.

TWENTY-FOUR

Phylum Chordata

INVERTEBRATE CHORDATES—TUNICATES

Introduction

The subphylum Urochordata includes a diversity of structural types and habits making a concise diagnosis impractical. The simplified classification in Table 24.1 may serve as a frame of reference for the introductory discussion that follows. The classification is that of Berrill (1950).

Table 24.1. Classification of Subphylum Urochordata

Class Ascidiacea	Class Thaliacea	Class Larvacea
Order Enterogona	Order Pyrosomida	Order Copelata
Order Pleurogona	Order Doliolida	
	Order Salpida	

Larval Forms

The relationship of tunicates to higher chordates is revealed mainly in the tadpolelike larvae of Ascidiacea and Thaliacea and in adults of the class Larvacea, which resemble modified larvae (Fig. 24.1A,B). Such forms possess a supporting rod, the *notochord*, in the tail and a tubular, dorsal nerve cord. These chordate traits are missing in adults of classes other than Larvacea.

Not all ascidians or thaliaceans pass through a tadpolelike larval phase; in some species of *Molgula* (Ascidiacea) and in salps, development is direct. In Ascidiacea the eggs and larval stage are the principal dispersive phases; adults are sessile. Adult thaliaceans and Larvacea, as well as their larval forms, are pelagic. A key differentiating the principal tunicate larval types follows.

---

1. Head with circular muscle bands; with a short, fleshy process, the ***cadophore***, (see Fig. 24.9A), emerging above the tail; notochordal axis with about 40 cells. Larva of Thaliacea (Doliolida).

1. Head without circular muscle bands; without a cadophore—2.

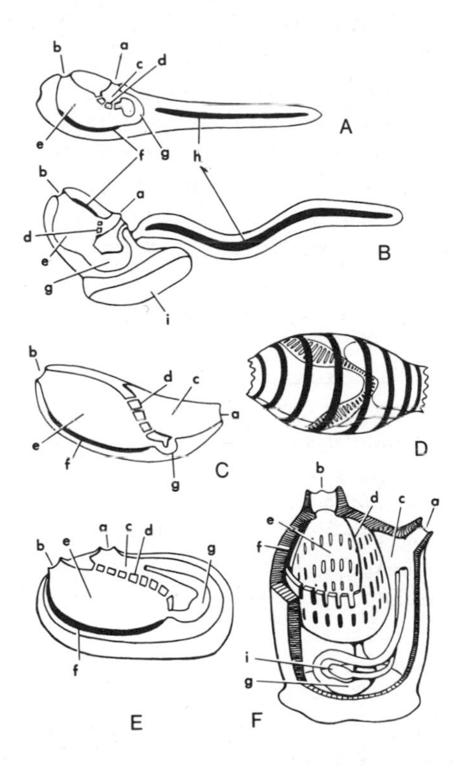

Fig. 24.1. Urochordata. All figures schematic. A, larval ascidian; B, adult larvacean; C, salp; D, doliolid; E, ascidian; F, same, dissected. Abbreviations are as follows: a, atrial siphon; b, branchial siphon; c, atrial chamber; d, gill partition; e, branchial chamber; f, endostyle; g, stomach; h, notochord; i, gonad.

2. Tail emerging terminally or ventrally; without gonads; notochordal axis with about 40 cells; Fig. 24.1A. Larva of Ascidiacea.

2. Tail emerging ventrally; with gonads; notochordal axis with 20 cells; Fig. 24.1B. Larvacea.

Adult Forms

Despite substantial diversity in external form and in the arrangement of internal parts, certain structural features may be followed through the tunicate classes (Fig. 24.1). The name "tunicate" derives from the external covering, the *tunic* or *test*, which may be tough and opaque and is frequently encrusted with debris in simple ascidians, or it may be soft, thin, and translucent. In some ascidians the test is impregnated with calcareous spicules. The composition of the test is highly unusual for an animal product, since cellulose is one of the characteristic ingredients. The tunic is secreted by a *mantle* which is the body wall proper.

There are, in most cases, two primary openings in the test. The incurrent opening, mouth, or *branchial siphon* opens into a pharynx or *branchial chamber* which is more or less separated from the *atrial* or *peribranchial chamber* by a partition pierced by rows of *stigmata* or gill slits. In Larvacea the peribranchial chamber is absent, and there are simply two gill openings in the wall of the pharynx leading directly to the exterior.

In simple ascidians, e.g., the various "sea squirts" (Fig. 24.1E,F), the branchial chamber may be viewed diagrammatically as an open basket, vaselike in shape, and suspended in a larger peribranchial chamber; the neck of the vase is the incurrent siphon. The atrial chamber, in turn, is pierced by the excurrent or *atrial siphon*. In thaliaceans (Fig. 24.1C,D) the branchial and atrial chambers may be essentially continuous when viewed from the side, as in salps, where the gill-bearing partition becomes a flat, perforated wall. The branchial and atrial openings are then at opposite ends of the long axis of a barrel- or spindle-shaped body. In ascidians, both openings may be close together, or the atrial siphon is lateral (actually posterior and dorsal) and the branchial siphon anterior and dorsal (Fig. 24.1E,F).

The pharynx is equipped with a grooved *endostyle* whose chief feature is its ciliated mucosa serving as the primary source of mucus which is directed laterally and dorsally to form a "rope" of entrapped plankton which then passes into the oesophagus. From here the food passes into the stomach, and then to the intestine, which opens into the atrial chamber. In Larvacea the rectum opens directly to the exterior. A circulatory system and a rather limited nervous system are present, though much reduced in some forms.

Tunicates are hermaphroditic, *Oikopleura dioica* being the single possible exception. Self-fertilization may occur, but in many genera it is blocked physiologically through mechanisms of sperm-ova incompatibility or by having the male and female gametes matured at different times.

The eggs may be shed and fertilized in the water, in which case the zygote normally develops into a pelagic, tadpolelike larva which subsequently attaches and metamorphoses into the adult. Most compound and some simple ascidians are viviparous, the fertilized eggs being retained in the atrial chamber until hatching, when the larvae escape by the atrial siphon. Among local species of simple ascidians this habit occurs in *Ascidia callosa*, *Polycarpa fibrosa*, *Boltenia echinata*, *Halocynthia pyriformis*, *Dendrodoa carnea*, *D. grossolaria*, *Molgula complanata*, and *M. citrina*. The larval period in tunicates is usually quite short, frequently lasting only a few hours. In the neotenic Larvacea, development is direct, since adults have an essentially larval form. In salps and a few ascidians the eggs are retained by the parent, develop directly to the adult form, and there is no free pelagic form.

Asexual reproduction by budding transforms the initial zooid into the colony of compound ascidians and pyrosomids. There is a complex alternation of asexual and sexual phases, following colony formation in salps and doliolids.

The tunicate feeding mechanism, as previously indicated, is geared to the utilization of planktonic food sources. In the case of Larvacea this filtering mechanism is so extraordinarily efficient that in many species the food items belong to the ultramicroscopic nannoplankton.

Sessile tunicates occur on a wide variety of substrata, including sand, rock, living plants, and animal exoskeletons. There are not parasitic forms. They range from the littoral to abyssal depths. Relatively few species of tunicates have penetrated estuarine situations with greatly reduced salinity. *Botryllus schlosseri*, *Perophora viridis*, *Molgula manhattensis*, and probably *Bostrichobranchus pilularis*, for example, are the only sessile forms reported from Chesapeake Bay; the first two species are only reported from the lower bay, but *Molgula* occurs at least as far north as the Patapsco River, where salinities drop in springtime below 10‰. Pelagic tunicates are chiefly oceanic, only a few forms entering coastal waters.

Systematic lists and distributional data are given in the subclass and ordinal sections that follow. Criteria for preliminary identification to these levels are given in Chapter Three.

Sessile and pelagic tunicates are considered in consecutive sections.

Class Ascidiacea
SESSILE TUNICATES

The sessile tunicates include both fixed species and others that are sessile but without firm attachment to a substratum. There is substantial variation in growth habit; since this variation crosses ordinal lines, it may be well to present the "diagnosis" in the form of an introductory key.

---

1. Colonial species; the zooids or individual members minute, usually shorter than 5 mm, embedded in a common matrix; frequently in more or less circular or oval "systems" with individual incurrent siphons but with a common excurrent opening; with or without embedded calcareous spicules; Fig. 24.3. Compound Ascidians.

1. Zooids solitary, usually larger than 5 mm, or colonial, stolonate species with minute individual zooids on separate branches—2.

2. Zooids minute, 4 mm or less, growing in vinelike colonies, the individual zooids on branches arising from a stolon; the zooids similar in structure to simple ascidians, i.e., with a compact, saclike body without an extensive postabdomen; 4 rows of stigmata; greenish; Fig. 24.2. *Perophora viridis.*

2. Zooids solitary, often aggregated but without a functional connection between individuals; zooids essentially saclike with incurrent and excurrent siphons relatively close together; Figs. 24.4–24.7. Simple Ascidians.

The species will be considered in the groups indicated, in keys following the systematic list.

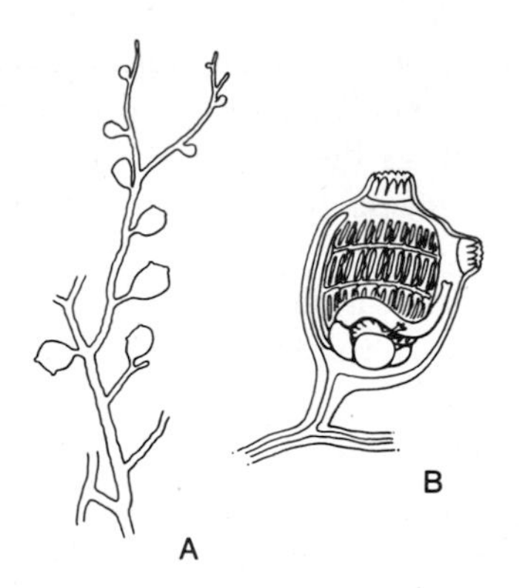

Fig. 24.2. Urochordata, Ascidiacea. *Perophora viridis.* A, colony form; B, individual zooid.

Systematic List and Distribution

The general classification follows Berrill (1950). Species lists are from Van Name (1945). The species are figured by Van Name (1945), and some of the common species have been figured in color by Miner (1950).

Order Enterogona		
Suborder Aplousobranchia		
Family Clavelinidae		
Eudistoma capsulatum	C, 39°	To 46 m
Distaplia clavata	B	40 to 62 m
Family Polyclinidae		
Amaroucium spitzbergense	B	15 to 154 m
Amaroucium glabrum	B+	Lit. to 457 m
Amaroucium pallidum	B+	Lit. to 862 m
Amaroucium stellatum	V+	8 to 13 m
Amaroucium pellucidum	V+	9 to 22 m
Amaroucium constellatum	V+	2 to 37 m
Synoicum pulmonaria	B	5 to 549 m
Family Didemnidae		
Didemnum albidum	B+	Lit. to 410 m
Didemnum candidum	V, 43°	Lit. to 64 m
Trididemnum tenerum	B+	33 to 137 m
Leptoclinides faeroensis	BV	>183 m
Lissoclinum aurem	B+	Shallow to 183 m
Suborder Phlebobranchia		
Family Cionidae		
Ciona intestinalis	B+	Lit. to 457 m
Family Perophoridae		
Perophora viridis	V+	Shallow to 27 m
Family Corellidae		
Corella borealis	A+	91 to 840 m
Chelyosoma macleayanum	B	1 to 565 m
Family Ascidiidae		
Ascidia prunum	B	5 to 366 m
Ascidia callosa	B	Lit. to 146 m
Ascidia obliqua	B	60 to 2196 m
Order Pleurogona		
Suborder Stolidobranchiata		
Family Styelidae		
Polycarpa fibrosa	B	1 to 1200 m
Dendrodoa pulchella	A, 44°	27 to 719 m
Dendrodoa grossularia	A+	Lit. to 600 m
Dendrodoa carnea	B, 40°	To 71 m
Styela coriacea	A, 43°	38 to 51 m
Styela partita	V+	Lit. to 27 m
Styela atlantica	V	113 to 726 m
Cnemidocarpa mollis	B+	5 to 274 m
Cnemidocarpa mortenseni	B	69 to 1065 m

Pelonaia corrugata	A, 43°	7 to 183 m
Botryllus schlosseri	BV	Lit. to 18 m
Botrylloides aureum	A, 42°	55 to 91 m
Family Pyuridae		
Boltenia ovifera	B+	7 to 494 m
Boltenia echinata	B+	Lit. to 296 m
Craterostigma singulare	B+	26 to 53 m
Halocynthia pyriformis	B	Lit. to 183 m
Family Molgulidae		
Molgula griffithsii	A+	55 to 732 m
Molgula siphonalis	A, 44°	Lit. to 146 m
Molgula citrina	B+	Lit. to 174 m
Molgula complanata	B+	Lit. to 549 m
Molgula manhattensis	V, 44°	Lit. to 29 m, eury.
Molgula provisionalis	B	Lit. to 5 m
Molgula robusta	B+	91 to 146 m
Molgula arenata	V+	5 to 46 m
Molgula lutulenta	V	123 to 260 m
Molgula retortiformis	B+	Lit. to 183 m
Bostrichobranchus pilularis	BV	1 to 220 m

Compound Ascidians. Including the families Clavelinidae, Polyclinidae, and Didemnidae in the order Enterogona, and the genera *Botryllus* and *Botrylloides*, family Styelidae in the order Pleurogona.

The identification of compound ascidians usually requires an examination of the individual members or zooids. This involves teasing apart a sample of the colony in alcohol with subsequent transfer to alcohol-glycerin mixture; the proportions are not critical. The tractability of different species to this treatment varies; in some cases whole zooids are easily obtained intact, in others they are not. The spicular content, if any, of the colony will probably become apparent in the course of the dissection. (Note, however, that in some species the spicules are scarce.) Some species also incorporate foreign particles such as sand grains in their tests. Hard particles should, therefore, be examined microscopically.

1. Colony with or without minute, calcareous, stellate spicules; colony broadly attached, usually thin, flat, never elevated; the zooids minute, <2.0 mm—2.

1. Without embedded spicules; colonies of various forms, sometimes clavate or capitate; the zooids usually >2.5 mm long plus a postabdomen—5.

2. Spicules needlelike, sometimes in crosses or clumps, 0.02–0.04 mm in diameter; spicules sometimes scarce; zooids <1.75 mm long with 3 rows of stigmata; colony whitish to yellow, 2 mm thick (rarely to 10 mm) by 60 mm in diameter; Fig. 24.3A. *Tridideinnum tenerum*.

2. Spicules burrlike or irregular; zooids with 4 rows of stigmata—3.

2. Spicules absent; zooids small, ca. 1.75 mm long, with about 9 rows of stigmata; colony in flat sheets to 100 mm diameter or, especially when growing on algae or grasses, in soft lobes; color extremely variable even in associated colonies; Fig. 24.3G,H. *Botryllus schlosseri*. Note also the very similar *Botrylloides aureum* reported from depths greater than 46 m north of 42°.

3. Zooids with sperm duct not coiled; spicules stellate or scarce and irregular in form, 0.03–0.05 mm in largest dimension; colony transparent to whitish or brownish, 3–6 mm thick by 30 mm diameter; Fig. 24.3B,C. *Lissoclinum aureum*.

3. Zooids with coiled sperm duct; spicules usually regular, stellate; Fig. 24.3D–F. *Didemnum*—4.

4. Spicules large, 0.05–0.08 mm; zooids 1.5–1.7 mm long; colony whitish to yellow or salmon, 2–3 mm thick by 100 mm diameter. *D. albidum*. This is the common Boreal species.

4. Spicules <0.05 mm, in boreal colonies <0.02 mm; zooids about 1.6 mm long, also smaller in Boreal colonies; whitish to yellow or red; 2–3 mm by 100 mm diameter. *D. candidum*. Northern populations with smaller spicules and zooids have been called *D. lutarum*.

Compound Ascidians without Spicules:

5. Zooids small, ca. 1.75 mm long; colony in flat sheets to 100 mm diameter or especially when growing on algae or seagrasses, in soft lobes; color extremely, variable even in associated colonies; Fig. 24.3G,H. *Botryllus schlosseri*. (Note also the very similar *Botrylloides aureum* reported from depths greater than 46 m north of 42°.)

5. Zooids larger, >2.5 mm long plus a postabdomen; colonies varied in form but usually not in flat sheets. Four genera with 9 local species key out here. These species are distinguished by details of zooid structure. While each species has a "typical" growth form, there is substantial variation in colony configuration. The species are compared in Table 24.42 and Fig. 24.3I–O.

Simple Ascidians. Including the families Cionidae, Corellidae, and Ascidiidae in the order Enterogona, and the families Styelidae (except *Botryllus* and *Botrylloides*), Pyuridae, and Molgulidae in the order Pleurogona.

Relatively few simple ascidians can be identified with full certainty on the basis of external characters alone. The examination of internal structures is not difficult. A simple dissection, requiring little more finesse than the peeling of a grape, exposes the internal parts. The test is cut along the midventral line (note left-right orientation in Figs. 24.5–24.7); the animal will be found lying more or less free inside, usually but loosely attached to the test except in the area of the siphons, where the dissection should be carried out so as to leave the siphons and adjacent test intact. When examined by transmitted light in liquid, the specimen can be oriented and the mantle opened by a full-length cut parallel to and close by the endostyle. The specimen may then be opened, as one would open a book fully to midpoint, exposing the internal anatomy and key traits.

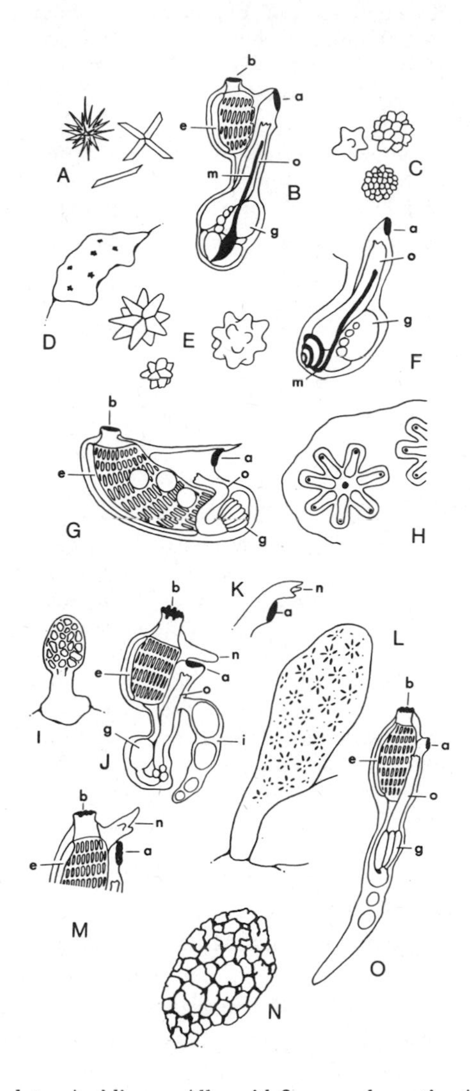

Fig. 24.3. Urochordata, Ascidiacea. All zooid figures schematic. A, *Trididemnum tenerum*, spicules; B, *Lissoclinum aurem*, zooid; C, same, spicules; D, *Didemnum*, colony form; E, same, spicules; F, same, zooid; G, *Botryllus schlosseri*, zooid; H, same, part of colony; I, *Distaplia clavata*, whole colony; J, same, zooid; K, *Synoicum pulmonaria*, atrial siphon; L, *Amaroucium stellatum*, colony; M, *Amaroucium glabrum*, part of zooid showing siphons and languet; N, *Amaroucium pellucidum*, colony; O, *Amaroucium pallidum*, zooid. Abbreviations are as follows: a, atrial siphon; b, branchial siphon; e, endostyle; g, stomach; i, incubatory pouch; m, sperm duct; n, languet; o, intestine.

Table 24.2. Comparison of Some Compound Ascidians without Spicules

Species	Colony Form	Colony Size, mm	Rows of Stigmata	Atrial Languet	Stomach Ridges	Color	Zooids in Systems	Sometimes Pedunculate	Fig. 24.3
Amaroucium									
spitzbergense	Small, round	28	4	Simple	4–5	Greenish, whitish	+	+	—
glabrum	Capitate to hemispherical	80	10–12	Lobed	12–15	Translucent, bluish gray Yellow-brown	+	+	M
constellatum	Turbinate, ovate, hemispherical	80	10–13	Lobed	22–25	Cream to reddish	+	+	—
pallidum	Globular to hemispherical	30	5–9	None	10–12	Whitish, yellowish	+	–	O
stellatum	Tabular, lobed	600	ca. 12	Simple	12	Whitish, bluish, reddish	+	–	L
pellucidum	Lobed to hemispherical, sand encrusted or infiltrated	200	12–18	Simple	?	Sand-colored	+	–	N
Synoicum									
pulmonaria	Spherical, pear-shaped	140	ca. 20	Trilobed	None	Whitish, yellowish	+	+	K
Eudistoma									
capsulatum	Round, oval	170	3	Simple	None	Brownish, reddish, violet	–	–	—
Distaplia									
clavata	Clavate or capitate	30	4	Simple	Numerous, weak	Whitish, yellowish	–	–*	I, J

* With incubatory pouch.

A preliminary key is used to sort the species into more manageable subgroups, four in number.

---

1. Branchial and atrial siphons dissimilar, with either 8 and 6 or 6 and 4 lobes, respectively; usually dull olive or brownish, frequently encrusted with foreign material—2.

1. Branchial and atrial siphons similar, 4-lobed or squarish, in some pyurids the lobes partly fused, the apertures then appearing subcircular or as a transverse cleft; frequently reddish in life, particularly about the siphons, the test encrusted or not—3.

2. Branchial siphon 8-lobed, atrial siphon 6-lobed; gonads unpaired, lying in or on the gut loop; branchial sac without folds, stigmata straight; Fig. 24.4. Suborder Phlebobranchia.

2. Branchial siphons 6-lobed, atrial siphon 4-lobed; gonads 1 or 2 on the mantle wall plus a large, saclike renal organ or "kidney" on the right wall; branchial sac without folds or with 5–7 folds on each side, the stigmata usually curved, in spirals; Fig. 24.7. Molgulidae.

3. Branchial sac with 5 or 6 folds on each side; the stigmata straight, or sometimes spiral on the folds (in *Boltenia* the stigmata are transverse to the longitudinal vessels); gonads 2–14, usually on the mantle wall, with one or more groups of "liver" tubules on the stomach wall; Fig. 24.6. Pyuridae.

3. Branchial sac without folds; siphons without lobes; stigmata strongly spiral; gonads 2, strongly lobed, one on each side on the mantle wall; a burrowing, sand-encrusted form; 11.5 mm; Fig. 24.6B. *Craterostigma singulare.* This is an aberrant member of Pyuridae.

3. Branchial sac with 4 folds on each side, sometimes indistinct or reduced to 1; stigmata straight; gonads 1–12 or more on the mantle wall; Fig. 24.5. Styelidae.

ORDER ENTEROGONA, SUBORDER PHLEBOBRANCHIA

---

1. Body form more or less oval in profile, the branchial siphon anterior, the atrial siphon dorsal; Fig. 24.4D—2.

1. Not so, either tall and cylindrical or low and disclike; Fig. 24.4E,F—3.

2. Gut loop and gonads on right side of branchial sac, the inner longitudinal vessels of the branchial sac without secondary papillae; test smooth and clean; 20 mm tall. *Corella borealis.*

2. Gut loop and gonads on left side of branchial sac, inner longitudinal vessels of branchial sac with secondary papillae; test sometimes encrusted; 45–50 mm tall; Fig. 24.4A–D. *Ascidia:* With 3 species of similar appearance, distinguished by differences in branchial structure: in *A. obliqua* papillae are only present at the intersections of vessels (Fig. 24.4A); in both *prunum* and *callosa* intermediate vessels are present, but the number and spacing of the internal longitudinal vessels differ as figured, *callosa*, Fig. 24.4B, *prunum*, Fig. 24.4C. Note also that *callosa* is viviparous.

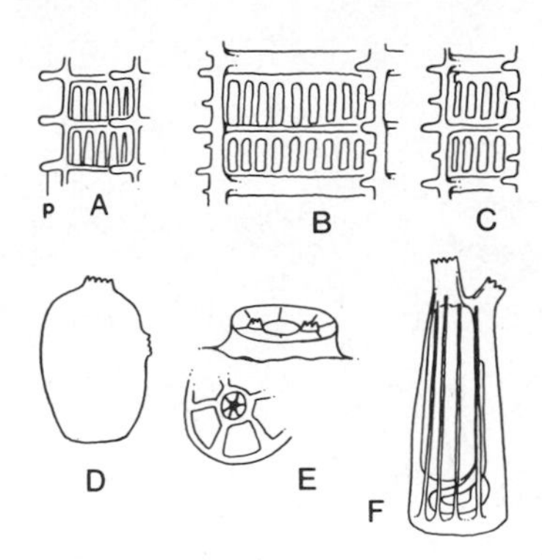

Fig. 24.4. Urochordata, Ascidiacea. A–C are details of branchial structure. A, *Ascidia obliqua;* B, *Ascidia callosa;* C, *Ascidia prunum;* D, *Ascidia* sp., zooid; E, *Chelyosoma macleayanum,* zooid viewed obliquely and from above; F, *Ciona intestinalis,* zooid. Papilla indicated by letter "p."

3. Disclike; the siphons surrounded by 6 wedge-shaped horny plates; 12 mm diameter; Fig. 24.4E. *Chelyosoma macleayanum.*

3. Tall and cylindrical or tapering upward; soft and gelatinous in appearance with 5–7 broad, muscular bands showing through the test; colorless to whitish or yellow to yellow-green; stomach, intestines, and gonads below (posterior to) the branchial sac; 120 mm tall; Fig. 24.4F. *Ciona intestinalis.*

ORDER PLEUROGONA, SUBORDER STOLIDOBRANCHIATA FAMILY STYELIDAE

---

1. With a single gonad on the right side; only one branchial fold clearly indicated; test apparently smooth, actually minutely wrinkled; body form, a depressed dome; in life resembling a drop of blood, pinkish to bright red; 11 mm diameter. Fig. 24.5A. *Dendrodoa carnea.* A somewhat similar form, *D. pulchella,* is yellow to brown or red; gonad with 2 or 3 branches; 16 mm diameter.

1. Not so; gonads 2 or more—2.

2. One gonad on each side, U-shaped; branchial sac without folds; body form elongate, sometimes wormlike, swelling basally, sand-encrusted; a burrowing species; 120 mm; Fig. 24.5B. *Pelonaia corrugata.*

2. With 1 or 2 gonads on each side, sinuous in shape; branchial sac with 4 well-developed folds; body form varying from a depressed dome to tall; exterior rough and wrinkled. *Styela: S. partita* has 2 gonads on each side; 40 mm (note a similar species in deep water, *S. atlantica*); Fig. 24.5C. *S. coriacea* has 1 gonad on each side; exterior with minute hemispherical tubercles; 20 mm.

2. With 3 gonads or with 4 or more on each side—3.

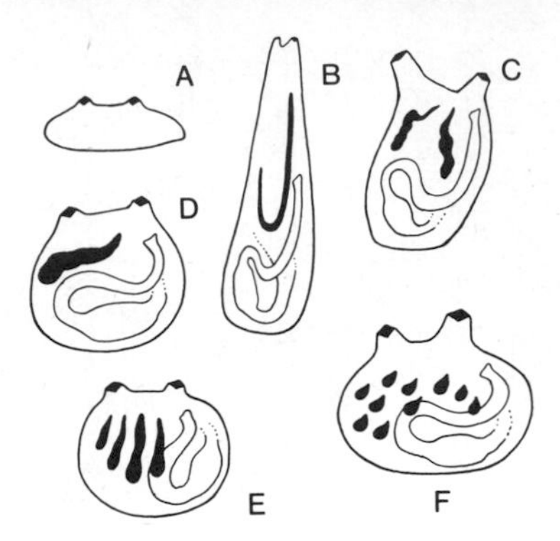

Fig. 24.5. Urochordata, Ascidiacea. All zooids viewed from left side with gonads shown in solid black (omitted from A). A, *Dendrodoa carnea;* B, *Pelonaia corrugata;* C, *Styela partita;* D, *Cnemidocarpa mortenseni;* E, *Cnemidocarpa mollis;* F, *Polycarpa fibrosa.*

3. With 3 gonads, 2 right and 1 left; branchial sac with 1 fold; body form round or oval, exterior usually clean and smooth, yellow to red; 19 mm; Fig. 24.5D. *Cnemidocarpa mortenseni.*

3. With 10 gonads, 6 right, 4 left; branchial sac with 4 slight folds; body form round or oval; exterior usually sand coated; a burrowing species; 14 mm; Fig. 24.5E. *Cnemidocarpa mollis.*

3. With up to 12 or more gonads on each side; branchial sac with 4 folds; body form round, more or less depressed, exterior rough, granular, encrusted; a burrower in mud; 18 mm; Fig. 24.5F. *Polycarpa fibrosa.* (Note: Several additional rare or deep water species of *Cnemidocarpa* and *Polycarpa* are known.)

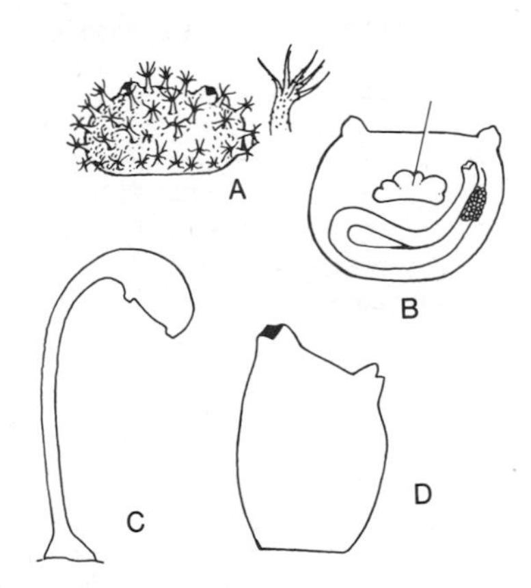

Fig. 24.6. Urochordata, Ascidiacea. Figure C viewed from right side, others from left. A, *Boltenia echinata,* with detail of individual spiny process; B, *Craterostigma singulare* with gonad indicated; C, *Boltenia ovifera;* D, *Halocynthia pyriformis.*

FAMILY PYURIDAE

1. Cactuslike, the test covered with spiny processes; no stalk; 15 mm; pinkish to deep salmon; Fig. 24.6A; a viviparous species. *Boltenia echinata.*
1. Pedunculate, the stem much longer than the body; body 90 mm, stem to 200 mm; whitish, yellowish, brownish, or red; Fig. 24.6C. *Boltenia ovifera.*
1. Test rounded, not pedunculate or with conspicuous spiny processes; covered with minute granules that are seen to be spiny under magnification, without encrustations; whitish, yellow to peach color or orange red; gonads 4–7 per side; 65 mm tall; Fig. 24.6D; frequently viviparous. *Halocynthia pyriformis.*

FAMILY MOLGULIDAE

1. With one gonad, bean-shaped; branchial sac without folds, stigmata spiral; an unattached burrowing form; 8 mm; Fig. 24.7A. *Bostrichobranchus pilularis.*

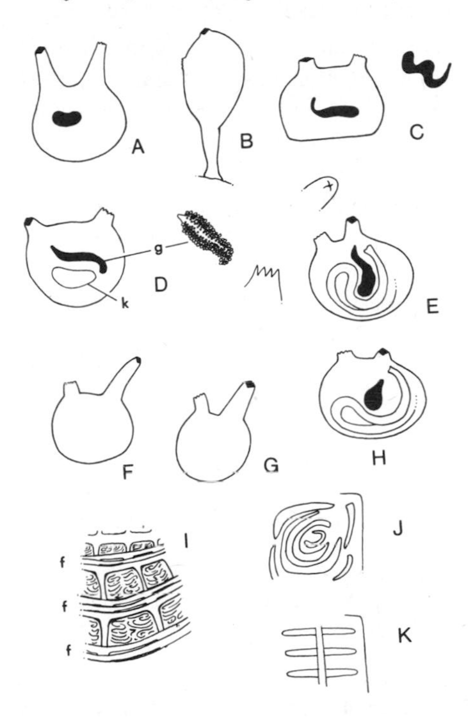

Fig. 24.7. Urochordata, Ascidiacea. A,C,D are viewed from the right side, all others from left, with gonads shown in solid black. A, *Bostrichobranchus pilularis;* B, *Molgula griffithsii;* C, *Molgula citrina*, with silhouette of left gonad shown at right; D, *Molgula arenata*, with detail of gonad at right; E, *Molgula robusta*, with details of branchial siphon (left) and contracted atrial siphon (above); F, *Molgula retortiformis;* G, *Molgula manhattensis;* H, *Molgula lutulenta;* I, detail of wall of pharynx showing folds; J, detail of wall of pharynx showing spiral stigmata; K, detail of pharynx wall showing simple stigmata. Abbreviations are as follows: g, gonad, k, kidney; f, folds.

1. With 2 gonads; branchial sac with 5–7 folds—2.

2. Pedunculate, the peduncle varying in length (rarely absent); siphons short, the atrial siphon anterior (on top), the branchial siphon dorsal (lateral); branchial sac with 5 folds on each side; body 25 mm; Fig. 24.7B. *Molgula griffithsii.*

2. Siphons both anterior; branchial sac with 6 or 7 folds; without a peduncle—3.

3. One or both gonads either U-, W-, or S-shaped:

- Both gonads U-shaped; 14.5 mm; test lightly attached to a substratum. *Molgula complanata.*
- Left gonad W- or S-shaped; 16 mm; test firmly attached; Fig. 24.7C. *Molgula citrina.*
  Both species are viviparous.

3. Not so—4.

4. Free, burrowing species; branchial folds, 6 on each side—5.

4. Attached species; branchial folds 6 or 7 on each side—6.

5. Siphons widely separated; gonads elongate; 20 mm; Fig. 24.7D. *Molgula arenata.*

5. Siphons close together; gonads elongate; 28 mm; Fig. 24.7E. *Molgula robusta.*

5. Siphons close together; gonads short; 15 mm; exterior of test with long, mosslike processes; Fig. 24.7H. *Molgula lutulenta.*

6. Branchial folds 7 on each side:

- Siphons close together; 24 mm. *Molgula siphonalis.*
- Siphons well separated; 75 mm; Fig. 24.7F. *Molgula retortiformis.*

6. Branchial folds 6 on each side:

- Siphons close together; 35 mm; Fig. 24.7G. *Molgula manhattensis.*
- Siphons well separated; 13 mm. *Molgula provisionalis.*

References

Barrington, E. J. W. 1965. The Biology of Hemichordata and Protochordata. Oliver and Boyd. Edinburgh and London.

Berrill, N. J. 1950. The Tunicata with an account of the British species. Ray Society.

Bullough, W. S. 1960. Practical Invertebrate Anatomy. Macmillan, London.

Grassé, P.-P., ed. 1948, Echinoderms, Stomochordes, Procordes. Tome XI. Traité de Zoologie.

Hopkinson, J. 1913. A bibliography of the Tunicata, 1469–1910. Royal Society.

Miner, R. W. 1950. Field Book of Seashore Life. G. P. Putnam's Sons.

Van Name, W. G. 1910. Compound ascidians of the coasts of New England and neighboring British Provinces. Proc. Boston Soc. nat. Hist. 34(11):339–424.

———. 1912. Simple ascidians of the coasts of New England and neighboring British Provinces. Proc. Boston Soc. nat. Hist. 34(13):439–619.

———. 1921. Ascidians of the West Indian region and Southeastern United States. Bull. Am. Mus. nat. Hist. 44:283–494.

———. 1945. The North and South American Ascidians. Bull. Am. Mus. nat. Hist. 84: vii + 476 pp.

Classes Thaliacea and Larvacea
PELAGIC TUNICATES

The classes Thaliacea and Larvacea are exclusively planktonic, quite different in appearance from the sessile Ascidiacea but presenting considerable diversity in their own right. The Larvacea resemble, superficially at least, pelagic ascidian larvae and are regarded by many students as neotenic forms, i.e., they become sexually mature but do not develop otherwise beyond an essentially larval state. The genera *Oikopleura* and *Fritillaria* are sometimes quantitatively important in the coastal plankton. Deevey (1952, 1956) reported *Oikopleura dioica* in concentrations of about 5000 per m^2 in Block Island and Long Island sounds. About 270,000 *Fritillaria borealis* were taken in a 10-minute tow of a meter net in Delaware Bay, according to Deevey (1960). Salps and doliolids (Thaliacea), though more typical of stations offshore, may also reach substantial concentrations in coastal waters, as witnessed by the taking of about a *bushel* of *Thalia democratica* in a ten minute, meter net tow in Delaware Bay (Deevey, 1960).

Pelagic tunicates are evidently less common in the Gulf of Maine than south of it. Bigelow (1926) regarded salps as one of the best indicators of tropical water in the gulf. They are chiefly epipelagic, though ranging to depths of greater than 1000 m. While some thaliacean genera and species are exclusively or primarily warm water forms, others range into high latitudes in Gulf Stream water in the eastern Atlantic. The occurrence of oceanic species inshore is dependent on unusual influxes of waters from the outer shelf or beyond. The Larvacea include Arctic and Boreal as well as warm water species. There are both winter-spring and summer-fall species, corresponding approximately to northern and southern centers of abundance.

Pelagic tunicates will be discussed in class and ordinal groups, but references for all groups are given at the conclusion of the account of Larvacea. Treatment of the higher categories of Thaliacea varies with different authors; the convention followed here recognizes three orders; Pyrosomida, Doliolida, and Salpida, of which the first will be discussed separately and the latter two collectively.

Class Thaliacea: Order Pyrosomida

Diagnosis. *Zooids in pelagic colonies consisting of numerous individuals embedded in a gelatinous, cartilaginous, or leathery tube closed at one end; colonies brilliantly luminescent* (*Fig. 24.8*).

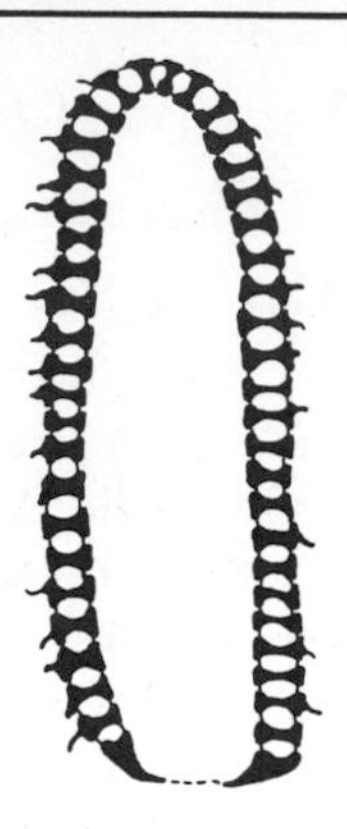

Fig. 24.8. Urochordata, Thaliacea. *Pyrosoma atlanticum*, schematic longitudinal section of colony.

Colonies of the bathypelagic *Pyrosoma spinosum* range in size exceptionally to 4 *meters* in the Indian Ocean; the usual size is about 500 mm. Large specimens of the more common western Atlantic species, *Pyrosoma atlanticum* (forms *giganteum*, *dipleurosoma*, and *intermedium*), reach about 550 mm with a diameter of about 70 mm and an aperture of 45 mm, and are nearly cylindrical. Individual zooids from such colonies are about 7.4 mm long. The test is yellowish, brownish, or greenish in color with a surface covered with rounded papillae plus long, fingerlike tentacles which may completely cover the surface but more commonly are less numerous and are scattered more or less at random. In preserved material the test wall is rigid and cartilaginous and about 6–8 mm thick. The shape of the colony varies in the different forms and may be long and slender or short and thick. Young individuals may be somewhat conical and nearly transparent and colorless.

The zooids are oriented with oral surface outermost. Currents generated by the individual zooids combine within the blind tubular chamber of the interior of the colony to produce a powerful jet which is further concentrated in passing through a constricted diaphragm at the open end. For descriptions of the individual zooids see Metcalf and Hopkins (1919), and Neumann (1935b).

Developmental details are incompletely known. Colonies are hermaphroditic with the individual zooids either protogynous or protandrous. Fertilization is internal. Young colonies develop by budding from primary zooids and are carried for a time in the colony until escaping via the tubular aperture.

Systematic List and Distribution

There is but one genus, and according to Metcalf and Hopkins (1919) the "species" interbreed (see also Neumann, 1935b). *Pyrosoma atlanticum* and

P. spinosum are usually presented as separate species, however, and the polymorphs of *atlanticum* have been variously described as "subspecies," "groups," or "forms." It seems pointless to attempt an explanation of this situation in terms of modern speciation theory. For further details see Metcalf and Hopkins (1919). There is scant information about the distribution of *Pyrosoma* in local coastal waters; this is a warm water form reported north to Cape Cod.

ORDERS DOLIOLIDA AND SALPIDA

Diagnosis. *Zooids solitary or joined in chainlike colonies, the individual zooids with a permanent, barrel-shaped, cylindrical, prismatic, or spindle-shaped test open at both ends; transparent; with complete or incomplete, hooplike muscle rings; polymorphic with sometimes quite different solitary and aggregate forms.*

The individual zooids are generally less than 50 mm long, but range in size in various species from about 1 mm in *Brooksia* to greater than 70 mm in *Iasis zonaria*. The test is thin, transparent, and semirigid; in locomotion the muscle bands compress the basically cylindrical test, forcing water out through the posterior or atrial opening; on relaxation of the musculature the test resumes its original shape. The atrial and branchial apertures open into large chambers. In doliolids a partition, the *gill lamella*, which is pierced by gill openings or *stigmata*, separates the two chambers, but in salps the chambers are essentially continuous, being separated only by an oblique *dorsal lamina* or gill bar.

There is a more or less opaque and sometimes luminescent visceral mass containing heart and digestive apparatus along the ventral body wall, and a cerebral ganglion and associated structures in the dorsal wall. The body plan is essentially bilaterally symmetrical.

Salps and doliolids are polymorphic, this variation being associated with metagenesis—the alternation of sexual and asexual generations. In doliolids the eggs are shed freely in the water and there is a pelagic larval stage. Salps are viviparous, the young being retained in the body of the parent. The life history of members of the two families shows other differences.

Among salps, individuals of the aggregate, colonial phase occur initially in chains which may number several hundred zooids. These individuals, termed *blastozooids* or "gonozooids," break loose in small groups when mature and eventually become solitary, breeding individuals. Salps are hermaphroditic but not self-fertilizing. Most species are presumably protogynous, the female elements maturing first; conversely, *Salpa cylindrica* is described as protandrous. Usually salps produce but a single egg, *Thetys vagina* and *Iasis zonaria* being exceptions. The zygote develops in the mother's body in association with a "diffusion placenta" and is released as a miniature salp; this *oozooid* has a ventral stolon along which the blastozooids are budded in chainlike sequence.

The situation in Doliolidae is substantially more complex, and the asexual phase is, itself, polymorphic. The oozooid begins life as a tailed larva somewhat similar to that of ascidian tunicates. However, the tail is relatively shorter; the "head" has circular muscle bands, and above the tail is a short, fleshy process called the *cadophore* (Fig. 24.9A). The larva metamorphoses gradually into a *nurse*, so called because with further development it becomes the focus of a colonial aggregation of zooids. These include two lateral and one median row of zooids which are budded from the ventral stolon. They subsequently separate from the stolon and become attached to the cadophore. The colony shows a division of labor which has been compared to that found in siphonophores. Thus the lateral rows of zooids, called *trophozooids* or *gastrozooids*, take over the feeding function and are permanent members of the colony. The median row of zooids, called *phorozooids*, lack gonads (as do gastrozooids); however, the phorozooids are liberated and carry buds that develop as true gonozooids. The phorozooids are themselves "nurses" in the sense that the gonozooids carried by them have not actually been budded from them but from the original oozooid. The gonozooids in turn become independent when mature and reproduce sexually, the offspring being the larvae that metamorphose into oozooids, and so on. For a more detailed account of this extraordinary life cycle see Berrill (1950) and the sources cited therein.

Systematic List and Distribution

This list is based on Metcalf (1918), Neumann (1935a), and the plankton surveys and regional lists cited in Chapter 3.

FAMILY DOLIOLIDAE

Neumann (1935a) listed 10 species in the Atlantic; some of these are now considered mere varieties, and the genus *Doliolum* has been subdivided with 3 genera in the western North Atlantic. A fourth genus, *Doliopsis*, is extralimital in tropical waters. The doliolids are all warm water forms but may be carried northward in the Gulf Stream as summer-fall visitors. *Doliolum nationalis* and *Dolioletta gegenbauri* appear to be the species most frequently reported in coastal waters, but *Doliolum denticulatum*, *Doliolina intermedium*, and *Doliolina mulleri* var. *krohni* are further possibilities; *tritonis* is regarded as a variety of *gegenbauri*.

FAMILY SALPIDAE

Former subgenera of *Salpa* have now been elevated to generic status with the species assigned to different genera, as follows.

Brooksia rostrata	—	*Salpa cylindrica*	B
Cyclosalpa floridana	—	*Salpa fusiformis*	B
Cyclosalpa pinnata	—	*Salpa maxima*	—
Cyclosalpa affinis	—	*Thalia democratica*	B
Iasis zonaria	B	*Thetys vagina*	B
Pegea confederata	—	*Traustedtia multitentaculata*	—

Those marked with a "B" have been recorded in the Gulf of Maine or off Nova Scotia. While virtually all of these species may occur on this coast, *Brooksia* and *Traustedtia* in particular are rare and the *Cyclosalpa* species may be confined to deep water. The salps are chiefly warm water forms, but they may be carried far north in the Gulf Stream and have been recorded in boreal and even subarctic latitudes in the eastern Atlantic. Bigelow (1926) found salps rare or sporadic visitors to the Gulf of Maine, usually in asexual swarms. The most frequently reported species in coastal waters south of Cape Cod are *fusiformis* and *democratica*, the latter probably being the most common species inshore.

Identification

The following keys are based in part on those of Harant and Vernieres (1938), Fraser (1947), and Metcalf (1918).

The families are separated as follows.

1. Branchial and atrial chambers separated by a partition penetrated by numerous gill slits or stigmata, Fig. 24.9. Doliolidae.
1. Branchial and atrial chambers essentially continuous, with an oblique gill bar. Salpidae, Fig. 24.10.

The often-cited criterion for separating these families on the basis of completeness or incompleteness of the muscle rings is untrustworthy, as the keys demonstrate.

FAMILY DOLIOLIDAE

Separate keys are required for the oozooids and for phorozooids and gonozooids; the initial separation may be accomplished as follows.

1. With 9 muscle bands and a cadophore; Fig. 24.9A. Oozoids. The nurses undergo progressive morphological specialization including the degeneration of the

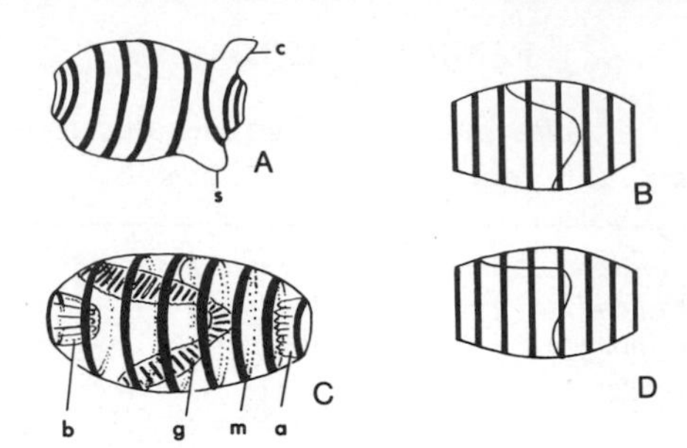

Fig. 24.9. Urochordata, Doliolidae. A, oozoid; B–D are phorozooids; B, *Dolioletta gengenbauri*, schematic, showing relationship of gill lamella and muscle bands for comparison with D; C, doliolid, showing atrial and branchial siphons, gill lamella, and muscle bands; D, *Doliolum nationalis*. Abbreviations are as follows: c, cadophore; g, gill lamella; m, muscle bands; s, stolon; a, atrial siphon; b, branchial siphon.

digestive tract, loss of gill slits, and in some species the second through eighth muscle bands become united in a single sheet.

1. With 8 muscle bands and no cadophore. Phorozooids and Gonozooids. Phorozooids lack gonads but have a vestigial stolon; gonozooids lack the stolon but have gonads.

Oozoids. This stage of *Doliolum nationalis* and *Doliolina mulleri* (var. *krohni*) is unknown.

---

1. Muscles II through VIII united in a single sheet. *Doliolum denticulatum.*

1. Muscles II through VIII separate—2.

2. Muscle bands narrower than ½ the spaces separating them; the muscles do not form a gradually increasing and diminishing series. *Doliolina intermedium.*

2. Muscle bands broader than ½ the interspaces; muscles both before and after the third band becoming progressively narrower. *Dolioletta gegenbauri.*

Phorozooids and Gonozooids. The species are compared in Table 24.3, simplified from Fraser (1947). The criteria include the general form of the digestive tract and the dorsal and ventral points of attachment of the gill lamella with reference to the adjacent muscle bands; cf. Figure 24.9B–D. Maximum sizes are indicated.

FAMILY SALPIDAE

Distinctive features of the musculature separate specimens greater than 3 mm long; note also differences in the shape of the test, as figured. *Traustedtia*, with 13 tentaclelike appendages protruding from the body, is omitted from

Table 24.3. **Doliolidae**

Species	Digestive Tract	Insertion of Gill Lamellae on Test		Length, mm
		Dorsal	Ventral	
intermedium	U-shaped	5−	4+	6
mulleri (v. *krohni*)	U/S-shaped	6+	5	7
denticulatum	Coiled	2+	3+	9
nationalis	Coiled	2+	4–5	4
gegenbauri	Coiled	3+	5−	17

the key. Aggregated and solitary forms are considered separately; the oozooids have a ventral stolon. Dimensions are maximum length; when a specific dimension is not given, the species may be presumed to be between 25 and 50 mm.

Solitary Forms

1. Digestive tract not forming a compact "nucleus." *Cyclosalpa:* In *floridana* the body rings are complete below except for the sixth, and weakly developed luminous organs are present; in *pinnata* and *affinis* the body rings are broken both above and below; *pinnata* has luminous organs, *affinis* does not. Cyclosalps reach lengths of 40–80 mm; Fig. 24.10A.

1. Digestive tract forming a compact "nucleus"—2.

2. Test with a pair of long protuberances; muscle rings broken or complete; Fig. 24.10B,C—3.

2. Test without a pair of long protuberances; muscle rings broken—4.

3. With 18–22 partly broken muscle rings behind the nerve ganglion; 220 mm; Fig. 24.10B. *Thetys vagina.*

3. Usually with 6 or 7 muscle rings, complete below except for the fifth; 25 mm; Fig. 24.10C. *Thalia democratica.*

4. Muscle rings parallel or nearly so—5.

4. Muscle rings not parallel—6.

5. With 9 or 10 muscles behind the nerve ganglion; 160 mm. *Salpa maxima.*

5. With 5 muscles behind the nerve ganglion; 65 mm; Fig. 24.10D. *Iasis zonaria.*

6. With 2 pairs of muscles forming a pair of X's; 120 mm; Fig. 24.10E. *Pegea confederata.*

6. With 9 or 10 muscle rings, including an X formed by the last pair and with the first 3 rings also touching dorsally; sometimes with spiny ridges; 80 mm; Fig. 24.10F. *Salpa fusiformis.*

6. With 9 muscle rings, the first 4 touching dorsally; Fig. 24.10G. *Salpa cylindrica.*

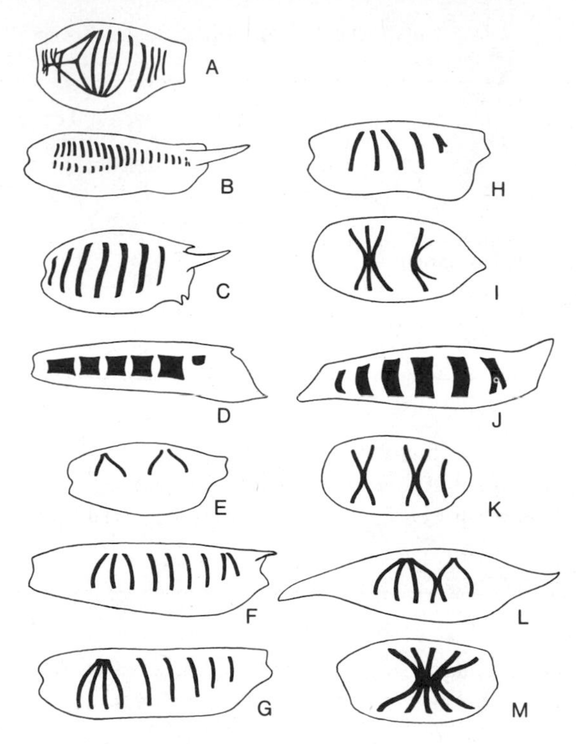

Fig. 24.10. Urochordata, Salpidae. Schematic figures outlining body form and muscle bands. All individuals are facing left and all are lateral views except I, K, M, which are dorsal. A–G are solitary forms. A, *Cyclosalpa* sp.; B, *Thetys vagina;* C, *Thalia democratica;* D, *Iasis zonaria;* E, *Pegea confederata;* F, *Salpa fusiformis;* G, *Salpa cylindrica;* H–M are aggregated forms; H, *Thetys vagina;* I, *Thalia democratica;* J, *Iasis zonaria;* K, *Pegea confederata;* L, *Salpa fusiformis;* M, *Salpa cylindrica.*

Aggregated Forms

1. Digestive tract forming a "nucleus"—2.

1. Digestive tract not forming a "nucleus." *Cyclosalpa* spp.: luminous organs in *pinnata* only; in *floridana* there are 3 body muscles, in *pinnata* and *affinis*, 4.

2. All muscle rings parallel, the last muscle band bifurcated below; 50 mm; Fig. 24.10J. *Iasis zonaria.*

2. Not all muscle rings parallel—3.

3. Muscle rings I and II, and III and IV forming pairs of well-separated X's; 120 mm; Fig. 24.10K. *Pegea confederata.*

3. Not so—4.

4. Muscles I to V converging dorsally; 12 mm; Fig. 24.10M. *Salpa cylindrica.*

4. Muscles I to IV converging dorsally—5.

4. Muscles I to III converging dorsally—6.

5. Muscles IV and V in contact laterally; 80 mm; Fig. 24.10L. *Salpa fusiformis.*
5. Muscles IV and V not in contact laterally; 150 mm. *Salpa maxima.*
6. Muscles I to III converging dorsally but not fused; 190 mm; Fig. 24.10H. *Thetys vagina.*
6. Muscles I to III touching dorsally; Fig. 24.10I. *Thalia democratica.*

Class Larvacea
APPENDICULARIANS (COPELATA)

Diagnosis. *Solitary, with a small trunk (1 to 9 mm), and a chordate tail ventrally attached to the body, constricted basally and emerging at right angles to the body axis, the tail usually 3–4 times the trunk length; trunk with gill slits opening directly to exterior. Appendicularians form a temporary, replaceable test or "house."*

Appendicularians have been described as permanently larval or neotenic animals, but they show such specialization that this description is really an oversimplification. For comparisons with the truly larval forms of Ascidiacea and Thaliacea see introductory key, page 595.

The gonads of appendicularians are large and conspicuous (Fig. 24.1B). Individuals are protandrous hermaphrodites, i.e., male elements are matured first. *Oikopleura dioica*, with sexes separate, is the sole known exception. Fertilization in all species is external and development proceeds directly without intervening forms.

The truly fantastic "house" is a vital element of the feeding mechanism, Figure 24.11F. It is secreted by a special oikoplastic epithelium on the anterior part of the trunk and is essentially a filtering mechanism for grading the size of planktonic food particles drawn into the test by the motions of the tail. The size of this house in different species ranges from that of a cherry to that of a baseball. In *Oikopleura longicauda* there are simple entrance and exit apertures whose size governs the dimensions of particles admitted. In other species the exit aperture is reduced and valvular, its function being that of an "escape hatch" for the animal. The filter mechanism consists of a pair of lateral windows fitted with a screen of crossed fibers plus an internal food trap. This system passes particles smaller than 30 microns, and oikopleurids feed mainly on the minute flagellates of the nannoplankton. Indeed, studies on *Oikopleura* gave the first clue to the importance of this class of ultrasmall plankters.

Systematic List and Distribution

The distribution of these small forms is imperfectly known. In some cases they are wide-ranging, and it is possible and perhaps likely that currently unrecorded species may be taken locally. In lieu of a formal check list the following systematic outline may serve as a frame of reference for the notes that follow.

Family Kowalewskaiidae, monogeneric and extralimital
Family Fritillariidae, *Tectillaria* extralimital; *Fritillaria* and *Appendicularia* in the local fauna
Family Oikopleuridae, *Oikopleura* and *Althoffia* locally plus several extralimital genera

---

Only *Oikopleura* and *Fritillaria* have been reported as residents in cold seas. At least five species of *Fritillaria* occur in the eastern North Atlantic, i.e., *borealis*, *venusta*, *tenella*, *gracilis*, and *aberrans*, with *borealis* apparently the most widely distributed form. Only *F. borealis*, *Oikopleura vanhoeffeni*, and *O. labradoriensis* have actually been reported in western North Atlantic waters. *F. borealis* and *O. labradoriensis* occur as winter-spring species on this coast at least as far south as Delaware Bay. The most likely warm water, summer-fall, species of *Oikopleura* on this coast is *O. dioica*. This species is widely distributed and is tolerant of a wide range of ecological conditions including temperatures as low as 3.2°C and salinities as low as 11.4‰. *O. longicauda*, *O. fusiformis*, and *Appendicularia sicula* have been recorded in the Gulf Stream and may appear inshore as accidental waifs. Additional warm water species reported in the approaches to western Europe include *O. parva*, *O. albicans*, *O. cophocerca*, and *Althoffia tumida*. See also introductory remarks, page 609.

Identification

---

1. Trunk relatively long, apparently tripartite; Fig. 24.11A,B. *Fritillaria: F. borealis* attains a trunk length of about 1.3 mm. For the identification of North Atlantic species, see Buckermann (1945).
1. Trunk oval—2.
2. Tail forked; trunk 0.8 mm, tail 2–3 times longer; Fig. 24.11C. *Appendicularia sicula*.
2. Tail pointed or rounded—3.
3. Tail tip with amphichordal cells; trunk 1.2 mm; Fig. 24.11D. *Althoffia tumida*.
3. Tail simple. *Oikopleura:* Distinctive patterns of *subnotochordal cells* in the tail musculature of the different species are figured. For additional criteria and formal keys see Buckermann (1945) and Harant and Vernieres (1938). *O. labradoriensis* attains a trunk length of 2.4 mm, with tail about 3.5–4 times longer; *longicauda*, *fusiformis*, and *dioica* grow to about half this size; *O. vanhoeffeni* is the largest form with a trunk length up to 9 mm.

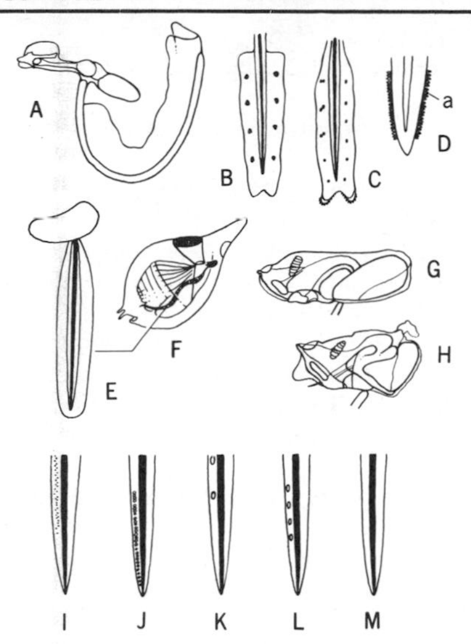

Fig. 24.11. Urochrodata, Larvacea. A, *Fritillaria*, whole animal; B, same, tail; C, *Appendicularia sicula;* tail; D, *Althoffia tumida*, tip of tail; E, *Oikopleura*, whole animal; F, "house" with animal indicated inside; G, *Oikopleura fusiformis*, trunk; H, *Oikopleura longicauda*, trunk; I–M are tail ends of *Oikopleura* species showing differential patterns of subnotochordial cells; I, *O. vanhoeffeni;* J, *O. labradoriensis;* K, *O. dioica;* L, *O. parva;* M, *O. fusiformis* and *longicauda*. Amphicordial cells indicated by letter "a."

References

Berrill, N. J. 1950. The Tunicata with an account of the British species. Ray Society.

Bigelow, H. 1926. Plankton of the offshore waters of the Gulf of Maine. Bull. Bur. Fish. Wash. (1924). xl(2):1–509.

Borgert, A. 1901. Die nordischen Dolioliden. Nord. Plank. Band. 2(3):1–4.

Brooks, W. K. 1908. The pelagic tunicata of the Gulf Stream. II. *Salpa floridana*. Carnegie Inst. Wash. Publ. 102:75–80.

———, and M. M. Metcalf. 1893. The genus *Salpa*. Mem. Johns Hopkins Univ. 2:1–371.

Buckermann, A. 1945. Appendicularia I–III, Sheet 7:1–8. Fiches d'identification du Zooplancton. Conseil International pour l'Exploration de la Mer.

Deevey, G. B. 1952. Quantity and composition of the zooplankton of Block Island Sound, 1949. *In* Hydrographic and biological studies of Block Island Sound. Bull. Bingham oceanogr. Coll. 13(3): 121–164.

———. 1956. Oceanography of Long Island Sound 1952–1954. V. Zooplankton. Bull. Bingham oceanogr. Coll. 15:62–112.

———. 1960. The zooplankton of the surface waters of the Delaware Bay Region. Bull. Bingham oceanogr. Coll. 17:5–52.

Fraser, J. H. 1947. Thaliacea I–II. Family Salpidae and family Doliolidae. Sheets 9, 10: 1–4, 1–4. Fiches d'identification du Zooplancton. Conseil International pour l'Exploration de la Mer.

Harant, H., and Paulette Vernieres. 1938. Appendiculaires et Thaliaces. Faune de France. 33(2):1–59.

Ihle, J. E. W. 1935. Tunicata. Desmomyaria. *In* Kukenthal and Krumbach's Handbuch der Zool. 5:401–544.

Metcalf, M. M. 1918. The Salpidae: a taxonomic study. Bull. U.S. natn. Mus. 100, 2(2):5–193.

———, and H. S. Hopkins. 1919. Pyrosoma—A taxonomic study based upon the collections of the United States Bureau of Fisheries and the United States National Museum. Bull. U.S. natn. Mus. 100, 2(3):195–275.

Miner, R. W. 1950. Field Book of Seashore Life. G. P. Putnam's Sons.

Neumann, G. 1935a. Doliolidae. *In* Bronn's Klassen und Ornungen des Tierreichs. Band 3 (Suppl. Tunicata), Abt. 2 Thaliacea 2 Buch: 1–67. Akad. Verlags.

———. 1935b. Pyrosomida (Tunicata). *In* Kukenthal and Krumbach's Handbuch der Zool. 5(2):226–323.

———. 1935c. Cyclomaria (Tunicata). *In* Kukenthal and Krumbach's Handbuch der Zool. 5(2):324–400.

General Index

(including taxonomic names above the level of genus)

*The *italic* page numbers indicate pages on which an illustration appears.

Systematic Index

(of generic and species names)

Generic names begin with a capital letter; species names are entirely in lower case. Common synonyms, not considered valid in this account, are given in parentheses; these are listed solely as an aid in using the literature. The list is not complete and is subject to almost constant revision. The citation *(g)* indicates that a generic synonym is given where the genus is the primary entry. Authors of original descriptions and the date of publication are given for most species. The reader is cautioned that this information has been taken largely from secondary sources, which sometimes neglect to give the date. Also, some sources do not follow the usual practice of placing the author's name in parentheses when the current generic name differs from the original. The somewhat archaic convention of appending author and date as a regular part of a species scientific name in general works has been abandoned by many authors. This practice obviously is important in formal works but has limited value for the general worker; such information is included here in deference to the few readers who made a point of suggesting its inclusion. Additional sources of confusion result from inconsistencies in the use of *i* or *ii* endings or of *a* for *um* or *us* and vice versa at the end of species names.

*The *italic* page numbers indicate pages on which an illustration appears.